Scientific Thought:
In Context

Scientific Thought: In Context

K. Lee Lerner & Brenda Wilmoth Lerner, Editors

VOLUME 2

CHEMISTRY TO SCIENCE PHILOSOPHY AND PRACTICE

GALE
CENGAGE Learning

Detroit • New York • San Francisco • New Haven, Conn • Waterville, Maine • London

Scientific Thought: In Context

K. Lee Lerner and Brenda Wilmoth Lerner, Editors

Project Editor: Elizabeth Manar

Editorial: Kathleen Edgar, Madeline Harris, Debra Kirby, Kristine Krapp, Kimberley McGrath

Production Technology: Paul Lewon

Rights Acquisition and Management: Mollika Basu, Margaret Abendroth, Ronald Montgomery, Kelly Quin, Edna Shy, Robyn Young

Composition and Electronic Capture: Evi Abou-El-Seoud, Mary Beth Trimper

Manufacturing: Wendy Blurton, Dorothy Maki

Imaging: Lezlie Light

Art Director: Jennifer Wahi

Product Management: Julia Furtaw, Janet Witalec

Indexing: Factiva, a Dow Jones & Reuters Company.

For product information and technology assistance, contact us at **Gale Customer Support, 1-800-877-4253.** For permission to use material from this text or product, submit all requests online at **www.cengage.com/permissions.** Further permissions questions can be emailed to **permissionrequest@cengage.com**

Cover photographs reproduced by permission of: © Digital Art/Corbis (DNA Strand); © Stapleton Collection/Corbis (*Tulipa* engraving by Basil Besler); © NASA/epa/Corbis (Hubble Telescope image of the spiral galaxy NGC 4013); © Bettman/Corbis (Chart showing the skeletons of five primates: human, gorilla, chimpanzee, orangutan, and gibbon. Georg Reimer, after Ernst Haechel's *The Battle of Evolution*, 1905); AP Photo/Ricardo Chaves/Zero Hora (A Brazilian scientist from PUC University studying a 220-million-year-old dinosaur fossil); and JLM Visuals (Radio telescopes, Socorra, New Mexico).

While every effort has been made to ensure the reliability of the information presented in this publication, Gale, a part of Cengage Learning, does not guarantee the accuracy of the data contained herein. Gale accepts no payment for listing; and inclusion in the publication of any organization, agency, institution, publication, service, or individual does not imply endorsement of the editors or publisher. Errors brought to the attention of the publisher and verified to the satisfaction of the publisher will be corrected in future editions.

LIBRARY OF CONGRESS CATALOGING-IN-PUBLICATION DATA

Scientific thought : in context / Brenda Wilmoth Lerner & K. Lee Lerner, editors.
   p. cm.
   Includes bibliographical references and index.
   ISBN 978-1-4144-0298-7 (set) -- ISBN 978-1-4144-0299-4 (vol. 1) -- ISBN 978-1-4144-0300-7 (vol. 2) -- ISBN 978-1-4144-0301-4 (vol. 3)
   1. Science--Philosophy. 2. Science--Methodology. I. Lerner, Brenda Wilmoth. II. Lerner, K. Lee.

Q175.S4238 2008
500--dc22
                                           2007051972

Gale
27500 Drake Rd.
Farmington Hills, MI, 48331-3535

ISBN-13: 978-1-4144-0298-7 (set)   ISBN-10: 1-4144-0298-8 (set)
ISBN-13: 978-1-4144-0299-4 (vol. 1)   ISBN-10: 1-4144-0299-6 (vol. 1)
ISBN-13: 978-1-4144-0300-7 (vol. 2)   ISBN-10: 1-4144-0300-3 (vol. 2)
ISBN-13: 978-1-4144-0301-4 (vol. 3)   ISBN-10: 1-4144-0301-1 (vol. 3)

This title is also available as an e-book.
ISBN-13: 978-1-4144-1085-2 (set) ISBN-10: 1-4144-1085-9 (set)
Contact your Gale, a part of Cengage Learning, sales representative for ordering information.

Printed in the United States of America
1 2 3 4 5 6 7 12 11 10 09 08

Contents

VOLUME 2

Contents

VOLUME 3

Advisors and Contributors

While compiling this volume, the editors relied upon the expertise and contributions of the following scientists, scholars, and researchers, who served as advisors and/or contributors for *Scientific Thought: In Context.*

Andrew Aberdein
Associate Professor
Florida Institute of Technology
Melbourne, Florida

Hanne Andersen
Associate Professor and Scholar
University of Aarhus
Aarhus, Denmark

Wilbur Applebaum
Professor Emeritus, Department of Humanities
Illinois Institute of Technology
Chicago, Illinois

William Arthur Atkins
Independent Scholar and Writer
Normal, Illinois

Ari Belenkiy
Mathematics Department
Bar-Ilan University
Ramat Gan, Israel

Julie Berwald
Geologist (Ocean Sciences) and Writer
Austin, Texas

John Burnham
Research Professor of History
Scholar in Residence, Medical Heritage Center

Ohio State University
Columbus, Ohio

William E. Burns
Independent Scholar
Washington, D.C.

Tamara E. Caraballo
Snohomish High School, Science Department
Snohomish, Washington

Philip Chaney, Ph.D., P.S.
Associate Professor, Geography
Auburn University
Auburn, Alabama

David F. Channell
Professor of Historical Studies
School of Arts & Humanities
University of Texas at Dallas
Dallas, Texas

Dennis W. Cheek
Vice President of Education
Ewing Marion Kauffman Foundation
Kansas City, Missouri

James Anthony Charles Corbett
Journalist
London, United Kingdom

Brian Dolan
Professor, History of Science and Medicine
University of California, San Francisco
San Francisco, California

Andrew Ede
Assistant Professor, History and Classics
University of Alberta
Edmonton, Alberta, Canada

Antonio Farina
Associate Professor, Embryology, Obstetrics, and Gynecology
University of Bologna
Bologna, Italy

Katrina Ford
Graduate Student, University of Auckland
Auckland, New Zealand

Larry Gilman
Electrical Engineer and Author
Independent Scholar and Writer
Sharon, Vermont

Amit Gupta
Independent Scholar and Journalist
Ahmedabad, India

William Hagan
Associate Professor of Chemistry
College of St. Rose
Albany, New York

Robert A. Hatch
*Interim Director, Center
for the Humanities*
Department of History
University of Florida
Gainesville, Florida

Tony Hawas
Writer and Journalist
Brisbane, Australia

Hal Hellman
*Independent Scholar
and Author*
Leonia, New Jersey

Matthew H. Hersch, J.D.
William Penn Fellow
History and Sociology of Science
University of Pennsylvania
Philadelphia, Pennsylvania

Brian D. Hoyle
Microbiologist and Author
Nova Scotia, Canada

Daniel Hudon
Lecturer in Physics
Boston University
Boston, Massachusetts

Alexandr Ioffe
Senior Scientist
Geological Institute, Russian
Academy of Sciences
Moscow, Russia

David T. King, Jr.
Professor, Department of Geology
Auburn University
Auburn, Alabama

Noretta Koertge
*Professor Emeritus, History
and Philosophy of Science*
Indiana University
Bloomington, Indiana

Kenneth T. LaPensee
*Epidemiologist and Medical
Policy Specialist*
Hampton, New Jersey

Christopher Lawrence
*Professor, Wellcome Trust
Centre for the History of
Medicine*
University College London
London, United Kingdom

**Adrienne Wilmoth Lerner,
J.D.**
Independent Research Scholar
Jacksonville, Florida

Eric v.d. Luft
*Lecturer, Center for Bioethics
and Humanities*
SUNY Upstate Medical University
Syracuse, New York

Lois N. Magner
Professor Emerita of History
Purdue University
West Lafayette, Indiana

David L. Morgan,
Assistant Professor, Physics
Eugene Lang College
The New School
New York, New York

Miriam C. Nagel
Independent Scholar
Avon, Connecticut

Lewis Pyenson
*Membre Correspondant,
Académie International
d'Histoire des Sciences (Paris)*
Research Professor, History, Center
for Louisiana Studies
University of Louisiana at Lafayette
Lafayette, Louisiana

Nicholas D. Pyenson
*Department of Integrative Biology
and Museum of Paleontology*
University of California, Berkeley

Giuseppina Ronzitti
Archives Henri Poincaré
Laboratoire de Philosophie et
d'Histoire des Sciences
Nancy, France

Anna Marie Eleanor Roos
*Research Associate, Wellcome
Unit for the History of
Medicine*
Oxford, United Kingdom

Gina Rumore
*Graduate Student, History
of Science and Technology*
University of Minnesota
Twin Cities, Minnesota

Concepcion Saenz-Cambra
Fulbright Research Scholar
Department of History
University of California
Santa Barbara, California

Martin Saltzman
Professor
Providence College
Providence, Rhode Island

James Satter
Independent Scholar
Minneapolis, Minnesota

Joachim Schummer
*Editor, HYLE: International
Journal for Philosophy of
Chemistry*
Heisenberg Fellow, Department
of Philosophy
University of Darmstadt, Germany

Angela Scobey
Independent Scholar
Mobile, Alabama

Sameer Shah
Graduate Student, UCLA
Los Angeles, California

Robert W. Smith
Professor, University of Alberta
Edmonton, Alberta, Canada

Matthew Stanley
*Institute for Advanced Study,
School of Historical Studies*
Princeton University
Princeton, New Jersey

Constance K. Stein
*Director of Cytogenetics/
Associate Professor*
SUNY Upstate Medical
University
Syracuse, New York

Florian Steger
*Institut fuer Geschichte und
Ethik der Medizin*
Friedrich-Alexander-Universität
Erlangen-Nuemberg, Germany

David J. Sturdy
Professor, School of History

University of Ulster, Coleraine
Coleraine, North Ireland, United
Kingdom

Edna Suárez
*Professor, National University
of México*

Visiting Scholar, Max Planck
Institute for the History of
Science
Berlin, Germany

Todd Timmons
Faculty Senate Chair

Associate Professor, Mathematics
and History of Science
University of Arkansas, Fort Smith
Fort Smith, Arkansas

Alain Touwaide
Historian of Science

Smithsonian Institution
Washington, D.C.

A. Bowdoin Van Riper
*Adjunct Professor, Social and
International Studies*

Southern Polytechnic State
University
Marietta, Georgia

Marco A. Viniegra Fernandez
*Graduate Student, History
of Science*

Harvard University
Cambridge, Massachusetts

Gary Weisel
*Associate Professor of
Physics*

Pennsylvania State University
Altoona, Pennsylvania

Melanie Barton Zoltán
Independent Scholar

Amherst, Massachusetts

Acknowledgments

The editors are grateful to the truly global group of distinguished scholars, researchers, and writers who contributed to *Scientific Thought: In Context*.

The editors also wish to thank the primary copyeditor, Amy Loerch, whose keen eyes and sound judgments greatly enhanced the quality and readability of the text.

The editors gratefully acknowledge and extend thanks to Janet Witalec and Julia Furtaw at Gale Cengage for their enduring faith in the project and for their sound content advice and flexibility. Without the able guidance and efforts of talented teams in rights and acquisition management and imaging at Gale, this book would not have been possible. We also thank Debra Kirby and her entire team of talented editors for their invaluable help in correcting copy. The editors also wish to acknowledge the contributions of Marcia Schiff at the Associated Press for her help in securing archival images.

Deep, sincere, and enduring gratitude is due project manager Elizabeth Manar who, over the years it took to write this book, showed nearly infinite patience. Especially during the rougher patches, when we mutually faced seemingly insurmountable obstacles, her intelligence, skill, passion for excellence, astute editing, and humanity were nothing short of inspirational.

Introduction

Written by a global array of experts in science, history, sociology, and law, *Scientific Thought: In Context* is intended to introduce to younger students (especially those just beginning their study of biology, chemistry, physics, earth science, etc.) insight into the power of science and a hint of the richness and complexity of scientific thought.

At the core of scientific thought is the assertion that the laws of physics and chemistry are the same throughout the universe. Although humans now understand some of the key laws that shape the cosmos, other laws—and the multitude of manifestations that evolve from them—remain shrouded in wonderful mystery to be peeled back by future generations. Such is the power of science's self-corrective mechanisms that, far above differences in culture and regardless of the language used to describe those laws or the social station of the eye viewing them (or whatever has evolved to be something akin to an eye in that part of the universe), ultimately the laws of science will be found to be the same.

These assertions regarding science and scientific thought do not, however, diminish value in the study of science as a cultural phenomena and manifestation. It is also true that students and other readers just beginning to explore the laws of biology and chemistry often have far less exposure to formal philosophical thought than to hard science. Accordingly, *Scientific Thought: In Context* is not intended to be a comprehensive history of science or to deeply explore the philosophy of science. Mindful of its audience, *Scientific Thought: In Context* only attempts to stab a toe, perhaps in some places to wade ankle deep, into the turbulent currents of intellectual thought surrounding the long history of scientific studies.

The editors and authors have intended only modest exploration—an audience appropriate glimpse—of the expanse of the philosophical waters. It would be inappropriate to force younger students to dive into the complexities of French narrative theory, evaluate the merits of ecofeminism, or suffer premature overexposure to elements of what can be an intellectually complex and acrimonious debate surrounding the tensions between science, philosophy, and cultural studies most often dubbed the "Science Wars." However, it is important for readers to know that vast oceans await. If this book allows students who are just beginning their studies of science and the history of science some insight into the diversity and rich complexity of the field, it will have achieved its primary objective.

Scientific Thought: In Context was designed to contain an extensive timeline as a chronological record of the evolution of science. The timeline serves as a valuable reference tool that allows readers to readily place events *in context* with scientific developments in other fields and historical milestones. Moreover, the timeline provides compelling reading in its own right. Starting with developments in ancient history, the timeline provides a narrative of the broad multicultural influences on the advancement of scientific thought.

To be sure, at its very best science transcends culture and unifies us. The deepest truths of science are knowable to all, regardless of age, gender, social condition, or cultural tradition.

Yet it is equally true that throughout the course of human history, science and society have advanced in a dynamic and mutual embrace, each influencing the other. While students learn of the grandeur and power of science and the scientific method, they should also know that scientific thought was (and is) often ignored in favor of cultural tradition, sifted through theological filters, or reduced to being a handmaiden to military tactics and weaponry. Scientific thought has been suppressed and swept from the philosophical stage during various periods in human history only to—by its virtue and strength as the most robust way to know the world—rekindle itself as a candle in the intellectual darkness.

It is important for all readers to understand that abundant evidence exists that sexism, racism, militarism, colonialism, and economic philosophy all influenced the course of human intellectual development, and the history of science is no exception. Studies of societal and cultural bias and prejudice also prove that selective picking and choosing of facts without scholarly discernment is, at best, unhelpful, and at worst—as in the case of Soviet Lysenkoism—deadly. Too readily does ignorance of science and scientific thought swell the ranks of book burning mobs, religious extremists, or cultish followers of pseudoscience.

With this in mind, the editors are indebted to a diversity of distinguished scholars for their generous contributions of time and compelling material. Given our intended audience, the selection and construction of entries was a balancing act of assessing what was interesting, informative, and intellectually understandable. Admittedly, the editors are scientists and not disposed to view science as a fashionable manifestation of political or popular culture. Accordingly, we do not agree with all of the philosophical and historical assertions made in this book. We have, however, encouraged scholars who represent highly diverse backgrounds and perspectives to express their views.

Ignorance, mysticism, and zealotry present true and grave dangers to science and the advancement of human rights and Enlightenment ideals. Within this context, the philosophical excesses on both sides of the "Science Wars" are trivial. Students not exposed to the intellectual heritage (and baggage) of scientific thought will be unable to make tempered and rational decisions regarding the appropriate application of scientific thought to modern issues.

Ideally, studies of the history of science from a cultural perspective can shed light upon previously overlooked contributors and their contributions and deepen appreciation for the commonality of science. Understanding the vulnerabilities of scientific thought, however historically transient, is also key to ensuring that science is advanced by intellectual ability and continues to offer humankind the best hope of a meritocracy of thought—where barriers of authority, class, wealth, religion, race or ethnicity might be cast aside.

K. Lee Lerner & Brenda Wilmoth Lerner, editors
Paris, France, December 2007

Primarily based in London and Paris, the Lerner & Lerner portfolio includes more than two dozen books and films that focus on science and science related issues.

The excerpt below comes from American physicist Richard Feynman's (1918–1988), *The Feynman Lectures on Physics.*

The things with which we concern ourselves in science appear in myriad forms, and with a multitude of attributes. For example, if we stand on the shore and look at the sea, we see the water, the waves breaking, the foam, the sloshing motion of the water, the sound, the air, the winds and the clouds, the sun and the blue sky, and light; there is sand and there are rocks of various hardness and permanence, color, and texture. There are animals and seaweed, hunger and disease, and to the observer on the beach; there may be even happiness and thought. Any other spot in nature has a similar variety of things and influences. It is always as complicated as that, no matter where it is. Curiosity demands that we ask questions, that we try to put things together and try to understand this multitude of aspects as perhaps resulting from the action of a relatively small number of elemental things and forces acting in an infinite variety of combinations....

A few hundred years ago, a method was devised to find partial answers to such questions. *Observation, reason,* and *experiment* make up what we call the *scientific method.*

What do we really mean by "understanding" something? We can imagine that this complicated array of moving things which constitutes "the world" is something like a great chess game being played by the gods, and we are observers of the game. We do not know what the rules of the game are; all we are allowed to do is to *watch* the playing. Of course, if we watch long enough, we may eventually catch on to a few of the rules....

—*Richard Feynman*

Editor's Note: Richard Feynman was awarded the 1965 Nobel Prize in Physics for his contributions to the advancement of quantum electrodynamics (QED). Dr. Feynman led an interesting life: As a young scientist he participated in the development of the first atomic bomb, and near the end of his life he was instrumental in determining the cause of the space shuttle *Challenger* disaster. Dr. Feynman played the bongo drums, loved diversity of culture, and expressed his views on the relationship of scientific thought to society through anecdotal stories contained in several books including *Surely You're Joking, Mr. Feynman.* He is considered by many to be one of the greatest teachers of physics.

About the *In Context* Series

Written by a global array of experts, yet aimed primarily at high school students and an interested general readership, the *In Context* series serves as an authoritative reference guide to essential concepts of science, the impacts of changes in scientific consensus, and the effects of science on social, political, and legal issues.

Cross-curricular in nature, *In Context* books align with, and support, national science standards and high school science curriculums across subjects in science and the humanities and facilitate science understanding important to higher achievement in the No Child Left Behind (NCLB) science testing. The inclusion of primary source documents and original essays written by leading experts serve the requirements of an increasing number of high school and international baccalaureate programs and are designed to provide additional insights on leading social issues, as well as spur critical thinking about the profound cultural connections of science.

In Context books also give special coverage to the impact of science on daily life, commerce, travel, and the future of industrialized and impoverished nations.

Each book in the series features entries with extensively developed words-to-know sections designed to facilitate understanding and increase both reading retention and the ability of students to understand reading in context without being overwhelmed by scientific terminology.

Entries are further designed to include standardized subheads that are specifically designed to present information related to the main focus of the book. Entries also include a listing of further resources (books, periodicals, Web sites, audio and visual media) and references to related entries.

In addition to maps, charts, tables and graphs, each *In Context* title has approximately 300 topic-related images that visually enrich the content. Each *In Context* title will also contain topic-specific timelines (a chronology of major events), a topic-specific glossary, a bibliography, and an index especially prepared to coordinate with the volume topic.

About This Book

The goal of *Scientific Thought: In Context* is to offer high-school and early college-age students insights into the essential facts and deeper cultural connections of scientific thought underpinning the study of modern science.

Because science increasingly plays a key role in shaping complex ethical and social debates, it is critical to have an understanding of the evolution of scientific thought. Arguably more critical than knowing the essential facts of science history (who discovered what and when) is the knowledge that advancements build upon earlier work and the realization that humanity for all our advances, has only recently grasped the most essential understanding of the nature and working of the cosmos.

In an attempt to enrich the reader's understanding of the mutually impacting relationship between science and culture, as space allows we have included primary sources that enhance the content of *In Context* entries. In keeping with the philosophy that much of the benefit from using primary sources derives from the reader's own process of inquiry, the contextual material introducing each primary source provides an unobtrusive introduction and springboard to critical thought.

General Structure

Scientific Thought: In Context is a collection of entries on diverse topics selected to provide insight into facets of scientific thought, especially touching on topics that relate to general science studies.

The articles in the book are meant to be understandable by anyone with a curiosity about science, and the first edition of *Scientific Thought: In Context* has been designed with ready reference in mind:

- Entries are arranged alphabetically, but key first words in each title help group subject areas such as physics or chemistry together for easier reference.
- The **chronology** (timeline) includes many of the most significant events in the history of science.
- An extensive **glossary** section provides readers with a ready reference for content-related terminology. In addition to defining terms within entries, specific "Words to Know" sidebars are placed within each entry.
- A **bibliography** section (citations of books, periodicals, Web sites, and audio and visual material) offers additional resources to those resources cited within each entry.
- A **comprehensive general index** guides the reader to topics and persons mentioned in the book.

Entry Structure

In Context entries are designed so that readers may navigate entries with ease. Toward that goal, entries are divided into easy-to-access sections:

- **Introduction**: A opening section designed to clearly identify the topic.
- **Words to Know** sidebar: Essential terms that enhance readability and critical understanding of entry content.
- Established but flexible **rubrics** customize content presentation and identify each section, enabling the reader to navigate entries with ease. Inside *Scientific Thought: In Context* entries readers will find a general scheme of organization. All entries contain a brief introduction, "Words to Know" sidebar, and then a section describing the essential history and scientific foundations of the topic. Sections titled "Impacts and Issues" or "Modern Cultural Connections" then interrelate key scientific, political, or social considerations related to the topic.
- More than 150 additional sidebars added by the editors enhance expert contributions by focusing on key areas, providing material for divergent studies or providing additional insights or context.
- If an entry contains a related primary source, it is appended to end of the author's text. Authors are not responsible for the selection or insertion of primary sources. Primary sources are designed to be related to the entry so as to stimulate critical thought, especially as to modern tangential connections to even the most ancient history of science.
- **Bibliography:** Citations of books, periodicals, Web sites, and audio and visual material used in preparation of the entry or that provide a stepping stone to further study.
- **"See also" references** clearly identify other content-related entries.

Scientific Thought: In Context Special Style Notes

Please note the following with regard to topics and entries included in *Scientific Thought: In Context*:

- Primary source selection and the composition of sidebars are not attributed to authors of signed entries to which the sidebars may be associated.
- Equations are, of course, often the most accurate and preferred language of science, and are essential to epidemiologists and medical statisticians. To better serve the intended audience of *Scientific Thought: In Context*, however, the editors attempted to minimize the inclusion of equations in favor of describing the elegance of thought or essential results such equations yield.
- A detailed understanding of biology and chemistry is neither assumed nor required for *Scientific Thought: In Context*. Accordingly, students and other readers should not be intimidated or deterred by the sometimes complex names of chemical molecules or biological classification. Where necessary, sufficient information regarding chemical structure or species classification is provided. If desired, more information can easily be obtained from any basic chemistry or biology reference.

Bibliography Citation Formats (How to cite articles and sources)

In Context titles adopt the following citation format:

Books

Feynman, Richard P. *The Feynman Lectures on Physics.* San Francisco, CA: Pearson/Addison-Wesley, 2006.

Feynman, Richard P., and Davies, Paul (preface). Edited by Robert B. Leighton and
 Matthew Sands *Six Easy Pieces: Essentials of Physics Explained by Its Most Brilliant
 Teacher.* New York: Basic Books, 2005.

Periodicals

Darwin, C.R., "Note on a Rock Seen on an Iceberg in 61° South Latitude." *Journal of
 the Geographical Society* 9 (March 1839): 528–529.
Dawkins, Richard. "Inferior Design." Review of *The Edge of Evolution: The Search for
 the Limits of Darwinism,* by Michael J. Behe. *New York Times* (July 1, 2007).
Elkin, Lynne Osman. "Rosalind Franklin and the Double Helix." *Physics Today* (February 2003): 63.

Web Sites

Aristotle. *Physics.* The Internet Classics Archive. http://classics.mit.edu/Aristotle/
 physics.html (accessed August 28, 2007).
Indiana State University School of Medicine. "Biochemistry of Nucleic Acids." http://
 www.indstate.edu/thcme/mwking/nucleic-acids.html (accessed January 24, 2008).
National Aeronautics and Space Administration (U.S.). "Cosmology: The Study of the
 Universe." September 26, 2006. http://map.gsfc.nasa.gov/m_uni.html (accessed
 February 6, 2008).
National Biological Information Infrastructure. "Botany." http://www.nbii.gov/
 portal/community/Communities/Plants,_Animals_&_Other_Organisms/
 Botany/ (accessed February 16, 2008).

Alternative Citation Formats

There are, however, alternative citation formats that may be useful to readers and examples of how to cite articles in often used alternative formats are shown below.

APA Style

Books: Kübler-Ross, Elizabeth. (1969) *On Death and Dying.* New York: Macmillan.
Excerpted in K. Lee Lerner and Brenda Wilmoth Lerner, eds. (2006) *Medicine, Health,
and Bioethics: Essential Primary Sources,* Farmington Hills, Mich.: Thomson Gale.

Periodicals: Venter, J. Craig, et al. (2001, February 16). "The Sequence of the Human Genome." *Science,* vol. 291, no. 5507, pp. 1304–51. Excerpted in K. Lee Lerner
and Brenda Wilmoth Lerner, eds. (2006) *Medicine, Health, and Bioethics: Essential
Primary Sources,* Farmington Hills, Mich.: Thomson Gale.

Web Sites: Johns Hopkins Hospital and Health System. "Patient Rights and Responsibilities." Retrieved January 14, 2006 from http://www.hopkinsmedicine.org/
patients/JHH/patient_rights.html. Excerpted in K. Lee Lerner and Brenda Wilmoth
Lerner, eds. (2006) *Medicine, Health, and Bioethics: Essential Primary Sources,* Farmington Hills, Mich.: Thomson Gale.

Chicago Style

Books: Kübler-Ross, Elizabeth. *On Death and Dying.* New York: Macmillan, 1969.
Excerpted in K. Lee Lerner and Brenda Wilmoth Lerner, eds. *Medicine, Health, and
Bioethics: Essential Primary Sources,* Farmington Hills, MI: Thomson Gale, 2006.

Periodicals: Venter, J. Craig, et al. "The Sequence of the Human Genome." *Sci-
ence* (2001): 291, 5507, 1304–1351. Excerpted in K. Lee Lerner and Brenda Wilmoth
Lerner, eds. *Medicine, Health, and Bioethics: Essential Primary Sources,* Farmington
Hills, MI: Thomson Gale, 2006.

Web Sites: *Johns Hopkins Hospital and Health System.* "Patient Rights and Responsibilities." <http://www.hopkinsmedicine.org/patients/JHH/patient_rights.html.> (accessed January 14, 2006). Excerpted in K. Lee Lerner and Brenda Wilmoth Lerner, eds. *Medicine, Health, and Bioethics: Essential Primary Sources*, Farmington Hills, MI: Thomson Gale, 2006.

MLA Style

Books: Kübler-Ross, Elizabeth. *On Death and Dying*, New York: Macmillan, 1969. Excerpted in K. Lee Lerner and Brenda Wilmoth Lerner, eds. *Medicine, Health, and Bioethics: Essential Primary Sources*, Farmington Hills, Mich.: Thomson Gale, 2006.

Periodicals: Venter, J. Craig, et al. "The Sequence of the Human Genome." *Science*, 291 (February 16, 2001): 5507, 1304–51. Excerpted in K. Lee Lerner and Brenda Wilmoth Lerner, eds. *Terrorism: Essential Primary Sources*, Farmington Hills, Mich.: Thomson Gale, 2006.

Web Sites: "Patient's Rights and Responsibilities." Johns Hopkins Hospital and Health System. January 14, 2006. <http://www.hopkinsmedicine.org/patients/JHH/patient_rights.html.> Excerpted in K. Lee Lerner and Brenda Wilmoth Lerner, eds. *Terrorism: Essential Primary Sources*, Farmington Hills, Mich.: Thomson Gale, 2006.

Turabian Style (Natural and Social Sciences)

Books: Kübler-Ross, Elizabeth. *On Death and Dying*, (New York: Macmillan, 1969). Excerpted in K. Lee Lerner and Brenda Wilmoth Lerner, eds. *Medicine, Health, and Bioethics: Essential Primary Sources*, (Farmington Hills, Mich.: Thomson Gale, 2006).

Periodicals: Venter, J. Craig, et al. "The Sequence of the Human Genome." *Science*, 291 (February 16, 2001): 5507, 1304–1351. Excerpted in K. Lee Lerner and Brenda Wilmoth Lerner, eds. *Medicine, Health, and Bioethics: Essential Primary Sources*, (Farmington Hills, Mich.: Thomson Gale, 2006).

Web Sites: Johns Hopkins Hospital and Health System. "Patient's Rights and Responsibilities." available from http://www.hopkinsmedicine.org/patients/JHH/patient_rights.html; accessed January 14, 2006. Excerpted in K. Lee Lerner and Brenda Wilmoth Lerner, eds. *Medicine, Health, and Bioethics: Essential Primary Sources*, (Farmington Hills, Mich.: Thomson Gale, 2006).

Using Primary Sources

The definition of what constitutes a primary source is often the subject of scholarly debate and interpretation. Although primary sources come from a wide spectrum of resources, they are united by the fact that they individually provide insight into the historical *milieu* (context and environment) during which they were produced. Primary sources include materials such as newspaper articles, press dispatches, autobiographies, essays, letters, diaries, speeches, song lyrics, posters, works of art—and, in the twenty-first century, web logs—that offer direct, first-hand insight or witness to events of their day.

Categories of primary sources include:

- Documents containing firsthand accounts of historic events by witnesses and participants. This category includes diary or journal entries, letters, e-mail, newspaper articles, interviews, memoirs, and testimony in legal proceedings.
- Documents or works representing the official views of both government leaders and leaders of other organizations. These include primary sources such as policy statements, speeches, interviews, press releases, government reports, and legislation.
- Works of art, including (but certainly not limited to) photographs, poems, and songs, including advertisements and reviews of those works that help establish an understanding of the cultural environment with regard to attitudes and perceptions of events.
- Secondary sources. In some cases, secondary sources or tertiary sources may be treated as primary sources. For example, if an entry written many years after an event, or to summarize an event, includes quotes, recollections, or retrospectives (accounts of the past) written by participants in the earlier event, the source can be considered a primary source.

Analysis of Primary Sources

The primary source material collected in this volume is not intended to provide a comprehensive or balanced overview of a topic or event. Rather, the primary sources are intended to generate interest and lay a foundation for further inquiry and study.

In order to properly analyze a primary source, readers should remain skeptical and develop probing questions about the source. Using historical documents requires that readers analyze them carefully and extract specific information. However, readers must also read "beyond the text" to garner larger clues about the social impact of the primary source.

In addition to providing information about their topics, primary sources may also supply a wealth of insight into their creator's viewpoint. For example, when reading a news article about an outbreak of disease, consider whether the reporter's words also indicate something about his or her origin, bias (an irrational disposition in favor of someone or something), prejudices (an irrational disposition against someone or something), or intended audience.

Students should remember that primary sources often contain information later proven to be false, or contain viewpoints and terms unacceptable to future generations. It is important to view the primary source within the historical and social context existing at its creation. If, for example, a newspaper article is written within hours or days of an event, later developments may reveal some assertions in the original article as false or misleading.

Test New Conclusions and Ideas

Whatever opinion or working hypothesis the reader forms, it is critical that they then test that hypothesis against other facts and sources related to the incident. For example, it might be wrong to conclude that factual mistakes are deliberate unless evidence can be produced of a pattern and practice of such mistakes with an intent to promote a false idea.

The difference between sound reasoning and pseudoscientific beliefs, preposterous conspiracy theories, or the birth of urban legends lies in the willingness to test new ideas against other sources rather than rest on one piece of evidence, such as a single primary source, that may contain errors. Sound reasoning requires that arguments and assertions guard against argument fallacies that utilize the following:

- false dilemmas (only two choices are given when in fact there are three or more options);
- arguments from ignorance (*argumentum ad ignorantiam*; because something is not known to be true, it is assumed to be false);
- possibilist fallacies (arguments in which "it could be" is usually followed by an unearned "therefore, it is." This is a favorite among conspiracy theorists, who attempt to demonstrate that a factual statement is true or false by establishing the possibility of its truth or falsity.);
- slippery slope arguments or fallacies (a series of increasingly dramatic consequences is drawn from an initial fact or idea);
- begging the question (the truth of the conclusion is assumed by the premises);
- straw man arguments (the arguer mischaracterizes an argument or theory and then attacks the merits of their own false representations);
- appeals to pity or force (the argument attempts to persuade people to agree by sympathy or force);
- prejudicial language (values or moral goodness, good and bad, are attached to certain arguments or facts);
- personal attacks (*ad hominem*; an attack on a person's character or circumstances);
- anecdotal or testimonial evidence (stories that are unsupported by impartial observation or data that is not reproducible);
- *post hoc* (after the fact) fallacies (because one thing follows another, it is held to cause the other);
- the fallacy of the appeal to authority (the argument rests upon the credentials of a person, not the evidence).

Despite the fact that some primary sources can contain false information or lead readers to false conclusions based on the "facts" presented, they remain an invaluable resource regarding past events. Primary sources allow readers and researchers to come as close as possible to understanding the perceptions and context of events and thus to more fully appreciate how and why misconceptions occur.

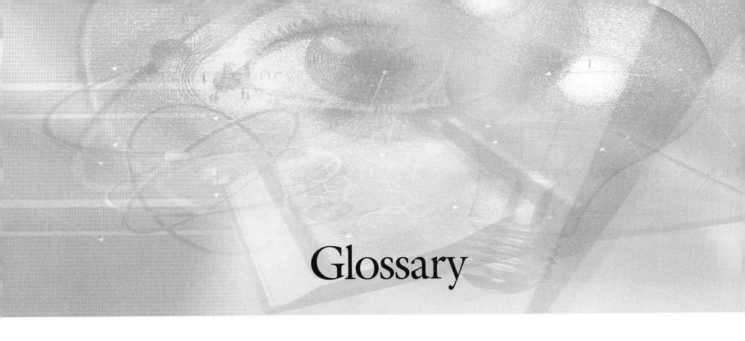

Glossary

A

ABERRANT: Deviation from some specified standard. In optics, the failure of a lens or mirror to focus an image perfectly at its intended focal point. In astronomy, the apparent displacement of an observed object due to the relative motion of the object and the telescope.

ABIOTIC: The portion of an ecosystem that is not living, such as water or soil.

ABSOLUTE MAGNITUDE: The apparent brightness of a star, measured in units of magnitudes, at a fixed distance of 10 parsecs.

ABSORPTION SPECTRUM: The spectrum formed when light passes through a cool gas.

ACADEMY: The world of university researchers, teachers, and students is often referred to as the academy. Specific academies, such as the U.S. National Academy of Sciences, are groups of experts asked to serve as a scientific advisory body, usually to a government.

ACCRATIELEON: A change in the velocity (either magnitude or direction) of an object.

ACCRETION DISK: When gas and dust are attracted to a heavy, spinning object such as a neutron star or black hole, they form a disk that rotates at right angles to the spin axis of the heavy object. Material from the innermost edge of the accretion disk continually falls into the heavy object.

ACETYLCHOLINE: One of many neurotransmitters employed by the human body and used throughout the nervous system. A neurotransmitter is a chemical released by nerve cells into a synapse—the narrow, fluid-filled space separating it from another nerve cell—in order to stimulate chemical activity in that other cell.

ACHROMATIC LENS: Achromatic lenses are built by forming lenses out of sandwiched parts that are made of different types of glass. The optical properties of each part of the lens compensate for or balance the properties of the others, reducing aberrations caused by other types of lenses. For example, simple glass lenses have the unwanted property of acting as prisms, that is, dividing white light up into rainbow-like spectra: This effect is termed chromatic aberration. Uncorrected lenses also suffer from spherical aberration, which is the focusing of different colors at different distances from the lens rather than at a single image plane.

ACQUIRED CHARACTERISTICS: Characteristics that are acquired by a plant or animal by interacting with its environment. These may include scars, lost limbs, calluses, strengthened or weakened muscles, and so forth. Although some acquired characteristics (e.g., low body weight due to malnutrition) can be inherited briefly and in part, it is now known that inheritance of acquired characteristics is not a significant factor in biological evolution.

ACUPUNCTURE: The Chinese practice of treating disease or pain by inserting very thin needles into specific sites in the body.

ADAPTIVE RADIATION: In evolutionary biology, adaptive radiation occurs when a species colonizes a new environment and rapidly evolves a wide range of new forms to adapt to niches or opportunities in that environment. Here "radiation" refers to the radiating or fanning-out of the species into ecological opportunities, not to nuclear radiation. A classic example of adaptive radiation is the history of cichlid fishes in the Great Rift Lakes of Africa. In Lake Victoria, for example, over 500 species of cichlid fish have evolved from a single ancestral species in the last 12,000 years or less.

ADDISON'S DISEASE: A disorder resulting from insufficient secretion of hormones from the adrenal cortex, named after the doctor who first characterized it, Englishman Thomas Addison (1793–1860). Its symptoms include weakness and bronze discoloration of the skin.

ADJUVANT THERAPY: Cancer treatments (radiation, chemotherapy) that are given after surgery in order to prevent recurrence of a tumor.

ADRENAL GLANDS: A pair of endocrine glands that sits atop the kidneys and releases hormones directly into the bloodstream. The adrenals are flattened, somewhat triangular bodies that, like other endocrine glands, receive a rich blood supply. The phrenic (from the diaphragm) and renal (from the kidney) arteries send many small branches to the adrenals, while a single large adrenal vein drains blood from the gland.

ADULT STEM CELL: A renewable and unspecialized cell found among specialized cells in a tissue or organ.

ALBEDO EFFECT: Albedo is the ability of a planet, moon, or other body in space to reflect light. Brighter objects have higher albedo, darker objects lower albedo. In climate science, the albedo effect is the influence of Earth's albedo on climate. Bright features such as ice caps tend to reflect solar energy into space, cooling the climate. Melting ice lowers Earth's albedo, making it a more efficient absorber of solar energy and tending to warm its climate.

ALCHEMY: The study of the reactions of chemicals in pre-modern times. It was often, but not always, directed by the goal of making gold. In a general sense, alchemy is perceived as the transmutation (transformation) of a common substance to something rare and valuable. Medieval alchemists are often portrayed as little more than quacks attempting to make gold from lead. This depiction is not entirely correct. To be sure, there were such characters, but for real alchemists, called adepts, the field was an almost divine mixture of science, mystery, and philosophy.

ALCOHOL: Any of the large number of molecules containing a hydroxyl (–OH) group bonded to a car-

bon atom to which only other carbon atoms or hydrogen atoms are bonded.

ALGORITHM: A mathematical relation between an observed quantity and a variable used in a step-by-step mathematical process to calculate a quantity.

ALLELE: Any of two or more alternative forms of a gene that occupy the same location on a chromosome.

ALLOGRAFT: Transplanted tissues or organs from donors of the same species.

ALLOTRANSPLANTATION: The transplantation of an organ to an individual from an animal of the same species, e.g., human to human. This is the most common form of transplantation. Xenotransplantation, on the other hand, is the transplantation of an organ from an animal of another species, e.g., a pig's heart valve into a human.

ALPHA PARTICLE: A positively charged nuclear particle that consists of two protons and two electrons; it is ejected at a high speed from disintegrating radioactive materials.

AMINO ACID: One of about two dozen relatively simple chemical compounds from which proteins are made, amino acids are the building blocks of proteins and serve many other functions in living organisms. They are nitrogen-containing organic compounds that consist of at least one acidic carboxyl group ($COOH$) and one amino group (NH_2). In alpha amino acids that are contained in the proteins found in cells, these two groups are both attached to a carbon atom, which also carries a hydrogen atom, plus a side chain known as the R group. The R group varies from one amino acid to another and gives each amino acid its distinctive properties. The prime function of DNA (deoxyribonucleic acid) is to carry the information needed to direct the proper sequential insertion of amino acids into protein chains during protein synthesis (translation). Although relatively simple compounds, amino acids can vary widely; to date, more than 80 different amino acids have been found in living organisms. Of these 80 amino acids, 22 are considered the precursors of animal proteins.

AMORPHOUS: A substance that lacks any well-defined structure. In earth science, substances made of randomly-organized atoms, such as glass. Crystals, whose atoms are organized in a definite pattern, are not amorphous; therefore, amorphous substances are sometimes said to be acrystalline (i.e., not-crystalline).

AMPLIFICATION: Increasing the strength of some signal such as the amount of electrical current passing through a transistor.

AMPUTATION: The surgical or accidental removal of a limb (arm or leg) or a distinct part of a limb (finger, toe, hand, foot, etc).

ANABOLISM: The process by which energy is used to build up complex molecules.

ANADROMOUS: Fish that migrate up streams and rivers to mate and produce offspring are said to be anadromous (from the Greek for "up-running"); all other fish are catadromous.

ANAEROBIC: Living or growing in an atmosphere lacking oxygen.

ANALOG: A process that is fluctuating or continually changing. In electronics, an analog signal is a base alternating current frequency that is modified, usually by amplification or varying of the frequency, in order to add information to the signal. Conventional forms of television and telephone transmissions use analog technology.

ANALOGOUS STRUCTURE: In evolutionary biology, an analogous structure is a part of a creature's anatomy that resembles a part of some other creature's anatomy, even though the two creatures are not related by evolutionary descent. The two structures evolved independently to perform the same function; for example, the eye of the squid and the human eye are analogous structures, as are birds' wings and bats' wings. Structures that resemble each other because of descent from a shared ancestor are homologous, not analogous.

ANALYSIS: In chemistry, the process of separating out the constituents of a complex chemical substance.

ANAPHYLAXIS: Any severe allergic reaction to a substance to which the body has been previously exposed. In the most severe form, the patient experiences anaphylactic shock, which may cause death through stopped breathing. Insect stings and penicillin are two common triggers for anaphylactic shock.

ANASTOMOSE: To make a connection between tubes in the body (blood vessels, digestive organs, or the like) that joins the mid-part of one tube to another, as opposed to connecting them end-to-end. Anastomosis may occur surgically or in the natural growth of a network of vessels. For example, the capillary beds that deliver blood to most of the body's tissues form an extensively anastomosed network.

ANEURYSM: A bulging or ballooning of part of a blood vessel. Aneurysms can burst, causing internal bleeding and severe damage, depending on the location or size of the break: a cerebral aneurysm can destroy part of the brain or cause death, while an aortic aneurysm is usually immediately fatal.

ANGULAR SIZE: The size of an object in the sky, measured in degrees.

ANIMALCULE: The term "animalcules," from Latin, means "tiny animals." The inventor of the microscope, Dutch scientist Anton van Leeuwenhoek (1632–1723), was the first person to observe microscopic animals and used this term to describe them. The word had previously been applied occasionally to any small animals, such as mice or insects.

ANTHROPOCENTRISM: The tendency to view human beings and their needs, values, and desires as all-important or central to the cosmos (from the Greek for "human-centered"). For example, the belief that Earth rotates on its axis so that humans can have a daily rest period would be anthropocentric.

ANTHROPOGENIC: Made by people or resulting from human activities. Usually used in the context of emissions that are produced as a result of human activities.

ANTIBIOTIC: Natural or synthetic compounds that kill or reduce populations of bacteria. There are a myriad of different antibiotics that act on different structural or biochemical components of bacteria. Antibiotics have no direct effect on viruses. Also, a specific drug, such as penicillin, that is used to fight infections caused by bacteria.

ANTIBODIES: Large, Y-shaped proteins found in blood that lock on to specific substances foreign to the body (antigens). Typical antigens are molecules found on the surfaces of viruses and bacteria. Some antibodies, when bound to antigens, act as flags targeting the antigen for attack by white blood cells; others combine with other blood molecules to attack the antigen directly. If a person's blood already contains antibodies for a particular antigen, the person's immune system attacks that antigen as soon as it appears. This is the basis of acquired immunity to specific viruses and bacteria, whether natural or instilled by immunization.

ANTIBODY: An antibody, or Y-shaped immunoglobulin, is a protein molecule found in the blood that is created by the immune system in response to the presence of an antigen (a foreign substance or particle). It marks foreign microorganisms in the body for destruction by other immune cells. There are a myriad of different antigens that are presented to the immune system. Hence, there are a myriad of antibodies that can be formed.

ANTIGEN: Any substance that the body considers foreign, such as a bacterial cell, that stimulates the body's immune system to produce antibodies. Antibodies, or Y-shaped immunoglobulins, are proteins that inactivate the antigen and help to remove it from the body. While antigens can be the source of infections from pathogenic (disease-causing) bacteria and viruses, organic molecules detrimental to the body from internal or environmental sources also act as antigens. Antigens are usually proteins or polysaccharides.

ANTIOXIDANTS: Antioxidants are substances, often found in foods and capable of being transferred to body cells, that interfere with oxidation reactions. These are reactions involving free radicals (free-floating charged molecules). Free radicals can damage cell chemistry by triggering unwanted chemical reactions; antioxidants reduce this damage.

ANTIPARTICLE: In particle physics, most fundamental particles having non-zero rest mass (e.g., neutron, proton, electron) have an antiparticle that is an almost perfect mirror-image of the particle. When a particle and its antiparticle meet, they annihilate each other, releasing all their energy in the form of photons. At the big bang, about 13.7 billion years ago, slightly more particles were produced than antiparticles. After all particle-antiparticle pairs annihilated each other, the remaining fraction of particles became all the ordinary matter seen in the universe today.

ANTISEPTIC: A substance that prevents or stops the growth and multiplication of pathogenic (disease-causing) microorganisms in or on living tissue. An antiseptic may kill a microorganism, but it does not necessarily have to. The treated microbes may only be weakened. The weaker, slower-growing microbes may then be more susceptible to the defense mechanisms of the host.

ANTISERUM (PLURAL: ANTISERUMS OR ANTISERA): A clear liquid blood fraction that contains certain antibodies against a specific agent, usually a bacterium or bacterial toxin, that is used to treat persons infected with that agent. Both antisera and antitoxins are means of proactively combating infections. The introduction of compounds to which the immune system responds is an attempt to build up protection against microorganisms or their toxins before the microbes actually invade the body. An antiserum is injected into the body to confer immunity against a pathogen (disease-causing organism) that is targeted by the antibody contained in the antiserum.

ANTITOXIN: An antidote to a toxin that neutralizes its poisonous effects. Both antisera and antitoxins are means of proactively combating infections. The introduction of compounds to which the immune system responds is an attempt to build up protection against microorganisms or their toxins before the microbes actually invade the body.

AORTA: The main artery of the body that arises from the left ventricle of the heart and runs down the body in front of the spine, supplying blood through branch arteries throughout the body.

APERIODICITY: A pattern or process that is not periodic; that is, it does not repeat itself after a fixed time interval or distance. Aperiodic geometric patterns or tilings were proposed in the 1960s and have since been found to be of importance in crystallography and the physics of solids.

APPARENT (OR RELATIVE) MAGNITUDE: The apparent brightness of a star as seen from Earth.

APPLIED GEOGRAPHY: Geographic knowledge applied to human activities such as settlement planning, agriculture, environmental protection, resource extraction, disaster relief, or military operations.

APPLIED MATHEMATICS: All mathematical knowledge used in physical science, economics, social science, and technology. Applied mathematics is often contrasted to pure mathematics, which is pursued without regard to any specific real-world use.

APPLIED SCIENTIFIC FIELD: An area of scientific knowledge that is used for some practical purpose—amusement, war, medicine, transport, communication, or other. Physics, chemistry, mathematics, biology, and other fields all have applied areas. Labeling some knowledge "applied" does not imply that knowledge that is not applied directly is useless: first, it satisfies the human desire to know, and second, the applied knowledge in each field would not exist or make sense without all the knowledge of the field, including that which is not directly applied.

APPRENTICE: A person serving at low wages in order to learn a skill, trade, or craft from an established practitioner. Apprenticeship was widespread in the European economy during the late Middle Ages.

ARCHAEOASTRONOMER: A scientist who studies the astronomical practices of ancient peoples, especially as reflected in the monuments and religious structures built by those peoples. For example, an archaeoastronomer might analyze the Great Pyramids or the megaliths of Stonehenge to discern their astronomical properties and the importance of those properties for their builders.

ARCHAEOASTRONOMY: The study of the astronomical practices of ancient peoples, especially as reflected in the monuments and religious structures built by those peoples.

ARCHETYPE: In the theories of Swiss psychologist Carl Jung (1875–1961), an image or event that is innately possessed by the minds of all human beings and that therefore recurs in stories and dreams as a basic, meaningful pattern across history and cultures. Examples of Jungian archetypes are birth and death, mother and father, the trickster and the wise woman, the bride and the hero.

AROMATIC: In organic chemistry, a compound whose molecular structure includes some variation of the benzene ring.

ARPANET: Early Department of Defense program, initiated by the Defense Advanced Research Projects Agency (DARPA), that led to the development of the Internet.

ARTERY: Any of the large elastic-walled blood vessels that carry blood away from the heart to other parts of the body.

ARTHAŚĀTRA: A prose work describing the principles of statecraft or political economy—how to run a country—that was written in India some time in the first few centuries AD. It is often compared to *The Prince* by Italian diplomat Niccolò Machiavelli (1469–1527), which also recommends ruthless measures for stabilizing centralized political power.

ASTROLABE: An instrument used throughout the Middle East and Europe from classical times through the Renaissance as an aid in observing star positions and calculating longitude and local time. It was used in astronomy, astrology, and navigation, and was a precursor of the slide rule. Astrolabes varied in design; typically, one consisted of a metal disk marked with lines conveying astronomical information and a center-mounted rotating pointer.

ASTROLOGY: The practice of studying the apparent motions of the planets, moon, sun, and stars in order to draw conclusions about human character (supposedly affected by sky patterns at the time of one's birth) or about the future of a person's affairs. Astrology has been practiced since at least about 3000 BC; European, Indian, and Chinese astrological systems have all been developed. Although astrology is still widely popular, it is not a form of science.

ASTRONOMICAL UNIT (AU): A unit of measurement equal to the average distance from Earth to the sun: 93 million miles (150 million km). Distances within the solar system are frequently expressed in AUs.

ASTRONOMY: The study of the physical and chemical properties of objects and matter outside Earth's atmosphere.

ATOM: The smallest particle in which an element can exist.

ATOMIC CLOCK: A device for keeping time based on natural oscillations within atoms.

ATOMIC NUMBER: The number of protons in the nucleus of an atom; the number that appears over the element symbol in the periodic table.

ATOMIC WEIGHT: A quantity indicating atomic mass that tells how much matter there is in something or how dense it is, rather than its weight. Atomic weight is expressed in units known as atomic mass units (amu).

ATTENUATED: A bacterium or virus that has been weakened, often used as the basis of a vaccine against the specific disease caused by the bacterium or virus.

AUGURS: Priests in ancient Rome who believed that they could discern the will of the gods by interpreting the flight patterns of birds.

AURICLE: Also known as pinna or external ear; the flap-like organ on either side of the head. Also, an atrium of a heart.

AUSPICES: In pre-Christian Roman religion, auspices were patterns of bird flight observed by priests called augurs and interpreted as revealing the will of the gods.

AUTAPOMORPHY: In evolutionary biology, an autapomorphy is a feature of a clade (a species or group of species evolved from a common ancestor) that is unique and original to that clade, that is, which was not inherited from earlier species. For example, hair is an autapomorphy of the clade of mammals; articulate speech is an autapomorphy of human beings.

AUTOGRAFT: A type of skin graft that uses tissue from another part of the patient's own body, and therefore has cells with the same genes, reducing the chances of complications associated with tissue rejection.

AUTOIMMUNE DISEASE: A disease in which the body's defense system attacks its own tissues and organs. Autoimmune diseases are conditions in which a person's immune system attacks the body's own cells, causing tissue destruction. Autoimmune diseases are classified as either general, in which the autoimmune reaction takes place simultaneously in a number of tissues, or organ specific, in which the autoimmune reaction targets a single organ. Autoimmunity is accepted as the cause of a wide range of disorders, and it is suspected to be responsible for many more. Among the most common diseases attributed to autoimmune disorders are rheumatoid arthritis, systemic lupus erythematosis (lupus), multiple sclerosis, myasthenia gravis, pernicious anemia, and scleroderma.

AUTOPSY: An examination, usually by dissection, of a body after death to determine the cause of death, extent of injuries, or other factors.

AXIOM: A mathematical statement accepted as true without being proven.

AZOIC THEORY: In the mid-nineteenth century, British naturalist Edward Forbes (1815–1854) proposed that the sea bottom was lifeless (azoic) at depths greater than 18,000 feet (5.5 km); he was mistaken. In geology, azoic rocks are those dating to the period before life existed on Earth.

B

B CELLS: Blood cells found in the vertebrate immune system that are responsible for making antibodies. There are several types of B cell, all with different roles in antibody production.

BACTERIA: Single-celled microorganisms that live in soil, water, plants, and animals. Their activities range from the development of disease to fermentation. They play a key role in the decay of organic matter and the cycling of nutrients. Some bacteria are agents of disease. Bacteria range in shape from spherical to rod-shaped to spiral. Different types of bacteria cause many sexually transmitted diseases, including syphilis, gonorrhea, and chlamydia. Bacteria also cause diseases ranging from typhoid to dysentery to tetanus. Bacterium is the singular form of bacteria.

BACTERIOLYTIC: Killing bacteria, usually by dissolving the cell membrane.

BACTERIOPHAGE: A bacteriophage, or phage, is a virus that infects a bacterial cell, taking over the host cell's genetic material, reproducing itself, and eventually destroying the bacterium. The word phage comes from the Greek word *phagein*, meaning "to eat."

A particular bacteriophage specifically infects one or a limited number of bacterial species.

BACTERIUM (PLURAL: BACTERIA): Any single-celled, microscopic, prokaryotic (nucleus-lacking) animal. Going by either total mass or species count, bacteria are Earth's dominant form of life.

BALMER SERIES: A hydrogen atom, if excited by added energy, releases photons with a variety of particular energies. When displayed as a spectrum, these particular energies show up as spikes or lines, which have been categorized into six groups or series. In 1885, Swiss mathematician Johann Balmer (1825–1898) produced a mathematical formula that predicted the locations of the lines in one particular group; this series was named the Balmer series in his honor.

BAND SPECTRUM: A representation of the mixture of different frequencies of light (or other vibrations) from a given source. All substances are made of atoms, and light is emitted from atoms when their electrons are excited (raised to higher energy levels) by receiving energy. When excited electrons lapse to lower energy levels, they emit light at particular frequencies, giving rise to spectral lines. In some substances, especially molecules, these lines can be so closely spaced as to appear like a continuous band rather than a series of spectral spikes. These spectra are termed band spectra.

BAND THEORY: In quantum physics, the physical theory that describes what bands are available in a substance and how electrons will occupy those bands. A band is a specific energy level that an electron may possess in a given atom or molecule. Band values vary from substance to substance and are often numerous.

BAROMETER: An instrument used to measure atmospheric pressure. A standard mercury barometer has a glass column about 30 inches (72.6 cm) long, closed at one end, with a mercury-filled reservoir. Mercury in the tube adjusts until the weight of the mercury column balances the atmospheric force exerted on the reservoir. High atmospheric pressure forces the mercury higher in the column. Low pressure allows the mercury to drop to a lower level in the column. An aneroid barometer uses a small, flexible metal box called an aneroid cell. The box is tightly sealed after some of the air is removed, so that small changes in external air pressure cause the cell to expand or contract.

BARYON: Subatomic particles that participate in strong force interactions. Baryons are composed of three quarks (or three antiquarks). Protons and neutrons are among the baryons.

BASE PAIR: In DNA molecules, the nitrogenous bases bind together in very restricted and specific ways to form base pairs between DNA strands. In DNA, adenine always bonds with thymine (A-T bond) on the opposite strand and cytosine always bonds with guanine to form a (C-G) base pair between strands of the DNA helix.

BASE PAIRS: In the long, twisted-ladder-shaped molecules used by all living things for passing on information to offspring—DNA and RNA—a base pair corresponds to a single rung in the ladder. Each half of each rung is a cluster of atoms called a base, so each rung is a pair of bases, a base pair. There are four bases in DNA, namely adenine, guanine, cytosine, and thymine. Adenine always bonds with thymine, and guanine with cytosine. The series of base pairs in a DNA molecule specifies a gene. Often a gene acts as a recipe for building a protein or may control the way other genes are used.

BEHAVIORALISM: The school of thought in political science that seeks to describe human group behavior mathematically.

BENTHOS: The plants and animals dwelling at the bottom of a body of water.

BENZENE: A ring-shaped hydrocarbon molecule that has long been used as an industrial solvent, that is, a liquid in which other substances can be dissolved. However, benzene is now known to be a potent cause of cancer and is not widely used.

BERNOULLI'S LAW: Named after Swiss mathematician David Bernoulli (1700–1782), a mathematical description of the fact that any increase in the velocity of a moving liquid decreases the pressure of that liquid.

BESTIARY: In medieval Europe, books devoted to describing a variety of animals. These precursors of later works in scientific natural history often drew moral lessons from the supposed behaviors of animals, such as the foresight exercised by groups of sleeping cranes in always posting a sentry.

BETA DECAY: Process by which a neutron in an atomic nucleus breaks apart into a proton and an electron.

BETA PARTICLE: An electron emitted by an atomic nucleus.

BIFURCATION: Splitting in two, especially of a physical structure such as a limb or blood vessel. The letter Y, for example, shows a bifurcation halfway up the stem.

BIG BANG: The beginning of the universe's expansion from a state of infinite or near infinite density and temperature to its present cool, dispersed state. It occurred about 13.7 billion years ago. It was not an explosion in empty space, like a bomb going off in a void, but is better characterized as an explosion of space itself.

BINARY: The binary number system uses only two digits, 0 and 1, and is basic to computer science. In astronomy, a binary star is a pair of stars orbiting each other and is sometimes called "a binary."

BINOMIAL NOMENCLATURE: System of naming plants and animals in which each species is given a two-part name, the first being the genus and the second being the species.

BIOCHEMICAL: A term used to define the chemical activity in living organisms.

BIODIVERSITY: Literally, "life diversity"; the number of different kinds of living things. The wide range of organisms—plants and animals—that exist within any given geographical region.

BIOENGINEERING: The term bioengineering is often used to denote the creation of artificial organs, organ parts, or tissue replacements that can be installed in the body to repair disease or injury. For example, the installation of a mitral heart valve is bioengineering.

BIOGEOGRAPHY: The study of the distribution and dispersal of plants and animals throughout the world.

BIOME: Well-defined terrestrial environment (e.g., desert, tundra, or tropical forest). The complex of living organisms found in an ecological region.

BIOMETRIC: The statistical analysis of biological data. In technology, biometric systems measure the distinctive properties of individual creatures, especially people, in order to identify them uniquely. Fingerprinting, retina or iris scanning, and computerized face recognition are biometric technologies.

BIOPIRACY: The exploitation of knowledge about biological systems—crop varieties, herbal medicines, or the like—originally created by an indigenous people. Biopirates profit from traditional biological know-how without compensating its creators.

BIOREMEDIATION: The use of living organisms to help repair damage such as that caused by oil spills.

BIOTECHNOLOGY: Any technique that uses parts of living organisms to create or modify products, plants, animals, or microorganisms for specific uses.

BIOTIC: Biotic refers to living organisms; for example, a material obtained from living organisms is a biotic material.

BIOTIC RESISTANCE HYPOTHESIS: The view that a native biota (system of plants and animals) inhabiting a region resists invasion by foreign species because the invaders are hindered by strong interactions with native species, including natural enemies (e.g., native plant-eaters). This theory is often contrasted to the natural-enemies hypothesis, which states that invasive species spread more readily because they do not encounter natural enemies.

BLACK BILE: The ancient Greek and Roman theory of medicine stated that four basic substances, the humors, exist in the human body, and that disease or exaggerated personality is caused by an imbalance among the humors. The four humors were blood, phlegm, yellow bile, and black bile; an excess of black bile was believed to cause melancholy and depression.

BLACK HOLE: A single point of infinitely small space containing the mass and gravity of a collapsed massive star. The gravity is so strong that light cannot escape.

BLACKBODY: An ideal emitter that radiates energy at the maximum possible rate per unit area at each wavelength for any given temperature. A blackbody also absorbs all the radiant energy incident on it; i.e., no energy is reflected or transmitted.

BLOODLETTING: Bloodletting, also known as phlebotomy, is the deliberate medical release of blood from the body, whether by allowing leeches to feed on the patient or by cutting a blood vessel. This technique was practiced by doctors from Greco-Roman times to the nineteenth century but was rarely of benefit and often harmful. However, there are certain rare medical conditions, such as hemochromatosis, in which the patient has excessive red blood cells and must undergo regular therapeutic bloodletting.

BOHR MODEL: Neils Bohr (1885–1962), a Danish physicist, proposed the Bohr model of the atom in 1913, at a time when physicists were still struggling to understand atomic structure. In the Bohr model of the atom, negatively charged electrons are bound to a positively charged (and much more massive) nucleus by electrostatic attraction, but prevented from falling into the nucleus by the fact that they can only exist in certain permitted energy states as described by quantum mechanics.

BOOLEAN ALGEBRA: A system that applies algebra to logic. Also called symbolic logic, it converts logic into mathematical symbols.

BOSON: According to the modern Standard Model of particle physics, all particles are either bosons or fermions. All forces are explained as exchanges of virtual bosons between fermions. Bosons have integer spin values; photons are bosons (spin = 1). The only particle predicted by the Standard Model that had not been experimentally observed as of early 2008 was the Higgs boson.

BOTANY: The branch of biology involving the study of plant life.

BOUNDED RATIONALITY: Economist and scientist Herbert Simon (1916–2001) won a Nobel Prize in Economics for his theory of bounded rationality, which challenged economists' traditional assumption that consumers and producers always make perfectly rational choices to maximize utility and profits. Instead, Simon said, they make choices whose rationality is limited by access to computational resources—that is, by lack of information about alternatives, time in which to ponder them, or education that would enable them to grasp a situation's complexities, for example.

BOVINE SPONGIFORM ENCEPHALOPATHY: A brain disease in cows caused by malformed proteins. Bovine means cow; spongiform refers to the spongy texture of the brains of cows that have the disease in advanced form; and encephalopathy means brain-disease. Eating tissue, especially nerve tissue, from cows infected by prions causes variant Creutzfeldt-Jakob disease in humans. Both bovine spongiform encephalopathy and Creutzfeldt-Jakob disease cause mental breakdown and eventual death.

BREWING: The practice of encouraging the growth of yeast—microscopic fungi—in mixtures of water and plant material to turn plant sugars into ethyl alcohol and carbon dioxide. After filtering and aging, the resulting mixture becomes an alcoholic beverage. If carbon dioxide is retained, it becomes bubbles, as in beer and champagne; if it is allowed to escape, non-carbonated alcoholic beverages such as wine are created.

C

C: In physics, for over a century, the italicized Roman letter c, sometimes written c_0, has stood for the speed of light in a vacuum, namely 670,616,629 miles per hour (1,079,252,848 km/h) or, as more commonly expressed in science, 2.99×10^7 m/s. This speed is constant, that is, no matter how fast a source and observer are moving with respect to each other, the observer will always measure the speed of light as *c*. In media such as air or glass, light travels more slowly, but still has the property of being constant for all observers.

CABBALISTIC: The Cabbalah (or Kabbalah) is an ancient Jewish tradition of biblical interpretation, still practiced by some Hasidic Jews. Any symbol or phrase having, or supposed to have, a secret, mystical sense is often described as cabbalistic.

CALCINATION: An old term used to describe the process of heating metals and other materials in air.

CALCULUS: The branch of mathematics that uses algebra to calculate changing quantities and motion.

CALORIC THEORY OF HEAT: In the 1780s, French scientist Antoine Lavoisier (1743–1794) proposed that heat is a substance resembling a fluid that can flow from one object to another. This fluid he named caloric. In the mid-1800s and beyond, this theory was replaced by the modern understanding of heat, which is that heat is the motion of large numbers of particles in a substance (atoms or molecules).

CALX: The powdery residue, generally of metal (metallic oxide), formed from the ore and oxygen from the air when a metal-bearing rock or mineral is heated to high temperature.

CAMBRIAN EXPLOSION: Relatively sudden evolution of a wide variety of multicellular forms of life at the beginning of the Cambrian period, about 530 million years ago, after billions of years during which Earth was inhabited almost entirely by single-celled organisms. A few forms of multicellular life did appear shortly before the Cambrian Explosion. The explosion may have been triggered by the ending of a major snowball Earth period or by the achievement of sufficiently high oxygen levels thanks to billions of years of oxygen production by algae (single-celled aquatic plants).

CAMERA OBSCURA: A dark room or box with a light-admitting hole that projects an image of the scene outside.

CAPACITY: In physics and electronics, capacity is a synonym for capacitance, the ability of a device to store electric charge.

CARAPACE: A protective covering or shield on an animal's back, e.g., the upper part of a turtle's shell.

CARBOHYDRATE: An organic compound present in the cells of all living organisms and a major organic nutrient for human beings; consists of carbon, hydrogen, and oxygen, and makes up sugar, starch, and cellulose.

CARCINOGENIC: The property of causing cancer. Common carcinogenic chemicals are asbestos, tobacco smoke, and benzene. Some viruses can also be carcinogenic.

CARD CATALOG: A device used for storing and retrieving information about the contents of a library. Each item in the library's collection—book, audio recording, or other—was identified on a small paperboard card. These cards were ordered alphabetically and stored in small drawers that could be browsed by a user looking for a particular item. The card recorded the physical location of the item, which the user then went and obtained. Since the 1980s, computer terminals have almost completely replaced card catalogs.

CARNOT CYCLE: A Carnot cycle—named after the French scientist who first identified such a cycle, Nicolas Carnot (1796–1832)—is a series of temperature-and-pressure states through which a system can be changed, returning eventually to its starting state. In moving through a Carnot cycle, a system does work on its environment, such as turning a shaft or moving a piston. All engines that turn heat into useful mechanical work are based on thermodynamic cycles like the Carnot cycle, but the Carnot cycle is the most efficient possible cycle and can only be approximated in a real machine.

CARTOGRAPHY: The science of mapmaking

CATABOLISM: The process by which large molecules are broken down into smaller ones with the release of energy.

CATALYST: Substance that speeds up a chemical process without actually changing the products of reaction.

CATASTROPHISM: School of thought in geology that holds that the rates of processes shaping Earth have varied greatly in the past, occasionally acting with violent suddenness (catastrophes). In nineteenth-century geology, two main approaches to explaining how Earth achieved its present geographic state, including its fossil record, seas, mountains, valleys, and the like, were debated by scientists. Some favored gradualism, the theory that slowly operating causes still visible in the environment, such as erosion, were completely responsible for Earth's geological history. Others favored catastrophism, the theory that sudden events (catastrophes) had caused geological upheavals and mass extinctions in the past. Modern science admits the reality of both gradual and catastrophic causes.

CATHARTIC: Emotional relief through the expression or experience of strong emotions has, since the early 1600s, been termed cathartic (from the Greek *katharsis*, for "cleansing"). The term is especially used to refer to a sensation of emotional cleansing achieved through the contemplation of drama and other arts.

CAUSAL: A system of physical objects and forces in which the state of the system at any given time defines the state of the system at all future times. Machines are, ideally, causal systems: Each operation of the machine forces the next state of the machine to come into being, and only one next state is possible. Physicists are divided over whether the universe as a whole is a causal system. Causality—cause-and-effect—definitely operates at larger physical scales, but may not describe the action of atomic-scale events, which can in turn influence larger-scale (macro) events, as when a subatomic particle causes a mutation that influences the course of evolution. Whether subatomic events are truly random or truly causal is debated in quantum physics.

CAUTERY: The sealing off of a broken blood vessel during surgery by applying heat. The purpose is to minimize bleeding. Electrocautery—the use of an electrical tool for cautery—is common in modern surgery.

CELESTIAL MECHANICS: The study of the motions of celestial bodies such as planets, moons, and stars, moving in gravitational fields of force.

CELL CYCLE: Complete sequence of steps that must be performed by a cell in order to replicate itself, as seen from mitotic event to mitotic event. Most of the cycle consists of a growth period in which the cell takes on mass and replicates its DNA. Arrest of the cell cycle is an important feature in the reproduction of many organisms, including humans.

CELL MEMBRANE: The outer membrane of a cell, which separates it from the environment. Also called a plasma membrane or plasmalemma.

CELL WALL: A tough outer covering that overlies the plasma membrane of bacteria and plant cells.

CELLULAR RESPIRATION: The oxidation of food within cells, involving consumption of oxygen and production of carbon dioxide, water, and chemical energy in the form of ATP (adenosine triphosphate).

CEPHEIDS: Variable stars that can be used as standard candles.

CEREBELLUM: Latin for "little brain," the cerebellum is responsible for coordinating movement and balance. It occupies a place partially tucked under the forebrain's cerebral hemispheres. The cerebellum is extensively folded into parallel folds, or folia, giving it an appearance of having irregular pleats. The cerebellum possesses right and left hemispheres that are connected with the spinal cord and forebrain.

CEREBRAL: Of, or having to do with, the brain.

CEREBRAL CORTEX: The external gray matter surrounding the brain, made up of layers of nerve cells and fibers; it is thought to process sensory information and impulses.

CEREBRUM: The folded mass of nerve cells that sits over the rest of the brain and coordinates voluntary movements. It is also essential for memory, learning, and sensing things.

CETACEANS: Cetacea or cetaceans are an order of sea-dwelling mammals that includes whales, dolphins, and porpoises. They are hairless, breathe through a blowhole, spend their whole lives in the water, and are among the most intelligent animals.

CHAIN REACTION: A reaction in which a substance needed to initiate a reaction is also produced as the result of that reaction. With specific regard to a nuclear fission reaction, neutrons released during the reaction cause an ongoing series of fission reactions.

CHANDRASEKHAR LIMIT: In the early 1930s, Indian-American physicist Subrahmanyan Chandrasekhar (1910–1995) calculated the upper limit on star mass that can be supported by electron degeneracy pressure (the quantum-mechanical prohibition on energy-level sharing by electrons) against further gravitational collapse. A star remnant heavier than about 1.44 times the mass of our sun is beyond the Chandrasekhar limit and must collapse to form either a neutron star or black hole. Chandrasekhar was awarded a Nobel Prize in physics in 1983.

CHARGE: Describes an object's ability to repel or attract other objects. Protons have positive charges, while electrons have negative charges. Like charges repel each other while opposite charges, such as protons and electrons, attract one another.

CHEMOTHERAPY: The treatment of a disease or condition with chemicals that have a specific effect on its cause, such as a microorganism or cancer cell.

CHI: Chi or qi (pronounced "chee") is the life force supposed by traditional Chinese medicine and philosophy. There is no scientific evidence for the existence of chi.

CHIP LOG: A device (now obsolete) for measuring the speed of a surface vessel relative to the water. It consists of a small wooden board or chip that is attached to a line. The chip is cast overboard, where it drags in the water and is left behind by the motion of the ship. A line runs out for a fixed period of time, measured using an hourglass or other timepiece, giving a distance and time measurement that can be used to compute a velocity.

CHIRALITY: Also termed handedness, the quality of having non-superimposable mirror images.

CHLOROFLUOROCARBONS (CFCs): Chemical compounds containing chlorine, fluorine, carbon, and oxygen. They are widely used in refrigeration and air-conditioning systems and are destructive of the ozone layer in Earth's stratosphere.

CHROMATIC ABERRATION: A defect in the lens of refracting telescopes that results in a false color or blurring around the image. The problem arises because light of different wavelengths is focused at different distances from the lens.

CHRONOMETER: Any device that measures the passage of time. The term is especially used for accurate timepieces built for special purposes, such as navigation.

CHYMICAL: An obsolete spelling of "chemical," or having to do with chemistry.

CIRCUMNAVIGATE: Traveling completely around an object. It is possible to circumnavigate an island, a continent, or the entire world. Earth was first circumnavigated in the early 1500s by Portuguese sailors.

CITATION: In scholarly or scientific publication, a reference to or quotation of another work. The correct attribution of citations—explaining exactly where quotations have come from—is an essential aspect of modern scientific communication because it enables interested readers to trace ideas to their sources.

CLADISTICS: A method of determining evolutionary relationships based on evolutionary divergence; it is now used in many evolutionary disciplines.

CLADOGRAM: A diagram, resulting from a cladistic analysis, that depicts a hypothetical branching sequence of lineages leading to the taxa under consideration. The points of branching within a cladogram are called nodes. All taxa occur at the endpoints of the cladogram.

CLASSIFICATION SYSTEM: Any method of assigning a collection of objects, terms, creatures, or other things to a set of named categories or groups. In science, the best-known classification system is used in biology, which assigns every living creature to a unique place in a hierarchy of categories: variety, species, genus, family, order, class, phylum, kingdom, domain.

CLEAN WATER ACT: Any law that seeks to control water pollution may be referred to as a "clean water act." In the United States, the Federal Water Pollution Control Act of 1948 and its later amendments (1965, 1972, 1977, 1987) are the major body clean-water acts.

CLEAVAGE: The splitting of a rock or crystal along a natural smooth plane; that is, the stone cleaves along the plane. In biology, the term refers to the splitting of a cell during reproduction.

CLIMAX COMMUNITY: A relatively stable ecosystem characterized by large, old trees, marking the last stages of ecological succession.

CLONAL SELECTION THEORY: The theory that explains the basic workings of the adaptive immune system, which is the network of blood cells in vertebrates that fights infection and can produce immunity to some pathogens (disease-causing organisms). According to the theory, numerous cells called B cells circulate in the blood. Some B cells, called memory B cells, produce antibodies for specific antigens—that is, they produce Y-shaped molecules (antibodies) that attach to identifying molecules (antigens) on the surfaces of pathogens. Each individual B cell can produce only one kind of antibody. When an antigen for which some B cell already makes the right antibody appears in the body, many genetically identical copies (clones) of that B cell are made. This greatly increases the number of that type of antibodies being manufactured and, usually, defeats the infection before it can become harmful.

CLONING: A technique of genetic engineering in which an offspring is produced asexually (without joining egg and sperm) and has the exact same genes as its donor organism.

CO_2 LASER: Lasers are devices that produce beams of coherent light, that is, light whose photons are all aligned and of the same frequency and travel in the same direction. Laser light can be produced from a variety of substances, one of which is carbon dioxide (CO_2). High power can be achieved affordably with a CO_2 laser, which emits ultraviolet light, so this type of laser is often used for high-power industrial applications such as welding. It can also be used in medicine.

COALSACK NEBULA: The Coalsack Nebula is a cloud of dark dust about 600 light years (3.5×10^{15} miles or 5.7×10^{15} km) from Earth and visible to the naked eye as a blot on the Milky Way (the part of our own galaxy that is visible in Earth's sky).

COASTAL: Of or having to do with a coast, that is, an edge of a landmass bordering an ocean.

CODON: A triplet of three letters representing bases of DNA. Also known as a triad; a group of three nucleotides in a DNA (deoxyribonucleic acid) or RNA (ribonucleic acid) molecule that during translation indirectly codes for the production of a single specific amino acid in a protein chain. More specifically, a section of DNA or RNA that codes for a single amino acid or for the termination of translation.

COHERENCE: In wave physics, coherence occurs when two or more waves—for example, light waves—traveling together have the same phase, that is, line up peak to peak and trough to trough.

COHORT: A group of people (or any species) sharing a common characteristic. Cohorts are identified and grouped in cohort studies to determine the frequency of diseases or disease outcomes over time.

COLLIDER (PARTICLE): A particle collider is a machine that accelerates (speeds up) atomic nuclei or subatomic particles such as protons to velocities approaching the speed of light. Particles are made to travel in either a straight line (in a linear collider) or a circle (in a ring collider) and eventually to collide with other particles, whether in a stationary target or a beam of particles traveling in the opposite direction. The goal is to produce high-energy collisions that spew out new particles whose interactions with surrounding detectors can be recorded. These particle showers reveal information about the physics of fundamental particles.

COMET: A small body composed of ice and dust that passes through the solar system, trailing a glowing plume or "tail" of gas as it nears the sun. Some comets follow a parabolic or hyperbolic path around the sun and disappear into deep space. Others follow highly elongated elliptical orbits and return at regular intervals. By convention, a short-period comet is one that returns at intervals of 200 years or less, and a long-period comet is one that returns at intervals greater than 200 years.

COMMENSALISM: In biology, a type of relationship between organisms that is intermediate between parasitism (or antibiosis) and symbiosis. In parasitism, one organism benefits and the other is harmed; in commensalism, one benefits and the other is neither harmed nor helped; in symbiosis, both organisms are helped.

COMMON RULE: A set of ethical guidelines for medical research in the United States that has been adopted by 18 federal agencies. All universities and hospitals hoping to receive U.S. government funding must adhere to the Common Rule guidelines, and most research adheres to them even if it is not federally funded. The Common Rule forbids certain kinds of human research altogether and requires meaningful, informed consent from subjects (with full freedom to decline participation) for those kinds of research that are permitted.

COMMUNITY: A collection of populations that interact with each other in the same geographic region.

COMPARATIVE ANATOMY: The study of the similarities and differences of the structure of living things.

COMPASS: A device for detecting the presence and direction of a magnetic field.

COMPETENCE: Competence, in medicine, refers to the ability of a structure to hold fluid without leaking. For example, an un-punctured stomach is competent. Restoring competence is of a goal of reconstructive surgery in the event of severe injuries.

COMPOUND MICROSCOPE: A multiple-lens microscope (two or more lenses housed in a long tube).

COMPOUND NUCLEUS REACTION: When an atomic nucleus is struck by a fast-moving particle, a nuclear reaction takes place. There are two types of reaction, namely, direct nucleus reaction and compound nucleus reaction. In a compound nucleus reaction, the particle is absorbed by the target nucleus and a new nucleus is formed; this new nucleus then proceeds to undergo radioactive decay (spontaneous breakdown).

CONCOMITANT: When two or more drugs are given together, as is the case in treating cancer or AIDS, the drugs are termed concomitant medications.

CONDENSATION: The process by which vapor molecules reform a liquid; a phase change from the gas state to the liquid state.

CONGENITAL MALFORMATIONS: Congenital features are those that people or other animals are born with; a congenital malformation or disorder is one that is present at birth.

CONJUGATE: In mathematics, the conjugate of a complex number is another number that has the same real part and a negative part of equal magnitude but opposite sign. The word "conjugate" has other technical uses in physics, chemistry, biology, and mathematics.

CONSERVATION BIOLOGY: The branch of biology that involves conserving rapidly vanishing wild animals, plants, and places.

CONTAGION: Transmission of an infectious disease from one person to another.

CONTINUOUS WAVE (CW): In physics, a continuous wave is one that oscillates regularly and continuously, without beginning, change, or ending. Absolutely continuous waves do not exist but are approximated, for engineering and science purposes, by waves that are unchanging for a relatively long period of time.

CONUNDRUM: Any difficult question, especially one expressed as a verbal puzzle.

CONVERGENT BOUNDARY: The boundary where two or more tectonic plates (large shell-like areas of rock floating on Earth's mantle, sometimes continental in size) drift against each other, edge to edge.

CONVERGENT EVOLUTION: The evolution of a similar feature in separate species is termed convergent evolution. For example, the shapes of sharks and dolphins evolved convergently; no shared or common ancestor of these two families of species had such a shape.

COPENHAGEN INTERPRETATION: A way of viewing the equations of quantum mechanics. Its earliest proponents were Danish physicist Neils Bohr (1885–1962) and German physicist Werner Heisenberg (1901–1976), who worked together on quantum mechanics in Copenhagen, Denmark, in the late 1920s. According to the Copenhagen interpretation, the wave function that describes every particle's probability of being found in any given place is collapsed by measurement—that is, it is reduced from an infinite number of possibilities to a single, definite state. Other interpretations of quantum mechanics say that the wave function does not collapse, but continues to hold all its values in a state of superposition (all-at-onceness or layered being). In this theory, an infinite number of co-existing universes are being continually generated by all the particle interactions in the universe. The majority of modern physicists today affirm the Copenhagen interpretation, but a growing minority affirms the many-worlds interpretation.

CORPUSCULARIANISM: A physical theory propounded by some medieval alchemists who said that all objects are made of invisibly small units or corpuscles similar to atoms. English physicist Isaac Newton (1643–1727) employed a version of corpuscularianism when he developed his theory of light as a stream of particles.

CORTISOL: A hormone produced in the cortex of the adrenal gland. Its artificial form, hydrocortisone, is used to relieve inflammation from allergic reactions, such as exposure to poison ivy, and the like.

COSMIC MICROWAVE BACKGROUND (CMB): A bath of low-intensity radio waves that shines from all parts of the sky. It is the afterglow of the big bang, the beginning of time and the universe in a state of extremely high density about 13.7 billion years ago. The CMB was predicted by big bang theory before being detected. Slight variations in its intensity from one part of the sky to another were later predicted by the inflation version of big bang theory and have been confirmed in recent years by satellite observations. The CMB is one of the most important pieces of evidence for the big bang.

COSMIC RAY: The fast-moving particles (protons, electrons, and helium nuclei) coming from outer space and moving near the speed of light with which Earth is constantly struck. Most cosmic rays originate near extremely heavy objects, such as black holes and neutron stars, that are accelerating nearby matter. Most of the matter accelerated near these objects gets sucked in, but some gets slung out: Some of these particles become cosmic rays.

COSMOGONY: The study of the origin of the solar system or (usually) of the universe as a whole. In modern science, cosmology (the study of the properties universe as a whole) includes cosmogony.

COSMOLOGY: The study of the universe as a whole, its nature, and the relations between its various parts.

COSMOS: All that exists, considered as a single, structured thing, from the Greek word meaning both "order" and "ornament." In modern usage, the word is a synonym for universe.

COULOMB FORCE: In the 1780s, French physicist Charles de Coulomb (1736–1806) announced a simple mathematical law describing the force of attraction or repulsion between two electrical charges: charges of opposite sign (positive, negative) exert a pull on each other, while charges of same sign exert a push against each other. This pull or push is termed the Coulomb force.

COVALENT: In chemistry, a covalent bond between two atoms is produced when the atoms share one or more electrons. The other major type of chemical bond is ionic, in which atoms lacking electrons or endowed with extra electrons (and thus having positive or negative charges, respectively) are attracted to each other.

CREATION SCIENCE: A form of creationism, the belief that life, or at least some life or features of life, was created miraculously rather than by natural evolutionary processes. Creation science was developed after the U.S. Supreme Court ruled in *Epperson v. Arkansas* (1968) that all U.S. state laws banning the teaching of evolutionary biology were unconstitutional. In response, creationists developed creation science, a pseudoscientific set of claims asserting that the Genesis story of creation, taken literally, is supported by scientific evidence (it is not). In 1987, the Supreme Court ruled in *Edwards v. Aguillard* that creation science is religious doctrine, not science, and cannot be taught in U.S. public schools because the First Amendment to the U.S. Constitution forbids the furtherance by government of any religion.

CRETINISM: An obsolete term for the symptoms of severe thyroid deficiency, including mental retardation.

CRITICAL MASS: The minimum amount of fissionable uranium or plutonium that is necessary to maintain a chain reaction.

CROSS SECTION: A slice, or slice-like view, that reveals the interior of an object. Cutting an apple in half, for example, reveals a cross section of the apple to view. In medicine, an interior view using "slices" can be constructed by an imaging instrument.

CROSS STAFF: A cross-shaped instrument for determining the angular height of the sun or other heavenly body above the horizon. The user places one eye at one end of the long stick and aligns it with the horizon, then shifts a shorter cross-piece along the long stick until its upper end is aligned with the object being observed. The angular height of the object can then be read off from markings on the long stick.

CRUSADES: A series of military expeditions undertaken by European nations, mostly against the area centered on Jerusalem, from about AD 1100 to about 1300. Religious motives were usually proclaimed for Crusades, hence their name (from "crucifix" or cross).

CRYSTAL: Naturally occurring solid composed of atoms or molecules arranged in an orderly pattern that repeats at regular intervals.

CRYSTALLINE: Having a regular arrangement of atoms or molecules; the normal state of solid matter.

CULTURAL REVOLUTION: A period of government-mandated social upheaval in China from 1966 to 1976. Its proclaimed purpose was to solidify the Communist Revolution by purging it of old habits, old culture, old ideas, and old customs. Religion was persecuted; science and education were severely set back; and several million deaths were caused, both by direct killings and disruption of agriculture and the economy.

CUNEIFORM: One of the earliest forms of writing. It was developed by the Sumerian civilization in the area now known as southern Iraq in about 3000 BC. Cuneiform was inscribed in clay bars or pads called tablets by pressing a sharpened reed into wet clay. The clay later hardened, making a permanent document (tens of thousands still exist). The reed's sharpened end made a series of wedge-shaped dents in the wet clay to denote characters of the script, hence the term "cuneiform" (from the Latin *cuneus*, wedge).

CYCLONE: An area of low pressure in which winds blow counterclockwise in the Northern Hemisphere and clockwise in the Southern Hemisphere.

CYTOKINES: A class of substances secreted by certain cells in order to signal to other cells. In the vertebrate immune systems, one of the functions of cytokines is to attract white blood cells to infected tissue in order to combat invading pathogens (e.g., viruses or bacteria).

CYTOPLASM: The semifluid substance inside a cell that surrounds the nucleus and other membrane-enclosed organelles.

CYTOTOXICITY: The toxicity (poisonousness) of a substance to living cells. Some medications, such as those used in chemotherapy for cancer, can only be given in certain maximum doses because of their cytotoxicity.

D

DARCY'S LAW: A mathematical statement of how liquids flow through sand and other porous (hole-filled) substances. It is named after French scientist Henry Darcy (1803–1858).

DARK ENERGY: In 1998, astronomers discovered that the universe is not only expanding, but expanding faster all the time (accelerating). Since energy is required to accelerate an object, it followed that some form of energy must be driving the accelerated expansion. Scientists do not know the nature of this energy and have not, as of 2008, been able to observe any source for it, and so have dubbed it "dark energy" until a better understanding is obtained. Dark energy makes up about 74% of the material of the universe.

DARK MATTER: Unseen matter that has a gravitational effect on the motions of galaxies within clusters of galaxies.

DATUM (PLURAL: DATA): A single number resulting from a measurement is a datum; the plural of datum is data.

DEAD RECKONING: A navigational method of estimating one's location after a period of movement, based on earlier knowledge of one's position and knowledge of the direction and rate of travel. By adding a change vector to a known position, a new position can be calculated. Highly precise dead reckoning is used in inertial-guidance systems, which are initialized with precise location information and then measure all changes in velocity that they undergo by measuring the forces accelerating them. Inertial guidance systems are used in many guided missiles, aircraft, and spacecraft.

DECAY RATE: The rate of the radioactive decay process for a given radioactive substance. Atoms of any particular radioactive type tend, on average, to undergo spontaneous fission (breaking up of their nuclei) at a certain rate, as measured in fissioning atoms per unit of time.

DEDUCTION: A form of logical reasoning. In deduction, one reasons from general laws or principles to particular facts. For example, knowing that a stone has been dropped at a certain time and place and knowing the law of gravitation, one can deduce when the stone will strike the ground. The complementary or reverse form of reasoning is induction, which reasons from particular observations to general principles. For example, after making measurements of how dropped stones accelerate, one might induce the law of gravitation.

DEEP TIME: Until the 1700s, few thinkers in the Western world considered the possibility that Earth (or the universe as a whole) might be extremely old relative to human history—say, many million or perhaps many billions of years old, rather than 10,000 or so years old. In the late 1700s and early 1800s, however, geologists found convincing evidence that Earth was extremely old. The phrase "deep time" became popular in the late twentieth century to describe the vast extent of past time.

DEFERENT: In the Ptolemaic system of astronomy, the circular path that the center of an epicycle follows around Earth. In this system, used in Europe from the second century AD until the Copernican revolution in the 1500s, the motions of the planets relative to those of the fixed stars were explained by compounded circular motion: that is, the planets were said to move in circles around points that moved in circles around Earth. A deferent was a circle traced around a point in space near Earth called the equant; around the point tracing the circle was another circle, the epicycle, along which the planet moved. The combination of these motions imparted a complex motion to planets that accounted for their movements in the sky.

DEISM: A form of monotheism (belief in one God) that rejects the idea that God might be directly involved in the workings of the universe. According to deism, God created the universe but has since then allowed it to unfold entirely according to its own internal laws.

DELAYED TYPE HYPERSENSITIVITY (DTH): A form of tissue damage that can result from the action of the body's immune system. An aspect of the immune system termed cell-mediated immunity increases the tendency of certain white blood cells (T lymphocytes) to attack pathogens. When these white blood cells attack cells in the body, often skin cells, the result is DTH. DTH can be triggered by viruses, fungi, bacteria, medications, or chemicals such as nickel, with results that vary from person to person.

DELTA: Triangular-shaped area where a river flows into an ocean or lake, depositing sand, mud, and other sediment it has carried along its flow.

DEMIURGE: In some ancient theological systems, a secondary supernatural being (inferior to God) who is responsible for creating the universe.

DEOXYRIBONUCLEIC ACID (DNA): A double-stranded molecule joined together by bonds between base pairs. The strands are akin to the sides of a ladder and the base pairs the steps of the ladder; the molecule as a whole is helical (shaped as a gently twisted ladder). DNA is the carrier of genetic information (encoded in the specific series of base pairs) for almost all organisms, though some viruses use ribonucleic acid (RNA).

DEPLETION REGION: In integrated electronic circuitry, devices that control the flow of current in response to a control voltage (transistors), much as a valve controls the flow of water through a faucet, are made by mixing contaminant substances into parts of a solid crystal of silicon or other semi-conducting materials. These contaminants either donate extra electrons to the crystal matrix (add a negative charge) or are short of electrons (add a positive charge). Where a portion of the crystal with extra electrons touches a portion lacking electrons, some of the extra electrons are pulled over from the negatively-charged area to the positively-charged area; this continues until electrical forces are balanced. When a balanced state is achieved, an in-between layer exists that is neither positively nor negatively charged and so lacks carriers for current. This area, depleted of charge carriers, is termed a depletion region.

DEPRESSION: In psychology and medicine, a mental disorder characterized by hopelessness, unhappiness, and kindred feelings that are more severe than circumstances would warrant (e.g., a person who is sad because a loved one has died is not suffering from depression, whereas a healthy, functional person who is so unhappy that they are considering suicide may be suffering from depression). In economics, a prolonged period of reduced economic activity, worse than a recession.

DEPTH OF FIELD: In photography, depth of field refers to the range of distance from the lens in which objects are in focus. A shallow depth of field means that only objects at or near a specific distance are in focus; a deep depth of field means that objects over a wide range of distances are in focus.

DETERMINATION: In biology, determination of sex is usually by genetic inheritance: in human beings, for example, males inherit one X chromosome and one Y chromosome, while females inherit two X chromosomes. (There are a number of exceptions to this rule, but this is usually the case.)

DETERMINISM: The notion that a known effect can be attributed with certainty to a known cause.

DETERMINISTIC SYSTEM: A deterministic or causal system is a system of physical objects in which the state of the system at any given time defines or determines the state of the system at all future times. Machines are, ideally, deterministic systems: Each operation of the machine forces the next state of the machine to come into being, and only one next state is possible. Physicists are divided over whether the universe as a whole is deterministic; large-scale events are deterministic, a quality we rely on in designing machines, but whether subatomic events are truly random or truly deterministic is debated in quantum physics.

DICHOTOMY: Any division into two clearly-separate categories. For example, to classify all cells as eukaryotic or prokaryotic is to make a dichotomy.

DIFFERENTIATE: With regard to cells, a change into a more specialized cell type.

DIFFERENTIATION: In developmental science, the process by which cells mature into specialized cell types, such as blood cells, muscle cells, brain cells, and sex cells.

DIFFRACTION GRATING: A device consisting of a surface into which are etched very fine, closely spaced grooves that cause different wavelengths of light to reflect or refract (bend) by different amounts.

DIGITAL: The opposite of analog; it is a way of showing the quantity of something directly as digits or numbers.

DIODE: An electronic device that permits current to flow through itself in one direction but not in the other.

DIPLOID: Having two sets of chromosomes.

DISSECTION: Cutting and separating the body along its natural cleavage lines to allow scientific examination.

DISTILLATION: The process of separating liquids from solids or from other liquids with different boiling points by a method of evaporation and condensation, so that each component in a mixture can be collected separately in its pure form.

DIVERGENT BOUNDARY: Where two or more tectonic plates (large shell-like areas of rock floating on Earth's mantle, sometimes continental in size) are forced apart by the creation of new crust from below (as along the Atlantic midocean ridge) or pulled apart by plate motion (as in the Great Rift Valley of Africa), the boundary between the two plates is termed a divergent boundary.

DIVINATION: Any attempt to foresee the future using magical or supernatural means. Many ancient religions practiced forms of divination, and some religions still do. Attempts to read future events in random or pseudorandom patterns such as the internal organs of sacrificed animals, the positions of tea-leaves, and so forth are all forms of divination.

DOMINANCE: In genetics, dominance is a quality of certain genes. Every cell possesses two versions of every gene, such as the gene for eye color. If the genes are alike (say, both are for blue eyes), then the character coded for by the gene will definitely appear in the organism. If the genes are unlike (say, one copy of the eye-color gene is for brown eyes and the other is for blue eyes), then the outcome will be determined by whether one of the genes is dominant. If one of the genes is dominant, it will dominate or force the outcome regardless of what the other gene is. In the case of eye color, the gene for brown eyes is dominant: a child will have brown eyes even if only one of its eye-color genes is a brown-eyes gene. To have blue eyes, a child must have two copies of the gene for blue eyes. The gene for blue eyes is an example of a recessive gene, as opposed to a dominant gene.

DOPANT: An impurity added to a semiconducting material.

DOPING: The act of adding impurities (dopants) to change semiconductor properties. In athletics, the illegal use of steroids, drugs, hormones, or techniques such as blood doping (injecting previously drawn red blood cells to increase oxygen) in order to enhance athletic performance.

DYNAMICAL SYSTEM: In mathematics, any equation or set of equations that describes the changing of a system over time—in particular, the location of a certain point in the system, such as the tip of a pendulum or a particle suspended in a swirling fluid.

E

ECLIPTIC: The plane in which the orbits of the planets and asteroids in our solar system lie.

ECOSYSTEM: Any natural unit or entity including living and non-living parts that interact to produce a stable system through cyclic exchange of materials.

EL NIÑO: A warming of the surface waters of the eastern equatorial Pacific that occurs at irregular intervals of 2–7 years, usually lasting 1–2 years. Along the west coast of South America, southerly winds promote the upwelling of cold, nutrient-rich water that sustains large fish populations, which then sustain abundant sea birds, whose droppings support the fertilizer industry. Near the end of each calendar year, a warm current of nutrient-pool tropical water replaces the cold, nutrient-rich surface water. Because this condition often occurs around Christmas, it was named El Niño (Spanish for boy child, referring to the Christ child). In most years the warming last only a few weeks or a month, after which the weather patterns return to normal and fishing improves. How-ever, when El Niño conditions last for many months, more extensive ocean warming occurs and economic results can be disastrous. El Niño has been linked to wetter, colder winters in the United States; drier, hotter summers in South America and Europe; and drought in Africa.

ELECTRIC FIELD: An invisible physical influence that exerts a force on an electric charge. All electric charges produce electric fields. Magnetic fields that are changing (getting weaker or stronger) also produce electric fields.

ELECTROCARDIOGRAPHY: When a heart contracts, electrical fields are created as ions (charged atoms) that flow in and out of muscle cells. These electrical fields can be sensed as voltages on the surface of the body. Electrocardiography is the recording of these changing voltages over time as the heart beats, measured at certain standard points on the skin. The resulting visual collection of what looks like squiggly lines is an electrocardiogram. Cardiologists can diagnose many disorders of the heart simply by looking at an electrocardiogram.

ELECTROCONVULSIVE SHOCK THERAPY: Also known as electroshock. A treatment in which an electrical current is made to flow briefly through the brain, triggering a seizure. It is used as a treatment for severe depression and for severe forms of some other mental illnesses.

ELECTROMAGNETIC FORCE: Electrically charged particles experience a force when they are in the presence of an electrical field, whether they are moving or stationary; they also experience a force in the presence of a magnetic field, if they are in motion. These forces are manifestations of a single fundamental force, the electromagnetic force.

ELECTROMAGNETIC RADIATION: Radiation that has properties of varying electric and magnetic waves and that travels through a vacuum with the speed of light. X rays, ultraviolet light, visible light, microwaves and radio waves are examples of electromagnetic radiation differing only in their wavelengths (or frequency).

ELECTROMAGNETIC SPECTRUM: The entire range of radiant energies or wave frequencies from the longest to the shortest wavelengths—the categorization of solar radiation. Satellite sensors collect this energy, but what the detectors capture is only a small portion of the entire electromagnetic spectrum. The spectrum usually is divided into seven sections: radio, microwave, infrared, visible, ultraviolet, x-ray, and gamma-ray radiation.

ELECTROMAGNETIC WAVE: Method of travel for radiant energy, so called because radiant energy has both magnetic and electrical properties. Electromagnetic waves are produced when electric charges change their motion. Whether the frequency is high or low, all electromagnetic waves travel at 186,411 miles/second (300,000,000 meters/second).

ELECTROMAGNETISM: A form of magnetic energy produced by the flow of an electric current through a metal core. Also, the study of electric and magnetic fields and their interaction with electric charges and currents.

ELECTRON: A subatomic particle having a charge of −1.

ELECTRON MICROSCOPE: A microscope that uses a beam of electrons to produce an image at very high magnification.

ELECTROPHORESIS: The use of electrical fields to separate molecules of different sizes. Electrophoresis is a form of chromatography.

ELECTROWEAK FORCE: In physics, the single unified force of which the electromagnetic force and weak interaction are both manifestations. The electromagnetic force is caused by an electromagnetic field; the weak interaction is involved in certain types of radioactive decay (e.g., beta decay of neutrons).

ELEMENT: Substance consisting of only one type of atom.

ELEMENTARY PARTICLE: A subatomic particle that cannot be broken down into any simpler particle.

ELLIPSE: A two-dimensional figure defined in relation to two points called foci. If A and B are the foci of an ellipse, and X is a point on that ellipse, then the sum AX + BX will be the same for every point X.

ELLIPSOID: An egg-like three-dimensional shape defined by rotating an ellipse in space. In geodesy (measurement of Earth or other planetary bodies), the ellipsoid is the idealized, simplified shape of the planet. Features of the actual planet's shape are defined by their distances from the ellipsoid.

EMBRYO: The earliest stage of animal development in the uterus before the animal is considered a fetus.

EMBRYONIC STEM CELL: A stem cell found in embryos about a week old. Descendants of one of these cells can become any kind of tissue. These cells can reproduce indefinitely in the laboratory.

EMISSION LINE: An atom or molecule emits photons (light particles) when its electrons jump from higher to lower electron levels. These photons, depending on the magnitude of the jumps possible in that atom or molecule, have certain specific energies. When displayed as a spectrum, these emissions, at certain favored frequencies, appear as spikes or vertical lines—emission lines.

EMPIRICISM: The philosophical doctrine, first developed in the 1600s and 1700s, that all knowledge arises from sense impressions—things seen, heard, felt, and so on. It opposes the view that some knowledge is available to the mind *a priori*, that is, by the very nature of thought.

ENDEMIC: Present in a particular area or among a particular group of people.

ENDOCRINE SYSTEM: The collection of glands—small, diverse organs located in the head and torso—that release substances into the blood termed hormones. Hormones influence growth, sex and reproduction, mood, and other aspects of physiology.

ENDOPLASMIC RETICULUM: The network of membranes that extends throughout the cell and is involved in protein synthesis and lipid metabolism.

ENDOTHERMIC: Reaction that absorbs heat from its surroundings as the reaction proceeds.

ENERGY: The ability to do work or transfer heat.

ENLIGHTENMENT: An eighteenth-century movement that advocated reason and scientific discovery as the bases of intellectual authority.

ENTROPY: Measure of the disorder of a system.

ENVIRONMENTAL GEOGRAPHY: A subdiscipline of geography that emphasizes the relationship of human beings to their environment, with an emphasis on spatial relationships and resource use.

ENVIRONMENTAL PROTECTION AGENCY (EPA): An agency of the U.S. federal government that was formed in 1970 and charged with defining and enforcing regulations that protect the environment (water, air, forests, shores, wetlands, wild animals, etc.). Some other nations have agencies with similar names or functions.

ENZYME: Any of numerous complex proteins produced by living cells that act as catalysts, speeding up the rate of chemical reactions in living organisms.

EPICYCLES: In the Ptolemaic system of astronomy used in European science from the second century

AD until the Copernican revolution in the 1500s, the motions of the planets relative to those of the fixed stars were explained in terms of compounded circular motion: That is, the planets were thought to move in circles around points that moved in circles. A deferent was a circle traced around a point in space near Earth called the equant; around the point tracing the circle was another circle, the epicycle, along which the planet moved. The combination of these motions imparted a complex motion to planets that accounted for their movements in the sky.

EPIDEMIC: Rapidly spreading outbreak of a contagious disease.

EPIDEMIOLOGY: The study of the causes, distribution, and control of disease in populations.

EPIGENESIS: The development of an animal or plant from a single fertilized egg cell. In modern biology, the term refers to this whole process of development, including the specialization of various tissues; formerly, it distinguished this theory of development from the theory of preformationism, which was the idea that all living things begin as extremely small copies of themselves and simply get bigger as they develop.

EQUANT: In the Ptolemaic system of astronomy the motions of the planets relative to those of the fixed stars were explained in terms of compounded circular motion: that is, the planets were thought to move in circles around points that moved in circles. A deferent was a circle traced around a point in space near Earth called the equant, an arbitrary point inside a planet's orbit.

EQUATOR: The line circling a planet that divides it into northern and southern hemispheres (halfspheres). An imaginary plane passing through the planet's center and at right angles to its axis of rotation will intersect the planet's surface along its equator.

EQUILIBRIUM: When the reactants and products of a chemical reaction are in a constant ratio. The forward reaction and the reverse reaction occur at the same rate when a system is in equilibrium.

EROSION: Processes (mechanical and chemical) responsible for the wearing away, loosening, and dissolving of materials, particularly of Earth's crust.

ERYTHROCYTES: Also known as red blood cells, erythrocytes contain the pigment hemoglobin; the characteristic red color of blood is due to the eryth-rocytes, which have the remarkable capacity to combine with and release oxygen. Human red blood cells contain a 33% solution of hemoglobin. Oxygen is transported to living tissues of the body as oxyhemoglobin in red blood cells. The human red blood cell is a biconcave disc with an average of about a 0.0003 inch (0.008 mm) diameter. Erythrocytes are the most common cell type in blood (with an average of about 5,500,000 per ml in men and 5,000,000 per ml in women). Newborn babies have an even greater number of erythrocytes, with as many as 7,000,000 per ml. Red blood cells are suspended in plasma, which is the straw colored liquid part of the blood. Human, and most mammalian, erythrocytes have nuclei while they develop in the bone marrow.

ESCAPE VELOCITY: The escape velocity of a planet, asteroid, moon, or other object is the minimum initial velocity that must be imparted to a projectile (e.g., a bullet) at its surface if that object is to leave the vicinity of the planet or other object completely and permanently. Escape velocity is calculated on the presumption (false, in the case of Earth and some other planets) that there is no atmosphere to get in the way.

ESTRUS: In many mammalian species, for example dogs, females go through a regular period of enhanced fertility and sexual willingness; this period is termed estrus.

ESTUARY: Lower end of a river where ocean tides meet the river's current.

ETHANOL: Compound of carbon, hydrogen, and oxygen (CH_3CH_2OH) that is a clear liquid at room temperature; also known as drinking alcohol or ethyl alcohol. Ethanol can be produced by biological or chemical processes from sugars and other feedstocks and can be burned as a fuel in many internal-combustion engines, either mixed with gasoline or in pure form. Several governments, most notably in Brazil and the United States, encourage the production of ethanol from corn, switchgrass, algae, or other crops to substitute for imported fossil fuels. Ethanol is criticized by some as being based on environmentally destructive agriculture, putting human populations into competition with automobiles for the produce of arable land and providing little more energy (depending on the manufacturing process used) than is required to produce it.

ETHER: Also spelled aether; the medium that was once believed to fill space and to be responsible for carrying light and other electromagnetic waves.

ETHOLOGIST: A scientist who studies human behavior in terms of its biological influences and consequences.

ETHOLOGY: The scientific and objective study of the behavior of animals in the wild rather than in captivity. Also, the study of the formation of human character.

EUCLIDEAN GEOMETRY: A type of geometry based on certain axioms originally stated by Greek mathematician Euclid (c.325–c.270 BC).

EUKARYOTE: Multicellular organism whose cells contain distinct nuclei and membrane-bound structures called organelles.

EVENT HORIZON: A spherical surface around any black hole that is a boundary separating the vicinity of the black hole from the rest of the universe. Light cannot travel from inside the event horizon to outside; therefore, events inside the event horizon cannot be seen on the outside (hence the term "event horizon," as one cannot see what is over a horizon).

EVOLUTION: The theory that all plants and animals developed gradually from earlier forms over a long period of time and that variations within a species are the result of adaptive traits passed on from generation to generation.

EXOGAMY: The practice of always marrying outside a certain group, such as one's own village.

EXONERATE: To prove someone innocent.

EXOTHERMIC: Reaction that gives off heat to the environment as the reaction proceeds.

EXPLANANDUM: A phenomenon, in science, for which an explanation is sought. The explanation is termed an explanans.

F

FALSIFIABILITY: A claim is falsifiable if there is, at least in principle, some way of proving that the claim is false. For example, the claim that gravity obeys an inverse-square law is falsifiable (has falsifiability): One can conduct tests and see what kind of law gravitation obeys and whether it is in fact an inverse-square law. However, the claim that all life is a dream, or that God created the universe five minutes ago complete with all our memories of a deeper past, is not falsifiable: All possible evidence could be part of the dream or could have been created along with everything else. Most scientists agree that although falsifiability is not enough to make a claim scientific, to be scientific a claim must be falsifiable.

FALSIFIABLE: A theory or claim is falsifiable if there are reasonably possible conditions under which the theory or claim can be proved false. To be scientific, an idea must be falsifiable (though it takes more than falsifiability to make an idea a scientific one).

FAT: A type of lipid, or chemical compound, used as a source of energy, to provide insulation, and to protect organs in an animal body.

FATTY ACIDS: Carboxylic acids found in fats and oils; there are many varieties of fatty acid. Chemically, all consist of a hydrocarbon chain terminated by a carboxyl group (CO_2H).

FERMENTATION: Chemical reaction in which enzymes break down complex organic compounds (for example, carbohydrates and sugars) into simpler ones (for example, ethyl alcohol).

FIBER: A long, threadlike material, often used in the manufacture of cloth. Also, in terms of composite fillers, a fiber is a filler with one long dimension. In nutrition, the portion of food that remains undigested by physiological processes but that is essential to facilitate digestion and intestinal health. Fiber is also described as bulk or roughage. Fiber may be soluble (able to dissolve in water) or insoluble, natural or synthetic.

FIBER OPTICS: The use of glass threads or fibers to carry light rays.

FIELD EFFECT TRANSISTOR (FET): A transistor is an electronic device that controls a current flow in response to a separate electrical signal, like a valve regulating the flow of water through a faucet. A field-effect transistor (FET) accomplishes this by using an electrical field to pinch a conducting volume of semiconductor crystal; the stronger the field, the narrower the conducting zone and the higher its resistance (i.e., less current flows). Most of the transistors used in modern integrated circuits such as microprocessor chips are FETs.

FISSION: Splitting or breaking apart. In biology, the division of an organism into two or more parts, which each produce a new organism. Nuclear fission is a process in which the nucleus of an atom splits, usually into two daughter nuclei, with the transformation of tremendous levels of nuclear energy into heat and light.

FLYBY: A type of mission in which an unmanned spacecraft passes close to a planet without landing or going into orbit.

FOCAL LENGTH: In optics, the distance between the center of a focusing mirror or lens is the focal length of that mirror or lens.

FOMITE: An object or a surface to which infectious microorganisms such as bacteria or viruses can adhere and be transmitted. Transmission is often by touch.

FOOD CHAIN: A sequence of organisms, each of which uses the next lower member of the sequence as a food source.

FORCE: Any external agent that causes a change in the motion of a free body or that causes stress in a fixed body.

FORENSIC SCIENCE: A multidisciplinary subject used for examining crime scenes and gathering evidence to be used in prosecution of offenders in a court of law. Forensic science techniques are also used to examine compliance with international agreements regarding weapons of mass destruction.

FORMALISM: A theory of the nature of mathematics proposed by German mathematician David Hilbert (1862–1943). According to formalism, mathematics does not refer to abstract objects or to physical reality, but only to certain logical forms—namely, the forms of mathematical statements themselves (hence the term formalism). Mathematics, on this view, is a system of statements whose only real content is their form. Pure formalism was ruled out by Gödel's incompleteness theorem in 1931.

FOSSIL: Hardened remains or traces of plant or animal life from a previous geological period preserved in Earth's crust.

FOUNDATIONAL CRISIS: In the early twentieth century, a vigorous debate arose among mathematicians over what the nature of mathematics itself is. This debate has been termed the foundational crisis.

FOUNDATIONAL THEORY: In mathematics, any philosophical explanation of the nature or basic foundations of mathematics itself.

FRACTAL: A fractal is a geometric pattern in which each part of the pattern has the same character as the total pattern. Some natural patterns, such as fern branching, have a nearly fractal character; approxi-mately fractal patterns are also used in some technological applications, such as cell-phone antennas.

FRAUNHOFER LINE: Light from the sun, displayed as a spectrum (a plot showing intensity or brightness over a range of frequencies), features a number of narrow notches. These notches or lines, named after German physicist Joseph von Fraunhofer (1787–1826), who discovered them in 1802, are frequencies of comparatively low intensity light that atoms in the atmosphere of the sun absorb. In effect, they are colors of light that are emitted from the sun's interior but that the sun's atmosphere filters out before they can be radiated into space.

FREQUENCY: The rate at which vibrations, oscillations, or events take place over time. In characterizing light, frequency is expressed in cycles per second or in hertz (Hz) to describe the number of wave crests that pass a given point in a given period of time.

FRONT: A boundary between two different air masses. The difference between two air masses is sometimes unnoticeable, but when the colliding air masses have very different temperatures and amounts of water in them, turbulent weather can erupt.

FUNDAMENTAL THEOREM OF CALCULUS: In calculus, the two fundamental operations are integration (finding the area under a curve) and differentiation (finding the slope of a curve at a single point). The fundamental theorem of calculus is that these two operations are inverses of each other: Integrating a function can be undone by differentiating the result, and vice versa.

FUSION: The process stars use to produce energy to support themselves against their own gravity. Nuclear fusion is the process by which two light atomic nuclei combine to form one heavier atomic nucleus. As an example, a proton (the nucleus of a hydrogen atom) and a neutron will, under the proper circumstances, combine to form a deuteron (the nucleus of an atom of "heavy" hydrogen). In general, the mass of the heavier product nucleus is less than the total mass of the two lighter nuclei. Nuclear fusion is the initial driving process of nucleosynthesis.

G

GALAXY: An isolated system of stars and gas held together by its own gravity.

GAMETE: A male or female reproductive cell.

GAMMA RAY: Short-wavelength, high-energy radiation formed either by the decay of radioactive elements or by nuclear reactions.

GENE: A segment of a DNA (deoxyribonucleic acid) molecule contained in the nucleus of a cell that acts as a kind of code for the production of some specific protein. Genes carry instructions for the formation, functioning, and transmission of specific traits from one generation to another.

GENERAL RELATIVITY: In 1915, German-American physicist Albert Einstein (1879–1955) announced a mathematical theory that described the nature of gravitation by appeal to the curvature of space and time. This theory, general relativity, has been one of the most successful scientific theories of all time and has passed many rigorous observational tests by successfully predicting the precise behaviors of various experimental setups and distant astronomical objects.

GENETIC ENGINEERING: The process of manipulating specific genes of an organism to produce or improve a product or to analyze the genes.

GENETIC FINGERPRINTING: The use of DNA from individual people or other creatures to tell them apart. Except for identical multiple births (e.g., identical triplets) and a few other exceptional cases, every multicellular organism has unique DNA.

GENETICALLY MODIFIED FOODS: Foods derived from plants or animals whose DNA has been altered technologically, usually by inserting genes.

GENOME: All of the genetic information for a cell or organism. The complete sequence of genes within a cell or virus.

GENOTYPE: The genetic information that a living thing inherits from its parents that affects its makeup, appearance, and function.

GENUS (PLURAL: GENERA): Biologists class living things into groups; for instance, all wolves are members of a single species. Wolves, coyotes, dingoes, and jackals, which are all closely related through evolutionary descent, are classed into a single group called a genus: in this case, the genus *Canis*. Genera (of which there are thousands) are lumped into families. From there on up, the classification system recognizes order, class, phylum, kingdom, and domain.

GEOCENTRIC: A geocentric model of the solar system places a stationary Earth at the center of the solar system, with the sun and planets orbiting Earth.

GEOCENTRISM: The discarded theory that a stationary Earth lies at the center of the universe (or solar system).

GEOCHRONOLOGY: The determination of the absolute ages of rocks and other natural deposits. The basic tool of geochronology is radiometric dating, which measures the amounts of radioactive elements and their daughter isotopes in samples. The relative proportions of these atomic varieties reveal the sample's age.

GEOGRAPHY: Geography is the scientific study of the surface of Earth, including the interactions of landscapes, bodies of water, the atmosphere, human beings, plants, and animals.

GEOID: A surface of constant gravitational potential around Earth; an averaged surface perpendicular to the force of gravity.

GEOLOGIC COLUMN: An idealized assemblage of all the rock layers deposited throughout Earth's history. Only fragments of the total column exist in any one location (with possible rare exceptions), due to uneven episodes of erosion that have removed rather than deposited layers; however, by comparing overlapping segments of the column found in different places, geologists can describe the entire column. Geochronology is used to assign absolute dates to the various layers of the column.

GEOLOGIC RECORD: Evidence of Earth's history left in rocks and sediments over thousands to billions of years. Events that can be inferred from the geological record include climate changes, biological evolution, continental drift, and asteroid impacts.

GEOMETRIC ISOMERS: Stereoisomers in molecules with restricted rotation about a bond.

GLAND: A small organ that produces a hormone, a substance released into the blood that affects behavior, sex and reproduction, growth, and other body functions. Some of the glands in the human body are the testes, ovaries, thyroid gland, and pituitary gland.

GLOBAL WARMING: Warming of Earth's atmosphere that results from an increase in the concentration of gases that store heat, such as carbon dioxide.

GLOBULAR CLUSTER: A spherical collection of stars that appears cloudy to the naked eye.

GLUCOSE: Also known as blood sugar; a simple sugar broken down in cells to produce energy.

GLUON: The elementary particle thought to be responsible for carrying the strong force (which binds together protons and neutrons in the atomic nucleus).

GLYCOLYSIS: A series of chemical reactions that takes place in cells by which glucose is converted into pyruvate.

GLYCOPROTEIN: A membrane-bound protein that has attached branching carbohydrates. These may function in cell-cell recognition, such as in human blood groups and immune system response, as well as in resisting compression of cells.

GNOMON: The pointer that casts a shadow on a sundial.

GRADUALISM: A model of evolution that assumes slow, steady rates of change. Charles Darwin's (1809–1882) original concept of evolution by natural selection assumed gradualism. Contrast with punctuated equilibrium.

GRAND UNIFIED THEORIES (GUTs): Systems of physical laws that describe the strong nuclear, electromagnetic, and weak nuclear forces in terms of a single underlying field. All such theories omit the force of gravity; it has proved difficult to formulate a theory uniting gravity to the other forces. A theory of everything (TOE) did not yet exist as of 2008.

GRANT PROPOSAL: A formal request for a sum of money (a grant) from a private foundation or a government agency. The authors must describe their qualifications, the question their research is designed to address and how it will address it, what results they think will probably be obtained, and what the importance of their proposed work will be. Usually more grant proposals are received than can be funded, so the process of obtaining grants is competitive.

GRAPH: A flat representation of information on a pair of coordinate axes at right angles to each other (i.e., Cartesian coordinates). Each axis represents a variable, and each point on the graph represents a pair of variable values. Marks on the graph make the relationship between the two variables visually apparent. Graphs are the most common method of representing quantitative information visually.

GRAPHIC USER INTERFACE (GUI): (Pronounced "gooey.") Visual computer displays that are responsive to human users. Input to the display from the user may be via touch-screen, graphics tablet, keyboard, mouse, voice command, eye movement, or the like. Software in the computer decides on an appropriate response and changes a graphic display accordingly: For example, in response to a mouse movement, a GUI may move a cursor, shift an icon or image, or show other changes.

GREENHOUSE EFFECT: The warming of Earth's atmosphere due to water vapor, carbon dioxide, and other gases in the atmosphere that trap heat radiated from Earth's surface.

GREENHOUSE GAS: A gaseous component of the atmosphere contributing to the greenhouse effect. Greenhouse gases are transparent to certain wavelengths of the sun's radiant energy, allowing them to penetrate deep into the atmosphere or all the way into Earth's surface. Greenhouse gases and clouds prevent some of infrared radiation from escaping, trapping the heat near Earth's surface where it warms the lower atmosphere. Alteration of this natural barrier of atmospheric gases can raise or lower the mean global temperature of Earth.

GROSS DOMESTIC PRODUCT (GDP): A measure of total economic activity, whether of a nation, group of nations, or the world: slightly different from gross national product (GNP; used by economists until the early 1990s). Defined as the total monetary value of all goods and services produced over a given period of time (usually one year). GDP's limitations have been pointed out by many economists. GDP is a bulk or aggregate statistic and does not take into account inequity: thus, a country's GDP might increase while 99% of its population got poorer, as long as the richest 1% grew sufficiently richer during the same period.

GUILD: In the economy of medieval Europe, an association of craftsmen (e.g., bakers, stonemasons) that controlled admission to the profession, prices, and other aspects of their trade.

GUILLOTINE: An execution device that drops a heavy blade on the neck of the condemned, cutting off the head. Versions of the guillotine were used beginning in the Middle Ages, but the device acquired fame and its present name during the French Revolution, when Joseph-Ignace Guillotin (1738–1814) proposed it as a humane alternative to the usual methods of execution, which often involved torture (e.g., breaking on the wheel, which involved tying a person spread-eagle on a large, spoked wheel and breaking all their bones with a hammer).

H

HADRON: Subatomic particles that are affected by the strong nuclear force; they include both baryons (three-quark particles such as protons, neutrons, etc.) and mesons (two-quark particles such as pions).

HALF-LIFE: The time it takes for half of the original atoms of a radioactive element to be transformed into the daughter product.

HANTAVIRUS: A class of viruses carried by rodents that can infect human beings.

HAPLOID: Having a single set of unpaired chromosomes.

HEAT SHIELD: Any device that protects another, more delicate device from heat. The term is most often used for the lens-shaped caps of special material placed on the leading surfaces of space vehicles designed to enter atmospheres at high speed. During re-entry, compression of air in front of the heat shield heats the air to high temperatures. Some of this heat is absorbed by the heat shield, while some is left behind in a trail of heated air or radiated away. Transforming kinetic energy (energy of motion) to heat energy in this way slows the spacecraft. A heat shield may simply re-radiate the heat it has absorbed (as is done by the ceramic tiles on the space shuttle) or it may also allow some of its material to ablate (vaporize), removing more heat.

HELIOCENTRIC: A heliocentric model of the solar system places the sun at the center of the universe, with the planets and Earth orbiting around it.

HELIOCENTRISM: The theory that the sun lies at the center of the universe (or solar system) and Earth is in motion around it.

HELPER T CELL: (Better known as a T helper cell or effector T cell.) A type of white blood cell that does not directly attack pathogens (invading organisms) but plays a supportive role in triggering and guiding attack by other white blood cells.

HELSINKI DECLARATION: In June 1965, in Helsinki, Finland, the World Medical Association adopted a statement of ethical guidelines for the conduct of research involving human beings. The Helsinki Declaration has since been revised a number of times to reflect advances in ethical understanding. The basic principle is respect for human autonomy, that is, the right to make medical decisions based on an accurate understanding of the situation and free from efforts at coercion or manipulation.

HEMATOLOGY: The medical study of blood (from the Greek *haima*, blood).

HEMOCOEL: Invertebrates (animals lacking a spinal column) do not have a circulatory system of tubular arteries, capillaries, and veins: Instead, they have a body cavity of complex shape, the hemocoel, which contains circulating fluid (including blood) that bathes the creature's organs and muscles, supplying oxygen and removing wastes.

HEMOCYTE: Any blood cell, especially in an invertebrate (animal lacking a spinal column).

HEMOGLOBIN: The protein pigment in red blood cells that transports oxygen to the tissues and carbon dioxide from them.

HEMOLYTIC ANEMIA: A type of anemia caused by destruction of red blood cells at a rate faster than that at which they can be produced.

HERBAL: Herbal remedies or medicines are those derived directly from edible or medicinal plants (herbs) with a minimum of processing: typically, an herbal remedy consists of the dried and possibly crushed or powdered leaves of the plant.

HEREDITY: The genetic link between successive generations of living things.

HERITABILITY: The quality of being able to be passed on from parent to offspring. The more heritable a trait or characteristic is, the more likely it is to be passed on. The random appearance of heritable traits and their sifting by natural selection (which is non-random) is the basis of biological evolution.

HERTZ (HZ): The hertz (Hz) is the unit of frequency, defined as the number of times that an event is repeated in 1 second. It is named after German radio engineer Heinrich Hertz (1857–1894).

HERTZSPRUNG-RUSSELL DIAGRAM: A graphic representation of information about different kinds of stars. It was designed around 1910 by Danish astronomer Ejnar Hertzsprung (1873–1967) and American astronomer Henry Norris Russell (1877–1957). Each star is plotted as a dot on the graph. The horizontal dimension of the graph is in color (whiter stars to the left, redder stars to the right) and the vertical dimension is brightness (brighter stars higher up, dimmer stars lower down). The Hertzsprung-Russell diagram shows that the stars form several distinct patches or groups; these are related to different stages in each star's life cycle and to the mass of the star.

HETEROTOPIC: A heterotopic cell or tissue is one that is growing in an inappropriate location; for example, in the disease known as heterotopic ossification, bone grows in soft body tissues.

HEURISTICS: In artificial intelligence, flexible rules that allow a system to learn.

HIERARCHY OF HUMAN NEEDS: Also known as Maslow's hierarchy of needs, proposed by American psychologist Abraham Maslow (1908–1970) in 1943. Maslow proposed that human needs can be organized in a hierarchy or pyramid, with people addressing higher-up needs only after satisfying lower-down needs. The levels of need in Maslow's hierarchy, from the bottom up, are physiological (food, sex, air, etc.), safety, love, esteem, and self-actualization (creativity, ethics, etc.). The accuracy of this way of viewing human need and motivation has been disputed by some other psychologists.

HIGGS BOSON: The only particle predicted by the current Standard Model of particle physics that has not yet (as of early 2008) been experimentally observed. The Higgs boson, if it exists, is responsible for endowing all matter with mass. Intense and expensive searches for evidence of the Higgs boson are ongoing at the CERN laboratory in Europe and elsewhere.

HOLE: In the physics of solid materials (solid-state physics), a place in a crystalline array of atoms where an electron is missing, creating a localized positive charge. In semiconductors, holes can travel through the crystal as electrons shift, moving the gap from one position to the next. A moving hole behaves exactly as if it were a positively-charged particle or charge carrier.

HOLISM: A style of thought that can be contrasted to reductionism. Reductionism assumes that the properties of any complex system—a human brain, an ecosystem, or other—can be reduced to the interactions of its component parts (hence "reductionism"). Holism asserts that at least some system properties arise out of the system as a whole (hence "holism") and are not possessed by, or reducible to, the properties of the system's isolated, individual parts.

HOMEOPATHY: An alternative medical system invented by German doctor Christian F.S. Hahnemann (1755–1843) in the early 1800s. In homeopathy, diseases are treated using extremely small doses of substances believed to cause symptoms resembling the disease itself. Some remedies marketed as homeopathic are actually herbal remedies (e.g., arnica gel for its anti-inflammatory properties) and contain significant quantities of their active ingredients; these remedies are not, strictly speaking, homeopathic. Homeopathy is pseudoscientific (i.e., appears scientific but is not): Controlled studies have shown that homeopathy is no better than the placebo effect, which is the tendency of patients to feel better if they believe they are receiving an effective treatment, regardless of the presence of any actual treatment.

HOMEOSTASIS: State of being in balance; the tendency of an organism to maintain constant internal conditions despite large changes in the external environment.

HOMOLOGOUS SERIES: A group of organic compounds that have similar structure and chemical properties.

HOMOLOGOUS STRUCTURE: In evolutionary biology, a homologous structure in an organism is one that resembles a structure in another because both organisms share an ancestor. That is, the similarity is inherited. For example, the leg and finger bones found in humans, whales, bats, and horses are, despite wide differences in shape and use, similar in number and basic pattern: They are homologous structures evolved from the limb bones of the most recent common ancestor of all those species.

HOMOLOGY: Two structures are considered homologous when they are inherited from a common ancestor that possessed the structure. This may be difficult to determine when the structure has been modified through descent.

HOMOPLASY: Similarity between independently evolved structures in two different organisms. For example, the resemblance in outline between a shark and a dolphin is a homoplasy: The two evolved this shape from ancestors that had completely different shapes.

HOPS: Dried flowers of the vine *Humulus lupulus*, which give beer its characteristic bitter flavor and aroma.

HORMONE: A chemical produced in living cells that is carried by the blood to organs and tissues in distant parts of the body, where it regulates various bodily functions.

HOROSCOPE: A forecast, produced by an astrologer, of a person's immediate or long-term future based on the apparent positions of the stars and planets at the time of the person's birth. Such forecasts have no scientific basis.

HOSPITALISM: A nineteenth-century term for the epidemic septic diseases of hospital wards.

HTML: Hypertext markup language (HTML) is a system of instructions that can be used to define the appearance and behavior of a hypertext document. Hypertext is computer-displayed text that not only conveys a direct message, but, if clicked on by a user, acts as a link to other information (e.g., image, sound, text). HTML is the basis of the World Wide Web of linked hypertext documents. The earliest version of HTML was developed by physicist Tim Berners-Lee (1955–) in 1991.

HUMAN GEOGRAPHY: The study of the relationship of human beings to their environment, especially as related to spatial patterns of human settlement (where people live, how densely they settle, how they exploit their environment, etc.).

HUMOR: The ancient Greek and Roman theory of physiology stated that four basic substances, the humors, exist in the human body, and that disease is caused by imbalance among the humors. The four humors were blood, phlegm, yellow bile, and black bile.

HUMORAL THEORY OF DISEASE: The ancient Greek and Roman theory of physiology stated that four basic substances, the humors, exist in the human body, and that disease is caused by imbalance among the humors. The four humors were blood, phlegm, yellow bile, and black bile. The theory proposed that not only disease but temperament or disposition—what psychologists now call personality—were determined by humoral balance.

HURRICANE: Large, rotating system of thunderstorms whose highest wind speed exceeds 74 mph (119 km/h). Globally, such storms are termed tropical cyclones. The word "hurricane" is often reserved for tropical cyclones in the Atlantic.

HYDROCARBON: A chemical containing only carbon and hydrogen. Hydrocarbons are of prime economic importance because they encompass the constituents of the major fossil fuels, petroleum and natural gas, as well as plastics, waxes, and oils. In urban pollution, these components—along with NO_x and sunlight—contribute to the formation of tropospheric ozone.

HYPERFINE RADIO EMISSION: In radio emissions, the presence of very similar emission frequencies (vibration rates of electromagnetic waves emitted by the atom) from a substance such as hydrogen or helium that appear as closely-spaced lines on a spectrum.

HYPERTEXT: Hypertext is computer-displayed text that not only conveys a direct message but, if clicked on by a user, acts as a link to other information, whether text, audio, or image.

HYPOTENUSE: The longest side of a right triangle, which is opposite the right angle.

HYPOTHESIS: An idea phrased in the form of a statement that can be tested by observation and/or experiment.

HYSTERIA: Any state of overexcited or exaggerated emotion. In psychology, the term once referred to a disorder characterized by attention-seeking, extreme emotionalism, and somaticization (conversion of psychological stresses into physical symptoms). For many centuries, hysteria was thought to be a disorder particular to women, hence its name, which comes from the Greek for uterus, *hystera*. In recent decades, psychiatrists have rejected the idea that there is any psychiatric disorder suffered only by women and no longer diagnose "hysteria." The symptoms once associated with hysteria are now associated with other diagnoses, such as somaticization disorder and histrionic personality disorder.

I

IATROCHEMISTRY: A school of European medical-scientific thought that flourished mostly in the 1500s. It was descended from alchemy and was, in some ways, a precursor of modern medical science: Iatrochemists taught that medicines should be based on alchemy (chemistry as it was then understood).

IATROGENIC: (Pronounced eye-at-roh-GEN-ik.) Any infection, injury, or other disease condition caused by medical treatment.

IGNEOUS: Any rock that has formed directly out of molten material, such as lava or granite. Most other rocks are sedimentary, that is, have formed by the accumulation and cementing together of small particles (sediments) deposited by wind or water; or they are metamorphic, that is, have started out sedimentary but then been partially melted before being allowed to resolidify.

IMMUNITY: The condition of being able to resist the effects of a particular disease.

IMMUNIZATION: The process by which resistance to an infectious disease is increased.

IMMUNOGLOBULIN: An immunoglobulin (also known as an antibody) is a large, Y-shaped protein found in blood that combines with a specific substance foreign to the body (antigen). Typical antigens are molecules found on the surfaces of viruses and bacteria. Some immunoglobulins, when bound to antigens, act as flags targeting the antigen for attack by a white blood cell; others combine with other blood molecules to attack the antigen directly. If a person's blood already contains immunoglobulins (antibodies) for a particular antigen, the person's immune system attacks that antigen as soon as it appears: This is the basis of acquired immunity to specific viruses and bacteria, whether natural or instilled by immunization.

IMMUNOLOGY: The study of how the body responds to foreign substances and fights off infection and other disease. Immunologists study the molecules, cells, and organs of the human body that participate in this response.

IMMUNOSUPPRESSIVE: An immunosuppressive drug is one that suppresses or decreases the functioning of the immune system. Such drugs are given with organ transplants in order to prevent the immune system from attacking and destroying (rejecting) the transplanted organ.

IN SILICO: The phrase "in silico" is a pseudo-Latin phrase coined to describe experiments that happen only in the form of computer simulations. "Silico" is a reference to the fact that computers perform calculations using circuits made mostly of silicon. The phrase is based on the phrases "in vivo" (said of experiments performed in a live subject) and "in vitro" (said of experiments performed in a glass container in the laboratory).

IN VITRO: In the laboratory or under laboratory conditions (e.g. in the bacteriological context, the cultivation of organisms in glass Petri dishes).

IN VIVO: In the bacteriological context, growing microorganisms in experimental animals.

INDEPENDENCE: In the mathematical field of probability and statistics, two experiments or observations, say A and B, are independent if the occurrence of a certain outcome for event A does not influence the probabilities for the possible outcomes of event B.

INDUCTION: A form of logical reasoning. In induction, one reasons from particular facts to general laws or principles. For example, by making measurements of how dropped stones accelerate, one might induce the law of gravitation. The complementary form of reasoning is deduction, which reasons from general principles to particular events. For example, knowing that a stone has been dropped at a certain time and place and knowing the law of gravitation, one can deduce when the stone will strike the ground.

INERTIA: A behavior exhibited by moving bodies. Aristotle defined it in the 300s BC as the tendency of moving bodies to come to rest and remain at rest. Galileo redefined it in the AD 1600s as the tendency of moving bodies to remain in motion unless acted on by an outside force.

INERTIAL REFERENCE FRAME: A system of coordinates (spatial measurements of height, length, and width) that is stationary with respect to a point that is moving at constant speed in a straight line. The special theory of relativity, announced by German physicist Albert Einstein (1879–1955) in 1905 and now a basic part of physics, describes the behavior of objects in inertial reference frames.

INFINITY: Infinity is a mathematical quantity greater than any particular number, symbolized by ∞. The concept was first developed in India about 1,600 years ago.

INFORMED CONSENT: An ethical and informational process in which a person learns about a procedure or clinical trial, including potential risks or benefits, before deciding to voluntarily participate in a study or undergo a particular procedure.

INFUSORIA: A group of minute organisms found in decomposing matter and stagnant water.

INOCULATION: Treatment with a vaccine to produce immunity to a disease. Inoculation produces immunity by triggering production of antibodies in the body that attach to virus particles or bacteria that cause the particular disease. If such disease organisms do appear in the body, the antibodies attach to the organisms, and the body attacks them at once, preventing infection.

INORGANIC: Composed of minerals that are not derived from living plants and animals.

INTEGRATE: In mathematics, to perform the act of integration on a function—interpreted geometrically—to find the area under the curve of the function. In electronics, to integrate is to create multiple electronic devices (transistors, capacitors, resistors, etc.) in the substance of a single crystal substrate, such as a chip of silicon. Devices made in this way are termed integrated circuits.

INTEGRATED CIRCUIT: A system of interconnected electronic components such as transistors and capacitors that have been built as part of a single, solid crystalline structure of silicon or some other semiconductor. Because the components are integral to (a continuous part of) a single solid object, they are said to form an integrated circuit.

INTERFERENCE: The interaction of two or more waves.

INTERFEROMETER: Any instrument that uses the phenomenon of optical (light) wave interference to make measurements. Interferometers may be used to make extremely fine measurements of distance between two objects. In astronomy, an array of two or more telescopes linked electronically so as to record differences between light waves arriving simultaneously at the various telescopes is called an interferometer.

INTERNAL REVIEW BOARD (IRB): Also known as an institutional review board. A committee of personnel in a hospital where research involving human subjects is performed. The IRB's mission is to make sure that such work strictly obeys laws governing research involving humans.

INTERNET: A vast worldwide conglomeration of linked computer networks. The most significant component of the Internet is the World Wide Web.

INTERTIDAL: Between the tides. The intertidal zone is the area of land along the shore that is covered by water when the tide is highest and exposed when it is lowest.

INTUITIONIST: Mathematicians who affirm the intuitionist theory of the meaning of mathematics, which is that a mathematical statement is true if the mathematician can intuit (directly perceive) the truth of that statement.

INVASION ECOLOGY: The study of the ecological consequences of the introduction of invasive species, that is, species that have evolved in different ecosystems.

ION: Removing or adding electrons to an atom creates an ion (a charged object very similar to an atom).

ION ENGINE: A form of reaction engine that propels a spacecraft by ejecting high-speed charged atoms (ions) opposite to the direction of desired acceleration rather than by ejecting a stream of hot gas, as is done by a conventional rocket. Ion propulsion is more energy-efficient than rocket propulsion but produces far smaller thrust and can only be used for slow maneuvers in space, such as orbital adjustment or mid-course correction.

IONIC: In chemistry, an ionic bond between two atoms is formed when the atoms are ionized (given a positive or negative charge by the addition or removal of electrons) and then brought close enough together to be bound by electrical attraction. In architectural history, Ionic or Ionian classic architecture is distinguished by its use of stone columns ornamented by scrollwork at the top.

IONIZING RADIATION: Any electromagnetic or particulate radiation capable of direct or indirect ion production in its passage through matter. In general use, radiation that can cause tissue damage or death.

ISOMERISM: The occurrence of chemical compounds that have the same composition (the same numbers of the same types of atoms) but different geometrical arrangements of those atoms and, as a result, different physical properties. Two substances with the same atomic ingredients but different structure are called isomers.

ISOMERS: Two substances with the same atomic ingredients but different structures. They have the same composition (the same numbers of the same types of atoms) but different geometrical arrangements of those atoms and, as a result, different physical properties.

ISOTACH: (Literally, "same-speed.") A line on a weather map that connect points of equal wind velocity. Crossing from one isotach to another corresponds to movement from a region of higher wind speed to lower, or vice versa.

ISOTHERM: (Literally, "same-heat.") A line on a map or other drawing that connects points of equal temperature. Crossing from one isotherm to another corresponds to movement from a higher temperature to a lower, or vice versa.

ISOTOPE: Two or more forms of the same element with the same number of protons (atomic number), but different numbers of neutrons (atomic mass), e.g. ^{233}U and ^{235}U are two isotopes of uranium; both have 92 protons, but ^{233}U has 141 neutrons and ^{235}U has 143 neutrons.

J

JET STREAM: Currents of high-speed air in the atmosphere. Jet streams form along the boundaries of global air masses where there is a significant difference in atmospheric temperature. The jet streams may be several hundred miles across and 1–2 miles (1.6–3.2 km) deep at an altitude of 8–12 miles (12.9–19.3 km). They generally move west to east and are strongest in the winter, with core wind speeds as high as 250 mph (402.4 km/h). Changes in the jet stream indicate changes in the motion of the atmosphere and weather.

JOULE (J), KILOJOULE (KJ): The joule (J) is the unit of work in physics and engineering. It is named after English physicist James Prescott Joule (1818–1889) and is defined as the work done by a force of 1 newton (N) moving through a distance of 1 meter (m): $1\,J = 1\,N \times m = 1\,kg\,\dot{s}m^2/s^2$.

JOURNEYMAN: In the European economy in the Middle Ages, a person who had completed their apprenticeship (learned the essentials of their specific trade) but had not yet achieved the status of a master craftsmen. Journeymen often traveled from job to job or worked briefly for various employers.

JUNCTION TRANSISTOR: A transistor is an electronic device that controls a current flow in response to a separate electrical signal, like a valve regulating the flow of water through a faucet. A junction transistor, also known as a sandwich transistor, consists of three pieces or regions of semiconducting crystal arranged in a sandwich. The outer layers consist of crystal that has been doped (slightly contaminated with atoms that donate electrons, endowing the crystal with negative charges); the middle layer consists of crystal that has been doped with atoms that lack electrons, endowing the crystal with positive charges. (A sandwich can also be made with positive and negative in reversed roles.) The ability of the junction transistor arises from the properties of the junctions between the positively and negatively doped silicon, hence the name.

K

KAMAL: An observational instrument used in the Middle East and China from the Middle Ages through the nineteenth century. Like the cross staff, its purpose was to allow measurement of the angle between a particular astronomical object (usually a star) and the horizon. A wooden card attached to a string is held at arm's length while the string is held in one's teeth: one then moves the card along the string until it is aligned with the star. Knots along the string are then counted to the position of the card, providing a reading for the angle.

KAON: Also known as a K-meson. A fundamental particle containing a single quark. It is a type of meson and is several times heavier than a pion.

KINETIC ENERGY: The energy due to the motion of an object.

KOCH'S POSTULATES: In 1884, German doctor Robert Koch (1843–1910) and scientist Freidrich Loeffler (1852–1915) announced a set of standards— later refined by Koch—that they said would have to be met before an organism could be declared the cause of a given infectious disease. First, the germ must be found in all diseased organisms but not in healthy ones; second, the germ must be capable of being isolated and grown in culture; third, a healthy organism inoculated with the germ must contract the disease; and fourth, the inoculated organism must supply germs that are identical with those isolated originally from diseased organisms. Koch's postulates are now known to be too strict; not all disease-causing organisms can be grown in culture, for example, and not all organisms colonized by a disease-causing organism necessarily become sick.

KREBS CYCLE: A set of biochemical reactions that occur in the mitochondria. It is the final common pathway for the oxidation of food molecules such as sugars and fatty acids. It is also the source of intermediates in biosynthetic pathways, providing carbon skeletons for the synthesis of amino acids, nucleotides, and other key molecules in the cell. The Krebs cycle is also known as the citric acid cycle and the tricarboxylic acid cycle. The Krebs cycle is a cycle because, during its course, it regenerates one of its key reactants.

K-T BOUNDARY: At the very end of the Cretaceous period and the beginning of the Tertiary period, 65.5 million years ago, a mass extinction wiped out 65–70% of all plant and animal species in a short time. The layer or boundary separating Cretaceous sediments from Tertiary sediments is termed the K-T boundary. This layer is rich in iridium, an element more common in meteorites than Earth rocks, which suggests that the extinction was caused by an asteroid striking Earth. The Chicxulub crater in the Yucatan Peninsula, Mexico, was probably caused by this impact.

L

LABILE: Easily changed or highly variable; the term is especially used in psychology, where it describes emotional states that are especially prone to change. For example, a person who is smiling one moment, crying the next, and smiling again soon after is displaying labile emotions.

LACTOBACCILLUS: A type of anaerobic (non-oxygen-using) bacterium that is found in the human gut and in nature and is also used in preparation of fermented foods such as yogurt, kimchee, and sauerkraut.

LAMARCKISM: The belief that acquired characteristics can be inherited, that is, that changes to an organism that happen during its life can be passed on to offspring.

LAMBDA: Lambda (λ or Λ) is the eleventh letter of the Greek alphabet. It is used throughout the physical sciences to denote certain objects or classes of objects: for example, in cosmology it denotes the cosmological constant; in biology it is a type of bacterium used in genetic research (lambda phage), etc.

LANDSCAPE: An area of land having a definite local character that includes any structures or changes created by human beings. Generally, a landscape supports a single ecosystem or part of one. The term has no strict definition in geography or biology.

LASER: Acronym for light amplification by stimulated emission of radiation; a device that uses the movement of atoms and molecules to produce intense light with a precisely defined wavelength.

LASER MEDIUM (ACTIVE MEDIUM): Also known as an active medium or gain medium. The physical substance—liquid, solid, or gas—inside a laser that produces optical gain (amplification) by converting energy from an outside source into coherent laser light.

LATENT ENERGY: Also called latent internal energy or latent heat. The latent energy of a physical system is energy stored in the system by virtue of its phase (liquid, solid, or gas). For example, water contains latent energy that is released during freezing (i.e., freezing water releases latent energy as heat).

LATITUDE: The angular distance north or south of Earth's equator measured in degrees.

LAW OF LARGE NUMBERS: In the mathematics of probability, the statement that a large number of random experiments of the same type will, over many trials, produce outcomes that average out to the expected value, that is, the value predicted for the random variable describing those experiments. For example, the number of times that heads appears when flipping a fair coin is described by a random variable with an expected value of 50%, i.e., heads and tails are equally likely to appear. For a small number of flips, heads may not appear on 50% of flips: indeed, for three flips, the closest heads can possibly appear to half is 1 out of 3 (33%) or 2 out of 3 (66%) of the time. But if many flips—hundreds or thousands—are tried, the percentage of heads will inevitably converge to 50%, as predicted by the law of large numbers.

LAW OF SUPERPOSITION: In geology, the law of superposition states that barring the overturning of rock layers by later processes, deeper layers are always older than shallower layers. The law holds because (apart from later disruptions) sediments laid down earlier must be on the bottom and later materials must be on the top, like ingredients being added to an open-faced sandwich.

LENS: A lens is any substance that transmits or refracts light. Gravity can act as a lens and bend light near very massive objects such as stars. Water can act as a lens to refract (bend) light. In common use (such as the lens found in eyeglasses), a lens is often made of glass or similar substance to specifically bend the path of light so as to either focus it on a certain point or cause it to diverge away from it's established path. A simple lens has one refracting element or body. A compound lens is an array of multiple lenses. A lens may also act to refract light in portions of the electromagnetic spectrum outside that of visible light (e.g., radio lens, microwave lens, ultraviolet lens, etc.) Also, an almost clear, biconvex structure in the eye that, along with the cornea, helps to focus light onto the retina. It can become infected with inflammation, for instance, when contact lenses are improperly used.

LEPTONS: A group of subatomic particles not composed of quarks that includes electrons, muons, and tau particles.

LIBRARY: A collection of documents or records, whether printed, digitized, or stored in any other medium. In modern usage, the ability to access information stored outside the library, such as on distant computers accessed through the Internet, is part of the definition of a library. In computer science, a library is a collection of related software entities accessed by one or more programs (e.g., a font library accessed by word processing programs and graphics programs).

LICENSE: An official government permission to carry on a certain activity. While science is not a licensed profession, medicine is. To practice as a doctor in the United States, one must possess a license granted by the state in which one has attended medical school. In most other countries, medical licensure is granted by the central government.

LIFESTYLE DRUG: Any medication that is used to treat a non-medical condition such as wrinkles or baldness. The term is sometimes applied to medications for non-life-threatening conditions such as impotence.

LIGATION: The surgical tying-off or pinching of a tube in the body so that it is closed. Ligation is used in surgery to prevent blood loss from major vessels; ligation of the fallopian tubes, termed tubal ligation, is a method of sterilizing women and other female mammals.

LIGHT YEAR: Despite its name, the light year is a unit of distance, not time. The distance traveled by light in a vacuum in a period of one year, equal to 5.878 trillion miles (9.4607×10^{12} km). The nearest star to our own sun, Proxima Centauri, is a little over four light years away.

LIMNOLOGY: The scientific study of lakes, ponds, and other freshwater bodies.

LINCEI: (Italian for "lynx") The short name for Accademia dei Lincei, a science academy in Italy that was founded in 1603 by Italian botanist Federico Cesi (1586–1630).

LIPID: A fat or oil; a chemical compound used as a source of energy, to provide insulation, and to protect organs in an animal body.

LIPOPROTEIN: A large molecule composed of a lipid, such as cholesterol, and a protein.

LITHOTOMY: (From the Greek *lithos*, stone.) Any surgical operation to remove a stonelike object from the body, such as a gallstone or stone in the bladder.

LITTLE ICE AGE: A cold period that lasted from about AD 1550 to about AD 1850 in Europe, North America, and Asia. This period was marked by rapid expansion of mountain glaciers, especially in the Alps, Norway, Ireland, and Alaska. There were three maxima, beginning about 1650, about 1770, and 1850, each separated by slight warming intervals.

LOBOTOMY: Also termed psychosurgery. An operation in which damage is deliberately caused to the prefrontal cortex of the brain, a part of the brain associated with conscious behavior, socializing, and complex intellectual activity. The purpose of the operation is to produce a more passive personality in persons either mentally ill or supposed by others to be mentally ill, however, its effects can include destruction of the personality. The operation was particularly popular in the United States from the 1930s through the 1970s. Today, lobotomy is often viewed as a violation of human integrity; in 1996, Norway announced that it was paying compensation to all living lobotomy victims.

LOGISTIC MAP: An equation that describes the growth or shrinkage of a hypothetical population of organisms. It is noted for its chaotic long-term behavior: That is, very slight differences in initial conditions lead to widely different populations at future times.

LONGITUDE: The angular distance from the Greenwich meridian (0 degree), along the equator. This can be measured either east or west to the 180th meridian (180 degrees) or 0 degree to 360 degrees W.

LORAN CHART: The Long Range Navigation (LORAN) system was set up after World War II

(1939–1945) to aid ocean surface navigation. Radio pulses from fixed stations are compared by LORAN receivers on board ships; the time differences between receipt of pulses from different stations can be used to calculate position. LORAN charts are maps that show where different pulse-timing differences will be observed on the surface, drawn as colored lines on the map.

LUMINOSITY: The amount of light given off by an object. In astronomy, brightness is the amount of light from a star as seen on Earth, while luminosity is the amount of light actually given off by the star. Closer stars look brighter even though they may or may not be more luminous. Luminosity is measured in units of absolute magnitude, given by a mathematical formula stating how bright a given star would look if located at a standard distance from Earth.

LYCEUM: The garden in Athens where Greek scientist and philosopher Aristotle (384–322 BC) taught his students.

LYMPH: A fluid that runs through the lymphatic vessels, lymph nodes, and other lymphatic organs.

LYMPHOCYTE: A type of white blood cell that functions as part of the lymphatic and immune systems by stimulating antibody formation to attack specific invading substances.

LYSENKOISM: A type of pseudoscience, named after its founder, Trofim Lysenko (1898–1976), that arose in the Soviet Union in the 1930s and destroyed Soviet biology for decades. Lysenkoists denounced modern evolutionary biology and genetics.

LYSOZYME: An enzyme, discovered by Alexander Fleming (1881–1955), that kills bacteria.

M

MAGELLANIC CLOUDS: Small, irregularly-shaped galaxies near our own galaxy, the Milky Way, visible in the night sky of the Southern Hemisphere. They are named after Portuguese explorer Ferdinand Magellan (1480–1521), who was the first European to note them.

MAGNETIC FIELD: The space around an electric current or a magnet in which a magnetic force can be observed. Stationary magnets or moving electric charges will experience a magnetic force—a physical push or pull—inside a magnetic field.

MAGNETIC RESONANCE IMAGING (MRI): A form of medical imaging developed in the 1970s. In MRI, the body is placed in a strong magnetic field that aligns the nuclei of the body's atoms. The nuclei are then struck with brief pulses of radio-frequency electromagnetic radiation. Radio energy re-radiated from the nuclei is recorded by detectors around the body and processed by a computer to yield a cross-sectional image or, in some devices, a three-dimensional image.

MAIN SEQUENCE: A range of star types that includes most stars. It appears as a wavy diagonal stripe on the Hertzsprung-Russell diagram, a color-versus-brightness graph on which each star is represented as a dot. Typically, a newly formed star appears on the main sequence, remains there until it has exhausted its original nuclear fuel, then migrates off the main sequence to become a red giant. Eventually it migrates again to become a white dwarf (also off the main sequence).

MAJOR HISTOCOMPATIBILITY COMPLEX (MHC): The proteins that protrude from the surface of a cell that identify the cell as "self." In humans, the proteins coded by the genes of the major histocompatibility complex (MHC) include human leukocyte antigens (HLA), as well as other proteins. HLA proteins are present on the surface of most of the body's cells and are important in helping the immune system distinguish "self" from "non-self" molecules, cells, and other objects.

MALIGNANT: Cancerous.

MARICULTURE: The raising of plants or animals in seawater, usually for food, whether in the actual ocean or in an enclosure of some kind. Kelp, shrimp, and pearl oysters are examples of maricultured species.

MARINE: Refers to the ocean.

MARINER: A sailor. Also, Mariner refers to any of a series of ten robotic spacecraft launched by the United States from 1962 to 1973. The Mariners visited Venus, Mars, and Mercury.

MARSUPIALS: A subclass of mammals. All marsupials bear their young at an early, almost embryonic, stage of development and then care for them inside a skin pouch called the marsupia. Opossums and kangaroos are typical marsupials.

MASS: Measure of the total amount of matter in an object. Also, an object's quantity of matter as shown by its gravitational pull on another object.

MASS EXTINCTION: A type of extinction event characterized by high levels (or rates) of species extinction in a geologically short period of time.

MASTER: In computer science and electrical engineering, a device that controls another device is sometimes termed a master. In this case, a device controlled by the master is termed a slave.

MATERIAL THEORY OF HEAT: Several material theories of heat were proposed in previous centuries. According to these theories of heat, heat is a material substance or fluid; when one body is heated and another cooled, this fluid is supposed to flow from the hot to the cool body. Since the nineteenth century it has been known that heat is actually the rapid, random motion of atoms and molecules, not a substance in its own right.

MATRILINEAL: Refers to descent through the mother. This descent may be cultural—for example, Jewishness is often defined as descent from a Jewish mother—or biological, as in the case of mitochrondrial DNA. Mitochondria are organelles found inside cells that have separate DNA from the rest of the cell; their DNA is descended from the egg cell supplied by the mother, since sperm contain far fewer mitochrondria and the few mitochondria they do contain do not enter the egg during fertilization.

MATRIX (PLURAL: MATRICES): In mathematics, a matrix is an array of numbers, variables, or equations arranged in rows and columns. A matrix is often a two-dimensional rectangular array like a chessboard, but matrices may also have more dimensions.

MATRIX MECHANICS: A form of quantum mechanics that provides a mathematical description of the physical laws governing the behavior of atomic- and subatomic-size objects. It was proposed in 1925 by German physicists Max Born (1882–1970), Werner Heisenberg (1901–1976), and Pascual Jordan (1902–1980).

MATTER: Anything that has mass and takes up space.

MAUNDER MINIMUM: A historic dip in the number of sunspots from about 1645 to 1715. During this period the European climate was unusually cold, but this may have been a coincidence. The Maunder minimum is named after English astronomer Edward Maunder (1851–1928), who identified its occurrence from historical records in 1893.

MAXWELL'S EQUATIONS: In 1861, Scottish physicist James Clerk Maxwell (1831–1879) published a set of equations describing the interactions of the electric and magnetic fields. These equations later came to be known as Maxwell's equations and are still basic to physics and engineering.

MEAN FREE PATH: The mean free path of a free-moving particle is the average distance it travels before striking an obstacle (e.g., another particle). The mean free path is longer in less-dense media such as diffuse gases.

MECHANICAL EQUIVALENT OF HEAT: The amount of work that must be done to produce a certain amount of heat. Mechanical work can be converted into heat by friction. Heat and mechanical work (force acting through distance) are interchangeable because heat consists of objects (particles) in motion and thus is essentially mechanical.

MECHANICAL PHILOSOPHY: Mechanical philosophy was a school of thought that prospered in the 1600s and asserted that the universe consists entirely of atoms in motion obeying mechanical laws (where the term "mechanical" refers to the motions and interactions of solid objects, not necessarily to the behavior of artificial machines).

MECHANICAL THEORY OF HEAT: Developed in the mid-nineteenth century and refined since, the theory (system of explanations) stating that heat is not a substance or fluid in its own right but consists entirely of mechanical motions, namely, the motions of particles. In a solid, these motions consist of oscillations or vibrations around a fixed point; in a gas or liquid, they consist of rapid straight-line movement interrupted frequently by collisions with other particles.

MEIOSIS: Process of cell division by which a diploid cell produces four haploid cells.

MENHIR: A large standing stone put in place by human beings, usually in the distant past. Menhirs had religious significance for their makers; often this religious meaning was combined with astronomical properties, such as alignment with the sun or moon at a solstice or other special time.

MESON: Subatomic particles composed of a quark and an antiquark; they participate in strong force interactions.

MESSENGER RNA (mRNA): A long, chainlike molecule that carries information in the cell (hence "messenger" RNA). mRNA is manufactured in the nucleus of the cell, where the chemical code of a section of DNA is copied to an mRNA molecule. This molecule is then transported out of the cell to one of the cell's ribosomes. A ribosome is a molecular factory that uses mRNA as a blueprint for making proteins. Once the mRNA has been used by the ribosome to make a protein, it is broken down by other chemicals in the cell.

METABOLISM: The sum of all the physiological (physical and chemical) processes that take place in a living organism by which the organism maintains life, including the breakdown of substances to provide energy for the organism.

METAMORPHIC: In geology, a sedimentary rock that has been partially melted and then allowed to re-solidify. This changes (metamorphoses) the character of the original rock.

METAPHYSICS: The branch of philosophy that is concerned with basic concepts such as time, space, being, reality, and the like. Especially in the twentieth century it has interacted closely with physics, which is also concerned with such questions.

METASTASIS: Spreading of a cancerous growth by shedding cells that grow in other locations.

METEOROLOGY: The science that deals with Earth's atmosphere and its phenomena and with weather and weather forecasting.

METRIC SYSTEM: A system of measurement used by all scientists and in common practice by almost every nation of the world except a few countries, such as the United States.

MICROAEROPHILIC: A type of bacteria that requires oxygen to live, but in relatively small amounts (less than 20% of the concentration found in the atmosphere). Several disease-causing organisms, such as the bacteria that cause Lyme disease, are microaerophilic.

MICROBE: Any microorganism.

MICROBIOLOGY: Branch of biology dealing with microscopic forms of life.

MICROFILM: Microfilms are miniature films used for photographing objects and documents. The images on these films cannot be seen without an optical aid, either in a form of a magnifying glass or a projector.

MICROORGANISMS: All fully-grown, independent living things too small to be seen with the naked eye. Individual cells in the tissues of multicellular organisms are not considered microorganisms, because they are not independent organisms.

MICROWAVE: Electromagnetic radiation with wavelengths between about .039 inches (1 mm) and 3.3 feet (1 m).

MILKY WAY: The galaxy in which our solar system is located.

MITOCHONDRIA: Cellular organelles, of round and elongated shapes, that are found in the cytoplasm and produce adenosine triphosphate (ATP) near intra-cellular sites where energy is needed. Shape, amount, and intra-cellular position of mitochondria are not fixed, and their movements inside cells are influenced by the cytoskeleton, usually being in close relationship with the energetic demands of each cell type. For instance, cells that have a high consumption of energy, such as muscular, neural, retinal, and gonadic cells present much greater amounts of mitochondria than those with a lower energetic demand, such as fibroblasts and lymphocytes. Their position in cells also varies, with larger concentrations of mitochondria near the intra-cellular areas of higher energy consumption.

MITOSIS: Process of cell division resulting in the formation of two daughter cells genetically identical to the parent cell.

MODERNISM: A style of thought, closely associated with art, literature, and architecture, that arose and flourished in the late nineteenth and early twentieth centuries. Modernists tended to assume that ongoing improvement of the human condition ("progress") was possible through logical thought, applied science, and the construction of rational, unornamented buildings.

MOLECULE: Two or more atoms chemically combined.

MOMENTUM: The mass of a moving object multiplied by its velocity.

MONISM: Any philosophy that denies the existence of duality or division in some aspect of being, such as between mind and body or nature and God. The term can refer to a philosophy or theology that denies the reality of all distinctions whatever.

MONOCHROMATIC LIGHT: Light can be treated, in many circumstances, as a wave traveling through space. Although the velocity of this wave is always the same in a given medium, its rate of vibration can vary. Light waves with rates of vibration (frequencies) in a certain range are visible to the naked eye and so are called visible light. In the visible range, we perceive different frequencies as distinct unmixed colors. Light of a single frequency, unmixed with other frequencies, is therefore termed monochromatic (single-colored) light.

MONOGENY: Monogeny, also called monogenesis, is a theory derived from literal readings of the biblical book of Genesis, which states that all human beings descended from a single original couple (e.g., Adam and Eve). Monogenesis is no longer a scientific the-ory. It is now known that human beings evolved as a population from populations of pre-human creatures. We do not descend from a literal First Couple and our genetic characteristics show that we have never experienced a population bottleneck as small as a single breeding pair.

MONOPHYLETIC: Term applied to a group of organisms that includes the most recent common ancestor of all of its members and all of the descendants of that most recent common ancestor. A monophyletic group is called a clade.

MOORE'S LAW: For about 40 years, the number of electronic components that can be manufactured on a single integrated circuit (microchip) at a certain cost has doubled every few years. This trend has been described as Moore's law since 1965, when it was identified by U.S. engineer Gordon Moore (1929–).

MORPHOGENESIS: In geology, the development of distinct landforms. In biology, it is an aspect of developmental biology, that is, the biology of the growth of the individual organism. Specifically, it is the development in the growing individual of formed organs and limbs, as distinct from the replication of individual cells or the differentiation of cells into various tissue types (nerve, muscle, epithelium, etc.).

MOTIVE POWER: The motive power of a device is the source of energy that makes it go. The term is usually applied to machines that actually move (e.g., vehicles).

MRI: MRI stands for magnetic resonance imaging, a form of medical imaging developed in the 1970s. In MRI, the body is placed in a strong magnetic field that aligns the nuclei of the body's atoms. The nuclei are then struck with brief pulses of radio-frequency electromagnetic radiation. Radio energy re-radiated from the nuclei is recorded by detectors around the body and processed by a computer to yield a cross-sectional image or, in some devices, a three-dimensional image.

MULSUM: A mixture of honey and wine served as an appetizer at formal meals in ancient Rome.

MULTIVARIATE ANALYSIS: In the mathematics of probability and statistics, multivariate (many-variable) analysis is the mathematical characterization of data using more than one random variable.

MULTIVERSE: Some physicists and cosmologists propose that the universe we observe is only one of an infinite number of universes, either superimposed in this space (in the case of the quantum multiverse) or, in other theories, located at distances so extreme that they cannot be observed. It is possible that both types of multiverse exist. Multiverse theories are suggested by the equations of physics, but not, so far, by observation. Direct observation of multiverses will presumably never be possible, but indirect evidence for or against their existence may someday be available.

MUST: Grape juice that is going to be fermented to make wine, or is in the process of being fermented.

MYCOLOGY: The branch of biology dealing with fungi.

MYOPIA: A synonym for near-sightedness. It is caused by elongation of the eyeball in the direction of vision so that images of far-away objects are focused inside the eye rather than on the retina (the light-sensitive surface at the back of the eye's interior).

MYXEDEMA: Underactivity of the thyroid gland causes a number of symptoms, including fluid swelling (edema) of the tissues under the skin. This imparts a waxy, smooth appearance to the skin. Either this swelling or the whole complex of thyroid insufficiency symptoms, including weight gain and mental retardation, can be termed myxedema. An obsolete (now offensive) term for the symptoms of congenital thyroid insufficiency is cretinism.

N

NANOTECHNOLOGY: Nanotechnology describes device components that are generally less than 100 nanometers (1/1,000,000 of a millimeter), thus making them on a molecular scale.

NANOTUBE: In materials science, a tube-shaped structure that is molecular in size. Such tubes rarely occur in nature but can be produced in large quantities in the laboratory and are increasingly being put to technological use. Carbon nanotubes, in particular, have useful electrical, optical, and mechanical properties, being extremely strong by weight and excellent conductors of electricity.

NATURAL PHILOSOPHY: A term first used in the late 1300s, "natural philosophy" signifies the study of natural objects and forces—what today is more likely to be termed physical science or physics.

NATURAL SELECTION: Also known as survival of the fittest; the natural process by which those organisms best adapted to their environment survive and pass their traits to offspring.

NATURALISM: Philosophical naturalism is the doctrine that nature is all that exists: There are no souls, spirits, or gods. Methodological naturalism is the practice of looking only for natural (physical) explanations for observable phenomena, while excluding the possibility that spirits, God, gods, or the like have caused phenomena. Methodological naturalism is basic to modern science, but does not require adherence to philosophical naturalism; that is, religious believers may, as scientists, be methodological naturalists without being philosophical naturalists.

NEBULA (PLURAL: NEBULAE): A patch of the night sky that appears cloudy to the naked eye.

NEBULAR THEORY: First proposed by Swedish scientist Emanuel Swedenborg (1688–1772) in 1734, the idea that the sun and its planets formed by condensation of matter from a cloud of dust or gas (a nebula). In more complex form, this remains the scientific explanation for the formation of all planetary systems.

NEUROLOGY: The scientific study of the nervous system, especially its structure, functions, and abnormalities.

NEURON: The basic cellular unit of the nervous system. The unique morphological and intercellular structure of the neuron is dedicated to the efficient and rapid transmission of neural signals. Within the neuron, the neural signal travels electrically. At the synapse, the gap between neurons, neural signals are conveyed chemically by a limited number of chemicals termed neurotransmitters. Specialized parts of the neuron facilitate the production, release, binding, and uptake of these neurotransmitters.

NEUROSURGERY: Any surgery involving the nervous system, but the term usually refers to surgery involving the brain or spinal cord.

NEUROTRANSMITTER: A chemical that transmits electrical impulses (information) between nerve cells or nerve and muscle cells.

NEUTRINO: A high-energy subatomic particle resulting from certain nuclear reactions that has no electrical charge and no mass, or such a small mass as to be undetectable.

NEUTRON: A subatomic particle with a mass of about one atomic mass unit and no electrical charge that is found in the nucleus of an atom.

NEUTRON STAR: An extremely dense object that remains after a massive star explodes in a supernova. Neutron stars have radii of about 6.2 miles (10 km) and masses of about 1.4 to 3.0 solar masses.

NEW SYNTHESIS: Also referred to as the neo-Darwinian synthesis, modern synthesis, or modern evolutionary synthesis. The combination of nineteenth-century Darwinian insights into natural selection operating on random heritable variations, as described by English naturalist Charles Darwin, (1809–1882), with mathematically rigorous descriptions of heredity based originally on the work of Gregor Mendel (1822–1884). This synthesis was achieved by a number of scientists from the 1910s through the 1940s.

NICHE: The portion of the environment that a species occupies, defined in terms of the conditions under which an organism can survive. It may be affected by the presence of other competing organisms.

NIHILISM: The attitude or doctrine that the universe as a whole, and human life in particular, are meaningless.

NODE: A place where the amplitude of vibration is zero. In botany, the point on the stem where leaves are attached; or the point of branching of the stem.

NOMENCLATURE: From the Latin *nomen*, name, and *clatura*, calling. The system of words used in a particular field of knowledge, scientific or otherwise, for naming objects of study. In science, Greek and Latin words are often used to construct words in nomenclature.

NOMINALIST: A person who affirms nominalism, the philosophical doctrine that universal concepts or abstractions (including mathematical abstractions) have no reality except as names (hence "nominalism," from the Latin *nomen*, for "name").

NONINVASIVE: In medicine, noninvasive practices are those that do not penetrate the body. They may employ natural orifices (e.g., mouth, anus) to place sensors or other devices inside the body but do not require breaking the skin. Ultrasound, x rays, MRI, and other imaging modalities are all noninvasive.

NONNATIVE: A nonnative species is one that has been suddenly and recently introduced to an ecosystem, especially by human action, whether deliberate or accidental.

NOSOCOMIAL INFECTION: A nosocomial infection is an infection that is acquired in a hospital. More precisely, the Centers for Disease Control in Atlanta, Georgia, defines a nosocomial infection as a localized infection or one that is widely spread throughout the body that results from an adverse reaction to an infectious microorganism or toxin that was not present at the time of admission to the hospital.

NOSOLOGY: Nosology is the scientific classification of diseases.

NOVA: A star that suddenly increases in light output and then fades away to its former obscure state within a few months or years.

NUCLEAR MAGNETIC RESONANCE SPECTROSCOPY: A highly sensitive spectroscopic method used to analyze atoms with inherent magnetic properties (known as nuclear spin).

NUCLEAR TRANSFER TECHNOLOGY: Nuclear transfer technology is that collection of laboratory methods that allows scientists to transfer the nucleus of one cell into another. This technology is used in some cloning methods (methods for artificially producing genetically identical organisms).

NUCLEIC ACID: A long, ladder-like molecule built of chained nucleotides (a type of chemical compound). RNA and DNA (ribonucleic acid and deoxyribonucleic acid), the molecules used by all living things to create offspring and run the chemical processes of cells, are nucleic acids.

NUCLEON: A nucleon is a proton or neutron. The term derives from the fact that all atomic nuclei are made solely of protons and neutrons.

NUCLEOSYNTHESIS: The process of building up the nuclei of atoms heavier than hydrogen. The most common hydrogen nucleus is a single particle (a proton); all other elements possess more than one proton and at least as many neutrons as protons. Nucleosynthesis began with the big bang, which produced hydrogen, helium, and some lithium; all later nucleosynthesis has occurred in the hearts of stars. All elements heavier than hydrogen of which Earth and humans are made were forged in stellar interiors by nucleosynthesis.

NUCLEOTIDE: The basic unit of a nucleic acid. It consists of a simple sugar, a phosphate group, and a nitrogen-containing base.

NUCLEUS: Any dense central structure can be termed a nucleus. In physics, the nucleus of the atom is the tiny, dense cluster of protons and neutrons that contains most of the mass of the atom and that (by the number of protons it contains) defines the chemical identity of the atom. In astronomy, the large, dense cluster of stars at the center of a galaxy is the galactic nucleus. In biology, the nucleus is a membrane-bounded organelle, found in eukaryotic cells, that contains the chromosomes and nucleolus. Intact eukaryotic cells are comprised of a nucleus and cytoplasm. A nuclear envelope encloses chromatin, the nucleolus, and a matrix, which together fill the nuclear space.

NUTRIENT CYCLE: The cycling of biologically important elements from one molecular form to another and back to the original form.

O

OBJECTIVE: Objective statements are reports of observations that omit emotions and opinions. Such statements can be verified by any observer who has suitable opportunities and equipment; they do not depend on the feelings or viewpoint of the original observer. Ideally, science is based on objective observations.

OEDIPAL COMPLEX: In the psychological theories of Austrian psychologist Sigmund Freud (1856–1939), an Oedipal (or Oedipus) complex is a stage of psychological development during which children view their father as an enemy because he supposedly competes for their mother's love. The complex is named after Oedipus, a character in Greek mythology who killed his father and married his mother (in both cases unknowingly). Freud's claims have been criticized by some philosophers of science as being nonscientific, since they are not subject to objective testing or falsification.

ONTOGENY: The development of an individual multicellular organism from its single-celled beginning (e.g., fertilized egg cell). It is often contrasted to phylogeny, the evolutionary history of a species, and sometimes confused with *ontology*, the philosophical study of the nature of being.

ONTOLOGICAL ARGUMENT: A philosophical argument for the existence of God. It was proposed in the Middle Ages and later famously articulated by French mathematician and philosopher René Descartes (1596–1650). "Ontological" means having to do with the nature of being. According to the ontological argument, God must be perfect, but a perfect being that did not exist would not be perfect; therefore, God must exist. Although forms of the ontological argument are still defended today, most philosophers and theologians have abandoned it as invalid because it attempts to establish a fact (God's existence) through appeal to a mere pattern of words. Persons responding to the ontological argument often point out that one could say "I have the idea of a perfect unicorn, and a perfect unicorn would have the attribute of existence, therefore my unicorn exists," even though there are, in fact, no unicorns.

OPEN REVIEW: In the process of checking scientific claims for publication, the reading of papers by experts whose identities are made known to the authors they are reviewing. This makes reviewers more accountable for their actions as well as giving them credit for their work.

OPERANT CONDITIONING: A type of conditioning or learning in which a person or animal learns to perform or not perform a particular behavior based on its positive rewards or negative consequences.

OPPORTUNISTIC INFECTIONS: Infections caused by viruses or bacteria that are normally not harmful, but which are made dangerous by failures of the immune system or preexisting infections.

OPTICAL ISOMERS: Stereoisomer with chiral or asymmetric centers.

OPTICAL PUMPING: The addition of energy to a laser medium (gas, liquid, or solid with appropriate properties) in order to stimulate the emission of coherent light (laser light). A laser medium may also be pumped using heat, electric current, or chemical reactions in the medium.

ORBIT: The path followed by a body (such as a planet) in its travel around another body (such as the sun).

ORBITAL: In an atom or molecule, a pattern of electron density or probable location that may be occupied by an electron. An orbital is not a path traced in space by an electron, as an orbit is traced by a planet or satellite in space: This is why the term "orbital" rather than "orbit" is used.

ORBITALS: An energy state in the atomic model that describes where an electron will likely be.

ORGANELLE: A membrane-bounded cellular "organ" that performs a specific set of functions within a eukaryotic cell.

ORGANIC: Made of or coming from living matter.

ORGANOMETALLIC: Any chemical compound containing a metal atom bound to an organic group (cluster of atoms containing carbon). Hemoglobin, which contains iron and transports oxygen in blood, is an example of an organometallic compound.

ORGANOTHERAPY: The treatment of human disease using extracts from animal tissues or organs.

OVUM (PLURAL: OVA.): A mature female sex cell produced in the ovaries.

OXYGEN CRISIS (OR OXYGEN CATASTROPHE): The great increase in the amount of free molecular oxygen

(O_2) that occurred about half a billion years ago, just before the diversification of multicellular life known as the Cambrian explosion. Only green plants (e.g., algae) produce O_2 in large quantities. O_2 combines readily with many other molecules and compounds and so does not last long in nature unless continually resupplied. Algae had been producing O_2 for over a billion years by the time of the oxygen crisis, but this oxygen had been absorbed by rocks. When the rocks could no longer absorb O_2 as fast as the algae produced it, the amount of O_2 in Earth's atmosphere increased dramatically.

OZONE LAYER: A layer of ozone that begins approximately 9.32 miles (15 km) above Earth and thins to an almost negligible amount at about 31.07 miles (50 km); it shields Earth from harmful ultraviolet radiation from the sun. The highest natural concentration of ozone (approximately 10 parts per million by volume) occurs in the stratosphere at approximately 15.53 miles (25 km) above Earth. The stratospheric ozone concentration changes throughout the year as stratospheric circulation changes with the seasons. Natural events such as volcanoes and solar flares can produce changes in ozone concentration, but man-made changes are of the greatest concern.

P

PANDEMIC: An outbreak of a disease affecting large numbers of people over a wide geographical area.

PARALLAX: The difference in direction or change of position of an object in the sky when it is viewed from two different points on Earth.

PARAPHYLETIC: Term applied to a group of organisms that includes the most recent common ancestor of all of its members, but not all of the descendants of that most recent common ancestor.

PARASITE: An organism that lives in or on a host organism and that gets its nourishment from that host.

PARTHENOGENESIS: The reproduction from an ovum (egg cell, in animals) without a sperm cell. In parthenogenesis, no male is necessary for reproduction. Many plants and a number of animals are capable of parthenogenesis.

PARTICLE ACCELERATOR: A device for accelerating subatomic particles to very high speeds for the purpose of studying the properties of matter at very high energies.

PATHOGEN: A disease-causing agent, such as a bacteria, virus, fungus, etc.

PATRILINEAL: Descent through the father. This descent may be cultural—for example, in Western culture, family names are usually inherited patrilineally—or biological, as in the case of genes carried on the Y chromosome (which male offspring derive only from the father).

PAULI EXCLUSION PRINCIPLE: In physics, the rule that no two fermions can have the same quantum number, first stated by Austrian physicist Wolfgang Pauli (1900–1958) in 1925. The exclusion principle accounts for the solidity of ordinary matter and the electron-shell structures of atoms, which in turn determine their chemical properties.

PELAGIC: Referring to the open oceans.

PENICILLIN: The first antibiotic, discovered by Sir Alexander Fleming (1881–1955), which is produced by a species of a mold microorganism.

PERIOD: The time it takes for a cyclical phenomenon to repeat. In pulsars, the time between two consecutive pulses.

PERIODIC LAW: Russian chemist Dmitri Mendeleyev (1834–1907) thought there to be a law of nature that would explain why there were regular repetitions of chemical properties when elements were arranged in order of atomic weight.

PERIODIC POTENTIAL: Periodic potential refers to the regular pattern of electric field potential inside a crystal, in which atoms are built up in recurring (periodic) structures like a grid or lattice.

PERMIAN-TRIASSIC EXTINCTION: Some 254.1 million years ago, at the end of the Permian period and the beginning of the Triassic period, the most severe mass extinction on Earth so far occurred. About 95% of all marine species and about 70% of all land-dwelling species disappeared. Various causes have been proposed. Some scientists theorize that several causes, coincidentally occurring near each other in time, caused the extinction.

PERTURBATIONS: In celestial mechanics, deviations of a planet's orbit from the shape predicted for it by Newton's laws.

PET: PET stands for positron emission tomography, a medical imaging technique developed in the 1970s. In PET, a chemical compound is prepared containing atoms of an unstable isotope of carbon, oxygen, nitrogen, or fluorine that emits a positron (antiparticle of an electron) when it decays. This chemical is designed to be absorbed by target tissues in the body. As the unstable atoms in the tissues decay, the positrons they emit encounter nearby electrons. These electron-positron pairs annihilate, as particle-antiparticle pairs always do when united, emitting two high-energy photons that leave the body in opposite directions. These photons are recorded by detectors surrounding the patient, and this information is processed by a computer to yield three-dimensional images of where radioactivity is occurring in the body. This images the tissues that have absorbed the chemical containing the unstable isotope. PET imaging is particularly used in cancer imaging and brain studies.

PEUTINGER TABLE: Also called the Tabula Peutingeriana, the Peutinger Table is a map copied in the 1200s from an original drawn in the 300s or 400s. The Peutinger Table is the only surviving ancient map showing the Roman system of public roads in Europe, northern Africa, and southern and eastern Asia. It is named not after the original artist, who is unknown, but after a sixteenth-century owner of the document, German politician Konrad Peutinger (1465–1547).

PHAGOCYTES: Phagocytes, also called white blood cells, are cells that circulate in the blood and can engulf and destroy disease organisms such as viruses and bacteria by dousing them with digestive enzymes.

PHASE: In the study of waves, a repeating or periodic wave's phase is the relationship of its pattern of peaks and valleys to a fixed reference (e.g. time).

PHASE TRANSITION: A substance that moves from one physical phase to another—gas to solid or liquid, solid to liquid or gas, liquid to solid or gas—is said to undergo a phase transition.

PHENETICS: In biology, the practice of grouping organisms based on numerical analysis of differences and similarities of anatomy (morphology); it is also termed numerical taxonomy and is an alternative to the more common method, cladistics.

PHENOTYPE: The visible characteristics or physical shape produced by a living thing's genotype.

PHILOSOPHER'S STONE: A material thought by alchemists to have the power to bring about the transmutation of metals.

PHILOSOPHICAL ELEMENTS OR PRINCIPLES: The underlying essential constituents of bodies that cannot be directly perceived.

PHLEBOTOMY: Phlebotomy, also known as bloodletting, is the deliberate medical release of blood from the body, whether by allowing leeches to feed on the patient or by cutting a blood vessel. This technique, which was practiced on a wide variety of patients in Western medicine from Greco-Roman times to the nineteenth century, was rarely of benefit and no doubt did far more harm than good. However, there are certain rare medical conditions, such as hemochromatosis, in which the patient has excessive red blood cells and must undergo regular therapeutic phlebotomy.

PHLEGM: The ancient Greek and Roman theory of physiology stated that four basic substances, the humors, exist in the human body, and that disease is caused by an imbalance among the humors. The four humors were blood, phlegm, yellow bile, and black bile; an excess of phlegm was believed to cause dull-wittedness and apathy.

PHOTOELECTRIC EFFECT: The phenomenon in which light falling upon certain metals stimulates the emission of electrons and changes light into electricity.

PHOTON: Smallest individual unit of electromagnetic radiation (light energy). These light particles are emitted by an atom as excess energy when that atom returns from an excited state (high energy) to its normal state. According to modern physics, it is not accurate to imagine any subatomic particle, including a photon, as a tiny, hard ball: Rather, a particle also has wave properties and is spread out through space. Thus, light has wave properties, despite being conveyed by photons.

PHYLOGENY: The evolutionary relationships among organisms; the patterns of lineage branching produced by the true evolutionary history of the organisms being considered.

PHYSICAL GEOGRAPHY: The subdiscipline of geography that concentrates on natural patterns and features of Earth's (or any other planet or moon's) surface, as opposed to patterns and features produced by human activity.

PHYSIOGNOMY: The pseudoscience of judging character from facial features or expressions, practiced from ancient Greek times through the nineteenth century. Physiognomists supposed that inborn vices or virtues could be detected in such features as closely-set eyes, low or high foreheads, bumps on the skull, and the like. Such beliefs were often recorded in literary descriptions of character, especially in nineteenth century writers.

PHYTOPLANKTON: Microscopic marine organisms (mostly algae and diatoms) that are responsible for most of the photosynthetic activity in the oceans.

PION: A pion or pi meson is a type of fundamental particle containing two quarks. There are three pions; all are involved in carrying the strong nuclear force.

PLACEBO: A false treatment that gives a patient the sense that a medicine is working when, in actuality, no medicine was actually received.

PLANCK'S CONSTANT: A fixed number (constant) that appears in quantum physics. Any photon can only carry some integer multiple (i.e., once, twice, three times, N times) of the fixed energy unit or quantum $E = hv$, where E is energy, v is frequency, and h is Planck's constant, whose value is approximately 6.626×10^{-34} Jśs. German physicist Max Planck (1858–1947) described the value of this constant in 1899 and in 1918 was awarded the Nobel Prize in physics.

PLANETARY ASPECTS: Also called astrological aspects or simply aspects, these are angles measured between planets, as seen from Earth, at a given moment. In astrology, these angles are imagined to be of importance in predicting the planetary influence on a human being's character and fate.

PLANKTON: Floating minute animal and plant life.

PLASMA: Matter in the form of electrically charged atomic particles that form when a gas becomes so hot that electrons break away from the atoms. Also, the colorless, liquid portion of the blood in which blood cells and other substances are suspended.

PLATE: Rigid parts of Earth's crust and part of Earth's upper mantle that move and adjoin each other along zones of seismic activity.

PLATONIC SOLIDS: The convex geometric shapes that have for their sides or facets identical polygons with sides of equal lengths. Polygons are flat shapes whose sides consist of straight line segments. Familiar polygons having sides with equal length are the square, pentagon, and equilateral triangle. Only five solid shapes having such shapes as sides or facets are possible, namely the five Platonic solids. The two most familiar Platonic solids are the cube and the tetrahedron (four-sided pyramid).

PLUTONISM: In modern geology, the hardening of magma far below the surface to form igneous rock; granite, for example, is formed by plutonism.

P-N JUNCTION: In electronics, devices of microscopic size can be made by adding small amounts of dopant or contaminant substances to portions of a semiconducting crystal, usually silicon. Some dopants lack electrons and so contribute positive charges to the crystal; others have extra electrons and so contribute negative charges. Positive dopants are termed p-type, negative dopants n-type. Where a p-type region abuts or contacts an n-type region, a p-n junction is created. Electrons from the n-type side of the junction are drawn to the positive charges on the p-type side until electrostatic forces balance out and prevent further charge movement. This creates a slab-shaped region of electric neutrality along the p-n junction. Since this region is depleted of charge carriers, it is termed a depletion zone and cannot carry current. Placing a voltage across the p-n junction that is positive on the p side narrows the depletion zone and allows current to flow through it; applying voltage that is negative on the p side widens the depletion zone so that current cannot flow. A p-n junction thus allows one-directional current flow. Used in this fashion, a p-n junction is termed a diode.

PNEUMA: In the Stoic philosophy of ancient Greece, a person's spirit or life-force. More generally, this word referred to the inhaled and exhaled breath. In modern medical terminology it is the root of lung-related terms such as pneumonia.

POISEUILLE'S CAPILLARY FLOW FORMULA: A mathematical statement relating the volume and velocity of liquid flow through a tube to the tube's diameter, the pressure at both ends, and viscosity (liquidity) of the fluid. It was created by French physician Jean Poiseuille (1799–1869) in the 1840s.

POLARIMETER: An instrument for measuring the polarization of light, that is, the way in which its transverse electromagnetic fields have been restricted to vibration in specific directions.

POLE STAR: The Pole Star is Polaris, a visible star that happens to be aligned with Earth's spin axis so that all the other stars appear to spin around Polaris while it remains fixed. Polaris is visible only in the Northern Hemisphere. In the Southern Hemisphere, no visible star happens to be aligned with Earth's axis, so there is no southern Pole Star.

POLIO: A disease (*poliomyelitis*) caused by a virus (poliovirus [PV]) that can result in muscle weakness, paralysis, or death.

POLYETHISM: In social insects such as ants and bees, the specialization of various individuals for different functions—queen, worker, drone, and so forth. Polyethism can be either morphological (individuals with different specializations have different anatomies) or temporal (individuals of different ages can be assigned different functions). Temporal polyethism is the main form of polyethism in honeybees.

POLYGENIC: In genetics, a trait in an organism that is controlled by more than one gene. Most complex traits are polygenic to the extent that they are gene-controlled. (Almost all gene expression is also influenced by environmental factors such as diet, stress, usage, and the like.)

POLYGENY: In evolutionary biology, the obsolete theory that different human races (also a questionable concept, biologically) had separate evolutionary origins. In genetics, polygeny occurs when a single trait in an organism is controlled by more than one gene, i.e., is polygenic.

POLYGYNY: Having more than one wife; a society where this pattern prevails is polygynous. An approximate synonym for polygyny is polygamy; however, polygamy usually describes having more than one wife in a context (e.g., U.S. law) where this is aberrant or criminal relationship, while polygyny tends to denote a socially accepted practice.

POLYMATH: Any person who is highly capable in several fields of knowledge—for example, physics, mathematics, and music.

POLYPHYLETIC: Term applied to a group of organisms that does not include the most recent common ancestor of those organisms; the ancestor does not possess the character shared by members of the group.

POLYSACCHARIDE: A molecule that consists of several sugar molecules linked together in a long chain of multiple monosaccharides.

POPULATION: A complete set of individuals, objects, or events that belong to some category.

POPULATION GENETICS: The study of changes in gene frequency (the number of organisms possessing a certain version of a gene) in populations. Gene frequencies are affected by random drift, natural selection, founder effect, and other factors.

POPULATION INVERSION: In physics, a population inversion occurs when a greater number of atoms or molecules in a system are in a high-energy state than are in a low-energy state. In a laser, high-energy atoms in a laser medium featuring population inversion emit the coherent photons of laser light.

PORTOLAN CHARTS: Navigational charts produced in Europe and elsewhere during the Middle Ages.

POSITIVISM: The philosophical doctrine that only science can produce valid knowledge. According to positivism, there is no other source of knowledge but science (e.g., religion, intuition, common sense, etc.). This view was first articulated by French philosopher August Comte (1798–1857).

POSITRON: The antiparticle of the electron. It has the same mass and spin as the electron, but its charge, though equal in magnitude, is opposite in sign to that of the electron.

POST HOC: Latin for "after this." The phrase is often used to label a common error in reasoning, namely the assumption that because event B happens after event A, it must have been caused by event A. The error is tempting because when A does actually cause B, B always does follow A; however, B may follow A even when it is not caused by A. In the latter case, the fact that B follows A may be coincidence or occur because both are caused by some third thing C. The corrective to the post-hoc fallacy is sometimes stated as "Correlation does not prove causation," i.e., just because things happen together, are correlated, it does not necessarily mean that one is causing the other.

POSTMODERNISM: A style of thought that tends to reject the belief that any qualities possessed by human beings are inherent in their nature, an idea that postmodernists call essentialism. It also emphasizes that all belief systems, including science, are shaped by relationships of "power," that is, the power held by some human beings over others.

POSTULATE: To postulate is to propose the existence of a thing or rule for the purpose of enabling a chain of reasoning. Generally, postulates are simple propositions that are self-evident or nearly so.

PRAGMATISM: The philosophical doctrine that holds that the truth-value of an idea relates to its practical usefulness; if adopting an idea leads to useful results, the idea can, according to pragmatism, be deemed "true," or at least more true than other ideas that lead to less-useful results.

PRECAMBRIAN: All geologic time before the beginning of the Paleozoic era. This includes about 90% of all geologic time and spans the time from the beginning of Earth, about 4.5 billion years ago, to 544 million years ago. Its name means "before Cambrian."

PRIME MERIDIAN: A meridian is a circle drawn around a planet and passing through its north and south poles; the prime meridian is the meridian arbitrarily (i.e., for no necessary reason) named as the 0th meridian, the one from which numbering of the other meridians shall begin. For example, on Earth, the prime meridian passes through Greenwich, England; on Mars, it passes through the center of crater Airy-0 (zero), selected for that purpose in 1969.

PRIMORDIAL CLOUD: In astronomy, any cloud of dust and gas from which a star, solar system, galaxy, or star cluster condenses under the mutual gravitational attraction of its particles. Often, the cloud from which our own solar system condensed is referred to as "the" primordial cloud.

PRION: A protein having correct chemical composition but a misfolded shape that is capable of inducing similar, normally-folded proteins to become misshapen. Prions are not alive but can cause infectious disease because, if ingested (e.g., by eating meat), they can trigger the formation of misshapen proteins in a previously uninfected organism. The body's chemical mechanisms for removing the normally-folded version of the protein do not work, so masses of the misfolded proteins accumulate in nerve tissues, eventually causing mental breakdown and death. In humans, infection by prions causes variant Creutzfeldt-Jakob disease.

PRISM: A triangular or wedge-shaped block of glass that breaks up light into its constituent colors.

PROBABILITY: The degree of chance that something might happen.

PROKARYOTE: An organism (usually single-celled) with a cell or cells that lack a true nucleus, which is a saclike structure that contains the cell's DNA and associated proteins. In prokaryotes, DNA floats freely about the interior of the cell. Bacteria and archaea are prokaryotes; single-celled or multicellular organisms that have a cell nucleus are called eukaryotes.

PROPER MOTION: The movement of stars across the sky caused by their relative motion to Earth.

PROTEIN: A complex chemical compound that consists of many amino acids attached to each other that are essential to the structure and functioning of all living cells.

PROTEOMICS: Proteomics is a discipline of microbiology and molecular biology that arose from the gene sequencing efforts that culminated with the publication of the sequence of the human genome in 2001. In addition to the human genome, sequences of disease-causing bacteria and other microorganisms continue to be deduced, although fundamental knowledge of the sequence of nucleotides that comprise deoxyribonucleic acid (DNA) reveals only a portion of the encoded protein structure. Proteins are an essential element of bacterial structure and function; for example, a variety of proteins enable a microorganism to establish and maintain an infection. Thus, knowledge of the three-dimensional structure and associations of proteins is vital for a full understanding of microorganism behavior and operation. Proteomics is an approach to unravel the structure and function of proteins.

PROTOCONTINENT: A landmass that is in the process of being developed by geological processes into a true or major continent.

PROTON: Particle found in a nucleus with a positive charge. The number of these gives the atomic number.

PROXIMATE: Nearby. For example, the closest star to our sun is Proxima Centauri, about 4.22 light years away.

PSEUDOSCIENCE: Arguments or ideas, often laced with scientific terminology or bizarre calculations, based on theories developed outside of the scientific method and thus not subject to scientific validation.

PSYCHIATRY: The branch of medicine dealing with the study of the mind and the diagnosis, treatment, and prevention of emotional and behavioral disorders.

PSYCHOANALYSIS: The method of analyzing psychic phenomenon and treating emotional disorders that involves treatment sessions during which the patient is encouraged to talk freely about personal experiences, especially about early childhood and dreams.

PSYCHOLOGY: The study of the mind, especially as it relates to behavior.

PSYCHROMETER: An instrument designed to measure dew point and relative humidity, consisting of two thermometers (one dry bulb and one wet bulb). The dew point and humidity levels are determined by drying the wet bulb (either by fanning or whirling the instrument) and comparing the difference between the wet and dry bulbs with preexisting calculations.

PULSAR: A rapidly rotating neutron star that has a strong magnetic field.

PULSED LASER: Lasers may either be operated continuously or in bursts. Those that are kept on for relatively long intervals are termed continuous lasers, and those that are turned on in brief flashes or bursts are termed pulsed lasers. Pulsed lasers can produce extremely brief pulses, on the order of a few femtoseconds (a femtosecond is a millionth of a billionth of a second, 1×10^{-15} second).

PUMPING: In lasers, the provision of energy to the laser medium; some of this energy is converted by the laser medium to coherent (phase-matched) laser light. Pumping may be accomplished using light, electric current, or chemical reactions.

PUNCTUATED EQUILIBRIUM: A model of evolution in which change occurs in relatively rapid bursts, followed by longer periods of stasis.

PURE MATHEMATICS: The pursuit and expansion of mathematical knowledge and technique without regard to whether the mathematics produced will ever have a technological application. Pure mathematics is often contrasted with applied mathematics. Often, pure mathematics has turned out to have unforeseen applications.

PURGE: In power politics, a campaign to remove (often violently) persons of certain ethnic backgrounds or beliefs from a government agency or other segment of the population. In medicine, to purge is to empty the bowels by taking a laxative.

PYRHELIOMETER: An instrument for measuring the intensity of solar radiation received at Earth's surface. The Eppley pyrheliometer measures not only the intensity but the duration of solar radiation, allowing the total insolation (exposure to the sun) at the instrument's location over a given period to be calculated.

Q

QUADRANT: One of the four regions in the Cartesian coordinate system formed by the intersection of the x- and y-axes.

QUANTUM: Something that exists in discrete units.

QUANTUM MECHANICS: A system of physical principles that arose in the early twentieth century to improve upon those developed earlier by Isaac Newton (1643–1727), specifically with respect to submicroscopic phenomena.

QUANTUM NUMBER: A variable in quantum mechanics that specifies the state of electrons in an atom.

QUANTUM PHYSICS: Quantum physics, also called quantum mechanics, is the science of the behavior of matter and energy at any scale where their quantum nature is significant (which is usually, though not always, the atomic or subatomic scale). The quantum nature of matter and energy refers to the restriction of changes to definite, sudden steps rather than smooth passage through a range of in-between values. Quantum physics began to be developed in the early twentieth century and continues to be one of the most vigorously investigated areas of physics.

QUANTUM TUNNELING: The appearance of a particle on the other side of a barrier (e.g., an electric field) that it does not have the energy to overcome. Tunneling occurs because every particle can be potentially observed at all points in the universe at any future time, only with much slighter probability at points far from its nominal location; if a particle is close to a barrier, then by the laws of quantum physics it can, with a certain probability, simply appear on the other side of the barrier without ever having energy enough to get over. It is as if a motionless marble sitting in a cup suddenly appeared on the table beside the cup—indeed, the probability of this event, although very small, is not, according to quantum physics, zero. Quantum tunneling is used in some microelectronic devices, such as tunnel diodes.

QUARANTINE: The practice of separating people who have been exposed to an infectious agent, but have not yet developed symptoms, from the general population. This can be done voluntarily or involuntarily by the authority of states and the federal Centers for Disease Control and Prevention.

QUARKS: Subatomic elementary particles that combine in threes to form other particles (neutrons, protons, and others) and are never observed alone. Quarks come in six varieties or flavors—the term is arbitrary and has nothing to do with the sense of taste—namely, charm, strange, top, bottom, up, and down.

QUASAR: A small, powerful source of energy that looks like a star but is believed to be the active nucleus of a distant galaxy.

R

RACEMIC: In optics, some transparent compounds rotate polarized light to the right (dextrorotatory compounds); others rotate it to the left (levorotatory compounds). A racemic substance is a mixture of equal parts of dextrorotatory and levorotatory compounds.

RADIATION: Energy transmitted in the form of electromagnetic waves or subatomic particles.

RADIATION THERAPY: Use of radioactive substances to kill cancer cells in the human body.

RADIATIVE FORCING: A change in the balance between incoming solar radiation and outgoing infrared radiation. Without any radiative forcing, solar radiation coming to Earth would continue to be approximately equal to the infrared radiation emitted from Earth. The addition of greenhouse gases traps an increased fraction of the infrared radiation, reradiating it back toward the surface and creating a warming influence (i.e., positive radiative forcing) because incoming solar radiation will exceed outgoing infrared radiation.

RADICAL THEORY: Also called the free-radical theory. In chemistry, the idea, first proposed in the 1950s, that aging is the result of accumulated biochemical damage to body cells from free radicals, charged atoms or molecules with a strong tendency to participate in chemical reactions (beneficial or not). The theory is not widely accepted among biologists studying aging, though there is agreement that free radicals can cause some disease conditions that progress with age.

RADIO GALAXY: A galaxy with an active nucleus that shines brightly in radio wavelengths.

RADIOACTIVE DECAY: The predictable manner in which a population of atoms of a radioactive element spontaneously disintegrates over time:

RADIOACTIVITY: The property possessed by some elements of spontaneously emitting energy in the form of particles or waves by disintegration of their atomic nuclei.

RADIOMETRIC DATING: The use of naturally occurring radioactive elements and their decay products to determine the absolute age of the rocks containing those elements.

RADIOMETRY: The measurement of optical electromagnetic radiation, which is defined as electromagnetic radiation in the frequency range of 3×10^{11} and 3×10^{16} Hz (cycles per second). Optical radiation includes visible light but is not limited to it.

RADIOSONDE: An instrument for collecting data in the atmosphere and then transmitting that data back to Earth by means of radio waves.

RANKING: An ordering of things or variables in some linear order according to some standard of comparison, first to last or highest to lowest. Non-parametric statistical analysis often deals with rank data such as letter grades in academic settings or place winners in sports.

RAREFACTION: A decrease in density. Also, a region of space with a lower-than-normal density. Particularly, a region of low density in a sound wave traveling through a gas such as air.

RATIONALIST PHILOSOPHERS: All philosophers who privilege or emphasize the use of reason in the quest for truth, while downplaying faith, experience, intuition, or other ways of knowing. Non-rationalist philosophers are not, in general, anti-reason, but see reason as one mode of mental activity leading to knowledge or wisdom, rather than as the only admissible one.

RECAPITULATION: In nineteenth-century and early twentieth-century evolutionary biology, the idea that animal embryos replay or recapitulate the evolutionary history of their ancestors (their phylogeny) during individual growth (ontogeny) was commonplace. The idea, often called simply recapitulation or the biogenetic law, was first popularized in 1866 by German biologist Ernst Haeckel (1834–1919) and is often summed up in the phrase "Ontogeny recapitulates phylogeny." However, it was soon discredited by biologists and is now known to be untrue, although in some cases certain features of ancestral development do appear as traces in embryonic development.

RECOMBINANT DNA: DNA that is cut using specific enzymes so that a gene or DNA sequence can be inserted.

REDSHIFT: The lengthening of the frequency of light waves toward the red end of the visible light spectrum as they travel away from an observer; most commonly used to describe movement of stars away from Earth.

REDUCTION: A process in which a chemical substance gives off oxygen or takes on electrons.

REDUCTIONISM: A style of thought that can be contrasted to holism. Reductionism assumes that the properties of any complex system—a human brain, an ecosystem, or other—can be reduced (hence "reductionism") to the interactions of its component parts. Holism asserts that at least some system properties arise out of the system as a whole (hence "holism") and are not possessed by, or reducible to, the properties of the system's isolated, individual parts.

REEF: A large ridge or mound-like structure within a body of water that is built by calcareous organisms such as corals, red algae, and bivalves. A barrier reef is a reef growing offshore from a land mass and separated by a lagoon or estuary, e.g., the Great Barrier Reef of Australia. A patch reef is a discontinuous reef growing in small areas, separated by bare areas of sand or debris, often part of a larger reef complex.

REFLECTING TELESCOPE: A telescope that uses mirrors to form a magnified image of distant objects rather than lenses (as is done in a refracting telescope).

REFORMATION: The Reformation was a movement for reform of the Roman Catholic Church that began in Germany in the early 1500s and led, eventually, to the establishment of Protestantism.

REFRACTING TELESCOPE: A telescope that uses lenses to form a magnified image of distant objects rather than mirrors (as is done in a reflecting telescope).

REFRACTION: The deflection from a straight line of a light ray or other energy beam when passing from one optical medium (such as air) to another (such as glass) in which its velocity is different.

RELATIVISM: In philosophy, the belief that moral and aesthetic values—beliefs about right/wrong and beauty, respectively—have no independent or ultimate truth or falsehood, but are defined relative to particular societies. A relativist will argue that what is wrong or ugly in one society may be right or lovely (or at least unimportant) in another society.

RELATIVITY THEORY: Relativity theory is actually a pair of theories, special relativity and general relativity, that were first proposed by German physicist Albert Einstein (1879–1955) in 1905 and 1915, respectively. Special relativity describes the effects of unaccelerated (constant-speed) straight-line motion on mass, length, and the flow of time; one consequence of special relativity theory is that mass and energy are interchangeable. General relativity describes accelerated and curved-line motion and the curvature of space by mass, which accounts for the force of gravity. Both theories are used throughout modern physics and astronomy and in some engineering applications (e.g., general relativity is applied in global positioning system [GPS] technology). Although relativity theory has been tested thousands of times against observations and experiments and found to be highly accurate, it has not yet (as of 2008) been united with quantum mechanics, a major challenge of modern physics.

RENAISSANCE: In the 1400s through the 1600s, a revival of interest in classical (Greek and Roman) art, philosophy, and literature triggered many changes in European culture; although this period is often termed the Renaissance, historians do not all agree that the period was actually distinct enough from preceding and following times to merit designation as a separate period. The term "Renaissance" (French for rebirth), which was popularized as a label for this period in the nineteenth century, implies that the immediately preceding period was senile or dead, which is not asserted by most modern historians.

RENORMALIZATION: A mathematical method in quantum physics first proposed in 1947 to deal with apparently absurd infinities in the equations of the theory known as quantum electrodynamics. In renormalization, infinities of opposite sign (positive and negative) are allowed to cancel each other out, producing a finite result that matches observation.

REPRODUCTIVE CLONING: A type of cloning that involves somatic cell transfer to induce pregnancy. This type of cloning requires a surrogate mother.

RESECTION: The surgical cutting out of part of an organ.

RESOLUTION: The ability of a sensor to detect objects of a specified size. The resolution of a satellite sensor or the images that it produces refers to the smallest object that can be detected.

RESOLVE: To examine a celestial object through a telescope and find that it is composed of stars.

RESONATOR: Any device that vibrates at a certain favored frequency (resonates). In acoustics and radio engineering, resonators are often hollow cavities that trap waves of a certain length.

RESTORATION ECOLOGY: The subdiscipline of ecology that studies how damaged ecosystems can be restored to something approximating their original condition.

RESTRICTION ENZYME: A special type of protein that can recognize and cut DNA at certain sequences of bases to help scientists separate out a specific gene. The enzymes are used to generate fragments of DNA that can be subsequently joined together to create new stretches of DNA.

RESURRECTIONIST: In England in the early nineteenth century, a resurrectionist or resurrection man was a person who illegally dug up recently buried corpses for sale to medical schools so that medical students might dissect them. The only legal source of corpses at that time was the execution of murderers, but executions had declined, so demand exceeded supply until 1832. In that year, Parliament passed the Anatomy Act, which allowed some other types of corpses to be used for medical dissection.

RETROGRADE MOTION: All planets in our solar system appear to move across the sky relative to the fixed stars. Most of the time they appear to move from west to east, opposite to the sun and moon; however, periodically, due to differences between Earth's orbital motion and theirs, they appear to switch direction and move from east to west. They do not actually reverse direction, but appear to do so. This reversed motion is termed retrograde motion.

RETROROCKET: A rocket fired against the direction of travel of a spacecraft orbiting a planet or moon in order to decrease its velocity and lower its orbit.

RETROSPECTIVE STUDIES: Also called retrospective cohort studies or historical studies. Medical studies in which researchers study the medical records of patients to determine the health influence of diet or some other factor that varies from one patient to another. Such a study is termed retrospective because the patients were not selected with a view to the study, but are studied in retrospect because they happen to be available.

RETROVIRUS: Viruses in which the genetic material consists of ribonucleic acid (RNA) instead of the usual deoxyribonucleic acid (DNA). Retroviruses produce an enzyme known as reverse transcriptase that can transform RNA into DNA, which can then be permanently integrated into the DNA of the infected host cells.

RIBONUCLEIC ACID (RNA): A long, twisted molecule that, similar to DNA (deoxyribonucleic acid), encodes information along its length as a series of molecular subunits (nucleotides). In most cells, information is copied off DNA onto RNA molecules, which are then transported to other parts of the cell and used as templates or blueprints for manufacturing proteins. The earliest forms of life probably used only RNA as their hereditary material; this hypothetical stage of life's history is termed the RNA world. DNA resembles a ladder, with two spiraling side-pieces linked by rungs; RNA resembles a piece of DNA cut lengthwise, a single side-piece with a series of half-rungs hanging off the side.

RIBOSOME: An organelle, that is, a small structure found inside cells. Information from DNA in the cell nucleus is copied to messenger RNA (mRNA) molecules and passed to the ribosomes, which use the mRNA as a blueprint for the manufacture of proteins.

RIBOSOMES: Organelles that play a key role in the manufacture of proteins. Found throughout the cell, ribosomes are composed of ribosomal ribonucleic acid (rRNA) and proteins. They are the sites of protein synthesis.

RIGHT TRIANGLE: A triangle that contains a 90°, or right, angle.

ROCKET: A machine that burns fuel and oxidizer in a combustion chamber and squirts the resulting hot, expanding gases out a nozzle at high speed. As described by Newton's Third Law of Motion, the force that accelerates the gas out of the nozzle is balanced by an equal and opposite force acting against the combustion chamber. This reactive force accelerates the rocket in the opposite direction to the gas.

RUBY: A precious crystalline stone, reddish and translucent when cut and polished, chemically consisting of aluminum oxide plus chromium (Al_2O_3Cr). It was used in some early lasers as the laser medium.

S

SALVARSAN: A synthetic agent used to cure syphilis, introduced in 1911.

SATISFICING: An economics term invented by Nobel Prize-winning economist and artificial intelligence researcher Herbert Simon (1916–2001). It refers to decision-making behavior that does not seek a perfect outcome or solution, but rather seeks a result that is merely satisfactory. Much real-world decision-making takes the form of satisficing rather than of perfect optimizing.

SCANNING ELECTRON MICROSCOPES (SEM): Devices that illuminate a target with a beam of electrons in order to produce highly magnified images of them. The name of the device refers to the fact that it scans a narrow electron beam back and forth over a sample to build up an image. Some scanning electron microscopes can image features as small as 1 nanometer (1 billionth of a meter).

SCANNING PROBE MICROSCOPE: Devices that produce highly magnified images of objects by moving a finely-pointed probe near their surface without touching it. The name of the device refers to the fact that it scans the probe back and forth over a sample to build up an image. Some scanning probe microscopes can image features as small as a single atom.

SCHIZOPHRENIA: A serious mental illness characterized by isolation from others and thought and emotional disturbances.

SCIENTIFIC JOURNAL: A periodical that contains articles that have been reviewed by expert scientists before being accepted for publication. Journals that publish opinions about science or that look scientific but have not been peer-reviewed by recognized scientists with relevant expertise are not scientific journals.

SCIENTIFIC METHOD: Collecting evidence meticulously and theorizing from it.

SCRIBE: In various ancient cultures (e.g., Egypt), a literate person employed to write and read documents for religious, legal, and business purposes. In ancient times the ability to read and write was a rare skill, and the scribe was a valued professional. More generally, any person who copies documents by hand in any historical period can be termed a scribe.

SEARCH ENGINE: Computer programs that search for specific strings of characters in memory. The most commonly used search engines are those that search the World Wide Web (e.g., Safari, Explorer, Firefox, Flock).

SEMICONDUCTOR: Substance, such as silicon or germanium, whose ability to carry electrical current is lower than that of a conductor (like metal) and higher than that of insulators (like rubber).

SENSITIVE DEPENDENCE ON INITIAL CONDITIONS: In a chaotic system, the future state of the system—what orbit a comet ends up following a million years from now, for example—can vary greatly depending on very slight changes in the original state of the system. The original conditions of the chaotic system are termed its initial conditions, and the future state is said to depend sensitively on the initial conditions.

SEPSIS: Also called bacteremia. A bacterial infection in the bloodstream or body tissues. This is a very broad term covering the presence of many types of microscopic disease-causing organisms. Closely related terms include septicemia and septic syndrome. According to the Society of Critical Care Medicine, severe sepsis affects about 750,000 people in the United States each year. However, it is predicted to rapidly rise to one million people by 2010 due to the aging U.S. population. Over the decade of the 1990s, the incident rate of sepsis increased over 91%.

SERUM: The clear, thin, sticky fluid of the blood that contains no cells or platelets.

SERUM SICKNESS: An allergic reaction to an antiserum, which is a preparation of the clear fraction of blood (serum) containing antibodies for a specific disease and administered as a vaccine.

SESSILE: Any animal that is rooted to one place. Barnacles, for example, have a mobile larval stage of life and a sessile adult stage of life.

SET: A collection of objects, physical or abstract.

SEVERE ACUTE RESPIRATORY SYNDROME (SARS): SARS is a severe respiratory infection caused by the SARS coronavirus. It has an unusually high mortality rate (death rate among those infected) for an infectious disease, almost 10%.

SHELL MODEL OF THE NUCLEUS: Also known simply as the shell model. A mathematical description of atomic nuclei in terms of the energy levels permitted by quantum mechanics to its neutrons and protons. The model was first proposed in 1949 and has been refined since.

SINGULARITY: In physics, a point in space-time where matter has infinite density and gravitational force is infinite—at least as described by the equations of general relativity. In mathematics, any point near which a function goes to infinity (that is, is undefined) may be termed a pole or singularity.

SOCIOPATHY: A psychological disorder characterized by extreme hostility to other persons and lack of conscience (the sense of right and wrong).

SOLAR MASS: A unit of mass measurement for stars; equivalent to the sun's mass.

SOLAR PROMINENCES: Loop- or arch-shaped masses of hot, luminous gas that often form above the sun's surface, shaped by intense magnetic fields. They may persist for days or weeks and span thousands of miles.

SOLID STATE PHYSICS: All physics devoted to the properties of solids—substances that retain a fixed shape unless subjected to sufficient force—fall under the heading of solid-state physics.

SOMATIC CELL: A cell that is part of the body (Greek, *soma*). Somatic cells are distinguished from germline cells (sperm and egg cells), which are directly involved in reproduction. All somatic cells except red blood cells (which contain no DNA) contain two copies of the organism's DNA; germline cells contain only one copy. When two germline cells unite, they produce a somatic cell (fertilized egg) with the usual two copies, which then divides and re-divides, duplicating its DNA at every step, as the organism grows.

SOUTHERN OCEAN: Also called the Antarctic Ocean. That part of the world ocean found south of 60° south latitude, ringing the continent of Antarctica.

SOY: Soy, also called soybean, is a legume native to Asia. Over 200 million tons of soybeans are harvested yearly worldwide. Processing of soybeans yields edible flour, oil, and curd (tofu). Soybean oil is also used in a wide variety of waxy or oily industrial products.

SPACE: The set of points that can be occupied by objects. Its properties are described by physicists in terms of geometry: space is not composed or made of any more fundamental material. In general relativity, space has three dimensions: that is, each point in space can be specified by three numbers representing distance measurements. In mathematics and computer science, all labeled measurements or quantities can be considered dimensions, and any collection of dimensions, however numerous or whatever their nature, can be termed a "space."

SPACE-TIME: The four-dimensional realm of space (which has three dimensions, often termed length, width, height) plus time (a fourth dimension). Time and space are neither independent nor interchangeable, but form a stretchy background, realm, or setting in which physical events occur. The term "stretchy" here refers to the fact that motions in space give rise to differences in time flow for moving and stationary observers: There is no one, unique rate of time flow throughout space, and no universal present moment. In particular, as described by special relativity, any stationary observer of a moving clock will see that clock as running slower than one to which they are attached. Here, "stationary" is a relative term because an observer attached to the "moving" clock will see exactly the same situation in reverse: Either observer's point of view can just as well be termed stationary (or moving) as that of the other. General relativity describes how space-time is curved by matter, giving rise to gravitation.

SPECIAL RELATIVITY: The theory proposed in 1905 by German physicist Albert Einstein (1879–1955) that describes the effects of straight-line, unaccelerated motion on time, distance, and mass. According to the theory, which has been validated by thousands of experiments, observers in constant, straight-line motion with respect to each other see that the other's mass is increased, their size in the direction of travel is decreased, and their clocks run slower. These effects are real, not illusory. The equivalence of mass and energy, which powers nuclear weapons and stars, follows from the theory. It is incorporated into the more complex and far-reaching theory of general relativity, proposed by Einstein in 1915 and also validated by many experiments.

SPECIATION: Speciation is the process by which new species arise. Although there are a variety of definitions, all involve an isolation event that separates an interbreeding population until the daughter lineages evolve in separate trajectories.

SPECIES: A group of living things that can breed together in the wild.

SPECIFIC HEAT: Also known as its specific heat capacity. The amount of heat energy required to raise one unit mass of that substance by a fixed unit of temperature (usually 1° Celsius). It takes more energy to heat substances that have high specific heats.

SPECTRAL LINE: A spectrum is a graph of the frequencies found in a mixture of waves, such as light waves. The graph line is higher at frequencies where more energy is present; when energy is concentrated in a narrow range of frequencies, the line draws a spike or vertical line. For example, a pure blue light appears as a spectral spike or line in the blue part of the visible light spectrum. Alternatively, lines can appear as narrow notches in an otherwise continuous-appearing spectrum. These notches are called absorption lines because they usually represent frequencies that have been absorbed after light was emitted by some source, such as the sun.

SPECTROSCOPE: An instrument used to measure the absorption, scattering, or emission patterns of electromagnetic radiation by atoms or molecules.

SPECTROSCOPY: Spectroscopy is the measurement of the absorption, scattering, or emission of electromagnetic radiation by atoms or molecules. The process separates the light of an object (generally, a star) into its component colors so that the various elements present within that object can be identified.

SPECTRUM: The dispersion of light into differential wave forms (e.g., blue light, red light).

SPERM: A mature male sex cell secreted in semen during male ejaculation.

SPHERICAL ABERRATION: The focusing of different colors at different distances from the lens rather than at a single image plane.

SPIRAL NEBULA: A distant, spiral-shaped collection of stars that appears cloudy without a telescope.

SPONTANEOUS GENERATION: Also known as abiogenesis; the incorrect theory that living things can be generated from nonliving things.

STABILITY: The quality or property of remaining in one state until forcefully disturbed. For example, a pyramid standing on its base is stable, but a pyramid balanced on its point is unstable. The distinction between stable and unstable systems appears throughout technology and the sciences.

STAGING: In rocketry, the practice of dividing a missile or spacecraft that is to be launched from the ground into separable components (stages). Each stage has its own fuel tanks and rocket motors. As a staged rocket ascends, it consumes the fuel in the tanks of its bottommost stage; when that stage is empty, it is dropped and the next stage is ignited. Each successive stage is much smaller than the previous one. The advantage of staging is that the weight of empty (now-useless) tanks does not need to be lofted into orbit.

STANDARD CANDLE: An astronomical object whose absolute magnitude (known luminosity) and apparent magnitude (observed brightness) are known, and which can therefore have its distance determined.

STANDARD MODEL: In particle physics, the theory that describes the fundamental particles that make up ordinary matter and the particles that are exchanged between them to create all of the fundamental forces except gravity. Gravity is not described by the Standard Model but by the theory of general relativity; scientists hope to someday modify the two so that they can be united into a single theory that describes all known phenomena.

STATISTICS: The branch of mathematics that deals with the analysis of large numbers of measurements (quantitative data).

STELLAR MAGNITUDE: The brightness of a star. The brightest stars in the sky are first magnitude, and dimmer stars have higher magnitude numbers.

STEM CELL: An unspecialized cell in an embryo or adult that has the potential to divide and grow into other types of specialized cells or tissues in the body.

Stem cells give rise to cells that have specialized forms and functions such as nerve or muscle cells.

STEREOCHEMISTRY: The branch of chemistry that deals with the way the three-dimensional shapes of molecules (stably bound clusters of atoms) affect their properties.

STEREOISOMER: Two molecules that contain the same numbers of the same types of atoms but differ in the spatial arrangement of those atoms are stereoisomers of each other.

STIMULATED EMISSION: The process of emitting light in which one photon stimulates the generation of a second photon, which is identical in all respects to the first.

STOICS: The Greek philosopher Zeno (334–262 BC) founded the Stoic school of philosophy in Athens, Greece, in the third century BC. The name "Stoic" arose from the fact that Zeno taught in the Stoa, a pillared, roofed walkway. The Stoics taught that happiness could be achieved by mastering passion (any form of strong feeling) and becoming virtuous.

STRANGE ATTRACTOR: In the mathematical description of systems, an attractor is a state that the system tends to approach over time. For example, a state of regular back-and-forth swinging is an attractor for a playground swing or pendulum that is being regularly pushed. A strange attractor is a state that does not repeat itself with such perfect regularity as a swinging pendulum; rather, the system may assume a never-repeating series of states that are close to the attractor but never simply converge to it.

STRATIGRAPHY: The branch of geology that deals with the layers of rocks or the objects embedded within those layers.

STREAMLINES: In the physics of flowing fluids, curves that show the paths followed by particles in different parts of the flow. For a fluid moving smoothly through a curved pipe, the streamlines would be lines inside the fluid that parallel the curvature of the pipe. In principle, an infinite number of streamlines could be drawn inside any body of flowing fluid.

STEREOISOMERISM: Two molecules that contain the same numbers of the same types of atoms but differ in the spatial arrangement of those atoms are stereoisomers of each other; the occurrence of stereoisomers is termed stereoisomerism.

STROMATOLITES: The bulb- or pillow-shaped rocky masses formed by some blue-green algae in shallow ocean water. Some fossil stromatolites are among the oldest traces of life on Earth; the oldest known stromatolites that were definitely formed by biological processes are 2.7 billion years old.

STRONG FORCE: In atomic physics, one of the four fundamental forces of nature (electromagnetic, weak, strong, gravitational). The strong force is repulsive at extremely close range (much smaller than an atomic nucleus) and attractive at longer ranges. Within the atomic nucleus, the attraction of the strong force between protons is stronger than the repulsive electromagnetic force between them (like charges repel); this is why atomic nuclei are not blown apart by the mutual electromagnetic repulsion of their positively-charged protons. On the other hand, protons and neutrons are kept from simply collapsing into each other by the repulsive effect of the strong force at extremely short range. At longer-than-nuclear ranges, the Coulomb force of electromagnetic attraction or repulsion is far stronger than the strong nuclear force, so it dominates outside the nucleus.

STRUCTURAL THEORY: In chemistry, the view, first put forward in the mid-nineteenth century by Russian chemist Aleksandr Butlerov (1828–1886), that the properties of a chemical substance arise both from the atoms that make it up and the way those atoms are structured to form molecules. That is, according to the structural theory, not only the composition but the shape of a molecule matters.

SUBDUCTION: Tectonic process that involves one plate being forced down into the mantle at an oceanic trench, where it eventually undergoes partial melting.

SUBLIMATION: Transformation of a solid to the gaseous state without passing through the liquid state.

SUBORBITAL: A rocket flight that passes above the atmosphere into space but does not achieve an orbit around Earth. The first American in space, Alan Shephard (1923–1998), made a suborbital flight on May 5, 1961, from Cape Canaveral space center in Florida (later Cape Kennedy) to a height of 116 miles (187 km), landing in the Atlantic 302 miles (486 km) from his launch point.

SUCCESSION: In ecology, succession refers to the replacement of certain plant and animal species by others in the process of regrowth over land disturbed by fire, logging, or other causes. The nature of succession in any given place depends on the local environment. In areas that are naturally forested, succession typically involves colonization by weeds, then fast-growing shrubs, then softwood trees such as aspens, and finally large, long-lived trees.

SUGAR: Any of several small carbohydrates, such as glucose, that are sweet to the taste.

SULCI: In medicine, the term "sulcus" (plural sulci) refers to one of the grooves or valleys in the surface of the brain formed by the folding of its surface. In astronomy, sulci are groove-like formations on the surface of a planet or moon.

SULFONAMIDES: Synthetic antibacterial agents produced in the 1930s.

SUNSPOT: A region on the surface (photosphere) of the sun that is temporarily cool and dark compared to surrounding areas.

SUPERCONTINENT: In geology, a supercontinent is formed when Earth's 14 or so major tectonic plates, which are in constant motion, happen to come together in a single, large landmass, leaving the rest of the planet covered with ocean. At least half a dozen supercontinents have formed during Earth's 4.5-billion year history.

SUPERFLUOUS: Anything which is unnecessary or more than needed for some purpose.

SUPPURATION: Formation and discharge of pus.

SYMBIOSIS: A pattern in which two or more organisms live in close connection with each other, often to the benefit of both or all organisms.

SYMPATHETIC NERVOUS SYSTEM: A collection of nerve cells that control a variety of internal functions when the body is exposed to stressful conditions.

SYMPLESIOMORPHY: In evolutionary biology, the occurrence of a similar trait in two or more taxa (groups of organisms) because those taxa all share a common ancestor from which they have inherited the trait.

SYNAPOMORPHY: An expressed character which, because it is shared by members of a classification taxa, is used to infer common ancestry.

SYNAPSE: Junction between nerve cells in the brain through which neurotransmitters travel and where the exchange of electrical or chemical information takes place.

SYNCHROTRON RADIATION: Radiation emitted by charged particles accelerated in magnetic fields to speeds near the speed of light, usually detected as radio wavelengths.

SYNODIC PERIOD: In astronomy, the time it takes for a planet or other astronomical object to reappear at the same point in the sky, relative to the position of the sun, is termed the synodic period of that object.

SYNTHESIS: The process of combining simpler chemicals to make a more complex compound.

SYPHILIS: An infectious sexually transmitted disease.

SYSTEMATICS: Field of biology that deals with the diversity of life. Systematics is usually divided into the two areas of phylogenetics and taxonomy.

T

T CELLS: A class of white blood cells distinguished by having the molecule known as the T-cell receptor on their surface. There are seven different varieties of T cell, all with distinct roles in the immune system.

TANGENT: A trigonometric function that represents the ratio of the opposite side of a right triangle to its adjacent side.

TAXONOMY: The science dealing with the identification, naming, and classification of plants and animals.

TECTONICS: In geology, large-scale processes that shape Earth's crust, such as those that build mountains and valleys. Plate tectonics is the theory that explains the large-scale features of Earth in terms of moving plates or rafts of rocky material floating about on the planet's liquid interior.

TELEMETRY: Telemetry is the process of making measurements from a remote location and transmitting those measurements to receiving equipment.

TERABYTE: In computer science, a unit of information, whether stored or transmitted, equal to one trillion bytes. Since each byte equals eight bits, where a bit is a single binary digit, i.e., a 0 or 1, a terabyte is 8 trillion bits.

TETRAVALENT: In chemistry, a tetravalent (four-valent) atom is one that has a valence of 4, that is, four electrons available for covalent (electron-sharing) bonding with other atoms. A tetravalent atom will bind with 4 other atoms; for example, carbon is tetravalent, so in methane, CH_4, it bonds covalently with four hydrogen atoms (which are monovalent).

THEORIES OF EVERYTHING (TOEs): Theories of everything are sought-after systems of physical law that would incorporate both quantum mechanics and general relativity into a single, coherent system of explanations. Presently, these two physical theories work extremely well in their own domains of application, but neither can be extended to all physical conditions.

THEORY: An explanation for some phenomenon that is based on observation, experimentation, and reasoning.

THERAPEUTIC CLONING: A type of cloning that involves somatic cell transfer to produce embryonic stem cells.

THERMOMETER: A device for obtaining temperature by measuring a temperature-dependent property, such as the height of a liquid in a sealed tube, and relating this to a numbered scale based on the rise and fall of the liquid.

THYMUS: The thymus is a lymphoid gland localized between the lungs in the anterior superior mediastinum (in the chest). Its cortex (i.e., external layer) is constituted by lymphatic tissue, with the internal portion containing lymphocytes. The thymus also has a thick reticular structure comprised of groups of granular cells enveloped by epithelial cells, known as Hassall's corpuscles. Much remains to be discovered about the thymus' physiological role and products.

TIME DILATION: When an object is stationary relative to an observer, any time-dependent process (e.g., the running of a clock) takes place at a certain rate as seen by the observer. As described by the theory of special relativity, when the object is made to move at a constant speed in a straight line with respect to the observer, clocks moving with the object are seen by the observer to run faster than when at rest. This slowing of time in moving reference frames, termed time dilation, is a real effect, not an illusion.

TOPOLOGY: A subfield of geometry. In particular, it is the mathematical study of shapes apart from any stretching or deformation that they may undergo.

TRANSCRIPTION: The transfer of genetic information from deoxyribonucleic acid (DNA) to ribonucleic acid (RNA). The process of transcription in prokaryotic cells such as bacteria differs from the process in eukaryotic cells (cells with a true nucleus), but the underlying result of both transcription processes is the same, which is to provide a template for the formation of proteins.

TRANSFER RNA (tRNA): In the molecular factories called ribosomes, where cells manufacture proteins, information is brought from DNA to the ribosome by messenger RNA. The ribosome then manufactures the protein according to the list of amino acids conveyed by the messenger RNA. Transfer RNA is a molecule that assists in this process, transporting one amino acid at a time to the growing chain of amino acids in the protein being manufactured by the ribosome.

TRANSFORM BOUNDARY: Where two or more tectonic plates (large shell-like areas of rock floating on Earth's mantle, sometimes continental in size) are sliding past each other edge to edge, the boundary between the two plates is termed a transform boundary. Earthquakes often occur along transform boundaries as the plates stick, build up strain, and slip suddenly.

TRANSFORMISM: In biology, the view that existing species have come about through the transformation of earlier living things. The term was once synonymous with evolutionism; today, it is obsolete, and "evolution" refers specifically to the theory of descent with modification as developed by English naturalist Charles Darwin (1809–1882) and greatly advanced by biologists over the last century.

TRANSFUSION: A technique used to replace blood lost during an accident, illness, or surgery.

TRANSISTOR: A solid-state electronic device that is used to control the flow of electricity in electronic equipment. It consists of a small block of semiconductor with at least three electrodes.

TRANSLATION: Translation is the process in which genetic information, carried by messenger RNA (mRNA), directs the synthesis of proteins from amino acids, whereby the primary structure of the protein is determined by the nucleotide sequence in the mRNA. Although there are some important differences between translation in bacteria and translation in eukaryotic cells, the overall process is similar.

TRANSMISSION ELECTRON MICROSCOPES (TEM): Devices that produce a highly magnified view of the structure of very thin samples of material. They do so by passing a beam of electrons through the material, which is then detected on the far side of the sample. Today, transmission electron microscopes are especially used in the examination of samples of metal and other solid materials whose microscopic properties are of interest.

TRANSMUTATION: In biology, changing in form or appearance, usually to a higher form of being. In physics, the conversion of one nuclide into another, as by neutron bombardment in a nuclear reactor. Also, in alchemy, the transformation of a base metal into gold.

TREPANATION: The surgical opening of a hole in the skull.

TRIAGE: The process of allocating limited medical resources in a situation of overwhelming need; life-and-death cases likely to benefit from treatment are the first to be treated.

TRIANGULATION: A means of navigation and direction finding based on the trigonometric principle that, for any triangle, when one side and two angles are known, the other two sides and angle can be calculated.

TRIGONOMETRIC FUNCTION: An angular function that can be described as the ratio of the sides of a right triangle to each other.

TROPHIC LEVEL: In biology, eating or ingestion of other creatures. The trophic level of a species is its place in the food chain; a creature that eats another is at a higher trophic level. For example, a frog is at a higher trophic level than a fly.

TRUE (OR GASEOUS) NEBULA: A large cloud of gas or dust in space.

TURBULENCE: Chaotic, rough flow or mixing in a gas or liquid. It is distinguished from smooth flow, which is called laminar flow because it tends to occur in neatly organized layers with smoothly ordered velocities.

U

ULTRASOUND: A form of energy that consists of waves traveling with frequencies higher than can be heard by humans. Also, a non-invasive diagnostic technique for imaging objects and the human body (often a fetus within a uterus) using ultrasound energy. The high-frequency sound waves produce a two-dimensional image of the body tissue by producing vibrations within it, allowing measurement and detection of abnormalities within the body.

UNCONSCIOUS: The part of the human mind of which one is not conscious, but which is actively involved in emotion, memory, the generation of ideas, and perception.

UNFALSIFIABLE: In the practice of science, a theory or claim is unfalsifiable if there are no reasonably possible conditions under which the theory or claim can be proved false. Claims that are unfalsifiable are not scientific because there is no objective way to check them.

UNIFICATION: In physics, the discovery of a single, consistent theoretical framework for describing phenomena that formerly had to be described by separate theories.

UNIFORM CIRCULAR MOTION: Motion in a perfect circle at a constant speed, thought by the ancient Greeks to be the most perfect form of motion and thus the form most natural for celestial bodies.

UNIFORMITARIANISM: Doctrine of geology promoted by English geologist Charles Lyell (1797–1895), asserting three theories: (1) actualism (uniform processes act throughout Earth's history), (2) gradualism (there is a slow, uniform rate of change throughout Earth's history), and (3) uniformity of state (Earth's conditions have always varied around a single, steady state). Uniformitarianism is often contrasted to catastrophism. Modern geology makes use of elements of both views, acknowledging that change can be drastic or slow, and that while some processes operate steadily over many millions of years, others (such as asteroid impacts) may happen rarely and cause catastrophic, sudden changes when they do.

UNIVERSALITY: The property of some mathematical representations of causal systems that certain overall properties of the system do not depend on its detailed properties.

UPLIFT: The process or result of raising a portion of Earth's crust through different tectonic mechanisms.

UREA: Chemical compound of carbon, hydrogen, nitrogen, and oxygen produced as waste by cells that break down protein or under laboratory conditions.

UTERUS: A pear-shaped, hollow muscular organ in which a fetus develops during pregnancy.

V

VACCINATION: The inoculation, or use of vaccines, to prevent specific diseases within humans and animals by producing an immunity to such diseases. The introduction of weakened or dead viruses or microorganisms into the body to create immunity by the production of specific antibodies.

VACUUM TUBE: A sealed glass tube in which the conduction of electricity takes place through a vacuum or gas.

VALENCE: The tendency of an atom to gain or lose electrons in reacting with other atoms.

VARIABLE (MATHEMATICAL): In mathematics, a quantity that can stand for anywhere from two to an infinite number of specific numerical values. A single specific number is termed a constant; variables vary, constants are constant. Variables are most commonly represented by italicized letters of the Latin alphabet, such as x or y.

VARIANT CREUTZFELDT-JAKOB DISEASE: A fatal degenerative disease of the central nervous system caused not by a living agent but by infectious, malformed proteins called prions. Variant Creutzfeldt-Jakob (vCJD) disease is a variety of Creutzfeldt-Jakob, first described in 1996, that tends to afflict young people and can be transmitted by blood transfusions or by eating the flesh of cattle infected with bovine spongiform encephalopathy ("mad cow disease").

VARIOLATION: Variolation was the pre-modern practice of deliberately infecting a person with smallpox in order to make them immune to a more serious form of the disease. It was dangerous, but did confer immunity on survivors.

VECTOR: Any agent, living or otherwise, that carries and transmits parasites and diseases. Also, an organism or chemical used to transport a gene into a new host cell.

VEGANISM: A lifestyle and/or belief system in which a person refuses to eat or use animal-based products, including meat, dairy products, leather, products tested on animals, etc. Veganism is often based upon ethical, moral, and or spiritual beliefs. Veganism is also a manifestation of concern about the rights of animals or concerns regarding mistreatment and killing of animals used in economic products or for commercial food production. Veganism is a healthful lifestyle choice for some, though it can also be based upon a mixture of scientific and pseudoscientific beliefs about the relationship of diet to human health.

VEIN: Vessel that transports blood to the heart.

VENTRICLE: A lower chamber of the heart from which blood is pumped into the arteries.

VIRTUAL PARTICLE: The Heisenberg uncertainty principle in quantum mechanics, named after its discoverer, German physicist Werner Heisenberg (1901–1976), states that it is impossible to know both the momentum and location of a particle at the same time with perfect precision. One of the consequences of this principle is that particles are permitted to pop into and out of existence for brief periods without violating the conservation of matter. These particles are termed virtual particles to distinguish them from permanent particles. Exchanges of virtual particles mediate the electromagnetic, weak, and strong forces.

VIRULENT: A virus or bacterium that is especially prone to cause infection.

VIRUS: A small, nonliving infectious agent that is a repository of nucleic acid. It consists of a core of genetic material (either deoxyribonucleic acid [DNA] or ribonucleic acid [RNA]) surrounded by a shell of protein and requires the presence of a living prokaryotic or eukaryotic cell for the replication of the nucleic acid. There are a number of different viruses that challenge the human immune system and that may produce disease in humans.

VIS VIVA: In the late 1600s, German scientist Gottfried Leibniz (1646–1716) proposed a theory of kinetic energy conservation that became known as *vis viva* (Latin for "living force"). The theory is now obsolete; kinetic energy is not, in fact, conserved, though momentum, a closely related quantity, is conserved.

VISCOSITY: The internal friction within a fluid that makes it resist flow.

VITALISM: The concept that compounds found within living organisms are somehow essentially different from those found in nonliving objects.

VITAMIN: A complex organic compound found naturally in plants and animals that the body needs in small amounts for normal growth and activity.

VIVISECTION: The dissection of a living animal. Until the twentieth century, it was practiced without anesthesia, so it was torturous for the subject animal. Vivisection is still practiced in medical research, but ethical guidelines require the anaesthetization of the animal so that the vivisection is painless.

VORTEX: A rotating column of a fluid such as air or water.

W

WAVELENGTH: The distance between one peak or trough of a wave (e.g., light wave) and the next corresponding peak or trough.

WAVE-PARTICLE DUALITY: The combination of wavelike and particle-like properties possessed by all physical objects, which is especially apparent for single subatomic particles such as electrons. Under some conditions, a particle behaves as a small object that can be definitively found in one place or another; under other conditions, a particle behaves as a wave that is spread throughout space and can add or subtract from similar waves in a way that would be impossible for a tiny, hard object that has a single, well-defined location at all times.

WEAK FORCE: In physics, the electroweak force is the single unified force of which the electromagnetic force and weak interaction are both manifestations. The weak interaction is involved in certain types of radioactive decay (e.g., beta decay of neutrons).

WEAK INTERACTION: In physics, the electroweak force is the single unified force of which the electromagnetic force and weak interaction are both manifestations. The weak interaction is involved in certain types of radioactive decay (e.g., beta decay of neutrons).

WHALING: The harvesting of whales for their products by individuals or commercial operations.

WMAP: The WMAP probe (Wilkinson Microwave Anisotropy Probe) is a satellite launched by NASA in 2001 and still operating as of 2008. Its mission was to measure slight variations in different directions (anisotropy) of the cosmic microwave background, a low-level bath of radio waves that is the afterglow of the big bang 13.7 billion years ago. The character of these variations has confirmed the expansion version of the big bang theory, helped pin down the age of the universe with unprecedented precision, helped characterize the amounts of dark matter and dark energy in the universe, and contributed to other advances in cosmology.

WORLD WIDE WEB: A collection of hypertext markup language documents that are linked to each other through the Internet. At least 10 billion (and possibly many more) pages are interlinked to form the World Wide Web, which has developed entirely since 1989.

X

XENOGRAFT: Tissues and organs used for transplantation that come from different animal species, like pigs or baboons.

Y

YEAST: A microorganism of the fungus family that promotes alcoholic fermentation and is also used as a leavening (fermentation) agent in baking.

YELLOW BILE: The ancient Greek and Roman theory of physiology stated that four basic substances, the humors, exist in the human body, and that disease is caused by an imbalance among the humors. The four humors were blood, phlegm, yellow bile, and black bile; an excess of yellow bile was believed to cause an angry, vindictive disposition.

Z

ZENO'S PARADOXES: A set of logical puzzles invented by the Greek philosopher Zeno (c.490–430 BC) that examines the nature of motion and appears to show that ordinary motions are impossible. The most famous is Zeno's apparent proof that a fast runner cannot overtake a slow tortoise: during every period of time, as the runner advances, the turtle advances too (though not as far), and it is apparently contradicted by logic that the runner should ever catch up. Today, we recognize that such a conclusion is false because an infinite series of finite numbers can add up to a finite sum. In this case, the infinite series of diminishing gains made by the tortoise add up to a finite number: The runner does indeed catch up to the tortoise in finite time and then pass it.

ZODIAC: In astrology, a strip or band of the visible heavens encircling Earth. It is centered on the ecliptic, which is that line in the sky along which the sun moves. The Zodiac is divided up into sectors named after various constellations (patterns of stars).

Chemistry: Biochemistry: The Chemistry of Life

■ Introduction

Biochemistry is the study of the basic molecules used by living things. Nucleic acids, proteins, lipids (fats), and carbohydrates all perform specific functions within and between cells that allow organisms to survive. Biochemists seek to understand how these substances are formed, what purposes they serve, and how they might be used elsewhere to make medicines or other products.

Biochemistry analyzes the structure and physical properties of biomolecules (those produced by living things) and the systems in which they are used. Many biochemists study a particular class of biomolecule such as pigment-producing compounds or molecules that form membranes and structure within cells. Other areas of study include the many chemicals of cellular energy pathways, hormones, and other regulatory molecules, the synthesis of proteins, and the makeup and function of DNA.

■ Historical Background and Scientific Foundations

Before the early nineteenth century, most chemists accepted the idea that only living organisms could produce biomolecules. Louis Pasteur popularized the principle of vitalism—the theory that some "vital force" present in living things was necessary to make biochemicals. However, in 1828 the German chemist Friedrich Wöhler (1800–1882) accidentally produced the biomolecule urea, a waste product of human metabolism, from the reaction of two different mineral compounds. This proved that the chemical processes within living beings could be understood within the same rules of the larger field of chemistry.

Much of the earliest work in biochemistry dealt with the fermentation of sugar into alcohol. Before yeast was recognized as a living organism, different researchers argued about its role in the process. Some thought it was totally uninvolved, others thought it was merely a catalyst— an agent that causes a chemical reaction but is not involved in the process itself. When Pasteur proved that natural fermentation does not take place without the presence of yeast, many scientists sought to determine how this happens. It was Eduard Buchner (1860–1917), a German chemist who proved that living yeast is not necessary to ferment sugar, but only the enzymes from within the yeast cells, a breakthrough for which he won the 1907 Nobel Prize in chemistry. This seemingly simple discovery led others to investigate the properties of enzymes and other proteins, eventually revealing the structures of these molecules.

Today, most biochemists accept the idea that the first biomolecules formed before any life was even present. In 1953 American chemists Stanley L. Miller (1930–2007) and Harold C. Urey (1893–1981) performed their famous experiment that showed that the chemical and weather conditions of the very young Earth could have produced organic biomolecules from nonliving mineral sources. Starting with water, methane, ammonia, and hydrogen, they added heat and electricity and allowed the experiment to circulate for several days. After a week, some of the simple compounds had formed amino acids, sugars, and lipids. Many of these biochemical units are able to combine to form more complex biopolymers, which could lead to the development of RNA, the genetic code of simple organisms. In other words, the first living things arose from nonliving molecules that now are known as what make up the basic units of life. It should be noted that no one has yet created living cells from an experiment of this type, but much research continues in this area.

The modern study of biochemistry centers around the isolation of new biomolecules from living organisms, learning about their structure and properties,

WORDS TO KNOW

AMINO ACID: One of about two dozen relatively simple chemical compounds from which proteins are made, amino acids are the building blocks of proteins and serve many other functions in living organisms. They are nitrogen-containing organic compounds that consist of at least one acidic carboxyl group (COOH) and one amino group (NH_2). In alpha amino acids that are contained in the proteins found in cells, these two groups are both attached to a carbon atom, which also carries a hydrogen atom, plus a side chain known as the R group. The R group varies from one amino acid to another and gives each amino acid its distinctive properties. The prime function of DNA (deoxyribonucleic acid) is to carry the information needed to direct the proper sequential insertion of amino acids into protein chain during protein synthesis (translation). Although relatively simple compounds, amino acids can vary widely, and to date more than 80 different amino acids have been found in living organisms. Of these 80 amino acids, 22 are considered the precursors of animal proteins.

ANTIBODIES: Large, Y-shaped proteins found in blood that lock on to specific substances foreign to the body (antigens). Typical antigens are molecules found on the surfaces of viruses and bacteria. Some antibodies, when bound to antigens, act as flags targeting the antigen for attack by white blood cells; others combine with other blood molecules to attack the antigen directly. If a person's blood already contains antibodies for a particular antigen, the person's immune system attacks that antigen as soon as it appears. This is the basis of acquired immunity to specific viruses and bacteria, whether natural or instilled by immunization.

CARBOHYDRATE: An organic compound present in the cells of all living organisms and a major organic nutrient for human beings; consists of carbon, hydrogen, and oxygen, and makes up sugar, starch, and cellulose.

CATALYST: Substance that speeds up a chemical process without actually changing the products of reaction.

DEOXYRIBONUCLEIC ACID (DNA): A double stranded molecule joined together by bonds between base pairs. The strands are akin to the sides of a ladder and the base pairs the steps of the ladder; the molecule as a whole is helical (shaped as a gently twisted ladder). DNA is the carrier of genetic information (encoded in the specific series of base pairs) for almost all organisms, though some viruses use ribonucelic acid (RNA).

ENZYME: Any of numerous complex proteins produced by living cells that act as catalysts, speeding up the rate of chemical reactions in living organisms.

GLUCOSE: Also known as blood sugar; a simple sugar broken down in cells to produce energy.

GLYCOLYSIS: A series of chemical reactions that takes place in cells by which glucose is converted into pyruvate.

GLYCOPROTEIN: A membrane-bound protein that has attached branching carbohydrates. These may function in cell-cell recognition, such as in human blood groups and immune system response, as well as in resisting compression of cells.

HORMONE: A chemical produced in living cells that is carried by the blood to organs and tissues in distant parts of the body, where it regulates various bodily functions.

KREBS CYCLE: A set of biochemical reactions that occur in the mitochondria. It is the final common pathway for the oxidation of food molecules such as sugars and fatty acids. It is also the source of intermediates in biosynthetic pathways, providing carbon skeletons for the synthesis of amino acids, nucleotides, and other key molecules in the cell. The Krebs cycle is also known as the citric acid cycle and the tricarboxylic acid cycle. The Krebs cycle is a cycle because, during its course, it regenerates one of its key reactants.

LIPID: A fat or oil; a chemical compound used as a source of energy, to provide insulation, and to protect organs in an animal body.

NUCLEIC ACID: A long, ladder-like molecule built of chained nucleotides (a type of chemical compound). RNA and DNA (ribonucleic acid and deoxyribonucleic acid), the molecules used by all living things to create offspring and run the chemical processes of cells, are nucleic acids.

PROTEIN: A complex chemical compound that consists of many amino acids attached to each other that are essential to the structure and functioning of all living cells.

---

understanding the complex cellular cycles and pathways in which they are used, and developing new uses for them. The four major types of molecules studied by biochemists are carbohydrates, lipids, proteins, and nucleic acids.

Some biomolecules are hybrids of these major classes or their components (e.g., high-density lipoproteins or HDL are hybrids of lipids and proteins, while chondroitin sulfate, a component of cartilage tissue and arthropod exoskeletons, is an amino-sugar chain). Each of these categories is very large, containing a vast number of very different biomolecules. Because of this great diversity, many biochemists specialize in a much smaller class of molecules, or even one particular molecule. Others might work to create better experimental methods or apply others' discoveries in new ways.

Friedrich Wöhler (1800–1882), the German chemist who discovered cyanates in 1828, showed that heating the inorganic compound ammonium cyanate gave urea. *SPL/Photo Researchers, Inc.*

Proteins

Many biochemists consider proteins the most important substances for study because they perform amazingly diverse tasks. For example, enzymes, the subject of so much early research, are proteins. Living things depend on enzymes to catalyze, or stimulate, important reactions that would not occur in their absence. Other proteins transport important materials around the cell or body, or store these materials for later consumption. Antibodies are specialized proteins produced by the immune system in response to a specific pathogen. They bind to disease-causing agents within the body, allowing immune system cells to recognize and destroy harmful intruders. Proteins perform many other functions within the body.

Most proteins are large, intricate molecules with chain-like structures that fold into a specific shape. Individual sections of the chain are made up of component amino acids. Though there are many kinds of amino acids, only 20 are used by all living things to produce thousands of different proteins. Biochemists have created several new amino acids that do not occur in any living organisms and have found important uses for them in industrial products.

Carbohydrates

Carbohydrates, the largest group of biomolecules, are dietary fuel—this is one of their major uses in living organisms. The most basic carbohydrates are monosaccharides, small simple sugar molecules. The most important of these is glucose, the body's major source of metabolic energy. More complex carbohydrates are used for longer-term storage of energy within the body, particularly in plants. These polysaccharides are also known as starches.

Very large carbohydrate molecules are usually not digested by humans, but still serve important tasks in many organisms. Cellulose gives structure to plants; it is the major component of wood and paper. Chitin, a glucose-based polymer, forms the exoskeleton in insects. Glycoproteins, or carbohydrates attached to amino acids, have important functions in the immune system and for blood clotting.

Lipids

Lipids are an important, although difficult to define, group of biochemicals. They are most commonly categorized as fats, but lipids include fats, oils, waxes, certain vitamins, some hormones, components of some cell membranes, steroids, fatty acids, and other related biomolecules. Until 1979 only two functions of lipids were known: energy storage and components of cell membranes. Then the lipid now known as platelet-activating factor was shown to help activate or mediate many immune responses. Following this discovery, other lipids were found to have important functions in sending signals within and between cells and transporting energy throughout the body.

Fatty acids are a class of lipids that are particularly important to human health. The body can produce most fatty acids, but those that must be consumed in the diet are called "essential fatty acids." Maintaining the correct balance between omega-3 and omega-6 fatty acids may help reduce the risk of cardiovascular disease such as high cholesterol, high blood pressure, heart attack, and stroke. Other studies have suggested that fatty acids might have a positive influence on many diseases such as diabetes, arthritis, and several types of cancer.

Nucleic Acids

Nucleic acids are most often associated with the large, important molecules they can form by joining together: DNA and RNA. However, nucleic acids have other

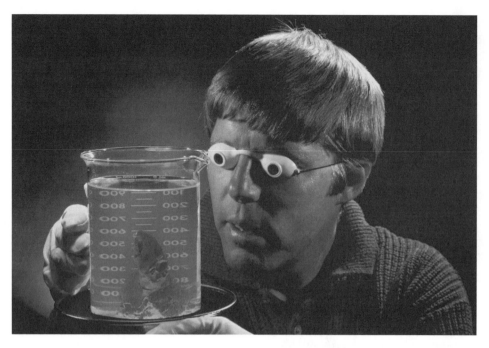

Dr. David Ward exposes cyanobacteria in ultraviolet light, causing chlorophyll to fluoresce. Its presence indicates photosynthesis. Such microbes released oxygen into the biosphere, allowing oxygen-based life to develop. *© Jonathan Blair/Corbis.*

important functions within the cell. They make up parts of coenzymes (molecules that carry material between enzymes), mediate and control various reactions, and store energy. When nucleic acids join with other base chemicals in a long chain, they form DNA or RNA. The specific sequence they form carries genetic information capable of replicating itself. DNA is often thought of as a set of instructions for making an organism. All of the necessary instructions are encoded in the way that the nucleic acids line up.

Cellular Respiration

Though many biochemists still study individual molecules, one of the most important features of modern biochemistry is the study of biochemical systems within the body or individual cells. Biochemical reactions often occur in a series, with the product of one step being used as material for the next. Many of these reactions are part of a cell's metabolism, which is vital for sustaining life.

Cellular respiration is the process by which cells obtain fuel, release its energy, and remove waste products. Almost all aerobic (oxygen-using) organisms, from yeast to humans, use the same reactions to fuel their life processes. One of these important reactions is glycolysis, which occurs in both plant and animal cells. Glycolysis transforms glucose, a relatively simple carbohydrate, into a molecule called pyruvate and a small amount of adenosine triphosphate (ATP), a high-energy molecule

used by cells. Because this pathway is used by so many different kinds of life, it is considered one of the most ancient; it was probably present in a very distant common ancestor of most living things.

In aerobic cells, a process known as the citric acid cycle or the Krebs cycle uses the pyruvate from glycolysis to produce more energy. It is the final system that changes lipids, carbohydrates, and fatty acids into energy for the cell. There are eight steps in the citric acid cycle, each providing products that are used in the next step. The final step provides a molecule that contains the right amount of carbon to start the cycle all over again. Enzymes within a cell's mitochondria catalyze each step, allowing the release of more energy. The citric acid cycle also produces molecules that are the raw material for a number of other biochemical processes.

Plant cells use biochemical systems to generate energy, too. Many plants generate their energy from the sun, a process known as photosynthesis. This complex biochemical system produces chemical energy from photons, the subatomic particles that make up light.

■ Modern Cultural Connections

Much of this increased knowledge about biochemicals and biochemical systems has been put to use in the field of medicine. Because enzymes are well understood, specific blood tests have been developed to detect abnormal enzyme levels that indicate disease. Liver enzyme tests

Dr. H.D. West, left, supervises two students who carry out an experiment in the biochemistry laboratory on September 18, 1956, at Meharry Medical College in Nashville, Tennessee, a college founded in 1876 for blacks and other people of color.   *AP Images.*

can determine how well the liver is working or how far liver disease may have progressed. Cardiac enzyme tests can confirm whether a patient has had a heart attack and how severe it might have been. Drug developers use biochemical research to design compounds that address the specific causes of diseases, rather than blindly searching vast numbers of chemicals for one that works.

As our knowledge of biochemistry has increased, doctors and researchers have become more able to address the causes of disease rather than simply treating the symptoms. Research continues into more obscure and complex biochemicals, yielding new medicines, antioxidants, and antibiotics. The ability to understand and develop these new molecules grew directly from the understanding of the most basic biomolecules. Recognizing that the chemistry of life can be studied objectively has led to the science of biochemistry as we recognize it today.

■ Primary Source Connection

In 2006, a blood doping scandal in cycling known as Operación Puerto (Operation Mountain Pass) became public, disqualifying several well-known cyclists before the start of the Tour de France. Investigators located a lab in Spain and a doctor who was providing cyclists with banned substances and stocking blood samples from several riders.

That same summer, American cyclist Floyd Landis staged one of the most impressive come-from-behind victories in the history of cycling to win the Tour de France. Cycling officials charged Landis of doping after Tour officials asserted that one of Landis's blood sample—taken after the tough mountain stage where he regained the lead—tested positive for high levels of testosterone.

As of May 2007, blood doping scandals continue to mar professional cycling. Operación Puerto implicated several of the top riders in the sport, forcing some into early retirement. Prominent riders were once again caught doping during the 2007 Tour de France, and former Tour champion Landis was stripped of his win after being found guilty of using banned performance enhancing substances. Several riders face legal prosecution in their home countries or the possibility of being stripped of past victories.

While blood doping has become a problem in many sports, professional cycling has garnered much of the spotlight for doping scandals in recent years. As tests for various doping products have become better, more accurate, and more thorough, more riders are being caught blood doping and using illegal performance-enhancing substances.

The following article about the doping scandals was written by Jerome Pugmire, a sports writer for the Associated Press.

Stanford University biochemistry scientist in his lab in Stanford, California, in 2005. He is researching what he calls "test-tube evolution," which allows for scientists to simultaneously screen millions of chemical compounds for their potential as drug targets. *AP Images/Paul Sakuma.*

TOUR DE FRANCE FACES LONG RIDE BACK AFTER DOPING SCANDALS

PARIS (AP)—Cyclists who have admitted using banned drugs say the Tour de France may need years to recover from the stigma of cheating, denial and lying that devastated the 2007 race.

[2006's] Tour was bad enough, with Floyd Landis' positive test coming days after the race. This time, doping rocked the 104-year-old institution to its core.

"I thought this year would have been better," former rider Frankie Andreu said. "Obviously it wasn't. So I'm not confident that even next year will be better."

French Sports Minister Roselyne Bachelot promised on Monday that the 2008 Tour will be "clean and renovated," likely with tougher doping sanctions, unannounced hotel room searches and other measures.

Patrice Clerc, head of Amaury Sport Organization that organizes the Tour, said next year's race will be the first step in rebuilding high-level cycling.

"The 2008 Tour will not be like the 2007 Tour," Clerc said. "I commit myself to that."

This time, fan favorite Alexandre Vinokourov, race leader Michael Rasmussen and Italian rider Cristian Moreni were all cited for doping or, in Rasmussen's case, for lying about his whereabouts while skipping tests.

German rider Patrik Sinkewitz also tested positive, except his test was from before the race and revealed during it.

"It's going to take five and 10 years until we have faith in the riders," Britain's David Millar said. "That's such a shame for the younger guys who are coming through and deserve it now because they're getting put in the same bracket."

Cynicism among some fans was clear.

"Tour of Transfusion," read one roadside banner.

Many now look to the new guard of young riders to stand up against doping. But will 24-year-olds like Tour winner Alberto Contador, Linus Gerdemann and Markus Fothen speak their minds?

Gerdemann, who won an Alpine stage on July 14, already has.

"We have to go that way, otherwise cycling is dead," Gerdemann said. "Everyone has to understand that this is the new way, and there are no other possibilities."

That won the approval of Millar, who like Andreu, used the performance enhancer EPO.

"It's going to take awhile to earn the trust," Andreu told The Associated Press.

After Sinkewitz's positive test for testosterone, two German television stations ended their coverage.

"It's important that riders have an opinion and say it," Fothen said. "So much silence. In the past was a generation that did things that were not good. Now we are a new generation. I can speak loud."

Andreu, a former teammate of seven-time Tour winner Lance Armstrong, wants Fothen to keep talking.

"It could be a generational thing because the guys grew up racing in the '90s fell into maybe taking stuff in order to perform," said Andreu, who admitted taking EPO in 1999.

Credit Agricole sporting director Roger Legeay says it will take more than youth.

"In 1998 they said we'd see a new generation," Legeay said. "In 2004 we'd see a new generation . . . so history repeats itself. Today we really have all the means at our disposal. Urine tests, medical records, DNA, random tests."

After clinching his Tour title at Saturday's time trial, Contador said he would take a DNA test, but only if asked.

"I'm innocent and I don't have to prove anything to anyone," Contador said. "Who should I have give my blood to? You?"

Contador never tested positive and there is no evidence tying him to blood-doping. Yet the fact he had to face questions reflects the current climate of suspicion.

One rider at this race, Germany's Erik Zabel, previously admitted taking EPO in 1996. Unlike Millar and Andreu, he has said hardly anything about doping.

"I said to him that we talk about it, that we should do an interview together," Millar said. "He's got to talk more about it. We can't just admit it and bury it."

The old guard like Zabel will soon be gone. Vinokourov and Landis may yet never ride the Tour again.

Millar accepts that fans may not start believing any time soon.

"They have every right not to," Millar said. "We expect a lot of our grand champions, and even when they do make mistakes, they don't face up to them. It's unfortunate, it's kind of a tragic twist."

Andreu remembers clearly the pressures to use banned drugs.

"You always wondered what the next guy was doing," Andreu said. "If you're trying to win the Tour de France and you think everybody else is doing stuff it becomes an arms race. And it might be a mysterious arms race because you never know, but you don't want to be caught out. So it becomes a game."

Jerome Pugmire

PUGMIRE, JEROME. "TOUR DE FRANCE FACES LONG RIDE BACK AFTER DOPING SCANDALS." *ASSOCIATED PRESS NEWSWIRE* (JULY 30, 2007).

SEE ALSO *Biology: Cell Biology; Biology: Genetics, DNA, and the Genetic Code; Biology: Sociobiology; Chemistry: Organic Chemistry.*

BIBLIOGRAPHY

Books

Berg, Jeremy M., John L. Tymoczko, and Lubert Stryer. *Biochemistry.* 5th ed. New York: W.H. Freeman, 2001.

Periodicals

Miller, Stanley L. "Production of Amino Acids under Possible Primitive Earth Conditions." *Science* 117 (May 15, 1953).

Pugmire, Jerome. "Tour de France Faces Long Ride Back after Doping Scandals." *Associated Press Newswire* (July 30, 2007).

Web Sites

Indiana State University School of Medicine. "Biochemistry of Nucleic Acids." http://www.indstate.edu/thcme/mwking/nucleic-acids.html (accessed January 24, 2008).

Lipid Library. "About Lipids." December 15, 2007. http://www.lipidlibrary.co.uk/lipids.html (accessed January 24, 2008).

Nobel Foundation. "Cell-Free Fermentation." http://www.nobelprize.org/nobel_prizes/chemistry/laureates/1907/buchner-lecture.pdf (accessed January 24, 2008).

United States National Library of Medicine. National Institutes of Health. MedLine Plus. "Carbohydrates." January 22, 2008. http://www.nlm.nih.gov/medlineplus/carbohydrates.html (accessed January 24, 2008).

University of Akron. Department of Chemistry. "Carbohydrates." http://ull.chemistry.uakron.edu/genobc/Chapter_17/ (accessed January 24, 2008).

University of Maryland Medical Center. "Omega-3 Fatty Acids." May 1, 2007. http://www.umm.edu/altmed/articles/omega-3-000316.htm#Uses (accessed January 24, 2008).

University of Texas Institute for Cellular and Molecular Biology "Glycolysis." July 18, 1998. http://biotech.icmb.utexas.edu/glycolysis/glycohome.html (accessed January 24, 2008).

Kenneth T. LaPensee

Chemistry: Chaos Theory

■ Introduction

Chaos theory is the study of mathematical systems that exhibit certain characteristic properties, one of which is extraordinarily erratic behavior. Examples of such systems include population growth, turbulent fluids, and the motion of the planets. Though chaotic systems had been recognized (but not defined) throughout human history, it was not until the 1970s that the mathematical tools existed to examine these sorts of complicated behaviors in a quantitative fashion.

Through intensive interdisciplinary work by an international set of researchers, chaos study has moved from a small group of interested practitioners into a worldwide phenomenon. Chaos theory has been applied to weather, populations, economics, turbulence, information theory, and neuroscience, among other topics. Chaos theory entered into public consciousness with the publication of colorful fractal pictures, popular science books, and public debates over its validity.

Although popular among select audiences, chaos was seen by others as an attack on the existing way of doing science, reductionism, and early practitioners met with a lot of resistance. Even today, some researchers are concerned about the use of chaos in both the natural and social sciences.

■ Historical Background and Scientific Foundations

Before the term "chaos" was used by mathematicians and scientists, chaotic phenomena had been observed in nature. In the middle of the seventeenth century, Isaac Newton (1642–1727) developed differential equations, which show how quantities change over time, and used them to describe the laws of planetary motion. Newton was able to solve the problem of determining the loca-

tion of a single planet orbiting the sun at a particular time in the future, the so-called "two-body problem," but extending this to a system of more than two bodies, such as the moon or other planets, was problematic. The equations of the conservation laws (such as the conservation of energy) that governed such complex motion could not be solved with simple algebraic methods. The many-body problem was more than theoretical: knowledge of the moon's position was important for celestial navigation. However, existing analytic methods were not powerful enough to solve it, and it was not until 1885 that progress was made.

That year, a mathematics professor in Sweden held a mathematical competition to honor King Oscar II's sixtieth birthday, asking entrants to address any of four pressing questions. One of them, tackled by Henri Poincaré (1854–1912), was the problem that Newton could not solve: the many-body problem. The question read something like this: Given a number of masses that obey Newton's law of gravitation, find the equations describing the position of the masses at any time.

Poincaré's paper, which for simplicity analyzed a three-body problem, derived a result that demonstrated the stability of the solar system (a many-body problem). In the future, he posited, the planets would not fly off into deep space. When this paper was being prepared for publication, however, several mathematical errors became apparent that invalidated this result. (In his defense, though, the point that Poincaré was trying to prove was so complex that it took vast energies just to perform the calculations, energies that would later be made easier by the computer.)

In the years since Newton's first attempt, however, mathematicians had come to understand that not all differential equations could be solved exactly, and mathematical tools had been developed to explain the qualitative properties of the solution. The judges, under the

WORDS TO KNOW

APERIODICITY: A pattern or process that is periodic and repeats itself after a fixed time interval or distance; patterns that are not periodic are aperiodic. Aperiodic geometric patterns or tilings were proposed in the 1960s and have since been found to be of importance in crystallography and the physics of solids.

BIFURCATION: Splitting in two, especially of a physical structure such as a limb or blood vessel. The letter Y, for example, shows a bifurcation halfway up the stem.

DETERMINISTIC SYSTEM: A deterministic or causal system is a system of physical objects in which the state of the system at any given time defines or determines the state of the system at all future times. Machines are, ideally, deterministic systems: each operation of the machine forces the next state of the machine to come into being, and only one next state is possible. Physicists are divided over whether the universe as a whole is deterministic; large-scale events are deterministic, a quality we rely on in designing machines, but whether subatomic events are truly random or truly deterministic is debated in quantum physics.

DYNAMICAL SYSTEM: In mathematics, any equation or set of equations that describes the changing of a system over time—in particular, the location of a certain point in the system, such as the tip of a pendulum or a particle suspended in a swirling fluid.

FRACTAL: A fractal is a geometric pattern in which each part of the pattern has the same character as the total pattern. Some natural patterns, such as fern branching, have a nearly fractal character; approximately fractal patterns are also used in some technological applications, such as cell-phone antennas.

LOGISTIC MAP: An equation that describes the growth or shrinkage of a hypothetical population of organisms. It is noted for its chaotic long-term behavior: that is, very slight differences in initial conditions lead to widely different populations at future times.

PHASE TRANSITION: A substance that moves from one physical phase to another—gas to solid or liquid, solid to liquid

or gas, liquid to solid or gas—is said to undergo a phase transition.

REDUCTIONISM: A style of thought that can be contrasted to holism. Reductionism assumes that the properties of any complex system—a human brain, an ecosystem, or other—can be reduced (hence "reductionism") to the interactions of its component parts. Holism asserts that at least some system properties arise out of the system as a whole (hence "holism") and are not possessed by, or reducible to, the properties of the system's isolated, individual parts.

SENSITIVE DEPENDENCE ON INITIAL CONDITIONS: In a chaotic system, the future state of the system—what orbit a comet ends up following a million years from now, for example—can vary greatly depending on very slight changes in the original state of the system. The original conditions of the chaotic system are termed its initial conditions, and the future state is said to depend sensitively on the initial conditions.

STRANGE ATTRACTOR: In the mathematical description of systems, an attractor is a state that the system tends to approach over time. For example, a state of regular back-and-forth swinging is an attractor for a playground swing or pendulum that is being regularly pushed. A strange attractor is a state that does not repeat itself with such perfect regularity as a swinging pendulum; rather, the system may assume a never-repeating series of states that are close to the attractor but never simply converge to it.

TOPOLOGY: A sub-field of geometry. In particular, it is the mathematical study of shapes apart from any stretching or deformation that they may undergo.

TURBULENCE: Chaotic, rough flow or mixing in a gas or liquid. It is distinguished from smooth flow, which is called laminar flow because it tends to occur in neatly organized layers with smoothly ordered velocities.

UNIVERSALITY: The property of some mathematical representations of causal systems that certain overall properties of the system do not depend on its detailed properties.

time constraints of selecting a winner before the birthday celebrations, saw the importance of Poincaré's paper and awarded him the prize.

A greatly corrected, revised, and lengthened version of the paper published in 1890, titled *Sur le probléme des trios corps et les equations de la dynamique,"* (On the Three-body Problem and the Equations of Motions; in modern terms, Equations of "Dynamics") found a different result from the first. Far from depicting a stable Newtonian universe, Poincaré's second analy-

sis of the problem showed chaotic behavior, describing a figure of curves that formed an intricate mesh so complex that he refused to attempt to draw it. His work, because of the impossibility of simple solutions, was largely qualitative and based on geometric reasoning. He also noted that a small difference in the initial position of the planets resulted in very large differences in the position of the bodies in the long run. A meteorologist would come to similar conclusions in the 1960s.

The First Route to Chaos

At the Massachusetts Institute of Technology in 1960, American meteorologist Edward Lorenz (1917–) programmed 12 weather-simulating equations into his vacuum-tube computer. While the equations governing the motion of air and water—both treated as fluids mathematically—had long been known, using them to predict weather changes had proved extremely difficult. These equations, like those Poincaré and Newton had studied before, described a dynamical system—a system that evolves over time.

In a dynamical system, both initial conditions (such as the positions and velocities of the planets at a particular time) and a set of rules (the differential equations governing motion) are used to calculate the state of the system in the future. For most dynamical systems, the future state of the system cannot be computed immediately: to do so, the state of the system has to be calculated for each moment from the beginning to the end point. Without a computer, this task was virtually impossible, but with his computer, Lorenz was able to calculate a series of numbers that showed various features of the simulated meteorological system, like temperature.

In the winter of 1962 Lorenz wanted to examine a particular sequence of numbers from his weather simulation in closer detail. When he ran it a second time, however, he got radically different results. This was unexpected because the weather system was deterministic; with the same initial conditions and the same equations, the system should have behaved exactly the same. Investigating the discrepancy, Lorenz discovered the cause of the strange behavior: he had run his second simulation with the initial condition 0.506, but the first calculation had used six digits, 0.506127.

In a dynamical system, Lorenz found, two very slightly different initial conditions (0.506 instead of 0.506127) led to radically different behavior in the long run—a conclusion that harked back to Poincaré's conclusion. The technical term for this conclusion, "sensitive dependence on initial conditions," was eventually condensed in popular shorthand to "the butterfly effect," which is based on the idea that even the flap of a butterfly's wings could produce a small atmospheric change that would create a chain of events resulting in a drastic change in weather patterns. Its practical result suggests that perfect weather prediction might be impossible. If the equations governing weather were anything like the equations Lorenz used to simulate the weather in his computer, the smallest discrepancy in the initial data would yield radically different results in the future state of the weather system.

To examine this effect further, Lorenz created a simpler three-equation system to describe atmospheric heat flow, known as convection. Like his previous 12-equation model, the three convection equations were deterministic, and for the most part the system was

periodic: it repeated its behavior after a set period of time. However, when the temperature difference between the top of the atmosphere and the bottom of the atmosphere was great enough, the system would show turbulence. (This is similar to a pot of water. As the bottom heats up, the temperature difference between the top of the water and bottom of the water grows ever larger, until eventually turbulence occurs and the water boils.)

The second model behaved similar to the first, displaying a similar sensitive dependence on initial conditions. Unlike Poincaré, however, Lorenz was able to use his computer to generate a graphical display of his system. Plotting the evolution of the convection system, he started with a particular initial condition, then traced a curve that looked like butterfly wings. Further investigation revealed that all initial conditions would trace a similar curve—they were "attracted" to the same general region. These curves defined a space that would come to be known as the Lorenz attractor, part of a larger class of mathematical objects known as strange attractors.

After Lorenz, work with physical systems led to the discovery of more strange attractors, including the Rössler attractor, which came out of the study of chemical reactions, and the Chua attractor, which resulted from the study of an electronic circuit. From Lorenz's pioneering work, it is clear that modern chaos theory would not have been possible without computers to perform millions of operations in a short period of time. But computers were more than simply calculating robots—they became electronic laboratories themselves, simulating natural phenomena.

Two More Routes to Chaos

Chaos theory did not develop with one person or in one place, but was the product of different scientists working in different places, solving different problems. In addition to Lorenz, two other figures in the 1960s established key sections of chaos theory's intellectual framework.

One of these was the American mathematician Stephen Smale (1930–) at the University of California at Berkeley, who became internationally famous for his study of topology (the study of the properties that remain unchanged as geometric figures are stretched and folded). In the 1960s, Smale turned his research from topology to dynamical systems. Early on, he theorized that stable systems—those in which a small perturbation would not change the overall outcome of the system—could not behave erratically. At this point, he was not aware of the Lorenz attractor, which was a structurally stable system that *did* behave erratically. (This is not surprising, since Lorenz's paper was published in a specialty journal, traditionally read only by meteorologists. Like many important scientific discoveries, the importance of his work was not immediately recognized.)

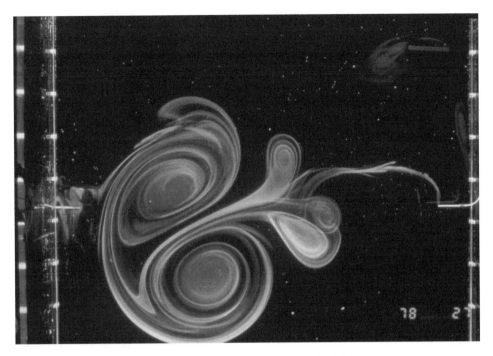

Chaotic systems: head-on collision of two dipolar vortices in a stratified fluid environment. The original vortices have exchanged a partner to form two new (mixed) dipoles which are moving at roughly right angles to the original direction of travel, that is, towards the top and bottom of the image. Dipolar vortices are relevant to turbulence in large-scale geophysical systems such as the atmosphere or oceans. Turbulence in fluid systems is one example of a chaotic system. *G. van Heijst & J. Flor/Photo Researchers, Inc.*

One of Smale's colleagues, however, told Smale that the Lorenz attractor refuted his conjecture. Forced to rethink his conclusion, Smale came up with a visual way to understand why his conjecture was incorrect. The horseshoe map was a topological version of sensitive dependence on initial conditions, a way to deform a system (say a square sheet of rubber) topologically through repeated squeezing and folding, so that any two points close to each other in the original system ended up arbitrarily far apart after enough folding and stretching.

Smale's work expanded the study of dynamical systems into the domain of topology and reshaped the disciplinary boundary between physics and mathematics. Smale himself said, "When I started my professional work in mathematics in 1960, which is not so long ago, modern mathematics in its entirety—in its entirety—was rejected by physicists, including the most avant-garde mathematical physicists ... By 1968 this had completely turned around."

The last figure setting the stage for an explosion of research in the 1970s was Belgian-French mathematical physicist David Ruelle (1935–). Located at a prestigious institute outside of Paris in 1971, he and mathematician Floris Takens (1940–) published "On the Nature of Turbulence," which suggested that the onset of turbulence in a fluid was caused by topological properties of the fluid equations themselves, known as the Navier-Stokes equations. Just as the equations Lorenz used to study convection gave rise to a strange attractor—the butterfly-shaped Lorenz attractor—the authors argued that the Navier-Stokes equations also gave rise to strange attractors, and the presence of these attractors was responsible for the onset of turbulence.

Since the early nineteenth century, the transition of a stable system into a turbulent one had been ill-understood. Some scientists have called it the greatest unsolved problem in classical physics. The prevailing theory before Ruelle and Takens was promoted by Russian physicist Lev Landau (1908–1968) in the 1940s.

His model argued that eddies formed in fluids will generate smaller eddies within them, and these smaller eddies will then generate even smaller eddies within them, *ad infinitum*. As more and more eddies are created, the fluid flow begins its transformation from being predictable and regular to exhibiting turbulence. The eddies, in Landau's theory, are created through small external disturbances to the fluid. For Landau, the transition to turbulence was based not on the fluid equations themselves, but on external noise influencing the system. Taking the opposite position, Ruelle and Takens argued that turbulence could be explained by the fluid equations alone.

At the beginning of the 1970s the study of fluid mechanics and the question of turbulence lay at the uneasy intersection between mathematics and engineering. Ruelle and Takens's work had two important consequences. First, it inaugurated the merger between the study of fluid mechanics and dynamical systems. Second, it suggested physical experiments that could test the validity of Ruelle's ideas. No longer would the study of these strange dynamical systems be confined to paper or computer simulations. With Lorenz, Smale, and Ruelle building the foundation, by the end of the 1960s the stage was set for chaos to enter the academic community.

An Era of Interdisciplinarity

Lorenz was a meteorologist, Smale a mathematician, and Ruelle a mathematical physicist, yet all three were working on similar things. The 1970s ushered in a period of interdisciplinarity, in which biologists, hydrodynamicists, meteorologists, mathematicians, and physicists, among others, would read each others' papers, attend the same conferences, and enter into work traditionally outside their own specialized fields. Divisions among scientific disciplines have never been so solid as to prevent interdisciplinary work, but chaos, acting as a common meeting ground, brought together scientific nomads. It was also during this decade that experimental work brought some credibility to chaos, but at the same time, practitioners were unable to pin down exactly what they were studying.

Scientists, when investigating natural phenomena, would often encounter phase transitions—points at which a system changed its character dramatically. One example was the onset of turbulence, but there are many more, such as the magnetization of a nonmagnet or the transformation of a conductor into a superconductor. A number of studies drew analogies between various types of phase transitions; mathematically their descriptions appeared similar. In 1973 American physicists Harry L. Swinney (1939–) and Jerry P. Gollub (1944–) were investigating phase transitions in fluids. They set up an inexpensive apparatus: two cylinders, one inside the other, with a fluid in between. As they began to rotate the inner cylinder (keeping the outer cylinder at rest), they studied the properties of fluid flow. Once they reached a certain threshold, they observed turbulence. The experiment was not new; it had been performed before in 1923. What was new was the data-collection method, using the deflection of a laser beam in the fluid to measure the scattering.

Swinney and Gollub expected to confirm the older theory of turbulence proposed by Laundau. What they found comported better with the ideas of Ruelle and Takens. A second similar and widely discussed experiment was performed by French physicist Albert J. Libchaber (1934–) in 1977. The study of turbulence and convection was beginning to become an interesting scientific topic again. Instead of being squarely in the purview of mathematics or engineering, it began to be

investigated by hydrodynamicists, plasma physicists, statistical physicists, thermodynamicists, and chemists. In the years between 1973 and 1977, a good number of conferences were held on turbulence. These acted simultaneously as sites where disciplines collided and where collaborations were forged.

James Yorke (1941–), an American mathematician, had stumbled across Lorenz's 1963 paper almost a decade after its publication and was enchanted by its conclusions. He sent a copy to Smale and made and distributed a number of other copies for his colleagues. He also cowrote a 1973 article in the widely read *American Mathematical Monthly* that brought the ideas to a new generation of mathematicians. The title of this article—"Period Three Implies Chaos"—marked the first time "chaos" was used in a technical sense.

Yorke's article made an important claim using a simple formula that was first used in 1838. The logistic equation is a model that predicts the population of a species over time, given information about how fast it reproduces and the maximum population sustainable in an environment. Yorke used the logistic equation to illustrate a powerful point: For most situations, the population of a species would remain the same after a long enough time had passed. However, after tweaking the conditions, he found that the population could eventually oscillate between two values—one year the population would be x, the next year y, the year after x, and so forth. Mathematically, this is called a bifurcation. It turns out the conditions could be tweaked so that the population would oscillate between 4, 8, 16, 32, etc., values. At a certain point, however, the system would become what Yorke called chaotic: The population would not oscillate among a fixed number of values, but rather, would never repeat itself.

Yorke used the logistic equation to illustrate a powerful mathematical theorem. Technically, it said that in any one-dimensional system (like the logistic equation), if a cycle of period 3 appears, then there would have to be cycles of every other period, as well as chaotic cycles. In a general sense this meant that even a simple equation can demonstrate some peculiar and unexpected behavior—chaos. More than that, in the midst of chaotic behavior, there can be pockets in which the system behaves nicely. In this way, chaos began to take on the meaning of its opposite: order. Even though the term chaos implied something erratic, it arose out of very simple systems, and even when chaos was observed, there seemed to be some kind of mathematical order within the chaos.

Universality in Chaos

In 1975 mathematical physicist Mitchell Feigenbaum (1944–) was a researcher at Los Alamos National Laboratory. After listening to a lecture by Smale, he began his work on the simple logistic map. He found that the bifurcations came at somewhat predictable intervals. In mathematical terms, they were converging geometrically. The convergent rate for the logistic map was calculated to be about 4.669. Feigenbaum discovered that the critical value of 4.669 occurred not just in the logistic map, but in a large class of equations—just as Yorke found that chaotic behavior occurred in a large class of equations.

He recalled in 1980 that "I spent a part of a day trying to fit the convergent rate value, 4.669, to the mathematical constants I knew. The task was fruitless, save for the fact that it made the number memorable." That indicated that this number was a new universal constant. Just as the ratio of the circumference to the diameter of any circle will always be π, the convergent rate for any of a large class of equations will always be about 4.669.

Feigenbaum believed he had discovered a new law of nature. If a natural phenomenon could be described mathematically by one of the large class of equations, then Feigenbaum's work was a quantitative way to discuss the route to phenomena exhibiting chaotic behavior. Through his work, the analogy made between hydrodynamics (the transition to turbulence) and phase transitions in physics was reinforced. The equations were all part of the same class.

As often occurs with historically important works, Feigenbaum had difficulty publishing—it was not quite mathematics to the mathematicians because it did not have a grand level of abstraction, nor was his claim rigorously proved enough for physicists. In spite of this, his ideas spread through lectures and conversation and generated excitement in an assortment of intellectual circles. In retrospect, scientists look back on this work as a watershed, bestowing a sense of legitimacy to the idea that chaos was present in many natural systems, and that this chaotic behavior could be explored quantitatively instead of qualitatively and geometrically.

Moreover, in subsequent analyses, he used mathematical tools (renormalization group methods) commonly used by physicists, thus making his chaos relevant for that community. As a result, a number of physicists in the late 1970s and early 1980s began to look at hydrodynamics and turbulence, when previously the problem had belonged in the domain of mathematicians and engineers; a disciplinary reorientation took place. Libchaber's 1977 experiment on turbulence, for example, was concerned with calculating Feigenbaum's constant. His experimental value and the theoretical value were dissimilar, but not incompatible, and later experiments conducted by others in the early 1980s brought a closer correspondence between the theoretical and experimental values. These experiments showed that Feigenbaum's constant, and chaos theory in general, had a role to play in the physical world.

Two important chaos meetings were held in the late 1970s. In 1977, the New York Academy of Sciences

IN CONTEXT: ICONIC IMAGES OF CHAOS

The Lorenz attractor is a graphical representation of the behavior of the convection equations. The logistic map is a discrete version of the logistic equation and is mathematically written: $x_{n+1} = rx(1 - x_n)$. We need not go into details about the mathematics here, but what should be clear is that this simple one-dimensional equation demonstrates some very complicated behavior.

As the value of r increases, the long-term behavior of the system changes. For small values of r, the system settles down to a single value. As r increases, something special happens: the system settles down not to one, but two values: It bifurcates.

As r increases further, the system settles down to four values, then eight, then sixteen, and so on. Eventually, when r reaches a critical value, the system never repeats itself, never settles down to a finite set of numbers. Note that the behavior of the equation becomes more and more erratic as r increases, but even in the chaotic regime, there are pockets of nonerratic behavior.

held a conference on bifurcation theory and applications in scientific disciplines. This meeting, the first of such a large scale, brought dozens of researchers together: economists, physicists, chemists, biologists, and others. A second meeting was held two years later, this time attended by hundreds of researchers. Chaos was becoming a more popular topic of study in a variety of fields. By 1990, one bibliography of chaos—in Chaos II—listed 117 books and 2,244 articles on the subject from a variety of disciplines, and, significantly, from many different countries too.

While chaos theory's audience was increasing, there was—and remains—no single accepted definition. As late as 1994 American mathematician Steven Strogatz (1959–) wrote that: "No definition of the term chaos is universally accepted yet, but almost everyone would agree on the three ingredients used in the following working definition: Chaos is aperiodic long-term behavior in a deterministic system that exhibits sensitive dependence on initial conditions."

"Aperiodic long-term behavior" simply means that the system does not settle down to a state in which nothing moves, or a state in which the system repeats itself over and over. Instead, the system should eventually lead to erratic behavior, like that in a Lorenz attractor.

Popularization: Chaos in Popular Consciousness

As chaos became a popular topic in the sciences, ideas about chaos theory also entered into the public domain.

IN CONTEXT: CHAOS IN NATURE

By the mid-1980s a wide-ranging set of systems had been shown to have exhibited chaotic behavior. More than mathematics and physics, even animals and humans fell under its purview as noted in an article in *Scientific American* in 1986:

> In the past few years a growing number of systems have been shown to exhibit randomness due to a simple chaotic attractor. Among them are the convection pattern of fluid heated in a small box, oscillating concentration levels in a stirred-chemical reaction, the beating of chicken-heart cells and a large number of chemical and mechanical oscillators. In addition, computer models of phenomena ranging from epidemics to the electricity of a nerve cell to stellar oscillations have been shown to possess this simple type of randomness. There are even experiments now under way that are searching for chaos in areas as disparate as brain waves and economics.

SOURCE: *Crutchfield, James P., et al. "Chaos."* Scientific American 54, *no. 12 (December 1986): 46–57.*

People learned about chaos through various channels, and one of these channels was fractals. Benoit Mandelbrot (1924–), a Polish-born, French-educated American mathematician, saw a universe full not of ordinary Euclidean geometry, in which figures were smooth, but full of "fractals," a term he coined in 1975.

Fractals are objects that display self-similarity at all scales: If a fractal is magnified, the resulting image will have properties similar to the original fractal. Because fractals are so intricate—they can be magnified over and over again—they are often described as beautiful. As a result of their intricacy, however, everyday concepts like distance and area are difficult to apply. When studying these figures, Mandelbrot devised a new way to measure an object's dimensions. With his definition, a line still has one dimension, a plane still has two, but fractals could possess a dimension in-between—a nonintegral dimension, such as a $^3/_2$ dimension.

Strange attractors such as the Lorenz attractor, it turns out, are not points, lines, or surfaces, but rather fractals with a dimension between 2 and 3. (A trajectory in the Lorenz attractor is infinitely long, but it is bounded by a finite volume—the butterfly shape.) Thus, understanding the properties of fractals and uncovering ways to study them could shed light on strange attractors—and consequently, on chaos.

Also important, however, is the explosion of fractals into popular culture. Mandelbrot's popular books *Fractals: Form, Chance, and Dimension* and *The Frac-*

tal Geometry of Nature, transported fractals to a wider audience, including a number of professionals. Ruelle exclaimed, "I have not spoken of the esthetic appeal of strange attractors. These systems of curves, these clouds of points suggest sometimes fireworks or galaxies, sometimes strange and disquieting vegetal proliferations. A realm lies there of forms to explore, and harmonies to discover."

The 1970s and 1980s fractal geometry was used to create art—illustrations of nature in addition to more abstract entities, known as Julia sets. Artists too found fodder in a new concept of space.

In addition to Mandelbrot's popular books and the widely reproduced pictures of fractals, American author James Gleick's (1954–) popular science book *Chaos: Making a New Science* also helped propel chaos theory into popular consciousness. Published in 1987, *Chaos* was an immediate bestseller.

Narrating the history of chaos as a paradigm shift, Gleick introduced chaos theory not only to the wider public, but also to numerous scientists. The impact of this book can be seen in the high number of citations it receives in both scientific and popular literature. (The *Science Citation Index,* which counts the number of times a particular book or article is cited in various research journals, puts the number at well over 1,000 articles.)

If Gleick introduced the concept of chaos, the publication of American novelist Michael Crichton's (1942–) book *Jurassic Park* in 1990 and the release of Steven Spielberg's (1946–) motion picture of the same name three years later put chaos theory center stage.

In the movie, the scientist Malcolm saw the dinosaur park as a physical system governed by chaos. His interpretation of chaos, however, was a theory of inherent unpredictability and disorder, an interpretation that is not quite accurate. Chaotic systems are often defined by deterministic equations—meaning that, in theory, at least, they are perfectly predictable. And, as noted before, one of the key features of chaos is the concept of order—not disorder. Regardless, *Jurassic Park,* Gleick's book, and captivating pictures of fractals, generated by many on their home computers, helped bring an interest in chaos to the wider public.

Chaos Outside the Traditional Sciences

Social scientists have long used statistics and mathematical models to better understand social phenomena; some hold the hope expressed in the introduction to *Chaos Theory in the Social Sciences: Foundations and Applications,* edited by L. Douglas Kiel and Euel Elliott, that chaos theory will be "a promising means for a convergence of the sciences that will serve to enhance understanding of both natural and social phenomena." Chaos theory, these scientists believe, may finally explain the complex phenomena observed in social systems.

Computer-generated fractal image derived from chaos theory. Fractals are patterns formed by repeating some simple process on an ever decreasing scale. This image was made by repeating mathematical operations upon data in the complex plane. Chaos theory may be used to model processes such as the weather and the beating of the heart.   *Mehau Kulyk/Photo Researchers, Inc.*

Beginning in the 1990s, political science, economics, sociology, and even in literary theory began to use chaos theory. Some of these fields rely heavily on its mathematical tools to analyze social data, while others capitalize on the metaphorical power of chaos. Political scientists, for example, have studied shifting public opinions and international conflict using some of the same mathematical tools used to study chaos. Some economists posit that because many simple economic systems show a sensitive dependence on initial conditions, the standard form of economics should be reevaluated, neoclassical economics. And literary theorist N. Katherine Hayles makes the case that since both science and the humanities are rooted in the same culture, when scientists were countenancing the implications of disorder, so too were fiction authors and literary theorists.

■ Modern Cultural Connections

In recent years, chaos has lost some of the excitement that fueled it earlier. Per Bak (1948–2002), a Danish theoretical physicist at Brookhaven National Labora-

tory, believed that chaos theory had run its course as early as 1985. Physicist J. Doyne Farmer (1952–) said, "After a while, though, I got pretty bored with chaos … [I] felt 'So what?' The basic theory had already been fleshed out. So there wasn't that excitement of being on the frontier, where things aren't understood." Chaos today can be seen not as a highly active area of scientific research, but instead as a conglomeration of mathematical tools used in a wide variety of disciplines.

Chaos, however, has taken on a new incarnation in recent years: complexity, which, like chaos, defies a universal definition. (One scientist compiled 31 distinct definitions.) However, most often it is described as the state between order and chaos. It is in this regime that complexity researchers believe self-organization occurs. In highly ordered systems, nothing novel can emerge. Chaotic systems are too erratic to have something structured emerge. It is at the intersection, some believe, that interesting and complex behavior emerges. Complexity theorists work in a number of different fields, including artificial intelligence, information theory, linguistics, chemistry, physiology, evolutionary biology, computer science, archeology, and network theory. In fact, in

BENOIT MANDELBROT (1924–), FATHER OF FRACTALS

Though the face of Benoit Mandelbrot (1924–) is not well known, the colorful fractal figures he pioneered are familiar to almost everyone. He is one of the most famous living mathematicians, having developed a geometry that would infiltrate a number of different arenas, including the study of turbulence, the stock market, physiology, and art.

Mandelbrot was born in Warsaw and raised in France. His father was a buyer and seller of clothes, his mother was a medical doctor, and his uncle, Szolem Mandelbrojt (1899–1983), was a Polish-born French mathematician at the Collège de France. While his uncle taught the younger Mandelbrot that mathematics was an honorable profession, he was also a purist who believed that mathematics and beauty were mutually exclusive.

Benoit Mandelbrot's mathematics education took place mainly in France, punctuated by brief stints at the California Institute of Technology and Princeton's Institute for Advanced Study. He was seriously affected by the existing political conditions, noting in an interview that "when I look back I see a pattern. For a long time that pattern was imposed by catastrophes, namely the fall of Poland and the occupation of France during the second world war. Those events dictated everything … Being raised under such hair-raising conditions can have a strong effect on someone's personality." In 1952 he obtained a doctorate in mathematical sciences at the University of Paris.

Mandelbrot joined the research team at the International Business Machines (IBM) Thomas J. Watson Research Center in New York in 1958, remaining there until 1987, when he joined the mathematics department at Yale University. At IBM, Mandelbrot reveled in the freedom afforded in his research, allowing him to move in directions that a university position would not encourage; his research included the flooding of the Nile River, cotton prices, and the geometric shape of coastlines. It was in these studies that the idea of fractals began to form, a concept Mandelbrot continued to expand and develop in the following years. One of the most famous fractals, called the Mandelbrot Set, is based on work done by French mathematicians Gaston Julia (1893–1978) and Pierre Fatou (1878–1929).

The complex Mandelbrot set is created through a simple mathematical transformation. Every point in the plane undergoes this transformation, and if the result is infinity, the point is shaded a particular color. If the result is not infinite, the point is shaded black. Interestingly, his uncle Szolem introduced him to this work in 1945, but Mandelbrot would not return to it until the 1970s. Mandelbrot retired from Yale in 2005.

1999, the prestigious journal *Science* devoted an entire issue to examining research done on complex systems in a variety of fields.

The major institution promoting the study of complex systems is the Santa Fe Institute (SFI) in New Mexico. George Cowan, founder of the SFI, and once the head of research at the national laboratory at Los Alamos, also advised President Ronald Reagan on the White House Science Council. During this time, he was made further aware of the interconnections between science and morality, economics, the environment, and other topics. Looking back, he said:

> The royal road to a Nobel Prize has generally been through the reductionist approach.… You look for the solution of some more or less idealized set of problems, somewhat divorced from the real world, and constrained sufficiently so that you can find a solution … [a]nd that leads to more and more fragmentation of science. Whereas the real world demands—though I hate the word—a more holistic approach.

A reductionist view, he believed, also leads to studying simple systems, instead of whole, messy, complicated systems.

With the rise of computer use and numerical simulation, complex behavior could be investigated. By the 1980s the complex behaviors Cowan was interested in were being investigated. These complex systems were not linear, in the sense that they could be broken down into smaller parts; what made them interesting was their global behavior created by the nonlinear interactions between individual parts. He took this belief and transformed it into the now world-famous interdisciplinary SFI, founded in 1984, eventually adding three Nobel laureates to its staff. This institute is now synonymous with the study of complexity. At the institute and elsewhere, the study of chaos theory was incorporated into a broader framework of "complexity."

Chaos and complexity: These terms hint at a new structure for science. Like Cowan, many see the sciences moving away from reductionism and linearity into something completely different. The term "paradigm shift" is often invoked by chaos researchers and popular writers to describe this transformation. In 1962, historian Thomas Kuhn (1922–1996) argued that science did not accumulate facts progressively. Instead, science is a cyclical process, punctuated by large-scale transformations known as "paradigm shifts." (Paradigms, loosely defined, are ways of understanding the world. Paradigm shifts are brought about by finding anomalies that cannot be explained in the current paradigm.) Resolving these anomalies led to a new way of understanding the world, a way in which the anomaly makes sense. It should be noted that historians of science have found much to praise and criticize with Kuhn's theory about how science operates.

IN CONTEXT: REDUCTIONISM VS NONREDUCTIONISM

In his review of a book by Stephen Wolfram titled *A New Kind of Science*, Nobel Prize–winning physicist Steven Weinberg noted that the increasing study of complexity in physics had led to a disciplinary division—those who believe that reductionism is the best way to study nature, and those who argue that complexity is better. Weinberg levies a common criticism at those favoring complexity: there is no set paradigm that complexity has produced.

There is a low-intensity culture war going on between scientists who specialize in free-floating theories [like chaos] and those (mostly particle physicists) who pursue the old reductionist dream of finding laws of nature that are not explained by anything else, but that lie at the roots of all chains of explanation. The conflict usually comes to public attention only when particle physicists are trying to get funding for a large new accelerator. Their opponents are exasperated when they hear talk about particle physicists searching for the fundamental laws of nature. They argue that the theories of heat or chaos or complexity or broken symmetry are equally fundamental, because the general principles of these theories do not depend on what kind of particles make up the systems to which they are applied. In return, particle physicists like me point out that, although these free-floating theories are interesting and important, they are not truly fundamental, because they may or may not apply to a given system; to justify applying one of these theories in a given context you have to be able to deduce the axioms of the theory in that context from the really fundamental laws of nature....

Lately particle physicists have been having trouble holding up their end of this debate. Progress toward a fundamental theory has been painfully slow for decades, largely because the great success of the "Standard Model" developed in the 1960s and 1970s has left us with fewer puzzles that could point to our next step. Scientists studying chaos and complexity also like to emphasize that their work is applicable to the rich variety of everyday life, where elementary particle physics has no direct relevance.

Scientists studying complexity are particularly exuberant these days. Some of them discover surprising similarities in the properties of very different complex phenomena, including stock market fluctuations, collapsing sand piles, and earthquakes ... But all this work has not come together in a general theory of complexity. No one knows how to judge which complex systems share the properties of other systems, or how in general to characterize what kinds of complexity make it extremely difficult to calculate the behavior of some large systems and not others. The scientists who work on these two different types of problem don't even seem to communicate very well with each other. Particle physicists like to say that the theory of complexity is the most exciting new thing in science in a generation, except that it has the one disadvantage of not existing.

SOURCE: *Weinberg, Steven. "Is the Universe a Computer?"* New York Review of Books *49, no. 16 (October 24, 2002).*

Researchers interested in chaos and complexity tout the beginnings of a paradigm shift. The old paradigm, reductionism, was on its way to being replaced, or at least supplanted by, nonreductionism, which sees the world as a system of many parts that interact in nonlinear ways. Many scientists are skeptical of this claim. They do not believe complexity is a paradigm shift in the Kuhnian sense, nor do they believe that the reductionist program has yielded all its secrets. Whether a paradigm shift, a fad, or something else, science has of recent been changing its flavor, and chaos and complexity have played a role.

■ Primary Source Connection

This patent describes an unusual application of chaos theory to a real-world problem: wrinkly clothes. The patent was issued October 1, 1996.

CHAOS WASHING MACHINE AND A METHOD OF WASHING THEREOF

■ Description of the Prior Arts

Washing machines presently employed use a pulsator or drum. The washing machines using the pulsator increase the washing power by irregular flow of washing water in the washing tank by repeatedly rotating the pulsator disposed in the bottom of the washing tank clockwise and counter-clockwise.

The washing machine using the drum increases the washing power by a head of laundry derived by rotating the drum itself in which the laundry and washing water are contained.

However, such washing machines using the pulsators have the disadvantage in that the laundry is wrinkled while being rotated together with the washing water, and further, it is difficult to obtain the higher washing effect due to limitations of the washing power that is dependent on the rotation power of water current.

On the other hand, washing machines using the drum have the disadvantage due to the difficulty of obtaining the higher washing effect because of the limited washing power that is dependent on the head of the laundry, as well as the laundry being wrinkled by the regular and reverse rotation of the drum.

Further, a problem exists in that the manufacturing cost increases because an additional program or hardware, must be installed in order to prevent the wrinkling of the laundry.

■ Summary of the Invention

It is an object of the present invention to provide a chaos washing machine which can improve the washing effect by using the random generation of water current in order to overcome the aforementioned defects.

It is another object of the present invention to provide a chaos washing machine which can reduce the wrinkling of the laundry by creating a random stream of water for short periods, random generation of the water stream and then producing a strong turbulent flow in the washing tank.

It is a further object of the present invention to provide a method of washing of a chaos washing machine which can improve a washing power and prevent the wrinkling of clothes by using the random generation of the water current.

These and other objects of the present invention are accomplished by means of a chaos washing machine which is composed of a first washing tank having a plurality of inducing holes for inducing the water current in which a detergent dissolved into the space where the laundry is contained, a second washing tank for enclosing the washing tank to be filled with water and a detergent, a water current fan and a fan motor creating a turbulent flow to water in which a detergent is dissolved and pushing the turbulent flow from the second washing tank through the inducing holes into the first washing tank and a washing tank motor for rotating the first washing tank.

[...]

As stated above, according to the chaos washing machine of using a convection flow of the present invention, the laundry is not wrinkled. Therefore, the present invention eliminates the need for mechanically and electrically means for sensing the wrinkle of the laundry or for proceeding the twist preventing pattern of the laundry such as the prior pulsator washing tank thereby easily accomplishing the structure of the washing machine and the design and the operation of the control program.

Further, according to the present invention, the laundry will be not rotated with the washing tank. Therefore, the present invention can reduce the damage created to the laundry that is created by being regular and reverse rotation of the laundry such as the prior pulsator or drum washing machines thereby extending the life of cloth and maintaining the cleanliness of clothes after washing.

Bo Wang

WANG, BO. "CHAOS WASHING MACHINE AND A METHOD OF WASHING THEREOF." *PATENT STORM.* HTTP://WWW. PATENTSTORM.US/PATENTS/5560230-FULLTEXT.HTML (ACCESSED NOVEMBER 21, 2007).

SEE ALSO *Physics: Aristotelian Physics; Physics: Articulation of Classical Physical Law; Physics: Newtonian Physics.*

BIBLIOGRAPHY

Books

Barrow-Green, June. *Poincaré and the Three-Body Problem.* Providence: American Mathematical Society, 1997.

Gleick, James. *Chaos: Making a New Science.* New York: Viking Press, 1987.

Hao, Bai-Lin, ed. *Chaos II.* Singapore: World Scientific, 1990.

Kiel, L. Douglas, and Euel Elliott, eds. *Chaos Theory in the Social Sciences: Foundations and Applications.* Ann Arbor: The University of Michigan Press, 1996.

Kauffman, Stuart A. "The Sciences of Complexity and 'Origins of Order.'" In *PSA: Proceedings of the Biennial Meeting of the Philosophical Science Association.* Vol. 2, 299–322. Chicago: University of Chicago Press.

Kuhn, Thomas. *The Structure of Scientific Revolutions.* Chicago: The University of Chicago Press, 1962.

Strogatz, Steven H. *Nonlinear Dynamics and Chaos: With Applications to Physics, Biology, Chemistry, and Engineering.* Cambridge: Westview Press, 1994.

Waldrop, M. Mitchell. *Complexity: The Emerging Science at the Edge of Order and Chaos.* New York: Simon & Schuster, 1992.

Wise, M. Norton, ed. *Growing Explanations: Historical Perspectives on Recent Science.* Durham: Duke University Press, 2004.

Periodicals

Aubin, David, and Amy Dahan Dalmedico. "Writing the History of Dynamical Systems and Chaos: Longue Durée and Revolution, Disciplines and Cultures." *Historia Mathematica* 29, no. 3 (2002): 273–339.

Crutchfield, James P., et al. "Chaos." *Scientific American* 54, no. 12 (December 1986): 46–57.

Feigenbaum, Mitchell J. "Universal Behavior in Nonlinear Systems." *Physica D: Nonlinear Phenomena* 7, nos. 1–3, (May 1983): 16–39.

Horgan, John. "From Complexity to Perplexity." *Scientific American* (June 1995): 104–109.

Jamieson, Valerie. "A Fractal Life." *New Scientist* (13 November 2004): 50–53.

Li, Tien-Yien, and James Yorke. "Period Three Implies Chaos." *American Mathematical Monthly* 82 (1975): 985–992.

Lorenz, Edward N. "Deterministic Nonperiodic Flow" *Journal of Atmospheric Sciences* 20, no. 2 (March 1963): 130–141.

Poincaré, Henri. "Sur le probléme des trios corps et les equations de la dynamique." *Acta Mathematica* 13 (1890): 1–270.

Ruelle, David, and Floris Takens. "On the Nature of Turbulence." *Communications in Mathematical Physics* 23, no. 4 (December 1971).

Shearer, Rhonda Roland. "Chaos Theory and Fractal Geometry: Their Potential Impact on the Future of Art." *Leonardo* 25, no. 2 (March 1992): 143–152.

Weinberg, Steven. "Is the Universe a Computer?" *New York Review of Books* 49, no. 16 (October 24, 2002).

Sameer Shah

Chemistry: Chemical Bonds

■ Introduction

Since the dawn of chemistry, scientists have sought to understand the forces that hold atoms together in a chemical bond. Since molecules (a name that means "little masses") are invisible to the naked eye, researchers have had to rely on indirect methods to unravel these structures. American physical chemist Gilbert Lewis

WORDS TO KNOW

COVALENT: In chemistry, a covalent bond between two atoms is produced when the atoms share one or more electrons. The other major type of chemical bond is an ionic bond, in which atoms lacking electrons or endowed with extra electrons (and thus having positive or negative charges, respectively) are attracted to each other.

IONIC: In chemistry, an ionic bond between two atoms is formed when the atoms are ionized (given a positive or negative charge by the addition or removal of electrons) and then brought close enough together to be bound by electrical attraction. In architectural history, Ionic or Ionian classic architecture is distinguished by its use of stone columns ornamented by scrollwork at the top.

ORBITAL: In an atom or molecule, a pattern of electron density or probable location that may be occupied by an electron. An orbital is not a path traced in space by an electron, as an orbit is traced by a planet or satellite in space: this is why the term "orbital" rather than "orbit" is used.

QUANTUM: Something that exists in discrete units.

VALENCE: The tendency of an atom to gain or lose electrons in reacting with other atoms.

(1875–1946) endowed molecular theory with a formidable array of concepts that continue to influence how structural chemistry is practiced and taught throughout the world. The consensus on the chemical bond that emerged during the early twentieth century has yielded powerful insights into many areas of science.

■ Historical Background and Scientific Foundations

An important inference about bonding was that some compounds are "ionic"—they form charged species when dissolved in a liquid (such as water). Throughout the nineteenth century, a small but distinguished group of scientists subscribed to this hypothesis. These ionists (as their opponents often called them) included, among others, the Prussian physicist Rudolf Clausius (1822–1888), and the British scientist Michael Faraday (1791–1867). In addition to his wide-ranging discoveries, Faraday introduced many of the terms that scientists continue to use today: ion (a charged particle), cation (a positively charged species) and anion (one bearing a negative charge). While his studies focused on high-temperature melts, they helped to bring ions into the common vocabulary of science.

During the nineteenth century, speculation and debate centered around the electrical interactions that could stabilize chemical compounds. These concepts acquired a new legitimacy in 1897 when the British physicist Sir Joseph J. Thomson (1856–1940) published the first of a series of experimental articles on the properties of electrons from gases subjected to a high voltage.

His discovery that cathode rays consisted of such particles (for which he eventually measured both the charge and the mass) led to new efforts to incorporate the electron into models for the chemical bond. As a physicist, Thomson relied on familiar principles (like

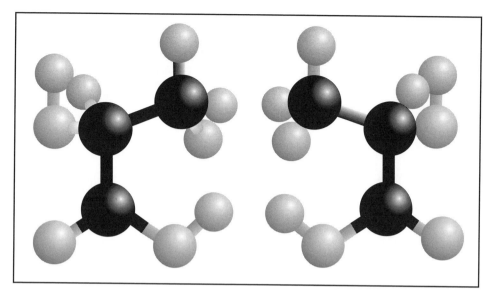

A Lewis diagram representation of water bonding (a two dimensional depiction of the basic orientation of water molecules). The profound influences of electronegativity, charge distribution, and geometry are ignored in this simplified depiction.   *Cengage Learning, Gale.*

electrostatic attraction) to explain the existence of molecules. His early theories treated all compounds as ionic and thus could not adequately account for the stability of molecules that did not dissociate into charged particles. In spite of such drawbacks, other scientists did embrace Thomson's ideas and developed variations of his electrostatic model.

The Science

One chemist who made a distinction between ionic and covalent compounds was Lewis, who published a seminal paper in 1916 entitled "The Atom and the Molecule." Like many scientists who also served as instructors, his research on the chemical bond began in 1902 as an effort to help his students understand the valence of different elements. Later, the advent of quantum mechanics drastically altered the theoretical interpretation of Lewis's model, but it remains part of the modern vocabulary of every chemist.

Lewis was quite unusual in that he spent over a decade developing his bonding theory before publishing it. This was partly because as an instructor at Harvard University he could find no chemist interested in working on such abstract concepts. By 1916, however, he had a large department of chemists at the University of California on whom he could rely for constructive feedback.

Another benefit of this delay was that researchers had a much better count of the number of electrons for each element, thanks to the spectroscopic data compiled by British physicist Henry Moseley (1887–1915). Lewis was able to incorporate this information into molecular pictures, called Lewis structures, that use dots to represent the number of valence electrons contributed by each

atom in a covalent bond. Each dot denotes an electron: In water (H_2O), the hydrogen contributes one electron, while the oxygen atom has six in its valence shell. Lewis assumed that these latter atoms also had two electrons in an inner core, which did not play a direct role in the bonding and thus were not shown explicitly. A noteworthy feature is that the oxygen has two pairs of valence electrons not used for bonding.

Lewis's real innovation was his treatment of covalent bonds as shared electron pairs. Therefore, the two dots that hydrogen and oxygen have in common denote a single bond in the water molecule, and the four electrons between the oxygen molecules in O_2 represent the much stronger double bond. He attributed the affinity of hydrogen (with only one electron) for other atoms as a natural tendency to complete its shell, achieving a configuration similar to that of helium (with two electrons). Likewise, oxygen in his scheme will usually form two single bonds or one double bond, so that it will acquire another two electrons, like neon. Lewis thus explained his bonding rules by noting that neon and helium belong to a group of elements known as noble gases, which have filled electron shells and are especially unreactive.

One of Lewis's most compelling achievements was that his 1916 publication solved several puzzles that had plagued chemists for generations, including the structure of ammonium chloride. Some chemists believed that NH_4Cl really consisted of two molecules: an ammonia part (NH_3) in close proximity to HCl. Other scientists believed that nitrogen in this molecule must have five "ionic" bonds to H^- and Cl^- anions.

Lewis cleverly solved this conundrum by suggesting that nitrogen may form four covalent bonds to

GILBERT NEWTON LEWIS (1875–1946)

Gilbert Newton Lewis (1875–1946) was born in the town of West Newton, Massachusetts, and was raised by progressive, independent parents. At the age of seven, his family moved to Nebraska, and Lewis was homeschooled until he attended high school in the city of Lincoln. His parents and teachers recognized his remarkable intelligence, and at the age of 16 he was admitted to the University of Nebraska.

Seeking a more cosmopolitan environment, he transferred to Harvard University, from which he received his bachelor's degree in chemistry in 1896. After a year of teaching at Phillips Academy in New Hampshire, Lewis returned to Harvard to begin research for his doctorate. His faculty advisor was Theodore Richards (1868–1928), a well-known American chemist who received a Nobel prize in 1914 for his precise atomic weight measurements.

Lewis earned his doctoral degree in 1899, then spent a year in the German laboratories of chemists Walther Nernst (1864–1941) in Göttingen and Russian-born Wilhelm Ostwald (1853–1932), then in Leipzig. By the time Lewis arrived, each of these scientists was an established figure in the emerging field of physical chemistry. The American student followed an educational route that was common at the end of the nineteenth and the beginning of the twentieth centuries: Aspiring scientists from the United States performed an apprenticeship at more established universities in Germany, then the world leader in science.

In 1901 Lewis returned to Harvard as an instructor in chemistry, only to leave after three years to work in a government laboratory in the Philippines. In 1905 he accepted an offer to join the faculty of the Massachusetts Institute of Technology, then in a period of expansion under the direction of American analytical and organic chemist Albert Noyes (1857–1941). He remained at MIT for seven years, focusing his research on the energies of chemical reactions.

Lewis was well-known enough by 1912 to be hired by the University of California in Berkeley to head its chemistry department. Under his leadership, he encouraged instructors to learn about recent developments in their discipline and to teach those concepts in the classroom. He also organized weekly meetings at which students and faculty members presented their research and discussed its significance. Such sessions were common in Europe, but rare within any American university at the time.

Lewis remained at the University of California for the rest of his career, continuing the research he had begun at MIT and Harvard. In later years he shifted his focus increasingly to the study of spectroscopy, developing new methods to characterize light. He also played an important role as the author of books that trained a generation of chemists in his concepts and models. He suffered a heart attack and died in 1946 while working in his Berkeley laboratory.

---

hydrogen, but also an ionic bond between NH_4^+ and Cl^-. Implicit in this interpretation is the central concept that a base can be defined as an electron-pair donor, and that an acid can be an electron-pair acceptor. However, a similar structure was proposed a decade earlier by Swiss chemist Alfred Werner (1866–1919), whose work was less widely read or discussed.

Among the early advocates of Lewis's model were his fellow chemists at the University of California. Wendell Latimer (1893–1955) published an important discussion in 1920, in which he analyzed the unusual behavior of water. Its most remarkable property is that water remains in the liquid state over such a wide range of temperatures: It melts at 32°F (0°C) and boils at 212°F (100°C) at normal atmospheric pressure and was therefore chosen as the basis of the Celsius (centigrade) scale. By contrast, the related compound hydrogen sulfide (H_2S) is liquid only over a 73°F (23°C) range and boils at −76°F (−60°C).

Latimer suggested that water's strange behavior might be due to the electron pairs on the oxygen, which could be attracted to the hydrogens from neighboring molecules. Our modern name for this interaction is hydrogen bonding, and it is widely believed to explain many of water's properties. Such bonds are much weaker than a covalent bond, but each water molecule can act as both a donor and acceptor. The high boiling point thus reflects the cumulative energy (in the form of heat) needed to transform liquid water into a gas, where such hydrogen bonds no longer exist.

■ Modern Cultural Connections

A fundamental defect of Lewis's approach was that he could not explain why a pair of atoms should form a shared-electron bond. Based on the laws of physics as they were known in 1916, the electrons should really fly apart, repelled by its electrical charges. Once again, he was painfully conscious of his dilemma, and stated that such rules needed to be "suspended" for his model to work. Other physicists criticized the implication that the electrons had to remain in one place for a bond to form, since they thought that it was more likely that the electrons would be in motion. In the end, a revolution in physics was necessary before a new justification of the shared pair would emerge.

During the late 1920s, the way scientists viewed electrons underwent dramatic change with the emergence of quantum mechanics. According to the interpretation

Typical Lewis diagrams for water and molecular oxygen.   *Cengage Learning, Gale.*

The reaction of ammonia (NH_3) with hydrogen to form ammonium (NH_4) a critical nitrogenous fertilizer for plants. Charge is not shown.   *Cengage Learning, Gale.*

of German physicist Max Born (1882–1970), the electron was no longer a particle fixed in space and time, but a wave for which you could only calculate the likelihood (or probability) of where it might be found. American chemist Robert Mulliken (1896–1986) thus introduced the term orbital to describe such regions of space.

Another important conclusion from quantum theory is that electrons only exist in certain energy states. In 1927 a pair of German physicists published the first such treatment of the hydrogen molecule. While Fritz London (1900–1954) and Walter Heitler (1904–1981) explicitly included a stabilizing attraction between the electrons and the nuclei, they also viewed the bond as an exchange between the atoms. Furthermore, they showed that the two electrons in hydrogen (H_2) reside within a true bonding level, which is lower in energy than the free atoms. The study by Heitler and London marked the first time scientists provided a compelling rationale for the pairing of electrons in the Lewis model.

This strategy developed by Heitler and London is known as valence bond theory. As its name implies, one characteristic was that this approach ignored the inner-shell (core) electrons and thus underestimated repulsive effects. However, valence bond theory can also be viewed as a segmented treatment, in which molecules are formed one bond at a time.

Other physicists and chemists adopted a more holistic approach, in which the bonds arise naturally from the interaction of the orbitals from each atom within a molecule. This strategy was usually called molecular orbital theory, and it was first applied in 1927 to the hydrogen molecule by the American physicist Edward Condon (1902–1974). Refinements were introduced by

the German physicist Friedrich Hund (1896–1997) and by Mulliken.

The advocates of molecular orbital theory, especially Mulliken, rightly claimed that they achieved more accurate predictions of molecular structure (such as the bond angles in NH_3 and H_2O). On the other hand, such calculations were extremely tedious (when everything was done with a pencil and paper), and the resulting image of bonding was much more abstract. Inevitably, this research was slowly accepted by the community of chemists, but most still embraced the graphic nature of the valence bond approach.

Almost a century after Lewis introduced the shared electron pair, this image remains a useful concept for every chemist. While scientists have frequently debated whether chemical architecture is best viewed through the lens of valence bond theory or molecular orbital methods, these approaches are now seen as complementary. Some daunting challenges remain, especially in applying these tools to the three-dimensional structure of proteins. Our understanding of the chemical bond has been greatly aided by theoretical modeling, and the next generation of scientists undoubtedly will apply such methods in exciting ways.

SEE ALSO   *Chemistry: Molecular Structure and Stereochemistry; Physics: Spectroscopy; Physics: The Quantum Hypothesis.*

BIBLIOGRAPHY

Books

Stranges, Anthony N. *Electrons and Valence: Development of the Theory, 1900–1925.* College Station, TX: Texas A&M University Press, 1982.

Periodicals

Hoffmann, R., S. Shaik, and P.C. Hiberty. "A Conversation on VB vs. MO Theory: A Never-Ending Rivalry?" *Accounts of Chemical Research* 36, no. 10 (September 5, 2003): 750–756.

Kohler, R.E., Jr. "The Origin of G.N. Lewis's Theory of the Shared Pair Bond." *Historical Studies in the Physical Sciences* 3 (1971): 342–376.

Latimer, W.M., and W.H. Rodebush. "Polarity and Ionization from the Standpoint of the Lewis Theory of Valence." *Journal of the American Chemical Society* 42 (1920): 1,419–1,433.

Lewis, G.N. "The Atom and the Molecule." *Journal of the American Chemical Society* 38 (1916): 762–785.

William J. Hagan

Chemistry: Chemical Reactions and the Conservation of Mass and Energy

■ Introduction

Chemical reactions involve molecules, the smallest units of matter that retain a substance's unique properties. Molecules, in turn, are composed of atoms, the smallest unit in which an element can exist. Molecules that represent a chemical combination of different atoms are called compounds. Molecules can also be composed of only one kind of atom; oxygen molecules, for example, have two oxygen atoms bonded together.

When two different molecules combine to make a new product, the reaction is called a synthesis reaction. In a decomposition reaction one molecule splits into smaller molecules. More commonly, chemical reactions involve two or more molecules swapping atoms in what are called replacement reactions. In every reaction the molecular reactant's atoms (or mass) are conserved. This means that the atoms that go into the reaction are the same as those that are found after the reaction in the molecules of the products.

All molecules have some degree of heat energy. Exothermic reactions increase heat energy in their environments; endothermic reactions absorb heat energy from their environment. Heat is a form of kinetic energy; "kinetic" comes from the Greek *kinetikos*, which means "in motion," because both the atoms in the molecules and the molecules themselves are moving. The degree of movement determines the temperature of the molecule; the collective total of the movements determines the heat content of the system.

In addition to having heat energy, molecules have stored or potential energy in the chemical bonds that hold them together. The potential energy of chemical bonds depends on the type of bond between the atoms of elements and on the unique characteristics of each element.

The total energy of the system, the kinetic energy of heat added to the potential energy stored in the chemical bonds at the beginning of a reaction, is the same as the total energy of the system at the end of the reaction. Like mass, the total energy involved in a chemical reaction is also conserved.

■ Historical Background and Scientific Foundations

Chemistry evolved from alchemy, a practice and philosophy that proposed, among other ideas, that the so-called base elements could be turned into gold. Alchemy combined the philosophical ideas of ancient Greeks with Arabic chemical arts. Greek philosophers introduced the term "element," but believed there were only four: earth, air, fire, and water. They considered atoms the essential particles of matter, but thought they were indivisible and didn't understand that they were the smallest units of chemical elements. Only in the seventeenth century did the modern concepts of elements, atoms, and molecules begin to emerge.

The breakthrough that led to modern chemistry came from the study of gases by British natural philosopher Robert Boyle (1627–1691). Until this point alchemists were concerned only with solids and liquids. Boyle's studies and writings inspired many others to investigate gases, including the one we now call oxygen, which was isolated in 1774 by British scientist Joseph Priestley (1733–1804).

Oxygen's discovery raised questions about what exactly it was, questions that were answered by French chemist Antoine Laurent Lavoisier (1743–1794). He established the modern concept of a chemical element and identified Priestley's discovery as an element, which he

WORDS TO KNOW

ATOM: The smallest particle in which an element can exist.

COVALENT: In chemistry, a covalent bond between two atoms is produced when the atoms share one or more electrons. The other major type of chemical bond is ionic, where atoms lacking electrons or endowed with extra electrons (and thus having positive or negative charges, respectively) are attracted to each other.

ELECTRON: A subatomic particle having a negative charge of −1.

ENDOTHERMIC: Reaction that absorbs heat from its surroundings as the reaction proceeds.

EQUILIBRIUM: When the reactants and products of a chemical reaction are in a constant ratio. The forward reaction and the reverse reactions occur at the same rate when a system is in equilibrium.

EXOTHERMIC: Reaction that gives off heat to the environment as the reaction proceeds.

INORGANIC: Composed of minerals that are not derived from living plants and animals.

ION: Removing or adding electrons to an atom creates an ion (a charged object very similar to an atom).

MOLECULE: Two or more atoms chemically combined.

NUCLEUS: Any dense central structure can be termed a nucleus. In physics, the nucleus of the atom is the tiny, dense cluster of protons and neutrons that contains most of the mass of the atom and that (by the number of protons it contains) defines the chemical identity of the atom. In astronomy, the large, dense cluster of stars at the center of a galaxy is the galactic nucleus. In biology, the nucleus is a membrane-bounded organelle, found in eukaryotic cells, that contains the chromosomes and nucleolus. Intact eukaryotic cells are comprised of a nucleus and cytoplasm. A nuclear envelope encloses chromatin, the nucleolus, and a matrix, which together fill the nuclear space.

ORGANIC: Made of or coming from living matter.

RADIOACTIVE DECAY: The predictable manner in which a population of atoms of a radioactive element spontaneously disintegrates over time.

Advances in chemistry happened quickly after Lavoisier's work was published. Swedish chemist Jöns Berzelius (1779–1848) set out to make systematic and very precise measurements of chemicals. He published a table of atomic weights in 1826 that still agrees with many of the values accepted today.

Electricity was discovered at about this time, and many chemists explored its effects. When English chemist Humphry Davy (1778–1829) passed a current through molten caustic soda (sodium hydroxide), a chemical favored for its reactive properties since ancient times, the result was pure elemental sodium. Humphry used the same method to isolate potassium.

The idea of ionic compounds (those that form ions [electrically charged particles] in solutions) emerged from Swedish chemist Svante Arrhenius's (1859–1927) studies of reactions in solutions that conduct electricity. He won a 1903 Nobel Prize for his contributions.

Like Berzelius, other nineteenth-century scientists saw vast differences between compounds that came from living beings and those that came from nonliving sources, although this distinction was challenged in 1828 when Friedrich Wöhler (1800–1882) was able to synthesize urea, a compound previously found only in the urine of most animals. This spurred the development of organic chemistry, the science of carbon compounds—whatever their source—as a separate discipline. Organic compounds, scientists learned, are based on the unique shape of the carbon atom, making their molecular bonds different from those in ionic compounds.

The questions of bonding were not easily answered until atomic structure was better understood. In 1896 French physicist Henri Becquerel (1852–1908) discovered radioactivity. The following year British physicist J.J. Thomson (1856–1940) discovered the electron, a negatively charged subatomic particle common to all atoms that is emitted during radioactive decay. Becquerel won the Nobel Prize in 1903 for his discovery; Thomson won the prize in 1906.

The radioactivity of certain atoms led to further speculation on the structure of atoms. Thomson's model, called the plum pudding model, suggested that electrons are scattered randomly through a positive mass, giving a neutral atom. That model was discarded in 1911 when New Zealand–born British physicist Ernest Rutherford (1871–1937) discovered atoms have a dense positive nucleus.

The final model of an atom was described by Danish physicist Niels Bohr (1885–1962) in 1913, work for which he won the Nobel Prize for physics in 1922. Bohr discovered that an atom's electrons surround the dense positive nucleus but can move only in very specific regions called shells or orbits. His basic ideas have been refined to a model that works well to explain chemical reactions.

named oxygen. Lavoisier published his ideas in 1789 in a treatise called *Traité élémentaire de chimie* (Elementary treatise on chemistry). Following Lavoisier, English meteorologist John Dalton (1766–1844) promoted a theory of atoms, chemical elements, molecules, and compounds.

Portrait of Monsieur Lavoisier and His Wife by Jacques-Louis David. (Antoine Laurent Lavoisier; 1743–1794). © *Bettmann/Corbis.*

Types of Chemical Reactions

Every chemical reaction involves the electrons in the outer orbits of atoms. In reactions involving ionic bonds, electrons are given or taken to make ions of opposite charge that attract to form a chemical bond. Ionic reactions occur between metals and nonmetals.

In reactions that involve carbon and nonmetal atoms, bonding is called "covalent," because electrons in the outer orbits (valence electrons) are shared in pairs in which the electrons are spinning oppositely, giving them north-south magnetic properties that make them attract. The bonding paired electrons are not always equally shared between atoms.

Collections of covalently bonded atoms can form molecules or ions. For example, the sulfate ion is made of four oxygen atoms clustered around a sulfur atom. It

Sir Humphry Davy, 1778–1829; English chemist who discovered several chemical elements (including sodium and potassium) and compounds, invented the miner's safety lamp, and became one of the greatest exponents of the scientific method. Plate 5 from Freeman, Samuel (1773–1857) *Lives of Eminent and Illustrious Englishmen, from Alfred the Great to the Latest Times,* vol. 8. Glasgow & Edinburgh, 1834–37. *HIP/Art Resource, NY.*

has a −2 charge because it has two extra electrons. Such ions can form ionic bonds with metals—one example is copper sulfate.

For a reaction to occur the atoms must collide in such a way that the valence electrons can be rearranged. Reactions usually happen in solutions in which molecules are moving randomly, making favorable collisions more probable. In addition, they must collide with enough force to break their existing bonds and collide at just the right angle to form new bonds. This takes energy, which is released when the product bonds are formed. Exothermic reactions release more energy in the formation of products than they use to break the bonds of reactants. If the reverse is true, the reaction is endothermic.

Many chemical reactions are reversible. At first the only reaction is the forward reaction but, as products accumulate in the system, the reverse reaction becomes more likely. In time, all observable properties, such as the color of the reaction mixture, appear to be constant. When the forward reaction is happening as fast as the reverse, a condition called equilibrium has been reached.

Equilibrium reactions were studied by French chemist Henry-Louis Le Chatelier (1850–1936). His observation, known still as Le Chatelier's principle, was that if an equilibrium reaction is disturbed by an ex-

ternal occurrence, either the forward or reverse reaction will increase until equilibrium is restored.

The speed at which a reaction proceeds depends on a number of factors, including what types of molecules are involved, their concentration, and the temperature of the reactants. Even with optimum concentration and temperature, the probability of collisions between atoms in molecules may not produce products directly. Very often some intermediate product forms first, which will then either form the products or go back to reform reactants. American chemist Henry Eyring (1901–1981) suggested the concept of intermediate products, which he called activated complexes. The energy needed to form the activated complex is called the activation energy.

In many reactions catalysts can aid in the formation of intermediate molecules that then form products faster. Catalysts generally work by lowering the activation energy required to make an intermediate product; they are not part of the final product. A catalyst speeds up a reaction but never produces a greater concentration of the products. However, speeding up the reaction can be very advantageous in some chemical reactions.

There are catalysts for a great many chemical reactions used in industrial processes. Catalysts, such as enzymes—proteins that are key to every living organism—are also continuously at work. Three American biochemists, Christian Anfinsen (1916–1992), Stanford Moore (1913–1982), and William Stein (1911–1980), shared the 1972 Nobel Prize for their contributions to the understanding of catalytic activity as it relates to RNA, an essential part of all living cells.

Catalysts

The observation that chemical reactions reach equilibrium and the fact that temperature and pressure can influence the rate of a reaction had a major impact on one famous chemical reaction: the Haber-Bosch process for synthesizing nitrogen and hydrogen into the soluble nitrogen product ammonia.

By manipulating Le Chatelier's principle, the German physical chemist Fritz Haber (1868–1934) found a catalyst that created favorable conditions for producing ammonia. Haber's process was scaled up to industrial production levels in 1908 by German chemist Carl Bosch (1874–1940), who worked for the German chemical company BASF.

The Haber-Bosch process is credited with prolonging World War I by increasing the German's ability to produce munitions and grow more food after the British navy tried to block Germany from receiving critical nitrogen compounds from Chile. It is also considered to have affected food production worldwide, thereby contributing to the rapid increase in world population since World War II. Haber won the Nobel Prize in 1918; Bosch was honored in 1931.

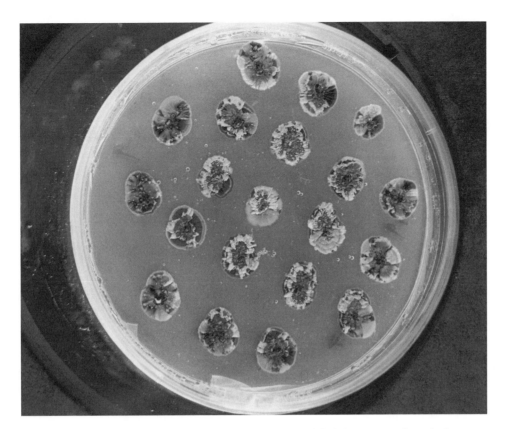

A petri dish containing colonies of recombinant (genetically modified) Streptomyces bacteria that may produce cellulase, which is able to ferment plant cellulose to produce ethanol for use as a fuel. The enzyme breaks down cellulose (the primary structural material in plants). Ethanol is added to gasoline to improve vehicle performance and reduce air pollution. Fuels produced by biological means (biofuels) are renewable, unlike fossil fuels. The technology has been developed by the National Renewable Energy Laboratory in Colorado. *NREI/US Department of Energy/Photo Researchers, Inc.*

Late-twentieth century advances in chemistry have considerably improved the quality of modern life. Unfortunately chemistry has also contributed to such problems as pesticide poisonings, landfills full of plastics that do not biodegrade, and air pollution.

Modern Cultural Connections

Chemical reactions happen continuously in every living organism. They are also at the heart of all chemical industries. The products of chemical reactions are everywhere.

One of the more noteworthy chemical reactions is the synthesis of ethanol as a renewable alternative fuel to supplement the use of petroleum-based fuels like gasoline. Even more noteworthy is the development of enzymes that make possible the conversion of cellulose (such as waste plant and wood fibers) into ethanol. Presently ethanol is produced from corn and sugar cane. Since both are also food crops, the production of ethanol for fuel is competing with world food supply demands.

Yet another advance in the movement toward green chemistry is the use of environmentally friendly materials and processes to manufacture everything from pharmaceuticals to plastics. The production of biodegradable plastics may help relieve the waste disposal problem that developed from the widespread use of many types of plastics. Research and development of chemical synthesis processes has also produced biocompatible plastics that can be used in life-saving medical devices. Other plastics are being developed and used in all parts of vehicles to reduce their weight and so save fuel.

SEE ALSO *Chemistry: Biochemistry: The Chemistry of Life; Chemistry: Chemical Bonds; Chemistry: Chemical Reactions and the Conservation of Mass and Energy; Chemistry: Molecular Structure and Stereochemistry; Chemistry: Organic Chemistry; Chemistry: States of Matter: Solids, Liquids, Gases, and Plasma.*

BIBLIOGRAPHY

Books

Keeler, James, and Peter Wolhers. *Why Chemical Reactions Happen.* Oxford: Oxford University Press, 2003.

Moore, Walter J. *Physical Chemistry.* Englewood Cliffs, New Jersey: Prentice Hall, 1979.

Morris, Richard. *The Last Sorcerers: The Path from Alchemy to the Periodic Table.* Washington, DC, Joseph Henry Press, 2003.

Weeks, Mary Elvira. *Discovery of the Elements.* Whitefish, Montana: Kessinger Publishing Company, 2003.

Miriam C. Nagel

Chemistry: Fermentation

■ Introduction

Fermentation is a biochemical process that is initiated by the actions of naturally occurring microorganisms acting on virtually any type of plant or animal product. It happens anywhere when the environmental conditions are right, with or without man's intervention. If fermentation is carried out under controlled conditions, it enriches the flavor and aroma of foods while adding to their food value and safe storage. It is a relatively easy, efficient, and low energy food enrichment and preservation process. Fermentation as a process was developed using native foods in villages to produce products that were adapted to individual cultures and traditions.

Fermentation of foods as part of cultures and traditions can be found in all parts of the world. It can be traced back to the beginning of recorded history. Bread, beer, wine, and cheese have been found as food staples back thousands of years. Soy sauce, sauerkraut, and yogurts are also fermentation products with a long history linked to different cultures.

Understanding the science behind fermentation did not happen until the microscope was invented, and scientists discovered the world of microorganisms. Fermentation is the decomposition of organic compounds into simpler compounds, which occurs when the right microorganisms are present along with the right conditions for their growth. All living organisms produce organic compounds. Today, organic compounds are also produced in chemical laboratories, some of them through fermentation processes.

The most common compounds associated with fermentation are sugars from fruits; the fermentation products are wines which contain alcohol. Fermented cereal grains are equally common and produce beer, also an alcoholic product. In a different fermentation process, cereal grains are used to produce leavened breads. The breads are leavened, expanded in volume, and made light in texture by carbon dioxide gas that is produced in the fermentation process. Alcohol is not a significant product in making bread, because the fermentation process is so short.

The microorganisms associated with fermentation are all non-green plants classified as bacteria, yeasts, and molds. Yeasts and bacteria are single-celled, with bacteria being smaller than yeasts. There are a great number of different varieties of yeasts, bacteria, and molds; not all of them good are for fermentation. Some microorganisms are actually harmful or produce toxic products.

Fermentation preserves foods, because it produces alcohol and acid products that block the growth of the harmful microorganisms. Bread and alcohol are produced with yeasts. Cheeses are produced with molds and bacteria, and the specific type of microorganism used depends on the type of cheese desired. Soy sauce is produced using molds at the start of a very long process that begins with soybeans.

■ Historical Background and Scientific Foundations

From the start of recorded history, artifacts have been found that indicate fermentation was being used to produce alcoholic drinks and fermented dairy products. The oldest known ancient wine jar that has been found dates back more than 7,000 years to about 5400 BC. A residue found in the jar was analyzed and identified as having come from wine.

The production of wines and beers was common in ancient countries in the Middle East. In Mesopotamia and Syria as far back as 2600 BC, brewing beer was a home industry. Women were the brewers and vendors. They were the local distillers and alchemists. A woman

WORDS TO KNOW

ALCOHOL: Any of the large number of molecules containing a hydroxyl (-OH) group bonded to a carbon atom to which only other carbon atoms or hydrogen atoms are bonded.

BACTERIA: Single-celled microorganisms that live in soil, water, plants, and animals. Their activities range from the development of disease to fermentation. They play a key role in the decay of organic matter and the cycling of nutrients. Some bacteria are agents of disease. Bacteria range in shape from spherical to rod-shaped to spiral. Different types of bacteria cause many sexually transmitted diseases, including syphilis, gonorrhea, and chlamydia. Bacteria also cause diseases ranging from typhoid to dysentery to tetanus. Bacterium is the singular form of bacteria.

FERMENTATION: Chemical reaction in which enzymes break down complex organic compounds (for example, carbohydrates and sugars) into simpler ones (for example, ethyl alcohol).

GENETIC ENGINEERING: The process of manipulating specific genes of an organism to produce or improve a product or to analyze the genes.

MICROAEROPHILIC: A type of bacteria that requires oxygen to live, but in relatively small amounts (less than 20% of the concentration found in the atmosphere). Several disease-causing organisms, such as the bacterium that cause Lyme disease, are microaerophilic.

MICROORGANISMS: All fully-grown, independent living things too small to be seen with the naked eye. Individual cells in the tissues of multicellular organisms are not considered microorganisms because they are not independent organisms.

SOY: Soy, also called soybean, is a legume native to Asia. Over 200 million tons of soybeans are harvested yearly worldwide. Processing of soybeans yields edible flour, oil, and curd (tofu). Soybean oil is also used in a wide variety of waxy or oily industrial products.

SPONTANEOUS GENERATION: Also known as abiogenesis; the incorrect theory that living things can be generated from nonliving things.

YEAST: A microorganism of the fungus family that promotes alcoholic fermentation and is also used as a leavening (fermentation) agent in baking.

Jose Challe examines a bottle of champagne by candlelight to check the state of its sediment. He is at work performing remuage, the process of agitating champagne bottles to loosen the sediment that accumulates during fermentation. © *Charles O'Rear/Corbis.*

efficient. By 1667 distillation methods had improved, with advances in glass making and distillation equipment. The properties of pure alcohol were observed. The liquid looked like water, but it burned and was a solvent for many things not easily mixed with water. The distillation of alcohol became very important to the alchemist, apothecary, and brewer in Europe during this period in history. Alcoholic distillates were called *aqua vitae*, or water of life.

The history of fermented milk products such as yogurt has been traced to early civilizations in the Middle East and Asia. The army of Mongolian leader Genghis Khan (c.1160–1227) was reputedly so strong and successful because of their diet, which was largely based on yogurt. In Northern Europe the recipe for Skyr (a fermented, creamy cheese that resembles yogurt) was allegedly brought to Iceland by the Vikings more than 1,100 years ago. It is still a national staple in Iceland.

Although the craft of fermenting food products was well advanced by the 1600s, the science behind the art

known as Maria the Jewess, an alchemist in Alexandria, Egypt, about AD 100, is said to have invented a three-armed still.

The distillation of wine and beer was carried out by the early Arabic cultures, but their methods were not

Eduard Buchner (1860–1917), the German chemist who won the Nobel prize for Chemistry for proving that alcohol fermentation is due to chemical processes in yeast. *Hulton Archive/Getty Images.*

was unknown. There were scientific advances in other areas such as those achieved by the Italian scientist Galileo Galilei (1564–1642) who had turned his telescope to the skies. Galileo also made a simple microscope but it was not very clear or powerful. A century later British scientist Robert Hooke (1635–1703) redesigned the microscope and discovered the wonders of seeing details on fleas that were never known before. Hooke published his design for a microscope and illustrations of magnified objects in 1664. His publication, *Micrographia*, caught the attention of a Dutchman, Antoni van Leeuwenhoek (1632–1723), who then made an even better microscope. Van Leeuwenhoek made a microscope that could clearly magnify over 200 times.

Van Leeuwenhoek was not only skilled in grinding lenses, he was very curious about the world around him as it was seen through his microscope. When he looked at pond water he discovered microorganisms. He wrote so clearly about this wondrous new microscopic world in 1674 that from his descriptions of microorganisms scientists can instantly identify them today.

Although the existence of microorganisms was discovered in the 1600s, no connection was made between microorganisms and fermentation until after French

scientist Louis Pasteur (1822–1895) became dean of science at Lille University in 1854. A brewer came to Pasteur with a problem. His beer was turning sour after fermentation. Pasteur conducted a very detailed scientific study of fermentation during which he discovered and proved that living yeast was necessary to produce beer. He also discovered that there are a great many microorganisms and that some of them spoiled foods. Only the correct type of yeast would make good beer, or wine. Fermentation processes entered the scientific era with Pasteur's discoveries.

German chemist Eduard Buchner (1860–1917) was studying the chemistry of fermentation in the 1890s when he discovered the living whole yeast was not necessary for fermentation. He received the Nobel Prize in 1907 for his research that led to the discovery that fermentation only needed something made by the yeast cells; that fermentation could be carried out without the whole yeast cells.

Buchner's discovery was advanced further by the work of British chemist Arthur Harden (1865–1940) and the work of Swedish chemist Hans Karl August Simon von Euler-Chelpin (1873–1964), whose investigations of the fermentation of sugars led to their discovery that the extracted yeast material that makes fermentation possible contains enzymes. It is the enzymes that are needed for fermentation. Enzymes are catalysts, complex organic compounds that make a reaction happen faster.

Shortly before Pasteur discovered the science of fermentation, in 1828 German chemist Friedrich Wöhler (1800–1882) synthesized the organic compound urea, a compound that is found in nature as a waste product of urine. The synthesis of organic compounds was now possible in the laboratory. Until Wöhler accidentally prepared urea, it was thought all organic compounds could only be made in living organisms. Seventeen years after Wöhler synthesized urea, in 1845, his pupil Hermann Kolbe (1818–1884) synthesized acetic acid entirely from constituent elements. Up until this time acetic acid had only been produced in vinegar through fermentation processes.

The synthesis of organic compounds in the laboratory increased greatly over the years, with an estimated 75,000 different compounds being synthesized by 1900. The synthesis of compounds other than alcohol using fermentation was not considered a cost-effective alternative until the 1920s, when efficient fermentation processes began to be developed to produce enzymes, the solvent acetone, another alcohol called butanol, acetic acid, lactic acid, and citric acid on an industrial scale. Along with the industrial production of chemicals came the industrialization of the fermented food products such as wine, beer, cheeses, yogurts, and breads.

Fermentation also played a role in one of the greatest discoveries in medicine during the twentieth century.

Although the use of natural materials to treat wounds also goes way back in history, it was not until Scottish biologist Alexander Fleming (1881–1955) observed a common mold had destroyed his culture of a disease bacteria that antibiotics were recognized as powerful tools to fight infections. In 1945 the Nobel Prize was given to Fleming, UK scientist Ernst Chain (1906–1979), and Australian Howard Walter Florey (1898–1968) for their discovery of penicillin and its curative effects.

The name antibiotic was first used by Russian-American microbiologist Selman Waksman (1888–1973) in 1941. Waksman led a research group at Rutgers University that was devoted to the discovery of a treatment for tuberculosis and other critical infectious diseases. Fermentation processes are used in the industrial preparation of antibiotics. Waksman received the Nobel Prize in 1952 for his discovery of streptomycin, the first antibiotic effective against tuberculosis.

The Science of the Art of Fermentation

Fermentation is a biochemical process in which complex organic molecules are broken down into smaller molecules. It is called a biochemical reaction because the reaction is catalyzed by enzymes produced by microorganisms. Catalyzed means the reaction speed is increased by the action of a substance that does not become part of the product of the reaction. Enzymes are naturally produced organic catalysts. Organic molecules all contain carbon and are generally large molecules. Until the mid-1800s scientists believed that only living organisms could produce these molecules, thus their name organic compounds.

Catalyzed reactions are described as stepwise reactions because they involve intermediate products. The intermediate products are called activated complexes. In some fermentations several steps are involved between the reacting molecules and the final products when more than one enzyme is involved in the reaction.

The microorganisms used in fermentations are specific to the particular fermentation reaction. For example, certain yeasts are used for the production of ethyl alcohol (commonly just called alcohol) while a specific bacteria is used for the production of the alcohol called butanol and a solvent called acetone that are produced together in one reaction. Not all microorganisms are useful for fermentation. Many microorganisms are involved in breaking down complex organic material into basic constituent molecules—they decompose the material entirely to waste products. In some such decomposition reactions the results are toxic.

The microorganisms used in fermentation are naturally found on plants, in the soil, or from natural fermentations. Pure cultures of useful microorganisms are now available commercially from specialized sources in many countries around the world. For such fermentation products as sourdough bread, a starter culture

is continued from one batch of bread to another by the baker.

Fermentation cultures include bacteria, yeasts, and molds. It is not uncommon for more than one type of microorganism to be involved in a fermentation reaction. Fermentation processes produce alcoholic or acidic environments that tend to inhibit the growth of undesirable microorganisms. That is why fermentation is considered a good method to preserve certain foods. Even the microorganisms that produce the enzymes for the specific reaction have a limited tolerance for alcohol, or acid, and so the fermentation reactions are self-limiting. In the case of alcohol production the yeast's limit of tolerance is at about an alcohol concentration of 12 percent.

The growth and activity of microorganisms depend on the temperature, pH, oxygen concentration, moisture content, and nutrients available. The temperature range varies with the microorganism, but it is generally between 41–104°F (5° and 40°C). The pH measures acidity of the solution on a scale of 0–14, with neutral being 7 and the lower the number the more acidic the solution. Above a pH of 7 the solution is alkaline. Bacteria generally do best in near neutral mixtures. Yeasts grow in slightly acidic conditions, and molds tolerate a wide range of pH from acid to alkaline.

Oxygen presence is a significant factor. While oxygen is needed for all metabolic activities, including those of microorganisms, some fermentation reactions have to be conducted without air present and are called anaerobic fermentations; others need air present and are called aerobic fermentations. In anaerobic reactions the oxygen needed for microorganism metabolism is obtained from other compounds present in the reaction mixture. There are also some reactions that grow in reduced amounts of atmospheric oxygen. These microorganisms are described as microaerophilic organisms, meaning they need just a little air.

Yeasts and bacteria are very small unicellular non-green organisms. Being non-green they cannot produce their own food. Yeasts and bacteria are only visible with a microscope as they are on the order of 0.00002 inches (0.005 mm) in diameter. Yeasts are irregularly oval, but bacteria have diverse shapes, including rod shapes. Yeasts were the first microorganisms to be studied as they are used to make wine, beer, and bread. The alcoholic products wine and beer are made in an anaerobic process. Alcohol is fermented in an aerobic process by bacteria to make vinegar. The acid in vinegar is acetic acid. If pure alcohol is used to make acetic acid the flavors that make vinegar tasty are missing.

Bacteria give sourdough bread its special taste. What are known as lactic acid bacteria are used to ferment milks in a microaerophilic fermentation process. The lactic acid bacteria are a diverse group of bacteria that are useful in acid food fermentations. Industrial

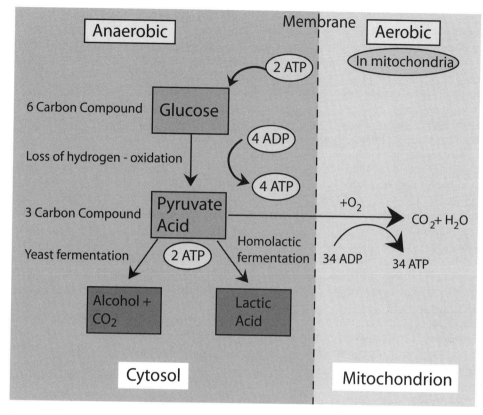

Anaerobic fermentation.   *Cengage Learning, Gale.*

production of lactic acid uses sugar from molasses, corn, or milk in a controlled fermentation process. Commercially-produced lactic acid is primarily used in foods, though a small amount is used in the production of other chemicals.

The fermentation of starch-containing grains using a particular bacteria strain to produce butyl alcohol and acetone was developed and patented by Russian-British chemist Chaim Weizmann (1874–1952). Acetone is a solvent used in the chemical industry and also in the manufacture of cordite. Weizmann's process became particularly important during World War I (1914–1918) because cordite is used as a propellant for cartridges and shells in wartime.

Yeasts and molds are both classified as fungi. Although molds are multi-cellular plants, they are microorganisms visible individually only with a microscope. They are also non-green but they may have other colors such as blue or black. Since they are non-green, they are also dependent on other organisms for food. Yeasts multiply by budding, but bacteria multiply by dividing. Molds grow by spreading filaments and reproduce by shedding spores into the air.

Antibiotics are mass produced in aerobic fermentation processes, using specifically cultured strains of microorganisms for maximum yield. Some molds are

useful in fermentation processes to produce antibiotics and add flavor to cheeses. Others are helpful in industrial processes to produce citric and other specific organic acids. But the vast majority of molds are spoilers of foods and, when they are in the air, can cause allergic reactions for some people.

Influence on Science and Society

The impact of Pasteur's discovery of the part microorganisms play in the fermentation process in the 1800s brought a centuries-old debate about spontaneous generation of organisms in certain matter to a public conclusion. Spontaneous generation seemed to explain the sudden appearance of organisms where there were no obvious parents for them. Maggots in decaying matter were an example of organisms that did not appear to descend from any living organisms of the same kind. Although the relation between flies and maggots was determined before Pasteur's time, the discovery of microorganisms by van Leeuwenhoek opened up a new mystery.

French naturalist Félix A. Pouchet (1800–1872) was a leader of the proponents of spontaneous generation at the time Pasteur was describing to colleagues his meticulous experiments. Pasteur was certain that he had proved the concept of spontaneous generation was wrong. The debate on whether spontaneous generation explained the presence of microorganisms in fermentation mixtures, or whether the organisms only appeared if there were invisible organisms present in the environment surrounding a culture medium, was very heated.

Pouchet challenged Pasteur to prove in a public demonstration that no organisms would grow inside a heated, then sealed, flask containing a culture medium—as Pasteur insisted would be the case. If no organisms were found when the flask was opened, Pouchet said he would admit he was wrong. Pouchet had tried a similar experiment on his own and found there were microorganisms growing.

Pasteur took his experiment to the appointed meeting with the Academy of Sciences, but Pouchet did not appear. However, the witnesses to Pasteur's experiment all saw that he was right. There were no microorganisms in the flasks when they were opened. The scientists who witnessed Pasteur's success attested to his results, and the debate of spontaneous generation was finally settled. It was recognized that Pouchet's experiment had been flawed because he did not completely sterilize his flask.

In current times, the influence of the development of the science of fermentation processes has vastly improved the yield and quality of products. Some traditional fermentation processes are steeped in local culture but are disappearing from common use. The Food and Agricultural Organization (FAO) of the United Nations has publicized the need for governments to create a supportive environment to preserve the heritage and culture of traditional fermentation processes around the world.

FAO points out that the introduction of western-style foods is displacing traditional fermented foods that have been staples for thousands of years. The nutritional value of fermented foods has been very important, especially in regions where the diet would otherwise be lacking in essential vitamins or other nutrients. Lack of proper nourishment has historically been a problem in underdeveloped countries. In places where fermented foods have been traditional staples, the disease rate from poor nutrition is lower. FAO promotes improving native fermentation practices for traditional foods to be even more effective as diet supplements. At the same time, FAO is encouraging protection of the culture and traditions that make each region unique.

Advances in the efficiency of fermentation processes have had a positive influence on the industrial level preparation of chemicals. The use of fermentation is becoming more cost-effective and it can be a green process for manufacturing some chemicals. Greener means the process is more environmentally friendly, a very important consideration.

■ Modern Cultural Connections

The development of the science of fermentation has had a profound impact on society. One of the greatest societal impacts started with Weizmann, who discovered a fermentation process to develop acetone from maize (corn) in 1916. The availability of acetone was critical to the British government's efforts in World War I, as acetone was needed to supply the troops with ammunition.

During the war the First Lord of the Admiralty asked Winston Churchill (1874–1965), Prime Minister of the United Kingdom from 1940–1945, to supply thirty thousand tons of acetone. Weizmann succeeded in scaling up his small laboratory production process to an industrial scale, a feat that later gave him the unofficial title of father of industrial fermentation.

The Second World War (1939–1945) provided another story of the impact of the science of fermentation. The recently discovered life-saving antibiotic penicillin was needed in large quantities. The task to mass-produce such a drug was an enormous challenge. This was met with a deep-tank fermentation process that was developed at a major drug manufacturing company to produce the quantities of the antibiotic needed. The antibiotic penicillin saved many lives that would have been lost to infections from war-related injuries. The commercial scale development of many antibiotics can be traced to that discovery.

One of the most ancient of fermentation products, alcohol, has become the center of efforts to replace fossil fuels with renewable energy sources. Considerable research is being devoted to the development of enzymes that will improve the fermentation process to produce ethanol (pure alcohol) for fuel from a variety of biomass sources. Corn, as well as corn stalks and other waste vegetative materials, is being used to produce ethanol. Although all sources for fermentation are not yet economical, considerable research is being conducted to make the fermentation of waste plant products practical. There is even an effort by a New Zealand firm to make ethanol from carbon monoxide using a bacterial fermentation process. The carbon monoxide is a waste gas from a steel mill process. Carbon monoxide is a very simple carbon compound, but it is a source of carbon, and carbon is the key to the production of ethanol.

Fermentation processes use microorganisms to reduce large organic molecules to smaller molecules. The basic process is primarily associated with the production of a desirable product, but it can also be used to reduce large undesirable organic molecules to small harmless ones. Microorganisms have long been known to reduce all kinds of waste materials naturally. Bacterial strains now have been developed that can digest oil to clean up oil spills.

In 2001, scientists at a U.S. National Laboratory developed microorganisms that can clean up coal to make it cleaner burning as a fuel. The bacteria they use have been adapted from geothermal locations in the South Pacific and North America. Development through altering the traits of microorganisms can include genetic engineering.

Genetic engineering of organisms is opening up new frontiers in the science of fermentation. Genetic engineering means altering the inherited traits of an organism to make it even more desirable for a specific trait. Not only microorganisms are genetically engineered today; certain plant crops used in fermentation are being genetically engineered for greater productivity, disease and pest resistance, or other desirable traits—though not without controversy.

SEE ALSO *Biomedicine and Health: Antibiotics and Antiseptics; Biomedicine and Health: Bacteriology; Biotechnology and the manipulation of Genes; Chemistry: Chemical Bonds; Chemistry: Fermentation: A Cultural Chemistry; Chemistry: Chemical Reactions and the Conservation of Mass and Energy; Chemistry: States of Matter: Solids, Liquids, Gases, and Plasma.*

BIBLIOGRAPHY
Books

De Kruif, Paul. *Microbe Hunters.* New York: Harcourt, Brace & World, Inc., 1954.

Moore, F.J. *A History of Chemistry.* New York and London: McGraw Hill Book Company, Inc., 1939.

Salzberg, Hugh W. *From Caveman to Chemist.* Washington, D.C., American Chemical Society, 1991.

Shreve, R. Norris. *Chemical Process Industries.* New York and London: McGraw Hill Book Company, 1967.

Taylor, F. Sherwood. *An Illustrated History of Science.* London: William Heinemann, 1956.

Web Sites

Chemical Heritage Foundation. http://www.chemheritge.org (accessed August 17, 2007).

Food and Agricultural Organization of the United Nations. http://www.fao.org (accessed August 17, 2007).

Nobel Prize Organization. http://nobelprize.org (accessed August 17, 2007).

Miriam C. Nagel

Chemistry: Fermentation: A Cultural Chemistry

■ Introduction

The term fermentation stems from the Latin *fervere*, "to boil." It originally referred to a substance whose properties changed through a bubbling or foaming process, like leavening bread or brewing beer. Although fermentation processes have been used by humans for at least 10,000 years, the chemistry behind them was not understood until the nineteenth century, when French chemist and microbiologist Louis Pasteur (1822–1895) proved that fermentation was caused by microorganisms.

Microbes like bacteria and fungi digest glucose, a 6-carbon sugar, and split it into two 3-carbon molecules called pyruvic acid. (This process, called glycolysis, is an anaerobic process, one that takes place without oxygen.) This releases electrons and generates adenosine triphosphate (ATP), an energy storage molecule that powers the cell. Depending on the microorganism involved, the production of pyruvic acid also releases ethyl alcohol, butyl alcohol, and lactic or citric acids. Alcohol is used to brew beer and preserve food; lactic acid is used to make cheese and yogurt. Fermentation also produces carbon dioxide, which makes the foam in beer, the bubbles in some wines, and the air that makes bread rise.

■ Historical Background and Scientific Foundations

The Ancient World

Fermentation and human history are intricately tied together. Our earliest historical records indicate that humans unknowingly domesticated wild yeasts to ferment food and drink, particularly *Saccharomyces cerevisiae* for wine production, *Saccharomyces carlsbergensis* for brewing beer, and "leaven"—most likely a mixture of milk bacteria (lactobacilli) and yeasts—for bread baking. Bakers saved a portion of leaven from each batch of dough as a "starter" for the next one. Each starter has a distinctive flavor, depending on the type and quality of wild yeast in the air. Some strains of yeast, such as those that give sourdough bread its distinctive taste, are cultivated commercially today.

Since grains and fruits ferment naturally, it is likely that the earliest civilizations knew how to make alcoholic beverages. Wild yeasts, for example, collect on the skin of grapes and other fruits, causing fermentation. Ancient pottery shards show the presence of tartrates, a byproduct of fermentation, indicating that the Chinese were brewing a drink made of rice, honey, and fruit as early as 7000 BC. Centuries later, they developed a process to break down complex sugars in rice to simple sugars like glucose, which were then fermented to make rice wine. The first stage in this process is carried out by a mold called *Aspergillus oryzae* followed by a sake strain of *S. cerevisiae*.

Wine was also being produced 8,000 years ago in Caucasus Mountain settlements, as well as in ancient Iran. Sumerians in ancient Babylonia (modern Turkey) considered beer a divine drink; they worshipped Ninkasi, the goddess of beer and alcohol, composing a hymn to her 4,000 years ago. In the Gilgamesh epic (c. 3000 BC), Gilgamesh's boon companion, the bestial Enkidu, was "civilized" by learning to eat bread and beer, the two main products of Babylonian fermentation:

They placed food in front of him,

They placed beer in front of him;

Enkidu knew nothing about eating bread for food,

And of drinking beer he had not been taught.

The harlot spoke to Enkidu, saying:

"Eat the food, Enkidu, it is the way one lives.

Drink the beer, as is the custom of the land."

Enkidu ate the food until he was sated,

He drank the beer—seven jugs!—and became expansive and sang with joy!

Winemaker Nontsikelelo Biyela, who had never even tasted wine before giving up her job to obtain a winemaking degree, tests wine in the fermentation room at a winery in Cape Town, South Africa. *Rodger Bosch/AFP/Getty Images.*

WORDS TO KNOW

BREWING: The practice of encouraging the growth of yeast—microscopic fungi—in mixtures of water and plant material to turn plant sugars into ethyl alcohol and carbon dioxide. After filtering and aging, the resulting mixture becomes an alcoholic beverage. If carbon dioxide is retained, it becomes bubbles, as in beer and champagne; if it is allowed to escape, non-carbonated alcoholic beverages such as wine are created.

DISTILLATION: The process of separating liquids from solids or from other liquids with different boiling points by a method of evaporation and condensation, so that each component in a mixture can be collected separately in its pure form.

ETHANOL: Compound of carbon, hydrogen, and oxygen (CH_3CH_2OH) that is a clear liquid at room temperature; also known as drinking alcohol or ethyl alcohol. Ethanol can be produced by biological or chemical processes from sugars and other feedstocks and can be burned as a fuel in many internal-combustion engines, either mixed with gasoline or in pure form. Several governments, most notably Brazil and the United States, encourage the production of ethanol from corn, switchgrass, algae, or other crops to substitute for imported fossil fuels. Ethanol is criticized by some as being based on environmentally destructive agriculture, putting human populations into competition with automobiles for the produce of arable land and providing little more energy (depending on the manufacturing process used) than is required to produce it.

GLYCOLYSIS: A series of chemical reactions that takes place in cells by which glucose is converted into pyruvate.

HOPS: Dried flowers of the vine *Humulus lupulus,* which give beer its characteristic bitter flavor and aroma.

LACTOBACCILLUS: A type of anaerobic (non-oxygen-using) bacterium that is found in the human gut and in nature and is also used in preparation of fermented foods such as yogurt, kimchee, and sauerkraut.

MULSUM: A mixture of honey and wine served as an appetizer at formal meals in ancient Rome.

MUST: Grape juice that is going to be fermented to make wine, or is in the process of being fermented.

MYCOLOGY: The branch of biology dealing with fungi.

YEAST: A microorganism of the fungus family that promotes alcoholic fermentation and is also used as a leavening (fermentation) agent in baking.

He was elated and his face glowed.

He splashed his shaggy body with water,

And rubbed himself with oil, and turned into a human...

The Babylonian ruler Hammurabi (d. 1750 BC), author of the oldest set of written laws, dictated beer rations for each social class, a normal worker receiving two liters per day.

Ancient Egyptian tomb paintings dating from 2400 BC show how they crushed barley and mixed it with water to make dried cakes. When reconstituted with water, an extract of the cakes was fermented in vessels to make a type of beer. Egyptian records from the same period also refer to the use of grapes for viniculture.

Wine Making in Antiquity

Brewing techniques eventually moved east to the Mediterranean, where wild grapes grew in abundance. The ancient Greeks and Etruscans considered wine an essential part of their diet, planting vineyards in their colonies near the Black Sea and in France and Spain. So revered was wine that the Greek religious cult of Bacchus saw drunkenness as a type of connection with the divine. Libations poured on tombs to honor the deceased

Jars of Mas Amiel's Maury wine, aging in the sun. The wine is made in the Roussillon (Northern Catalonia) wine region of France.   © *Owen Franken/Corbis.*

involved a mixture of olive oil and wine. Carthage, an ancient Phoenician civilization in North Africa, developed such advanced viniculture that books were written about it; the city became wealthy exporting wine to the nascent Roman republic.

After the Romans conquered Carthage in 241 BC, the Roman Senate decreed that Carthaginian wine treatises be translated into Latin. The writer and politician Cato the Elder (234–149 BC) wrote an agricultural treatise named *De Agri Cultura* that incorporated Carthaginian techniques. The Romans planted vineyards as far north as Germany in steep river valleys such as the Mosel and Ahr. The valleys protected the grapes from the wind and kept them warm, and the perpendicular sunshine hit the ripening vines with greater heat, producing sweet juice that is better fermented by yeast. Wine was exported to Roman provinces and was particularly popular when sweetened with honey as a beverage called mulsum. Because distillation had not yet been discovered, wine was the drink with the highest alcohol content; it was often diluted with water.

Archaeological ruins at Pompeii and other ancient sites provide evidence of the Roman wine-making process. After the grapes were harvested, they were crushed to produce the first premium pressing of juice. The resulting grape mash was then put through a simple wine press for a second extraction used to make cheaper wine. The final residue from the pressing was mixed with water and given to the farm laborers.

The juice that resulted from pressing, known as must, then fermented either outside or in pitch-covered clay jars that were buried underground. (Modern wineries use stainless steel vats, adding sugar and yeast rather than relying on wild strains from the atmosphere). As Romans liked sweet wines, grapes were often left to ripen on the vine as long as possible until the first frosts of autumn, so the sugar could be concentrated, a technique still used today. Sometimes the must was boiled to concentrate sugars, or spices were added for flavor or to mask a sour vinegary wine. Wines were stored in two-handled tapered clay jugs called amphora, which typically litter most Roman archaeological sites. To extend the wine's life (though most was drunk within the year to prevent spoilage), the Romans also stored it in oak barrels to age the wine and give it added flavor.

The History of Beer Brewing

In colder parts of the Roman Empire where grape cultivation was difficult, beer brewing predominated, usually by making successive extracts from a batch of brown malt.

Malting is the process of immersing barley (or wheat) in water from 55–60°F (12.7–15.5°C) for 40 to 50 hours; its volume increases and the grain begins to sprout, causing its root sheaths, called chits, to break through the husk. As the barley germinates, the seed embryo secretes a hormone called gibberellic acid, which initiates the production of an enzyme called alpha-amylase; this in turn both converts the barley starch into fermentable sugars that the embryo can use as food and produces carbon dioxide. Large amounts of malting barley have to be turned and aerated to dissipate the gases and heat emitted during germination.

The malt is then dried to remove most of the moisture in a process called kilning or roasting, which reduces enzyme activity. Successively wetting and kilning malts imparts different colors and flavors to the finished beer. A high proportion of chocolate malt (a dark-brown roast) mixed with roasted and ungerminated barley is used to make stouts and porters.

Next the malt is mixed with hot water (144–162° F/ 62–72°C)—and sometimes other grains—in a process called mashing. This completes the conversion of starch into fermentable sugar. The aqueous part of the mixture, called wort, is separated, filtered, and then boiled for 60 to 90 minutes. The excess grains can be used for cattle feed.

By the eighth century brewers had discovered that hops can be added to the wort to impart a tangy, slightly bitter flavor. The female flowers of hop plants contain a chemical called humulone that gives bitters and pilsners their characteristic flavor. Interestingly, recent research has revealed that humulone is an angiogenesis inhibitor—a substance that prevents the growth of new blood vessels. Since cancer tumors can only grow and spread if they make new blood supplies, humulone may be useful in chemotherapy, and there is scientific speculation that beer consumption may help prevent cancer, specifically prostate cancer.

The wort is then filtered again, cooled, and stored in a collecting tank where yeast is added to begin fermentation. Before brewers understood yeast's role in fermentation, a successful brew hinged on the chance of attracting wild yeasts. A bad brew was often blamed on the influence of magic or witches.

In traditional brewing, a first fermentation took a week for ale and more than three weeks to make a lager. A secondary fermentation, in which extra sugar or actively fermenting wort was added, purged the beer of unwanted bitter compounds, a process that could take from a week to three months.

Ancient and medieval brewers waited for the yeast to rise to the top of the brewing vessel, then skimmed it off in a process called top fermentation. The first extract was the strongest, giving the best beer, and the third extract was the poorest, producing what was called small beer. Top-fermented beers, typical of British beers, were known as ales. Another fermenting process, in which the yeast sank to the bottom of the brewing vat, produced lager, a beer that become popular in Germany. As brewing was primarily a winter occupation, the term comes from the German *lagern* "to store." Though in the past the yeast that was skimmed off or sank was discarded, modern companies in England and Australia concentrate it into a salty extract called Marmite or Vegemite that is used as a sandwich spread and nutritional supplement, as the yeast provides a high quantity of B vitamins.

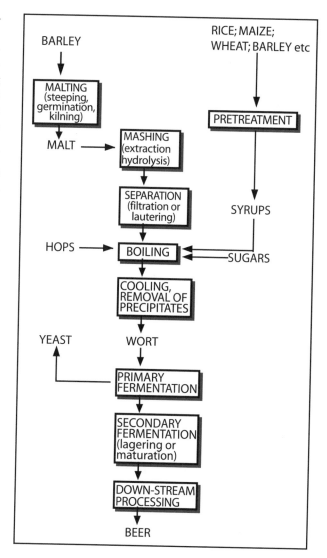

Beer brewing. *Cengage Learning, Gale.*

After the Roman Empire fell in the fifth century AD, brewing technologies were preserved in Christian monasteries. Monks had originally started making their own sacramental wine and brewed beer as a nutritional alternative to meals. This was important during their many fasts, since drinking liquid was not considered to break the fast. The excess was sold to raise money for the monastery.

Some religious houses, in fact, made brewing a big business: eleven out of twelve monastic houses in medieval Yorkshire, for example, had brew houses on their premises. It is not only likely that monks discovered the advantages of adding hops to beer, but medieval monasteries such as those in St. Gall in Switzerland also added secret proprietary mixtures of herbs to their brews to widen their public appeal. The beer industry became so important in Germany that the first food purity laws, the *Reinheitsgebot* of 1516, regulated beer purity, requiring that it only contained water, malted barley, malted wheat, and hops.

Other Domestic Fermentation

Outside of the monasteries, brewing, like baking, was primarily a female occupation. Because of this, taverns and alehouses were often run by female alewives; the labor shortage that followed the Black Plague (1347) only increased this trend. Women in the medieval era also used other fermentation techniques to pickle food and to make cheese and yogurt.

In these processes, salting to control the growth of the microbes is combined with fermentation to pickle food. When cucumbers are pickled, for example, microbes turn carbohydrates into acid, and their color changes from bright to olive green. To stop microbial activity, pickles are stored in a brine solution of 8–10% for the first week, with the salt concentration increased by 1% a week until it stands at 16%. The pickles, then cured, have the excess salt leached from them by soaking them in warm water for 10–14 hours.

Cheese-making dates back to ancient Babylonia, and it is likely its production was accidental, as milk was transported in the stomachs of young animals slaughtered for food. Lactobacilli in the milk combined with the sun's warmth permitted the milk sugars to ferment. The acid in the stomach lining, called rennet, caused milk to curdle. Swaying the stomach bag broke up the solid curds, producing the liquid whey. As the curds set in a soft gel, the cheese was drained, salted, and packaged to make soft cheese. For hard cheese, the cheese would be baked to further drain the whey and make it more solid. Molds were used to drive out water and shape the final product. As cheese ages, microbes (usually lactobacilli) typically break down the milk protein (called casein) into amino acids and reduce milk fats into simpler fatty acids; both molecules impart flavor and texture.

Yogurt, or fermented milk, was made simply by allowing milk to heat in the sun. The milk's sugar was fermented by natural lactobacillus to produce lactic acid, which reacted with the milk protein to thicken and solidify the milk.

Though in ancient Egypt the secret of cheese-making was kept by temple priests, in ancient Greece and Rome cheese was part of the daily diet, as it was a way to store otherwise perishable milk year-round. In the Middle Ages, monks in North Yorkshire abbeys made Wensleydale cheese from ewe's milk; cheddar cheese was invented near England's Cheddar Gorge in 1500.

Louis Pasteur Discovers the Fermentation Process

Though fermentation was used throughout history to make a variety of foodstuffs, the biological and chemical

mechanisms behind it were not revealed until the nineteenth century. No one knew, for example, why beer and wine would suddenly sour during production. And, as the Industrial Revolution commercialized beer and wine production on a vast scale, a "bad batch" could mean great financial losses.

Louis Pasteur, dean of sciences at the University of Lille, was asked for help by an industrialist who was producing beer from beet sugar. The batches continually were going sour. By examining the brewer's samples under the microscope, Pasteur realized that correctly aged beer contained spherical yeast, but sour batches contained yeasts that were elongated. Pasteur also noted that the desirable globe-shaped yeasts produced alcohol, but the rod-shaped yeasts produced lactic acid, making the batches go sour. These findings made him realize 1) yeasts were responsible for fermentation, 2) different types of yeasts made different byproducts, and 3) yeasts did not need oxygen to metabolize. Anaerobic respiration (without oxygen) to this day is sometimes referred to as the "Pasteur effect."

Pasteur advised the industrialist that after the "good" yeasts did their work and fermented to produce alcohol, the liquid be heated gently to 122° F (50° C) to kill the "bad yeast" and prevent souring as the beer aged. This process, now called pasteurization, is used today in milk, cheese, yogurt, beer, and wine production. In 1876 Pasteur published the results of his findings, the *Études sur la Bière* (Studies on beer). His work led to a scientific understanding of the brewing and winemaking processes. Pasteur modestly wrote in the preface of his book:

Louis Pasteur's (1822–1895) experiment showing that fermentation and putrefaction are caused by air-borne organisms. *Image Select/Art Resource, NY.*

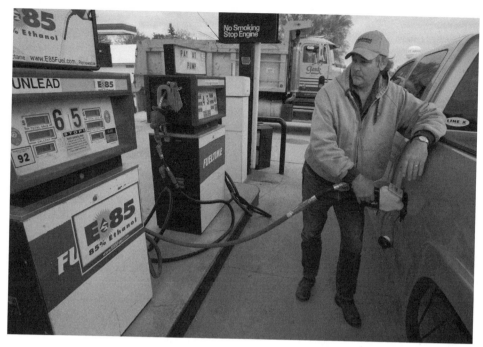

A corn and soybean farmer fills his truck with E-85 fuel, a blend of 85 percent grain alcohol distilled from corn and 15 percent gas, at a gas station.   *Charlie Neibergall/AP Images.*

I need not hazard any prediction concerning the advantages likely to accrue to the brewing industry from the adoption of such a process of brewing as my study of the subject has enabled me to devise, and from an application of the novel facts upon which this process is founded. Time is the best appraiser of scientific work, and I am not unaware that an industrial discovery rarely produces all its fruits in the hands of its first inventor.

From his research, Pasteur discovered that sour batches could also be caused by microbes other than yeast, such as molds or bacteria. Realizing that each type of fermentation was the result of a specific type of microbe, he applied this conclusion to disease, suggesting diseases were also microbe or "germ"-specific. This insight is known as the germ theory of disease, and Pasteur went on to develop vaccines for anthrax, chicken cholera, and rabies based on this principle.

◼ Modern Cultural Connections

After Pasteur made his discoveries, production of the industrial solvents butyl and ethyl alcohol became the most important industrial fermentations. In the 1960s, however, chemical engineers devised nonbiological ways to produce alcohol and acetone that relied on synthetic starting materials derived from then-cheap and plentiful petroleum. Growing concerns in the 1980s about petroleum conservation led to renewed interest in producing alcohols from microbial fermentation, using cellulose and starches from agricultural wastes that would otherwise go to a landfill. Fermentations using microbes also do not produce toxic byproducts, as do chemical processes.

Ethanol, fuel for automobiles produced by microbial fermentation of cornstarch, has also become a topic of current scientific and political interest. In 1999 Brazil mandated that by 2003 all new cars in the country had to run on E85, a fuel that was 85% ethyl alcohol and 15% gasoline. American, Japanese, and European manufacturers began to produce cars that were E85 compatible as well.

Molecular biologists, in turn, have genetically engineered bacteria, yeasts, and even some mammalian cells to produce ethyl alcohol more efficiently via fermentation in giant vessels called bioreactors. The bioreactors provide appropriate nutrients, stirrers, temperature, and pH (level of acidity or alkalinity). Genetically modified yeasts are also being used in modern brewing to shorten the time it takes to ferment the wort. Transgenic barleys have been developed with genes that encode for certain desired enzymes like alpha-amylase.

Industrial fermentation with fungi such as yeasts is also used to produce a large variety of other commercial products as well. Other products of industrial mycology include modern antibiotics like penicillin or streptomycin; food coatings, such as pullulan, a complex sugar produced from glucose by yeast; and vitamin B_2 from brewer's yeast. Citric acid, used in foods, detergent, and pharamaceutical applications, is produced industrially from *Aspergillus niger*. From bread and beer to fuel, humans have depended on microbial fermentation to create a wide variety of products necessary to their well-being.

SEE ALSO *Chemistry: Biochemistry: The Chemistry of Life; Chemistry: Fermentation.*

BIBLIOGRAPHY

Books

Bennett, Judith M. *Ale, Beer, and Brewsters in England: Women's Work in a Changing World, 1300–1600.* New York: Oxford University Press, 1996.

Bigelis, Ramunas. "Fungal Fermentation: Industrial." In *Encyclopedia of Life Sciences.* New York: John Wiley and Sons, Ltd., 2001.

Epic of Gilgamesh. Translated by Maureen Gallery Kovacs. Stanford: Stanford University Press, 1990.

Johnson, Hugh. *Vintage: The Story of Wine.* New York: Simon & Schuster, 1989.

Pasteur, Louis. *Louis Pasteur's Studies on Fermentation: The Diseases of Beer, Their Causes, and the Means of Preventing Them.* MacMillan & Co., 1879. Reprint Edition, BeerBooks.com, 2005.

Sandler, Merton, and Roger Pinder. *Wine: A Scientific Exploration.* London: Taylor & Francis, 2002.

Stansbury, P.F., A. Whitaker, and S.J. Hall. *Principles of Fermentation Technology.* New York: Butterworth-Heinemann, 1997.

Periodicals

Brown, William. "Description of the Buildings of Twelve Small Yorkshire Priories at the Reforma-tion." *Yorkshire Archaeological Journal* ix (1886), 197–215.

Linko, Matti, Auli Haikara, Anneli Ritala, and Merja Penttil. "Recent Advances in the Malting and Brewing Industry." *Journal of Biotechnology* 65, nos. 2–3 (October 27, 1998): 85–98.

Pausch, Mark H., Donald R. Kirsch, and Sanford J. Silverman. "Saccharomyces cerevisiae: Applications." In *Encyclopedia of Life Sciences.* New York: John Wiley and Sons, Ltd. (2001).

Rossiter, J.J. "Wine and Oil Processing at Roman Farms in Italy." *Phoenix* 35, no. 4. (Winter 1981): 345–361.

Shimamura M., et al. "Inhibition of Angiogenesis by Humulone, a Bitter Acid from Beer Hop." *Biochemical and Biophysical Research Communications* 289, no. 1 (November 23, 2001): 220–224.

Web Sites

Encyclopedia Romana. "Wine and Rome." http://penelope.uchicago.edu/~grout/encyclopaedia_romana/wine/wine.html (accessed October 22, 2007).

National Health Museum: Access Excellence. "Microbial Fermentations: Changed the Course of Human History." http://www.accessexcellence.org/LC/SS/ferm_background.html (accessed October 22, 2007).

New Scientist.com. "World's Earliest Tipple Discovered in China." http://www.newscientist.com/article/dn6759.html accessed October 22, 2007).

Anna Marie Eleanor Roos

Chemistry: Molecular Structure and Stereochemistry

■ Introduction

A daunting hurdle for early chemists was the recognition that molecules have a three-dimensional shape, a concept known today as stereochemistry. This article will focus on the tetrahedral geometry of carbon and the crucial role played in its development by two young Europeans during the second half of the nineteenth century. At a time when some scientists still doubted the very existence of atoms, or whether molecular architecture could ever be known, French chemist Joseph-Achille Le Bel (1847–1930) and Dutch physical chemist Jacobus van't Hoff (1852–1911) independently introduced this novel idea. The rapid (and almost universal) adoption of the tetrahedron effected a quiet revolution as chemists tacitly employed such structures for the depiction of the spatial arrangement in carbon. The van't Hoff–Le Bel hypothesis has dominated stereochemistry and has thus provided scientists with a powerful tool for the prediction of molecular properties.

■ Historical Background and Scientific Foundations

One of the great systematizers of the first half of the nineteenth century was the Swedish chemist Jöns Berzelius (1779–1848). Not surprisingly, he was one of the early advocates of the atomic theory, a notion first proposed in 1808 by the British chemist and meteorologist John Dalton (1766–1844), which stated that elements consist of tiny particles that combine in specific ratios to form molecules. While the Scandinavian enthusiastically endorsed the central tenants of Dalton's hypothesis, however, he suspected that many of the latter's formulas were inaccurate.

From his conviction that the errors arose from insufficient data on the relative weights of the chemical elements, Berzelius carried out precise measurements of volume and mass changes during reactions. He compiled extensive tables of atomic weights that he published during the second decade of the nineteenth century (and revised throughout his career). Of equal importance, he introduced the convention of letters and numbers that chemists continue to use in their formulas. For example, Berzelius denoted water as H_2O (indicating two parts hydrogen for each oxygen) and ammonia as NH_3 (conveying a 3:1 ratio of hydrogen to nitrogen).

Chemists during this period became quite adept at the art of analysis. Among the most prolific in the characterization of substances from animals and plants was the German chemist Justus von Liebig (1803–1873), who created one of the first teaching laboratories in Europe. As the volume of analytical data expanded, many scientists were deeply puzzled by substances known as isomers, which shared the same chemical formula but exhibited different physical properties. In 1828, for example, another German chemist, Friedrich Wöhler (1800–1882) found the transformation of ammonium cyanate into urea (named for its presence in urine). Both had the same composition (two nitrogens, four hydrogens, and a carbon), but were obviously isomers: urea had a melt temperature of 271°F (133°C), but ammonium cyanate underwent decomposition on heating.

Two Frenchmen made important discoveries during the first half of the nineteenth century that had important implications for molecular structure. Physicist Jean-Baptiste Biot (1774–1862) devoted a large portion of his prolific career to the study of polarized light, a term that refers to the parallel orientation after a beam of light passes through a crystal. He carried out innovative measurements of its interaction with solids, liquids, and gases. In 1815 Biot found that several natural oils (including turpentine and lemon) had the ability to rotate light, and in 1828 reported that tartaric acid (from grapes) had

WORDS TO KNOW

CHIRALITY: Also termed handedness, the quality of having non-superimposable mirror images.

GEOMETRIC ISOMERS: Stereoisomers in molecules with restricted rotation about a bond.

ISOMERS: Two substances with the same atomic ingredients but different structures. They have the same composition (the same numbers of the same types of atoms) but different geometrical arrangements of those atoms and, as a result, different physical properties.

OPTICAL ISOMERS: Stereoisomer with chiral or asymmetric centers.

ORBITALS: An energy state in the atomic model that describes where an electron will likely be.

POLARIMETER: An instrument for measuring the polarization of light, that is, the way in which its transverse electromagnetic fields have been restricted to vibration in specific directions.

PROTEIN: A complex chemical compound that consists of many amino acids attached to each other that are essential to the structure and functioning of all living cells.

RACEMIC: In optics, some transparent compounds rotate polarized light to the right (dextrorotatory compounds); others rotate it to the left (levorotatory compounds). A racemic substance is a mixture of equal parts dextrorotatory and levorotatory compounds.

STEREOCHEMISTRY: The branch of chemistry that deals with the way the three-dimensional shapes of molecules (stably bound clusters of atoms) affect their properties.

STEREOISOMER: Two molecules that contain the same numbers of the same types of atoms but differ in the spatial arrangement of those atoms are stereoisomers of each other.

VALENCE: The tendency of an atom to gain or lose electrons in reacting with other atoms.

the same property. He even developed improved devices (called polarimeters) to measure solutions and suggested that chemists could use the optical rotation as a proxy for the concentration of compounds in solution.

One chemist who took Biot's recommendation to heart was French chemist and microbiologist Louis Pasteur (1822–1895). Like many Europeans in wine-producing regions, he was intrigued by the physical and chemical changes that occurred during the fermentation process. For his research as a candidate for a doctoral degree, Pasteur focused on a chemical from wine sediments known as racemic acid. Curiously, this substance had the same elemental composition as Biot's

tartaric acid, but did not affect polarized light. During the 1840s, most chemists regarded optical rotation as an exotic technique with no practical value, but Pasteur, with the help of his elder countryman, dramatically altered that perception.

Pasteur was able to draw from a rich tradition in the study of crystals that had existed for nearly seven decades. In 1781, the French mineralogist René-Just Haüy (1743–1822) first noticed how calcium carbonate (calcite) breaks when he accidentally dropped a specimen. He subsequently carried out extensive geometric analyses to learn how such minerals might form. Another French chemist, who sought to extend Haüy's methods to chemical compounds, was Auguste Laurent (1808–1853), who shared a laboratory with Pasteur at the Normal School in Paris. Laurent suggested that Pasteur focus his studies on the shape of the wine crystals, and also gave him important practical advice that ensured the experiment's success.

When Pasteur checked the solid under a microscope, he was shocked to find two types of crystals (in right- and left-handed forms). Fortunately, Pasteur was able to prove his own adage that "In the field of observation, chance favors the prepared mind." Pasteur knew that Biot had reported a similar phenomenon in quartz. Working alone in the laboratory, he carefully separated the two materials and found (using Biot's own polarimeter) that each rotated light in opposite directions. Upon publishing this remarkable data in 1848, he proposed a mirror-image relationship at the molecular level, but the primitive state of structural chemistry prevented Pasteur from any more definitive conclusions.

An essential element in the structure of carbon-containing compounds is the number of linkages that each forms. The English chemist William Odling (1829–1921) was the first to recognize (in 1855) that carbon could form four bonds to other atoms. However, it was the German chemist August Kekulé (1829–1896) who first assigned the formula CH_4 to methane (a gas produced by microbes living in marshes) and assigned double and triple bonds to other molecular structures. In 1857 Kekulé introduced the term "valence" to describe the number of links that an atom formed. He played an especially prominent role in understanding molecular structures, and emphasized that chemical formulas could be a predictive tool.

One of Kekulé's champions was the Russian chemist Alexander Butlerov (1828–1886). He wrote more clearly than the German, who often buried important concepts in obscure footnotes. Many scientists thus learned about the Kekulé valences through Butlerov's publications. In an 1861 article, Butlerov also expressed a distinction between the chemical structures inferred from experiments and the absolute arrangement of atoms in the true molecule. One year later, he published a provocative discussion of ethane (C_2H_6), in

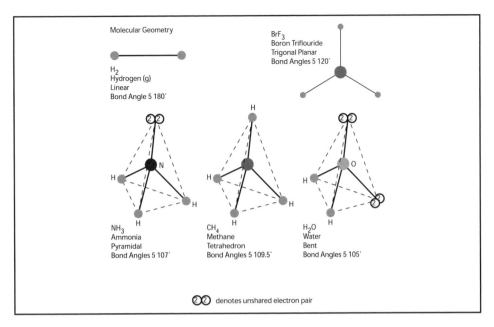

Molecular geometry.   *Illustration by Argosy. Cengage Learning, Gale.*

which he proposed a tetrahedral geometry for carbon. However, he made no effort to apply this insight to more complex compounds with other atoms in place of hydrogen.

The Science

In 1873 the German chemist Johannes Wislicenus (1835–1902) had just made the puzzling observation that lactic acid, a compound found in sour milk ($C_3H_6O_3$), did not rotate polarized light, whereas another form of lactic acid from meat did. Van't Hoff suggested that molecules with four different groups at the central carbon could exist as a pair of mirror-image structures (often called optical isomers). According to this hypothesis, sour milk contained a mixture of the two isomers, such that the effects would cancel and no rotation would be observed with polarized light. By contrast, meat should contain just one, although van't Hoff could not specify which. He also included a list of compounds that are not expected to rotate polarized light: $CH_3(CH_2)_nCH_3$, $CH_3(CH_2)_nCO_2H$, and $CH_3(CH_2)_nCH_2OH$ (where n is any positive integer). He noted that none of these formulas possessed an asymmetric carbon, and thus would show no optical effect.

Le Bel and van't Hoff published their analyses almost simultaneously. While both discussed lactic acid, Le Bel adopted a more abstract (and therefore more general) treatment. For example, he was intrigued by the symmetry of methane (CH_4) and how the geometry contrasted with other derivatives. While van't Hoff focused entirely on the prediction of isomers (thus his model was therefore more likely to be used by chemists),

Le Bel explained the same phenomena in the language of physics. He also concluded his article with some prescient observations on the stereochemistry of chemical reactions. Specifically, he made a bold prediction: if one started with a compound that lacked an asymmetric carbon, then the products would contain equal quantities of the mirror-image forms. Le Bel advanced an even bolder prediction when he suggested that a chemist could influence the stereochemical outcome of a reaction by employing substances with a known handedness or even by shining polarized light on the solution.

Both men also described the structures of compounds containing carbon-carbon (C=C) double bonds. They implicitly recognized that this unit was inherently rigid, which would give rise to distinct isomers, but the atoms with single bonds were free to rotate. Furthermore, van't Hoff suggested that some observations could be explained by the presence of two different geometries about the central double bond.

A noteworthy factor in Le Bel and van't Hoff's presentations is the striking difference in their use of illustrations. Le Bel employed a convention, borrowed from the British chemist Alexander Crum Brown (1838–1922), in which solid lines denoted the connections between atoms. While he referred repeatedly to carbon's tetrahedral geometry, nowhere did he provide a true perspective drawing. By contrast, van't Hoff dramatically illustrated his article with drawings of the mirror-image pair of isomers containing a true tetrahedron, as well as possible geometries for maleic and fumaric acids. Chemists were able to follow his arguments more easily than Le Bel's.

H:Ö:H   :Ö::Ö:

Figure showing the mirror image relationship between the
isomeric conformations of lactic acid.   *Cengage Learning, Gale.*

Portrait of Joseph Achille Le Bel (1847–1930), French organic
chemist who showed that carbon compounds which are
chemically identical can be structurally different. In 1874 Le Bel
and J. van't Hoff independently published results showing that
the bonds formed by carbon radiated in three dimensions rather
than all lying in a plane. This allows different arrangements of
atoms within a molecule. The various forms of molecules can
have different properties; Le Bel had discovered this effect of
stereochemistry because of the way that different forms polarized
light. Le Bel spent the rest of his career between his family's
factory and a private laboratory.   *SPL/Photo Researchers, Inc.*

Influences on Science and Society

At first van't Hoff and Le Bel's publications were virtually ignored by the chemical community. Not surprisingly, such a chilly reception impelled each to pursue
alternate projects. Le Bel turned to botany, while van't
Hoff studied reaction rates and the properties of membranes (winning the first Nobel Prize for chemistry in

1901). Fortunately, the concept of a tetrahedral carbon
atom gained a few patrons among the German chemists,
especially Wiscilenus and Victor Meyer (1848–1897).
Significantly, the former arranged for a German translation and promoted its discussion at scientific meetings.
While van't Hoff became the public spokesman for the
model, he always gave equal credit to the absent Le Bel.
Meyer not only employed the tetrahedral carbon in his
own textbook, he also introduced the term "stereochemistry" to refer to the three-dimensional molecular
shapes. Finally, the British physicist John Strutt, Lord
Rayleigh (1842–1919), coined the equally important
word "chirality" from the Greek for right- and left-
handedness.

Wiscilenus in particular played an often unrecog-
nized role in the early dissemination of van't Hoff and
Le Bel's hypothesis. As an administrator in the universities at Wurzburg and Leipzig, he could lend his professional prestige to the tetrahedral carbon. Just four
years after van't Hoff and Le Bel's article appeared,
Wiscilenus began teaching it to his students as part of
the Wurzburg chemistry curriculum. A decade later he
also published the first of a series of research articles in
which he described ten independent lines of evidence to
assign the structures of maleic acid and fumaric acid.

For example, maleic acid can be converted into its
isomer (but not the reverse), suggesting that the oxy-
gens are more stable when the atoms are further apart.
This brilliant tour de force was replete with three-dimensional diagrams in which he relied extensively on the
concept of tetrahedral carbon. Wiscilenus also coined
the term geometric isomers, which chemists still use to
refer to compounds that differ only in the arrangement
of groups about a double bond.

Another German chemist who relied explicitly on
the tetrahedral carbon was Emil Fischer (1852–1919).
He is best known for his extensive studies of sugars,
which began in 1884 and spanned two decades. He
brought order to this complex subject by characterizing
the numerous asymmetric carbons in these compounds.
He also made significant contributions to biotechnology
and the life sciences when he examined the behavior of
different sugars in the presence of yeasts, demonstrating
(among other findings) that the sugar mannose was digested by eleven varieties of yeast, but its stereoisomers
were not affected at all. In related research, Fischer introduced his vivid lock-and-key analogy to describe the
remarkable specificity of sugar-protein interactions.

$$H:N:H \quad + \quad H \quad = \quad H:N:H$$

(with H below the N on the left, and H above and below the N on the right)

Figure showing the structures of fumaric acid (on the left) and maleic acid (on the right), as these compounds are known today. These molecular conformations are of Cis-Trans isomers. *Cengage Learning, Gale.*

Fischer and his contemporaries recognized that they could only hypothesize whether or not two compounds were similar in their structure. Chemists at the end of the nineteenth century lacked the technology to map the true arrangement of atoms in space, and so Fischer made an educated guess for an isomer of tartaric acid as a standard. However, an x-ray study in 1951 headed by Dutch physicist Johannes Bijvoet (1892–1980) finally proved Fischer correct.

Pasteur occupied a unique position in discussions of chirality, because of his pioneering studies of optical rotation and crystalline form. While he publicly stressed the connection between molecular asymmetry and living processes, he simultaneously pursued private experiments in search of a nonbiological basis for the selection. Among other factors, he explored magnetic fields and sunlight reflected through a mirror, without observing any effect on the stereochemistry. Late in his career, Pasteur speculated in 1883 that life on our planet might be governed by a cosmic asymmetry, which had previously eluded scientists. Some historians have even suggested that he may have deferred this research for fear of alienating his royal patron, Napoleon III.

Chemists would learn from spectroscopic studies and other physical properties that the hydrogens in methane really do conform to the geometry predicted by van't Hoff and Le Bel. In developing his model for bonding in methane, American chemist Linus Pauling (1901–1994) assumed that he could ignore the inner electrons of carbon. Furthermore, to get the right arrangement, he devised the notion of hybrid orbitals (functions that define the location and energy of an electron), which had spatial properties in between those of the pure atomic orbitals. Pauling showed in 1931 that the tetrahedral geometry for the bonds in methane was consistent with the rules of quantum mechanics, at a time when other scientists were focused on much simpler molecules.

Pauling also popularized a way of portraying certain molecules as a combination of different structures in order to predict bond angles, charge distribution, and other physical properties. He also established the convention of using a double-headed arrow to depict these molecules. By drawing three structures in this manner, he and other chemists sought to convey the concept that the real molecule was actually in between these resonance forms. In this case, it is known from other studies of BF_3 that it has a trigonal planar geometry with equivalent B-F bonds.

The concept of resonance had been used in molecular theory since the inception of quantum mechanics, especially in London and in Heitler's treatment of hydrogen. Pauling, however, did more than any other scientist to demonstrate the advantages of this depiction. While some chemists misinterpreted resonance as an oscillation, it nevertheless remains an important tool for representing molecules.

The first support for van't Hoff and Le Bel's postulate of rotation about single bonds came from several American chemists during the 1930's. Henry Eyring (1901–1981), who modeled the rates of reactions, tried to estimate how fast the carbon-carbon axis moved in ethane (C_2H_6). He calculated that the bond must be twisting at a very rapid rate: 7×10^{13} times per second at room temperature. While many scientists accepted this value, George Kistiakowsky (1900–1982) and his colleagues used more direct experimental measurements to determine a rotation that was 300 times slower. However, this revised estimate remains fast, justifying the assumptions of van't Hoff and Le Bel.

■ Modern Cultural Connections

The concept of stereochemistry had profound effects on biology. Beginning in the late 1930s, Pauling embarked on an extended study of proteins, which serve such diverse functions as building materials, catalysts (which speed up reactions), and transport vehicles. For 15 years he bombarded proteins with x rays to determine the relative atomic positions, and also built models to interpret the data. In 1951 he finally published the first of a

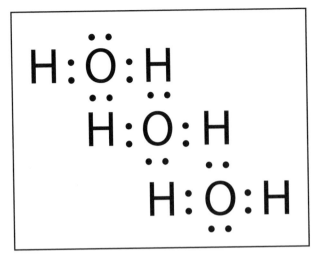

Linus Pauling's (1901–1994) depiction of BF₃ as an average of three contributing forms, showing charge and bonding distribution. *Cengage Learning, Gale.*

series of research articles in which he provided explicit evidence that parts of these proteins exhibit a helical twist (like a corkscrew), held in place through hydrogen bonds. Although he studied silk, muscles, tendons, gelatin, hair, feathers, and horn, he found that the same principles apply to all proteins. Importantly, the helix is also a direct consequence of asymmetric carbons, which all possess the same chirality in the backbone of the structure. For these (and many other) achievements, Pauling received the Nobel Prize for chemistry in 1954.

Nearly all molecules found in living systems exhibit chirality. One rancorous debate during recent decades has centered on the vexing question of how we ended up with one set of optical isomers. Of course, attempts to simulate the formation of such compounds on the early Earth have typically started with carbon in the form of CH_4, CO, or CO_2. Because these gases all lack an asymmetric center, the products of the reactions always contain equal amounts of the two isomers (what chemists now call a racemic mixture), just as Le Bel had predicted. For example, when the American chemist Stanley Miller (1930–2007) published his first such synthesis from CH_4, NH_3, H_2, and H_2O subjected to simulated lightning, he reported that the amino acids (the building blocks of proteins) did not rotate polarized light.

Some geologists, such as the American Robert Hazen (1948–) of the Carnegie Institution, have suggested that life might have begun on mineral surfaces. He has done experiments on Haüy's calcite, in which different faces have the ability to bind amino acids of one chirality, which (in principle) could join together to form small proteins. Once the first organisms appeared, evolution would undoubtedly have favored those with one type of handedness. However, Hazen's model is also deeply rooted in the assumption that the origin of life (and chirality) was a very local phenomenon, perhaps occurring on the edge of a mineral that happened to stick out of the primordial mud in a tidal lagoon.

Other scientists have suggested that a more fundamental cosmic asymmetry played a role. For example, American chemist William Bonner (1919–) of Stanford University has proposed that a spinning neutron star remnant from a supernova may have passed near our solar system early in its history and bathed our neighborhood in circularly polarized sunlight that could have induced the formation or destruction of specific stereoisomers. He and his collaborators even carried out laboratory simulations using racemic mixtures of amino acids, which can be transformed after irradiation into a product that has a preponderance of one structure.

Surprising support for this unearthly notion has come from the analysis of a meteorite that fell in Australia in 1969, but which represents debris from the earliest period of our solar system. Some meteorites are especially rich in carbon-containing compounds, in contrast to the stony meteorites—from which they are completely absent. Planetary geologists have suggested that the ingredients for life could have been brought to our planet via this process. In 1997, the American chemist John Cronin (1937–) of Arizona State University reported results on some especially hardy amino acids. Remarkably, these compounds are non-racemic, indicating an extraterrestrial selection in the distant past. In the intervening years, the suite of such amino acids

Two time Nobel Prize winner Linus Pauling (1901–1994) stands at the blackboard in his chemistry laboratory.   *© Bettman/Corbis.*

has grown substantially, and they consistently reveal a bias toward the same stereochemistry found in modern proteins. Cronin believes that the ubiquitous preference for one type of chirality is most consistent with a flux of polarized light, such as from a neutron star or through light scattering by magnetically aligned interstellar dust.

Chirality has been a compelling issue not only in the study of life's origins, but also in the design and efficacy of drugs. Because biological molecules have an inherent handedness, many pharmaceuticals have asymmetric carbons that cause them to bind to specific sites in your body. In general, only one isomer is effective, while its mirror image may even be toxic. A striking example is thalidomide, which was sold throughout Europe during the 1950s to treat depression and, later, nausea. The drug was withdrawn when physicians reported severe birth defects, which some scientists attributed to one of the mirror-image forms based on studies in rats. In 1998, however, the U.S. government approved its limited use to treat the side effects of leprosy, and other applications may be pending. This drug continues to be sold as racemic mixture, partly because of its lower cost,

but also because the isomers are rapidly interconverted inside the human body.

In conclusion, the concept of molecular handedness has had a profound influence on numerous applications in science and medicine. From the shape of a protein to the possible extraterrestrial origins of chirality, the tetrahedral carbon has become an integral part of modern chemistry. The tentative hypothesis of van't Hoff and Le Bel that was introduced in 1874 now rests on an ever-expanding body of data. Chemists will undoubtedly build on this tradition with new advances in medicine and other materials.

SEE ALSO *Chemistry: Biochemistry: The Chemistry of Life; Chemistry: Chemical Bonds; Chemistry: Fermentation; Chemistry: Organic Chemistry.*

BIBLIOGRAPHY

Books

Ramberg, P.J. *Chemical Structure, Spatial Arrangement: The Early History of Stereochemistry, 1874–1914.* Burlington, VT: Ashgate Publishing Company, 2003.

Periodicals

Chyba, C.F. "Origins of Life: A Left-Handed Solar System?" *Nature* 389, no. 6648 (1997): 234–235.

Farley, J., and G.L. Geison. "Science, Politics and Spontaneous Generation in Nineteenth-Century France: The Pasteur-Pouchet Debate." *Bulletin of the History of Medicine* 48 (1974): 161–198.

Hazen, R.M. "Life's Rocky Start." *Scientific American*. 284, no. 4 (2001): 76–85.

Palladino, P. "Stereochemistry and the Nature of Life: Mechanist, Vitalist, and Evolutionary Perspectives." *Isis* 81 (1990): 44–67.

Snelders, H.A.M. "The Reception of J.H. van't Hoff's Theory of the Asymmetric Carbon Atom." *Journal of Chemical Education* 51, no. 1 (January 1974): 2–7.

William J. Hagan

Chemistry: Organic Chemistry

■ Introduction

Organic chemistry is the branch of chemistry that focuses on the properties and reactions of compounds that contain carbon atoms. The carbon atom is unique because it is the only element that can bond to itself, forming chains that can contain hundreds of atoms. Carbon can also combine with a wide variety of other elements.

■ Historical Background and Scientific Foundations

More than a million carbon compounds have been discovered, and new ones are constantly being synthesized. One reason for this large number is isomerism, particularly structural isomerism, first recognized in 1830 by the Swedish chemist Jöns Jakob Berzelius (1779–1848). In this arrangement, two or more compounds contain the same number of carbon and hydrogen atoms arranged in different ways to form unique compounds with distinct chemical and physical properties. A compound formed from five carbons and twelve hydrogen atoms ($C_5 H_{12}$) will, for example, produce three unique compounds. Other forms of isomerism distinguished by the spatial arrangements of their atoms were discovered later in the nineteenth century.

In the eighteenth and early part of the nineteenth centuries, when many carbon-based compounds were first isolated, they had only been extracted from plants and animals. For example, formic acid had been isolated from ants, salicylic acid had been isolated from willow bark, and urea had been found in urine. This entrenched the idea that organic molecules must be created by (or by the actions of) a living organism. This idea also fit with the doctrine known as vitalism, a theory that developed in reaction to the rise of the mechanistic approach to life being merely a physical process.

The Influence of Liebig and Wöhler

Two German chemists, Justus von Liebig (1803–1873) and Friedrich Wöhler (1800–1882), were responsible for the emergence of organic chemistry in the early nineteenth century. Their quantitative analytical methods helped establish the constitution of newly isolated and synthesized carbon compounds. Both were inspiring teachers who established laboratory work as the basic model for chemical education, teaching students who came from all over Europe and America. The pupils then emulated their methods when they returned home to train the next generation of chemists.

Liebig, the most prominent chemist in nineteenth-century Europe, studied chemistry at the Universities of Bonn and Erlangen, where he obtained his PhD in 1822. After further study in Paris, he chaired the chemistry department at the University of Giessen before moving to the University of Munich, where he spent the remainder of his life. Liebig's major contributions were the development of new methods for the quick and precise measurement of the quantities of carbon, hydrogen, and nitrogen in organic compounds. This allowed Liebig and his students to identify a host of new organic compounds. Much of his work after 1840 was related to agricultural and biological chemistry, including a study of fermentation and methods for increasing soil fertility and yields through the use of artificial fertilizers.

The Synthesis of Urea and the Demise of Vitalism

Wöhler studied medicine at the Universities of Marburg and Heidelberg, obtaining his medical degree in 1823. He went on to study with Berzelius in Stockholm and to hold positions in technical schools in Berlin and

WORDS TO KNOW

AROMATIC: In organic chemistry, a compound whose molecular structure includes some variation of the benzene ring.

BENZENE: A ring-shaped hydrocarbon molecule that has long been used an industrial solvent, that is, a liquid in which other substances can be dissolved. However, benzene is now known to be a potent cause of cancer and is less used.

HOMOLOGOUS SERIES: A group of organic compounds that have similar structure and chemical properties.

HYDROCARBON: A chemical containing only carbon and hydrogen. Hydrocarbons are of prime economic importance because they encompass the constituents of the major fossil fuels, petroleum and natural gas, as well as plastics, waxes, and oils. In urban pollution, these components—along with NO_x and sunlight—contribute to the formation of tropospheric ozone.

INORGANIC: Composed of minerals that are not derived from living plants and animals.

ISOMERISM: The occurrence of chemical compounds that have the same composition (the same numbers of the same types of atoms) but different geometrical arrangements of those atoms and, as a result, different physical properties. Two substances with the same atomic ingredients but different structure are called isomers.

ORGANOMETALLIC: Any chemical compound containing a metal atom bound to an organic group (cluster of atoms containing carbon). Hemoglobin, which contains iron and transports oxygen in blood, is an example of an organometallic compound.

RADICAL THEORY: Also called the free-radical theory. In chemistry, the idea, first proposed in the 1950s, that aging is the result of accumulated biochemical damage to body cells from free radicals, charged atoms or molecules with a strong tendency to participate in chemical reactions (beneficial

or not). The theory is not widely accepted among biologists studying ageing, though there is agreement that free radicals can cause some disease conditions that progress with age.

STEREOISOMERISM: Two molecules that contain the same numbers of the same types of atoms but differ in the spatial arrangement of those atoms are stereoisomers of each other; the occurrence of stereoisomers is termed stereoisomerism.

STRUCTURAL THEORY: In chemistry, the view, first put forward in the mid-nineteenth century by Russian chemist Aleksandr Butlerov (1828–1886), that the properties of a chemical substance arise both from the atoms that make it up and the way those atoms are structured to form molecules. That is, according to the structural theory, not only the composition but the shape of a molecule matters.

SYNTHESIS: The process of combining simpler chemicals to make a more complex compound.

TETRAVALENT: In chemistry, a tetravalent (four-valent) atom is one that has a valence of 4, that is, four electrons available for covalent (electron-sharing) bonding with other atoms. A tetravalent atom will bind with 4 other atoms; for example, carbon is tetravalent, so in methane, CH_4, it bonds covalently with four hydrogen atoms (which are monovalent).

UREA: Chemical compound of carbon, hydrogen, nitrogen, and oxygen produced as waste by cells that break down protein or under laboratory conditions.

VALENCE: The tendency of an atom to gain or lose electrons in reacting with other atoms.

VITALISM: The concept that compounds found within living organisms are somehow essentially different from those found in nonliving objects.

Kassel. In 1836 he was appointed professor of chemistry in the medical faculty of Göttingen University, a position he held for the rest of his life. Wöhler's major research interests were in inorganic chemistry, where he isolated the elements boron, silicon, aluminum, cerium, and just missed being the discoverer of vanadium and niobium. Like Liebig, he also established a school of research and teaching, but is best known for synthesizing urea.

In 1828 Wöhler synthesized the organic compound urea in the laboratory using the inorganic compound ammonium cyanate. Urea had previously been found only in urine—that is, from a biological source. While this dealt a blow to vitalism, it did not fully spell its demise. Skeptics believed that compounds associated with

living organisms were produced by a "vital force" not available to the chemist. This idea persisted until 1844 when Hermann Kolbe (1818–1884) proved definitively that organic compounds could be produced under laboratory conditions by synthesizing acetic acid from the simple inorganic compounds carbon disulfide and chlorine.

Structural Theory and Its Development

In the early part of the nineteenth century a new problem emerged as chemists tried to classify and bring some order to the ever-increasing number of organic substances. This began a period of confusion and controversy that would last several decades, until the development of the structural theory of organic chemistry.

Analytical methods pioneered by both Liebig and Wöhler were able to determine the content of organic molecules, but did not show how their elements were arranged. The first attempt to solve the problem was known as radical theory. Berzelius proposed a way to understand the formation of inorganic compounds by assuming that some elements had a positive charge and others had a negative charge. Scientists knew that opposite electrical charges would attract; thus the formation of sodium chloride could be explained by assuming that sodium is positively charged and chlorine is negatively charged.

Radical theory was pioneered in the 1830s by Liebig and the French chemist Jean-Baptiste André Dumas (1800–1884). Radicals were thought to be a stable group of elements that were joined together to produce an electropositive group that was joined to an electronegative atom to form a compound. In the case of organic compounds the radical would contain carbon, hydrogen, and various other atoms in combination with an electronegative inorganic partner. Thus ethyl alcohol was represented as $C_2H_4 H_2O$.

The electropositive radical was capable of transformation but should remain intact. This was demonstrated by Liebig and Wöhler in the case of benzaldehyde, which was extracted from almonds. The fact that compounds made from benzaldehyde all contained the radical $C_{14}H_{10}O_2$ provided evidence for radical theory.

Radical theory lost its hold on organic chemistry, however, when scientists realized it was possible to substitute one atom for another in what was assumed to be a "stable" radical. This substitution was further shown by Dumas to be an atom of a different charge. For example, in some compounds the substitution of an electronegative chlorine for electropositive hydrogen did not significantly change the character of the product. Dumas was able to convert acetic acid to trichloroacetic acid, and the product still remained an acid.

Radical theory gave way in the 1840s to what became known as type theory. It was developed by two of Dumas's students, Auguste Laurent (1808–1853) and French chemist Charles Frédéric Gerhardt (1816–1856), who proposed that organic (and even inorganic) substances were derived from simple molecules by substitution. Alexander Williamson (1824–1904) experimentally proved the existence of the water type with his synthesis in 1850 of diethyl ether from the potassium salt of ethyl alcohol and ethyl iodide.

In 1845 Laurent also introduced the concept of homologous series—a group of similar compounds that differed by a single unit. The first three members of the alkane series, for example, are methane (CH_4), ethane (C_2H_6), and propane (C_3H_8); each succeeding molecule contains an additional CH_2 group. The German chemist Hermann Kopp (1817–1892) showed that each additional CH_2 increases the boiling point of each successive group member by a fixed number of degrees.

Three other types were proposed by Gerhardt in 1853, the ammonia (NH_3), hydrogen (H_2), and hydrogen chloride (HCl) types. These types together with the water type could be used to classify the already large number of organic compounds into four distinct groupings. Gerhardt wrote "By exchanging their hydrogens among certain groups, these types give rise to acids, to alcohols, to ethers, to hydrides, to radicals, to organic chlorides, to acetones, to alkalis." Type theory was thus an advance—but because it still allowed multiple classifications for the same molecule, major problems remained to preoccupy chemists in the latter part of the nineteenth century.

Valences

The concept of valence (the combining power of an element) slowly developed in the period from 1850 to 1870 through the work of the English chemist Edward Frankland (1825–1899) and the German August Kekulé (1829–1896). Frankland had begun his chemical studies in London and continued them in Germany with chemist Robert Wilhelm Bunsen (1811–1899) at Marburg. Frankland's view of valence was derived from the earlier radical theory; Kekulé's was based on type theory.

Frankland attempted to synthesize the hydrocarbon radical ethyl (C_2H_5) by reacting ethyl iodide with zinc. This produced not the desired result but butane, with diethyl zinc, the first example of an organometallic compound, as a byproduct. What was crucial about this discovery was that the diethyl zinc (C_2H_5)2Zn always contained twice as much ethyl as zinc. Frankland went on to show that in other organometallic compounds that a definite combining power existed between the organic portion and the metal. In 1852 Frankland proposed that inorganic and organic compounds both had a combining power that came to be known as valence. Frankland also recognized multiple valences and gave examples where the valence was 3 or 5 such as in PCl_3 & PCl_5.

August Kekulé

Friedrich August Kekulé (1829–1896) was born in Darmstadt Germany, a descendent of a noble Bohemian family from Stradonitz, a city near Prague. Kekulé had studied architecture at Geissen, but because of Liebig's influence he changed to chemistry, with additional studies in biological classification. Kekulé obtained his doctorate in Paris in 1851.

As a student in Paris, Kekulé had studied with Dumas and Gerhardt and was naturally drawn to type theory. Further study in London from 1854 to 1855 brought him into contact with two very original thinkers: Williamson and William Odling (1829–1921), chemists who studied chemical structure to understand chemical properties. Kekulé hypothesized that carbon was tetravalent, that is, could combine with four

August Kekulé (1829–1896), German organic chemist, circa 1885. In 1865, Kekulé published his theory of the structure of the benzene ring as ring of six carbon atoms attached by double and single bonds alternately. *Oxford Science Archive/Heritage-Images/The Image Works.*

substituents to form organic molecules. His approach was purely mechanical and may have been influenced by his initial interest in architecture. In 1858 Kekulé proposed that carbon could form chains by using some of its valences to bond to other carbon atoms. Kekulé's representations are not the familiar structural formulas we use today. Those were established by Scottish chemist Archibald Scott Couper (1831–1892) in 1858.

Couper had studied in Paris with the French chemist Charles-Adolphe Wurtz (1817–1884). He had ideas similar to those of Kekulé, but due to a series of mishaps his paper was published after Kekulé's, and thus he failed to receive the credit he should have. In contrast to the conservative and cautious Kekulé, Couper drew formulas for molecules that used dotted lines to represent the bonds between carbon atoms.

The graphical structural formulas that are used today were introduced by Alexander Crum Brown (1838–1922) in 1861. Initially he wrote the elements using the letters such as C, H, O, etc., with circles around them, as Dalton had done in his exposition of the atomic the-

ory in the first decade of the nineteenth century. The circles were connected with solid lines, with the number of lines equal to the valence—four in the case of carbon. The circles were eventually dropped to create the structural formulas we still use today.

Stereochemistry and Aromaticity

The idea that molecules are three dimensional was not realized until the latter part of the nineteenth century. Discoveries made by French scientist Louis Pasteur (1822–1895), Dutch chemist Jacobus van't Hoff (1852–1911), and French chemist Joseph-Achille Le Bel (1847–1930) provided the keys to understanding the three-dimensional nature of many organic molecules.

Pasteur is best known for his microbiological research, but his initial training was in chemistry. In his doctoral research Pasteur studied the puzzling differences between two forms of the same compound: tartaric acid. Pasteur found that natural tartaric acid, isolated from the fermentation of grapes, rotated the polarization plane of light that passed through it. Racemic acid, a synthesized product that had an identical chemical formula, however, did not.

The French physicist Jean-Baptiste Biot (1774–1862) had observed this same phenomenon in 1815 with several organic liquids, including oil of turpentine. Salts of tartaric acid had been studied by other chemists as well, in particular the German chemist Eilhard Mitscherlich (1794–1863), who reported in 1844 that the physical properties of the sodium-ammonium salts of the optically active tartaric acid and the racemic acid were the same in every respect except their interaction with plane-polarized light.

Pasteur found that a comprehensive study of different tartrates showed that the optically active tartrates had an asymmetrical crystalline shape. All the racemic acid crystals were symmetrical. One particular compound, the sodium ammonium double salt of tartaric acid, produced very large well-defined crystals. Using a microscope and tweezers Pasteur was able to isolate two forms of the crystals. When subjected to plane polarized light, one bent the light to the right and the other to the left. The rotation was identical for both except the direction. If mixed in equal amounts there was no rotation at all. This phenomenon came to be known as stereoisomerism. Pasteur proposed that it was possible in nature for molecules to be asymmetric; by extension an association was made by Pasteur between asymmetry and life.

Pasteur and Kekulé had speculated that stereoisomerism might be caused by either the tetravalency of carbon itself or the way carbon is oriented in space as a result. This question was answered in 1874 by van't Hoff and Le Bel, independently of each other.

At this time van't Hoff was a professor in the veterinary school in Utrecht, Holland. His initial proposal, written in Dutch, attracted little attention. An extended form, *La chimie dans l'espace* (Chemistry in space) published in French in 1875, attracted much interest and controversy. He explained stereoisomerism by proposing that the four carbon valences were on the apexes of a tetrahedron. Four different substituents bonded to the central carbon atom could produce two structures that were mirror images of each other; this would then produce the asymmetry in carbon compounds.

Thus two mirror images could exist that were identical in all their properties except for the way they affected polarized light. These mirror images came to be known as enantiomers. Pasteur found that if a mixture contains equal amounts of each type, they cancel each other; this is why racemic acid did not affect polarized light and the natural tartaric acid did. Any mixture with equal numbers of enantiomers is now known as a racemic mixture.

Le Bel's explanation was similar, but he started with a carbon that had four identical groups and looked at what happened with successively substituted atoms. With four different substituents, an asymmetry is produced in which the mirror image is no longer superimposable on the original. Molecules that have more than one center of asymmetry (that is, more than one stereogenic center) will have 2^n stereoisomers, where n is the number of sterogenic centers. In organic chemistry, each carbon atom to which four other atoms (or groups of atoms) are bonded constitutes a stereogenic center. Not every carbon atom in an organic molecule is necessarily a stereogenic center.

An example of how this rule applies is the glucose molecule ($C_6H_{12}O_6$), whose stereoisomers were first studied by German chemist Emil Fischer (1852–1919). Each glucose molecule has three centers of asymmetry, which allows the atoms to combine into $2^n = 8$ (eight) unique compounds, which Fischer identified. One of the major achievements of twentieth-century synthetic organic chemistry has been to develop methods to produce particular stereoisomers when multiple routes are possible.

One major structural problem that remained to be solved was the structure of benzene (C_6H_6) and aromatic compounds in general. Benzene was a formidable challenge because its unusual properties could not be based on conventional ideas of structure. Many chemists had tried and failed to solve the benzene problem before Kekulé devised the first rational structure in 1865.

In a story of dubious authenticity, Kekulé is said to have had a daydream in which he envisioned a snake catching its tail; this led him to visualize benzene's hexagonal structure, in which six carbon atoms alternate in single and double bonds. This model predicted that four unique isomers existed in disubstituted benzene molecules (benzene molecules containing two substituted atoms or groups of atoms), but only three were

known. Kekulé proposed that one of these isomers has two forms that differ only by the locations of the single and double bonds. If an equilibrium exists between these two forms, all the carbon atoms become equivalent. This explanation became the basis of aromatic chemistry until the development of molecular orbital theory in the 1930s.

The Birth of the Synthetic Organic Chemical Industry

Organic compounds such as ethyl alcohol, found in beer and wine, and acetic acid, used in vinegar, have been produced since antiquity. Other naturally occurring organic compounds have been used for millennia as dyes and medicinal agents. But the production of organic substances in manufacturing plants did not being in earnest until the 1850s. The synthetic organic chemical industry began when a process was developed in the late-eighteenth century to produce gas from coal. This produced coal tar, a waste product that was considered a nuisance as it had only limited use, mainly as a water-proofing agent.

Organic chemists began to examine the constituents of coal tar in earnest in the 1840s. One of the primary investigators was the German chemist August Wilhelm Hofmann (1818–1892). A student of Liebig, he became the first professor of chemistry at the newly founded Royal College of Chemistry in London in 1844. Hofmann had begun to analyze coal tar in Liebig's laboratory and continued this work when he moved to London.

Hofmann found that coal tar's major components were aromatic hydrocarbons, a class of compounds based on the benzene molecule, first isolated by Michael Faraday (1791–1867) in 1825 from compressed oil gas. Hofmann and his students isolated at least 20 different substances from coal tar, the most important being aniline ($C_6H_5NH_2$), an organic analog of ammonia, and phenol (C_6H_5OH), which was used as one of the first antiseptics.

In 1856 William Henry Perkin (1838–1907), one of Hofmann's students at the Royal College of Chemistry, tried to synthesize quinine, a drug used to treat malaria. At that time quinine could only be extracted from the bark of the cinchona, a tree that was native to the Dutch East Indies. Hofmann thought that quinine might be the aromatic hydrocarbon naphthalene. Perkin took a different approach and used an impure form of an aniline derivative called allyl toluidine and treated it with potassium dichromate. The reaction failed to produce quinine, and Perkin repeated the experiment using aniline itself.

This produced a brown substance which, when dissolved in alcohol, produced a vivid purple solution. Perkin was only 18 at the time, but he had the insight to see that this might be useful as a dye, especially since purple is a color with few natural sources. When Perkin's samples were sent to a dyer, he found that it dyed silk a subdued purple that became known as mauve.

In partnership with his father and brother, Perkin built a plant to manufacture this synthetic dye on a commercial scale. Perkin's genius was his ability to convert a laboratory process into a commercial product. His success led many English competitors to develop their own lines of aniline-based dyes. Using various derivatives of aniline and reaction conditions, the whole spectrum of colors became available, eliminating the need for natural dyes and dooming that industry.

Britain became the center of the synthetic organic chemical industry for the next two decades. The dye industry, in particular, became a magnet for many highly trained German organic chemists. The experience they gained in Britain had two distinct consequences. First, German chemists became masters of laboratory synthesis, developing synthetic replacements for the natural red dye alizarin and blue dye indigo. Second, Germany became the center of the organic chemical industry, founding of such companies as BASF, AGFA, Bayer, and Hoechst. Partnerships that developed between academia and the industry made Germany the world leader in chemical production by 1914. This became a problem for the United States during World War I, when exports of German organic chemicals were stopped by the British blockade.

Although synthetic dyes had been developed, pharmaceuticals were still derived mainly from natural sources. One key breakthrough was the synthesis of salicylic acid by Kolbe in 1853. A natural product derived from willow bark, salicylic acid was known for its pain-relieving abilities. Although one of Kolbe's students converted the laboratory synthesis into a commercial product, salicylic acid taken internally produced unwanted side effects. The acetyl derivative prepared in 1897 by a chemist working for the Bayer Company in Germany was found to produce the same pain relief without the side effects. This compound was sold as aspirin beginning in 1899 and has remained a staple product for over a century.

In the latter part of the nineteenth century, German physician and bacteriologist Robert Koch's (1843–1910) postulates proved the validity of the germ theory of disease and ushered in a more rational approach to the treatment of bacterial infections. Although synthetic dyes had previously been used to stain cells for medical studies, scientists such as German medical researcher Paul Ehrlich (1854–1915) thought that some dyes might actually destroy bacteria also. The dye methylene blue, for example, killed the parasite associated with malaria.

Ehrlich believed that a drug must be linked in a rational and systematic manner to its target. Ehrlich's best-known work was on syphilis, which is caused by a spirochete, a type of bacterium. Ehrlich had been studying organic compounds that contained arsenic as a treatment against certain tropical diseases, only to find that most of these, although somewhat effective, had toxic side effects. Based on his knowledge of organic chemistry, Ehrlich synthesized a series of organo-arsenic compounds, going through 605 before the next one, marketed under the trade name Salvarsan, proved successful. An even more effective compound called Neosalvarsan was produced in 1912; this remained the standard treatment until the beginning of the antibiotic era in the late 1940s.

Because little was known about how drugs actually worked, very few breakthroughs occurred between 1920 and 1940. If a drug proved effective it was purely by chance in most cases, not by design. A good example of this was the discovery of sulfa drugs by the Bayer division of I.G. Farben in Germany. Thousands of compounds were synthesized by Bayer chemists before one, whose structure was similar to a class of dyes known as azo compounds, showed any medicinal potential. The drug Prontosil was patented in 1932 and was the first synthetic antibacterial effective against streptococcal and staphylococcal bacteria (but not against enterobacteria).

What made Prontosil effective was that it decomposed *in vivo* to a compound called sulfanilamide, a relatively simple compound first synthesized in 1908. Thousands of sulfanilamide analogs were created in the 1930s, and several proved effective. This was the only antibacterial therapy available in any quantity through World War II. Penicillin, an antibiotic first used in 1942, was much more effective, but it was difficult to produce in large quantities, and much of what was made was intended for military use.

The era after 1945 marked the golden age of organic chemistry and the discovery of new drugs. New synthetic techniques and, most importantly, a revolution in instrumentation that included nuclear magnetic resonance, mass spectrometry, and various types of infrared, visible, and ultraviolet spectroscopy made it possible to determine the structures of both natural and synthetic products rapidly. Tremendous strides were made in finding medicinal agents to deal with a variety of conditions.

Industrial Advances

Prior to 1939 insecticides were mainly inorganic in nature and posed many health hazards, especially when used in the agricultural sector. In 1939 the organic insecticide DDT (dichloro-diphenyl-trichloroethane) was found to be very effective against many pests, especially those that carry diseases such as malaria. Spraying DDT saved an enormous number of lives during World War II because of its ability to prevent outbreaks of malaria and other tropical diseases. Various other chlorinated hydrocarbon pesticides were introduced when effectiveness and safety concerns made the use of DDT less desirable. These second-generation insecticides such as Chlordane, Lindane, and Dieldrin also had their problems and were replaced with an entirely new class of organophosphates.

In the post-World War II era, organic herbicides came to be an important addition to agriculture. One of the earliest commercial products, 2,4-D (2,4-dichlorophenoxyacetic acid), highly effective against broad-leaf plants, improved agricultural yields. Agent orange is a 50-50 mixture of 2,4-D and 2,4,5,T (2,4,5-trichlorophenoxyacetic acid) that was used extensively as a defoliant during the Vietnam war, sometimes creating serious health effects in those who came into contact with it. Trace amounts of dioxin, a highly carcinogenic material, are produced in the manufacture of 2,4,5-T. The manufacture of 2,4-D does not produce dioxin and is still used to control unwanted vegetation. Many new types of herbicides have been produced since then that are different in structure than the chloroacetic acids, and many are sold for use in both agriculture and for the home gardener.

Petroleum and Petrochemicals

The manufacture of chemicals from petroleum in the nineteenth century was spurred by the increasing need for new fuel sources to replace plant oils and animal tallow. The discovery that coal could be converted into a gas was promising, but mass distribution required large facilities for production and an infrastructure for distribution. A more convenient fuel was needed.

In the mid-nineteenth century a process devised by Scottish chemist James Young (1811–1883) heated certain forms of coal to produce a liquid that, after distillation and purification, proved to be an excellent illuminant. Sold under the trade name kerosene, it became the dominant method of lighting worldwide for the balance of the nineteenth and early part of the twentieth century.

In the United States, kerosene was mass-produced in plants using the Young process. However, the large amounts of industrial waste produced, the need to import special coal, and the required royalty payments led to the search for an alternative.

Crude oil seeps had been found in various parts of the United States, including one in Titusville, Pennsylvania. A New York-based business syndicate thought that it might be possible to drill into the earth, pump this crude oil to the surface, then distill it into a product similar to kerosene. The process worked, and after petroleum was found in other states and other countries, oil eventually became the world's dominant energy source.

German chemists had pioneered the production of petrochemicals from both coal and natural gas (methane) from 1900 to 1930, devising processes to produce phenol, ethylene, ethylene oxide, acetone, vinyl acetate, and vinyl chloride. In the 1920s American chemists began to use the byproducts of the crude-oil refining process to produce many of the same products that German chemists had been able to make from coal.

Many of the products produced from petroleum form the basis of synthetic macromolecules. For example, ethylene can be converted directly into polyethylene, or transformed into styrene to produce polystyrene or vinyl chloride to produce polyvinyl chloride. Oil and natural gas have proved to be some of the most valuable of all organic substances for producing either other organic materials or intermediates. The petrochemical industry became the leading chemical industry in the United States after 1945; it remains an important part of the economy.

Organic Chemistry in the Twentieth Century

While the focus in the nineteenth century was on the synthesis of new molecules and the structure of natural products, research in the twentieth century centered on the reactions of organic molecules. By studying the processes involved in converting a reactant to a product, scientists hoped to follow the same path using a different reactant. A major step in this direction occurred with the development of the Lewis-Langmuir theory of covalent bonding, based upon the concept of the electron pair as the basis of the chemical bond as formulated by American physical chemist Gilbert N. Lewis (1875–1946) and Irving Langmuir (1881–1967).

This began the era of the electronic interpretation of reaction mechanisms. For a new generation of chemists, particularly those in Great Britain such as Robert Robinson (1886–1975) and Christopher Ingold (1893–1970), the Lewis-Langmuir theory became a way to understand the unique structure of molecules such as benzene and how reactions produced certain products and not others.

The tools of physical chemistry, such as thermodynamics and kinetics (the measurement of the rates of reactions) were used to validate electronic interpretation. This hybrid combination of organic and physical chemistry became known as physical organic chemistry, a term coined by the American physical chemist Louis Hammett (1894–1987) for his 1940 pioneering textbook in the new type of study.

Many significant investigations had been performed before 1940 by British and American chemists, including the British investigators Arthur Lapworth (1872–1941) and Kennedy J.P. Orton (1872–1930) and the Americans James Bryant Conant (1893–1978), Howard Lucas (1885–1963), and Frank C. Whitmore (1887–1947) among others. After 1945 American chemists became the preeminent practitioners of physical organic chemistry. Many reactions whose mechanisms had remained a

Professor Derek Barton (1918–1998), organic chemist, in his laboratory at the Imperial College of Science in London. He was jointly awarded the 1969 Nobel Prize for chemistry for his work on the geometry of natural products. In 1950 he showed that organic molecules could be assigned a preferred conformation. *© Hulton-Deutsch Collection/Corbis.*

mystery for decades were revealed by applying the techniques pioneered by individuals in the decades and centuries before.

■ Modern Cultural Connections

In the latter part of the twentieth century organic chemistry focused increasingly on the chemistry of life and biochemical processes. This usually involved developing methods to synthesize complex molecules, especially those with many centers for optical isomerism, and the structural elucidation of natural products found to be active against various conditions such as cancer.

Organic chemistry provides a host of useful products that have made the world more colorful with synthetic dyes, improved public health through pesticides and drugs, and increased the crop yields with fertilizers

and pesticides. Synthetic fibers and engineered materials are also the products of organic chemistry.

SEE ALSO *Chemistry: Biochemistry: The Chemistry of Life; Chemistry: Chemical Bonds; Chemistry: Chemical Reactions and the Conservation of Mass and Energy; Chemistry: Molecular Structure and Stereochemistry; Chemistry: The Practice of Alchemy; Chemistry: States of Matter: Solids, Liquids, Gases, and Plasma.*

BIBLIOGRAPHY

Books

Aftalion, Fred. *A History of the International Chemical Industry.* Translated by Otto Theodor Benfey. Philadelphia: University of Pennsylvania Press, 1991.

Brock, William H. *The Chemical Tree: A History of Chemistry.* New York: W.W. Norton, 2000.

Haber, L.F. *The Chemical Industry During the Nineteenth Century: Study of the Economic Aspect of Applied Chemistry in Europe and North America.* London: Oxford University Press, 1958.

Ihde, Aaron J. *The Development of Modern Chemistry.* New York: Dover, 1984.

Spitz, Peter H. *Petrochemicals: The Rise of an Industry.* New York: John Wiley & Sons, 1988.

Tarbell, Dean S., and Ann Tracy Tarbell, *Essays on the History of Organic Chemistry in the United States, 1875–1955.* Nashville: Folio Publishers, 1986.

Martin Saltzman

Chemistry: States of Matter: Solids, Liquids, Gases, and Plasma

■ Introduction

There are three common states of matter on Earth: solid, liquid, and gas. State defines a physical property of matter. The defining characteristics that determine state include the number and chemical makeup of the molecules and how a particular collection of them is represented by such physical properties as their collective volume and shape, their reaction to temperature, and their reaction to pressure.

The solid state of matter is characterized by a fixed shape that is dependent on the number and characteristics of the specific atoms and molecules that make up the solid. The solid state resists change in shape due to external pressure. To a very small degree relative to the other states, the density (mass/volume) of solids is temperature dependent. Solids can be crystalline, which means they have very orderly structure to the particles that make up the solid, or the solid can be amorphous and have no symmetry in the arrangement of the particles that make up the solid.

In the liquid state, matter will flow to take the shape of any container. The density of liquids is significantly dependent on temperature, but confined liquids resist change due to pressure. The gaseous state also will take the shape of a container but differs from liquids in that it will expand to fill any container so that if the container size is increased, the gas will expand to fill it. Temperature, pressure, and the number of molecules are all important parameters in defining the volume of a gas. Different gases are always completely miscible (mixable), but different liquids are not always miscible.

The term phase is used in certain contexts rather than state to describe the physical properties of matter. Phase is a more restricting term that is commonly used in scientific studies, but it is not often used in general conversation. Where phase differs from state can best be understood in describing liquid mixtures such as oil and water. Each physically and chemically different and separable liquid represents a separate phase. The individual entities, like the mixture, are in the liquid state.

Plasma is called the fourth s ate of matter. It is not common on Earth's surface, but 90% of the matter in space is in the plasma state. Naturally occurring plasma does make up the ionosphere, a layer of low-density charged particles in the upper atmosphere of Earth. The plasma state is similar to the gaseous state, except it does not contain atoms and molecules. Instead, plasmas consist of subatomic particles such as negatively charged electrons and positive ions.

WORDS TO KNOW

AMORPHOUS: A substance that lacks any well-defined structure. In earth science, substances made of randomly-organized atoms, such as glass. Crystals, whose atoms are organized in a definite pattern, are not amorphous; therefore, amorphous substances are sometimes said to be acrystalline (i.e., not-crystalline).

CRYSTALLINE: Having a regular arrangement of atoms or molecules; the normal state of solid matter.

PHASE: The term phase is sometimes used rather than state to describe the physical properties of matter. It differs in that each physically and chemically different material, for example in a mixture of water and oil, may be in a different phase but the same state.

PLASMA: Matter in the form of electrically charged atomic particles that form when a gas becomes so hot that electrons break away from the atoms.

■ Historical Background and Scientific Foundations

The terms solid and liquid are English versions of Latin words for the terms. The term gas has a history. It can be traced back to the Renaissance period (1400–1650) when Jan Baptist van Helmont (1579–1644) gave the name gas to what is now known to be carbon dioxide. Until that time gases were called airs or spirits or vapors. Studies of gases became a significant part of the evolution of modern chemistry. The Ideal Gas Laws that are used by chemists today emerged from a series of studies that followed van Helmont's famous work.

British scientist Robert Boyle (1627–1691) experimented with gases and published his observations in 1660 in a treatise he titled *The Spring of the Air*. Boyle observed that the product of the pressure on a gas times the volume of the gas remains constant at a constant temperature for any trapped gas. This fact is now called Boyle's Law.

A fascination with gases in general and with the recently-discovered, lighter-than-air gas hydrogen in particular led French chemist Jacques Charles (1746–1823) to recognize in 1787 that the volume of a fixed weight of gas is directly proportional to its temperature at constant pressure. Known as Charles' Law today, this is the second of the gas laws.

Another French scientist, Joseph Louis Gay-Lussac (1778–1850), experimented with the reactions of particular gases and found that they always reacted in volume ratios of small whole numbers. He announced his finding in 1808. Gay-Lussac's discovery is now known as the law of combining volumes.

Gay-Lussac's discovery is not one of the Ideal Gas Laws, but it was part of the background that led Italian physicist Amadeo Avagadro (1776–1856) to suggest in 1811 that the volumes of gases are determined by the number of molecules of gas present. Avogadro's hypothesis was also inspired by the ideas promoting the theory of atoms and molecules by British scientist John Dalton (1766–1844) that were published in 1808. Avogadro's major contribution was to recognize that it is the number of molecules, not atoms, that determines the volume of a gas at a given temperature and pressure.

Molecules are made of atoms. The number of atoms in a molecule varies with the nature of the particular molecule. Oxygen has two atoms in a molecule. Water vapor has three, and methane, commonly found in decaying material, has four.

Combining all the individual contributions, the Ideal Gas Laws state that the pressure times the volume of a gas will equal the number of molecules times the temperature with a constant multiplied to relate the various units used in the measurements. This is stated mathematically as: $PV = nRT$, where n is the number of molecules and R is the constant.

Solids and liquids are the condensed states of matter. Physicists conduct studies of condensed matter, particularly at the atomic and molecular level. Condensed Matter Physics is also known as Solid State Physics. Liquids are included in the study of Solid State Physics under the heading of Soft Condensed Matter. Interest in crystalline solids has been documented to prehistoric times in collections of native crystals that have holes and engravings made in them. Ancient Greeks also studied crystals.

The discovery of the structure of crystals was made accidentally by a French monk, René-Just Haüy (1743–1822), when he dropped a fine calcite crystal from a crystal collection of the Treasurer of France. Upon examination of the pieces, Haüy noted the consistent regular and repeating shapes of all the pieces. In 1817, he published the story of this discovery and his later work on crystals in a number of books that included *Traité de Cristallographie* (Treatise on Crystallography). Haüy's work is considered to be the beginning of the important science of crystallography, which has led to major advances in understanding the chemical structure of crystalline solids.

When x rays were discovered, the study of crystals expanded. German physicist Max von Laue (1879–1960) received the Nobel Prize in 1914, and the father-son team of British physicists Sir William Henry Bragg (1862–1942) and William Lawrence Bragg (1890–1971) got the Nobel Prize in 1915 for their contributions to x-ray crystallography. Through advances in these early studies, the study of crystalline molecules has provided structural information on a great many compounds that include proteins and polymers.

There is no similar "Eureka!" moment related to the history of the liquid state of matter. However, the fluidity and solvent properties of liquids have had considerable importance in the advancement of all sciences, as in life science, physics, and chemistry.

The Science

At about the same time Avogadro was presenting his hypothesis and John Dalton was promoting the theory of atoms and molecules, British botanist Robert Brown (1773–1858) was observing through his microscope the random and ceaseless motions of pollen grains suspended in water. This observation in 1827 became known as Brownian motion. Brown did not attempt to explain the motion, but in 1905 German-American physicist Albert Einstein (1879–1955) suggested the motion was the result of random thermal motions of the liquid molecules. Einstein's work and the studies that produced the Ideal Gas Laws led to the Kinetic Molecular Theory, which explains states of matters and changes of state.

Gases are described as composed of randomly moving molecules with perfectly elastic collisions as they bound off the walls of their confining container. In liquids the molecules have equally random motions, but the molecules are tumbling and interacting in close proximity to each other so they flow to fill a container. The viscosity of the liquid depends on the type of molecules and how they interact and also on the temperature of the fluid. Solids have random thermal vibrations, the degree of which depends on the nature of the molecules and the temperature of the solid. The molecules of solids vibrate in place. They do not move about in space like liquid molecules; that is why a solid has a fixed shape.

Changing state involves the addition (or subtraction) of heat to change the degree of motion of the molecules, that is to change the temperature. Starting with a solid that is a pure substance, adding heat makes the molecular vibrations increase. At the melting temperature adding heat no longer makes the solid warmer but instead, the heat causes the bonds between molecules to break. When enough heat has been added to break all the intermolecular bonds, the solid melts and becomes a liquid. The amount of heat needed to melt a pure solid is called the heat of fusion for that solid.

A similar increase in motion and increase in temperature occurs as a liquid is heated until the liquid molecules reach the temperature where the energy absorbed by the molecules has increased their velocity to the point the molecules can escape the weak bonding that kept them close as a liquid. At the point of vaporization, the temperature again does not raise because all the energy added is used to cause molecules to escape

from the surface of the liquid until the liquid is completely converted to a gas. Once a gas, adding heat again increases the temperature. The heat needed to convert a pure liquid to a gas is called the heat of vaporization.

At very high temperatures, such as those in lightening or on the sun, the molecules of the gas are converted to atoms, and the atoms are converted to subatomic particles that are positively and negatively charged. At that point the gas has become plasma.

Influences on Science and Society

The gaseous, liquid, and solid states of matter all have had significant roles in the development of modern chemistry and physics. Starting with Robert Boyle's experiments with gases that led to his publishing *The Spring of the Air* treatise in 1660, research with gases significantly influenced the development of modern theories on chemical elements, atoms, and molecules.

British scientist Joseph Priestley (1733–1804) made a key discovery when he isolated oxygen, a gas he called dephlogisticated air, in keeping with the theory of that time that a mysterious substance called phlogiston was in all substances. The confusion of exactly what Priestley had discovered was solved by French chemist Antoine Laurent Lavoisier (1743–1794) who named the gas oxygen and called it an element, an element being by his definition a unique chemical substance. Priestley did not agree with Lavoisier. The existing concept of elements was very limited, but Lavoisier gained popularity among scientists and the discovery of more elements accelerated.

Dry ice in water looks smoky, but the effect is caused by sublimation, the ability to transform from a solid directly to a gas. *© Steve Klaver/Star Ledger/Corbis.*

Gases also figured in the advancement of the theory that matter is composed of atoms and molecules. Although this theory was not widely accepted at the time Dalton and Avogadro were promoting their ideas, it gained momentum with the growing popularity of Lavoisier's ideas. Unfortunately, his career as a leader in the advancement of modern chemistry was abruptly ended. Lavoisier was not only a chemist but he was also wealthy. His family's aristocratic status caused his death at the guillotine during the French Revolution (1789–1799).

Liquids played a part in the advancement of the concept that matter is made of atoms with Robert Brown's observation in 1827 of pollen grains moving in a liquid, albeit the influence was slow in coming. Even though evidence was gathering in the support of the existence of atoms and molecules, the concepts were not universally accepted until after Einstein suggested that the motion of the liquid molecules caused Brownian motion. When Einstein presented his ideas in 1905, theoretical physics was not a well-organized science. Einstein's studies attracted other key scientists to advance the theory of atoms and molecules, as well as the science of physics.

The solid state has been particularly influential as a result of Einstein's contributions and the advances in physics that have led to the development of atomic energy. The development of the atomic bomb and the applications of nuclear energy are all encompassed in a scientific discipline that is now called Solid State Physics.

■ Modern Cultural Connections

Austrian botanist Friedrich Reinitzer (1857–1927) found a solid, crystalline, cholesterol-based substance that did not melt all at once, as a pure solid is expected to melt. The substance appeared to have two melting points. At 229°F (145°C) the crystals melted to a cloudy liquid. With further heating, to 288°F (178°C), the substance changed to a clear liquid. Reinitzer asked German physicist Otto Lehmann (1855–1922), an expert in crystal optics, if he could help explain the unusual melting of the cholesterol-based crystals. Lehmann determined the cloudy phase represented a new state of matter, which he called liquid crystals. Researchers following the discovery of the cholesterol-based liquid crystals found that other substances, often rod-shaped molecules, also formed liquid crystals.

The curious new state of matter was investigated in the 1960s by French physicist Pierre-Gilles de Gennes (1932–2007). He received the Nobel Prize in 1991 for his contributions to the science of liquid crystals. Today, liquid crystals are used in electronic information displays for everything from cell phones to flat panel displays for computers and TVs. These liquid crystal displays are commonly called LCDs. The development of

these applications for liquid crystals was made possible by de Gennes' work.

Competing with liquid crystal displays for TVs are plasma displays that use xenon gas and neon gas trapped in tiny cells in which the gases are converted to plasma through high voltage applied to electrodes that are attached to the cells. Plasma displays have the advantage of a wider angle view than is available with a liquid crystal display.

Man-made plasma is also used in fluorescent lighting and in neon signs. Scientists are working on plasma reactors that may some day produce energy on a very large scale by using the same fusion process that is used in the sun.

SEE ALSO *Chemistry: Chemical Bonds; Chemistry: Chemical Reactions and the Conservation of Mass and Energy; Physics: Fundamental Forces and the Synthesis of Theory; Physics: Maxwell's Equations, Light and the Electromagnetic Spectrum; Physics: Newtonian Physics.*

BIBLIOGRAPHY

Books

Moore, F.J. *History of Chemistry.* New York: McGraw-Hill Book Company, Inc., 1939.

Moore, Walter J. *Physical Chemistry.* Englewood Cliffs, NJ: Prentice Hall, 1979.

Morris, Richard. *The Last Sorcerers: The Path from Alchemy to the Periodic Table.* Washington, D.C.: Joseph Henry Press, 2003.

Tabor, D. *Gases, Liquids, and Solids.* Baltimore, MD: Penguin Books, Inc. 1969.

Web Sites

Purdue Department of Chemistry. "General Chemistry Help: States of Matter." http://www.chem.purdue.edu/gchelp/atoms/states.html (accessed August 18, 2007).

Chemical Heritage Foundation. http://www.chemheritage.org/ (accessed August 18, 2007).

Department of Physics, Brown University. http://www.physics.brown.edu (accessed August 18, 2007).

Nobel Prize Organization, Physics Prize Information. http://nobelprize.org/nobel_prizes/physics/ (accessed August 18, 2007).

Princeton Plasma Physics Laboratory: Fusion Energy Educational Web Site. http://fusedweb.pppl.gov/FusEdWeb-home.html (accessed August 18, 2007).

Science Week: History of Physics: Einstein and Brownian Motion. http://scienceweek.com/2005/sw050318-1.htm (accessed August 18, 2007).

Miriam C. Nagel

Chemistry: The Periodic Table

■ Introduction

How did scientists come up with the idea for the periodic table? How has it changed as we learn more about atomic structure? And exactly why does it summarize so much useful chemical information?

The ancient Greek philosopher Aristotle (384–322 BC), British natural scientist Robert Boyle (1627–1691), and French chemist Antoine Laurent Lavoisier (1743–1794) all played important roles in this story, but the scientist who discovered the periodic system was Dmitry Ivanovich Mendeleyev (1834–1907). Ironically, he had no idea of the true basis of his table. He thought the chemical characteristics of an element depended on its atomic weight, although we now know that its key concept is atomic number. Yet following the reasoning processes of people like Mendeleyev, even when their result is not exactly the right answer, explains much about scientific progress.

■ Historical Background and Scientific Foundations

The Greek Theory of Four Elements

Modern science has its roots in the way ancient Greek philosophers such as Aristotle thought about the natural world. Earlier civilizations like the Babylonians had made careful astronomical observations; ancient Egyptians developed elaborate metallurgical techniques for refining copper and making alloys of gold. To this practical knowledge, the Greeks applied a curiosity about the underlying causes of natural phenomena.

Aristotle tried to find the fundamental elements of materials, such as solids, liquids, and gases. In *On Generation and Corruption*, written in 350 BC, he expressed his goal this way: "we are looking for the 'originative sources' of perceptible bodies" to understand the origins of their properties and how changes in their appearances take place. He assumed that there would be only a few, although we now know there are over a hundred chemical elements.

Aristotle postulated that there were four elements: earth, air, fire, and water. There were also two pairs of contrary qualities: hot/cold and wet/dry. Each element was associated with two of these fundamental qualities. So, for example, fire was associated with hotness and dryness. Its opposite, water, was cold and wet. Adding water to fire would temper its hot, dry qualities; the result would be something in between. Aristotle emphasized that these four elements were usually not encountered in pure form. For example, both metals and sand were "earths," but a ductile metal such as gold probably was a combination of the pure element earth with the pure element water. Blood was a liquid, an indication of the presence of water, but because of its role in sustaining life, it probably contained an unusual amount of elemental fire.

Aristotle's theory seems terribly simplistic and fanciful today, but it dominated scientific thought for centuries. Nevertheless, his attempt to explain the multiplicity of chemical materials in terms of a limited number of active ingredients still guides research today. The Aristotelian account, however, also allowed for the possibility that one material could transmute (change) into another.

Consider the Aristotelian account of what today we would call the water cycle. When the sun shines on water, the cold, wet water is turned into hot, wet air, or steam. If heat is removed, we once again get water. On a very cold, dry day (rarely encountered in the Mediterranean), the water might be transformed into ice, a material that has the coldness and dryness characteris-

Engraving of a bust of Greek philosopher Democritus (460–370 BC), who believed all matter is made up of various individual elements he called atoma or atomon (from which the English word atom is derived) or "indivisible units." These units could be combined in many ways with other types of units to create different substances. *Getty Images.*

tic of earth. This idea of impressing qualities on matter formed the philosophical basis for alchemy, an early scientific quest to turn base metals, such as lead, into gold.

Alchemical Principles

As the Roman Empire disintegrated, much ancient learning and science was lost. For example, the great library in Alexandria that had been started by one of Aristotle's students was burned on several occasions and eventually destroyed. As the Arab successors of Mohammed (570–632) occupied much of the Roman and Persian empires, interest in natural philosophy shifted eastward and became centered in Islamic cities such as Baghdad. Important Greek texts were translated into Syriac and Arabic. Ancient texts that alluded to the legendary Philosopher's stone and the possibility of transmuting base metals into gold were rediscovered.

Alchemy led to many important chemical discoveries. Early alchemists designed elaborate glassware and learned how to distill liquids. They synthesized hydrochloric,

WORDS TO KNOW

ALCHEMY: The study of the reactions of chemicals in pre-modern times. It was often, but not always, directed by the goal of making gold. In a general sense, alchemy is perceived as the transmutation (or, transformation) of a common substance to something rare and valuable. Medieval alchemists are often portrayed as little more than quacks attempting to make gold from lead. This depiction is not entirely correct. To be sure, there were such characters, but for real alchemists, called adepts, the field was an almost divine mixture of science, mystery, and philosophy.

ATOMIC NUMBER: The number of protons in the nucleus of an atom; the number that appears over the element symbol in the periodic table.

ATOMIC WEIGHT: A quantity indicating atomic mass that tells how much matter there is in something or how dense it is, rather than its weight. Atomic weight is expressed in units known as atomic mass units (amu).

CALCINATION: An old term used to describe the process of heating metals and other materials in air.

CALX: The powdery residue, generally of metal (metallic oxide), formed from the ore and oxygen from the air when a metal-bearing rock or mineral is heated to high temperature.

GENERATIVE ELEMENTS OR PRINCIPLES: The basic, active ingredients from which everything was thought to originate.

PERIODIC LAW: Russian chemist Dmitri Ivanovich Mendeleyev (1834–1907) thought there to be a law of nature that would explain why there were regular repetitions of chemical properties when elements were arranged in order of atomic weight.

PHILOSOPHER'S STONE: A material thought by alchemists to have the power to bring about the transmutation of metals.

PHILOSOPHICAL ELEMENTS OR PRINCIPLES: The underlying essential constituents of bodies that cannot be directly perceived.

QUANTUM NUMBER: A variable in quantum mechanics that specifies the state of electrons in an atom.

TRANSMUTATION: The conversion of one nuclide into another, as by neutron bombardment in a nuclear reactor. Also, in alchemy, the transformation of a base metal into gold.

VALENCE: The tendency of an atom to gain or lose electrons in reacting with other atoms.

sulfuric, and nitric acids and used them to dissolve metals and to react with soda, potash, and lime.

Early alchemists admired Aristotle and continued to work within his framework of four elements. But the Aristotelian qualities of hot & cold, wet & dry seemed far removed from the properties of the new chemicals they discovered. So they postulated three additional fundamental principles: sulfur, salt, and mercury.

Like Aristotle's elements, these, they believed, were never encountered in pure form. "Philosophical sulfur" was not the yellow powder used to fumigate unhealthy rooms, but it was responsible for the combustibility of materials. "Salt" was the principle that underlay the reactivity of acids (when acids are neutralized they form ordinary salts) and "philosophical mercury" accounted for metals' malleability and shiny surfaces.

Not surprisingly, ordinary mercury was a popular starting material in attempts to make gold. It is easy to laugh at stories of alchemists mixing up mercury, lead, and some bright yellow material such as saffron, burying it in manure, and then warming the whole ensemble in an oven for weeks. Yet there was a method behind this apparent madness.

It was widely believed that metals "grew" in mines, and gold nuggets were often found embedded in ore containing copper. Perhaps, they reasoned, copper or other metals were just immature forms of gold. If they could only speed up the growth process, alchemists reasoned, maybe they could duplicate in the laboratory what nature did more slowly in the bowels of the earth. Manure appeared to be a likely catalyst. After all, little insects and maggots were formed in manure, they believed, which gave it strong generative powers.

As Muslim conquerors moved across North Africa and entered Spain in AD 711, they brought scientific writings in Arabic, both translated texts of the ancient Greeks and innovations of their Arab successors. In the twelfth and thirteenth centuries, many of these writings were translated into Latin, often by Jewish scholars. The influx of these preserved and rediscovered Greek texts, along with Arabic contributions to natural philosophy, accelerated the rebirth of learning in European universities. The German-Swiss physician and alchemist Paracelsus (1493–1541) was particularly taken with the three new alchemical principles and promoted them as part of the new doctrines that would eventually supplant the old ancient Greek dogmas.

Paracelsus, sometimes called "the Luther of physicians," was quite a radical. He once burned copies of Aristotle's and the Greek physician Galen's (129–c.216) books in a public bonfire at the University of Basel to show how dangerously outdated their theories were. Paracelsus introduced the concept of chemical medicines and recommended a mercury compound for the treatment of syphilis. (Luckily only small portions of this poison were prescribed!)

Georg Ernst Stahl (1660–1734) was a German chemist and physician who developed the phlogiston theory of combustion. Phlogiston was thought to be a combustible substance found in flammable objects. Stahl thought that air was needed to absorb the phlogiston, as combustion only occurs in air. While the theory explained many of the observations associated with combustion, it was overturned by Antoine Laurent Lavoisier (1743–1794) in the 1770s. *SPL/Photo Researchers, Inc.*

The Phlogiston Theory

Nearly 200 years after Paracelsus, German chemist and physician Georg Ernst Stahl (1660–1734) modified the three-principle theory and used it to systematize a whole series of important reactions. He renamed the sulfurous principle "phlogiston" (from the Greek *phlogistos* for "flammable") and claimed that it was contained in every combustible material, and forcibly expelled in all combustion reactions producing heat and light. To explain why combustion won't occur unless there is lots of air around, Stahl claimed that it was necessary to carry away the phlogiston.

We now know that Stahl had it just backward—when materials such as charcoal or sulfur burn, they are not decomposing. Instead, something from the air is added. The air is a passive receptacle for something that is given off; it supplies oxygen, which combines with whatever is burning. The heat and light given off is energy produced by the rapid oxidation reaction.

Yet Stahl successfully used the phlogiston theory to systematize some important chemical reactions—a good example of how scientists can make progress even when their starting assumptions are wrong. Stahl's first insight was to note the parallels between what happens when metals are heated in air and turned into a powder (a process he called "calcination") and ordinary combustion, such as when wood burns. Stahl noted that in both cases heat and light might be given off (perhaps you may have seen the sparks when iron filings are sprinkled into a flame) and that air was required to keep the reaction going. (He would have said phlogiston was escaping while we know that oxygen is being combined, but we also recognize the parallel between the combustion of metals and nonmetals. Both are oxidation reactions.)

Then Stahl used his system to describe what happened when metallic ores were reduced to metals. He noted that copper miners often mixed the minerals they took from the ground with charcoal and heated them while stirring the molten results with a green stick. Stahl reasoned that the metallic ores had lost their phlogiston. They were powdery just like the metal calxes that were the products of calcination. To turn them into shiny metals, one needed to add phlogiston. But he was convinced that charcoal was a rich source of phlogiston because it burned so well. So although they didn't realize it, the miners were transferring phlogiston from the charcoal into the metal ore and thus producing metals. Since the miners thought of the charcoal as only a fuel, not as an ingredient in a chemical reaction, they sometimes didn't add enough charcoal to reduce all of the metal present in the ore. So Stahl explained to them the necessity of doing so.

Stahl spoke of the reactions above as analyses and syntheses. In calcination, a metal was analyzed (separated) into a metallic powder and phlogiston. What the miners were doing was synthesizing the metal by combining the metallic ore with phlogiston that came from the charcoal. One of Stahl's students proposed the axiom that every analysis should be followed by a synthesis if one wanted to prove that one understood the composition of a material. Stahl certainly felt that he had proved that metals were compounds, not elements. Lavoisier was soon to use the same methods of analysis and synthesis to prove him wrong!

Lavoisier's Scientific Criterion for Chemical Elements

Aristotle thought of elements as "originative principles." Nearly 2,000 years later in *The Sceptical Chymist* (1661), the Irish chemist Robert Boyle defined elements as "certain primitive and simple, or perfectly unmingled bodies; which not being made of any other bodies, or of one another, are the ingredients of which all those called perfectly mixt bodies are immediately compounded, and into which they are ultimately resolved."

But how could chemists determine exactly what these primitive and simple materials were? Aristotle had proposed four elements: earth, air, fire, and water; the alchemists added salt, sulfur, and mercury. Yet none of these so-called elements could be put in a bottle or used in experiments. (Remember that the alchemists' third element was "philosophical mercury," not the quicksilver that was used in barometers.)

It was the French chemist and public servant Antoine Laurent Lavoisier (1743–1794) who finally provided a workable definition of a chemical element in 1789. Since it was impossible to reach agreement about the invisible constituents of bodies, Lavoisier proposed in *Traité élémentaire de chimie* (*Elements of Chemistry;* 1789) that: "we apply the term elements, or principles of bodies, to express our idea of the last point which analysis is capable of reaching, we must admit, as elements, all the substances into which we are capable, by any means, to reduce bodies by decomposition."

Lavoisier recognized that the list of elements might change as chemists figured out how to further decompose complex materials. Thus he tentatively included lime and magnesia on his list, but noted that they might not be simple substances. (We now know they were oxides of calcium and magnesium.) Still another step was required before chemists could unambiguously agree on a list of elements. Stahl was dead by the time Lavoisier proposed his definition, but had he been alive, he might have argued that metals were not elements—they contained phlogiston, which was then freed upon heating. So phlogistonists considered a metal calx to be a simpler substance than a metal.

Lavoisier provided a direct way to show that metals were indeed elements. He weighed the sample of metal, heated it in air, and then weighed the powdery product. Since the calx was heavier, something must have been added in calcination. Through more elaborate experiments Lavoisier isolated the gas from the atmosphere that was absorbed during both calcination and combustion and called it oxygen, from the Greek *oxys,* which means acidic, or sharp.

Chemists rapidly accepted Lavoisier's definition of an element and adopted the helpful strategy of tracking the weights of products and ingredients. When Lavoisier published his *Elements of Chemistry*, there were 33 elements on his list, including the common metals, sulfur, arsenic, phosphorus, oxygen, and nitrogen. Soon hydrogen and chlorine were discovered by means of electrolysis; by the time Mendeleyev began his textbook, there were over 60, each with a set of distinctive chemical properties.

Assigning Atomic Weights

Throughout history, natural philosophers and chemists speculated about the existence of atoms. Because both Aristotle and (much later) French mathematician

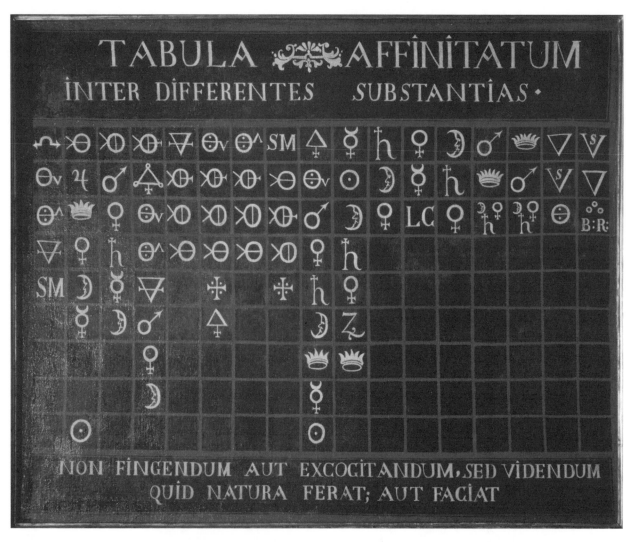

Tabula Affinitatum, a table of the affinities (the elements' ability to react and displace other elements in a compound) between the different substances, of Pietro Leopoldo (1747–1792), archduke of Tuscany. Elaborated by Torbern Olof Bergman (1735–1784), Swedish chemist, who developed a mineral classification scheme based on chemical characteristics. Eighteenth century. *SEF/Art Resource, NY.*

René Descartes (1596–1650) thought that a vacuum was physically impossible, they both rejected Democritus's (c.460–370 BC) theory of atoms in a void. However, virtually everyone was willing to entertain the idea that matter might be composed of tiny, invisible "corpuscles." Yet as was the case with elements before Lavoisier, it seemed impossible to find out anything specific about these particles. It was John Dalton (1766–1844), a British meteorologist and schoolteacher, who came up with the crucial inferences that made modern atomic theory viable.

In a book called *A New System of Chemical Philosophy* (1808), Dalton noted that it was well established that when elements combined to make a new compound, they almost always did so in the same proportion by weight. (Alloys were the exception.) So it seemed rea-

sonable to assume that the compound always contains the same ratio of atoms. Dalton put it this way:

> Therefore we may conclude that the ultimate particles of all homogeneous bodies are perfectly alike in weight, figure, &c. In other words, every particle of water is like every other particle of water; every particle of hydrogen is like every other particle of hydrogen, &c.

Knowing the formula for a compound, Dalton posited, would allow him to calculate the relative weights of the atoms. But how could he determine the formula? To solve this problem Dalton took the bold step of postulating rules for making conjectures about the formulae for compounds:

John Dalton's (1766–1844) table of elements of 1803 illustrating his atomic theory. The atoms of different elements were represented by symbols and their relative atomic weights compared to hydrogen by the numbers shown. © *Bettmann/Corbis*.

The following general rules may be adopted as guides in all our investigations respecting chemical synthesis.

1st. When only one combination of two bodies can be obtained, it must be presumed to be a binary one, unless some other cause appear to the contrary.

2d. When two combinations are observed, they must be presumed to be a binary and a ternary.

For example, if carbon and oxygen form only one compound, Dalton assumed it contains only two atoms. Hence its formula should be CO. So if a sample of carbon monoxide (using our name for it) consists of 3 grams of carbon combined with 4 grams of oxygen, we presume the ratio of their atomic weights is 3 to 4. If, however, carbon and oxygen form two different compounds (which is actually the case), then one of the combinations is CO, while the other consists of three atoms. Hence its formula should be either CO_2 or C_2O. If it analyzes into 3 grams of carbon for every 8 grams of oxygen, then we know the correct formula is what we now call carbon dioxide.

This example illustrates the way Dalton reasoned, but it makes the job of coming up with a comprehensive, consistent system of relative atomic weights look much easier than it actually was. First of all, the experimental data on combining weights were not very accurate. Secondly, Dalton got off on the wrong foot because since only one compound of hydrogen and oxygen was known at the time, his rules led him to assign water the formula HO instead of H_2O, and when he set the atomic weight of hydrogen as 1, oxygen came out to be half of its correct value!

Nearly 50 years later, at an 1860 conference in Karlsruhe, Germany, Italian chemist Stanislao Cannizzaro (1826–1910) finally presented a method for determining atomic weights that all chemists found acceptable. Attending that conference was a young Russian chemist studying in Germany. His name was Dimitri Ivanovich Mendeleyev (1834–1907).

Mendeleyev's Periodic Table and Periodic Law

Mendeleyev was born in a small Siberian town 1,300 miles (2,092 km) from Moscow in 1834. After his father died, Mendeleyev and his mother hitchhiked to the capital city so he could go to the University of Moscow. Unable to get a scholarship, they traveled another 400 miles (644 km) east to St. Petersburg. Soon after graduating first in his class, Mendeleyev went for postgraduate studies at Heidelberg University in Germany.

Because Germany at that time was home to many famous chemists, while Russia was something of a scientific backwater, when Mendeleyev returned to Russia to take up an academic position at the University of St. Petersburg, one of the first things he did was to write a 500-page chemistry textbook. It was a huge success both financially and intellectually, and Mendeleyev began to write another to cover the chemistry of all the elements.

He began the manuscript intending to write a chapter that discussed the chemical properties of each element. But it soon became apparent that he couldn't do so for each of the 60-odd elements known at the time. Furthermore, he needed some rational way to order the discussions. At the Karlsruhe Conference, Mendeleyev had become convinced that a key property of an element was its atomic weight. So he made a note card for each element and started arranging them on his desk in order of atomic weight, then grouped those that had similar chemical properties. (Legend has it that Mendeleyev liked to play the card game solitaire, in which cards end up in rows organized by both suit and numerical order, and that this experience helped him make his discovery.)

Others before Mendeleyev had noticed that some elements could be grouped into families. The German chemist Johann Wolfgang Döbereiner (1780–1849) had even pointed out that in triads such as lithium (7),

Dimitri Ivanovich Mendeleyev (1834–1907), a Russian chemist and professor in St. Petersburg, Russia, from 1867–1890. He devised the Periodic Table for classifying elements by their atomic weight. *Getty Images.*

sodium (23), and potassium (39), the atomic weight of the middle member was about halfway between the atomic weights of the extremes. In a similar vein, British chemist John Newlands (1837–1898) proposed the law of octaves because he noted that, when arranged by atomic weight, the elements' chemical properties repeated themselves in periods of eight—just like the notes in a musical scale. Few paid much attention to this pattern, however, in part because it broke down at elements of higher atomic weight.

Mendeleyev was convinced that there was a periodic law that governed the properties of elements. Because mass played a central role in Isaac Newton's (1642–1727) laws, Mendeleyev believed that atomic weight determined each element's chemical properties, and that it was no coincidence that there were periodic similarities among them. (We now know that indeed it is not a coincidence—each row on the periodic table represents a filled shell of electrons; in Mendeleyev's time, however, no one knew anything about the atomic structure.) He was so convinced that a natural law underlay his table that he made several bold predictions based on it.

First, he predicted the existence of three new elements based on gaps in his table. Mendeleyev was convinced that there would be two new elements between zinc and arsenic because the properties of arsenic showed that it should be in the phosphorus family. Furthermore the jump in atomic weight was unusually large. Using the known properties of the missing elements' neighbors, Mendeleyev predicted not only their atomic weights, but their densities and melting points as well. When gallium was discovered a few years later, its actual properties were extremely close to Mendeleyev's predictions. When other gap elements were discovered in short order, chemists all over the world were convinced that the periodic table reflected a deep truth about the world.

Mendeleyev also used his table to predict that some current experimental values for atomic weight, such as the 128 initially assigned to tellurium (Te), were incorrect. If he had arranged the elements in that row strictly according to increasing atomic weight, iodine (I) with atomic weight 127 would have come before tellurium (128). But Mendeleyev knew that because of its chemical properties, iodine belonged in the same horizontal row with fluorine, chlorine, and bromine, while tellurium was more like its neighbor to the left, selenium. So he placed these elements where they belonged, based on their chemistry, and challenged chemists to redetermine the values of atomic weights.

In many cases, Mendeleyev's prediction that an atomic weight was a little off was correct. But the so-called reversed pair of tellurium and iodine remained an anomaly. Look up today's values for the atomic weights of tellurium (127.6) and iodine (126.904), remembering that we now put families such as fluorine, chlorine, bromine, and iodine in vertical columns while Mendeleyev put them in horizontal rows.

Chemists continued to find new elements. Some filled gaps in Mendeleyev's table but others didn't seem to fit at all. Helium was first discovered through spectrographic analysis of sunlight and later isolated from uranium ore. (We now know that uranium emits alpha particles, which are helium nuclei.) At first Mendeleyev didn't think there was a place for helium in his table, but after other inert gases such as argon and neon were found in the atmosphere, he simply inserted a whole new family into the table. These gases formed a useful buffer between the highly reactive alkali metals such as lithium and sodium and the halogen gases such as fluorine and chlorine gases; because they do not form compounds he assigned them a zero valence. But Mendeleyev was never able to figure out what to do with the so-called rare-earth metals, now called lanthanides.

According to their atomic weights, these metals should all come immediately after barium. But the problem was that each of these 14 elements have the same valence (+3) and very similar chemical properties.

This violated the principle of periodicity on which Mendeleyev's table was based. He eventually concluded that the rare-earth metals "broke" the periodic law.

On the modern periodic table, both the lanthanides and actinides are displayed somewhat like a postscript below the table—even though according to atomic number they should appear above! Mendeleyev's contemporaries had no idea of how the rare earths could fit into the pattern represented by the periodic table. We now know that the quantum number theory predicts their existence, and they do fit smoothly into the modern theory of the elements that is based on atomic number.

By the end of the nineteenth century, the very concept of chemical element that had been developed by Boyle, Lavoisier, and Dalton and that underlay Mendeleyev's system was to undergo profound changes. With the discovery of radioactivity, scientists learned that atoms were not the smallest pieces of matter. And the discovery of isotopes proved that not all the atoms of a given element were alike. Yet despite these revolutionary changes, the basic shape of the periodic table remained intact. Let us see how this was possible.

From Atomic Weight to Atomic Number

Mendeleyev first learned about cathode rays, x rays, and radioactivity a few years before his death in 1907. However, he was not very receptive to the news, in part because reports of glowing fluorescent screens caused by invisible rays reminded him of the false claims spiritualists at that time made about occult phenomena. Mendeleyev had, in fact, debunked the practices of several Russian mediums, who claimed that they could communicate with the dead, uncovering their deceptive and unscrupulous methods in the process.

Conclusive evidence soon emerged, however, that atoms consisted of positive particles and negative electrons. The New Zealand-born British physicist Ernest Rutherford (1871–1937) further showed in 1911 that the positive particles were concentrated in a nucleus. In 1913 his British student Frederick Soddy (1877–1956) found that not all atoms of a pure element were identical. For example, although a material then called mesothorium (because it was found in thorium ores), had an atomic weight of 228, its chemical properties were identical to those of radium, whose atomic weight was 226. (We now know that thorium undergoes radioactive decay to produce radium.) Soddy called atoms that were chemically identical but had different atomic weights "isotopes." Using the mass spectrograph, scientists were able to separate all sorts of elements into their constituent isotopes.

These discoveries forced chemists to revise their definition of chemical elements. Boyle had stated that elements were the simplest substances out of which all materials were made, but it was now clear that the fundamental building blocks of matter were subatomic particles such as protons and electrons. Lavoisier had

			K = 39	Rb = 85	Cs = 133	—	—
			Ca = 40	Sr = 87	Ba = 137	—	—
			—	?Yt = 88?	?Di = 138?	Er = 178?	—
			Ti = 48?	Zr = 90	Ce = 140?	?La = 180?	Th = 231
			V = 51	Nb = 94	—	Ta = 182	—
			Cr = 52	Mo = 96	—	W = 184	U = 240
			Mn = 55	—	—	—	—
			Fe = 56	Ru = 104	—	Os = 195?	—
			Co = 59	Rh = 104	—	Ir = 197	—
			Ni = 59	Pd = 106	—	Pt = 198?	—
H = 1	Li = 7	Na = 23	Cu = 63	Ag = 108	—	Au = 199?	—
	Be = 9,4	Mg = 24	Zn = 65	Cd = 112	—	Hg = 200	—
	B = 11	Al = 27,3	—	In = 113	—	Tl = 204	—
	C = 12	Si = 28	—	Sn = 118	—	Pb = 207	—
	N = 14	P = 31	As = 75	Sb = 122	—	Bi = 208	—
	O = 16	S = 32	Se = 78	Te = 125?	—	—	—
	F = 19	Cl = 35,5	Br = 80	J = 127	—	—	—

Typische Elemente

Russian chemist Dmitri Ivanovich Mendeleyev's (1834–1907) periodic table of 1869. This is an early form of the periodic table, with similar chemical elements arranged horizontally. Chemical symbols and the atomic weights are used. The final version (1871) had the elements arranged in the familiar vertical columns or groups. *SPL/Photo Researchers, Inc.*

British physicist Francis Aston (1877–1945) demonstrated that all of an element's atoms did not have the same atomic weight, thus leading to a change in defining the identity of an element based on its atomic number instead of its atomic weight. © *Library of Congress.*

defined elements as the last products of analysis, but scientists could now divide samples of what had been thought to be homogeneous materials into isotopes.

In 1920, in fact, British physicist Francis Aston (1877–1945) reported that gaseous neon consisted of Ne-20, Ne-21, and Ne-22 while chlorine consisted mostly of Cl-35 and Cl-37 with traces of Cl-36 and Cl-38. John Dalton would have been shocked to learn that all of an element's atoms do NOT have the same atomic weight. And if elements' identities did not depend on their atomic weights, then order by atomic weight could hardly provide the foundation for Mendeleyev's periodic table.

The solution, which may seem obvious today, is to define chemical elements by their atomic number. Chemical properties depend on the nucleus's positive charge and the corresponding number of electrons around it. Atoms of the same atomic number have different atomic weights if they have a different number of neutrons in their nuclei. Mendeleyev's periodic table worked as well as it did because there is a very close correspondence between order by atomic weight and order

by atomic number. Sometimes the order is not the same (remember the reversed order of tellurium and iodine) but because nuclei that have many more neutrons than protons are unstable, Mendeleyev's idea of ordering elements by weight turned out to be a very close approximation of the true ordering principle: atomic number.

Some chemists balked at accepting this new definition, arguing that each isotope should be considered to be a separate element. After all, a uranium-235 atom is much more highly radioactive than a uranium-238 atom. There are also chemically significant differences between ordinary hydrogen H-1 and deuterium H-2. In fact deuterium oxide (so-called heavy water) is poisonous to animals. Following this logic, atomic weight would have remained the defining property.

The chemical community, however, wisely decided to use the atomic number. Can you imagine how complicated the periodic table would be if each individual isotope got its own box? Yet even with all this new information about atoms, one major puzzling feature of the periodic table remained: its periodicity.

Why do chemical properties repeat after every eight, or in some cases 18, elements? Why does the numerically ordered string of elements wrap so that if sodium (Na) is placed under lithium, where it obviously belongs, then chlorine automatically takes its place under fluorine? Why are some periods longer than others? And what about those pesky rare earths, where the valence remained +3 even as the atomic number kept increasing? The answers were provided in the 1920s by quantum theory.

Quantum Numbers and the Periodic Table

When heated, each element gives off light with a characteristic spectrum. (If you sprinkle ordinary table salt into a blue flame you will see the characteristic reddish-orange color of sodium.) When this emitted light is analyzed with a spectroscope, the spectrum breaks into sharp lines. Because sunlight contained lines that did not correspond to any element known on Earth, spectroscopers realized that the sun contained a previously unknown element, which they named helium, from the Greek *helios*, for sun.

But why would only certain specific wavelengths be emitted? This and many other questions led physicists such as Niels Bohr (1885–1962) to conclude that the electrons surrounding the nucleus occupied orbits at certain discrete distances. When an individual atom absorbed energy, say from a flame, an electron in one of those shells jumped up to a new orbit that was farther away from the nucleus. When it eventually fell, it produced light of a single wavelength, determined by the energy difference between the two orbits. In the case of a very simple atom such as hydrogen, Bohr was even able to calculate from his theory what the spectrum of excited hydrogen should be.

According to quantum theory, the orbits available to any electron in an atom are described by integers called quantum numbers. In an unexcited atom (one that is not being bombarded by an outside energy source), the electrons gravitate to the orbits of lowest energy. But only two electrons can occupy a given orbit. If we apply the principles of quantum mechanics to the periodic table, we find that there can only be two elements in the first row because there is only space for two electrons in the first shell.

Quantum mechanics, however, tells us that there can be eight electrons in the second shell because it contains four available orbits—and we indeed find eight elements in the second row of the periodic table. Lithium (atomic number 3) has two inner electrons and one electron in its outer shell. In chemical reactions it may lose that electron and become a Li^{+1} ion. Fluorine (atomic number 9) has two inner electrons and seven outer electrons. Because a full outer shell is an especially stable arrangement (that is why the rare gases are inert), in reactions fluorine atoms often gain an electron and become negative F^{-1} ions.

As we move to higher numbers of shells, describing the number of orbitals and their relative energies becomes more complicated, but the existence of horizontal families of elements (like the lanthanides and actinides) is predicted from the theory. All lanthanides have three electrons in their outermost shell (hence the +3 valence) but they differ in the number of electrons in the shell just beneath. Hence as the atomic number increases there is little variation in the properties of the outside electrons, which are those that typically enter into chemical bonding.

Modern Cultural Connections

Look on the walls of chemistry classrooms or laboratories in any country of the world and you will see a table of roughly 100 elements arranged to look something like a bed. Boxes containing the letters H (hydrogen) and He (helium) are stacked like bedposts on top of columns labeled as I and VIII, respectively. The atomic number of each element tells how many protons are in the nucleus; the Roman numeral of the column is an indication of the valence (or combining power) of the element. Since lithium is in column I (under hydrogen) but beryllium is in column II, then we expect the formula of lithium chloride to be LiCl while beryllium chloride is BeCl2. We also expect family resemblances between elements in the same column. For example, in column II a couple of squares under Be we find Ca for calcium and, right under it, Sr for strontium. Knowing that calcium is an important component of bones, we aren't surprised to hear that its neighbor strontium is also found in the skeleton. This family relationship explains why during the period of atomic bomb testing, scientists were very concerned that a radioactive isotope of strontium (called Sr-90) might end up in milk just like calcium and be absorbed into the bones of growing children.

Primary Source Connection

A World made by Atomes

Small Atomes of themselves a World may make,

As being subtle, and of every shape:

And as they dance about, fit places finde,

Such Formes as best agree, make every kinde.

For when we build a house of Bricke, and Stone, [5]

We lay them even, every one by one:

And when we finde a gap that's big, or small,

We seeke out Stones, to fit that place withall.

For when not fit, too big, or little be,

They fall away, and cannot stay we see. [10]

So Atomes, as they dance, finde places fit,

They there remaine, lye close, and fast will sticke.

Those that unfit, the rest that rove about,

Do never leave, untill they thrust them out.

Thus by their severall Motions, and their Formes, [15]

As severall work-men serve each others turnes.

And thus, by chance, may a New World create:

Or else predestined to worke my Fate.

Margaret Cavendish

CAVENDISH, MARGARET. *THE ATOMIC POEMS OF MARGARET (LUCAS) CAVENDISH, DUCHESS OF NEWCASTLE, FROM HER POEMS, AND FANCIES, 1653.* TRANSCRIBED AND EDITED BY LEIGH TILLMAN PARTINGTON. ATLANTA: WOMEN WRITERS RESOURCE PROJECT, EMORY UNIVERSITY, 1996.

SEE ALSO *Chemistry: Chemical Bonds; Chemistry: Chemical Reactions and the Conservation of Mass and Energy; Chemistry: Molecular Structure and Stereochemistry; Chemistry: States of Matter: Solids, Liquids, Gases, and Plasma; Chemistry: The Practice of Alchemy.*

BIBLIOGRAPHY

Books

Ball, Philip. *The Elements: A Very Short Introduction.* Oxford and New York: Oxford University Press, 2004.

———. *The Ingredients: A Guided Tour of the Elements.* Oxford and New York: Oxford University Press, 2002.

Boyle, Robert. *The sceptical chymist; or, Chymico-physical doubts & paradoxes, touching the experiments whereby vulgar spagirists are wont to endeavour to evince their salt, sulphur and mercury, to be the true principles of things. To which in this edition are subjoyn'd divers experiments and notes about the producibleness of chymical principles.* Oxford: 1680.

Brock, William H. *The Norton History of Chemistry.* New York and London: W.W. Norton & Co., 1992.

Cavendish, Margaret. *The Atomic Poems of Margaret (Lucas) Cavendish, Duchess of Newcastle, from her Poems, and Fancies, 1653.* Transcribed and edited by Leigh Tillman Partington. Atlanta: Women Writers Resource Project, Emory University, 1996.

Dalton, John. *A New System of Chemical Philosophy.* New York: Philosophical Library, 1964.

Gordin, Michael D. *A Well-Ordered Thing: Dmitrii Mendeleev and the Shadow of the Periodic Table.* New York: Basic Books, 2004.

Lavoisier, Antoine Laurent. *Elements of Chemistry.* New York: Dover Publications, 1965.

Mendeleyev, Dimitri. *Principles of Chemistry by D. Mendeléeff.* Translated from the Russian by George Kamensky. 6th ed. Edited by T.A. Lawson. London, New York: Longmans, Green, and Co., 1897.

Morris, Richard. *The Last Sorcerers: The Path from Alchemy to the Periodic Table.* Washington, DC: Joseph Henry Press, 2003.

Strathern, Paul. *Mendeleyev's Dream: The Quest for the Elements.* New York: St. Martin's Press, 2001.

Web Sites

AUS-e-TUTE. "History of the Periodic Table of the Elements." http://www.ausetute.com.au/pthistor.html (accessed June 8, 2006).

Chemical Heritage Foundation. "Chemical Achievers: The Human Face of the Chemical Sciences." http://www.chemheritage.org/classroom/chemach/index.html/ (accessed June 8, 2006).

Royal Society of Chemistry. "History of the Periodic Table." http://www.chemsoc.org/networks/learnnet/periodictable/pre16/develop/mendeleev.htm (accessed June 8, 2006).

Noretta Koertge

Chemistry: The Practice of Alchemy

■ Introduction

Alchemy was a form of early chemistry that aimed to transmute base metals into gold and to create an elixir of longevity or eternal life. Alchemy emerged in China in the fourth century BC and developed over centuries, laying a foundation for the modern science of chemistry. Social conventions, economics, and royal patronage affected alchemy's long history and development.

■ Historical Background and Scientific Foundations

Chinese Alchemy

The earliest record of alchemy is from China in the fourth century BC, and its practice was associated with the religion and philosophy of Taoism created by the sage Lao Tzu. The tao is a form of prime matter that is infinitely changeable, and the matter that the tao composes is described in terms of opposites. Yin is the passive female element that is cool and dark, and yang is the hot and light male element. The interaction and opposition between yin and yang was thought to produce the elements. Earth was the central element, while others were thought to be opposing pairs, such as fire and water, metal and wood. Taoist alchemists thought that by changing the proportions of these elements it was possible to transform matter.

Metals in particular were considered a product of chi (the life force that governs the body). Deep within the earth, chi was thought to create and incubate metals until they reached maturity. The most mature and perfect metal was gold. A form of drinkable or potable gold was considered by Chinese alchemists to be an elixir of immortal life, containing infinite chi and life-energy.

In their work, Chinese alchemists discovered processes such as metallic jewelry making, metallurgy, the development of furnaces, the making of alloys (including mosaic gold), and the development of gunpowder. Some of these ideas or processes may have reached the west via India, Greece, or the Arab world as early as the fourth century BC via trade or the cultural interactions in warfare. Alexander the Great (356–323 BC) was pushing eastward from Greece in his conquests at about the same time that Chinese alchemy began to flourish.

By 144 BC, Chinese alchemy was only openly practiced under royal patronage. Historical accounts note that in 135 BC alchemy was performed at the imperial Chinese courts by Li Shao-chün (c. second century BC). He had the reputation of being able to predict future events, transforming cinnabar powder (mercuric sulfide) into the golden elixir of life, and creating life-extending golden cups and bowls for drinking and eating. Li Shao-chün also advised the emperor to worship the Stove, a demi-goddess clothed in red garments with her head done up in a knot on the top of her head. She was the divinity responsible for cooking and brewing, as well as alchemy.

Although Chinese alchemy had as it primary goal the extension of life, the production of gold to create wealth was also of interest. In 56 BC, Liu Hsiang (77–6 BC), a courtier and scholar, attempted to make alchemical gold with the express patronage of the emperor, the resources of the Chinese imperial treasury, and the "recipes of immortality" provided by legendary predecessors. His utter failure may have lead to the enactment of laws forbidding alchemical practice in the Han dynasty from AD 9 until the third century and the eventual decline of alchemy in China.

WORDS TO KNOW

ALCHEMY: The study of the reactions of chemicals in pre-modern times. It was often, but not always, directed by the goal of making gold. In a general sense, alchemy is perceived as the transmutation (or, transformation) of a common substance to something rare and valuable. Medieval alchemists are often portrayed as little more than quacks attempting to make gold from lead. This depiction is not entirely correct. To be sure, there were such characters, but for real alchemists, called adepts, the field was an almost divine mixture of science, mystery, and philosophy.

CHI: Chi or qi (pronounced "chee") is the life force supposed by traditional Chinese medicine and philosophy. There is no scientific evidence for the existence of chi.

CHYMICAL: An obsolete spelling of "chemical," of or having to do with chemistry.

CONDENSATION: The process by which vapor molecules reform a liquid; a phase change from the gas state to the liquid state.

IATROCHEMISTRY: A school of European medical-scientific thought that flourished mostly in the 1500s. It was descended from alchemy and was, in some ways, a precursor of modern medical science: iatrochemists taught that medicines should be based on alchemy (chemistry as it was then understood).

NOSOLOGY: Nosology is the scientific classification of diseases.

RAREFACTION: A decrease in density. Also, a region of space with a lower-than-normal density. Particularly, a region of low density in a sound wave traveling through a gas such as air.

SUBLIMATION: Transformation of a solid to the gaseous state without passing through the liquid state.

TRANSMUTATION: The conversion of one nuclide into another, as by neutron bombardment in a nuclear reactor. Also, in alchemy, the transformation of a base metal into gold.

The Ancient Greeks

While the Chinese alchemists were searching for the elixir of life and developing practical applications, the ancient Greek philosophers were searching for the elemental principles of matter. Their initial inquiries took place at Miletus on the coast of Ionia in the sixth century BC. Thales of Miletus (c.624–546 BC) taught that "all things are water," as water could be in the form of solids (ice), liquid, and gas (steam). Thales also thought that Earth was a flat disc floating on an infinite ocean.

Earth was formed out of the oceans by having dirt piled up on top of it, similar to the silting he had observed in Egypt at the delta of the Nile River.

His colleague Anaximenes (c.585–c.528 BC) had similar beliefs, thinking all things came from air, and using examples of condensation and rarefaction (a decrease in air density and pressure as a soundwave passes) to explain material changes of state. We also think that Thales was the teacher of Anaximander (c.610 BC–c.546 BC). Anaximander also thought all of creation could be derived from a natural substance, not an element like water or air, but something called the apeiron or the boundless, a sort of a primal stuff. Anaximander stated that the apeiron was "the source of things that are—it is that from in which the coming to be of things takes place, and to which it returns when they perish."

Though some ancient Greek philosophers asserted the world was made of a homogeneous substance, like water, or air, or the apieron, how was it possible then to account for diversity and change in the natural world? From the sixth to the third centuries BC, the main preoccupation of philosophers in the Greek world was to explain change, but reconcile it with the idea that the universe also possesses constant eternal qualities. This debate and analysis of what is changing and what is constant led to the first ideas of elements and atoms.

In his attempts to explain change, the philosopher Empedocles (c.495 BC–c.435 BC) dismissed the idea of the apeiron. He stated instead that there are four elements (earth, air, water, and fire) that compose everything, and that their interactions cause change, just as all colors can be created by mixing three primary colors in appropriate proportions.

Other philosophers such as Leucippus (c. fifth century BC) and his pupil Democritus (460–370 BC) also believed that matter was made of component parts, but they postulated that these particles were characterless, infinitely small bits that interacted randomly in empty space to produce matter. These particles were called atoms from the Greek word *atomon*, meaning that which cannot be divided. Democritus thought the universe worked in a mechanical matter, driven by vibrations or the impacts of atoms making and breaking down matter. Though these atoms could not be directly observed, the atomists believed their existence was inferred through their effects in the natural world; for example, a wet piece of cloth became dry in the sunshine because the water atoms left the cloth, though individually they were invisible.

A century after the atomists, Aristotle (384–322 BC) incorporated the work of Empedocles into his own philosophy. Aristotle's formation of the rough equivalent of a polytechnic university, called the Lyceum, in Athens; his influence as tutor to Alexander the Great; and his vast array of philosophical writings made his ideas extremely important in the ancient and medieval world. While rejecting the work of the atomists and their idea of empty space (a vacuum), Aristotle added his own qualities of hot, dry, cold, and wet to Empedocles's elements of earth (cold and dry), air (hot and wet), water (cold and wet), and fire (hot and dry). Aristotle also promoted the idea, previously proposed by Chinese alchemists, that metals incubated in the wombs of the earth.

Metallurgy and copper smelting were also known in ancient Greece from the Bronze Age, as were the use of substances in dyeing, painting, pottery, and the production of perfume and cosmetics. By the fourth century AD, in the Hellenistic period, knowledge of these techniques had spread to Alexandria in Egypt, the capital city of Alexander the Great's empire. The Alexandrian alchemist Zosimus (c. fifth century BC) was using techniques of distillation, the vaporization of solids by heat (sublimation), as well as filtration; stills and condensers were also part of his laboratory equipment. This basic chemical knowledge, combined with the idea that metals could grow, meant that alchemists thought that they could replicate the process of metal growth. They also thought they could accelerate and alter the growth process. The goal was to reduce metals to their prime matter and then impose upon them the qualities of gold. As alchemy took shape, the power of imposing qualities was believed to reside in what became known as the philosophers' stone.

Arabs and Alchemy

After the Islamic Arabs conquered Alexandria in the seventh century, the secrets of alchemy were transferred to Baghdad and Damascus. The eight century Arab alchemist Jabir ibn Hayyan (721–815) believed the secret of the stone and the creation of gold lay in the Aristotelian four elements. Jabir was said by later medieval philosophers to have been a court physician practicing in what is now Iraq. It is also likely Jabir was a member of the Sufis, a type of Islamic mystic who sought salvation via contemplation or prayer; he seems to have made alchemy and the search for the philosophers' stone part of his quest.

Since elements could be changed based on shared qualities with other elements—e.g., fire could become air via heat or become earth by drying out—Jabir believed the transmutation of matter, and the subsequent creation of gold from base metals, was a possibility.

Reasoning that the transmutation of one metal into another was affected by the rearrangement of its qualities, Jabir believed the change was mediated by a substance or elixir (*al-iksir* in Arabic). This elixir, or the philosophers' stone, was thought to be a dry powder composed of a substance called carmot. In the search for the stone, Jabir did many experiments with substances that could dissolve and thus "transmute" matter. He most likely invented *aqua regia*, a mix of nitric acid

(distilled from saltpeter) and hydrochloric acid (distilled from salt), which is one of the few substances that can dissolve and purify gold. It is also possible he discovered citric and tartaric acids (from dregs left from wine-making). His term alkali is still used in modern chemistry.

Jabir also introduced a theory that was influential in alchemy, postulating that metals were mixtures of mercury, sulfur, and arsenic, except for gold, which was made up of sulfur and mercury alone. He theorized that gold contained the most mercury and the least sulfur, so other metals could be made into gold if their mercury content could be increased.

Medieval and Early Modern Alchemy in the West

As the Arabs expanded their empire into Spain and Portugal in the eighth century, they took their knowledge of alchemy with them, introducing it to the Christian West. The first Crusade (1095–1099) also introduced Christians to Muslim science. Jabir's treatises on chemistry, the *Kitab al-Kimya* and *Kitab al-Sab'een*, were translated into Latin by the twelfth century with modifications and additions. The translation of the *Kitab al-Kimya*, which was published by Robert of Chester in 1144, became known as *The Book of the Composition of Alchemy*; it was the first alchemical work to appear in Latin in Europe. Robert of Chester also translated the Koran from Arabic to Latin. The medieval philosopher Gerard of Cremona (c.1114–1187) also translated some of Jabir's works.

Allegorical woodcut captioned *Alchemical Allegory of Putrefaction*, as some alchemists believed decay was the first stage in the medieval science of transmuting base metals into gold. *Time Life Pictures/Getty Images.*

A later fourteenth-century Spanish alchemist named himself Geber, the Latin version of Jabir, to take advantage of Jabir's reputation. Some older scholarly works translated Jabir's name into the Latinate as Geber, so many historians refer to the medieval writer as pseudo-Geber to avoid confusion. The medieval Geber, or pseudo-Geber, based his works, such as *The Summit of Perfection*, on Arabic alchemy but modified them to include a form of atomism called corpuscularianism. Corpuscularianism was much like atomism, but postulated that the minute particles of matter continued to be divisible, unlike the indivisible atoms. The fact that matter could be divisible infinitely meant that transmutation of base metals into gold was even more likely. The medieval Geber also included Christian allusions in his work, postulating that finding the philosophers' stone was like going on a Christian pilgrimage and finding it was the grail of salvation of the soul.

The mixture of mysticism, science, and often inherent greed intrinsic to alchemy gave it a checkered reputation in medieval and early modern Europe. Though alchemy was seen as a spiritual quest by select adepts (trained and skilled persons), knowledge of alchemy was often confined to the wealthy (it cost money to have a laboratory). Alchemy was also tied to the rights of a sovereign to whatever precious metals were found (or created) within his domain and the right to mint money. When King Henry IV of England (1367–1413) passed a law that forbade the making of gold or silver via alchemy in 1402, the idea was not necessarily to completely forbid the practice, but to insure the king's right to grant gold-making monopolies. In other words, gold-making was one of the king's privileges.

As the king controlled the mining and production of precious metals and coinage, a fraudulent alchemist would be committing a direct offence against the royal court. A special method of executions for fraudulent gold-makers was derived by Duke Friedrick of Württemberg (1557–1608). Alchemists were hung on gold-plated gallows under the inscription, "I had it in mind to fix mercury, but I find it was reversed, and I've been fixed." Alchemists also could be imprisoned by ruling nobles if they were thought to be successful or potentially successful at making gold or silver, so they would not sell their secrets to a noble rival.

Paracelsus and the Use of Alchemy in Medicine

Amidst the economic and political concerns of alchemy, one individual by the name of Theophrastus Bombastus von Hohenheim (1493–1541) wished to use alchemy to improve medicine. Calling himself Paracelsus ("beyond Celsus") as a means to claim superiority over the Roman physician Celsus (c.25 BC–AD 50), he studied chemical reactions, tested the healing powers of chemicals inherent in plants, and believed that illness was caused by

The Alchemist at Work, a woodcut illustration from *Coelum Philosophorum* by Philippus Ulstadium, showing two alchemists working on an experiment. © *Stapleton Collection/Corbis.*

alchemical imbalances within the body. Paracelsus was the founder of iatrochemistry, or medical chemistry.

Paracelsus' emphasis on iatrochemistry was in direct opposition to the medical theory of his day, which was dominated by the doctrine of the bodily fluids or humors postulated by the ancient Roman physician Galen (129–c.216). Galen stated that good health relied on the balance of four humors or bodily fluids, defined as phlegm, blood, yellow bile, and black bile. The job of the humors was to nourish the body, as well as to provide the material for sperm and, in pregnancy, for the fetus.

Humoral balance was also influenced by one's complexion or temperament. There were four basic complexions, each caused by the dominance of one humor. The sanguinous personality resulted from the predominance of the blood and had a lively and cheerful temperament. Sanguine people also tended to have florid complexions from an excess of blood. The melancholy personality resulted from a surfeit of black bile, and melancholics were thought to be dark in skin and hair tone and prone to depression and worry. The phlegmatic person, dominated by the phlegm humor, was calm, slow, and prone to watery swellings in the body. Cholerics, with a surfeit of yellow bile, were energetic and quickly prone to anger.

If a person was too emotional, it was thought he had an "overactive" heart due to too much blood. Humoral balance could then be restored by therapeutic bloodletting via leeches or lancet. The doctor perforated the vein, and sometimes many shallow cuts were made. When the patient felt faint, and was considered to be calmer to the purgation of the excess humor, the bleeding was stopped. Bleeding was also done if another humor was too predominant, as the pure humor blood contained a smaller amount of the other humors. Humoral balance could also be achieved via diet or herbal remedies, using a treatment of opposites. For instance, if there was an overabundance of cold and moist phlegm, the physician would give the patient remedies associated with hot and dry yellow bile.

Paracelsus sought to overthrow Galenic theory and replace it with an alchemical model. He saw the growth

A physician performs bloodletting on his patient. © *Bettmann/ Corbis.*

of plants and animals, the ripening of fruit, the fermentation of wine and beer, and the digestion of food as essentially alchemical processes and was the first to use the word "chemistry" to describe them. Jabir had posited that mercury and sulfur were basic elemental principles; that metals were made of sulfur and mercury, with sulfur being considered a combustible substance; and that mercury was the predominant element in gold. Paracelsus added salt to this dyad of chemical elements, creating a *tria prima*, or three-principle, model of matter that could explain all transmutations. While Paracelsus had no desire to dispose of the Aristotelian schema of earth, air, water, and fire, he did feel that they were purely "spiritual in nature and only crude approximations of the objects by which we call these names."

Paracelsus also attempted to classify chemicals according how they performed in chemical reactions, and he insisted on the purity of a chemical and an exact quantity when giving it as medicine, which was a new concept and standard for medicine. He also did the best contemporary nosology (medical description) of the venereal disease syphilis in his day and promoted curing the disease with mercury, which in limited doses can be effective. He also realized that silicosis, a lung disease of miners, was due to inhaling vapors in mines, not the result of mountain spirits seeking their revenge. Paracelsus also insisted upon empirical observation to create his chemical medicaments, stating "what a doctor needs is not eloquence or knowledge of language of books, but a profound knowledge of Nature and her works."

Paracelsus' innovations did not mean that he was a modern doctor. He believed in astrological medicine and posited deep connections between the microcosm (little world) of the body and the macrocosm (larger universe) in health and disease. He thought that the shape of plants indicated their healing properties; for instance he wrote that an orchid shaped like a testicle or a mandrake root shaped like a small man could cure sexual dysfunction.

Robert Boyle, Chymistry and the Decline of Alchemy

Although Paracelsus used what we would now consider pseudo-scientific theories, it is best to understand him on his own terms. His ideas about chemistry reflected the state of chemical experimentation in the early modern period. Some historians have suggested that by the sixteenth century, any sort of chemistry should be referred to as chymistry to reflect its transitional status between what we would consider alchemy and chemistry. In other words, though early modern chymists (those who practiced from 1500–1700) attempted to transmute metals into gold, considered an alchemical practice, they also performed other experiments involving mass balance (determining which materials enter and leave a

ALTERIVS NON SIT, QVI SVVS ESSE POTEST.

LAVS DEO, PAX VIVIS, REQVIES ÆTERNA SEPVLTIS.

OMNE DONVM PERFECTVM À DEO, IMPERF. À DIABO.

AVREOLVS PHILIPPVS THEOPHRAST.

Engraved portrait, c. 1567, of alchemist and physician Paracelsus. *© Stapleton Collection/Corbis.*

system) or crystallographic analysis that would be considered more closely related to modern chemistry.

Robert Boyle (1627–1691), the seventeenth-century natural philosopher, would certainly be considered a chymist. A son of the extremely wealthy Earl of Cork, Boyle had the time and resources to build his own chymical laboratory and supply it with equipment. His subsequent work on vacuum pumps, the discovery of the relation of volume and pressure named Boyle's Law, the classification of alkalis and acids, and the use of pH indicators would be considered chemical research today, but the context of these discoveries was sometimes alchemical in origin.

Boyle is best known for subjecting Aristotle's theory of the elements, as well as Paracelsus' *tria prima* to analysis and thorough criticism. However, Boyle was

not against Jabir's idea that sulfur and mercury could be separated from metals; he believed that the philosophers' stone existed; and thought he had succeeded in the alchemical transformation of gold to silver.

On the other hand, Boyle did not think that sulfur and mercury were truly elemental principles, as he was an advocate of corpuscularianism, and he had little patience for the secrecy of most alchemical practice. Boyle believed that all alchemical experiments should be done rigorously, with keen empiricism, and that their results should be shared with other chymical practitioners so the experiments could be repeated and their results confirmed. It is little surprise that Boyle himself was one of the key early members of England's Royal Society, founded in 1660, the oldest national scientific society in the world that had a regular journal, the *Philosophical*

Transactions, recording the results of the experiments of its members.

Boyle published *The Sceptical Chymist* (1661) to criticize poor experimentation and systems of the elements such as Aristotle's and Paracelsus', which he saw as based only on theory and not on empirical proof. In his work, he did a series of experiments by fire analysis, in which he subjected samples to intense heat and subsequent distillation to reveal their elemental principles. He found that none of the theoretical frameworks of the elements explained what he was seeing in the laboratory. For instance, gold remains gold no matter how much it is heated. It does not break down into earth, air, water, fire, salt, sulfur, or mercury. When blood was observed during fire analysis, it seemed to give five substances: phlegm, spirit (vapor), oil, salt, and earth. These inconsistent results fostered Boyle's assertion that the chymist should not adopt theories that claimed that universal elements are present in all bodies.

Boyle thought all matter was made up of corpuscles, which were themselves made up of atoms. The different combinations of atoms resulted in different material forms. Though it was difficult to apply his corpuscular theory to laboratory practice, and ultimately corpuscularianism fell by the wayside, Boyle contributed greatly to the transformation of alchemy into a chemistry we would recognize today. His emphasis on precise experimentation, his questioning of the elemental system of the alchemists, his discovery of chemical laws, and his insistence that the results of chemical research should be shared openly made chemical practice a necessary and respected part of scientific inquiry. The Age of Alchemy was on the decline.

■ Modern Cultural Connections

In 1980, Nobel Prize-winning American chemist Glenn Seaborg (1912–1999) transmuted microscopic amounts of bismuth into gold. The altered atoms lasted only seconds and required a significant input of energy, thus negating any potential economic benefit. However, modern chemistry and physics had at last achieved the ancient alchemists' goal.

While alchemy is no longer considered a science, it remains part of various cultural traditions. Ayurvedic and traditional Chinese medicines still apply some ancient alchemical principles. Traditional alchemy is often a subject of contemporary fantasy fiction. The legend of the philosopher's stone is featured prominently in the popular *Harry Potter* book and movie series.

SEE ALSO *Astronomy and Cosmology: Cosmology; Physics: Aristotelian Physics; Science Philosophy and Practice: Pseudoscience and Popular Misconceptions.*

BIBLIOGRAPHY

Books

Boyle, Robert. *The Sceptical Chymist.* New York: Dover, 2003.

Levere, Trevor H. *Transforming Matter: A History of Chemistry from Alchemy to the Buckyball.* Baltimore: Johns Hopkins University Press, 2001.

Martin, Sean. *Alchemy and Alchemists.* Harpenden, UK: Pocket Essentials, 2001.

Morris, Richard. *Last Sorcerers: The Path from Alchemy to the Periodic Table.* Washington, DC: National Academies Press, 2003.

Periodicals

Debus, Allen G. "Fire Analysis and the Elements in the Sixteenth and Seventeenth Centuries," *Annals of Science* 23, 2 (June 1967): 127–147.

Dubs, Homer H. "The Beginnings of Alchemy," *Isis* 38, 1/2 (November 1947): 62–86.

Principe, Lawrence M., and William R. Newman. "Alchemy vs. Chymistry: The Etymological Origins of a Historiographic Mistake," *Early Science and Medicine* 3 (1998): 32–65.

Smith, Pamela H. "Alchemy as a Language of Mediation at the Habsburg Court," *Isis* 85, 1 (March 1994): 1–25.

Anna Marie Eleanor Roos

Computer Science: Artificial Intelligence

■ Introduction

Artificial intelligence (AI) is a branch of computer science that seeks to build machines that carry out tasks which, when performed by humans, require intelligence.

AI techniques made possible some self-guided navigation by the twin Mars Exploration Rovers (on Mars since early 2004), allowing the robot rovers to explore more of Mars than if they were steered entirely by commands radioed from Earth. AI programs for chess-playing and limited recognition of faces, text, images, and spoken words are commonplace today in industrial, military, and domestic computer systems. However, AI has so far not produced machines capable of handling everyday human language, translating between human languages fluently, and performing most other intelligent tasks that are routine for human beings. Whether this failure arises from the fundamental nature of thought or if AI research has simply not yet progressed far enough to build such systems is one of the many philosophical and technological debates that rage around AI. Other questions often raised in the context of AI include the following: What is thinking? Whatever it is, can machines do it? Why or why not? How would we know? Are people machines?

■ Historical Background and Scientific Foundations

In the seventeenth century, German mathematician Gottfried Leibniz (1646–1716) invented the binary number system, the foundation of digital computation. He expressed the belief that his system of notation could be used to formalize all knowledge: to every relevant object in the world, he said, one would assign a "determined characteristic number," as he called it, and by manipulating these numbers one could resolve any question. "If someone would doubt my results," Leibniz wrote, "I would say to him, 'Let us calculate, Sir,' and thus by taking pen and ink, we should settle the question."

Leibniz did not anticipate decision-making by mechanical means rather than by pen and ink—that is, artificial intelligence—but a century later British mathematician George Boole (1815–1864) described the logical rules, today called Boolean algebra, by which true-or-false statements can be identified with binary numbers and manipulated with mathematical rigor. At about the same time, the idea of mechanized calculation was advanced by British philosopher and engineer Charles Babbage (1791–1871), who designed what he called an Analytic Engine to perform mathematical and logical operations. His work was supported by Lady Ada Byron (1815–1852), who wrote the world's first computer program for the Analytic Engine (which was never completed). The possibility of AI was immediately apparent, even at this early period: In one letter, Ada Byron asked Boole if his Analytic Engine would "think."

World War II (1939–1945) boosted the development of digital computers for code-breaking and other military purposes. By the late 1940s, mathematicians and philosophers had real computers to ponder and addressed the question of artificial intelligence with fresh clarity. In 1950 British mathematician Alan Turing (1912–1954) published one of the most famous papers in computer-science history, "Computing Machinery and Intelligence." In it he took up the already old question "Can machines think?" and proposed that it could be answered by the now-famous Turing test, which he called the imitation game. The imitation game would work as follows: if a human interrogator communicating with both a human being and a computer in another room, say by exchanging typed messages, could not reliably tell which was the human and which the computer, even after an extended exchange—that is, if the computer could imitate unrestricted human conversation—then it would be reasonable to say that the computer thinks and is intelligent.

By 1950 early electronic computers could manipulate numbers at blinding speed. It looked as if Leibniz's old dream of reducing all thinking to computation was about to be realized. All that remained, some scientists thought, was to code the rules of human thought (human thought was assumed to depend on hidden rules) in binary form, supply the computer with a mass of digitally encoded facts about how the world works, and run some programs. In this heady atmosphere, many over-optimistic claims were made about progress in AI.

For example, AI pioneer Herbert Simon (1916–2001) predicted in 1957 that "within ten years a digital computer will be the world's chess champion." (A computer did not beat the world chess champion until 1997, 30 years behind schedule.) In 1965 Simon predicted that "Machines will be capable, within twenty years, of doing any work that a man can do." (They still cannot.) In

IN CONTEXT: WHAT IS INTELLIGENCE?

Jobs that seem hard to people, like handling large sets of numbers rapidly, are easy for computers, while things that seem easy to most people, like having a conversation or cleaning house, are hard—extremely hard—for computers.

The reason is that tasks like chess and arithmetic can be performed by applying a few strictly defined rules to pieces of coded information. In contrast, most daily activities of human beings are richly connected with a physical world full of endlessly various objects, persons, and meanings. The sheer number of facts that any normal person knows, and the number of ways in which they apply those facts in performing a typical task of daily life, including speech, is simply too large for even a modern computer to handle (even assuming that human intelligence can be understood in terms of applying rules to facts, which is a matter of dispute). A human being washing dishes by hand must deal simultaneously with the mechanical properties of arms and hands, caked-on food, grease, water, soap, scrubbers, utensils of scores of different shapes, plastics, metals, and ceramics, and so on. They must start with a disorganized heap of dirty tableware and finish with a pile of properly draining or towel-dried dishes, all acceptably clean, while not injuring themselves or flooding the kitchen and breaking a dish only rarely. To describe such a job in terms of chess-like rules that a computer can follow has turned out to be far more difficult than the pioneers of artificial intelligence thought in the 1950s and 1960s.

1968, Stanley Kubrick's hit movie *2001: A Space Odyssey* depicted a conversational computer, the HAL-9000, as being a reality in 2001. The film was not meant to be pure fiction: AI expert Marvin Minsky (1927–), hired as a technical consultant for the film, assured the director that the existence of computers like HAL by 2001 was a sure thing. As of 2007 no such computer was even close to being constructed. As one AI textbook put it in 2005, "The problem of over-promising has been a recurring one for the field of AI."

Speculation about thinking machines was common in the fiction of the mid-twentieth century, especially after the publication of Czech writer Karl Capek's play *Rossum's Universal Robots,* which introduced the word "robot" into the English language in 1923. Long before computers were commonplace in actual life, they saturated the popular imagination through hundreds of representations in science fiction—mostly sinister. During the 1950s and 1960s, popular interest in machine intelligence became pervasive, rivaled only by fear of the

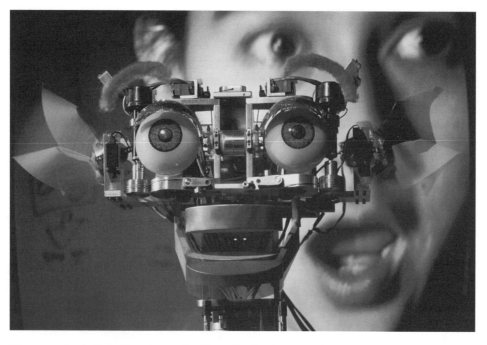

Kismet, a robot built in Massachusetts Institute of Technology's Artificial Intelligence Lab, can engage humans in simple social interactions, possibly leading to communication with human caregivers using infant-like expressions and vocalizations. Here, the robot shows a surprise reaction, similar to that of its creator, who is seen on a video monitor behind Kismet. *© George Steinmetz/Corbis.*

atomic bomb. Computers and robots starred in such hit movies as *The Day the Earth Stood Still* (1951), *Forbidden Planet* (1956), and *2001: A Space Odyssey* (1968). The press often presented startling claims that some new computerized device could think, learn, compose music, or perform some other human task. In 1970, for example, *Life* magazine announced excitedly that a turtle-like machine called Shaky, which could navigate a simple indoor environment, was the "first electronic person," and promised its readers that by 1985 at the latest we would "have a machine with the general intelligence of an average human being."

The Science

AI research restricted to specific problems such as pattern identification, question-answering, navigation, and the like is often called "weak" AI. AI research and philosophy concerned with the possibility of producing artificial minds comparable (or superior) to human minds is called "strong" AI. Strong AI has produced few real-world results; weak AI has produced many.

AI can also be divided into what is sometimes called "good old-fashioned artificial intelligence" (GO-FAI) and neural networks. GOFAI seeks to produce computer programs that apply symbolic rules to coded information in order to make decisions about how to manipulate objects, identify the content of sounds or images, steer vehicles, aim weapons, or the like. These programs run on ordinary digital computers. The other basic approach to AI is the connectionist or neural network approach. This seeks to mimic the way animal nervous systems achieve intelligent behavior, by linking together anywhere from dozens to billions of separate nerve cells—neurons—into a network. A neural network may be produced by building electronic neurons and linking them, or by simulating such a network on a regular digital computer (the more common approach). Neural networks may also be combined with GOFAI, rule-based type systems to create hybrid systems.

The mathematical theory behind AI is complex, but as a simplification it can be said that there are three basic elements to the handling of information in AI systems. The first is knowledge representation. Knowledge representation involves ways of organizing and annotating factual information in computer memory. The second element is searching, the means by which a computer sifts through a database (such as a list of possible disease diagnoses) or calculated alternatives (such as chessboard positions) to find a solution to a problem. The set of all possible solutions is often called a "problem space." Problem spaces are often too large to search completely, so AI programmers seek rules that govern the search process and make it more efficient. These rules are called heuristics.

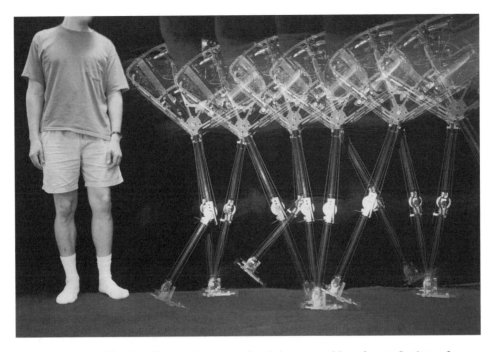

A student in the Artificial Intelligence Department Leg Laboratory at Massachusetts Institute of Technology is shown running a test of his PhD research project "Spring Flamingo." It is a computer-controlled hydraulic walking machine. Researchers believe that advancements in AI and biotechnology will help to increase mobility for amputees. *© George Steinmetz/Corbis.*

State space search is a class of search methods used widely in AI. State space search defines a problem as a "space" in a mathematical, not a physical sense, that is, as a linked network of possible states or nodes. Operators (computational functions) allow transitions from each state to others. The resulting network of states is called a graph. Nodes or states that correspond to acceptable solutions to the problem are called goal states. The mathematical discipline known as graph theory is used to reason about ways of searching state spaces for goal states. For example, the problem of a robot rover seeking a route through a field of boulders might be cast as a state space search. Acceptable goals would be points on the far side of the field, attained with the rover undamaged, untrapped, and standing with an acceptable degree of tilt. Intermediate positions of the rover in the boulder field might be symbolized as nodes in the graph: operators would correspond to maneuvers between intermediate positions. (This is not necessarily how autonomous rover navigation is actually calculated, but serves as an illustrative example.)

Specialized problem-solving languages have been developed for handling heuristics, state space searches, and other AI operations. These languages include LISP (for "list processing") and PROLOG (for *programmation en logique*, French for "programming in logic").

■ Modern Cultural Connections

Programs exploiting techniques developed in the AI field are now commonplace in commercial and military applications. Hundreds of thousands of industrial robots are used worldwide on assembly lines (more in Japan than any other country). Neural-network programs running on conventional digital computers are used by credit-card companies to search for anomalous card uses that might signal identity theft. Governments use such programs to search the Internet for behaviors that they consider illicit, including terrorism or forbidden politics; for example, China, named as a routine abuser of human rights by the U.S. State Department and nongovernmental human-rights organizations such as Amnesty International, began research on a nationwide digital surveillance project called Golden Shield in 2001. Golden Shield depends heavily on artificial intelligence techniques to recognize faces in video recorded by surveillance cameras and to decipher recorded telephone conversations. The system will, according to the International Centre for Human Rights and Democratic Development, be "effectively applying artificial intelligence routines to data analysis via complex algorithms, which enable automatic recognition and tracking. Such automation not only widens the surveillance net, it narrows the mesh."

As with other technologies, AI can cause social stresses even when there is no malicious intent. The RAND Corporation, a private strategic think-tank often hired by the U.S. military, reported in 2001 that "[t]he increasing sophistication of robotics, coupled with software advances (e.g., in artificial intelligence and speech understanding) removes jobs from the marketplace, both in low-skilled, entry-level positions and more sophisticated specialties." Military applications for AI are now occurring, including the partially self-guided weapons termed "smart bombs" and autonomous navigation by unmanned airplanes and submarines. Military research programs are directed toward the eventual development of fully autonomous weapons that would function without direct human supervision.

AI systems are also finding application in medicine, industry, game-playing, and numerous other areas. Wherever computers are interacting with complex environments, AI techniques are often being applied, usually in inconspicuous or built-in ways that are not apparent to users.

In the culture at large, AI participates prominently in unresolved debates about human nature, determinism, free will, and morality. Some thinkers argue that human beings are only "lumbering robots" (Richard Dawkins, 1941–) programmed by their DNA, and that the human brain is a carbon-based neural net comprised of billions of neural mini-robots, not essentially different from any other computer (Daniel Dennett, 1942–). In popular culture, Captain Jean-Luc Picard informs watchers of the TV drama *Star Trek: The Next Generation* that human beings "are machines—just machines of a different type." Some, such as physicist Roger Penrose (1931–) and philosophers John Searl (1932–) and Hubert Dreyfus (1929–), argue that the strong-AI equation of mechanical and human thought is based on fallacious assumptions about thinking, information, and physical systems.

SEE ALSO *Computer Science: Artificial Intelligence and Economics; Computer Science: Information Science and the Rise of the Internet; Computer Science: The Computer.*

BIBLIOGRAPHY

Books

Anderson, Alan Ross, ed. *Minds and Machines.* Englewood Cliffs, NJ: Prentice-Hall, 1964.

Crevier, Daniel. *AI: The Tumultuous History of the Search for Artificial Intelligence.* New York: Basic Books, 1993.

Dreyfus, Hubert L. *What Computers Can't Do: A Critique of Artificial Reason.* Cambridge, MA: MIT Press, 1992.

Padhy, N.P. *Artificial Intelligence and Intelligent Systems.* New Delhi, India: Oxford University Press, 2005.

Penrose, Roger. *The Emperor's New Mind: Concerning Computers, Minds, and the Laws of Physics.* New York: Oxford University Press, 1989.

Roland, Alex. *Strategic Computing: DARPA and the Quest for Machine Intelligence, 1983–1993.* Cambridge, MA: MIT Press, 2002.

Scientific American. *Understanding Artificial Intelligence.* New York: Warner Books, 2002.

Von Neumann, John. *The Computer and the Brain.* New Haven, CT: Yale University Press, 1958.

Larry Gilman

Computer Science: Artificial Intelligence and Economics: The Rise of Formalism and Behaviorism

■ Introduction

Starting the 1930s, American scientist Herbert Simon (1916–2001) has had a major influence on political science, economics, management science, cognitive psychology, the philosophy of science, and artificial intelligence (AI). Simon won the 1978 Nobel Prize in economics for his theory of bounded rationality, which challenged the traditional assumption in economics that consumers and producers always make perfectly rational choices to maximize utility and profits. Simon helped transform the social sciences into the behavioral sciences and was a leading advocate of behavioralism, the attempt in political science to describe group behavior mathematically. He also sought to extend his idea of mind as a physical symbol-system to the process of scientific discovery, thus contributing, albeit controversially, to the philosophy of science.

■ Historical Background and Scientific Foundations

Herbert Simon was born and grew up in Milwaukee, Wisconsin. His father, Arthur Carl Simon, an electrical engineer and patent attorney, immigrated to the United States in 1903; his mother, Edna Simon (born Merkel), was an American pianist of third-generation German and Czech descent. Simon had one brother, Clarence, five years older. Simon attributed his early interest in books and music to the influence of his mother. He camped, canoed, and hiked extensively in Wisconsin and formed lifelong interests in chess, classical piano, and beetles. Conversation around the family dinner table often involved politics and science. He was educated in the public school system of Milwaukee.

His mother's younger brother, Harold Merkel, who had studied economics at the University of Wisconsin, lived briefly with the family. He died young, but left his personal library of economics and psychology texts in the home. It was through these books that Simon became aware of the social sciences. Simon joined his high-school debate club, where he found himself defending unpopular causes such as free trade and disarmament. This moved him to read several textbooks on economics.

Simon entered the University of Chicago in 1933, determined to bring mathematical rigor to the social sciences. The university was loosely organized by today's standards; Simon's graduate transcript records that he formally attended only one course, boxing, in which he earned a "B." His studies in symbolic logic, statistics, advanced math, physics, economics, and political science were self-guided. In 1936, a term-paper project stimulated his interest in decision-making in organizations. His paper helped him obtain a research assistantship studying the internal operations of municipal administrations. This led to Simon being appointed to the directorship of a research group at the University of California, Berkeley, from 1939 to 1942, performing research on the administration of state relief programs. While in California he was enrolled as a doctoral student at the University of Chicago, taking his exams by mail. His work on state relief programs provided the material of his doctoral dissertation. He received his Ph.D. in political science in 1942.

In 1947, he published a revised form of his Ph.D. dissertation as *Administrative Behavior*, a landmark work in administration theory. By the late 1950s, *Administrative Behavior* had become standard reading in college courses on public administration, organizational sociology, and business education.

Early and Middle Career

After receiving his doctorate in 1942, Simon joined the political science faculty of the Illinois Institute of Technology in Chicago. Living in Chicago enabled Simon to participate regularly in the staff seminars of the Cowles Commission for Research in Economics at the University of Chicago. The Cowles Commission was a private foundation devoted to integrating mathematical techniques with economics. The Cowles seminars introduced Simon to mathematical economic theories, including new econometric (economy-measurement) techniques and John Maynard Keynes's general theory.

In 1949, the Carnegie Institute of Technology (which merged with Mellon Institute of Industrial Research in 1967 to form Carnegie Mellon University) was given a $6 million endowment by William Larimer Mellon to found its Graduate School of Industrial Administration (GSIA, renamed the David A. Tepper School of Business in 2004). Simon left the Illinois Institute of Technology to become the GSIA's first faculty member.

For the first five years of the GSIA's existence, Simon was a direct participant in all team research projects at the GSIA. Funding came mostly from contracts with the RAND Corporation, the Ford Foundation, the U.S. Air Force, and the U.S. Navy. RAND (an

acronym derived from "Research and Development") had been set up in 1946 to provide scientific expertise to the U.S. military, but was spun off as an independent, nonprofit think-tank in 1948. It remained influential in setting defense policy, including nuclear deterrence policy. In 1952, Simon was asked by several RAND scientists to participate in a social-psychology study of a simulated air-defense direction center. At RAND's facility in Santa Monica, Simon met Allen Newell (1927–1992). Newell and Simon found they had much in common, and together sought to use information-processing concepts to analyze the way air-defense personnel operated their machines. RAND exposed Simon to the most advanced computer technology of the day, which was used in its air-defense study for printing out maps—a startling accomplishment at the time. Computer-generated maps alerted Simon to the fact that computers can be used to handle non-numerical information.

His work with Newell moved Simon to consider more carefully the analogy of human brain to digital computer. He already conceived of the human mind (in its problem-solving capacity) primarily as a logic machine, as a device that took premises (claims of fact) and processed them to reach conclusions: this idea now began to take the specific form that the mind is like a computer that takes programming and data and processes them to produce output.

In November 1954, Newell heard a talk on a computerized pattern-recognition system. In the following months, Newell wrote a paper about the prospects for computer chess and left RAND briefly for Pittsburgh, where he intended to collaborate with Simon on a computer program that would play chess. In the summer of 1955, Newell moved back to RAND but continued commuting to Pittsburgh to work with Simon. The two eventually wrote a program called Logic Theorist, designed to prove mathematical theorems.

Logic Theorist was a stunning success, finding proofs for thirty-eight of the first fifty-two theorems of Alfred North Whitehead and Bertrand Russell's magisterial *Principia Mathematica* (1910–1913). Newell and Simon presented their results at the pivotal Dartmouth Summer Research Conference on Artificial Intelligence of 1956, where the term "artificial intelligence" was first used. Wishing to generalize the success of Logic Theorist, the two men next produced a program called General Problem Solver. The program never achieved its most ambitious goals, but the effort to produce it was formative of Simon and Newell's ideas about artificial intelligence.

Later Career

Given Simon and Newell's formalism, that is, their belief that all thought is symbol manipulation, it followed for them that the institutional boundary between computation and psychology reflects nothing real: both are ultimately ways of studying physical symbol systems. The digital computer might therefore not only model or mimic the human mind but work just as the human mind does. Psychologists, for their part, were deeply impressed by Logic Theorist and General Problem Solver, the apparent advent of the "thinking machines" long forecast by science fiction. Several landmark books in the late 1950s and 1960s signaled the cognitive revolution in psychology that overthrew S-R (stimulus-response) behaviorism, then the prevailing psychological orthodoxy. The use of computational models became commonplace—although never uncontroversial—in the study of cognitive psychology, and remains so today. Simon and Newell continued to contribute to this field, stimulating and incorporating the results of painstaking experimental work with human volunteers. For example, in the 1970s and 1980s their computer models of mental processes achieved close matches between program performance and human eye and finger movements and certain other behaviors.

Simon extended his ideas about intelligence as a formalistic problem-solving procedure guided by heuristic learning rules to the philosophy of science. Scientific discovery, he maintained, being a cognitive problem-solving process, could be modeled by computer programs. Beginning in the 1960s he repeatedly published on this theme.

■ Modern Cultural Connections

Administrative Theory

Prior to the 1930s, economics concerned itself mostly with the behavior of total markets rather than the behavior of individuals or organizations. However, interest in organization theory was stimulated by the growth of large companies through merger and monopoly and, within companies, the growth of managerial structures distinct from executive and productive operations. Simon was influenced by and contributed to this new area of study. His *Administrative Behavior* (1947) became one of the century's most important works in public administration and political science.

His approach to the study of administrative behavior had philosophical roots. While studying at the University of Chicago in the late 1930s, he adopted logical positivism under the influence of philosopher Rudolf Carnap and A.J. Ayer's *Language, Truth and Logic* (1936). Central to logical positivism was the verification theory of meaning, which asserts that a statement only has meaning if it is empirically verifiable. One version or offshoot of verification theory is operationalism, the claim that to understand something, one must understand the operations or procedures by which that something is brought into being. Simon brought his

operationalist viewpoint to the study of organizational behaviors. In *Administrative Behavior*, he promoted an operational model of human decision-making in organizations. Simon's model assumed essentially rational behavior (Simon, like most other economists, excluded the possibility of truly irrational behavior as a significant factor), but only within limits. The organizational decision maker, Simon said, is constantly faced with multiple means-and-ends strategies associated with various costs and consequences. Achievement of ends with minimal cost is the definition of "rationality." Simon's most important innovation was his observation that decision makers never approach perfect rationality in real life because of (1) inherently limited knowledge and cognitive ability and (2) contextual constraints such as limited resources. This is the principle Simon called "subjective rationality" in *Administrative Behavior* but later referred to as "bounded rationality."

Simon modeled the company or organization itself as a network of cooperating, boundedly rational decision makers. He rejected the classical view that a company can be effectively treated as omniscient, perfectly rational, and profit-maximizing.

Economics and Political Science

In the 1940s, Simon extended his theory of the bounded rationality organizational decision maker to microeconomics. The prevalent view among economic theorists, then as now, was the neoclassical rationality postulate, namely, the assertion that consumers rationally maximize their utility (the benefit they experience from their expenditures) while producers rationally maximize their profits. Simon viewed these postulates as unrealistic: all economic actors, like the intracorporate decision makers modeled in *Administrative Behavior*, apply limited powers of reasoning to limited information in constraining contexts, so optimization—always achieving the best of all possible outcomes—cannot be assumed. In a 1956 paper in *Psychological Review*, he invented the term *satisficing*, a combination of *satisfying* and *sufficing*. Satisficing is the selection and pursuit of a course of action that, the actor hopes, will be "good enough" to achieve a specific goal but which may not be optimal. Simon proposed that the neoclassical maximizing rationality postulate be replaced with the postulate that decision makers engage in satisficing behavior: consumers and entrepreneurs or corporations are satisficers, not optimizers.

Simon did not succeed in dislodging rational choice theory from economic theory. Rather, a mixture of competing rational-choice theories has remained dominant and theories emphasize bounded rationality have grown up in competition. Simon's bounded-rationality thesis contributed to the insitutionalist school of economics, which focuses on bounded rationality and adaptation in economic behavior.

Contributions to AI and Cognitive Psychology

In 1955, Simon and Newell realized that before they could build an intelligent machine they would need to invent an appropriate programming language. Working with RAND programmer Cliff Shaw, they developed a list-processing programming language called Information Processing Language (an important conceptual precursor of the computer language LISP, LIst Processor, formally specified in 1958 and still extensively used today in AI research). Information Processing Language was essentially a tool for managing the allocation of limited memory space. In December 1955, Simon tested his program concept by assigning logical tasks written on cards to human volunteers (including his three children, then ages 13, 11, and 9). Obeying the instructions on the cards, the volunteers enacted the program, solving a test problem. On the strength of this test, Simon announced to his mathematical modeling class in January 1956 that he, Newell, and Simon had invented a "thinking machine."

In 1956, using their new programming tools, Simon and Newell wrote Logic Theorist (also called Logic Theory Machine), a computer program designed to discover proofs for theorems in symbolic logic. The program stored axioms and already-proved theorems (including those proved by itself) in symbolic form. Supplied with a logical expression, it sought to prove the expression by combining its stored axioms and theorems. This is the program that was able to prove thirty-eight of the first fifty-two theorems of the *Principia Mathematica*.

Next, Newell and Simon moved on to a new, more ambitious project, General Problem Solver (GPS). GPS, they hoped, would show that a heuristic (learning-based) rather than an algorithmic (rule-based) approach to problem solving was not only at the core of human reasoning but could be replicated in machines. Algorithms are deterministic recipes or rule systems for carrying out tasks, while heuristic rules or heuristics are loosely defined rules that allow approach to solutions through trial and error. Working on GPS allowed Newell and Simon to develop their view of physical symbol systems. They claimed that hierarchic symbolic manipulation was the process used by humans, consciously or otherwise, in all problem-solving, and argued that human thought could be abstracted from neurology: also that thought, being symbolic, could be ported from one calculating mechanism to another.

All intelligent behavior, according to this central doctrine, is the behaving of sufficiently complex "physical symbol systems." Symbolic structures are constructed as in physical systems: "thinking" occurs when symbols are modified, duplicated, created, or otherwise transformed by physical processes corresponding to logical rules. The physical system, on this view—brains, computers, or cards being carried about a classroom by volunteers—is irrelevant: distinct physical systems are equivalent if they instantiate equivalent symbol systems. Not only can an appropriate physical symbol system reproduce any intelligent behavior, but, conversely, Simon and Newell claimed, any intelligent behavior must be found to be the product of some physical symbol system.

Newell and Simon's formalist view has shaped one of the two basic approaches in AI research for over half a century. The main alternative to the formalist approach has sought to realize intelligent behaviors in neural networks—devices employing numerous interconnected simple units (analogous to neurons) rather than physical symbol systems. Due to the slow progress of symbolic AI in the many years since Logic Theorist's success, the AI trend in the 1990s and 2000s has been toward the neural-network or connectionist approach.

Following Logic Theorist and General Problem Solver, a field of computational psychology sprang up and has continued in the decades since, though never without controversy. This field and AI overlap in personnel, techniques, and disputes; the main difference is that AI is oriented toward the manufacture of intelligent systems, while computational psychology is oriented toward understanding the human mind as an intelligent machine.

Newell and Simon's "thinking machine" programs were intended as psychological simulations from the beginning—realizations in software of exactly what goes on in the human mind (allegedly) during problem-solving. Later cognitive-simulation work by D. Marr, J.R. Anderson, Z.W. Pylyshyn, and others has derived directly from Newell and Simon's cognitive-simulation approach. Although not an unqualified success, GPS established theoretical ideas that have proved useful in analyzing some forms of problem-solving in the half-century since: specifically, structuring one's search of a problem-space using goals and sub-goals, conducting one's search using heuristic (flexible) rules or operators, and controlling processing using information about preconditions and results of operators.

Although cognitive models of the Newell-Simon type have produced impressive matches to empirical observations of human problem-solving behavior in some experiments, they have also been criticized by some other psychologists. For example, they argue that Simon and Newell went too far in assuming that *all* problem solving can be treated as sequential search in a state-space. Some other psychologists have rejected the premise that the human mind is fundamentally a symbol-processing system at all, although it may engage in symbol processing as a special-case behavior (e.g., when doing mental arithmetic, proving symbolic logic theorems, or counting out chess positions). Even some who

IN CONTEXT: THE DREYFUS DISPUTE

Hubert Dreyfus (1929–), a philosophy professor at the University of California, Berkeley, for most of his career, has been a vocal critic of formalist AI and particularly of Simon and Newell. In 1965, while a consultant for RAND, Dreyfus wrote a critical analysis of AI that was published in 1972 as the book *What Computer's Can't Do: A Critique of Artificial Reason.* (An updated version was released in 1992.)

Dreyfus acknowledged the value of the work done by Simon, Newell, Marvin Minsky, and their AI colleagues on list structures, database organization, and the like, and affirmed the theoretical possibility of simulating human thought in a supercomputer that models the physics of the entire human brain at the molecular level. However, he criticized Simon, Newell, and others for making certain assumptions and accused them of chronic exaggeration and overpromising about the achievements and future of AI. For example, in a talk given in 1957, Simon predicted that within a decade a computer would be the world chess champion, a computer would discover an important new theorem in mathematics, and that most new psychological theories would be in the form of computer programs. Dreyfus noted in 1972 that Simon's first two predictions had not been fulfilled over 10 years after having been made, and were in fact not even close to being fulfilled. He identified this failure as a special case of what he said was an overall pattern in AI research: "early, dramatic success based on the easy performance of simple tasks, or low-quality work on complex tasks, and then diminishing returns, disenchantment, and, in some cases, pessimism" (*What Computers Can't Do*).

Simon and Newell, Dreyfus argued, made bad predictions because they made four faulty assumptions: (1) the biological assumption that the brain processes information by discrete operations at some level, (2) the psychological assumption that the mind works primarily by applying formal rules to bits of symbolic information, (3) the epistemological assumption that all knowledge can take the form of logical statements, and (4) the knowledge assumption that the world can be completely modeled as a collection of context-free bits of information. Since all four assumptions were (Dreyfus argued) false, promises such as Simon's 1965 claim that "machines will be capable, within twenty years, of doing any work that a man can do" (*The Shape of Automation for Men and Management*), were doomed. Tempers flared on both sides of the debate. Dreyfus was reviled by the AI community and responded heatedly.

In the long run, some of Dreyfus's pessimistic forecasts have proved correct. A chess computer (IBM's custom-built Deep Blue) did not beat the human world chess champion, Garry Kasparov, until 1997, 40 years after Simon's original prediction. Moreover, it did not do so by mimicking human chess-playing strategies, but partly by evaluating about 18 billion board positions per move, far more than the 100 to 200 positions explicitly evaluated per move by a top human player. Yet Simon would not concede that Deep Blue beat Kasparov primarily by counting out billions of boards per move rather than by applying heuristic rules of the type Simon had been insisting since the 1950s were the basis of human chess-playing. As of 2008, programs using more advanced heuristics (e.g., Deep Junior, Deep Fritz) and running on less powerful hardware were still evaluating on the order of 1.5 billion positions per move; despite 50 years of research, it had still not been possible to model human chess-playing directly in software. Computers play excellent chess, but they do not play it like human beings.

support the assumptions of formalist computational psychology contend that the seriality (one-step-at-a-time nature) of Simon and Newell's approach is a drawback, and that human cognition has an inherently parallel nature, with many things happening at once rather than one thing happening after another.

By the early 2000s, Simon's 1967 prediction of fully human-equivalent AI capability by 1986 looked particularly quaint: AI had still not produced computers, for example, with any significant ability to converse in natural language. Computers still could not do most of the things that human beings can do. AI's most impressive demonstrations so far have been in chess-playing and autonomous vehicle navigation.

However, the importance of Simon's contributions to AI and numerous other fields is indisputable. His insistence on mathematical rigor and empirical verification, combined with his passion for considering problem-solving and decision-making in all its forms, has helped shape and reshape several fields of thought.

SEE ALSO *Computer Science: Artificial Intelligence; Computer Science: The Computer.*

BIBLIOGRAPHY

Books

Boden, Margaret. *Computer Models of Mind: Computational Approaches in Theoretical Psychology.* New York: Cambridge University Press, 1988.

Crevier, Daniel. *AI: The Tumultuous History of the Search for Artificial Intelligence.* New York: Basic Books, 1993.

Dreyfus, Hubert. *What Computers Still Can't Do: A Critique of Artificial Reason.* Cambridge, MA: MIT Press, 1992.

Gardner, Howard. *The Mind's New Science: A History of the Cognitive Revolution.* New York: Basic Books, 1985.

McCorduck, Pamela. *Machines Who Think.* Natick, MA: A.K. Peters, Ltd., 2004.

Simon, Herbert. *Administrative Behavior: A Study of Decision-Making Processes in Administrative Organizations.* 3rd ed. New York: Macmillan, 1976.
———. *Models of My Life.* New York: Basic Books, 1991.
———. *The Sciences of the Artificial.* 1st paperback ed. Cambridge, MA: MIT Press, 1970.
———. *The Shape of Automation for Men and Management.* New York: Harper & Row, 1965.

Periodicals

Bendor, Jonathan. "Herbert A. Simon: Political Scientist." *Annual Review of Political Science* 6 (2003): 433–471.
Feigenbaum, Edward A. "Herbert A. Simon, 1916–2001." *Science* 291 (2001): 2107.

Simon, Herbert. "The Information-processing Theory of Mind." *American Psychologist* 50 (1995): 507-508.
———. "Rational Choice and the Structure of the Environment." *Psychological Review* 63 (1956): 129–138.
———. "Theories of Decision Making in Economics." *American Economic Review* 49 (1954): 223–283.

Web Sites

Simon, Herbert. "Autobiography." 1978. http://www.nobelprize.org/nobel_prizes/economics/laureates/1978 (accessed January 5, 2008).

Larry Gilman

Computer Science: Information Science and the Rise of the Internet

■ Introduction

First emerging as an academic discipline in the 1960s, information science is the field of knowledge concerned with the storage, organization, search, and retrieval of information stored in all kinds of documents, including computer records. Because of the information explosion of the twentieth century, including millions of printed books, newspapers, and articles as well as trillions of terabytes on the Internet, it would be impossible to access most human knowledge without the tools devised by information science.

■ Historical Background and Scientific Foundations

The Ancient Period

Written language first developed in the Middle East in 3000 BC. By 2500 BC, Sumerian scribes were collecting large numbers of documents in central locations. The first documents listing other documents appear at about this time, showing the first effort to use an information technology system to manage the information itself. The first library in the world was probably assembled by the Assyrian king Ashurbanipal (685–627 BC), who authorized his chief scribe to collect every tablet in the kingdom—by force, if necessary—and bring it to the capital city to be added to his collection. (Documents were literally "tablets" at this time, small rectangles of solid clay with marks pressed into their surfaces by sharpened reeds.)

The first Chinese library was assembled about 1400 BC; the Egyptian Pharaoh Ramses II built one in Thebes about 1225 BC; and in India manuscript collections appeared about 1000 BC. The destruction of libraries was also routine. Ashurbanipal's library was destroyed when the Assyrian empire collapsed. In 213 BC the new emperor of China, Shi Huangdi, ordered every book in China destroyed so that he could replace them with books composed according to his own ideas.

The greatest library of ancient times, founded about 300 BC, was the Great Library at Alexandria, a city in Egypt named after the Greek conqueror Alexander the Great (356–323 BC). The Greek ruler of Egypt at that time, Ptolemy I (367–283 BC) built a royal library and offered special incentives, including money and free housing, to attract scholars from surrounding countries.

By 47 BC the collection contained about 700,000 written works, vastly more than any other collection in the world. This library also featured the world's first system of classification. Rather than works being mixed together in an indiscriminate jumble, they were segregated into different rooms based on their subject matter. The scholar Callimachus (305–240 BC) created a 120-volume catalog listing the contents of the entire library as it existed in his day. Tragically, the Great Library at Alexandria was destroyed. The exact date of the disaster and its cause are not certain; the library was probably burned, partly or entirely, on more than one occasion.

Throughout the Middle Ages, European written culture was kept alive mostly by the efforts of Christian monks in scattered monasteries, memorizing and copying the scriptures onto parchment. Even the largest collections of documents at this time did not number above the hundreds, and there was little need for special systems to manage information.

The Era of Print

The great turn toward the modern information era came about 1450, when German craftsman Johannes Gutenberg (1400–1468) invented the printing press and movable type. In this system, individual letters

WORDS TO KNOW

HTML: Hypertext markup language (HTML) is a system of instructions that can be used to define the appearance and behavior of a hypertext document. Hypertext is computer-displayed text that not only conveys a direct message, but, if clicked on by a user, acts as a link to other information (e.g., image, sound, text). HTML is the basis of the World Wide Web of linked hypertext documents. The earliest version of HTML was developed by physicist Tim Berners-Lee (1955–) in 1991.

HYPERTEXT: Hypertext is computer-displayed text that not only conveys a direct message but, if clicked on by a user, acts as a link to other information, whether text, audio, or image.

INTERNET: A vast worldwide conglomeration of linked computer networks. The most significant component of the Internet is the World Wide Web.

MICROFILM: Microfilms are miniature films used for photographing objects and documents. The images on these films cannot be seen without an optical aid, either in a form of a magnifying glass or a projector.

SCRIBE: In various ancient cultures (e.g., Egypt), a literate person employed to write and read documents for religious, legal, and business purposes. In ancient times the ability to read and write was a rare skill, and the scribe was a valued professional. More generally, any person who copies documents by hand in any historical period can be termed a scribe.

SEARCH ENGINE: Computer programs that search for specific strings of characters in memory. The most commonly used search engines are those that search the World Wide Web (e.g., Safari, Explorer, Firefox, Flock).

WORLD WIDE WEB: A collection of hypertext markup language documents that are linked to each other through the Internet. At least 10 billion (and possibly many more) pages are interlinked to form the World Wide Web, which has developed entirely since 1989.

---

are mass-produced, then arranged together in a block or plate to form a mirror image of a page of text. The plate is then wetted with ink and pressed against a blank sheet of paper. Gutenberg's method allowed for the rapid creation of new plates, making it possible to manufacture books far more cheaply and rapidly than ever before. Printing presses quickly swept across Europe, transforming its culture; printing presses remained the world's primary information technology for the next 600 years.

Over the next few centuries, printed books became steadily cheaper and more numerous. As knowledge expanded and book collections swelled, the dream of mechanized organization of the world's knowledge first surfaced. In 1532, Giulio Camillo (1480–1544), a Venetian, designed a curious device called the Theater of Memory. This consisted of a small room in which the user could sit, surrounded by an array of small windows covered by shutters. The shutters could be opened by a geared mechanism to reveal words and images. Although Camillo could not mechanize the processing of information, only display it, he seems to have anticipated a form of the "windows" principle by which most modern users interact with computers.

Camillo's Theater of Memory was never completed and no similar devices were built. However, it is the first recorded attempt to mechanize access to all knowledge, one that would eventually be fulfilled in some measure by the Internet.

Long before the Internet and computer technology emerged, however, the need to master large amounts of printed information was a continuous problem. In 1668 the Scientific Revolution generated so many journals and papers that English bishop John Wilkins (1614–1672) proposed a radical solution: a "Universal Language" that would place most of the world's knowledge into 40 categories, each with a unique four-letter name.

Wilkins argued that natural language was inadequate to handle the quantity and diversity of all known facts, an anticipation of modern computer languages that handle the quadrillions of bits of information available today. At first Wilkins's plan was received enthusiastically by scholars, just as 300 years later ARPANET, the prototype of the Internet, would be built for researchers. However, Wilkins's scheme was too arbitrary and rigid to fulfill its promise, and it was soon forgotten.

The information problem continued to grow. From the invention of the printing press to 1500, Europe manufactured about 20 million books; in the following century, it manufactured ten times as many. Demand grew for a compact, affordable way to access knowledge in many fields at once.

This demand was met by the invention of the encyclopedia. The first modern encyclopedia was edited by Denis Diderot (1713–1784), whose *Encyclopédie, ou dictionnaire raisonné des sciences, des arts et des métiers* (Encyclopedia, or a systematic dictionary of the sciences, the arts, and the professions) appeared in installments from 1751 to 1772. This was a revolutionary technology and recognized as such: The Pope and the kings of England and France all condemned it for giving the masses easy access to technical knowledge. The French government ordered publication of the work to stop and literally imprisoned 6,000 volumes in the Bastille. Even in the modern world repressive governments, such as

Ptolemy I Soter inaugurates the Great Library at Alexandria.   *Mary Evans/Photo Researchers, Inc.*

those of China, Iran, and Cuba, try to control access to the World Wide Web.

Throughout the eighteenth and nineteenth centuries, the organization of most libraries remained crude. As at the Great Library of Alexandria, libraries' contents were listed in specialized books called catalogs. At the beginning of the twentieth century, however, a new library technology appeared—the card catalog, a cabinet filled with small, alphabetically ordered drawers holding paper cards, one card for each book in the library. Other new schemes of library organization were also proposed. The most successful, at least in the United States (where it is still used in thousands of libraries) was that proposed by Melvil Dewey (1851–1931). Today, like physical card catalogs, Dewey's and other cataloging systems have been transferred to computers.

Mechanical Dreams, Library Science, and Rise of the Internet

The idea of combining text, photographs, microfilm, television, and other mechanical aids to produce a sort of twentieth-century Theater of Memory was the dream of Paul Otlet (1868–1944), who established the professional field of document tracking and organizing that gave rise to information science in the late 1940s. Years before electronic computers were invented, Otlet promoted the idea of a universal knowledge network that could be viewed on screen via what he called an "electric telescope." Sitting at a special circular desk packed with apparatus, the user would have access to essentially all the recorded knowledge of the world, a "universal book."

Although Otlet's vision of mechanized universal access was impractical, his insights into the relationships between texts was not. In the 1930s he invented the term "links" to name references in one text to other texts and described the resulting set of relationships between books, articles, and other information sources as a "web." He also saw the need not only for an elaborate information-handling machinery, but for some type of search and retrieval system—functions supplied today by search engines.

There was little interest in Otlet's ideas, however, especially as World War II (1939–1945) loomed. After

Computer Science: Information Science and the Rise of the Internet

the war, American engineer Vannevar Bush (1890–1974) proposed a system he called the Memex, widely hailed as the immediate conceptual forebear of the Internet. In a popular 1945 *Atlantic Monthly* article, "As We May Think," Bush proposed a workstation-type desk packed with apparatuses. Information would be stored on reels of microfilm; images and text would be projected upward onto desktop viewing screens. A built-in camera would provide the photographic equivalent of a scanner, allowing the user to add texts and images to the Memex's microfilm memory. Crucially, and recalling Otlet's notion of "links" (though there is no evidence that Bush knew of Otlet's work), Bush proposed that the Memex would allow its user to make notes linking one document to another to form "associative trails," anticipating the hyperlinks that tie the World Wide Web together.

Scholars have noted that, unlike the Internet, Memex's goal was to *contain* all information relevant to a given researcher, not to link to a network of outside sources. It was to be, in effect, a miniaturized private library plus "associative trails," not a terminal or node in a larger network. Its closest parallel is perhaps an MP3 player or iPod, not a personal computer linked to the Internet. Although Memex was never built, Bush's idea generated excitement, talk, and fresh awareness of the possibility of mechanizing access to information—just

as the digital electronic computer was in the process of being invented.

A decade later, in the mid-1950s, in an apparently unrelated development, American library science student Eugene Garfield (1925–) devised a method of citation ranking—a way to measure the influence of individual articles in scientific journals. The more a paper is cited, the more important it is deemed to be, and the higher a ranking it is given. Soon, Garfield's method became a standard fixture of scientific research, helping scientists cope with the overwhelming and ever-growing number of publications in their field by telling them which were the most important. The Science Citation index, a privately-produced database of citation rankings, is purchased by all university libraries for use by researchers. The significance of citation ranking is that its basic method was adopted half a century later as the core method of the search engine Google, which ranks Web pages rather than scientific papers and does so automatically (see sidebar). Once again, a technique developed for scholars broke ground that would eventually be used by many millions of users.

The electronic digital computer matured and became a commercial product (though still a large, rare, and expensive one) in the 1950s and 1960s. Most universities and large businesses acquired computers. Individual computers could now be linked to each other

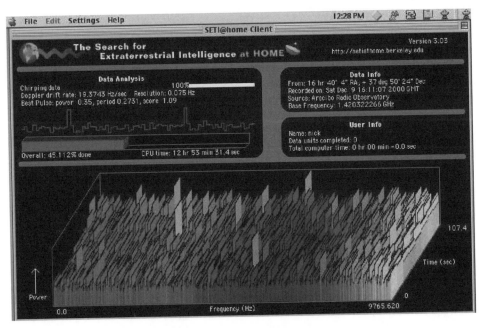

An image from a computer monitor shows the screen display of The Search for Extraterrestrial Intelligence at Home program. For SETI and now other projects, researchers are increasingly turning to a technique called distributed computing to complete in days or weeks computing tasks that would normally take months or years. By combining the idle processing power of thousands, even millions of personal computers on the Internet, they can form a virtual machine more powerful than even the world's fastest supercomputers. *AP Images.*

SCIENTIFIC THOUGHT: IN CONTEXT

electronically through the telephone system or other channels. In 1962, American computer scientist J.C.R. Licklider (1915–1990) and his colleagues described a vision of a network of computers that would allow users to exchange information with each other. Hired by the U.S. government's Advanced Research Projects Agency in 1963, Licklider set about orchestrating the creation of such a network, the Advanced Research Projects Agency Network (ARPANET). ARPANET would link universities and other institutions where researchers sponsored by ARPA worked. In 1969, the first elements of the system became functional—four computers at four American universities. By 1973 there were 40 nodes, and by 1981 there were 213, including international links, mostly to Europe. In the 1970s and early 1980s other, independent networks were built, but during the 1980s these were merged or interconnected with each other. In the late 1980s the first Internet service provider companies came into being, selling access to the growing Internet.

But there is more to today's Internet than a system of computers exchanging data: What has made the Internet useful to hundreds of millions of people is the World Wide Web. Early ARPANET and other computers featured a screen and a keyboard. Users typing at the keyboard saw letters appear on the screen, which might be entered as commands or messages. Messages from other network users or computers would also appear as lines of text. There were no windows—just the single screen, with a growing stack of lines of text—lines that would eventually scroll up out of view beyond the top of the screen. Users had to know specific computer commands to use the system. In the 1970s through the 1990s, several ways of organizing information on the screen and interacting with the computer were invented that would revolutionize this clunky standard interface. Collectively, these made possible the World Wide Web. The Web should not be confused with the Internet. The Internet is a communications network: the Web is a collection of software applications running on the computers connected to that network. In practice, the Web is a mass of several tens of billions of Web pages structured by hypertext markup language (HTML) and linked to each other by hyperlinks. Web pages are visually rich, and users interact with them either by typing in text or by clicking on words or images using a mouse to control a pointer.

These ways of presenting information and interacting with computers may seem obvious today, but were not always so. The mouse was the 1963 invention of American inventor Douglas Engelbart (1925–), a fan of Vannevar Bush's 1945 "As We May Think" essay. Hypertext was invented by Ted Nelson (1937–) in 1968 and first demonstrated publicly by Engelbart in the same year. The windows-type screen environment was developed by Xerox researchers in the mid 1970s. In

IN CONTEXT: DID AL GORE INVENT THE INTERNET?

During the presidential campaign season of 1999–2000, strange headlines appeared asserting that candidate Al Gore (1948–), a Democrat and Vice President of the United States from 1993 to 2001, had claimed in a TV interview to be the "inventor of the Internet." The story appeared in dozens of opinion columns and editorials. Many pundits described Gore as "delusional." The Associated Press ran a story headlined, "Republicans pounce on Gore's claim that he created the Internet" (March 11, 1999).

Gore must have been crazy, the commentators said: The Internet is not any one person's invention. If anyone could lay claim to that title of the Internet's inventor it might have been J.C.R. Licklider (1915–1990), who in the 1960s created the military-funded network of university computers called ARPANET, which eventually evolved into the Internet. Or perhaps the credit might be shared with Tim Berners-Lee, who in 1991 invented hypertext markup language (HTML), the system of coding that makes all Web pages possible.

The accusation against Gore was, however, fundamentally flawed. What he said in the TV interview was not that he had "invented" the Internet or was "the father" of the Internet, as many outlets wrongly reported, but this:

"During my service in the United States Congress, I took the initiative in creating the Internet."

Was this claim correct? As a matter of Internet history, it is: In the 1980s, Gore was the leader of bipartisan Congressional efforts to construct the physical backbone of what would become the Internet in the 1990s. He coined the term "information superhighway" and proposed legislation to fund the construction of high-speed, transcontinental data links. Senator Newt Gingrich, a Republican and otherwise a political enemy of Gore's, worked with him in the 1980s to lay the groundwork for the Internet. In 2000 Gingrich said, "in all fairness, Gore is the person who, in the Congress, most systematically worked to make sure that we got to the Internet." (C-SPAN, September 1, 2000).

Reports of Gore's delusions of grandeur were incorrect. Al Gore did not invent the Internet—and Al Gore never claimed to have invented the Internet.

Gore lost the 2000 presidential election to George W. Bush, but later won the 2007 Nobel Peace Prize, shared with the United Nations Intergovernmental Panel on Climate Change, for his groundbreaking work bringing attention to the problem of global climate change, most notably with the film *An Inconvenient Truth.*

1990, English physicist Tim Berners-Lee fused graphic user interfaces, HTML—a program he wrote himself to manage HTML documents on-screen (i.e., a Web browser)—and the Internet to produce the beginnings

IN CONTEXT: SEARCH ENGINES

Search engines such as Ask.com, Google, and Yahoo!Search are the spark of life in the Internet. Without such tools, looking for a specific piece of information on the Internet would be like trying to take a sip of water from an open fire hydrant.

A user of a search engine inputs a search term or phrase, and the search engine displays a number of Web pages chosen from the several tens of billions of Web pages that exist. The first step in enabling the engine to do this is to "crawl" the Web, that is, to automatically visit as many Web pages as possible—all of them, ideally—and index their contents. Indexing means making a list of words, each accompanied by a list of pointers to places where the word occurs. In a book, the pointer points to a page number; on the Internet, it points to a Web page address.

But simple indexing is not enough. Pages must also be ranked, that is, ordered from the ones that the user will probably find most interesting to those they will probably find least interesting.

Search engine companies use many secret methods for ranking Web pages, as well as other tricks. The basic method used by Google is known, since the mathematical formula was made public in 1999. For every page on the Web, Google calculates a number that says how important that page is: This number is called a page rank. When you perform a search, pages with high ranks are more likely to appear on your screen. The Google formula calculates a rank for a Web page—call it page Z—by looking at the pages that link to Z. For each page linking to Z, a number is calculated by dividing its own page rank by the number of its out-going links. Z's rank is then given as the sum of all these fractions (modified by an arbitrary number called the damping factor).

In this system, the page rank of every page depends on the page rank of many other pages, which in turn depend on the page ranks of other pages, and so on. In practice, powerful computers owned by search-engine companies are continually re-calculating page rankings for the entire Web as it grows.

of the World Wide Web. Indeed, there was a brief time when the World Wide Web included only a single computer, Berners-Lee's own.

■ Modern Cultural Connections

The invention of the World Wide Web in 1990 was possible because the Internet already existed. The creation of the Internet depended on artificial languages developed for computers and elaborate methods of organizing information inside computer memories. These techniques, in turn, can in part be traced to methods developed by the fields of documentation and information science in the early twentieth century. Without those techniques, which allowed scientists and engineers to trace what was important to them in the millions of pages of printed technical matter already being produced, it would have been difficult to invent modern computers originally.

Since 1990, Berners-Lee's desktop Web has indeed become worldwide, transforming every aspect of life in the industrialized countries from courtship to business. Techniques borrowed directly from library science are at the heart of the search engines (Google, Yahoo!Search, etc.) that make the Web functional in practice, and are therefore a lynchpin of the Internet-driven aspect of the modern global economy. All technologies have a potentially dark side, however, and the Internet and Web are no exception: the same information-management techniques of listing, searching, and ranking that can be used to connect users with information that is relevant to them can be run by corporations and governments in reverse, tracing Internet usage backward to individual people, who may then be discriminated against, spied upon, arrested, or even killed. Legitimate law enforcement, such tracking child-pornography traffickers and terrorists, makes use of essentially the same set of tools.

Despite their extreme popularity, it is difficult to point any overall improvement in society that can be attributed directly to the Internet or Web. In the industrialized countries, science literacy has increased only slightly, if at all, since the Internet exploded into daily life. Worldwide, military dictatorships remain in control despite millions of Internet users in their countries. In China, which has more Internet users than any other country except the United States (over 50 million), all Internet traffic enters and leaves the country through a cluster of supercomputers owned and operated by the government. This allows government agents to scan the content of all Web pages, blogs, e-mails, and other international Internet exchanges, looking for—and punishing the users of—content the government disapproves of for political reasons. Since the late 1990s, at least several dozen people have been jailed in China for using the Internet for forbidden political purposes; several have been tortured, according to the human rights organization Amnesty International. In 2006, the Internet search-engine company Google, submitting to Chinese government demands, agreed to voluntarily block access by users inside China to international sites banned by the Chinese government.

In the United States, a minority of eligible voters still participates in national elections, and the amount of time spent daily reading books (the only form in which most people can still have in-depth, prolonged encoun-

Chinese youth use computers at an Internet café in Beijing. *AP Images/Greg Baker.*

ters with fiction or nonfiction works) is declining. Simply making trillions of searchable terabytes available to people has not, by and large, made them freer, happier, or better educated. Furthermore, the Internet itself, according to some experts, is beset by deep organizational flaws, such as the one-way nature of hyperlinks. Flawed or shallow information, hoaxes, spam, propaganda, pornography, and advertising make up a large fraction—perhaps a majority—of what is available on the Internet. Old problems have migrated to the new medium, and the new medium has given rise to new problems.

Nevertheless, it can be argued that the Internet has had a net positive effect in at least some areas. For scholars and scientists, its original users, it is an essential tool for sharing large quantities of complex data rapidly. Some individuals have been empowered by the Internet to become more politically active, generally educated, or economically independent.

SEE ALSO *Computer Science: Artificial Intelligence; Computer Science: The Computer.*

BIBLIOGRAPHY

Books

Albarran, Alan B., and David H. Goff, eds. *Understanding the Web: Social, Political, and Economic Dimensions of the Internet.* Ames, IA: Iowa State University Press, 2000.

Arms, William Y. *Digital Libraries.* Cambridge, MA: M.I.T. Press, 2000.

Ceruzzi, Paul E. *A History of Modern Computing.* 2nd ed. Cambridge, MA: The M.I.T. Press, 2003.

Katz, James E., and Ronald E. Rice. *Social Consequences of Internet Use.* Cambridge, MA: The M.I.T. Press, 2002.

Lilley, Dorothy B., and Ronald W. Trice. *A History of Information Science, 1945–1985.* San Diego, CA: Academic Press, Inc., 1989.

Wright, Alex. *Glut: Mastering Information Through the Ages.* Washington, DC: Joseph Henry Press, 2007.

Periodicals

Dalbello, Marija. "Is There a Text in This Library? History of the Book and Digital Continuity." *Journal of Education for Library and Information Sciences* 43 (2002): 11–19.

Veith, Richard H. "Memex at 60: Internet or iPod?" *Journal of the American Society for Information Science and Technology* 57 (2006): 1233–1242.

Larry Gilman

Computer Science: Microchip Technology

■ Introduction

Microchips—also called silicon chips, integrated circuits, and several other terms—are small, thin, rectangular chips or tiles of a crystalline semiconductor, usually silicon, that have been layered with large numbers of microscopic transistors and other electronic devices. These devices are a part of the chip's crystal structure, that is, integral to it—hence the term "integrated circuit." An integrated circuit may contain billions of individual devices but is one solid object.

WORDS TO KNOW

INTEGRATED CIRCUIT: A system of interconnected electronic components such as transistors and capacitors that have been built as part of a single, solid crystalline structure of silicon or some other semiconductor. Because the components are integral to (a continuous part of) a single solid object, they are said to form an integrated circuit.

MOORE'S LAW: For about 40 years, the number of electronic components that can be manufactured on a single integrated circuit (microchip) at a certain cost has doubled every few years. This trend has been described as Moore's law since 1965, when it was identified by U.S. engineer Gordon Moore (1929–).

SEMICONDUCTOR: Substance, such as silicon or germanium, whose ability to carry electrical current is lower than that of a conductor (like metal) and higher than that of insulators (like rubber).

TRANSISTOR: A solid-state electronic device that is used to control the flow of electricity in electronic equipment. It consists of a small block of semiconductor with at least three electrodes.

The prefix "micro" refers not to the chip itself, although a typical microchip is quite small—a centimeter or less on a side—but to the microscopic components it contains. The microchip has made it possible to miniaturize computers, communications devices, controllers, and hundreds of other devices. Since 1971, whole computer CPUs (central processing units) have been placed on microchips. These affordable, highly complex devices—microprocessors—have been the basis of the computer revolution.

By 2008, at least 5 billion microchips were being manufactured every year in the United States alone, and many more were being manufactured globally. Microchips and computers are now used in scientific instruments, military weapons, personal entertainment devices, communications devices, vehicles, computers, and many other applications, and are an important part of the global economy. In 2007, the global semiconductor industry sold about $256 billion worth of microchips. The social effects of cheap computation have been profound, though not as overwhelming as computer enthusiasts have repeatedly predicted.

■ Historical Background and Scientific Foundations

Before the chip, electronics depended on the three-electrode vacuum tube, which was invented in 1907 by American inventor Lee De Forest (1873–1961). This device allowed the amplification of variations in a current (e.g., an audio signal) by using the signal to be amplified to control the flow of a more powerful current, somewhat like a small amount of force applied to wiggling a faucet valve can produce a pattern of identical wiggles in a more forceful flow of water. Such tubes are still called "valves" in British English for this reason. The vacuum tube was the beginning of modern electronics and made possible the invention of sensitive two-way radio, television, and electronic computers. It was, however, fragile, bulky, power-hungry, expensive, and prone to breakdown. A smaller, less wasteful, more reliable, cheaper alternative could, some scientists speculated in the 1930s, be made out of solid materials. Such a device would require no vacuum, no fragile glass bulb, and no glowing-hot filaments of wire. In 1947, scientists at Bell Laboratories in the United States built the first crude device of this kind. The new device, a transistor, did the same job as the vacuum tube but had none of its disadvantages.

For years, transistors were manufactured as separate (discrete) devices and wired together into circuits. Although a vast improvement over vacuum tubes, such circuits were still bulky and fragile. In 1958, the microchip was conceived independently, but at about the same time, by U.S. engineers Jack Kilby (1923–) and Robert Noyce (1927–1990). A microchip or integrated circuit has all the advantages of a discrete transistor circuit but is even smaller, more efficient, and more reliable. In 1962, microchips were used in the guidance computer of the U.S. Minuteman missile, a nuclear-tipped intercontinental ballistic missile intended to be launched from underground silos in the American Midwest. The

U.S. government also funded early microchip mass-production facilities as part of its Apollo moon-rocket program, for which it required lightweight digital computers. The Apollo command and lunar modules each had microchip-based computers with 32-kilobyte memories, that is, memories capable of storing 32,000 bytes. (A byte is eight binary digits, 0s or 1s, also called bits.)

Microchip progress was rapid, because profits were enormous: The industrialized world seemed to have in insatiable appetite for ever-more-complex electronic devices, as it still does. These could be made affordable, portable, and reliable only through microchips. A little over a decade after the first integrated circuit was tested, in 1971, Texas Instruments placed a calculator on a chip, the first commercial microprocessor. By the end of the decade, many manufacturers were making chips, and the number of transistors and other devices packed onto each chip was rising quickly.

By the early 2000s, a typical desktop computer contained around a million times more memory than the Apollo computers and performed calculations thousands of times faster. The contrast between an Apollo command-module computer costing millions of dollars in 1968 and a far-more-powerful desktop computer costing $2,000 or less in the early 2000s reflected the

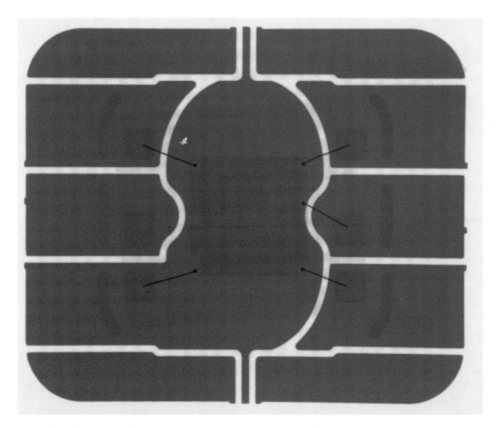

Microchips, like the one shown here, are also used in various credit cards. Such microchips are located in the corner of the credit card, often under a gold or silver covering. The chip electronically encodes the information required to carry out a monetary transaction. *Gusto/Photo Researchers, Inc.*

IN CONTEXT: NANOTECHNOLOGY

Nanotechnology extends on advances in microelectronics during the last decades of the twentieth century. The miniaturization of electrical components greatly increased the utility and portability of computers, imaging equipment, microphones, etc. Indeed, the production and wide use of now commonplace devices such as personal computers and cell phones was absolutely dependent on advances in microtechnology.

Despite these fundamental advances there remain real physical constraints (e.g., microchip design limitations) to further miniaturization based upon conventional engineering principles. Nanotechnologies intend to revolutionize components and manufacturing techniques to overcome these fundamental limitations. In addition, there are classes of biosensors and feedback control devices that require nanotechnology because—despite advances in microtechnology—present components remain too large or slow.

rapid changes in microchip technology in that interval, one of the most remarkable success stories in the history of technology.

For about 40 years, the number of electronic components that could be put on an individual microchip at a certain cost has doubled every few years. This trend has been described as Moore's Law since 1965, when U.S. engineer Gordon Moore (1929–) identified it. During those decades, engineers and physicists have continually striven to make electronic components smaller so that more could be fit on each microchip. (Simply making chips larger to fit more components on them would not have worked, since the time needed for signals to travel across a sprawling chip would slow its operation.) Since the early 1990s, however, designers have been warning that miniaturization is becoming steadily more difficult as the dimensions of transistors and other integrated devices approach the atomic scale, where quantum uncertainty will inevitably render traditional electronic designs unreliable. In 2005, an industry review of semiconductor technology found that the limits of the silicon-based microchip may be reached by about 2020. They agreed that alternative technologies, such as quantum computing or biology-based approaches, all have flaws, and that there is not yet any clear successor to pick up in 2020 where silicon leaves off.

Manufacture of a microchip begins with the growth in a factory of a pure, single crystal of silicon or other semiconducting element. A semiconductor is a substance whose resistance to electrical current is between that of a conductive metal and that of an insulating material such as glass (silicon dioxide, SiO_2). This large, cylindrical crystal is then sawed into disc-shaped wafers 4–12 inches (10–30 cm) across and only 0.01–0.024 inches (0.025–0.061 cm) thick. One side of each wafer is polished and then processed to produce upon it dozens of identical microchips. These are separated after the wafer is processed, placed in tiny protective boxes called packages, and connected electrically to the outside world by metal pins protruding from the packages. In the early 2000s, manufacturers began packing multiple microprocessors onto each chip so that the processors could work in parallel, speeding computation. In 2006, the first chips to contain over 1 billion transistors appeared; in 2008, the number jumped to 2 billion. This was about 50,000 times the number of transistors in the earliest microprocessors of the 1970s.

To produce a microchip requires massive factories that cost billions of dollars and must be retooled every few years as technology advances. The basics of the microchip fabrication process, however, have remained the same for decades—by bombarding the surface of the silicon wafer with atoms of various elements, impurities termed dopants can be introduced into the wafer's crystalline structure. These atoms have different properties from the silicon atoms around them and so populate the crystal either with extra electrons or with "holes," gaps in the crystal's electron structure that behave almost like positively-charged electrons.

Microscopically precise patterns of p-type (positively-doped, hole-rich) silicon and n-type (negatively-doped, electron-rich) silicon are projected optically onto a light-sensitive chemical coating on the wafer (a photoresist). Other chemicals etch away the parts of the photoresist that have not been exposed to the light, leaving a minutely patterned layer. The surface of the wafer is then bombarded by dopants, which only enter the crystal where it is not protected by the photoresist. Metal wires and new layers of doped silicon can be added by similar processes. Dozens of photoresist, etching, and deposition stages are used to build up the three-dimensional structure of a modern microchip. By crafting appropriately shaped p-type and n-type regions of crystal and covering them with multiple, interleaved layers of SiO_2, polycrystalline silicon (silicon comprised of small, jumbled crystals), and metal strips to conduct current from one place to another, a microchip can be endowed with millions or billions of interconnected, microscopic transistors.

■ Modern Cultural Connections

Since their appearance, microchips have transformed much of human society. They are now found in computers, guided missiles, "smart" bombs, satellites for communications or scientific exploration, hand-held communications devices, televisions, aircraft, spacecraft, and motor vehicles. Without microchips, such familiar devices as the personal computer, cell phone, personal digital assistant, calculator, Global Positioning System, and video game would not exist. As chip complexity increases and cost decreases thanks to improvements in manufacturing techniques, new applications for chips are constantly being found.

It would, then, be difficult to name a department of human activity that has not been affected by the microchip. However, its effects have not been as revolutionary as predicted or supposed by forecasters and futurologists. For example, efforts to replace the printed paper book with electronic texts (e-books) downloaded to computers or other chip-based viewing devices have repeatedly failed; most of the world's people still live in poverty and do not have access to sufficient food, clean water, or medical care, much less to a computer; most e-mail carried over the Internet is unwanted junk mail (spam); despite early predictions of a "paperless office," per capita paper consumption has risen, not fallen, since the advent of the microchip; studies have found that persons who spend more than a short amount of each day surfing the Web are more likely to suffer depression, probably as a result of decreased time spent with family and friends; and by 2007, experts estimated that up to 4 million Americans were behaviorally addicted to Internet pornography, with another 35 million viewing it regularly. Nor, on the other hand, despite intensive use of computers by governments to spy on their citizens both at home and abroad, have computers yet made it possible to produce an all-knowing dictatorship as imagined in science fiction. The microchip has produced few entirely new pastimes or economic activities. It has tended to modify existing patterns of human activity—personal, political, military, and economic—but not to transform them out of all recognition or to eliminate them.

Technologically, the great challenge as of the early 2000s was the likely upcoming death of Moore's Law, at least as regards the silicon microchip. Improvements to silicon technology, such as the breakthrough power-conserving technologies to reduce unwanted microchip heating announced by Intel and IBM in 2007, spurred hope of extending silicon's run, though not indefinitely. Quantum computing and other novel techniques were being intensively researched by governments and industries, but all still had to cross major technological hurdles before they could rival silicon's cheapness, speed, and device density.

SEE ALSO *Computer Science: Artificial Intelligence;*
Computer Science: Information Science and the Rise
of the Internet; Computer Science: The Computer.

IN CONTEXT: MICROCHIP TECHNOLOGIES

Microchip-based or enabled technologies stir both the imagination and controversy.

National identity cards are not creations of the twenty-first century. The Nazis used them, and, under apartheid, the South African government required blacks and "coloureds" to carry them at all times. ("Black" denotes only Black Africans, whereas coloured was a separate classification under the apartheid system denoting mixed race, including Indians. The term still has strong cultural connections, and use is still common, including in self-description, but it does not carry the now derogatory connotations of the U.S. term.) Under Nazi rule and apartheid, the cards listed name, residence, and work information; if found in an area to which the bearer was denied access, they were subject to arrest. Accordingly, national ID cards inspire distrust and fear among many. In an age of terrorism and identity fraud, however, some countries are considering using microchip enabled identity cards.

Many countries currently use national identity cards, including most European nations. As technology has progressed, their functions have evolved as well. Taiwan, a country that has used national ID cards since 1947, continued their use from the Japanese colonial government. Taiwan's cards also act as a police record, and the law mandates that they be carried at all times. Issues surrounding inclusion of fingerprint data stirred heated debate as citizens feared loss of privacy data.

In the summer of 2005, shortly after the terrorist attacks on the London, England, subway system, the British Parliament reopened the debate on national identity cards. During World War II the United Kingdom implemented a national ID card system, but it ended the program in 1952. Proponents believe that the cards would help thwart terrorism because every person entering, working in, or living in the country would be required to have one. They would increase the possibility of identifying terrorists before an attack could be carried out. Opponents argue that they cannot guarantee stopping terrorism and could facilitate the quarantining of individuals based on family lineage, ethnic background, or country of origin. Critics contend that they do not want their personal data compiled into a database that could possibly be seen by a computer hacker, nor do they want the government to have large files of their personal information. Also, individuals fear that the cards could make their movements and financial transactions too easy to track. Despite concerns, in February 2006, the British Parliament passed the Identity Cards Act. Registry will be mandatory when applying for documents like a passport, but individuals will not have to carry them at all times. Additionally, the cards will be recognized travel documents in the European Union, and they will contain a microchip holding a set of fingerprints as well as facial and iris scans.

BIBLIOGRAPHY

Books

Reid, T.R. *The Chip: How Two Americans Invented the Microchip and Launched a Revolution.* New York: Random House, 2001.

Yechuri, Sitaramarao S. *Microchips: A Simple Introduction.* Arlington, TX: Yechuri Software, 2004.

Periodicals

Macilwain, Colin. "Silicon Down to the Wire." *Nature* 436 (2005): 22-23.

Williams, Eric, et al. "The 1.7 Kilogram Microchip: Energy and Material Use in the Production of Semiconductor Devices." *Environmental Science and Technology* 36 (2002): 5,504–5,510.

Larry Gilman

Computer Science: The Computer

■ Introduction

Computers are machines that process information. Most modern computers are digital, meaning that they manipulate symbols according to logical rules. The basic symbols used in most digital computers are 0 and 1, which are grouped to designate numbers, words, colors, and the like. Early computer designs, dating from the nineteenth century, were purely mechanical devices. In the 1950s electronic computers came into wide use, although only by large organizations. Computers gradually became a consumer commodity, however, and by the early 2000s hundreds of millions were in use around the world; many millions more were embedded in products such as automobiles, ovens, telephones, personal digital assistants, music players, and toys. Computers, now essential to science and engineering, have effected deep transformations in almost every aspect of society. Utopian predictions of a world freed from drudgery, poverty, political oppression, and other ills by the computer, however, have not been fulfilled.

■ Historical Background and Scientific Foundations

For most of human history, computation was performed with some kind of aid: fingers, piles of pebbles, marks on clay or paper, abaci, and the like. The physical parts of these aids always symbolize or stand for quantities: a finger on the left hand, for instance, can stand for a 1, a finger on the right hand for a 5; beads on one wire of an abacus can stand for 1s, the beads on the next wire for 10s. Mechanical aids eased the burden on human memory and freed people to transform the information they stored with addition, multiplication, and other numerical systems.

WORDS TO KNOW

ALGORITHM: A mathematical relation between an observed quantity and a variable used in a step-by-step mathematical process to calculate a quantity.

ANALOG: A process that is fluctuating or continually changing. In electronics, an analog signal is a base alternating current frequency that is modified, usually by amplification or varying of the frequency, in order to add information to the signal. Conventional forms of television and telephone transmissions use analog technology.

BINARY: The binary number system uses only two digits, 0 and 1, and is basic to computer science. In astronomy, a binary star is a pair of stars orbiting each other and is sometimes called "a binary."

BOOLEAN ALGEBRA: A system that applies algebra to logic. Also called symbolic logic, it converts logic into mathematical symbols.

DIGITAL: The opposite of analog, it is a way of showing the quantity of something directly as digits or numbers.

DOPANT: An impurity added to a semiconducting material.

SEMICONDUCTOR: Substance, such as silicon or germanium, whose ability to carry electrical current is lower than that of a conductor (like metal) and higher than that of insulators (like rubber).

TRANSISTOR: A solid-state electronic device that is used to control the flow of electricity in electronic equipment. It consists of a small block of semiconductor with at least three electrodes.

VACUUM TUBE: A sealed glass tube in which the conduction of electricity takes place through a vacuum or gas.

A more complex class of devices began to be built in ancient times—mechanical analog computers. Around the end of the second century BC for example, the Greeks produced a sophisticated geared device known as the Antikythera mechanism, which was used to calculate lunar and solar eclipses as well as other astronomical movements. In the seventh or eighth century AD, Islamic scientists produced the first astrolabes, circular metal calculators used to predict the positions of the sun, moon, and stars that were used for navigation and for astrology.

Another early calculator was invented by Scottish scientist John Napier (1550–1617) in the late sixteenth century. "Napier's bones" were a set of marked ivory bars that sped up calculations by substituting addition and subtraction for multiplication and addition. After Napier invented logarithms in the early 1600s, the first slide rules were developed, using marked sliding sticks to record logarithmic relationships between numbers. The modern slide rule, invented in the 1850s, was standard in science and engineering work until the invention of the handheld electronic calculator in the 1960s.

Gottfried Wilhelm von Leibniz (1646–1716) might be called the world's first computer scientist. He invented binary (base 2) arithmetic, which uses only the numerals 0 and 1, unlike decimal or base 10 arithmetic, which uses the numerals 0 through 9. In 1673 he built a new type of calculator that could do multiplication; mechanical calculators that evolved from this design were used until the 1970s.

Another important step was taken by British mathematician George Boole (1815–1864) in 1854, when he described a system of logical rules, today called Boolean algebra, by which true-or-false statements could be handled with mathematical rigor. The 1s and 0s of Leibniz's binary arithmetic can be identified with the true and false statements of Boolean algebra. Since any two-state device—an on-off switch, for instance—can store the value of a Boolean variable, Boole's new algebra paved the way for mechanized general logic, not just arithmetic.

In the mid-1800s English mathematician Charles Babbage (1791–1871) conceived a revolutionary design that he called the Analytical Engine. If completed, it would have possessed all the basic features of a modern computer: Babbage had designed a memory, or "store," of a thousand 50-digit numbers, the equivalent of about 20 kilobytes—more memory than some electronic digital computers had in the 1970s. These numbers were to be processed in a central processing unit called the "mill." Finally, the machine would have been programmed with instructions on punch cards, using technology borrowed from the Jacquard loom.

Due a complex series of misfortunes, this brilliant invention was never built, and Babbage was largely forgotten after his death. It was not until the 1930s that the

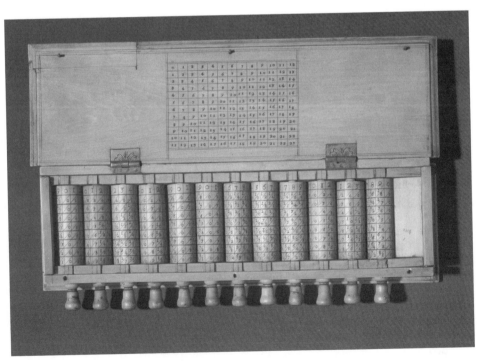

A calculating device created by Scottish mathematician John Napier consists of cylinders inscribed with multiplication tables also known as "Napier's bones." © *Visual Arts Library, London/Alamy.*

A 1946 photo of the ENIAC computer shows technicians in the process of programming the computer to solve a hydrodynamical problem. *Time & Life Pictures/Getty Images.*

concept of the programmable computer was approached again. By this time, mechanical on-off switches (relays) driven by electricity had been perfected for use in the telephone system; vacuum tubes, which can be also used as on-off switches, had been developed for radio. Both devices were adapted for use in digital computers.

In 1930 American engineer Vannevar Bush (1890–1974) built an analog computer called the Differential Analyzer. Although it could not perform logical operations on symbols as digital computers do, it allowed mechanical or electrical quantities (cam motions, voltages, etc.) to behave like particular mathematical functions. By measuring how these quantities behave, it could solve various mathematical problems, including differential equations. One advanced differential analyzer built by Bush in the late 1930s weighed 100 tons and contained some 2,000 vacuum tubes and 200 miles (322 km) of wiring. These analyzers were used to compute firing tables for artillery gunners during World War II (1939–1945).

But the future of computing did not lie with analog computers. In the late 1930s American engineer George R. Stibitz (1904–1995) and colleagues built a small digital computer based on telephone relays and Boolean algebra. Harvard mathematician Howard Aiken (1900–1973), collaborating with International Business Machines Corp. (IBM), began building the IBM Au-

tomatic Sequence Controlled Calculator in 1939, a relay-based computer that could be programmed with instructions coded on a reel of paper tape with small holes punched in it. In the same year mathematician John Atanasoff (1903–1995) and Clifford Berry (1918–1963) designed (but did not finish) a fully electronic vacuum tube computer that stored information on a rotating drum covered with small charge-storage devices called capacitors, a forerunner of the modern hard drive.

Increasingly sophisticated "second-generation" electronic computers built during the 1940s and early 1950s gradually eliminated all mechanical parts. Many of these advances were driven by military needs; England's Colossus, for example, was devoted to cracking the secret German military code Enigma. Individual computers were bult for specific projects, each with a unique name: Colossus, ENIAC, EDSAC.

The shift to standardized commercial units began with the UNIVAC I (Universal Automatic Computer I). The first one, bought by the U.S. Census Bureau in 1951, filled a large room. Nineteen were sold to government and industry from 1951 to 1954. Other large standardized computers were soon on the market, dominated for many years by IBM.

Until they were replaced by transistors, computers ran on vacuum tubes—sealed glass cylinders containing

The U.S. Census Bureau begins tabulating results of the 1954 Census of Business with the UNIVAC, a giant electronic computing system. *© Bettmann/Corbis.*

electrical components but almost no air. Each tube acted as an amplifier (in analog devices) or as an on-off switch (in digital devices). Vacuum tubes use a lot of energy and break down frequently, factors that limit computer size. Transistors, invented in the late 1940s, are small power-sipping devices made of solid crystal. The first all-transistor computer was the TRADIC (transistor digital computer), a special-purpose computer built in 1955 and installed in B-52 Stratofortress bombers. In 1957 IBM announced that it would replace vacuum tubes with transistors in all its computers.

In these "third-generation" computers, the transistors were small wired cylinders. These were made largely obsolete by the integrated circuit, a small solid tile or chip of semiconducting material (e.g., silicon) in which multiple transistors, resistors, capacitors, and wires have been created on layers and regions of the chip by depositing elements called dopants. The first commercial chip was produced in 1961; it contained four transistors and stored one bit ("binary digit," a 0 or 1). By 2007 some chips contained over a billion transistors. For example, the Dual-Core Intel Itanium 2 processor released in 2006 contained 1.72 billion transistors.

In 1970 an entire computer was put onto a single chip, and the microprocessor was born. This led to the development of handheld calculators the following year

and the debut of personal computers in 1974. Since then, the history of computing has largely been the evolution of increasingly powerful microprocessor chips at lower prices.

Starting in the 1980s, networked computers—arranged to exchange data over phone lines or other electronic channels—led to the development of the Internet and World Wide Web. In the 2000s, computing power, storage capacity, and connectivity continued to skyrocket, expanding the ways in which computers could be used, from scientific calculation to personal entertainment.

The Modern Digital Computer

Modern digital computers are based on binary arithmetic, which symbolizes all numbers as combinations of 0 and 1, such as 0110 for the number 8. Computers represent each 1 or 0 with an on-or-off electrical signal. Feeding these signals into devices produces other electrical signals according to the rules of Boolean algebra. All the millions of tasks carried out by modern computers, including graphics, sound, game playing, and the like, are built on Boolean bit-level operations.

The bit is the logical basis of computing; the transistor is its on-off switch. A small group of transistors can be linked together into a device called a flip-flop, which stores the value of a single bit. Other combinations of transistors take bits and perform Boolean operations on them. Millions of transistors are hidden inside most modern computers, many of them changing state (flipping from 0 to 1 or back again) billions of times per second. Market pressures have forced engineers to figure out how to make transistors smaller, faster, and more energy efficient. As of 2007 prototype transistors could change state almost a trillion times a second; others were only 3 nanometers wide.

Computers interface with humans through screens, printers, speakers, pads, keyboards, and mice. They communicate with other computers on high-speed communications channels, receive data from special sensors, and operate machinery using a wide array of actuators. As with computing hardware, there is intense market pressure to produce cheaper, lighter, more flexible, more impressive interface devices.

Computer Science

Computer science is the field of knowledge devoted to designing more efficient computer architectures (arrangements of hardware and software) and more useful methods, called algorithms, for tasks such as searching memory, identifying patterns, encrypting or decrypting data, and solving mathematical problems. Programs that perform specific tasks when executed by a computer are as essential to computer function as the hardware itself, making software production a major industry.

One rule of computer science is described by Moore's Law, which was first posited by Intel Corporation cofounder Gordon Moore (1929–) in 1965. In its modern form, the law says that the number of transistors on a single chip doubles every two years. Graphs of cost versus memory, processing power, or pixels (for digital cameras) show that the law has held approximately true for over 40 years. Driven by human ingenuity, motivated by the desire for profit, and working with the limits imposed by the laws of physics, engineers have found clever ways to make transistors smaller and stuff more of them into every square millimeter of chip surface.

Obviously the trend described by Moore's Law cannot continue forever. Some experts believe that device miniaturization will be limited by the size of the atom in as little as 10 or 20 years After that, the methods used to make computers will have to change fundamentally if more computing power is to be packed into the same tiny space.

Several options beckon, all in the research stage today. One of the likeliest candidates is quantum computing, which exploits the properties of matter and energy at the atomic scale. Quantum computing enthusiasts hope to store multiple bits in single atoms and to teach bits stored in groups of atoms to perform simultaneous calculations as interlinked wholes, breaking out of the one-step-at-a-time limitations of Boolean logic. The underlying physics is real, but whether computers based on quantum bits or "cubits" can be produced remains to be seen.

■ Modern Cultural Connections

Scientists and engineers were among the first to use computers and they rely on them more than ever today. Many scientific tasks can only be accomplished by solving equations, often by numerical methods—that is, the painstaking manipulation of many numbers. Given the extreme tedium, time, and expense of carrying out such calculations by hand, mechanized computing is an obvious need for science. Modern computers not only store data and communicate scientific results, but make

IN CONTEXT: ASSEMBLING THE TREE OF LIFE

In 1859 English biologist Charles Darwin (1809–1882) proposed a theory that all living things are related through an unbroken web of descent. He identified a natural process to explain how living things adapt to their environments— natural selection. Scientists have now turned to computers to better understand the patterns of evolution.

By the early 2000s large amounts of genetic code from thousands of different organisms had been deciphered and made available in computer databases such as GenBank, which is maintained by the U.S. National Center for Biotechnology Information. By 2005 GenBank contained at least partial DNA information for about 6% of all known animal and plant species. Using computers to compare gene sequences from different species, it is possible to tell how closely the species are related and to describe how all living species have diverged from their common ancestors over the hundreds of millions of years.

In 2002 the National Science Foundation launched a project called Assembling the Tree of Life to fund computer research to trace life's family tree. Three years later, scientists proved that by knowing only fragments of genomes (a species' complete DNA complement), it is possible to reconstruct subtrees—clusters of twigs, as it were—if not the whole tree, from the bottom up. This is good news because most genomes are not yet known in full.

Genomic evolutionary analysis is growing rapidly. "The algorithms are just not keeping pace with the data that are available," biologist Keith Crandall said in 2005, "so anyone with a better mousetrap is going to have a huge impact."

many scientific projects possible that would not exist otherwise. Space probes, for example, could not return data from distant moons and planets without onboard computers.

Impact on Society

The earliest practical computers were gigantic and extremely expensive, affordable only to government agencies and a few large private corporations. Starting in 1971, however, handheld microprocessor-based electronic calculators made the slide rule obsolete. But this still did not put general-purpose computing power on the desks of ordinary citizens. That only began to happen in the mid-1970s, when personal computers—computers intended for home use by individuals—first

appeared. The most famous was the Altair 8800, a build-it-yourself computer originally shipped with no screen or keyboard. Although it sold only a few thousand units, it is now credited with jump-starting the market. In 1977 the Apple II became the first mass-marketed personal computer, selling millions of units. After 1981 Apple computers were eclipsed by the IBM and compatible machines made by other manufacturers. In 1983 *Time* magazine declared the personal computer its Person of the Year for 1982.

In the 1980s personal computers became commonplace in American and European homes. They also appeared in schools, but early claims that computers would replace human teachers turned out to be false. In the early 1990s a network of university and government computers called ARPANET (after its sponsoring organization, the Advanced Research Projects Agency of the U.S. Department of Defense) evolved rapidly into the Internet, which now connects virtually all computer users to each other in a global web offering literally millions of cultural and scientific connections.

Computers now permeate all aspects of science, engineering, and manufacturing. They analyze genetic data, communicate with and control probes in deep space, analyze data returned by those probes for clues to the fundamental structure of the universe, and much more. Almost all scientific and engineering work that requires numerical calculation now involves computers.

Computers have become pervasive in industrialized countries. In 2003, for example, about 63% of American adults could access the Internet either at home or work, and this was not even the highest access rate in the world. At that time, 148 million Americans used the Internet and 80% of American Internet users were connected via high-speed broadband connections. By 2005 the average U.S. Internet user spent 3 hours online daily, compared to 1.5 hours watching television. The social effects of all this Internet use have been hailed by enthusiasts, assailed by critics, and studied by sociologists with somewhat uncertain results.

Some experts argue that communicating through the Web encourages the formation of communities, but the analogy with face-to-face interaction is questionable, especially since some studies show that people often find time for the Internet by spending less time with family and friends.

However, scientific results as of 2007 were contradictory. Some studies showed that Internet usage to be socially isolating, others did not. Indeed, both positive and negative claims about the impact of the Internet on society may have been exaggerated. Many important aspects of life, including people's emotional lives, close relationships, and choices about how to spend large blocks of time, turn out to be fairly stable and to resist sudden change even in the face of radical technological innovations like the Internet. For example, despite predictions by Microsoft CEO Bill Gates and others in the late 1990s that computerized books (e-books) were about to replace the printed variety, sales of e-books remain slight compared to paper ($20 million versus $25 billion).

There is less controversy, however, about the economic benefits of the Internet, which has produced a new class of professionals able to work from home all or part of the time, exchanging documents with employers rather than taking themselves physically to workplaces. Internet access, including broadband, is associated with higher personal income; poorer people are less likely to have Internet access and so are less likely to experience its economic benefits, a phenomenon known as the "digital divide."

SEE ALSO   *Computer Science: Artificial Intelligence; Computer Science: Information Science and the Rise of the Internet; Computer Science: Microchip Technology; Physics: Semiconductors.*

BIBLIOGRAPHY

Books

Campbell-Kelly, Martin, and William Aspray. *Computer: A History of the Information Machine.* Boulder, CO: Westview Press, 2004.

Ceruzzi, Paul E. *History of Modern Computing.* Cambridge, MA: MIT Press, 2003.

Ritchie, David. *Computer Pioneers: The Making of the Modern Computer.* New York: Simon and Schuster, 1986.

Shurkin, Joel. *Engines of the Mind: A History of the Computer.* New York: W.W. Norton & Company, 1984.

Periodicals

Buchanan, Bruce G. "A (Very) Brief History of Artificial Intelligence." *AI Magazine* (Winter 2005): 53–60.

Gorder, Pam Frost. "Computing Life's Family Tree." *Computing in Science and Engineering* 7 (2005): 3–6.

Larry Gilman

Earth Science: Atmospheric Science

■ Introduction

Atmospheric sciences are a group of disciplines that comprise the formal study of the atmosphere that envelopes Earth, from just above the ground all the way to near space, where a vacuum exists. Traditionally it is split into three fields of study that developed in approximately the following historical sequence: *meteorology*, which studies the motion and phenomena of the troposphere, also seeks to predict the weather and explain the various processes involved in weather and atmospheric phenomena; *climatology*, the study of the atmosphere at a given place over multiple years or greater periods of time; and *aerology* (or aeronomy), sometimes viewed as a subfield of meteorology, which focuses mainly on chemical and physical reactions that occur within the five major atmospheric layers.

■ Historical Background and Scientific Foundations

From ancient times, man has been keenly aware of the power of nature and the need to better understand the atmosphere and mitigate the impact of hurricanes, tornadoes, drought, torrential downpours, blizzards, and other types of severe weather. The ancient Greeks were the first to write about weather and climate, producing three key texts known today: *On Airs, Waters, and Places* by Hippocrates (460–375 BC) which among other things discusses the impact of climate on human health; *De Ventis* by Theophrastus (372–287 BC), an extensive discussion of the origin and function of winds; and Aristotle's (384–322 BC) *Meteorologica*, the first textbook on meteorology (which included everything between Earth and the moon). Aristotle discussed a wide range of weather phenomena including winds, clouds, rain, snow, hail, thunder, lightning, and rainbows, but time

quickly showed that virtually all of his explanations were incorrect, based as they were on reason with no basis in experimentation. Arabic scholars over the next several centuries added to the knowledge of the Greek philosophers and began to keep records of weather phenomena in their areas.

European interest in finding new lands, people, and riches in the sixteenth century led to the discovery of Arabic books and translations of ancient Greek texts on meteorology. Unfortunately, these voyages of discovery also generated fanciful and wildly exaggerated accounts of weather conditions, climate, and natural phenomena that would negatively influence scientific understanding of Earth's atmosphere for several centuries.

Galileo Galilei (1564–1642) invented a simple thermometer called a thermoscope in 1592 that estimated temperature change by measuring the expansion of air. Despite its pleasing aesthetics, it had no scale and was unreliable, although replicas of this device are still available today. Galileo's device was further improved into a clinical thermometer by fellow Italian Santorio Santorii (1561–1636) in 1612.

Evangelista Torricelli of Italy (1608–1647), an assistant to Galileo, invented a mercury barometer in 1643. The device consisted of a glass tube 3.9 feet (1.2 m) long, filled with mercury and inverted into a dish. Atmospheric pressure exerted on the mercury controlled the height of the mercury and created a vacuum at the top of the column. Torricelli's barometer was also used as a model for the creation of a liquid-in-glass thermometer that was invented in Florence, Italy, sometime in the mid-seventeenth century.

In his book *Les Météores*, René Descartes (1596–1650) reworked many of Aristotle's ideas from the *Meteorologica*; and while it perpetuated many false notions, it moved meteorology onto a somewhat firmer footing with an emphasis on the need for hypotheses and experiments. Perhaps most importantly, in 1662 Irish

WORDS TO KNOW

ALBEDO EFFECT: Albedo is the ability of a planet, moon, or other body in space to reflect light. Brighter objects have higher albedo, darker objects lower albedo. In climate science, the albedo effect is the influence of Earth's albedo on climate. Bright features such as ice caps tend to reflect solar energy into space, cooling the climate. Melting ice lowers Earth's albedo, making it a more efficient absorber of solar energy and tending to warm its climate.

BAROMETER: An instrument used to measure atmospheric pressure. A standard mercury barometer has a glass column about 30 inches (72.6 cm) long, closed at one end, with a mercury-filled reservoir. Mercury in the tube adjusts until the weight of the mercury column balances the atmospheric force exerted on the reservoir. High atmospheric pressure forces the mercury higher in the column. Low pressure allows the mercury to drop to a lower level in the column. An aneroid barometer uses a small, flexible metal box called an aneroid cell. The box is tightly sealed after some of the air is removed, so that small changes in external air pressure cause the cell to expand or contract.

CHLOROFLUOROCARBONS: Chemical compounds containing chlorine, fluorine, carbon, and oxygen. They are widely used in refrigeration and air-conditioning systems and are destructive of the ozone layer in Earth's stratosphere.

FRONT: A boundary between two different air masses. The difference between two air masses is sometimes unnoticeable, but when the colliding air masses have very different temperatures and amounts of water in them, turbulent weather can erupt.

HURRICANE: Large, rotating system of thunderstorms whose highest wind speed exceeds 74 mph (119 km/h). Globally, such storms are termed tropical cyclones: The word "hurricane" is often reserved for tropical cyclones in the Atlantic.

ISOTHERM: Literally, "same-heat." A line on a map or other drawing that connects points of equal temperature. Crossing from one isotherm to another corresponds to movement from a higher temperature to a lower, or vice versa.

JET STREAM: Currents of high-speed air in the atmosphere. Jet streams form along the boundaries of global air masses where there is a significant difference in atmospheric temperature. The jet streams may be several hundred miles across and 1–2 miles (1.6–3.2 km) deep at an altitude of 8–12 miles (12.9–19.3 km). They generally move west to east, and are strongest in the winter with core wind speeds as high as 250 mph (402.4 km/h). Changes in the jet stream indicate changes in the motion of the atmosphere and weather.

METEOROLOGY: The science that deals with Earth's atmosphere and its phenomena and with weather and weather forecasting.

OZONE LAYER: The layer of ozone that begins approximately 9.32 miles (15 km) above Earth and thins to an almost negligible amount at about 31.07 miles (50 km), shields Earth from harmful ultraviolet radiation from the sun. The highest natural concentration of ozone (approximately 10 parts per million by volume) occurs in the stratosphere at approximately 15.53 miles (25 km) above Earth. The stratospheric ozone concentration changes throughout the year as stratospheric circulation changes with the seasons. Natural events such as volcanoes and solar flares can produce changes in ozone concentration, but man-made changes are of the greatest concern.

PSYCHROMETER: An instrument designed to measure dew point and relative humidity, consisting of two thermometers (one dry bulb and one wet bulb). The dew point and humidity levels are determined by drying the wet bulb (either by fanning or whirling the instrument) and comparing the difference between the wet and dry bulbs with preexisting calculations.

PYRHELIOMETER: An instrument for measuring the intensity of solar radiation received at Earth's surface. The Eppley pyrheliometer measures not only the intensity but the duration of solar radiation, allowing the total insolation (exposure to the sun) at the instrument's location over a given period to be calculated.

RADIOMETRY: The measurement of optical electromagnetic radiation, which is defined as electromagnetic radiation in the frequency range of 3×10^{11} and 3×10^{16} Hz (cycles per second). Optical radiation includes visible light but is not limited to it.

RADIOSONDE: An instrument for collecting data in the atmosphere and then transmitting that data back to Earth by means of radio waves.

chemist and physicist Robert Boyle (1627–1691) formulated "Boyle's Law," the basic mathematical relationship between pressure and volume of a gas. In Paris, meanwhile, the government started to collect daily weather data. While the records from these early years are not standardized or complete, it set a global precedent regarding their importance.

Eighteenth-century technological advances in instrumentation allowed much more precise measurements of atmospheric data and encouraged systematic data collection on weather, climate, and the atmosphere, including the use of kites to measure temperatures above Earth's surface, building on the pioneering work of American scientist Benjamin Franklin's (1706–1790)

Evangelista Torricelli (1608–1647), who invented the barometer. © *Bettmann/Corbis.*

study of lightning in 1752. English scientist Sir Edmond Halley (1656–1742), for whom Halley's comet is named, used wind data from ships' logs to construct the first meteorological map in 1668. He focused on the tropical surface winds, called the trade winds. The map was a significant aid to ocean navigation and signaled the beginning of a much more systematic understanding of the dynamic nature of the atmosphere.

Halley knew that warm air rises near the equator and that cooler air moves downward over Earth's surface from its northernmost regions, but he thought that air simply "followed the sun" as it moved from east to west across the sky. Some 50 years later amateur meteorologist and member of the Royal Society George Hadley (1685–1768) was mystified as to why winds in the Northern Hemisphere, which he thought should flow in a straightforward northerly direction, had a pronounced westerly flow. He published information about the trade winds from various sources in his 1735 book, *Concerning the Cause of the General Trade-Winds.*

He argued that sunlight in the equatorial zone strikes Earth at almost a right angle, and therefore each unit area of the surface receives more sunlight than areas farther north. Since warm air is less dense than cold air at the same pressure, this warmer air must rise

and be replaced by cooler air flowing toward the equator from higher latitudes. Applying Newton's first law of motion, Hadley attributed the result to the rotation of Earth. Later study would show that this was incorrect, since in a rotating system it is angular momentum that is conserved, not linear momentum, but his general description moved the study of the atmosphere forward significantly. Temperature readings became more standardized with the introduction of the centigrade or Celsius scale in 1736, named after its inventor, astronomer Anders Celsius (1701–1744) of Sweden.

English chemist Henry Cavendish (1731–1810) discovered hydrogen as a component of the atmosphere and settled a running scientific dispute as to whether the composition of the air varied from place to place. Conducting over 60 experiments throughout what is today central London, Cavendish established that there was virtually no difference in the composition of the air in various locations. He calculated a concentration for oxygen in the atmosphere at 20.83%, a number that is only 0.12% removed from today's accepted value of 20.95%. He went on to show that there was no variation between air on the ground and that a few thousand feet up. Fellow Englishman John Dalton (1766–1844), the founder of modern chemistry, made pioneering studies that discerned correctly that no chemical reaction was involved in the evaporation of water into the atmosphere—a subject quite controversial at the time.

The atmospheric sciences began to develop more rapidly in the early nineteenth century, as did the physical and biological sciences. Weather data such as cloud types, temperature, atmospheric pressure, wind velocity, precipitation and other basic measures were now collected systematically in many places around the world, including New Haven, Connecticut, which has the longest continuous sequence of records in the United States, dating back to 1779. Some weather-observation stations in England have records that go back over 325 years, but most weather stations have continuous records that span only the last century or so. The United States War Department established a national weather service in 1870, at first using the army signal office to collect daily weather reports via telegraph from about 500 observers across the nation. The weather service was moved to the Department of Agriculture in 1891 and renamed the National Weather Bureau.

Alexander von Humboldt (1769–1859), a leading scientist of the day, used data from various places to construct the first world map of mean annual temperatures in 1816, using isotherms to connect identical temperature readings. Just the year before, much of North America and Europe experienced unusually cold weather as a result of the explosion of Mount Tambora in Indonesia in April 1815. The eruption sent 3.5 million cubic feet (100,000 cubic meters) of volcanic debris into the upper atmosphere, where it shrouded the entire globe.

By 1827 Prussian physicist and meteorologist Heinrich Wilhelm Dove (1803–1879) used weather maps and other information to develop his laws of storms that laid the foundation for weather prediction. Among his other important observations were that tropical cyclones rotate counterclockwise in the Northern Hemisphere and clockwise in the Southern Hemisphere. He investigated the effects of climate on the growth of plants. He was also fascinated by the manner in which heat is distributed over Earth's surface. Dove believed that low-pressure storm systems formed as polar and equatorial air masses met. Other refinements in this period included the invention of the psychrometer in 1825, the invention of the pyrheliometer to measure insolation in 1837, and the systematic use of balloons to collect measurements from varied levels of the lower atmosphere.

Map making that employed atmospheric data developed rapidly during this century. William Redfield (1789–1857) created the first weather map in the United States in 1831. He was drawn into the study of meteorology when he observed firsthand the damage of the "Great September Gale" of 1821 in southern New England. He determined from observations he made at the time and studying the aftermath in Connecticut that it was a "progressive whirlwind" (what we now call a hurricane). He noted that trees in Connecticut had fallen toward the northwest in the eastern part of the state and toward the southeast in the western part. From this data and other information, Redfield correctly deduced that the wind patterns were in a counterclockwise direction. He also concluded that low-pressure systems were caused by a spinning air mass that moved air from the center to the periphery of the storm, creating a calm spot in the middle. Since small mountains seemed to stop a hurricane's advance, he and other amateur meteorologists of the time wrongly concluded that hurricanes extended only a mile or so into the atmosphere, a belief that would only disappear after the laws of physics derived in the 1920s and 1930s laid them to rest. Redfield's many scientific interests led to his election in 1848 as the first president of the American Association for the Advancement of Science.

Fellow American meteorologist James Pollard Espy (1785–1860) proposed an alternate theory to Redfield's, arguing that air just above Earth's surface rises due to heat it receives from the sun; as it does so, it takes water vapor with it. (Today we realize that the air is not heated directly by the sun but rather that Earth's atmosphere acts as a reflector, trapping heat from the sun near Earth's surface, the albedo effect.) As this column of heated air rises and barometric pressure decreases, the air expands and becomes less dense. This causes the water vapor in the air to condense, which leads to the formation of a cloud. The condensation process releases heat that provides a mechanism by which more moist air can be drawn up into the storm. Espy's ideas linked barometric pressure and temperature and showed how the two concepts are related to low-pressure system formation.

German geographer Heinrich Berghaus (1797–1884) and his nephew Hermann Berghaus (1828–1890) developed a world map of precipitation in 1845 with their widely read *Physikalischer Atlas.* Heinrich Wilhelm Dove published maps of mean monthly temperatures in 1848 and Austrian geographer Alexander Supan (1847–1920) created a map of world temperature regions in 1879. Supan divided the world into five major climatic zones and created 34 climatic provinces within them. This was the first major taxonomy of climate zones and greatly advanced the field by bringing some order to the accumulating data. The Frenchman E. Renou drew the first map of mean pressure for Western Europe in 1862.

With the discovery of the stratosphere in 1902 and the ozone layer in 1913, the twentieth century witnessed a rapid acceleration in scientific understanding of the atmosphere as its different layers were clearly delineated.

Norwegian meteorologist Vilhelm Bjerknes (1862–1951), who was trained as a mathematical physicist, advanced a theory of atmospheric circulation that included both hydrodynamics and thermodynamics. His ideas were later influential in oceanography as well. Bjerknes concentrated on seven variables—air pressure, temperature, density, water vapor content, and three factors related to wind. Drawing on his extensive knowledge of physics, he knew that all seven had accompanying physical laws that could account for changes within them. He coupled this knowledge with a detailed and systematic approach to drawing weather maps.

With regard to wind for example, he created streamlines (lines of constant direction) and isotachs (lines of constant wind speed) from which vertical wind speeds could be determined using graphical calculus. By 1918 Bjerknes and his colleagues began to articulate a polar front theory regarding the weather in the Scandanavian countries, adapting the word "front" from its military use in World War I (1914–1918), because he viewed weather as a struggle between warm and cold currents in a global conflict. Bjerknes's influential approach to meteorology, which treated it as a product of the movement of distinct air masses and fronts that form on the boundaries between them, has become the basis for modern weather prediction. Bjerknes and his colleagues founded the Bergen School of Meteorology in 1917.

In addition to contributing terminology, World War I led to several technological innovations that benefited the study of the atmosphere, most notably aircraft. Predicting weather patterns and climate was essential to battlefield strategy not only due to the combatants' active air corps, but also to ensure that large-scale gas attacks did not go awry and lead to gassing one's own troops.

IN CONTEXT: THE GALVESTON HURRICANE OF 1900

Prior to Hurricane Katrina's devastation of New Orleans and a significant stretch of the Gulf Coast in 2005, the worst natural disaster in the history of the United States was the Galveston, Texas, hurricane of 1900. The city was nearly destroyed; the death toll ranged between 6,000 and 12,000 people.

Galveston, at the turn of the century, was on its way to becoming a significant American metropolis. But the developers who built on Galveston Island, a low sandbar in the Gulf of Mexico, had ignored reports of an 1841 storm that had flooded the island.

Around August 27, 1900, a storm began forming in the Cape Verde Islands in the North Atlantic. Heading westward, the storm struck Cuba the first week of September, dumping 24.34 inches (87 cm) of rain. A Havana observatory issued a statement predicting that the storm would likely intensify once it headed toward Florida. Meteorologists at the U.S. Weather Bureau in Washington, D.C., however, previously frustrated by "alarmist" reports they received from Cuba about the severity of West Indian storms, discounted the alarm.

By September 5 the storm was gradually heading northward. Weather Bureau forecasters predicted that it would move northward and threaten both Florida and the East Coast, ignoring Cuban forecasters' speculation that the storm would move northwestward into the Gulf of Mexico instead. On September 6 the Weather Bureau erroneously reported that the storm was 150 miles northwest of Key West and as late as that afternoon warned fishermen in New Jersey to remain in port. Meanwhile the storm was getting larger, faster, and heading toward Texas.

A ship, the *Louisiana,* ran into the storm in the Gulf early in the afternoon of the sixth and clocked wind speeds of 100 miles per hour (161 km/h). The very next day, the steamship *Pensacola* was nearly destroyed by the storm, but managed to remain afloat. Unfortunately, neither vessel was equipped with a wireless transmitter, so no warnings could be sent.

By the morning of the sixth it was clear that the storm was not going to hit Florida, and all U.S. Weather Bureau offices along the Gulf of Mexico were ordered to hoist storm warnings. There was no effort to indicate the possible severity of the storm, a standard practice at the time to avoid alarming the public or crippling business interests.

On Friday, September 7, Isaac M. Cline, chief of the local Weather Bureau office in Galveston, awoke to the sound of heavy breakers on the beach and confusing information from headquarters. Monitoring the breakers throughout the day and into the next, he found the interval between them lengthening, a sign that a large storm was likely brewing out in the Gulf. On the other hand, the sky was clear; neither high cirrus clouds nor the brick-dust sky that Cline believed were always associated with hurricanes were visible. The barometer was falling, however, and by Saturday morning the sea had risen high enough that the seaward end of a number of city streets were flooded. Cline telegraphed Washington, D.C., of this news, noting that opposing winds were now in play that he had not seen before. Still, no hurricane warning was issued from his office to the citizens of Galveston—such warnings had to be issued from Washington.

Cline did not realize the severity of the situation in part because he accepted that a storm surge like the one that swept the Ganges River basin in 1876, killing over 100,000 people, was impossible in Galveston because of the way that waters shoaled as they approached its coastline. He also thought that Galveston Bay would absorb any serious flooding, an argument he had earlier used to rebut the city's need for an expensive sea wall. At the same time, Cline seemed to ignore the 1841 inundation of Galveston Island and the complete havoc wrought by two separate hurricanes that struck the port of Indianola, 150 miles (240 km) southwest of the city, in 1875 and 1886; the last one devastated the city so severely it was never rebuilt.

Whether Cline ultimately issued a warning sometime in late morning or early afternoon Saturday is still a matter of dispute. No survivors remembered any warning being given. The storm struck the city full force on Saturday afternoon; Cline himself recorded that by 7:30 PM the water rose as much as 4 feet (1.2 m) within a few seconds, quickly reaching a level of about 10 feet (3 m) above the ground where his own home stood. At least another 5-foot (1.5-m) rise occurred within the next hour, coupled with winds as high as 140 mph (225 km/h). Cline's wife Cora, with 31 others who had sought refuge at his home, perished. Cline and his three children clung to debris for over three hours before the waters subsided enough to find high ground.

No one knew of the disaster until well into the following day. Even then, early reports of the dead, missing, and injured were viewed in Houston as impossibly high. Subsequently, Clara Barton, founder of the American Red Cross and a veteran of Civil War battlefield carnage, reported she had never seen anything like it. Corpses were burned in funeral pyres.

The storm's devastation left profound changes in its wake. The city began construction of a 17-foot- (5-m-) high, 6-mile- (10-km-) long seawall and started the painstaking process of raising the city to as much as 17 feet (5 m) above preexisting levels. From the 1920s to 1961, the seawall was extended still farther to its current length of 10 miles (16 km). These precautions proved their worth, substantially reducing damage from violent storms.

Meteorologists quickly figured out that the Germans had a major advantage because winds on the Western front made it easier for them to glide safely back to the German lines in the event of engine failure, while Allied planes were afforded no such wind assistance, and always had to be sure that they knew how far they were from troop lines so that if their engine failed they would be safely behind their own lines. Bjerknes seized on the public's fascination with aircraft to advocate for a global network of weather observatories and trans-Atlantic travel, a dream fulfilled in the 1930s.

By 1925 several nations were engaged in systematic collection of atmospheric data from aircraft. Radiosondes were employed for the first time in 1928 to capitalize on advances in balloons, radiometry, and knowledge of wind currents. This practice continues as the United Nations' World Meteorological Organization (WMO) coordinates the daily launch of hundreds of meteorological balloons to study the upper atmosphere and improve local weather forecasting.

The use of rockets to study the atmosphere followed the pioneering work of American inventor Robert H. Goddard (1882–1945), who launched a scientific rocket in the early 1930s with an attached camera and a barometer to collect data high up in the atmosphere. The Meteorological Rocket Network (MRN) was launched at North American missile ranges in 1959 under supervision of the military and later expanded to a global, nonmilitary system that embraces some 60 sites worldwide today. The MRN has roughly doubled the amount of the atmosphere that can be subjected to relatively thorough study. The United States Weather Bureau, as it was called in the 1930s, conducted large-scale analyses of air masses and adopted much of the work of the Bergen School of Meteorology. The bureau also funded the creation of more elaborate communication systems to facilitate weather forecasting.

World War II (1939–1945) spurred further developments in the study of the atmosphere, as the movement of troops, supplies, and tactics were strongly dependent on accurate weather forecasting and prediction of major climatic changes. The U.S. armed forces, for example, employed some 8,000 weather officers. During the battle for the Marshall Islands in 1942, when American naval task forces realized their vulnerability to Japanese counterattacks, they used weather fronts as natural barriers to detection, following the front when possible toward safe harbor. The U.S. Navy flew an aircraft directly into the eye of a hurricane for the first time in 1943 to collect scientific data and better understand how to protect troops and ships from these large natural phenomena. The chief weather forecaster for the Royal Air Force in Britain, Norwegian-born Sverre Petterssen (1898–1974) built on emerging knowledge of the jet stream and put it to use in nightly bombing runs over Germany, thereby conserving bomber fuel and extending the planes' range. The use of long-range weather forecasting (up to five days with reasonable accuracy) led to successful Allied forecasts for amphibious landings in both North Africa and on the beaches of Normandy.

Radar was another technological advance in World War II that turned out to have profound implications for atmospheric sciences. Radar today is used for the short-term prediction of rain or severe weather at ground locations and to study severe storms, precipitation development, and the spatial structure of precipitation patterns. Television weather reports routinely show live Doppler radar pictures of precipitation. High atmospheric radar is used to investigate winds and the thermal structure of the upper atmosphere.

A third boon from World War II was the extensive number of ex-military aircraft that were purchased by or donated to university researchers, increasing their ability to track storms and take measurements in the lower atmosphere. This included the American Thunderstorm Project, where, by flying through developing thunderstorms, scientists developed a much more accurate understanding of their formation, dynamics, and life cycle.

Monitoring atmospheric data and phenomena from space began in 1960 when the United States launched the first meteorological satellite, the Television and Infrared Observation Satellite (TIROS) I, which transmitted images of large-scale weather systems. The TIROS series of satellites and their successors continue to provide improved detection of developing major weather patterns, including the successful warning of Hurricane Carla in 1963 that permitted an early evacuation and saved many lives. Data from these types of satellites was supplemented in the United States by the deployment of a national Doppler radar network in 1990 that provides even more fine-grained tracking of storm systems. Efforts similar to those in the United States have spread to many countries including Russia, Japan, China, India, and the countries of Europe.

The National Science Foundation (NSF) established the National Center for Atmospheric Research at the University of Colorado at Boulder in 1960. It became a focal point for the entire community of researchers across the United States and for visiting scholars from many other nations. NSF provided substantial research funding and support for graduate education in atmospheric studies from the late 1950s onwards, and also gave the field its present name of atmospheric sciences. The American Meteorological Society picked up this designation in 1962 when it retitled its lead journal the *Journal of Atmospheric Sciences* and launched a new *Journal of Applied Meteorology* to affirm that the study of atmosphere is about much more than just weather prediction.

Military developments also helped advance atmospheric sciences during the 20 years a number of nations practiced open-air nuclear weapons testing. The release of radioactive particles high into the atmosphere required atmospheric scientists to track them; this generated new understanding of upper atmospheric wind patterns. This work contributed to the construction of the 1963 Nuclear Test Ban Treaty between the United States and Soviet Union, ending above-ground nuclear tests by these two nations. It also established the clear role of the upper atmosphere's influence on weather in the troposphere, an insight derived in large part from one of Bjerknes's former students, Swedish-American

An isolated supercell thunderstorm threatens south-central Kansas on June 5, 2004. The flying saucer-shaped severe storm produced baseball-sized hail. *Jim Reed/Getty Images.*

meteorologist Carl-Gustaf Rossby (1898–1957) at the Massachusetts Institute of Technology.

The U.S. National Climate Program Act of 1978 was a significant development for atmospheric sciences globally. It included a number of elements that greatly advanced scientific understanding of the atmosphere, including provisions for developing new methods to improve climate forecasts, gathering global data on a continuing basis, systems for disseminating climatological data and information, mechanisms for climate-related studies, experimental climate forecast centers, increasing international cooperation in climatology, and a robust program of research and development in atmospheric studies. Increased cooperation among nations led to the Global Atmospheric Research Programme (GARP) in the 1960s and 1970s.

John von Neumann (1903–1957), the Hungarian-born American mathematician and contributor to ENIAC, the first computer, was granted government funding to attempt to model the weather beginning in 1946. He drew upon earlier work by British physicist Lewis Fry Richardson (1881–1953) who developed a way to evaluate analytical equations describing the weather. Although the first computer weather predictions from this work in 1949 and 1951 were crude by today's standards, von Neuman's pioneering efforts attracted technically oriented people into meteorology to create more sophisticated models that progressively improved weather prediction and better understanding of the possible impact of human activities on the planet's atmosphere.

American mathematician Edward N. Lorenz (1917–) showed mathematically that a very small change in initial conditions within the atmosphere could have dramatic effects on later events. This led to the rigorous application of chaos theory to meteorology. Today's models for weather and global climate change employ the power of supercomputers and retrieve and utilize millions of data points and various simulations. Despite the sophistication of these models, Earth's atmosphere and the billions of interactions within it that occur on a daily basis are still vastly more complicated than the most advanced computer model of this dynamic system.

Coordinating global atmospheric information among nations and organizations is the responsibility of the World Meteorological Organization, an official agency of the United Nations. Since 1966 this global network has successfully identified and tracked every tropical storm, as well as all other major weather events.

Twentieth-century atmospheric sciences developed in close concert with the disciplines associated with oceanography. Wallace S. Broecker (1931–) studied the circulating currents of the northern Atlantic and discovered that they convey heat from tropical waters to the shores of northwest Europe. Broecker's "conveyor belt" dramatically influences the weather patterns of northern Europe and results in a much warmer climate that would be possible without this warm oceanic circulation system.

In the 1920s British physicist and statistican Sir Gilbert Walker (1868–1958), head of the Indian

Meteorological Survey, discovered an inverse relationship between the air pressure over the Indian Ocean and over the Pacific Ocean. He called this the southern oscillation. Scientists linked this phenomenon to one well known to fisherman off the coast of Peru, who for centuries had noticed the ocean generally warmed around Christmas, a phenomenon they called *El Niño* (Spanish for "little boy") in reference to the celebration of the birth of Jesus. This rise in water temperature along the equator in the Pacific Ocean was linked to heavier seasonal rains in South America. At UCLA, Norwegian-born American meteorologist Jacob Bjerknes (son of Vilhelm) tied the two together and christened them El Niño southern oscillation (ENSO). *La Niña* (little girl) or *El Viejo* (old man) is the opposite end of the oscillation when the Pacific Ocean water at the equator turns cooler than normal. Today scientists know that ENSO and La Niña affect both climate and fish populations in each hemisphere. Each oscillation lasts about five years, with the most intense temperature difference amounting to about 50°F (10°C). Climatic data gathered from the study of tree rings documents the existence of this oscillation over the past 750 years. Steeper changes in temperature observed in the past decade have led some scientists to speculate that this is further evidence of global warming.

Aerology, the third branch of the atmospheric sciences, has broadened and deepened considerably in the twentieth century with advances in chemistry and physics and aided by increased use of sophisticated monitoring instruments on the ground and in near space. The Swedish chemist Svante Arrhenius (1859–1927) presented a paper to the Stockholm Physical Society in 1895 in which he argued that widespread combustion of fossil fuels (e.g., coal and petroleum) would lead to global warming. Smog in cities within Great Britain was identified as a major health concern in 1905 and coal and chemical combustion were linked to the deaths of thousands of people between 1850–1960, including a particularly bad episode of London-type smog that led to over 4,000 deaths in London in 1952. The burgeoning city of Los Angeles had to contend with photochemical smog as early as the late 1940s although it was not described scientifically until 1951 by Dutch chemist Arie Haagen-Smit (1900–1977), who produced ozone in a laboratory, simulating a daily occurrence in the Los Angeles basin.

■ Modern Cultural Connections

In 1970 Paul Crutzen (1933–) reported that nitrogen oxides released as waste products from soil bacteria rise all the way into the stratosphere. There they are broken apart by sunlight, a chemical reaction that depletes the amount of protective ozone (a special form of isotopic

A satellite image taken by the U.S. National Oceanic & Atmospheric Administration (NOAA) shows Hurricane Katrina in the Gulf of Mexico on August 28, 2005. Authorities in New Orleans, Louisiana, ordered hundreds of thousands of residents to flee as Katrina strengthened into one of the strongest storms ever seen and barreled toward the vulnerable U.S. Gulf Coast city. Although Katrina weakened slightly before landfall, the low barometric pressures produced the highest storm surge ever recorded. At the time this picture was taken, Katrina was a Category 5 hurricane, with catastrophic winds of 175 mph (284 km/h) according to the U.S. National Hurricane Center in Miami. *© NOAA/CNP/Corbis.*

oxygen that helps shield Earth's surface from ultraviolet solar radiation) in the atmosphere. His report was not widely accepted within the scientific community until Mario Molina (1943–) and Sherwood Rowland (1927–), working at the University of California at Irvine reported in 1974 that chlorofluorocarbon (CFC) gases accelerate the decay of the ozone layer. Invented in 1928, CFCs were in wide use by the early 1970s as refrigerant gases in air conditioners and in aerosol cans. The 1995 Nobel Prize–winning research of these three chemists from Holland, Mexico, and the United States respectively, coupled with observations from James Anderson (1944–) that the concentration of chlorine oxide in the upper atmosphere above Texas was much higher than predicted, led to a flurry of research.

The ozone layer, discovered in 1913, was first measured over the Antarctic by a British survey team in

Water from a levee along the Inner Harbor Navigational Canal pours into the city after the structure broke under the force of Hurricane Katrina in 2005. *Vincent Laforet/AP Images.*

1957. By the 1970s it was clear that it was diminishing; by 1985 measurements indicated a layer 100 times thinner than in the 1970s. In 1982, Anderson and a team from Harvard University used a U2 aircraft to measure the ozone layer and reported the presence of a hole over Antarctica, with a similar hole discovered over the Arctic. Informed speculation suggested that both were caused by the widespread use of CFCs. In light of the growing scientific concern and the work of many environmental groups, governments around the globe agreed on a series of international protocols named after the city in which they were negotiated. The Montreal Protocols of 1987, 1990, and 1992 phased out the use of CFCs in most nations of the world, and mandated the substitution of new substances; China and India have yet to sign the agreement. The other major discovery in the 1980s and 1990s was that Earth appeared to be warming; this effect was also suspected to be caused by human activity.

The study of the atmosphere over the centuries demonstrates a number of important themes. Foremost is the role that technology has played. Second, military needs and funding have given scientists new tools and support for much basic and applied work. Third, controversies among scientists and a multidisciplinary approach have been critical to acquiring better knowledge. Finally, advances in atmospheric sciences have come from scientists and nonscientists around the world, not just a single nation or region. It is truly an international

science and important to the future of the planet we call home.

Atmospheric science today is a highly complex set of disciplines working together to elucidate the highly convoluted manner in which gas, radiative, aerosol, meteorological, cloud, transport, and surface processes combine their effects moment by moment in Earth's atmosphere. Comparative study of nearby planets with atmospheres has sharpened our understanding of our own. Land masses, oceans and seas, the sun, and human activity all contribute to the dynamic system of Earth's atmosphere and it is clear that the composition of the atmosphere has changed dramatically over time. While many major advances in our understanding of this gaseous envelope that surrounds Earth have occurred, there are still vastly more mysteries to be solved.

■ Primary Source Connection

Ozone is generally found both in the troposphere, which extends 5 to 9 miles (8 to 14.5 km) above sea level, and in the stratosphere, which extends between 10 and 31 miles (17 and 50 km) in altitude. In the troposphere, ozone is considered a pollutant because it contributes to the formation of smog. In the stratosphere, ozone occurs naturally and acts as an important protective shield against harmful radiation from the sun. Stratospheric ozone absorbs much of the ultraviolet energy at

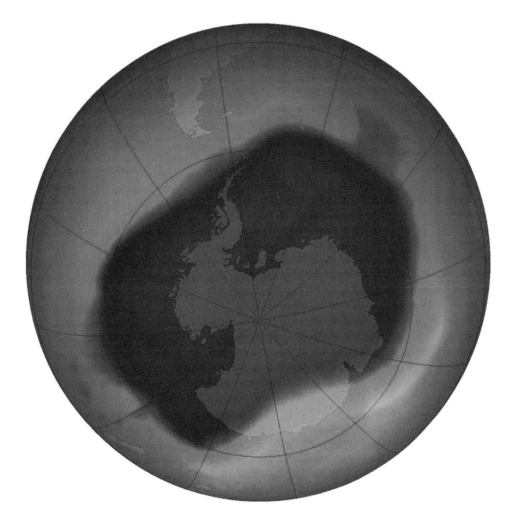

This image provided by NASA was compiled by the Ozone Monitoring Instrument on NASA's Aura satellite from September 21–30, 2006; the average area of the ozone hole was the largest ever observed, at 10.6 million square miles government scientists said Thursday Oct. 19, 2006. In this image, from September 24, the Antarctic ozone hole was equal to the record single-day largest area of 11.4 million square miles, reached on September 9, 2000. The so-called hole is a region where there is severe depletion of the layer of ozone (a form of oxygen) in the upper atmosphere that protects life on Earth by blocking the sun's ultraviolet rays. *NASA/AP Images.*

wavelengths between 240 and 320 nm that impinges on Earth. This ultraviolet radiation is responsible for sunburn, skin cancer, damage to vegetation, and higher rates of genetic mutation in many invertebrate animals.

Ozone is formed in the stratosphere by sunlight. Oxygen molecules are generally found as two oxygen atoms joined together, symbolized as O_2. When ultraviolet energy from the sun strikes an oxygen molecule, it can dissociate the two atoms of oxygen, O. Oxygen atoms are extremely reactive and quickly join with a molecule of oxygen to form the ozone molecule (O_3).

Ozone can be destroyed naturally when an atom of oxygen joins with ozone to form O_4. The O_4 molecule then splits to form two molecules of oxygen, O_2.

However, this reaction occurs too slowly to account for the concentrations of ozone that are actually found in the stratosphere. Research has shown that other gases have the ability to destroy ozone as well. In particular, manmade chemicals that contain chlorine, known as chlorofluorocarbons or CFCs, are particularly effective at destroying ozone. CFCs are used in refrigeration and cooling systems, aerosols and solvents. Also, gases containing bromine and nitrogen oxides are involved in the breakdown of ozone. Nitrogen oxides result from the burning of fossil fuels.

In the following article, the authors discuss holes in the protective ozone layer, their potential causes, and the possibility of closing them.

GOOD NEWS AND A PUZZLE: EARTH'S OZONE LAYER APPEARS TO BE ON THE ROAD TO RECOVERY.

May 26, 2006: Think of the ozone layer as Earth's sunglasses, protecting life on the surface from the harmful glare of the sun's strongest ultraviolet rays, which can cause skin cancer and other maladies.

People were understandably alarmed, then, in the 1980s when scientists noticed that manmade chemicals in the atmosphere were destroying this layer. Governments quickly enacted an international treaty, called the Montreal Protocol, to ban ozone-destroying gases such as CFCs then found in aerosol cans and air conditioners.

Today, almost 20 years later, reports continue of large ozone holes opening over Antarctica, allowing dangerous UV rays through to Earth's surface. Indeed, the 2005 ozone hole was one of the biggest ever, spanning 24 million sq km in area, nearly the size of North America.

Listening to this news, you might suppose that little progress has been made. You'd be wrong.

While the ozone hole over Antarctica continues to open wide, the ozone layer around the rest of the planet seems to be on the mend. For the last 9 years, worldwide ozone has remained roughly constant, halting the decline first noticed in the 1980s.

The question is *why?* Is the Montreal Protocol responsible? Or is some other process at work?

It's a complicated question. CFCs are not the only things that can influence the ozone layer; sunspots, volcanoes and weather also play a role. Ultraviolet rays from sunspots boost the ozone layer, while sulfurous gases emitted by some volcanoes can weaken it. Cold air in the stratosphere can either weaken or boost the ozone layer, depending on altitude and latitude. These processes and others are laid out in a review just published in the May 4th issue of Nature: "The search for signs of recovery of the ozone layer" by Elizabeth Weatherhead and Signe Andersen.

Sorting out cause and effect is difficult, but a group of NASA and university researchers may have made some headway. Their new study, entitled "Attribution of recovery in lower-stratospheric ozone," was just accepted for publication in the Journal of Geophysical Research. It concludes that about half of the recent trend is due to CFC reductions.

Lead author Eun-Su Yang of the Georgia Institute of Technology explains: "We measured ozone concentrations at different altitudes using satellites, balloons and instruments on the ground. Then we compared our measurements with computer predictions of ozone recovery, [calculated from real, measured reductions in CFCs]." Their calculations took into account the known behavior of the sunspot cycle (which peaked in 2001),

seasonal changes in the ozone layer, and Quasi-Biennial Oscillations, a type of stratospheric wind pattern known to affect ozone.

What they found is both good news and a puzzle.

The good news: In the upper stratosphere (above roughly 18 km), ozone recovery can be explained almost entirely by CFC reductions. "Up there, the Montreal Protocol seems to be working," says co-author Mike Newchurch of the Global Hydrology and Climate Center in Huntsville, Alabama.

The puzzle: In the lower stratosphere (between 10 and 18 km) ozone has recovered even better than changes in CFCs alone would predict. Something else must be affecting the trend at these lower altitudes.

The "something else" could be atmospheric wind patterns. "Winds carry ozone from the equator where it is made to higher latitudes where it is destroyed. Changing wind patterns affect the balance of ozone and could be boosting the recovery below 18 km," says Newchurch. This explanation seems to offer the best fit to the computer model of Yang et al. The jury is still out, however; other sources of natural or manmade variability may yet prove to be the cause of the lower-stratosphere's bonus ozone.

Whatever the explanation, if the trend continues, the global ozone layer should be restored to 1980 levels sometime between 2030 and 2070. By then even the Antarctic ozone hole might close—for good.

Patrick L. Barry
Tony Phillips

BARRY, PATRICK L., AND TONY PHILLIPS. "GOOD NEWS AND A PUZZLE: EARTH'S OZONE LAYER APPEARS TO BE ON THE ROAD TO RECOVERY." *SCIENCE@NASA* (MAY 26, 2006).

SEE ALSO *Chemistry: Chaos Theory; Earth Science: Climate Change; Earth Science: Oceanography and Water Science.*

BIBLIOGRAPHY

Books

Anderson, Katharine. *Predicting the Weather: Victorians and the Science of Meteorology.* Chicago: University of Chicago Press, 2005.

Buderi, Robert. *The Invention that Changed the World: How a Small Group of Radar Pioneers Won the Second World War and Launched a Technological Revolution.* New York: Simon & Schuster, 1996.

Cagin, Seth, and Phillip Dray. *Between Earth and Sky: How CFCs Changed Our World and Endangered the Ozone Layer.* New York: Pantheon, 1993.

Fleming, James Rodger. *Meteorology in America, 1800–1870.* Baltimore, MD: Johns Hopkins University Press, 1990.

Fleming, James Rodger, ed. *Historical Essays on Meteorology, 1919–1995: The Diamond Anniversary*

History Volume of the American Meteorological Society. Boston, MA: American Meteorological Society, 1996.

Friedman, Robert Marc. *Appropriating the Weather: Vilhelm Bjerknes and the Construction of a Modern Meteorology*. Ithaca, NY: Cornell University Press, 1989.

Hughes, Patrick. *A Century of Weather Service: A History of the Birth and Growth of the National Weather Service, 1870–1970*. New York: Gordon and Breach, 1970.

Kutzbach, Gisela. *The Thermal Theory of Cyclones: A History of Meteorological Thought in the Nineteenth Century*. Boston: American Meteorological Society, 1979.

Larson, Erik. *Isaac's Storm: A Man, a Time, and the Deadliest Hurricane in History*. New York: Crown Publishers, 1999.

Nash, J. Madeleine. *El Niño: Unlocking the Secrets of the Master Weather-Maker*. New York: Warner Books, 2002.

Nebeker, Frederik. *Calculating the Weather: Meteorology in the 20th Century*. San Diego: Academic Press, 1995.

Williams, James Thaxter. *The History of Weather*. Commack, NY: Nova Science Publishers, 1999.

Periodicals

Barry, Patrick L., and Tony Phillips. "Good News and a Puzzle: Earth's Ozone Layer Appears to Be on the Road to Recovery." *Science@NASA* (May 26, 2006).

Mazuzan, George T. "Up, Up, and Away: The Reinvigoration of Meteorology in the United States: 1958–1962." *Bulletin of the American Meteorological Society* 69 (1988): 1152–1163.

Smith, W.L., et al. "The Meteorological Satellite— Overview of 25 Years of Operation." *Science* 231 (January 31, 1986): 455–462.

Dennis Cheek
Kim Cheek

Earth Science: Climate Change

■ Introduction

Climate is the average weather of a region over time. Climates are shaped by a global machinery of ocean currents, winds, forests, ice caps, mountain ranges, bacteria, planetary orbital motions, human activities, and many other factors.

Scientists are sure that global climate change is happening and are almost certain that it is mostly anthropogenic (caused by humans). Since the 1990s, most scientific and public debate has turned from the question of whether climate change is real to the question of what should be done about it. However, some doubters or skeptics, including many political commentators and a relatively small number of scientists, have claimed since the 1980s that global climate is either not happening or, if it is happening, is not caused primarily by human beings. Criticism of mainstream science continues to be voiced. However, there is consensus among the great majority of scientists studying climate that present-day global climate change is real, significant, and primarily human-caused.

The consensus is that Earth's climate is changing as a result of human activities that have added large quantities of certain gases to the atmosphere, especially carbon dioxide (CO_2) and methane (CH_4). The well-being or survival of hundreds of millions of people may soon be threatened by rising sea levels, disrupted food production, extreme weather, and emergent diseases; a large fraction of the world's species of animals and plants may go extinct in the next century or two as a result of climate change and other human pressures on the environment. Estimates of the money costs of climate change over the next century range in the scores of trillions of dollars.

■ Historical Background and Scientific Foundations

The idea that Earth's atmosphere might act as a one-way valve for solar energy, letting light in but not letting heat out, was first suggested in the early 1800s. In 1824, French scientist Joseph Fourier (1768–1830) described the greenhouse effect accurately, using the scientific language of his day, when he wrote that "the temperature [of Earth] can be augmented by the interposition of the atmosphere, because heat in the state of light finds less resistance in penetrating the air, than in repassing into the air when converted into non-luminous heat."

In 1895, Swedish chemist Svante Arrhenius (1859–1927) suggested that changes in atmospheric CO_2 concentrations could change Earth's climate. He estimated that doubling CO_2 would increase average global temperature by 9°F (5°C). This was not far wrong, by today's standards: In 2007, the United Nations' Intergovernmental Panel on Climate Change (IPCC) said that the result of doubling CO_2 would most likely be a 5.4°F (3°C) increase in global temperature. In 1908, Arrhenius was the first scientist to suggest the possibility of an anthropogenic greenhouse effect. Human beings, he suggested, by burning fossil fuels such as coal and so increasing the amount of CO_2 in the atmosphere, might warm Earth's climate. Unable to foretell the huge increase in fossil-fuel use that was about to occur in the twentieth century, Arrhenius suggested that a greenhouse effect might become noticeable in 3,000 years; in fact, it was detectable by the 1990s, less than 100 years later.

In the 1930s, English inventor Guy Stewart Callendar (1898–1964) estimated that doubling CO_2 would cause 3.6°F (2°C) of global warming and theorized correctly that warming would be greater in the polar regions. This is being observed today: The Arctic and the circum-Antarctic region (though not central Antarctica) are warming about twice as fast as the rest of the world. Cooling in central Antarctica may be caused by increased snowfall there due to warmer temperatures over the Antarctic Ocean, or to the strengthening of the circumpolar vortex (circular wind blowing around Antarctica) due to the loss of heat-absorbing ozone over the continent. In any case, Callendar and some other researchers of that time were mistaken in their belief that anthropogenic global warming was already, in the early twentieth century, clearly detectable in the climate record.

By the mid-1950s, scientific understanding of Earth's climate system was advancing rapidly and the possibility of anthropogenic climate change was widely discussed by scientists. However, nobody had yet found a way to make the precise measurements of atmospheric greenhouse gases. Such measurements would show whether humans were actually increasing the amounts of such gases in the atmosphere: perhaps, some scientists theorized, the oceans were absorbing CO_2 as fast as we were releasing it. In 1958, American scientist Charles David Keeling (1928–2005) developed sensitive new instruments to measure atmospheric CO_2 and began operating them on the summit of Mauna Loa volcano in Hawaii. After just a few years, his data showed a clear result: Although CO_2 fell each summer as green plants grew in the Northern Hemisphere (where most of the world's land is), it rose again in the autumn and winter—and it always rose farther than it fell. The result was an upward-tilted zigzag line showing a steady increase in atmospheric CO_2. Since atmospheric CO_2 was indeed increasing, an anthropogenic greenhouse effect might be occurring.

Keeling's chart of rising CO_2, now known as the Keeling curve, has become an icon of global warming. His measurements and thousands of similar ones, including measurements of air samples trapped in ancient ice layers in Greenland and Antarctica, show beyond doubt that human activities have raised atmospheric CO_2 from about 280 parts per million in 1750 to 383 parts per million as of November 2007, a 36.8% increase. (Parts per million refers to the number of molecules in a mixture; for example, 280 parts per million CO_2 means that out of every 1 million air molecules, 280 are CO_2.) Atmospheric CO_2 continues to increase.

Through the 1970s, however, even knowing about the Keeling curve, scientists were still uncertain about whether Earth was about to experience global cooling or global warming. Aerosols—small solid or liquid particles suspended in the air—tend to cool Earth while greenhouse gases tend to warm it, and both are added to the atmosphere by burning fuels. It seemed possible that the cooling effect of aerosols might be balanc-

WORDS TO KNOW

ANTHROPOGENIC: Made by people or resulting from human activities. Usually used in the context of emissions that are produced as a result of human activities.

GREENHOUSE EFFECT: The warming of Earth's atmosphere due to water vapor, carbon dioxide, and other gases in the atmosphere that trap heat radiated from Earth's surface.

GREENHOUSE GAS: A gaseous component of the atmosphere contributing to the greenhouse effect. Greenhouse gases are transparent to certain wavelengths of the sun's radiant energy, allowing them to penetrate deep into the atmosphere or all the way into Earth's surface. Greenhouse gases and clouds prevent some infrared radiation from escaping, trapping the heat near Earth's surface where it warms the lower atmosphere. Alteration of this natural barrier of atmospheric gases can raise or lower the mean global temperature of Earth.

RADIATIVE FORCING: A change in the balance between incoming solar radiation and outgoing infrared radiation. Without any radiative forcing, solar radiation coming to Earth would continue to be approximately equal to the infrared radiation emitted from Earth. The addition of greenhouse gases traps an increased fraction of the infrared radiation, reradiating it back toward the surface and creating a warming influence (i.e., positive radiative forcing) because incoming solar radiation will exceed outgoing infrared radiation.

ing the warming effect of increased CO_2 and methane (CH_4), the second-most-important greenhouse gas. Perhaps, scientists also speculated, the amount of energy from the sun was changing, or as-yet-unknown natural processes were changing climate. Although a few sensational articles in news magazines proclaimed that the world was on the verge of a new Ice Age, the opinion held by the great majority of scientists was that they did not yet know enough to say what was going to happen to Earth's climate in the near future.

Meanwhile, the Keeling curve continued to climb. Surface temperature measurements from thousands of weather stations showed warming trends in most parts of the world, and data from hundreds of tide gauges showed that sea levels were rising. Layered cylinders of muck from ocean bottoms and of ancient snow layers from deep inside the ice caps of Greenland and Antarctica

Industrial smoke, dense with soot, pollutes the environment, contributing to global warming. The factory burns hydrocarbons to make carbon black, a pigment used in inks and paints, and a substance used by the tire industry for vulcanized rubber.
© *Andrew Holbrooke/Corbis.*

began to reveal more about paleoclimate (climate before the beginning of instrumental records), hinting that relatively small changes to radiative forcing, such as slight changes in Earth's orbit, might trigger large climate changes. The first mathematical descriptions of the climate mechanism, called climate models, were developed in the 1960s. Although crude by today's standards, they predicted several degrees of warming over the next century, now known to be a reasonably correct result. Starting in the 1970s, satellites began to monitor Earth's temperature from outer space, supplying a flood of new, independent information about climate.

By the mid-1980s, the doubts of most scientists had been answered: The world was indeed warming. What was more, it would continue to warm, and human beings were the primary cause. Scientists warned that global warming would cause a host of other climate changes, many dangerous or costly, from rising seas to shifting rainfall patterns. It became increasingly clear that the question of global warming was more than a matter of abstract curiosity: All human life ultimately depends on farming, and all farming ultimately depends on climate.

Also, hundreds of millions of people live within a meter or so of sea level: rising waters might force their resettlement, if that were possible.

In 1985, a United Nations (UN) scientific conference in Austria agreed that significant human-caused global warming was probably about to occur. In 1987, the tenth congress of the UN's World Meteorological Organization recommended ongoing, long-term assessment of climate change by an international group of scientists. The new group, the Intergovernmental Panel on Climate Change (IPCC), was formed in 1988 and given the job of reporting on the scientific community's understanding of climate so that decision-makers could make informed decisions. The IPCC's reports, issued every two to five years, have been influential in the global discussion of climate change. In 2007 the organization shared a Nobel Peace Prize with former U.S. vice president and climate-change activist Al Gore (1948–). The first IPCC Assessment Report was issued in 1990 and the fourth in 2007. A fifth report is due around 2012.

The IPCC's 1990 report advised that global warming was probably happening and might cause many problems. Mildly alarmed—there were still many doubts and uncertainties—almost all the world's nations sent representatives to a climate summit in Rio de Janeiro, Brazil, in 1992. There, a treaty addressing the problem of global climate change was negotiated. This treaty, the United Nations Framework Convention on Climate Change (UNFCCC), did not place binding obligations on any countries but did acknowledge the reality of anthropogenic global climate change. Under the UNFCCC, industrialized countries made a nonbinding commitment to reduce their greenhouse-gas emissions and to help poorer countries reduce theirs as well.

Throughout the 1990s, a minority of scientists challenged the basis of global climate change theory. Some proposed that the globe was not warming significantly at all; others argued that warming might be occurring, but was due to natural processes, not anthropogenic greenhouse gases. Those doubting that Earth was warming drew attention to potential problems such as urban heat islands (see sidebar). The great majority of climate scientists believe today that the urban-heat-island objection has been answered by adjusting temperature readings downward: Even the adjusted readings show that the world is warming. Another scientific objection was the discrepancy between satellite measurements of Earth's temperature and ground-based (thermometer) measurements: The satellite data seemed to show that Earth was, if anything, cooling rather than warming. Careful re-analysis in the early 2000s showed that the satellite data had been misinterpreted: When correctly used, it also showed warming. The U.S. government announced in 2006 that the dispute had been resolved: "This significant discrepancy no longer exists because errors in the satellite and radiosonde data have

Oceanographers test flow rates aboard the National Oceanic and Atmospheric Administration ship *Malcolm Baldridge* in Miami in January 1996, after the ship completed a year-long scientific expedition to gather critical data on the ocean's role in global climate change.   *AP Images/Alan Diaz.*

IN CONTEXT: URBAN HEAT ISLANDS

Urban heat islands are areas of increased warmth in and near built-up areas such as cities and suburbs. As areas become covered with buildings and pavements, natural ground surfaces—such as moist soil, vegetation, and open water—are replaced with stony or tarry surfaces that absorb solar energy effectively and are therefore warmer. Such surfaces also dry more quickly after precipitation, reducing evaporation, which has a cooling effect in green areas. The result is that built-up or urban areas are warmer than country or rural areas.

Some skeptics of the reality of global warming have argued that the world only appears to have become warmer because too many temperature-measuring stations have been located near urban areas. As urban areas have grown over the last century, temperatures at urban or near-urban measuring stations have risen, creating an illusion of global warming. According to these critics, scientists have being measuring too many temperatures in places that have warmed for reasons that have nothing to do with global climate change.

Climatologists agree, however, that temperature data must be adjusted for urban heat-island effects. Such adjustments are standard practice in climate measurement. Today, the claim that global warming is an illusion created by expanding urban heat islands has no scientific standing. The reality of global warming has been established by numerous, independent, convergent lines of evidence, including over 540 million separate readings of ground and sea surface temperature and over 30 years of satellite measurements scanning the whole surface of Earth. Global warming does not necessarily proceed smoothly: A single year, or even a run of years, may show no warming or even cooling, while the long-term trend proceeds upward.

been identified and corrected. New data sets have also been developed that do not show such discrepancies" (Climate Change Science Program, 2006).

■ Modern Cultural Connections

Despite the efforts of a small number of scientists and a large number of political commentators to cast doubt on the reality of human-caused nature of climate change, a scientific consensus on climate change emerged starting in the late 1980s. Computerized climate models became more complex and realistic every year, the Keeling curve continued to climb, global average temperature continued to rise, and paleoclimate studies showed that, thanks to human activity, there is now more CO_2 in the air than there had been for at least 800,000 years. Earth, climatologists confirmed, was probably warmer by the late 1990s than it had been for at least 1,100 years, maybe far longer.

The countries that had signed the UNFCCC in 1992, including the world's then-largest greenhouse polluter, the United States, held regular meetings in the following years to discuss the changing science of climate change and to plan counter-action. These meetings resulted in a protocol or add-on to the UNFCCC, the Kyoto Protocol of 1997. The Kyoto Protocol was rejected by U.S. leaders and those of a few other countries—Australia did not sign the protocol until December 2007—but it was affirmed by all other signatories of the UNFCCC. Under Kyoto, industrialized countries made binding promises to reduce their own greenhouse emissions and to help developing countries do the same.

Kyoto was controversial. The United States refused to commit to the protocol because it did not require rapidly developing nations such as China to reduce their emissions. In any case, Kyoto did serve as a beginning for international action on climate change, establishing mechanisms for carbon emissions trading and formalizing the commitment of most industrial nations to reduce their emissions. Kyoto was not meant to be the final word on climate action; signers agreed from the beginning to replace Kyoto with an updated agreement starting in 2012.

In 2001 and in 2007, the IPCC released its third and fourth Assessment Reports on climate change. The 2007 report, prepared by over 2,500 scientists and economists appointed by scores of governments, declared that global warming was "unequivocal" (i.e., definite) and that there was at least a 90% probability that human beings were the main cause. It also said that climate change could be mitigated (made less severe) at fairly low cost if prompt action were taken to reduce greenhouse emissions. The report had unprecedented impact on world opinion on climate change, creating a heightened sense of urgency. After the 2007 IPCC report, voices doubting the reality of climate change were more marginalized than ever. As of early 2008, however, the United States and China, the world's largest greenhouse polluters, both still remained unfriendly to the idea of binding emissions limits, making the future of efforts to control global climate change uncertain.

Despite spreading recognition of the climate-change issue, public understanding of the causes of global warming has remained low. In 1994, 57% of the public thought that it was "definitely true" or "probably true" that the greenhouse effect is caused by a hole in Earth's atmosphere; in 2000, the number was 54%, not significantly different. (Global warming is caused by greenhouse gases added to the atmosphere, not by a hole in the atmosphere.) In 2001, only 15% of U.S. citizens could correctly identify the burning of fossil fuels as the primary cause of global warming, a tie with Brazil and significantly less than in Mexico, where 26% of respondents (still a relatively small number) could answer correctly. The public is also confused about whether a scientific consensus exists on global warming: as of 2007, depending on how the question was worded, the number of Americans who think that scientists have reached agreement that human-released carbon dioxide is the major cause of global warming varied from a third to slightly over 60%.

However, international polls in 2006 showed that large percentages of the populations of many countries consider global warming a serious threat. The highest percentages found were in South Korea (where 96% of those polled believed that climate change is either a "critical threat" or "important but not critical" threat), Australia (95%), and Mexico (93%). The Ukraine had the lowest level of concern, about 66%—still a solid majority. In the United States, 87% saw the threat of global warming as either critical or important.

■ Primary Source Connection

The following article was written by Peter N. Spotts, a science and technology writer for the *Christian Science Monitor*. Founded in 1908, the *Christian Science Monitor* is an international newspaper based in Boston, Massachusetts. This article predicts that, on the heels of the release of the IPCC "Climate Change 2007" report, the upcoming United Nations Climate Change Conference could result in an international agreement for restricting greenhouse-gas emissions. In fact, during the conference held December 3–14, 2007, in Bali, Indonesia, a "Bali roadmap" for a future agreement on climate change was adopted by the member nations in attendance.

CLIMATE REPORT A KEY TO WORLD EMISSIONS AGREEMENT IN BALI

Despite concern among scientists that politics have watered it down in distillation, the synthesis is expected to add urgency to next month's emissions meeting in Indonesia.

Sometimes warnings pack more punch when they come in a concentrated form—and at the right moment.

That's the hope United Nations officials have expressed after the weekend release of the last of four reports this year on global warming and options for trying to bring it under control.

The report reflects rising scientific confidence—and remaining uncertainties—in describing current and projected effects of global warming. That, plus the report's condensed size and terse talking points, virtually ensure it will play a key role in adding urgency to negotiations that begin on Dec. 3 in Nasu Dua, on the island of Bali in Indonesia.

The aim of next month's meeting is to gain consensus on a formal framework for reaching a new emissions-reduction pact over the next two years. A new pact would pick up where the 1997 Kyoto Protocol leaves off. Currently it only requires industrial countries to reduce their greenhouse-gas emissions by an average 5.5 percent below 1990 levels between 2008 and 2012.

Over the weekend, delegates from 140 nations meeting in Valencia, Spain, adopted the Intergovernmental Panel on Climate Change's (IPCC) "synthesis report." It's a thin volume distilled from three larger tomes the UN-sponsored group of scientists, economists, and other experts released earlier this year. It draws no new conclusions. But it does insist that some effects of global warming, such as sea-level rise, are inevitable and will continue for centuries, even if all heat-trapping greenhouse-gas emissions stopped tomorrow. Heading off the worst of the effects means moving aggressively to curb rising greenhouse gases, mainly carbon dioxide from burning fossil fuels, the report suggests.

The message "could not be simpler," says UN Secretary-General Ban Ki Moon. "Global, sweeping, concerted action is needed now; there is no time to waste."

The synthesis report reiterates that the warming of Earth's climate is "unequivocal" and that the scientists involved express "very high confidence" that human activities have warmed the climate since 1750. It also indicates that human influence on climate has contributed to rising sea levels, shifting storm tracks, increasing temperature extremes, and raising the risk of heat waves, droughts, and heavy rains.

The synthesis report uses sea levels to illustrate what many see as unavoidable long-term effects, depending on the rate of warming.

Even the most aggressive scenario to curb greenhouse-gas emissions—with emissions peaking by 2015 and falling to between 50 and 80 percent of 2000 levels by 2050—would still warm the planet enough to ensure that over the next millennium, global average sea level would rise by up to 4.6 feet. The least aggressive scenario, which yields the largest warming, would raise sea levels by up to 12 feet. These increases come merely from heating the oceans, which expand when warmed. The scenarios don't take into account meltwater that icecaps in Greenland or Antarctica would contribute as the global average temperatures rise.

The concentration of carbon dioxide in the atmosphere needed to hold sea-level rise to a minimum "is basically where we are right now," says Ronald Stouffer, a researcher and a member of the synthesis report's core writing team. But global average temperatures today do not yet reflect "in any way, shape, or form, the amount of carbon dioxide and other greenhouse gases in the atmosphere," he says, because of the inertia in the climate system.

Trends in carbon-dioxide emissions hint at the tough job that awaits negotiators heading for Nasu Dua, Indonesia next month. A team of government and university scientists from Australia, the US, Britain, France, and Austria reported in late October that between 2000 and 2006, carbon dioxide concentrations in the atmosphere grew at the fastest rate since monitoring began in 1959. Some two-thirds of the increase comes from industrial emissions and deforestation, the researchers note. But, they add, another 18 percent can be traced to oceans and plants, which are becoming less efficient at soaking up CO2. Computer models the IPCC uses to track Earth's natural carbon cycle have projected a slowdown in CO2 uptake by oceans and plants. According to the team, the slowdown is larger and is coming earlier than models project. CO2 concentrations are at their highest

level in at least 650,000 years and likely the last 2 million years, the team noted.

Criticism that the IPCC process is too political often comes from conservative groups. They argue that the worriers have hijacked the IPCC process, leading to a litany of gloomy scenarios.

However, concern about politics and the IPCC process also comes from some scientists, who argue that because the IPCC operates by consensus among the political delegations who must approve the reports the scientists produce, the reports may understate the challenges humanity faces from global warming.

The reports, which appear every five to six years, represent a snapshot of the science that is now about two years old, notes Dominique Bachelet, an associate professor in the biological and ecological engineering department at Oregon State University in Corvallis, Ore. "The climate is changing so fast," and while the authors are writing and assembling the reports, "science is moving on."

The synthesis report and its progenitors serve as a highly useful baseline, she says, "but it's a conservative baseline."

Despite the challenges, the UN's Mr. Ban says he remains optimistic that countries can agree. "I'm encouraged by the level of political will. I look forward to China and the US to play a more constructive role" at Bali. "Both can lead."

Peter N. Spotts

SPOTTS, PETER N. "CLIMATE REPORT A KEY TO WORLD EMISSIONS AGREEMENT IN BALI." *CHRISTIAN SCIENCE MONITOR* (NOVEMBER 19, 2007).

SEE ALSO *Earth Science: Atmospheric Science.*

BIBLIOGRAPHY
Books
Weart, Spencer. *The Discovery of Global Warming.* Cambridge, MA: Harvard University Press, 2004.

Periodicals
Begley, Sharon. "Global-Warming Deniers: A Well-Funded Machine." *Newsweek* (March 18, 2007).
Brechin, Steven R. "Comparative Public Opinion and Knowledge on Global Climatic Change and the Kyoto Protocol: The U.S. Versus the World?" *The International Journal of Sociology and Public Policy* 23 (2003): 106–135.
Hopking, Michael. "Climate Skeptics Switch Focus to Economics." *Nature* 445 (2007): 582–583.
Mahlman, J. D. "Science and Nonscience Concerning Human-Caused Climate Warming." *Annual Review of Energy and the Environment* 23 (1998): 83–105.

Mooney, Chris. "Blinded By Science: How 'Balanced' Coverage Lets the Scientific Fringe Hijack Reality." *Columbia Journalism Review* (November/December 2004).

Nisbet, Matthew C., and Teresa Myers. "Twenty Years of Public Opinion about Global Warming." *Public Opinion Quarterly* 71 (2007): 444–470.

Oreskes, Naomi. "The Scientific Consensus on Climate Change." *Science* 306 (2004): 1686.

Sample, Ian. "Scientists Offered Cash to Dispute Climate Study." *The Guardian [UK]* (February 2, 2007).

Spotts, Peter N. "Climate Report a Key to World Emissions Agreement in Bali." *Christian Science Monitor* (November 19, 2007).

Web Sites

Chicago Council on Global Affairs and WorldPublicOpinion.org. "Poll Finds Worldwide Agreement that Climate Change Is a Threat." March 13, 2007. http://www.worldpublicopinion.org/pipa/articles/home_page/329.php (accessed January 22, 2008).

Climate Change Science Program (U.S. Government). "Temperature Trends in the Lower Atmosphere: Steps for Understanding and Reconciling Differences." April 2006. http://www.climatescience.gov/Library/sap/sap1-1/finalreport/default.htm (accessed January 22, 2008).

Larry Gilman

Earth Science: Exploration

■ Introduction

Motivations for exploration have ranged from a nation's quest for territory and power to an individual's desire for religious fulfillment, economic prosperity, or scientific discovery. From the ancient Phoenicians to the robotic Mars missions, sociocultural factors have mingled with the desire for knowledge in voyages of discovery.

■ Historical Background and Scientific Foundations

Although prehistoric indigenous peoples colonized the American continent, Asia, and the Arctic tundra, the first written account of exploration comes from ancient Egypt. The walls of the tomb of Harkhuf (fl. 2290–2270 BC) described a trading journey in which he went to Nubia "with three hundred asses laden with incense, ebony, heknu, grain, panthers. . . ivory, [throwsticks], and every good product." Over time, Egypt became a crossroads of trade for Mediterranean and African "good products," exporting them to Cyprus, Crete, Greece, Syro-Palestine, Punt, and Nubia.

Phoenician explorers in the first millennium also sailed the Mediterranean in search of trade. Phoenicia's rough and hilly climate meant that it could not sustain itself on domestic agriculture alone, so it exported cedar wood, glass, and murex (a beautiful and expensive purple dye obtained from snails). With limited goods to offer, the Phoenicians became masters at copying other people's arts and selling them at lower prices, using their adaptive skills to create sailing ships that used both Greek keels and Egyptian ribs and cross-braces. The Phoenicians were the first people to use the pole star for navigation and to sail by tacking—sailing diagonally left and right to catch prevailing winds and move in a desired direction.

With their ships and navigational techniques, the Phoenicians left the Mediterranean, and under the leadership of Himlico (fl. 5th c. BC) went to Britain in 500 BC to trade for tin. He and his crew may also have visited the southwest coast of India and Sri Lanka; the Greek historian Herodotus (c.484–430/420 BC) records that in 600 BC they sailed around Africa, hugging the coast as they went, since navigating in open waters was still perilous. The Phoenicians also founded colonies, such as Cades (now Cadiz) on the coast of Spain, and Carthage in North Africa. Controlling the passage between the eastern and western Mediterranean, Carthaginians traded olives and olive oil, wine, wheat, linen, cotton, and expensive spices from Asia. By the fourth century BC, the ancient Greeks challenged the Phoenician's naval dominance, and in 146 BC the

Romans destroyed Carthage in the Third Punic War, ending Phoenician exploration.

The ancient Greeks had long practiced *apoikia,* or colonization, since the dry rocky soil of their islands could not support a large population. Between 650–600 BC settlers populated points as far afield as the Black Sea, Sicily, southern Italy (known as *Magna Graecia* or Greater Greece), and the coasts of France and Spain. It was little wonder that the ancient poet Homer's tales of discovery and conquest, the *Odyssey* and the *Iliad,* were centrally important in Greek culture.

The Greeks also began extensive trade with the Near East, exporting olive oil and wine. They imported not only luxuries such as spices, but Eastern myths and artistic motifs as well. Figures of gorgons, sphinxes, and griffins appeared on their famous red- or black-painted pottery, and Greek statuary adopted the monumental style of Egyptian sculpture. The Greeks also incorporated near-Eastern philosophy into their science. Scholars believe that the Greek philosopher Pythagoras (c.580–c.500 BC) traveled to Egypt and Mesopotamia, bringing back their methods in early geometry and astronomy to his philosophical school based in Croton in southern Italy. Pythagoras recognized that the laws of nature were written mathematically, and devoted his school to the search for empirical verification of such laws via experiments in music and sound as well as number theory.

In addition to their travels to the Near East, Greek explorers also seem to have explored the Atlantic. Pytheas of Marseilles (380–310 BC) seems to have reached Britain either via naval voyage from Marseilles, or by crossing France on foot to Bordeaux and sailing from there in the 320s BC. Pytheas also seemed aware of Ireland, as well as the Hebrides and Orkneys, and may have even reached Iceland and the Gulf of Finland.

Like Phoenicia and ancient Greece, the growth and consolidation of the Roman Empire was built on exploration and colonization. Successful Roman military conquest of France, Britain, Spain, North Africa, and Germany quelled most border disturbances by the reign of Augustus (63 BC–AD 14). Increasing economic prosperity and the subsequent demand for luxury goods prompted voyages to Southeast Asia and Africa. In 145 BC Greek historian and explorer Polybius (c.200–c.118 BC) was sent to investigate the west coast of Africa south of Gibraltar by the Roman general Scipio Africanus (236–183 BC). Polybius reached Senegal, where he saw crocodiles and hippopotamuses swimming in the river. Later Roman emperors tried to repeat this success and sent several expeditions to the northern oceans, known to be rich in fish, but the loss of several ships due to storms ended their efforts.

Roman contributions to geography were also significant. Claudius Ptolemy's (AD c.90–c.168) *Geographia* (AD 150) was one of the first works to utilize map projections with a grid of latitude and longitude. An earlier map of the Roman Empire was made by Marcus Agrippa (c.63–12 BC), a Roman war hero during the reign of Augustus. He summoned leading geographers to Rome and provided them with the archives of field and coastal coordinates gathered by the Roman army and navy. The resulting map was 60 feet (18 m) long, was either painted or inlaid in stone, and occupied a public colonnade in Rome. The map illustrated the famous Roman highway system, and was accompanied by a geographical commentary by Agrippa that provided the physical dimensions of provinces, the lengths of important rivers, and the distances between cities.

Such geographical accomplishments meant that wealthier Romans became some of the first tourists. Romans toured the Parthenon, took boat rides down the Nile, visited the Pyramids, and were led by professional tour guides called *mystagogi* (those who show sacred places to foreigners). In their own manner of cultural exploration and assimilation, they also bought souvenirs such as painted glass vials showing the lighthouses of Alexandria and miniature statues of Apollo, and they watched floor shows where Egyptian priests fed pet crocodiles and polished their teeth.

The Fall of Rome to the Renaissance

After the fall of Rome's western empire in the fifth century, European expansion and colonization quickly came to an end. Roman contributions to geography and cartography did, however, survive in the Byzantine (eastern) region of the empire, where they were adopted by Islamic scholars. The rise of Islam in the seventh century and the founding of the capital of Baghdad in the eighth century by Abbasid Caliph Harun al-Rashid (c.766–809), led to renewed trade with the East Indies, Persia, and China.

The most remarkable Islamic geographer was Ibn Battuta (1304–c.1368) who over the course of 30 years visited the lands of every Muslim ruler in his time, traveling 75,000 miles (120,700 km) in the interior of Africa, India, China, and what is now Turkey. He was a keen empiricist (observer), and recorded his observations, providing a unique ethnographic glimpse into the culture of the medieval East. He described geography, social and religious customs, and even food. In a book detailing his travels, translated by H.A.R. Gibb, he describes the importance of the betel plant (containing a mild stimulant similar to caffeine) in India:

> Betel-trees are grown like vines on cane trellises or else trained up coco-palms. They have no fruit and are only grown for their leaves. The Indians have a high opinion of betel, and if a man visits a friend and the latter gives him five leaves of it, you would think he had given him the world, especially if he is a prince or notable. A gift of betel

is a far greater honour than a gift of gold and silver. It is used in the following way: First one takes areca-nuts, which are like nutmegs, crushes them into small bits and chews them. Then the betel leaves are taken, a little chalk is put on them, and they are chewed with the areca-nuts.

By the ninth century the Vikings were the primary explorers of the early Middle Ages, making voyages to trade iron ore for glass and woolens from Ireland and Russia. In the process, they developed impressive ships with strong keels that could handle rough northern waters. They journeyed from Scandinavia into Britain, Greenland, and likely Iceland around AD 1000, and also mounted expeditions of conquest, creating fortified harbors in Ireland from which they could raid farther south and east into England and France.

Normandy ("land of the Norsemen") was a Danish colony in the tenth century. In England, the Vikings demanded extortionate bribes, called "Danegeld," from the Anglo-Saxons; the Vikings ruled England briefly from 1016 until 1042, when the Anglo-Saxons regained control for an even shorter period. Excavated ruins and Nordic sagas show that Vikings like Leif Eriksson (fl. 11th century) reached North America (Newfoundland) 500 years before Columbus. In 1960 a Scandinavian settlement was discovered in L'Anse aux Meadows, Newfoundland, proving that Vikings had indeed been in North America.

By the late twelfth century, there was a renewed emphasis on European exploration. The Crusades brought Westerners into contact with the Middle East, and the rise of cities encouraged exploratory voyages to increase trade. Venetian adventurer and merchant Marco Polo's (c.1254–1324) account of his visit to China was enormously popular, as was Sir John Mandeville's (fl. 14th century) more fanciful book *The Voyage and Travels of Sir John Mandeville, Knight*. This gave rise to the legend of the kingdom of Prester John, a mythical Christian king in India or central Asia who would aid Crusaders in their fight against Islam and whose realm contained a fountain of youth.

Such travel tales made the Portuguese realize the wealth to be gained via trade with the East, and they devoted their energies in the fifteenth century to determining the most efficient sea routes around Africa to India and China. Under the leadership of Henry the Navigator (1394–1460), the Portuguese government mapped the coast of Mauritania, discovered the Azores, and mounted several expeditions to Africa and Asia in their sleek caravels (single-stern ships). Bartolomeu Dias (1450–1500) circumnavigated Africa around the Cape of Good Hope. In 1498 Vasco da Gama (1460–1524) became the first sailor to travel from Portugal to India, where he attempted to secure Calcutta under Portuguese dominion. The opening of a sea route to India created a monopoly for the Portuguese.

Other European powers who wanted to cash in on the lucrative trade with India and China attempted to find a westward route. To this end, the Spanish government backed Christopher Columbus (1451–1506) who, motivated by a combination of missionary zeal and desire for personal glory, discovered Hispaniola, the Bahamas, and Cuba, although he thought at first that he had reached the Chinese mainland or another point in Asia.

The encounter between the Old and New Worlds, while a triumph of technology and navigation, was an environmental catastrophe. New European species wiped out native flora and fauna, and European diseases like smallpox nearly destroyed entire races of Native Americans. The strange peoples, plants, and animals brought back to Western Europe from the New World engendered a renewed intellectual appreciation for the "wonders of nature." Scholars have speculated that such encounters spurred a desire to observe and classify nature, which became part of the basis for the Scientific Revolution in the seventeenth century.

Illustration showing Christopher Columbus at Isla Margarita, an island that he discovered in 1498 on one of his trips to the Americas. The image appeared in *The Narrative and Critical History of America* edited by Justin Winsor in 1886. © *Visual Arts Library (London)/Alamy.*

Voyages to the New World continued in the sixteenth and seventeenth centuries. The Pacific Ocean was discovered by Spanish conquistador Vasco Nuñez de Balboa (1475–1519) in 1513. Portuguese navigator Ferdinand Magellan (c.1480–1521), launched an expedition in 1519 that became the first to circumnavigate the globe. (Magellan himself was killed in the Philippines in 1521, and did not live to see the mission completed.) The Aztec and Incan Empires were discovered and conquered by Spanish explorers Hernán Cortés (1485–1547) and Francisco Pizarro (1471–1541), respectively. Cortés and Pizarro were driven by a lust for gold, territory, and a hope of finding *el Dorado* ("the gilded one") a variation on the legend of Prester John. Their expeditions also had far-reaching economic effects. The opening of wildly productive silver mines in Peru by Spanish conquistadors using Native American slave labor brought so much bullion to Western Europe, that the last part of the sixteenth century saw tremendous inflation in the prices of food and durable goods.

From the Enlightenment to the Modern Era

The encounters between the Old and New Worlds led to some of the first comparative anthropological and sociological studies of culture, primarily in the eighteenth century. French philosopher Charles-Louis Montesquieu in his *Spirit of the Laws* (1748) discussed the effect of climate, primarily heat and cold on the human body and on the intellectual outlooks of society. He espoused "geographical determinism": the belief that certain climates were superior to others. The temperate climate of France, he said, was ideal; races living closer to the equator were "hot-blooded," and those in northern countries "stiff." Montesquieu also claimed hot southern or eastern climates led to despotisms (absolute rule/ tyranny), and moderate western climates to constitutional monarchies such as those in England and France.

As western Europe began to colonize Africa and the Middle East in the late eighteenth and nineteenth centuries, their policy makers relied on theories such as Montesquieu's to impose governments on native peoples who were "under the sway" of their "inferior" climates.

European colonialism also gave impetus to exploration of those areas of the world that remained unknown to the West, such as Australia, New Zealand, and the African interior. The voyages of Captain James Cook (1728–1779) and Sir Joseph Banks (1743–1820) to Australasia and the Pacific led to a program of economic botany in the British Empire. Indigenous plants were sent back to the Royal Botanic Gardens at Kew, investigated for possible uses, and then exported to colonial plantations. In this manner, tea plantations using Chinese plants were established in the British colony of India, so Britain did not have to pay to import Chinese tea.

The expansion of British whaling ships and seal traders into the Arctic and Antarctic led to an interest in

One of the early explorers of Antarctica was Robert Scott (1868–1912). Here, men, sleds, and dogs tackle the snow and ice during Scott's Terra Nova Antarctic Expedition (1910–1913), with the ship *Terra Nova* in the background. Scott died during the expedition. *Time & Life Pictures/Getty Images.*

reaching the North and South Poles. In 1828 Sir William Parry (1790–1855) nearly reached the North Pole, a goal that was finally accomplished by American explorer Robert Peary (1856–1920) in 1909. The first scientific expedition to survey Antarctica was led by British explorer James Ross (1800–1862) in 1839–1843, which further whetted appetites to explore the region. In 1895 the Sixth International Geographic meeting in London adopted a resolution that "the exploration of the Antarctic Regions is the greatest piece of geographical exploration still to be undertaken." Though the Congress recommended that "this work should be undertaken before the close of the century," it was not until 1911 that Norwegian explorer Roald Amundsen (1872–1928) reached the pole. An expedition by Ernest Shackleton (1874–1922) subsequently made heroic yet tragic efforts to cross the continent from 1914–1917.

The Antarctic Treaty of 1959 prohibits any one from staking territorial claims on Antarctica, designating it for scientific purposes only. Presently, the Antarctica research station is taking ice core samples to monitor global warming; it is also the darkest place on Earth and thus ideal for astronomical research.

■ Modern Cultural Connections

Those who plumb the depths of the sea and engage in manned or robotic missions to the stars are the last explorers. Using seismometers at the bottom of the ocean, robotic divers, and deep sea submarines, scientists in the Deep Ocean Exploration Institute at Woods Hole Oceanographic Institution investigate the planetary forces and phenomena that generate earthquakes and tsunamis, support communities of life of the ocean floor that hold key clues to the evolution, and that forge large offshore mineral, oil, and gas deposits. Under the leadership of Dr. Bob Ballard (1942–), Woods Hole was also responsible for the discovery of the shipwrecked RMS *Titanic* in 1985.

As does deep ocean exploration, space exploration has scientific, economic, and geopolitical facets. Space exploration began in the late 1950s with the launch of the Soviet Union's *Sputnik* satellite on October 4, 1957, the first manmade object to orbit Earth. Responding to the Cold War–induced "space race," the United States made the first manned moon landing by *Apollo 11* on July 20, 1969.

After the first 20 years of exploration, NASA's budgetary cuts shifted emphasis to renewable space hardware (such as the Space Shuttle and instrumentation such as the Hubble Telescope). The end of the Cold War also saw partnerships in space with the former Soviet Union, exemplified by the International Space Station with entrepreneurs such as Sir Richard Branson (1950–) of Virgin Group, Ltd., forming the Spaceship Company in 2005 to promote space tourism, build suborbital spaceships, and launch aircraft. The price for a two-hour space ride is projected to be $200,000. NASA has advocated manned lunar and Mars missions after 2010 not only due to gains in scientific knowledge, but out of pragmatic concerns. With the onset of global climate change, humans may be forced to colonize space to ensure their survival.

SEE ALSO *Earth Science: Geography; Earth Science: Navigation.*

BIBLIOGRAPHY

Books

Campbell, Mary Baine. *Wonder & Science: Imagining Worlds in Early Modern Europe.* Ithaca: Cornell University Press, 2004.

Chartrand, René, Mark Harrison, et al. *The Vikings: Voyagers of Discovery and Plunder.* Botley, Oxford: Osprey Publishing, 2006.

Cohler, Anne M., Basia Carolyn Miller, and Harold Samuel Stone, eds. *Montesquieu: The Spirit of the Laws.* Cambridge Texts in the History of Political Thought. Cambridge: Cambridge University Press, 1989.

Fara, Patricia. *Sex, Botany, and Empire: The Story of Carl Linnaeus and Joseph Banks.* New York: Columbia University Press, 1994.

Gibb, H.A.R., ed. *Ibn Battuta: Travels in Asia and Africa 1325–1345.* New York: Routledge and Kegan Paul, 2004.

Green, Peter. *Ancient Greece: A Concise History.* London: Thames and Hudson, 1973.

Landis, Marilyn J. *Antarctica: Exploring the Extreme: 400 Years of Adventure.* Chicago: Chicago Review Press, 2003.

Marincola, John M., ed. *Herodotus: The Histories.* New York: Penguin, 2003.

Perrottet, Tony. *Pagan Holiday: On the Trail of Ancient Roman Tourists.* Westminster, MD: Random House, 2003.

Rawlinson, George. *Phoenicia: History of a Civilization.* New York: I.B. Tauris, 2005.

Roller, Duane W. *Through the Pillars of Herakles: Greco-Roman Exploration of the Atlantic.* London: Routledge, 2006.

Russell, Peter E. *Prince Henry "The Navigator": A Life.* New Haven: Yale University Press, 2000.

Schiebinger, Londa. *Plants and Empire: Colonial Bioprospecting in the Atlantic World.* Cambridge: Harvard University Press, 2007.

Van Pelt, Michel. *Space Tourism: Adventures in Earth Orbit and Beyond.* New York: Springer, 2005.

Wertheim, Margaret. *Pythagoras' Trousers: God, Physics, and the Gender Wars.* New York: W.W. Norton and Company, 1997.

Web Sites

National Aeronautics and Space Administration. "NASA Archives." http://www.nasa.gov/topics/history/history_interactive_archive_1.html (accessed January 21, 2008).

Anna Marie Eleanor Roos

Earth Science: Geodesy

■ Introduction

Geodesy is the science of measuring the shape and size of a planet or moon. Earth itself, because we have easy access to it, is by far the most accurately measured planetary body. Many measurements of Earth's shape and of the precise locations and motions of points fixed to its surface are made using satellites, which, since the late-1990s, have allowed differences as small as 0.4 inches (1 cm) to be measured from space. The motions of continents, the melting of ice caps, changes in Earth's surface wrought by earthquakes, and other phenomena are now routinely detected from space using geodetic methods. Since precision navigation and positioning are increasingly important in today's world, the continuous updating and improvement of geodetic information is crucial in many fields.

WORDS TO KNOW

ELLIPSOID: An egg-like three-dimensional shape defined by rotating an ellipse in space. In geodesy (measurement of Earth or other planetary bodies), the ellipsoid is the idealized, simplified shape of the planet. Features of the actual planet's shape are defined by their distances from the ellipsoid.

GEOID: A surface of constant gravitational potential around Earth; an averaged surface perpendicular to the force of gravity.

TRIANGULATION: A means of navigation and direction finding based on the trigonometric principle that, for any triangle, when one side and two angles are known, the other two sides and triangles can be calculated.

■ Historical Background and Scientific Foundations

The philosophers of ancient Greece were aware that Earth is round. They were also the first people to measure its size. About 250 BC, the Greek philosopher Eratosthenes (384–322 BC) estimated the size of Earth by measuring (with an assistant's help) the lengths of the shadow cast by a stick in Alexandria, Egypt, at the same time of the same day that the sun stood directly overhead in the city of Syrene several hundred miles away, casting no shadows. Applying trigonometry to this information, Eratosthenes calculated that Earth's equatorial circumference (the distance around the planet along the equator) is 25,000 miles (40,233 km)—remarkably close to the true value of 24,901 miles (40,075 km).

Many centuries passed before additional significant fundamental progress in geodesy occurred. Around 1600, as European trade, empire, and warfare expanded and physical science began to advance on many fronts, modern geodesy was born with the widespread application of careful triangulation. Triangulation is the determination of the location of a reasonably fixed point (say, a marker set into bedrock) by measuring the angles to two other points of known location, often termed control points. By applying trigonometry to these measurements, the location of the fixed point can be mapped. The newly mapped point can then be used as a control point for a new triangulation. In this way, networks of triangulation can be spread over entire countries or, ultimately, continents.

In 1660, the Royal Society in London was formed to promote scientific investigation. In 1666, the French formed a rival organization, L'Academie Royale des Sciences. The two bodies soon became embroiled in a patriotic dispute about the shape of Earth. The French maintained that Earth is a prolate spheroid, that is, an almost spherical object lengthened in the direction of its axis (North to South), somewhat like an egg. The English maintained that it is an oblate spheroid, that is, an almost spherical object flattened at the poles and broadened at the Equator, somewhat like a tomato. In the 1730s, to settle the dispute, L'Academie Royales des Sciences mounted expeditions to points near the Equator in South America and relative near the North Pole in northern Scandinavia to measure the curvature of Earth's surface in each region. If Earth is prolate, it must be flatter near the equator; if it is oblate, it must be flatter near the pole. The expeditions showed that Earth is, in fact, oblate. This shape is caused by the rotation of Earth on its axis, which causes an apparent centrifugal force that pulls outward from the spin axis (e.g., along the equator). The fact that such measurements could even be made shows how advanced the science of geodesy had become by that time.

In the 1900s mathematical and instrumental advances allowed more precise geodesy. The meter was defined as one 1/10,000,000 of the distance from the equator to the North Pole through Paris. However, geodetic networks of triangulation remained national affairs: Each national measurement network was a separate creation, so there was no way to precisely relate points in different countries, much less in different continents. This was amended in the first two-thirds of the twentieth century, using precise measurements of local gravity (Earth's surface gravity varies slightly from place

to place), observations of the stars, and more precise instruments for conventional triangulation.

In the mid-1980s, geodesists began to use satellite-based geodetic measurement systems. These included laser ranging (bouncing laser light from a satellite off the surface of Earth), very long baseline interferometry (synthetic aperture radar), and the Global Positioning System, a constellation of satellites exchanging radio signals that allows a receiver anywhere on Earth's surface to calculate its position to high accuracy.

Modern geodesy uses several basic concepts to describe the shape of Earth (or any other planet or moon). First is the reference ellipsoid. This is an idealized geometric figure, a perfectly symmetrical oblate spheroid. The reference ellipsoid does not reflect that actual bumpiness of Earth's surface. Its purpose is to give geodesists a reference or standard in comparison to which the variations of Earth's actual shape can be specified.

Second is the geoid. This is also an abstract surface—it does not correspond exactly to the surface of the land or sea, except at certain points—but has a more detailed physical basis than the reference ellipsoid. At sea, the geoid is the surface that would coincide with the surface of the world ocean if it were unaffected by tides, currents, or the like: over land, it is the surface where the sea surface would be found if it were able to send fingers of water to every point. It is, in essence, the idealized global sea level. The geoid is not symmetrical because Earth's gravity field is not symmetrical, but varies from

Azimuth marker. An azimuth is defined as an angle, usually measured in degrees, between a reference plane and a point. In mapping, directions (the azimuths) from a central point are kept. The direction is usually due north, but any direction is possible. *Phil Degginger/Alamy.*

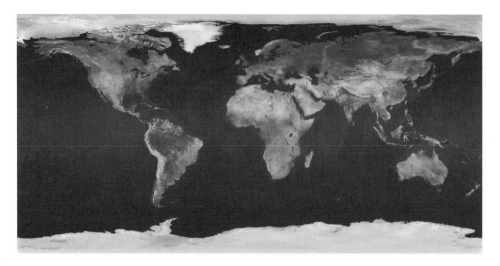

The use of satellites has helped researchers learn more about Earth. In this image, a composite of many different satellite images, Earth is shown as it would appear from space, with no clouds or sea ice present. © *Worldspec/NASA/Alamy.*

place to place: Where gravity is stronger, ideal sea level would be higher (more water would be drawn to such spots), and the geoid is higher.

Third, two bodies of measurements are specified with respect to the geoid: These measurements define the locations of certain well-characterized points on Earth's surface. Each body of measurements is called a datum. The vertical datum is a set of points with precisely known heights, and the horizontal datum is a set of points with precisely known latitude (north-south location), longitude (east-west location), or both. The two datums are updated every so many years to compensate for movements of Earth's crust. In the United States, datum points are marked by round brass plates set into bedrock and stamped with identifying information. These plates may often be seen on mountaintops. The datums supply a foundation for measuring other points or changes in the landscape. For example, the vertical datum gives a standard against which surveyors can measure subsidence of land, such as is occurring in the vicinity of New Orleans, Louisiana. Or, the horizontal datum allows surveyors to monitor land slippage in the vicinity of the San Andreas fault in California.

Because the geoid is defined as mean (average) local sea level (or what mean sea level would be, if the sea were present), which is in turn affected by how strong Earth's gravity happens to be in a given area, geodesists are interested in measuring variations in Earth's gravity field. In 2002, a pair of U.S. satellites was launched into Earth's orbit on a Russian rocket, forming the Gravity Recovery and Climate Experiment (GRACE). The two GRACE satellites fly about 137 miles (220 km) apart and continually measure the distance between them to high precision using lasers. As the leading satellite passes above an area of higher gravity, it dips slightly, increas-

ing the distance between the satellites: As the second satellite enters the high-gravity area, it dips too and the distance between them decreases. As the satellites move out of the high-gravity area they gain altitude again, first one and then the other. Scientists have been able to build up a detailed gravity map of Earth from these continual slight changes in distance between the two satellites. This information has allowed geodetic measurements of both stable and time-varying features of Earth's mass distribution with unprecedented accuracy. As of 2008, the mission continued.

■ Modern Cultural Connections

The Global Positioning System is now incorporated in many automotive navigation systems and in millions of miniature devices for surveying, wilderness orienteering, land-sea rescue guidance, and other purposes, including weapons targeting.

The GRACE satellite has performed types of environmental measurement that were hitherto impossible. For example, GRACE has enabled researchers to measure the amount of groundwater contained in the Congo River basin, to discover a large meteor crater hidden under the Antarctic ice, to assist in measuring a subtle gravitational effect called frame dragging (a test of general relativity), and to measure the rate of ice loss from the Greenland and Antarctic ice caps. These ice caps are so large—Greenland, with only a tenth as much ice as Antarctica, contains about 596,000 cubic miles (2.5 million cubic km) of ice—that their gravitational pull can be detected from space. Further, so fast are these giant masses of ice melting—Greenland lost about 19 cubic miles (80 cubic km) of ice per year from

Geologist Bolivar Caceres places Differential Global Positioning System equipment on the Cotopaxi volcano in Ecuador, January 3, 2004. Ecuadorean and French scientists are studying the effects that the volcano's reactivation and global warming can have upon the volcano's icecap. *Jorge Vinueza/AFP/ Getty Images.*

1997 to 2003—that the gravitational change due to this loss can also be detected.

The melting rates of the Greenland and Antarctic ice sheets are significant because melting of land-based ice sheets raises sea level. The degree to which global climate change will raise sea level by melting ice has been a matter of scientific debate, but satellite geodesy data from GRACE and other missions is helping reduce this uncertainty. In 2006, scientists announced that GRACE data had confirmed that Greenland is losing ice at a surprisingly fast and accelerating rate and that Antarctica is losing ice as well. These data were widely reported in the mass media, helping communicate to the public the reality and seriousness of global climate change. Rising sea levels may inundate some coastal settlements over the next century and increase the vulnerability of others to storms. The amount of sea-level rise will determine the amount of damage and displacement of populations that is caused.

■ Primary Source Connection

Scottish scientist James Hutton (1726–1797) wrote his *Theory of Earth* in four volumes. Hutton's work included the assertion that Earth's core was hot and that its surface moved and was weathered over time. His theory of deep time claimed Earth was much older than the few thousand years asserted by most of his contemporaries.

Hutton's work thus laid the foundation for several modern geologic principles.

THEORY OF THE EARTH

CHAPTER I.

THEORY of the EARTH; or an Investigation of the Laws observable in the Composition, Dissolution, and Restoration, of Land upon the Globe.

SECTION I.

Prospect of the Subject to be treated.

When we trace the parts of which this terrestrial system is composed, and when we view the general connection of those several parts, the whole presents a machine of a peculiar construction by which it is adapted to a certain end. We perceive a fabric, erected in wisdom, to obtain a purpose worthy of the power that is apparent in the production of it.

We know little of the earth's internal parts, or of the materials which compose it at any considerable depth below the surface. But upon the surface of this globe, the more inert matter is replenished with plants, and with animal and intellectual beings.

Where so many living creatures are to ply their respective powers, in pursuing the end for which they were intended, we are not to look for nature in a quiescent state; matter itself must be in motion, and the scenes of life a continued or repeated series of agitations and events.

This globe of the earth is a habitable world; and on its fitness for this purpose, our sense of wisdom in its formation must depend. To judge of this point, we must keep in view, not only the end, but the means also by which that end is obtained. These are, the form of the whole, the materials of which it is composed, and the several powers which concur, counteract, or balance one another, in procuring the general result.

The form and constitution of the mass are not more evidently calculated for the purpose of this earth as a habitable world, than are the various substances of which that complicated body is composed. Soft and hard parts variously combine to form a medium consistence, adapted to the use of plants and animals; wet and dry are properly mixed for nutrition, or the support of those growing bodies; and hot and cold produce a temperature or climate no less required than a soil: Insomuch, that there is not any particular, respecting either the qualities of the materials, or the construction of the machine, more obvious to our perception, than are the presence and efficacy of design and intelligence in the power that conducts the work.

In taking this view of things, where ends and means are made the object of attention, we may hope to find a principle upon which the comparative importance of parts in the system of nature may be estimated, and also a rule for selecting the object of our inquiries. Under this direction, science may find a fit subject of investigation in every particular, whether of form, quality, or active power, that presents itself in this system of motion and of life; and which, without a proper attention to this character of the system, might appear anomalous and incomprehensible.

It is not only by seeing those general operations of the globe which depend upon its peculiar construction as a machine, but also by perceiving how far the particulars, in the construction of that machine, depend upon the general operations of the globe, that we are enabled to understand the constitution of this earth as a thing formed by design. We shall thus also be led to acknowledge an order, not unworthy of Divine wisdom, in a subject which, in another view, has appeared as the work of chance, or as absolute disorder and confusion.

To acquire a general or comprehensive view of this mechanism of the globe, by which it is adapted to the purpose of being a habitable world, it is necessary to distinguish three different bodies which compose the whole. These are, a solid body of earth, an aqueous body of sea, and an elastic fluid of air.

It is the proper shape and disposition of these three bodies that form this globe into a habitable world; and it is the manner in which these constituent bodies are adjusted to each other, and the laws of action by which they are maintained in their proper qualities and respective departments, that form the Theory of the machine which we are now to examine.

Let us begin with some general sketch of the particulars now mentioned.

1st, There is a central body in the globe. This body supports those parts which come to be more immediately exposed to our view, or which may be examined by our sense and observation. This first part is commonly supposed to be solid and inert; but such a conclusion is only mere conjecture; and we shall afterwards find occasion, perhaps, to form another judgment in relation to this subject, after we have examined strictly, upon scientific principles, what appears upon the surface, and have formed conclusions concerning that which must have been transacted in some more central part.

2dly, We find a fluid body of water. This, by gravitation, is reduced to a spherical form, and by the centrifugal force of the earth's rotation, is become oblate. The purpose of this fluid body is essential in the constitution of the world; for, besides affording the means of life and motion to a multifarious race of animals, it is the source of growth and circulation to the organized bodies of this earth, in being the receptacle of the rivers, and the fountain of our vapours.

3dly, We have an irregular body of land raised above the level of the ocean. This, no doubt, is the smallest portion of the globe; but it is the part to us by far most interesting. It is upon the surface of this part that plants are made to grow; consequently, it is by virtue of this land that animal life, as well as vegetation, is sustained in this world.

Lastly, We have a surrounding body of atmosphere, which completes the globe. This vital fluid is no less necessary, in the constitution of the world, than are the other parts; for there is hardly an operation upon the surface of the earth, that is not conducted or promoted by its means. It is a necessary condition for the sustenance of fire; it is the breath of life to animals; it is at least an instrument in vegetation; and, while it contributes to give fertility and health to things that grow, it is employed in preventing noxious effects from such as go into corruption. In short, it is the proper means of circulation for the matter of this world, by raising up the water of the ocean, and pouring it forth upon the surface of the earth.

Such is the mechanism of the globe: Let us now mention some of those powers by which motion is produced, and activity procured to the mere machine.

First, There is the progressive force, or moving power, by which this planetary body, if solely actuated, would depart continually from the path which it now pursues, and thus be for ever removed from its end, whether as a planetary body, or as a globe sustaining plants and animals, which may be termed a living world.

But this moving body is also actuated by gravitation, which inclines it directly to the central body of the sun. Thus it is made to revolve about that luminary, and to preserve its path.

It is also upon the same principles, that each particular part upon the surface of this globe, is alternately exposed to the influence of light and darkness, in the diurnal rotation of the earth, as well as in its annual revolution. In this manner are produced the vicissitudes of night and day, so variable in the different latitudes from the equator to the pole, and so beautifully calculated to equalise the benefits of light, so variously distributed in the different regions of the globe.

Gravitation, and the vis infita of matter, thus form the first two powers distinguishable in the operations of our system, and wisely adapted to the purpose for which they are employed.

We next observe the influence of light and heat, of cold and condensation. It is by means of these two powers that the various operations of this living world are more immediately transacted; although the other powers are no less required, in order to produce or modify these great agents in the economy of life, and system of our changing things.

We do not now inquire into the nature of those powers, or investigate the laws of light and heat, of cold and condemnation, by which the various purposes of this world are accomplished; we are only to mention those effects which are made sensible to the common understanding of mankind, and which necessarily imply a power that is employed. Thus, it is by the operation of those powers that the varieties of season in spring and autumn are obtained, that we are blessed with the vicissitudes of summer's heat and winter's cold, and that we possess the benefit of artificial light and culinary fire.

We are thus bountifully provided with the necessaries of life; we are supplied with things conducive to the growth and preservation of our animal nature, and with fit subjects to employ and to nourish our intellectual powers.

There are other actuating powers employed in the operations of this globe, which we are little more than able to enumerate; such are those of electricity, magnetism, and subterraneous heat or mineral fire.

Powers of such magnitude or force, are not to be supposed useless in a machine contrived surely not without wisdom; but they are mentioned here chiefly on account of their general effect; and it is sufficient to have named powers, of which the actual existence is well known, but of which the proper use in the constitution of the world is still obscure. The laws of electricity and magnetism have been well examined by philosophers; but the purposes of those powers in the economy of the globe have not been discovered. Subterraneous fire, again, although the most conspicuous in the operations of this world, and often examined by philosophers, is a power which has been still less understood, whether with regard to its efficient or final cause. It has hitherto appeared more like the accident of natural things, than the inherent property of the mineral region. It is in this last light, however, that I wish to exhibit it, as a great power acting a material part in the operations of the globe, and as an essential part in the constitution of this world.

We have thus surveyed the machine in general, with those moving powers, by which its operations, diversified almost ad infinitum, are performed. Let us now confine our view, more particularly, to that part of the machine on which we dwell, that so we may consider the natural consequences of those operations which, being within our view, we are better qualified to examine.

This subject is important to the human race, to the possessor of this world, to the intelligent being Man, who foresees events to come, and who, in contemplating his future interest, is led to inquire concerning causes, in order that he may judge of events which otherwise he could not know.

If, in pursuing this object, we employ our skill in research, not in forming vain conjectures; and if data are to be found, on which Science may form just conclusions, we should not long remain in ignorance with respect to the natural history of this earth, a subject on which hitherto opinion only, and not evidence, has decided: For in no subject, perhaps, is there naturally less defect of evidence, although philosophers, led by prejudice, or misguided by false theory, may have neglected to employ that light by which they should have seen the system of this world.

But to proceed in pursuing a little farther our general or preparatory ideas. A solid body of land could not have answered the purpose of a habitable world; for, a soil is necessary to the growth of plants; and a soil is nothing but the materials collected from the destruction of the solid land. Therefore, the surface of this land, inhabited by man, and covered with plants and animals, is made by nature to decay, in dissolving from that hard and, compact state in which it is found below the soil; and this soil is necessarily washed away, by the continual circulation of the water, running from the summits of the mountains towards the general receptacle of that fluid. The heights of our land are thus levelled with the shores; our fertile plains are formed from the ruins of the mountains; and those travelling materials are still pursued by the moving water, and propelled along the inclined surface of the earth. These moveable materials, delivered into the sea, cannot, for a long continuance, rest upon the shore; for, by the agitation of the winds, the tides and currents, every moveable thing is carried farther and farther along the shelving bottom of the sea, towards the unfathomable regions of the ocean.

James Hutton

HUTTON, JAMES. *THEORY OF THE EARTH*, VOLUME 1 (OF 4) EDINBURGH: 1795. AVAILABLE AT *PROJECT GUTENBERG*, JULY 9, 2004. HTTP://WWW.GUTENBERG.ORG/ FILES/12861/12861-H/12861-H.HTM (ACCESSED MARCH 5, 2008).

SEE ALSO *Astronomy and Cosmology: Setting the Cosmic Calendar: Arguing the Age of the Cosmos and Earth.*

BIBLIOGRAPHY

Periodicals

Gordon, Richard G., and Seth Stein. "Global Tectonics and Space Geodesy." *Science* 256 (1992): 333–341.

Heiskanen, W.A. "New Era of Geodesy." *Science* 121 (1955): 48–50.

Meade, Charles, and David T. Sandwell. "Synthetic Aperture Radar for Geodesy." *Science* 273 (1996): 1181–1182.

Smalley, R., Jr. "Space Geodetic Evidence for Rapid Strain Rates in the New Madrid Seismic Zone of Central USA." *Nature* 435 (2005): 1088–1090.

Web Sites

National Ocean Service (US). "Welcome to Geodesy." March 8, 2005. http://oceanservice.noaa.gov/ education/kits/geodesy/welcome.html (accessed February 9, 2008).

Larry Gilman

Earth Science: Geography

■ Introduction

Many people think of geography simply as the study of maps and places. Although the science of geography requires understanding these things, the field is much broader than that. Geography is the study of Earth's surface and the way it affects people, the environment, politics, and other events or physical features.

The science of geography has three main branches: Physical geography is the study of Earth, its physical features, landforms, weather, soils, oceans, water, and attributes such as size and shape, both in the present and the past. Human geography is the study of people and their distribution, particularly with respect to culture, religion, politics, economics, health, demographics, and other characteristics. Environmental geography examines how people interact with the environment in relation to the space they occupy. Each branch overlaps the others in some ways, but differs in the kinds of questions it seeks to answer and in the focus of its work.

■ Historical Background and Scientific Foundations

As a science, geography relies heavily on the use of maps to show and explain the concepts it studies in the places they occur. Without maps, geographers would have a difficult time expressing and analyzing the data they collect. Cartography is the ancient art, and recent science, of mapmaking. People have used maps for thousands of years; the oldest ones known come from the Middle East, where the earliest civilizations began. One map of a Babylonian city shows hills, rivers or canals, and a specific plot of land with its size and owner labeled on it. Another ancient map from Turkey shows a city plan with individual buildings and roads, as well as a volcano. The advanced nature of these maps, showing both human-made and natural features in different scales, indicates that the idea of maps was not new when they were drawn. The Babylonians also produced a world map, showing a circular world surrounded by an ocean.

The Greeks and Romans were advanced cartographers. Ancient Greek geographers were interested in both physical and human geography, though those terms were not in use at the time. The poet Homer (fl. 8th or 9th century BC) was famous as a geographer for describing the shape of Earth as he knew it: land surrounded by sea. Later, Pythagoras (580–500 BC) advanced the idea that Earth was spherical; Aristotle (384–322 BC) verified this with experiments. The Roman historian Herodotus (c.484–c.430 BC) included a great deal of geographical information in his writing. His descriptions of places such as Egypt are so accurate that they have been used to guide archaeological expeditions. A famous passage from Herodotus speculates that the reason for the annual Nile River floods must be melting snow far upstream, although he was unsure where the snows fell since Africa is so hot. He was also interested in the people of the world and described their customs, though it should be noted that some of his work is based on observations and interviews, while other parts were mostly rumor and speculation.

One of the most important uses of maps in the ancient world was for travel. The Romans built an extensive network of roads that allowed armies and commerce to move through their huge empire. By the fifth century AD, a map known today as the Peutinger Table had been developed that showed a great number of roads with towns, cities, and inns along them. Remarkably, the map stretches from Ireland, Spain, and Morocco in the west to Sri Lanka and perhaps even China in the east. Though the dimension representing longitude is compressed, many features are easily recognizable even today. The original ancient map does not survive, but a medieval copy was discovered and published in the sixteenth century. The Romans were also skilled surveyors and made maps to plan colonies and towns. One of the largest city maps was of Rome, carved on marble by order of the Emperor Septimius Severus (AD 146–211). It showed the plans of temples and public buildings like theaters and even simple homes and apartment buildings.

The study of geography reached its most advanced state in the ancient world in the book *Geography* by Ptolemy (AD c.90–c.168), a Romanized Greek-Egyptian geographer. Ptolemy set out to record and update all the knowledge of the world that he was able to obtain. He stands out because he did not base his ideas on a philosophical ideal, but instead used data to shape his

A portion of a full-size facsimile of German cartographer Martin Waldseemüller's 1507 world map is shown. It was this map, divided into 12 panels covering a total of 36 square feet, that the name "America" first appeared. *AP Images.*

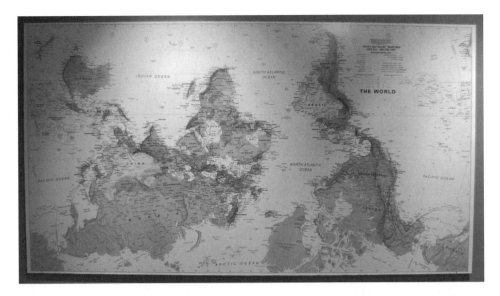

A world map drawn to a different perspective is seen at a Boston Public Library map exhibit titled "Faces and Places," which offers a lesson of geography, history, economics, and politics through the study of maps, which include the world, various countries, and old Boston. *Patricia McDonnell/AP Images.*

conclusions. Ptolemy was the first to use a latitude and longitude grid to position features on his map, though its proportions were flawed since it was impossible for the Romans to measure longitude accurately. Ptolemy's work, lost or ignored for almost 1,000 years, was rediscovered in Europe during the late Middle Ages and provided a basis for the maps during the Renaissance and the age of exploration.

During the Middle Ages, the study of geography (and science in general) declined in Europe. However, people of other cultures at the time were studying Earth, using principles of geography to map and understand it. As early as the fifth century BC, the Chinese wrote about the geography of their empire. They described the physical characteristics of the land, rivers, lakes, and other features. They were also interested in agricultural crops, goods produced, and trades practiced in each particular area. Later, Chinese geographers drew maps with accurate scales and included geography as an important part of their imperial bureaucracy. Islamic geographers in the Middle East preserved the works of the Greek geographers and produced detailed works that included maps of the *hajj*, or yearly religious pilgrimage to Mecca.

The field of geography grew from cartography as scholars sought to understand more about the world and wanted to locate their data in specific places. In the nineteenth century, scientists such as the German Alexander von Humboldt (1769–1859) traveled widely to study the physical features of Earth, transforming the study of geography from a literary or philosophical pursuit to one of rigorous scientific inquiry. Humboldt insisted on using the scientific method and inductive reasoning to understand the phenomena he observed, including plants, weather, and agriculture.

The study of human geography was also taking shape at this time. German geographer Carl Ritter (1779–1859) is considered the father of the field. More of an academic than an explorer like Humboldt, Ritter concentrated his work on human habits and settlements. He was also concerned with how people used, changed, and adapted to particular areas, climates, or habitats. Humboldt, too, was interested in human geography, investigating rates of disease with respect to temperature.

■ Modern Cultural Connections

In the second half of the twentieth century, the study of geography became more systematic. Instead of simply collecting data, geographers began to analyze its meaning and connect it with other data and ideas. They also began to develop theories about the nature of their knowledge and to challenge themselves with new models and ways of thinking. Where their predecessors concentrated on facts and measurements, present-day geographers are concerned with the study of systems and the way individual factors influence one another. With these foundations, the modern study of geography has expanded into a vast field with numerous individual areas of study.

The study of geography is important because it helps us comprehend the world, its people, and the environment. At its most basic, physical geography is the knowledge of where things are located. It also attempts to understand what features are located in what

General view of the rubble of Margala Towers in Islamabad, Pakistan, after an earthquake measuring 7.6 on the Richter scale struck in the early hours of October 8, 2005, killing more than 30 people and trapping dozens in the apartment complex. Using principles of applied geography, scientists now encourage building structures that sustain less damage during earth-quakes in areas where they frequently strike. *AP Images.*

places, and in what relation to one another. Sometimes these features are not obvious to an uninformed observer. Beyond plotting the location and properties of mountains and rivers, physical geography also includes things that are important to our changing world and environment. Locations and frequency of wildfires; soil types and erosion rates; the sizes, locations, and effects of earthquakes; frequency and severity of floods, all of these topics and more are relevant to our lives, especially when viewed in the context of global climate change.

Human geography studies the distribution of people, their effect on the environment, and vice versa. It includes the physical presentation of social sciences such as economics, sociology, linguistics, politics, and demographics. Human geographers might examine how the environment influences disease in a community, for example, or why some people practice a traditional religion while others converted to a different faith. Cultural geography is a broad type of human geography that looks at many aspects of a culture, for example language, religion, food, and the economy, and how these things vary by location. Some human geographers are guided by philosophical principles; these include Marxist geographers who study the relationships between the perceived exploitation of workers and the places they inhabit, and feminist geographers who study gender equality based

on location. All categories of human geography study people and how their lives are affected by the spaces they inhabit, although the focus of each subdiscipline may be very different.

The third branch, environmental geography, is the study of the way people use and perceive space in the natural world. It links physical and human geography because it deals with aspects of both disciplines. It is easy to understand how people might cut away hills for a highway or fill in coastal areas to create more land for building, but environmental geography addresses more than this. It is most heavily concerned with how human activity influences natural patterns like weather and water cycles. For example, an environmental geographer might be interested in a new water table forming under Las Vegas, where none had existed before the rapid population growth of the past 50 years. To understand how it developed, geographers would look at its physical location, extent, composition, and other characteristics. They would also study the population growth rate and the way that residents use water resources to determine how this influences the environment. Geographers would be equally interested in the ways that the physical environment in that location influences human behavior.

One of the most important developments in the field is the idea of "applied geography." Going beyond research, applied geography uses science to solve problems. For example, determining past wildfire patterns might help predict future outbreaks, or the study of earthquake damage might reveal new techniques for construction or urban planning. Applied geography also helps explain the distribution of natural resources and how to manage them effectively.

From the development of the earliest maps to the most recent achievements, the science of geography will continue to influence the growth and development of the world's societies, environments, and people.

SEE ALSO   *Earth Science: Atmospheric Science; Earth Science: Climate Change; Earth Science: Geodesy; Earth Science: Geologic Ages and Dating Techniques; Earth Science: Gradualism and Catastrophism; Earth Science: Oceanography and Water Science; Earth Science: Plate Tectonics: The Unifying Theory of Geology.*

BIBLIOGRAPHY

Books

Gregory, Kenneth J. *The Changing Nature of Physical Geography.* New York: Oxford University Press, 2000.

Mitchell, Don. *Cultural Geography: A Critical Introduction.* Boston: Blackwell, 2000.

Periodicals

Pattison, William D. "The Four Traditions of Geography." *Journal of Geography* 63, no. 5 (May 1964): 211-216.

Web Sites

Newberry Library: The Hermon Dunlap Smith Center for the History of Cartography. "Cartographic Images of the World on the Eve of the Discoveries." http://www.newberry.org/smith/slidesets/ss08.html (accessed November 13, 2007).

Valparaiso University. Department of Geography and Meteorology. "Von Humboldt and Ritter." October 14, 1996. http://www.valpo.edu/geomet/geo/courses/geo466/topics/humboldt.html (accessed November 13, 2007).

René Nougayrède
Kenneth T. LaPensee

Earth Science: Geologic Ages and Dating Techniques

■ Introduction

Earth is about 4.5 billion years old. Geologists divide this age into major and minor units of time that describe the kinds of geological processes and life forms that existed in them. Earth's geologic record was formed by constant change, just like those that occur routinely today. Though some events were catastrophic, much of Earth's geology was influenced by normal weather, erosion, and other processes spread over very long geologic ages. Accurate dating of the geologic ages is fundamental to the study of geology and paleontology, and provides important context to the life sciences, meteorology, oceanography, geophysics, and hydrology.

■ Historical Background and Scientific Foundations

In the mid-seventeenth century, James Ussher (1581–1656), the Archbishop of Ireland, compiled a chronology of Earth by adding up the generations named in the Bible. He determined that Earth was created the night before October 23, 4004 BC. This would make the world about 5,650 years old in Ussher's day and about 6,000 years old now.

Although Ussher also based his calculations on a painstaking analysis of many other literary sources as well as the Bible, his was not a scientific investigation. His chronology represented the common belief among Christians of his time that biblical events, including the creation account in Genesis, happened exactly as they were written. By the nineteenth century it had become a popular opinion among scientists and scholars that Earth was created in a single event and that its short history was altered only by the great biblical flood.

Ussher's chronology was widely accepted at the time that early geologic investigators began their work.

In the 1750s, however, Giovanni Arduino (1714–1795), an Italian professor studying mining and surveying, began to realize that different kinds of rocks had been deposited at different times in history. He divided the many different kinds of rocks that he studied into four broad categories: Rocks from the primary age, consisting of igneous or metamorphic rocks at the cores of mountains, were the first to be deposited. Rocks from the secondary age were sedimentary layers deposited on the sides of mountains, on top of primary layers. Deposits from the tertiary age were made up of hills of gravel and sand, i.e., broken pieces of rocks that were formed in earlier ages. The quaternary age is the current era, in which rocks and soils are still being deposited.

Arduino's work did not create an absolute time scale, but more of a relative chronology, a sequence of events without distinct time spans imposed on it. He described which deposits had to be older than others by understanding that sedimentary rocks, gravels, and sands are made of older rocks that have been broken apart to make up new materials. Although Arduino's chronology was broad and relative, it is still generally accepted as correct and has been incorporated into modern methods of geologic dating.

The nineteenth century was an important period in the study of geology. Many important discoveries, particularly in Britain, laid the foundation for a scientific investigation of Earth. A Scottish scholar and farmer named James Hutton (1726–1797) introduced many of the fundamental geological concepts that led to a scientific understanding of Earth's age and development.

A curious man, Hutton sought to understand how the geological features that he saw around him had come to be. He recognized that for one type of rock to intrude on another, the first rock formation must have been molten. This violated the prevailing theory that most rocks were deposited from the waters of the biblical flood. Hutton called his theory that all rocks were

WORDS TO KNOW

CAMBRIAN EXPLOSION: Relatively sudden evolution of a wide variety of multicellular forms of life at the beginning of the Cambrian period, about 530 million years ago, after billions of years during which Earth was inhabited almost entirely by single-celled organisms. A few forms of multicellular life did appear shortly before the Cambrian Explosion. The explosion may have been triggered by the ending of a major snowball Earth period or by the achievement of sufficiently high oxygen levels thanks to billions of years of oxygen production by algae (single-celled aquatic plants).

GEOLOGIC RECORD: Evidence of Earth's history left in rocks and sediments over thousands to billions of years. Events that can be inferred from the geological record include climate changes, biological evolution, continental drift, and asteroid impacts.

HALF-LIFE: The time it takes for half of the original atoms of a radioactive element to be transformed into the daughter product.

IGNEOUS: Any rock that has formed directly out of molten material, such as lava or granite.

K-T BOUNDARY: At the very end of the Cretaceous period and the beginning of the Tertiary period, 65.5 million years ago, a mass extinction wiped out 65–70% of all plant and animal species in a short time. The layer or boundary separating Cretaceous sediments from Tertiary sediments is termed the K-T boundary. This layer is rich in iridium, an element more common in meteorites than Earth rocks, which suggests that the extinction was caused by an asteroid striking Earth. The Chicxulub crater in the Yucatan Peninsula, Mexico, is probably the crater caused by this impact.

LAW OF SUPERPOSITION: In geology, the law of superposition states that barring the overturning of rock layers by later processes, deeper layers are always older than shallower layers. The law holds because (apart from later disruptions) sediments laid down earlier must be on the bottom and later materials must be on the top.

METAMORPHIC: In geology, a sedimentary rock that has been partially melted and then allowed to resolidify. This changes (metamorphoses) the character of the original rock.

OXYGEN CRISIS (OR OXYGEN CATASTROPHE): The great increase in the amount of free molecular oxygen (O_2) that occurred about half a billion years ago, just before the diversification of multicellular life known as the Cambrian explosion. Only green plants (e.g., algae) produce O_2 in large quantities. O_2 combines readily with many other molecules and compounds and so does not last long in nature unless continually resupplied. Algae had been producing O_2 for over a billion years by the time of the oxygen crisis, but this oxygen had been absorbed by rocks. When

the rocks could no longer absorb O_2 as fast as the algae produced it, the amount of O_2 in Earth's atmosphere increased dramatically.

PERMIAN-TRIASSIC EXTINCTION: Some 254.1 million years ago, at the end of the Permian period and the beginning of the Triassic period, the most severe mass extinction on Earth so far occurred. About 95% of all marine species and about 70% of all land-dwelling species disappeared. Various causes have been proposed. Some scientists theorize that several causes, coincidentally occurring near each other in time, caused the extinction.

PLUTONISM: In modern geology, the hardening of magma far below the surface to form igneous rock; granite, for example, is formed by plutonism.

PRECAMBRIAN: All geologic time before the beginning of the Paleozoic era. This includes about 90% of all geologic time and spans the time from the beginning of Earth, about 4.5 billion years ago, to 544 million years ago. Its name means "before Cambrian."

PROTOCONTINENT: A landmass that is in the process of being developed by geological processes into a true or major continent.

RADIOACTIVE DECAY: The predictable manner in which a population of atoms of a radioactive element spontaneously disintegrates over time.

RADIOMETRIC DATING: The use of naturally occurring radioactive elements and their decay products to determine the absolute age of the rocks containing those elements.

SUPERCONTINENT: In geology, a supercontinent is formed when Earth's 14 or so major tectonic plates, which are in constant motion, happen to come together in a single, large landmass, leaving the rest of the planet covered with ocean. At least half a dozen supercontinents have formed during Earth's 4.5-billion year history.

UNIFORMITARIANISM: Doctrine of geology promoted by English geologist Charles Lyell (1797–1895), asserting three theories: (1) actualism (uniform processes acting throughout Earth's history), (2) gradualism (slow, uniform rate of change throughout Earth's history), and (3) uniformity of state (Earth's conditions have always varied around a single, steady state). Uniformitarianism is often contrasted to catastrophism. Modern geology makes use of elements of both views, acknowledging that change can be drastic or slow, and that while some processes operate steadily over many millions of years, others (such as asteroid impacts) may happen rarely and cause catastrophic, sudden changes when they do.

the result of volcanic activity "Plutonism," after the Roman god of the underworld. Scientists know today that only igneous rocks are formed from a molten state, and only sedimentary rocks are deposited under bodies of water.

One enduring idea that James Hutton introduced, however, was the principle of uniformitarianism. He argued that processes such as volcanism, weather, and erosion that happen today have been occurring at the same rate and intensity throughout Earth's history. This led him to realize that Earth must be much older than the 6,000 years calculated by Ussher, an important breakthrough in the study of Earth's geologic ages, since it means that natural laws can be determined and expected to apply in the same ways over time.

The law of superposition had been formulated in the seventeenth century by Danish geologist Nicolaus Steno (1638–1686); Hutton applied this principle when studying Siccar Point, a place where upturned shale layers were covered with a horizontal layer of sandstone. Since the sandstone must have formed after the shale solidified, the sandstone must be younger than the shale. This law is central to understanding geologic dating, because it can be applied to individual layers, kinds of fossils, or even whole geologic ages: Newer material is deposited on top of older. Therefore, since fossils of trilobites are found in layers far below those of birds, it can be concluded that trilobites lived in an age long before birds appeared.

The law of superposition ties in tightly with the law of floral and faunal succession, which states that different kinds of plants and animals (flora and fauna) occurred in a specific and identifiable order wherever they are found. The English geologist William Smith (1769–1839) developed this law while working as a surveyor for mining and canal-building companies in the late 1700s and early 1800s. By descending into mines, Smith was able to study the rock layers, or strata, noting their composition and the kinds of fossils preserved in them. This led him to conclude that the order of strata and fossils was consistent even in different places. Though all layers were not visible in one place, Smith, in his job as a wandering surveyor, strung data from different locations together. This allowed him to establish a relative geological chronology and create his greatest work, a geological map of England. However, the development of an absolute chronology relating the distant past to the present was not yet possible.

In the nineteenth century, using the work of Hutton, Smith, Arduino, and others, it became possible to divide geologic time into eons, eras, periods, and epochs based on the kinds of rock present, the layers they fell into, and the kinds of life present in them. Though geologists and paleontologists could put most of the ages of Earth's past in order, they still had to estimate how long ago those events happened and how long they

lasted. The development of a measured absolute time scale would have to wait until the twentieth century.

Currently, the International Commission on Stratigraphy recognizes four eons and more than 70 eras, periods, and epochs within them. Eons are usually agreed to be the largest division of time on the geologic scale. Until recently, the three most distant eons, the Hadean, the Archean, and the Proterozoic, were collectively referred to as the Precambrian eon. Use of this term has since been officially discontinued, but still occurs occasionally, especially in older texts.

The Hadean eon began 4.5 billion years ago, when Earth formed from debris as the solar system coalesced, and ended about 3.8 billion years ago. At the beginning of this eon, Earth was molten. As it cooled, it separated into layers, producing the core, mantle, and crust that exist today. A favored scientific theory holds that the moon was also formed at this time, when a massive object the size of Mars collided with Earth. Shortly thereafter, Earth was pelted with meteorites during the

Geologists examine the walls of a trench for evidence of ancient earthquakes. Earthquakes appear as noticeable disruptions in the stratigraphy. By examing the wall, scientists can estimate when earthquakes occurred. © *Roger Ressmeyer/Corbis.*

late heavy bombardment, increasing the environment's hostility to life. Chemical analyses have supported the theory that oceans formed during the Hadean eon, although because of atmospheric conditions they were more acidic than they are today. There is no evidence of life from the Hadean eon, and it is so distant in time that only a few scattered samples of minerals, and no complete rock formations, survive.

The Archean eon, extending from 3.8 to 2.5 billion years ago, was characterized by extreme geologic activity, which formed the first protocontinents. Because radioactive materials were present in amounts far greater than today, the heat from their decay drove volcanism and other geologic activity at extraordinary rates. Geologists do not agree on whether plate tectonics were active during the Archean eon or whether Earth's crust was even made up of plates at this time. However, smaller land masses were developing and being destroyed and rocks were forming at much higher temperatures than are possible today. The atmosphere was radically different, as well, containing very little oxygen. Certain kinds of iron ores formed that would not have been possible had oxygen been present in the atmosphere. Most modern life forms could not have survived in the oxygen-poor atmosphere of the Archean eon.

Despite the lack of oxygen, the first kinds of life arose in the Archean: anaerobic bacteria that formed into mats, columns, or cones called stromatolites. Many different kinds of fossilized stromatolites have been found in different rock formations around the world, giving geologists a great deal of information about the Archean eon. They show that different kinds of bacteria lived together in an ecosystem, that some bacteria used photosynthesis to generate energy from the sun while others relied on different sources, and that areas of both shallow and deep water were present. Although the fossil evidence from the Archean is limited, all the life forms discovered so far have been single-celled prokaryotes that lack a nucleus.

The Archean was followed by the Proterozoic, occurring between 2.5 billion and 545 million years ago. During this eon life began to transform into types that we recognize today, changing Earth along with it. Shallow seas formed and the atmosphere began to change as well. During the Paleoproterozoic era (the earliest part of the Proterozoic eon) an event known as the oxygen catastrophe occurred: a relatively sudden increase in the amount of available oxygen which was the result of a complex chain of events. Since the Archean eon, early bacteria had been excreting oxygen as a waste product. Initially, most of the oxygen was consumed in the oxidation of minerals and metals such as iron. As the amount of unoxidized iron began to decrease, the amount of oxygen in the atmosphere increased. This poisoned some types of anaerobic Archean bacteria, but spurred others to use oxygen in their metabolism, a much more efficient way of processing energy. Aerobic organisms became dominant in the Proterozoic eon.

The Mesoproterozoic era (the middle part of the Proterozoic eon) saw the development of eukaryotes, single-celled organisms with a nucleus. During the end of the Neoproterozoic era (the most recent part of the eon), in a division known as the Ediacaran period, the earliest complex multicellular organisms appeared. These soft-bodied creatures appear to have lived on the bottom of shallow seas, not unlike modern corals or sponges. They were diverse in size, structural complexity, shape, and symmetry. The Ediacaran period is the most recently recognized of all the eons, eras, and periods, named for the Ediacara area in Australia, where many of the fossils have been found.

Alongside the rapidly changing life forms of the Proterozoic eon, significant geological processes were occurring. The supercontinent called Rodinia formed at the end of the Stenian period in the Mesoproterozoic. The first ice ages occurred during the Proterozoic era. At times ice may have covered Earth entirely, a hypothesis known as the "snowball earth" theory.

The end of the Proterozoic is marked by a dramatic event in the fossil record known as the Cambrian explosion. At this time, a remarkable increase in the numbers and types of species is seen, as well as the first hard-bodied animals, i.e., those with shells. The Cambrian period marks the beginning of the Phanerozoic eon, the eon of "visible life" spanning the past 545 million years and continuing today. During this time, life evolved from the simplest sponges, jellyfish, and worms to include almost everything we can think of that is alive today.

Geological periods during the Phanerozoic are divided into smaller epochs based on changes in the kinds of life that appear in the fossil record. The larger number of fossilized species present and the relatively short period of time since their deposit allow this more precise dating. The largest divisions of the Phanerozoic eon are the Paleozoic, Mesozoic, and Cenozoic eras. Each lasted for millions of years and each is broadly characterized by the degree of development that the life within it has undergone.

The Paleozoic is divided into the Cambrian, Ordovician, Silurian, Devonian, Carboniferous (which is sometimes divided into the Mississippian and Pennsylvanian eras) and Permian periods. Each of these is further divided into several epochs, some named for places where their major characteristics were discovered, others simply divided into early, middle, and late epochs. During the Paleozoic era, insects, plants, the first vertebrate animals, amphibians, reptiles, fish, sharks, and corals all appeared. Often, it is the changes in the kinds of animals and plants that are used to decide boundaries between the different periods.

Despite the emphasis on life in describing the various ages of the Paleozoic, geologic processes were still

View of a linear accelerator, used as part of an accelerator mass spectrometer (AMS) for radiocarbon dating. This device is capable of counting the relatively few carbon-14 atoms in a sample being radiocarbon dated. The carbon-12 content may be assessed using conventional mass spectrometry. The proportion of carbon-14 to carbon-12 atoms in the sample may be used to determine the radiocarbon age of an organic object. This is then adjusted by various corrections to give the true age. Photographed at Oxford University. *James King-Holmes/Photo Researchers, Inc.*

under way. Supercontinents formed and broke apart, several ice ages advanced and retreated, temperatures fluctuated, and sea levels rose and fell. These diverse processes influenced the many changes in life that are recorded in the fossils of the era—coal deposits in Europe laid down during the Carboniferous period are one of its more famous features. At the end of the Paleozoic era, a disastrous event known as the Permian-Triassic extinction led to the destruction of almost all Paleozoic species. Though there have been efforts to link this extinction to a meteorite impact, no convincing evidence of a large enough collision during this time period has been found.

Dinosaurs appeared during the Mesozoic era. The names of the periods in the Mesozoic era may sound familiar: Triassic, Jurassic, and Cretaceous. During this 180 million-year era, all the familiar dinosaurs such as triceratops, tyrannosaurus, stegosaurus, diplodocus, and apatosaurus flourished at different times. Some modern animals have ancestors that first appeared during the Mesozoic era, including birds, crocodiles, and mammals. Plants continued to develop, and the first flowering plants appeared.

The end of the Mesozoic era can be seen clearly in some rock layers. Known as the K-T (Cretaceous-Tertiary) boundary, this dark line of sediment is rich in the element iridium. Another massive extinction of species occurred at this time, possibly because of one or more meteorite impacts along with a period of intense volcanic activity. This would have decreased the amount of sunlight reaching Earth's surface, killing plants and, eventually, animals. Not all geologists and paleontologists are convinced that the K-T extinction was a catastrophic event; some argue that it occurred over a few million years after slower climate changes.

The Cenozoic era, the current era of geologic time, is divided into the Paleogene and Neogene periods, and further into the Paleocene, Eocene, Oligocene, Miocene, Pliocene, Pleistocene, and Holocene epochs. During the Cenozoic, the supercontinent of Gondwana broke apart, and the continents reached their current positions. Several ice ages occurred, and the poles became ice-covered. The first mammals began to flourish in the Paleocene; the first apes appeared in the Miocene; and the first human ancestors in the Pliocene. Modern humans, along with large animals such as mammoths and wooly rhinoceroses, appeared in the Pleistocene. The Holocene epoch, currently ongoing, began with the end of the last ice age, less than 10,000 years ago.

Though this vast span of time was largely understood by the end of the nineteenth century, geologists, paleontologists, and scientists of other disciplines were still curious about Earth's absolute age, using different approaches to tackle the problem. In the 1860s William Thomson (1824–1907), more commonly known as Lord Kelvin, applied his theories of thermodynamics to determine Earth's age. He surmised that Earth was between 20 and 40 million years old by calculating the time it should take for it to cool from a liquid to a solid. Though his calculations and some of his assumptions were correct, he failed to account for heat added by radioactivity. Around the turn of the twentieth century, Irish geologist John Joly (1857–1933) estimated Earth's age by analyzing the salt content of the seas. He then assumed that the oceans had started off as freshwater, and that all the salt had washed into them from the land. This relied on the assumption that the rate of salt coming into the oceans was constant and that no salt had ever been removed from the seas. By this calculation he arrived at an age of about 100 million years.

Scientists needed a method that relied on something measurable over Earth's entire lifespan. In rocks older than about 600 million years, it becomes impossible to use fossils to calculate their age because very few, if any, exist in these rocks. There are, however, a number of naturally radioactive elements that have been decaying since the formation of Earth. With the discovery of radiation and the calculation of half-lives in the twentieth century it finally became possible to determine the age of Earth's oldest rocks.

Radioactive decay is the spontaneous change in the nucleus of an element by the escape of a proton or neutron. Once a particle escapes the nucleus of an atom, it becomes a different isotope of the same element, or sometimes a different element altogether. The ratio of the original parent element to the daughter element produced by decay determines how long the element has been decaying. The half-life of an isotope is the amount of time it takes for half of the sample to decay. In 1906, New Zealand-born British physicist Ernest Rutherford (1871–1937) discovered that uranium and thorium decayed into isotopes of lead. By 1907 Bertram Boltwood (1870–1927), an American chemist studying radioactive materials, had calculated the age of certain rocks based on analysis of their radioactivity.

Radiometric dating, a well-regarded way to establish the age of rocks, is still based on the same principles laid out by Rutherford and Boltwood. It assumes that the half-lives of elements do not change over time, and that the sample has not been contaminated by the addition or removal of radioactive material. Zirconium crystals are usually analyzed because they trap uranium in their structure. Analyzing the decay of uranium to lead is useful because the half-life of uranium^{238} is 700 million years. Even longer dates can be measured with potassium-to-argon decay, with a half-life of 1.3 billion years, and rubidium-to-strontium decay, with a half-life of 50 billion years. Carbon-14 dating is useful for measuring very short ages on the geologic time scale. With a half-life of 5,730 years, carbon-14 decay is useful for measuring dates up to about 70,000 years. This makes the method particularly useful for dating samples from the Holocene and late Pleistocene epochs.

■ Modern Cultural Connections

Radiometric dating is the key to developing and understanding an absolute time scale of Earth and its geologic ages. When geological events, rock formations, and individual species can be placed accurately in time, it becomes possible to understand their relationships

to each other and to events and circumstances present today. Many scientific disciplines rely on an understanding of the geological past to make accurate observations and predictions. Some of these sciences, like meteorology, hydrology, and oceanography have important roles to play in understanding and possibly mitigating the effects of global climate change and population growth. By studying climate changes in the past or uncovering the reasons for mass extinctions, it might be possible to foresee disasters and figure out how to avert them.

SEE ALSO *Biology: Miller-Urey Experiment; Earth Science: Climate Change; Earth Science: Gradualism and Catastrophism; Earth Science: Plate Tectonics: The Unifying Theory of Geology.*

BIBLIOGRAPHY

Books

Gould, Stephen J. *Wonderful Life: The Burgess Shale and the Nature of History.* New York: W.W. Norton & Company, 1989.

Schopf, J. William. *Cradle of Life: The Discovery of Earth's Earliest Fossils.* Princeton, NJ: Princeton University Press, 1999.

Shimer, John A. *This Changing Earth: An Introduction to Geology.* New York: Harper & Row, 1968.

Weiner, Jonathan. *Planet Earth.* New York: Bantam Books, 1986.

Periodicals

Jensen, Soren. "The Proterozoic and Earliest Cambrian Trace Fossil Record; Patterns, Problems and Perspectives." *Integrative and Comparative Biology* 43, no. 1 (February 2003): 219–228.

Knoll, Andrew H., and Sean B. Carroll. "Early Animal Evolution: Emerging Views from Comparative Biology and Geology." *Science* 284, no. 5,423 (June 25, 1999): 2,129–2,137.

Pope, Kevin O., Steven L. D'hondt, and Charles R. Marshall. "Meteorite Impact and the Mass Extinction of Species at the Cretaceous/Tertiary Boundary." *Proceedings of the National Academy of Sciences of the United States of America* 95, no. 19 (September 15, 1998): 11028–11029.

Web Sites

American Museum of Natural History. "James Hutton: The Founder of Modern Geology." http://www.amnh.org/education/resources/rfl/web/essay-books/earth/p_hutton.html (accessed October 9, 2007).

BBC News. "Geological Time Gets a New Period." May 17, 2004. http://news.bbc.co.uk/2/hi/

science/nature/3721481.stm (accessed November 22, 2007).

University of Maryland, Department of Geology. "GEOL 102, Historical Geololgy—The Archean Eon I: The Earliest Rocks." January 10, 2007. http://www.geol.umd.edu/~tholtz/G102/102arch1.htm (accessed October 11, 2007).

University of Texas at Dallas, Department of Geosciences. "When Did Plate Tectonics Begin?" October 4, 2006. http://www.utdallas.edu/~rjstern/PlateTectonicsStart (accessed October 11, 2007).

David King Jr.

Earth Science: Gradualism and Catastrophism

■ Introduction

Gradualism and catastrophism were schools of thought in the earth sciences that explained the major features of Earth's surface and life's history by appealing to different sorts of causes. Gradualists explained geological features as the result of slowly acting processes such as erosion, while catastrophists argued that Earth had been shaped mainly by a series of violent events or catastrophes, whether over a relatively short time (6,000 to 10,000 years) or over many millions of years. In the early nineteenth century, gradualism seemed to win out completely over catastrophism, but in the late twentieth century scientists discovered that catastrophic events have also played a major role in Earth's history. For example, Earth's moon was probably formed by the collision of the early Earth with a Mars-sized object, and it is now known that the history of life has been repeatedly shaped by impacts of asteroids or comets that wiped out large numbers of species, allowing new ones to evolve—including ourselves. Today, the dispute between gradualism and catastrophism is largely over, as scientists recognize that both gradual and sudden processes have shaped the surface of Earth and the course of biological evolution.

■ Historical Background and Scientific Foundations

In Europe during the Middle Ages and Renaissance, theories about Earth's history and of the origin of life tended to follow a literal reading of the Bible, especially the book of Genesis, which describes the creation of Earth and a world-covering flood survived by Noah and the other passengers on his ark. Genealogies in the Bible that linked one generation to another and gave lifespans in years seemed to show that Earth could be no more

WORDS TO KNOW

CATASTROPHISM: School of thought in geology which holds that the rates of processes shaping Earth have varied greatly in the past, occasionally acting with violent suddenness (catastrophes).

EROSION: Processes (mechanical and chemical) responsible for the wearing away, loosening, and dissolving of materials, particularly of Earth's crust.

GRADUALISM: A model of evolution that assumes slow, steady rates of change. Charles Darwin's original concept of evolution by natural selection assumed gradualism. Contrast with punctuated equilibrium.

UNIFORMITARIANISM: Doctrine of geology promoted by English geologist Charles Lyell (1797–1895), asserting three theories: (1) actualism (uniform processes acting throughout Earth's history), (2) gradualism (slow, uniform rate of change throughout Earth's history), and (3) uniformity of state (Earth's conditions have always varied around a single, steady state). Uniformitarianism is often contrasted to catastrophism. Modern geology makes use of elements of both views, acknowledging that change can be drastic or slow, and that while some processes operate steadily over many millions of years, others (such as asteroid impacts) may happen rarely and cause catastrophic, sudden changes when they do.

UPLIFT: The process or result of raising a portion of the Earth's crust through different tectonic mechanisms.

than about 10,000 years old, as young-Earth creationists still believe today. Given such a short span of time, it seemed reasonable to suppose that Earth's features had either been created as they are today or shaped mainly by brief, violent events—catastrophes. As for plants and animals, they were believed to have all appeared suddenly, as described in Genesis.

In the eighteenth and early nineteenth centuries, scientists began to gather evidence that Earth is far older than 10,000 or so years—perhaps millions or thousands of millions of years old. (We now know that it formed about 4.5 billion years ago; the universe itself formed about 13.7 billion years ago.) Some scientists, such as the French naturalist Georges Cuvier (1773–1838), formulated a non-biblical, scientific version of catastrophism that accommodated this new vision of an ancient Earth. Cuvier showed in the 1790s that some species of animals known only from fossils were truly extinct, overturning earlier assumptions that all animals present at the Creation must still survive somewhere on Earth. Moreover, he argued that extinctions occurred in waves caused by great disasters or catastrophes. Other theorists, such as German geologist Abraham Werner (1749–1817), proposed that most features of Earth's crust could be explained by the recession of waters from a single, great, world-covering flood, possibly the biblical Noachian flood. These were the doctrines of the catastrophist school of thought.

However, in the late 1780s the Scottish geologist James Hutton (1726–1797) proposed a different view. He suggested that Earth's history was cyclic, with uplift (raising of land) and erosion (wearing away of land) going on continuously, uniformly, and gradually. Erosion removed particles and washed them to places where they were deposited, eventually forming layers of sedimentary rock; uplift caused by Earth's internal heat raised these sedimentary layers so that they could be eroded once again. Uplift, erosion, deposition; uplift, erosion, deposition; an eternal cycle. He proposed that in Earth's cyclic system there was "no vestige of a beginning,—no prospect of an end." Time, he taught, "is to nature endless and as nothing" (quoted in Gould, 1987).

For decades, both catastrophism and Hutton's slowly cycling world had scientific supporters. But in 1830 through 1833, Scottish geologist Charles Lyell (1797–1875) published his persuasive work *Principles of Geology*, in which he developed Hutton's ideas further. Lyell argued for a strict form of what was quickly labeled uniformitarianism, the doctrine that all past events in Earth's history can be explained by causes still in operation today: volcanic eruptions, erosion and deposition of sediments, uplift. Like Hutton, Lyell asserted that Earth had always looked much as did today and always would, as eternally-operating causes recycled its surface. Moreover, Lyell taught, these causes were gradual, not catastrophic: They worked at a slow, steady pace through deep time.

Lyell's view triumphed, and by the end of the nineteenth century catastrophism had been abandoned, thereafter to be reviled as superstitious nonsense. So strong was this rejection that later writers of histories and textbooks often mistakenly spoke of catastrophist

scientists such as Cuvier as biblical literalists who clung to an untenable reading of sacred texts rather than looking at the evidence of geology. This is an inaccurate depiction of Cuvier and his colleagues, who were skilled scientists. But throughout much of the twentieth century, catastrophist ideas were literally unthinkable in the earth sciences. Fossil evidence accumulated showing that Cuvier had been right about episodes of mass extinction in the deep past, but it was taught that gradual causes such as continental drift, not catastrophes, must be responsible for such changes.

This changed dramatically beginning in 1980, when Walter Alvarez (1940–) and his father, Luis Alvarez (1911–1988), published evidence showing that the great mass extinction that occurred at the geological moment dividing the Cretaceous and Tertiary periods, known as the K-T extinction event, was caused at least in part by the impact of a comet or asteroid some 6 miles (about 10 km) in diameter. Shock waves and dust clouds from the impact had, the Alvarezes argued, killed off the last of the dinosaurs along with 65-70% of all other plant and animal species alive at that time.

Evidence for the Alvarez hypothesis came not only from the sudden nature of the extinctions, as shown in the fossil record, but from a thin layer of particles found all over the world at the boundary between the Cretaceous and Tertiary sediments, dust that must have been deposited at the time of the extinctions, 65 million years ago. The Alvarezes showed that this dust is rich in iridium, an element more common in meteorites than on Earth. The best explanation for this iridium-rich layer was the impact of a gigantic asteroid. The Alvarezes even identified the impact crater, the 40-mile-wide (70 km) Chicxulub crater on the Yucatan peninsula in Mexico, which had been discovered by petroleum geologist Glen Penfield in 1978. The crater was the right size and age to have caused the K-T extinction.

Some scientists found the new theory distasteful because it was a revival of catastrophism. However, evidence mounted, and Earth scientists were forced to re-admit catastrophic events into their toolkit of explanatory possibilities. In 2007, the editors of the journal *Nature*, looking back on the whole episode, said that the effect of the Alvarez hypothesis was to "change how scientists and others see the world, and reintroduce catastrophism to the Earth sciences." Later, scientists would marshal evidence that the Chicxulub impact probably only finished off a process of extinction that had already begun due to vast volcanic eruptions in the area that is now India.

Today, gradualism or uniformitarianism blends with catastrophism as it is recognized that catastrophes, though occasional, may still occur, just as they did in the past. Catastrophes can be seen as one of Lyell's ever-present causes working over long spans of time. This was demonstrated spectacularly in 1994, when comet Shoemaker-Levy 9 collided with Jupiter, releasing at

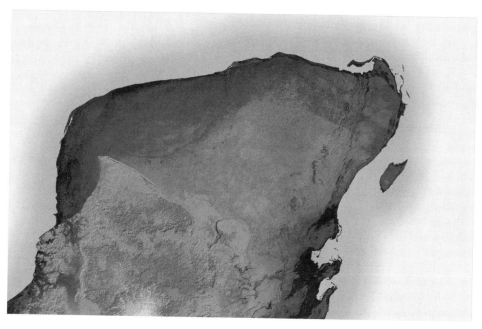

This NASA image is a high resolution topographic map of the Yucatan Peninsula released on March 7, 2003, in Washington, D.C., and created with data collected by the *Endeavor* Space Shuttle Radar Topography Mission, conducted from February 11 to 22, 2000, which made detailed measurements of 80% of Earth's landmass. In the upper left portion of the peninsula, a faint dark arc is visible, indicating the remnants of the Chicxulub impact crater. The crater was caused by a cataclysmic asteroid impact which, scientist theorize, may have caused the Cretatious-Tertiary Extinction of most life on Earth, including the dinosaurs, 65 million years ago. *NASA/Getty Images.*

least 750 times more energy than all the nuclear weapons on Earth exploded at once. Telescopes revealed the white flash of the impact and the dark scars that lingered in Jupiter's atmosphere for days, a dramatic proof that cosmic catastrophes can still happen.

■ Modern Cultural Connections

Catastrophism, especially in the form of cosmic collisions, is now taken seriously as a possible threat to human life, mostly as a result of the discovery that such collisions have occurred repeatedly in Earth's past. In the 1980s and 1990s, a number of projects were begun by astronomers to discover and track asteroids that might impact Earth someday; many were commenced even before Shoemaker-Levy hit Jupiter. In 1992, the U.S. Congress ordered the National Aeronautics and Space Administration (NASA) to identify 90% of large asteroids that approach Earth closely by 2002. This goal was later revised to include the discovery of at least 90% of all near-Earth objects with diameters greater than .62 miles (1 km) by 2008. In 2005, Congress ordered NASA to detect 90% or more of near-Earth objects with diameters of 490 feet (150 meters) or more by 2020. In early 2008, NASA announced that it would meet the goal of detecting most 1-km objects by the end of that year.

Various schemes have been proposed for deflecting any asteroid found to be on a catastrophic collision course with Earth, including exploding nuclear weapons near its surface, darkening part of its surface so that light pressure changes its course, using a spacecraft's gravitational attraction to an asteroid to tow it slightly off course, and more. The farther away an asteroid is when deflection efforts begin, the smaller the deflection that would be required to make it miss Earth, thus preventing a repeat of the disaster that extinguished the dinosaurs.

SEE ALSO *Biology: Evolutionary Theory; Biology: Paleontology.*

BIBLIOGRAPHY

Books

Gould, Stephen Jay. *Time's Arrow, Time's Cycle.* Cambridge, MA: Harvard University Press, 1987.
Rudwick, Martin J.S. *Georges Cuvier, Fossil Bones, and Geological Catastrophes: New Translations and Interpretations of the Primary Texts.* Chicago: University of Chicago Press, 1997.

Periodicals

The Editors of *Nature.* "The Big Splash." *Nature* 449 (2007): 1–2.

Larry Gilman

Earth Science: Navigation

■ Introduction

Maritime navigation began when vessels from ancient Greece, China, and Phoenicia first ventured along the coastline, lining up landmarks (such as a near outcrop of rock against a distant point on land) to plot a course. Heeding the flight paths of birds, wind direction, Pole Star, and the path of the sun allowed sailors to make ever more ambitious voyages in open waters. The Chinese and Europeans independently invented the compass around the eleventh century, and European sailors also developed navigational instruments such as the cross-staff and quadrant in the twelfth and thirteenth centuries, making it possible for medieval mariners to find their latitude (the angular distance between an imaginary line around a heavenly body parallel to its equator and Earth's equator).

While determining latitude was relatively straightforward, figuring out longitude was much trickier. Englishman John Harrison's (1693–1776) invention of the chronometer, a reliable clock whose spring mechanism was not affected by the pitching of the waves, permitted an accurate reading of local time, making determination of longitude possible. This allowed sailors to determine how far north or south of the equator they were, as well as east or west of the prime meridian, the imaginary half circle running pole to pole though Greenwich, England. In other words, a ship could now determine its coordinates on Earth's map grid.

By the end of the twentieth century, radio technology was beginning to be used to determine directions, and in 1904, time signals were sent to ships via radio to check their chronometers for errors. Radar was installed for the first time on an American warship in 1937, and in the twentieth century, satellite navigation systems and the development of global positioning systems (GPS) permitted ships and air traffic to navigate the oceans and the atmosphere.

■ Historical Background and Scientific Foundations

Early Navigation

Early navigators stayed in sight of land, steering from point to point and creating coastal charts using dead reckoning—an estimate of current position based on a previously determined position. Sailors based their calculations on their known speed, elapsed time, and the direction of their course via the formula Time = Distance / Speed. The elapsed time between two points was computed via counting or an hourglass, and speed was determined by an instrument called a chip log, a wooden board, reel, and attached log line with uniformly spaced knots. The board was tossed overboard and floated while the ship moved past while the log line ran out for a fixed period of time. The ship's speed was indicated by the length of the line (and number of knots) that had passed over the stern. In areas like ancient Greece, which are dominated by chains of islands, navigating with fixed land reference points was fairly reliable.

The Role of the Pole Star

In open waters however, it was also necessary to steer by the sun and the stars. The sun's path or ecliptic through the sky gave early Phoenician and Arab sailors their direction. The Pole Star, or Polaris, always remains in a fixed position in the sky, and its distance above the horizon is a direct measure of terrestrial latitude. So if one is at the equator, the Pole Star will be zero degrees above the horizon; at the North Pole it is 90 degrees directly overhead.

The ancient Greeks, Phoenecians, and Arabs likely used one or two finger widths to measure Polaris's distance from the horizon. To return home, ancient sailors only had to sail south or north to bring Polaris to the

WORDS TO KNOW

ASTROLABE: An instrument used throughout the Middle East and Europe from classical times through the Renaissance as an aid in observing star positions and calculating longitude and local time. It was used in astronomy, astrology, and navigation, and was a precursor of the slide rule. Astrolabes varied in design; typically, one consisted of a metal disk marked with lines conveying astronomical information and a center-mounted rotating pointer.

CHIP LOG: A device (now obsolete) for measuring the speed of a surface vessel relative to the water. It consists of a small wooden board or chip that is attached to a line. The chip is cast overboard, where it drags in the water and is left behind by the motion of the ship. A line runs out for a fixed period of time, measured using an hourglass or other timepiece, giving a distance and time measurement that can be used to compute a velocity.

CHRONOMETER: Any device that measures the passage of time. The term is especially used for accurate timepieces built for special purposes, such as navigation.

COMPASS: A device for detecting the presence and direction of a magnetic field.

CROSS STAFF: A cross-shaped instrument for determining the angular height of the sun or other heavenly body above the horizon. The user places an eye at one end of the long stick and aligns it with the horizon, then shifts a shorter crosspiece along the long stick until its upper end is aligned with the object being observed. The angular height of the object can then be read off from markings on the long stick.

DEAD RECKONING: A navigational method of estimating one's location after a period of movement, based on earlier knowledge of one's position and knowledge of the direction and rate of travel. By adding a change vector to a known position, a new position can be calculated. Highly precise dead reckoning is used in inertial-guidance systems, which are initialized with precise location information and then measure all changes in velocity that they undergo by measuring the forces accelerating them. Inertial guidance systems are used in many guided missiles, aircraft, and spacecraft.

EQUATOR: The line circling a planet that divides it into northern and southern hemispheres (half-spheres). An imaginary plane passing through the planet's center and at right angles to its axis of rotation will intersect the planet's surface along its equator.

KAMAL: An observational instrument used in the Middle East and China from the Middle Ages through the nineteenth century. Like the cross staff, its purpose was to allow measurement of the angle between a particular astronomical object (usually a star) and the horizon. A wooden card attached to a string is held at arm's length while the string is held in one's teeth: one then moves the card along the string until it is aligned with the star. Knots along the string are then counted to the position of the card, providing a reading for the angle.

LATITUDE: The angular distance north or south of Earth's equator measured in degrees.

LONGITUDE: The angular distance from the Greenwich meridian (0 degree), along the equator. This can be measured either east or west to the 180th meridian (180 degrees) or 0 degree to 360 degrees W.

LORAN CHART: The Long Range Navigation (LORAN) system was set up after World War II to aid ocean surface navigation. Radio pulses from fixed stations are compared by LORAN receivers on board ships; the time differences between receipt of pulses from different stations can be used to calculate position. LORAN charts are maps that show where different pulse-timing differences will be observed on the surface, drawn as colored lines on the map.

POLE STAR: The Pole Star is Polaris, a visible star that happens to be aligned with Earth's spin axis so that all the other stars appear to spin around Polaris while it remains fixed. Polaris is visible only in the Northern Hemisphere. In the Southern Hemisphere, no visible star happens to be aligned with Earth's axis, so there is no southern Pole Star.

PORTOLAN CHARTS: Navigational charts produced in Europe and elsewhere during the Middle Ages.

PRIME MERIDIAN: A meridian is a circle drawn around a planet and passing through its north and south poles; the prime meridian is the meridian arbitrarily (i.e., for no necessary reason) named as the 0th meridian, the one from which numbering of the other meridians shall begin. For example, on Earth, the prime meridian passes through Greenwich, England; on Mars, it passes through the center of crater Airy-0 (zero), selected for that purpose in 1969.

QUADRANT: One of the four regions in the Cartesian coordinate system formed by the intersection of the x- and y-axes.

altitude of their home port, and then turn as appropriate to keep Polaris at a constant angle, "sailing down" the latitude.

By the eighth century, Arabs used an instrument called a *kamal*, a rectangular plate attached to a string, to make their observations of latitude. The sailor moved the plate closer or farther from his face until the distance between the Pole Star and the horizon corresponded to the plate's top and bottom. The plate's distance from the face was measured by the string, with the length marked by tying a knot. Soon, Arab sailors created almanacs of different ports that recorded which knot on a *kamal* corresponded to the height of the Pole Star for each port they visited.

The Arabs also invented the astrolabe approximately in the first century AD. Adopted by medieval European navigators, the astrolabe was a simple metal disc engraved with marks to measure the height of a star or the sun over the horizon at different latitudes. A more complex planispheric astrolabe is a two-dimensional model of the skies as they appear in relation to Earth, including a *rete* or star pointer to indicate stellar positions. It was used to tell time via solar altitude, to calculate the duration of the night and day, and of course, to track the stars.

Another popular medieval tool for determining latitude was the cross-staff, an adaptation of the kamal. It was a T-shaped device with a base and slidable crossbar to measure the sun's or star's height. The base was held to the eye, while the sliding top was pulled back until the desired celestial object was at the top and the horizon at the bottom. This made the navigator look like an archer taking aim at the sky, the origin of the phrase "shooting the stars." In the seventeenth century the cross-staff was embellished with mirrors and prisms for celestial observations at night, developing into the sextant.

The Compass

Chinese texts indicate that the compass was used in marine navigation by the eleventh century, perhaps as much as a century ahead of its development in Europe. (The first European compass was described by Peter Peregrinus [fl. thirteenth century] in a book about magnets in 1269, but it may have been used before this date.) The compass not only permitted the development of more accurate nautical charts (called portolan charts), but accounted for increasing economic prosperity for the people of the Mediterranean area in the late thirteenth century. Italian city-states were particularly engaged with

Italian explorer Amerigo Vespucci (1451–1512) is depicted using an astrolabe to chart the stars for navigation. *The Art Archive/Musée des Arts Africains et Océaniens/Gianni Dagli Orti.*

Near East trade. The compass extended the ability of the Venetians, Florentines, and Genoese to embark on trade voyages during the winter months when the sky was cloudy and it was difficult to navigate by the sun and stars. Mediterranean voyages around Spain to England and the Netherlands also became safer. The sponsoring nations' ensuing prosperity may have contributed to the cultural and artistic achievements of the Renaissance, which were funded by monied merchants such as the Medici.

Improved navigational techniques also led to a renaissance of exploration by the Portuguese in the fifteenth century under the leadership of Henry the Navigator (1394–1460), prince of Portugal. He founded institutes dedicated to cartography, where maps that the Portuguese created of the African coastlines were held as state secrets. By the sixteenth century, the Spanish, English, and French were using improved instruments to explore the New World. Christopher Columbus (1451–1506), however, in his attempt to find a western passage to the Orient mainly used dead reckoning—little wonder he did not reach his initial goal and discovered Cuba instead.

Finding Longitude

Although it was relatively easy to find latitude at sea with navigational instruments by the mid eighteenth-century, it was still not possible to discern a ship's longitude. To figure out longitude at sea, one has to know time at both the home port (or another place where the longitude is known), and time aboard ship. The differences in time are used to calculate geographical position. Earth takes 24 hours to revolve 360 degrees, so one hour is 1/24 of the revolution—and 1/24 of 360 degrees is 15 degrees. Each hour's difference from the home port is therefore 15 degrees of longitude to the east or west, depending on which way the ship is sailing.

The secret of finding longitude was in making a clock that was accurate at sea, something that had not been possible in an era of pendulum clocks—the bob was simply useless in a ship rolling and pitching on the waves. The damp sea air also meant that traditional lubricants would not work on the metal parts of clocks, and their wooden components would swell.

In 1707 several of British Admiral Sir Cloudesley Shovell's (1650–1707) ships, on a return voyage from France, miscalculated their position in a deep fog and were wrecked on rocks near the Scilly Islands. The resulting deaths of 1,400 men, including Shovell, prompted the British government to sponsor the longitude prize in 1714. The act promised a prize of £20,000 to anyone who could solve the longitude problem to an accuracy of ½ degree.

John Harrison (1693–1776), a skilled instrument maker, devised a series of brass and steel chronometers that would not rust or unduly expand and contract with temperature, enclosed in tropical wood (that self-lubricated) and utilized a spring mechanism rather than a pendulum. The last and best clock (1759) also had a caged roller bearing and ran independently of outside gravitational force, which was perfect for a moving ship. One of his instruments, made into a pocket watch, was carried by James Cook (1728–1779) on his voyages from the tropics to the Antarctic; it was never off by more than 8 seconds per day. Harrison's chronometers finally allowed the calculation of accurate longitude at sea.

Navigation by Radio and Radar

After Harrison's invention, nineteenth-century navigation instruments such as the marine chronometer and sextants were improved and refined. The advent of electronics technology in the twentieth century, particularly radio, raised navigation to a new level. Radio technology maintains links between ships (and, eventually, aircraft) and known locations on Earth via the use of electromagnetic radiation, which has wavelengths that are longer than visible light. Bad weather, distance, and darkness thus do not interfere with radio direction finders on aircraft and marine vessels.

The ship or plane receives synchronized pulse signals from transmitting stations, often located in lighthouses or radio control towers. By measuring the time difference between the arrivals of signals from a pair of these transmitting stations, a curved line of position can be plotted on a LORAN (long-range navigation) chart. A second set of signals from another pair of stations produces another curved line of position, and crossing the two lines gives a fixed position for the ship or aircraft.

In the 1930s Scottish physicist Robert Alexander Watson-Watt (1892–1973), Arnold F. Wilkins, and Alan Blumlein (1903–1942) developed radar systems as part of the war effort to detect incoming German aircraft. Radar developed out of a British belief that some sort of visible death ray could be used as an offensive weapon against the Nazis. While ultimately that line of inquiry came to naught, British Post Office engineers did notice that an airplane flying through an experimental high frequency beam caused the beam to jump on an oscilloscope. Watson-Watt and Wilkins concluded it would be possible to develop a form of aircraft-detection system, in which a plane would return a radio signal aimed at it back to its source.

Between 1935 and 1937 an experimental radar system called Chain Home proved that it could detect enemy aircraft. Radar uses wavelengths that are shorter than radio but longer than light waves; from a single landmark it can give a direct indication of distance. Radar does this by measuring the time lag between the plane or boat sending and receiving signal pulses to the landmark. The distance to the reflecting surface of

Mike Ferguson, chief economist for the state of Idaho, is the author of *GPS Land Navigator,* a complete guide book to using global positioning equipment. The strongest message in his new book is that the GPS user must be well-versed in the electronic wonder as well as carry and understand compasses and topographic maps.   *AP Images/Troy Maben.*

the landmark is given in terms of the velocity of light. It was used to great effect in the Battle of Britain in 1940 to guide the Royal Air Force (RAF) pilots toward the incoming German bombers. Radar was also installed on British and American warships.

Radar technology, developed in wartime, became a mainstay of commercial navigation, forming the basis of radiation-contact guidance equipment called the Doppler system. In this type of radar system, continuous waves rather than pulses are transmitted from the plane or ship to the landmark. There is a measurable shift in frequency between reflected and transmitted waves. Called a Doppler shift, this can be quantified to measure the distance between the ship or plane and the landmark.

Satellite Technology

The launch of the Soviet satellite *Sputnik* in 1957 gave American scientists the inspiration for the global

positioning system (GPS). Dr. Richard Kershner (1913–1982), from the Johns Hopkins Applied Physics Laboratory, was monitoring *Sputnik*'s radio transmissions. He noticed that due to the Doppler shift, he measured a different frequency as *Sputnik* approached than as it traveled away from them. By measuring the Doppler shift, they could pinpoint where the satellite was in its orbit.

Using this principle, the U.S. Navy began to develop a satellite navigation system in 1960; there are presently 31 broadcasting satellites in the global positioning system. Though administered by the U.S. Department of Defense, the GPS system is free for civilian use and navigation. Using Doppler shift calculations, GPS receivers calculate their position by measuring the distance between themselves and three or more GPS satellites. The increasingly low cost of GPS chips means that GPS satellite signals are used to power remote door openers, automotive navigation systems in cars, cell phones, systems that track vehicles carrying dangerous goods, and the location of utility lines. Several governments are studying how the loss of GPS signals (for example from a downed satellite) would affect industry and communications infrastructure.

■ Modern Cultural Connections

Navigation tells us where we are and where we are going. Our sense of place now has a global context, and using navigational systems we have made the first strides into outer space. From dead reckoning to GPS, navigation permits the transmission of goods and knowledge, and its increasing efficiency means the expansion of our world and our greater freedom to travel and learn.

SEE ALSO   *Earth Science: Exploration.*

BIBLIOGRAPHY

Books

Alexander, Robert Charles. *The Inventor of Stereo: The Life and Works of Alan Dower Blumlein.* Oxford: Focal Press, 1999.

Feinburg, Richard. *Polynesian Seafaring and Navigation: Ocean Travel in Anutan Culture and Society.* Kent State University Press, 1988.

Sobel, Dava. *Longitude: The True Story of a Lone Genius Who Solved the Greatest Scientific Problem of His Time.* New York: Penguin, 1996.

Taylor, E.G.R. *The Haven-Finding Art: A History of Navigation from Odysseus to Captain Cook.* New York: Abelard-Schuman, 1957.

Periodicals

Draper, Charles S. "Navigation—From Canoes to Spaceships." *Proceedings of the American Philosophical Society* 104, no. 2 (April 19, 1960): 113–123.

Finney, Ben. "Colonizing an Island World." *Transactions of the American Philosophical Society* 86, no. 5 (1996): 71–116.

Hsu, Mei-Ling. "Chinese Marine Cartography: Sea-Charts of Pre-Modern China." *Imago Mundi* 40 (1988): 96-112.

Lane, Frederic C. "The Economic Meaning of the Invention of the Compass." *The American Historical Review* 68, no. 3 (April 1963): 605–617.

Web Sites

Radiocommunications Agency. "Study into the Impact of UK Commercial and Domestic Services Resulting from the Loss of GPS Signals." http://www.ofcom.org.uk/static/archive/ra/topics/research/topics/other/gpsreport/gps-report.pdf (accessed October 4, 2007).

Anna Marie Eleanor Roos

Earth Science: Oceanography and Water Science

■ Introduction

Water is a fundamental substance on Earth, covering approximately two-thirds of the surface of the planet. In addition, water has unique chemical properties that make it crucial to numerous reactions that contribute to a significant portion of the biological, geological, and physical processes on Earth. The study of water and its role in the processes on Earth is called water science.

Water science is an applied scientific field encompassing the study of the behavior of all forms of water. This includes both fresh and saline waters as well as gaseous, liquid, and solid forms of water. Some of the larger areas of water science research include hydrology, the study of the movement of water; limnology, the study of lakes and rivers; and oceanography, the study of oceans. Within each of these fields, the biological, physical, and chemical processes of water are further subdivided into areas of research.

Oceanography is one of the larger subjects within the field of water science. Approximately 97% of all water on Earth is marine (related to the oceans). The oceans control weather and climate throughout the planet, drive numerous economies via fisheries, trade, and recreation, and act as a major reservoir for nearly every biogeochemical cycle as well as economically important minerals and petrochemicals. As in all fields of water science, oceanographers specialize in understanding the chemical, physical, geological, ecological, and biological processes that influence and are influenced by the ocean.

■ Historical Background and Scientific Foundations

Although it was not a formalized subject, the concepts involved in water science probably originated with the advent of permanent human settlements and the development of agriculture. As early as 4000 BC, Egyptian farmers studied the flow of water in order to divert the Nile River to water crops. They began to understand how to build dams and levies to curtail flooding. The building of dams was also practiced in ancient Assyria, Mesopotamia, and China, and the use of these water-controlling structures contributed greatly to the growth of these civilizations.

The study of the oceans originated with the great mariners of the ancient world. The Polynesians may have begun sailing the Pacific more than 20,000 years ago. Phoenicians likely sailed throughout the Mediterranean Ocean by 2000 BC. Much of this early exploration was associated with trade and the discovery of new resources. However while sailing the oceans, ancient sailors accumulated knowledge of currents, tides, geography, and the distribution of fish and other marine organisms.

By 750 BC aqueducts, human-made structures built specifically to carry water from natural sources to cities for public use, were constructed in ancient India, Persia, Assyria, and Egypt, as well as Italy, where the Romans showed themselves to be masters of hydrology.

Greek philosopher Aristotle (384–322 BC) is credited with the first organized study of marine biology. Using his philosophy based in observation, induction, and reason, he identified nearly 200 species of marine invertebrates and fish. He also accurately identified whales and dolphins as mammals. Aristotle and other Greek naturalists also correctly established many of the processes involved in the hydrologic cycle.

In the fifteenth century AD, Europeans began a great exploration of the world's oceans. The Portuguese, Dutch, English, and Spanish launched large naval fleets and made numerous advancements in understanding the oceans. Much of the development involved solving practical problems associated with building boats that could sail faster, producing tools that could navigate more accurately, and creating maps of the oceans.

In the 1600s and 1700s, significant work on the hydrologic cycle was accomplished in Europe. French scientists established that rainfall could account for stream flow. Additional work by English naturalists showed that the quantity of water that evaporated from the Mediterranean was related to the amount of rainfall that occurred in the surrounding lands.

English chemist and mathematician John Dalton (1766–1844) completely described the hydrologic cycle in the 1800s. Throughout the rest of the century, European mathematicians continued to apply physical laws to the behavior of water and established some of the fundamental principles of water science: Bernoulli's law, Darcy's law, and Poiseulle's capillary flow formula.

Between 1872 and 1876, the Royal Society of London funded the first major scientific exploration of the oceans. The society outfitted a war ship called the HMS *Challenger* to accommodate scientific research. The ship sailed more than 68,000 mi (109,435 km) collecting data on the biology and physics of every ocean except the Indian. The collections from the expeditions fill 50 volumes and required nearly two decades to analyze.

In the early 1800s, the U.S. government recognized the need for an increased understanding of the oceans to defend its coastlines and improve the safety of fishermen and sailors. They established the Naval Depot of Charts and Instruments in 1830 and the Fish Commission in 1871. Soon after, the first oceanographic institutions were established in Woods Hole on Cape Cod in Massachusetts: the Marine Biological Laboratory and Woods Hole Oceanographic Institute. Both of these facilities still play extremely active roles in the fields of oceanography and water science today.

One of the major fields of water science, hydrology, became well-established as a research entity in the early part of the twentieth century. In 1922, the International Union of Geodesy and Geophysics established the Section of Scientific Hydrology, and in 1930 the American Geophysical Union added the Hydrology Section, creating formal forums for the exchange of ideas concerning hydrological research. Both of these associations continue to advance communication between scientists working in hydrology today.

A second major field in water science, limnology, which is the study of lakes, also became formalized in the early part of the twentieth century. The Limnology Society of America (LSA) was established in 1936 to facilitate the exchange of information on aquaculture. LSA merged with the Oceanographic Society of the Pacific to form the American Society of Limnology and Oceanography (ASLO) in 1948. ASLO, and its journal *Limnology and Oceanography,* continue to provide a forum for the exchange of ideas between all fields of water science.

In the twentieth century, universities and governments became major drivers in water science research,

WORDS TO KNOW

APPLIED SCIENTIFIC FIELD: An area of scientific knowledge that is used for some practical purpose—amusement, war, medicine, transport, communication, or other. Physics, chemistry, mathematics, biology, and other fields all have applied areas. Labeling some knowledge "applied" does not imply that knowledge that is not applied directly is useless: first, because it satisfies the human desire to know, and second, because the applied knowledge in each field would not exist or make sense without all the knowledge of the field, including that which is not directly applied.

BERNOULLI'S LAW: Named after Swiss mathematician David Bernoulli (1700–1782), a mathematical description of the fact that any increase in the velocity of a moving liquid decreases the pressure of that liquid.

DARCY'S LAW: A mathematical statement of how liquids flow through sand and other porous (hole-filled) substances. It is named after French scientist Henry Darcy (1803–1858).

GROSS DOMESTIC PRODUCT (GDP): A measure of total economic activity, whether of a nation, group of nations, or the world: slightly different from Gross National Product (GNP, used by economists until the early 1990s). Defined as the total monetary value of all goods and services produced over a given period of time (usually one year). GDP's limitations have been pointed out by many economists. GDP is a bulk or aggregate statistic and does not take into account inequity: thus, a country's GDP might increase while 99% of its population got poorer, as long as the richest 1% grew sufficiently richer during the same period.

MARINE: Refers to the ocean.

MARINERS: A sailor. Also, Mariner refers to any of a series of ten robotic spacecraft launched by the United States from 1962 to 1973. The Mariners visited Venus, Mars, and Mercury.

and oceanography in particular. Primary researchers at universities and government institutions compete for domestic and international grants and fellowships to fund research. The governments of many industrialized countries support fleets of research vessels, ocean-viewing satellites, and other sophisticated equipment used to study the way that water behaves on Earth.

Scientific and Cultural Preconceptions

Water science and oceanography both have their roots in day-to-day activities. Because water is fundamental to life, some of the first priorities of any civilization are

Jacques Cousteau (1910–1997) made significant contributions to the study of the oceans. *AP Images.*

providing freshwater to the public and removing sewage from areas of dense population. Fishing is a basic method for providing a high-protein food source to a population. Farmers channel the flow of water to irrigate their crops. Trade and travel both rely on waterways. Rivers have long been used as a way to move products and people. Ancient people explored the ocean as a means of discovering new resources and used aquatic vessels to defend their coastlines. Although knowledge of the processes involved with the behavior of water and the oceans has been gathered since the beginning of civilization, water science and oceanography were not truly formal academic subjects prior to the twentieth century.

As populations grew during the industrial revolution, the need for water management brought about some formalization to water science. During this time significant effort was put into building structures to prevent flooding and to aid in irrigation. For example, more than 200 dams were built in England at the end of the nineteenth century. In conjunction with the Works Progress Administration (WPA) in the 1930s, numerous dams were built throughout the United States, block-

ing entire rivers. Hydrology and water science emerged along with the technology that supported these water management projects.

The onset of World War II in 1939 brought about research into technologies to support naval and marine soldiers. In particular, amphibious assaults on Europe and the Pacific Islands by the United States drove increased ability to predict wave and tidal conditions. With the development of submarine warfare, the necessity to map the seafloor and features of ocean basins such as magnetic fields were developed. This served as the germination of the academic field of oceanography.

Water science became an academic research field in the second half of the twentieth century, however public knowledge of this field was rather limited for many more decades. The efforts of environmental activists in the 1970s led to the passage of numerous laws that brought water science and oceanographic research into the public eye. These laws include the Marine Mammal Protection Act of 1972, the Endangered Species Act of 1973, and the Clean Water Act of 1977.

At the end of the millennium, water science and oceanography were pulled even more firmly into the public sphere. With the advent of interest in climate change, the entertainment industry turned toward the environment, and in particular the romantic notions associated with oceans and waterways for source material. Numerous films like *Jaws* (1975), *Whale Rider* (2002), *Finding Nemo* (2003), and *Happy Feet* (2006) depict water-related themes. Of major significance to water science, the documentary, *An Inconvenient Truth,* incorporates conclusions from water science as evidence for climate change. Discussing the career and climate change work of former Vice President Al Gore, who wrote a book by the same title, the film won an Academy Award. Gore went on to share the Nobel Peace Prize with the Intergovernmental Panel on Climate Change in 2007.

The Science

Both water science and the sub-discipline oceanography are large, applied fields of scientific study. Some of the fields of scientific research in water science include hydrology, hydrogeology, and limnology. Aquaculture is an area of research that straddles water science and oceanography because farming of aquatic crops occurs in both fresh and marine waters. Biological oceanography, chemical oceanography, physical oceanography, marine geology, and remote sensing all fall within the discipline of oceanography. This list of sub-disciplines is by no means comprehensive, but rather represents a significant cross-section of the more active fields of research. Each of these more specialized fields of research are, in and of themselves, extremely complex fields of study. Many of these fields borrow tools from other fields of water science, making water science extremely interdisciplinary and interactive.

Hydrology is a branch of engineering focusing on the study of the physical properties of freshwater and how it moves through lakes, rivers, and aquifers. Hydrologists often use computer models to study the flow of water and then use these models to predict the ways that changes to the environment will affect water supplies and water flows. Hydrogeology is a branch of hydrology concerned with the distribution of freshwater on Earth. Hydrogeologists study the ways that geological features affect groundwater flow and storage. Both hydrologists and hydrogeologists are concerned with pollution and the ways that contaminants can affect freshwater supplies.

Limnology is the study of the chemical, physical, geological, and biological processes that affect freshwater lakes, rivers, aquifers, and wetlands. In addition, the study of saline lakes falls under the domain of limnology. Limnologists, like hydrologists, are also concerned with pollution of natural waters. They study the ways in which humans impact the interactions between the living and non-living elements of aquatic environments.

Aquaculture is the farming of animals and plants under controlled conditions in marine or freshwater environments. Research into aquaculture revolves around managing environmental factors to ensure that crops grow quickly and in good health. A major area of research involves developing methods to reduce the pollution generated by aquatic farms. Some of the more economically valuable aquaculture crops include catfish, salmon, oysters, shrimp, crawfish, and kelp.

One of the major goals of biological oceanography is to understand the distribution of marine organisms in the ocean. Numerous factors influence why certain organisms are located in one location and not another. These factors include ocean temperature, dissolved chemicals called nutrients found in the water, ocean currents, and available sunlight. In turn, marine organisms influence the oceans themselves by consuming nutrients and releasing waste products. Research into global warming has focused on marine phytoplankton, which absorb carbon dioxide from the environment in quantities equal to that of all terrestrial plants.

Chemical oceanographers study the chemicals that are dissolved in ocean waters. Variations in the chemical composition of ocean waters result from the influence of weather patterns and atmospheric interactions, runoff from coastlines, dissolution of minerals from the seafloor, and the metabolic processes of biological organisms.

Physical characteristics in the ocean include temperature, salinity, density, and the ability to transmit light and sound. These fundamental properties control ocean currents, wave forces, and the amount of energy absorbed and released by the ocean. These processes, which are studied by physical oceanographers, further affect weather patterns and climate change throughout Earth.

IN CONTEXT: POLYNESIANS, AN ANCIENT SEAFARING PEOPLE

The Polynesians were probably the earliest, and certainly the most ambitious, civilization to navigate the oceans. The area over which the ancient Polynesians sailed covered 26 million mi (66 million km) of open ocean in the western Pacific. Although estimates vary, anthropologists generally agree that these ancient seafaring people settled many of the islands in the Pacific 30,000 years ago. By 20,000 years ago, they had colonized what is today the Philippines and by 2,500 years ago, the Polynesian culture was firmly established on the islands of Tonga, Samoa, the Marquesas, and Tahiti.

The Polynesians developed sophisticated technologies in order to sail the immense distances between islands. They built large boats with dual-hulls that could carry up to 100 people. They developed methods of storing food, water, and seeds for their great voyages. They accumulated a knowledge of the flight patterns of birds and an understanding of changes in the temperature, salinity, and color of ocean water to help find land. There is evidence that Polynesians also created stick charts to map the locations of islands. Bamboo sticks were tied together to represent the motion of water. Straight sticks indicated currents and curved pieces of bamboo showed the way that waves bend around islands. Knots and shells tied to junctions of the sticks indicated the locations of islands.

The Polynesian seafarers are responsible for colonizing Hawaii between AD 450 and 600. This accomplishment required sailing more than 2,000 mi (5,080 km) across the Pacific, from the southern hemisphere to the northern hemisphere, where the navigational stars are completely different. In addition, Hawaii lies to the north of the doldrums, a part of the ocean where the wind often ceases and where Polynesian travelers must have had to paddle with oars. This feat likely qualifies the ancient Polynesians as some of the greatest oceanographers of all time.

Marine geologists study the geological features of the ocean. One of the major focuses of marine geologists is the regions where tectonic plates meet at spreading centers in the deep ocean. In these places clues to the composition of the inner Earth can be found. Understanding how the movements of these plates occur aids in earthquake prediction. A second major focus of marine geology is the study of the chemical and physical properties of sediments on the sea floor. Understanding these properties provides insight into Earth's climactic record and the location of economically important resources such as minerals and fossil fuels.

Remote sensing is a technique in which measurements of an object or process are recorded from a distance

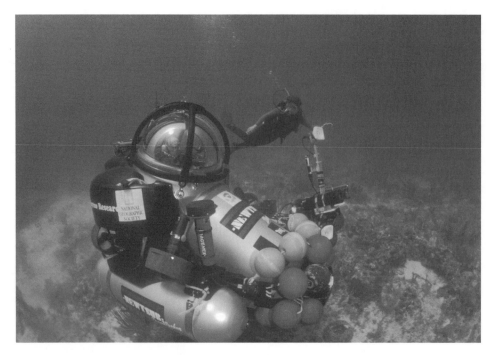

Researchers use various devices, like this one-person submersible, to get a closer look at their subjects.   © *Stephen Frink Collection/Alamy.*

using an electronic instrument. Often electromagnetic energy is directed at objects, such as the surface of the ocean or a lake. The energy interacts with the object and an instrument collects the resulting electromagnetic signal. Researchers analyze these signals to determine the characteristics of the surface. In water science, remote sensing is used to map geological formations surrounding and within bodies of water, to locate objects at the bottom of the ocean, to detect the distribution of microscopic organisms found in high densities, and to track the flow of various water bodies. Special radio tags can be attached to sharks and fish and are used to track migrations using satellites. Because aquatic environments are often difficult to access, the field of remote sensing has played an important role in furthering understanding of the oceans and freshwater bodies.

Influences on Science and Society

Water science and oceanography influence other fields of science greatly. Because water influences so many different processes on Earth, the study of water affects the study of nearly every other environmental science field. In particular, the connections between ocean circulation and climate have greatly influenced the pattern of scientific research in the 1990s and the early part of the twenty-first century.

Growing concern about changes in the global climate has compounded the impact of water science on environmental science in general. The Intergovernmental Panel on Climate Change estimates that the increase

in carbon dioxide in the atmosphere between 1905 and 2005 is responsible for a rise in temperature on both land and in the ocean of approximately 1.5°F (0.87°C). The effects of this increase in temperature on the ocean

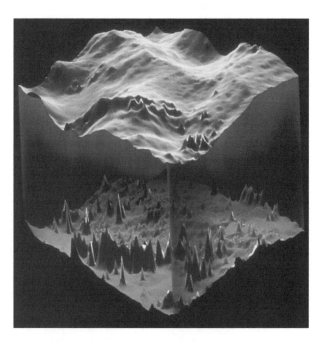

The surface of an ocean reflects the topography of the ocean floor, and a precise picture of the surface can now be drawn with the aid of satellite photography by the *ERS-1* satellite, launched in 1991.   © *Alain Nogues/Corbis Sygma.*

In downtown Manaus, Brazil, favelas reach down to the river that resembles a sewer in places. Hygiene and sanitation are non-existent, waterborne diseases are rife. As of 2006, an estimated one billion people do not have routine access to safe potable water. *© Collart Herve/Corbis Sygma.*

are intimately tied to the circulation patterns in the ocean and thus to weather patterns.

An example of the influence of global climate and ocean circulation illustrates this connection. Ocean circulation follows what has been called the great ocean conveyor belt. Under this paradigm, salty cold water in the North Atlantic, called North Atlantic Deep Water, sinks to the bottom of the ocean and moves into the southern hemisphere. It eventually wells up to the surface on the eastern borders of oceans and near the equator. As the water sinks, it gives up heat to the atmosphere. In regions where the water rises to the surface, heat is lost from the atmosphere as it warms the deep, cold water. These, and other, connections between ocean circulation and atmospheric temperature patterns greatly influence weather and climate.

Nearly all climate models predict that changes in atmospheric temperature will increase the rate of ice melt from the ice sheets in Greenland, which are located in the North Atlantic Ocean. If these ice sheets melt, the formation of North Atlantic Deep Water, which drives the great ocean conveyor belt, will cease. This will effectively halt the current pattern of circulation in the world's oceans. Understanding the effects of such

a massive paradigm shift to the oceans and links to hydrologic patterns throughout the planet influence the research being done by numerous oceanographers.

Understanding connections between changes to global weather patterns and climate are being studied by water science researchers as well as meteorologists, ecologists, and economists. Scientists are trying to assess the effects of climate change, not only on water processes, but on how these processes may be linked to processes in other parts of the environment as well.

Because water is so fundamental to basic human needs, water science impacts nearly every aspect of society. Changes to the environment often refer to changes to the flow or quality of water in a certain region. Pollution by industry, especially because of the burning of fossil fuels, has affected numerous water processes. Researchers in the field of water science continue to follow these changes and predict their effects. Some examples spanning the spectrum of water issues, from rainwater, to coastal erosion, to water pollution, follow.

Acid rain results when airborne pollutants change the chemistry of rain, making it more acidic. Acid rain was a serious problem in the eastern United States and in the Black Forest in Germany, destroying large swaths

IN CONTEXT: REMOTE SENSING

Remote sensing is being used by researchers at the University of Stanford to understand the swimming patterns of one of the ocean's top predators: the great white shark. A type of tag called a pop-up tag is attached to the animal's dorsal fin. This tag remains attached to the shark for several months and then it detaches from the shark, pops to the surface of the ocean, and sends a signal to a satellite. Researchers then download the tag's signal at their lab in California and analyze the results.

The tags that researchers attach to the sharks record the depth of the shark in the ocean, the temperature of the water, and the light intensity. Using these three measurements along with satellite maps of the surface temperature in the ocean, scientists can reconstruct the path that the shark follows.

Results from the tags indicate that the sharks remain near the shore of California in the fall, where they can easily prey on young seals living in rookeries. During this time they rarely dive deeper than 90 feet (27 m). In the winter the sharks leave the islands and migrate out into the Pacific Ocean. Some of them travel as far as the Hawaiian Islands, a distance of more than 2,000 mi (5,080 km). Others travel out into the Pacific Ocean south of Baja California far from any land and they remain there for several months. The reason for this migration is unknown.

The tags also showed that the sharks prefer to swim at two different depths during their migrations. They were most often within 15 ft (4.6 m) of the surface or at a depth of about 1,000 ft (305 m). They could occasionally dive as deep as 2,000 ft (610 m) below the surface. The temperature range over which the sharks swim varies greatly from as high as 75°F (24°C) to as cold as 45°F (7°C). Remote sensing has made research in extreme environments, like the open ocean, and on extreme creatures, like the ocean's top predator, much more feasible.

of old growth forests in the 1970s and 1980s. In 1990, as part of the Clean Air Act, the U.S. Environmental Protection Agency introduced a program to reduce the emissions of the pollutants that cause acid rain. This program has been generally successful.

More people live near coastlines than anywhere else on land. Urban development impedes natural patterns of erosion and deposition. Rainwater, which is the fundamental driver of coastal processes, is unable to slowly soak through impervious cover like concrete and cement. Instead it collects into rivulets and rushes into the ocean through small areas where the ground is uncovered. This causes erosion of the beachfront, destroying not only native ecosystems but often beach houses and other development. In 1972, the U.S. government enacted the Coastal Zone Management Act to more effectively regulate development in coastal regions.

The threat to coastlines is further intensified with the prediction of sea level rising because of climate change. Increased global temperatures are predicted to cause a sea level increase of up to 3.5 ft (1.07 m) in the next century. This sea level rise threatens the people living in places built below the current sea level, such as the Netherlands, Bangladesh, and Florida in the United States.

In some parts of the world, the supply of water itself does not meet the demands of the people in the area. In the western United States, the water systems are failing under the enormous growth in population in California and Nevada. The Colorado River, which is the major water source in the region, no longer flows out at the Sea of Cortez as all the water is consumed for agriculture to supply populations before it reaches the ocean. In the Middle East, one of the significant issues for debate in the peace process involves water rights to the Jordan River.

The food harvested from aquatic environments makes up a significant portion of people's diets throughout the world. In particular, many people enjoy eating top predators such as tuna, swordfish, shark, and filefish. Industrial waste, sewage, and rainwater runoff often contain high quantities of heavy metals, such as lead and mercury. These metals enter the marine food chain when microorganisms absorb the metals from seawater. Large, predatory fish eat many smaller prey and therefore ingest great quantities of heavy metals. The resulting bioaccumulation of heavy metals has contaminated most top marine predators in the oceans. In particular, the U.S. Food and Drug Administration advises that people in sensitive categories—women of child-bearing age and small children—avoid eating several species of marine fish.

■ Modern Cultural Connections

The growing concern about the effects of climate change on Earth and its impacts on the processes that affect Earth's systems have had a strong impact on society. Following the documentary, *An Inconvenient Truth,* the public became more aware of the ways in which water is intimately tied to all aspects of the planet's climate. The Intergovernmental Panel on Climate Change (IPCC) issued its Fourth Assessment Report in 2007, which details probable effects of climate change on the planet. A significant number of these effects are related to oceanic processes and other water issues.

As a result of increased temperatures, glaciers are expected to retreat in most high altitudes and at the poles. This will cause a rise in sea-level, threatening people living in low-lying areas. This type of ocean intrusion would be devastating, both socially and economically. In particular, the large-scale displacement of populations has the potential to result in regional conflict.

The pattern of precipitation is expected to change as the climate warms. Such changes will result in drought in some places. For example, the Amazon Basin is already experiencing its worst drought in 100 years. This threatens not only the ecosystem, but also the source of food for the people in the region. In other places, like Europe, flooding is expected to occur because of excessive rain.

Most climate predictions indicate that an increased frequency of tropical cyclones in the North Atlantic may result from increases in ocean surface temperature. The intensity of cyclones is also expected to increase. These events can be disastrous, not only for the people affected by the cyclones, but also for insurers, reinsurers, and banks who help those affected recover in the wake of intense weather events.

Some economists have attempted to estimate the costs associated with climate change. The *Stern Review*, published in 2006, estimated that even under the most conservative of scenarios, the costs to the United States were a loss of 1% of the gross domestic product (GDP) initially, followed by large declines when the temperature increases by 5°F (2.8°C). An increase in wind speed of just 5–10% results in hurricane damage increases costing 0.13% of the U.S. GDP. In the United Kingdom, costs associated with flooding are expected to increase between 0.1 to 0.3% of the GDP.

The complexities of the interaction of water processes and climate change are significant and multifaceted, and are greatly enhanced by the impact of global climate change. The effects on society are yet to be seen, but it is certain that continued study in water science and oceanography are crucial to understanding, and perhaps mitigating, these impacts.

SEE ALSO *Earth Science: Atmospheric Science; Earth Science: Climate Change; Earth Science: Exploration; Earth Science: Geography; Earth Science: Navigation; Earth Science: Plate Tectonics: The Unifying Theory of Geology.*

BIBLIOGRAPHY

Books

Garrison, Tom. *Oceanography: An Invitation to Marine Science.* 5th ed. Stamford, CT: Thompson/Brooks Cole, 2004.

Prager, Ellen, and Sylvia A. Earle. *The Oceans.* New York: McGraw-Hill, 2000.

Web Sites

American Geophysical Union. "Home Page." 2008. http://agu.org (accessed January 24, 2008).

American Society of Limnology and Oceanography. "Home Page." http://www.aslo.org/ (accessed January 24, 2008).

The Boston Museum of Science. "Oceans Alive: The Living Sea." 1998. http://www.mos.org/oceans/life/index.html (accessed March 2, 2007).

Environmental Literacy Council. "Water." January 8, 2007. http://www.enviroliteracy.org/category.php/14.html (accessed January 24, 2008).

HM Treasury. "Stern Review Final Report." http://www.hm-treasury.gov.uk/independent_reviews/stern_review_economics_climate_change/stern_review_report.cfm (accessed January 29, 2008).

Intergovernmental Panel on Climate Change. "Intergovernmental Panel on Climate Change Home Page." http://www.ipcc.ch/ (accessed January 30, 2008).

International Union of Geodesy and Geophysics. "International Association of Hydrological Sciences (IAHS)." December 27, 2007. http://www.iugg.org/associations/iahs.html (accessed January 29, 2008).

Marine Biological Laboratories. "Home Page." http://www.mbl.edu (accessed January 29, 2008).

Natural History Museum London. "The HMS Challenger Expedition." March 8, 2007. http://www.nhm.ac.uk/nature-online/science-of-natural-history/expeditions-collecting/fathom-challengervoyage/the-hms-challenger-expedition–18721876.html (accessed January 29, 2008).

Scripps Institution of Oceanography. "Global Discoveries for Tomorrow's World." http://sio.ucsd.edu/ (accessed January 29, 2008).

Tagging of Pacific Predators. "White Sharks." 2007. http://www.topp.org/species/white_shark (accessed January 29, 2008).

Woods Hole Oceanographic Institution. http://www.whoi.edu/ (accessed March 2, 2007).

Julie Berwald

Earth Science: Plate Tectonics: The Unifying Theory of Geology

■ Introduction

Plate tectonics is the unifying theory of geology, the framework into which are fitted all other explanations of large-scale geological phenomena, such as earthquakes, volcanoes, and the existence of ocean basins and continents. Plate tectonics describes and explains the movements of lithospheric plates, which are large areas of rocky crust, like fragments of eggshell thousands of miles across, that float and drift on the asthenosphere (the molten or malleable upper layer of Earth's mantle). Some of these moving plates bear the continents, while others underlie the ocean basins. Given the great expanse of geologic time—Earth is over 4.5 billion years old—even small plate velocities of inches or centimeters per year can radically rearranged the continents and oceans of Earth over many millions of years. Many lines of evidence show that Earth's continents have clustered together and then drifted apart again repeatedly during its history.

"Tectonics" is a geology term referring to all large-scale processes that shape planetary crusts, including that of Earth; plate tectonics is thus not the only possible kind of tectonics. As of 2008, Earth was the only body in the solar system definitely known to have plate tectonics.

■ Historical Background and Scientific Foundations

The interior bulk of Earth, which has an average radius (distance from center to surface) of 3,960 miles (6,370 km), is divided into core, mantle, and crust. The core is a ball at the center of the planet consisting mostly of iron, nickel, and sulfur: It has a radius of about 1,860 miles (3,000 km). It is surrounded by the mantle which is rocky, rather than metallic, and about 1,800 miles (2,900 km) thick. The mantle, in turn, is topped by the crust, whose outer face is the surface of Earth. The crust is divided into the areas of oceanic and continental crusts: Oceans overlie most oceanic crust, while most of the surface overlying continental crust is above sea level. The oceanic crust is thin (3-4.3 mi [5-7 km]), basaltic (<50% SiO_2), dense, and young (<250 million years old). In contrast, the continental crust is thick (18.6-40 mi [30-65 km]), granitic (>60% SiO_2), light, and old (250–3,700 million years old). SiO_2 Silicon dioxide (SiO_2) is known as the mineral quartz when it is in a crystalline form or glass when it is an amorphous form; it is a relatively light mineral, and the continents float on the mantle because they contain so much SiO_2. Oceanic crust eventually sinks into the mantle as it ages and thickens (accumulating solidified mantle material on its lower surface), since it does not contain enough SiO_2. Oceanic crust that sinks or is forced down into the mantle is continually replaced by fresh oceanic crust along certain ridges that zig-zag across the ocean floors, traversing Earth like the seams on a baseball.

The outer crust is subdivided into seven major (e.g., North American, South American, Pacific) and about a dozen minor lithospheric plates. These lithospheric plates, composed of crust and the outer layer of the mantle, contain varying proportions of oceanic and continental crust.

The visible continents and the oceanic floors, both a part of the lithospheric plates upon which they ride, shift position slowly over time in response to the geothermal forces generated by radioactive decay and residual heat left over from the formation of Earth. This heat causes the soft and liquid material in the mantle to turn over like the contents of a boiling pot, only more slowly; the lithospheric or tectonic plates shift about on the surface of this rolling mantle material like islands of scum jostling about on the surface of boiling soup. The resulting plate velocities are usually measured in centimeters per year.

Containing both crust and the upper region of the mantle, lithospheric plates are approximately 60 miles (approximately 100 km) thick: They are thicker under continental crust than under oceanic crust. Miles below the surface, a sudden change in composition from crust material to mantle pyriditite is termed the Mohorovičić discontinuity (or simply the Moho for easier pronunciation) by geologists in honor of the man who discovered it in 1910, Croation seismologist Andrija Mohorovičić (1857–1936).

Plate tectonic boundaries are regions where the edges of lithospheric plates meet. There are three types of plate tectonic boundary, namely, divergent, convergent, and transform. Divergent boundaries describe areas under tension where plates are pushed apart by magma upwelling from the mantle. New oceanic crust is scrolled out from both sides of these ridgelike boundaries, aging and thickening as it slowly moves away from its origin; such a boundary runs all the way down the middle of the Atlantic from the Arctic to the Antarctic. As the twin sheets of oceanic crust expand from the mid-ocean ridge, the Atlantic sea-floor spreads and the Americas get farther away from Africa and Europe.

Convergent boundaries are sites of collision between plates. This results either in crustal uplift (e.g., when continental crust collides with continental crust, squeezing up mountains) or in subduction, where the edge of an oceanic plate is driven downward into, and ultimately dissolved in, the molten mantle. Because Earth must remain the same size—no significant amount of matter is being added to or removed from it—its surface area is constant, so destruction of crust at subduction zones is exactly balanced, over time, by creation of new crust at divergent boundaries. Most creation and destruction of crust affects only oceanic crust, which is why continental crust is, on average, far older than oceanic crust—much of it almost as old as Earth itself.

Transform boundaries are areas such as the San Andreas fault in California, where the edges of tectonic plates slide along each other. The jerky stick-and-slip motion of the rubbing plate edges causes earthquakes. At triple points where three plates converge (e.g., where the Philippine sea plate merges into the North American and Pacific plate subduction zone), the situation becomes more complex. Also, mid-plate stresses can exist due to forces acting on plate boundaries and may result in bowing, fracturing, or mid-plate earthquakes such as the 1812 New Madrid earthquake, which devastated part of what is now the state of Missouri.

When oceanic crust collides with oceanic crust, both plates may subduct, V-ing downward to form an oceanic trench up to 36,000 feet (10,973 m) deep (as in

WORDS TO KNOW

CONVERGENT BOUNDARY: The boundary where two or more tectonic plates (large shell-like areas of rock floating on Earth's mantle, sometimes continental in size) drift against each other, edge to edge.

DIVERGENT BOUNDARY: Where two or more tectonic plates (large shell-like areas of rock floating on Earth's mantle, sometimes continental in size) are forced apart by the creation of new crust from below (as along the Atlantic midocean ridge) or pulled apart by plate motion (as in the Great Rift Valley of Africa), the boundary between the two plates is termed a divergent boundary.

PLATE: Rigid parts of Earth's crust and part of Earth's upper mantle that move and adjoin each other along zones of seismic activity.

SUBDUCTION: Tectonic process that involves one plate being forced down into the mantle at an oceanic trench, where it eventually undergoes partial melting.

TECTONICS: In geology, large-scale processes that shape Earth's crust, such as those that build mountains and valleys. Plate tectonics is the theory that explains the large-scale features of Earth in terms of moving plates or rafts of rocky material floating about on the planet's liquid interior.

TRANSFORM BOUNDARY: Where two or more tectonic plates (large shell-like areas of rock floating on Earth's mantle, sometimes continental in size) are sliding past each other edge to edge, the boundary between the two plates is termed a transform boundary. Earthquakes often occur along transform boundaries as the plates stick, build up strain, and slip suddenly.

the Marianas trench in the western Pacific, the deepest point in the world ocean). As the oceanic crusts subduct, material may be scraped off and clump up to form an accretion prism and oceanic island arcs.

When oceanic crust collides with the less-dense, more-buoyant continental crust, the oceanic crust subducts under the lighter continental crust and the continental crust may be wrinkled under compression to form mountain chains (e.g., the Andes). The subducting oceanic crust melts as it penetrates deeper into the asthenosphere, and rising molten material and gases from the melted crust may contribute to the formation of volcanic arcs such as those found along the Pacific Rim.

Although lithospheric plates move very slowly, the plates have tremendous mass. Accordingly, at collision, each lithospheric plate carries tremendous momentum (the mathematical product of velocity and mass) and the kinetic energy to drive subduction. Ultimately, all

A unique, nearly complete fossilized skeleton of a prehistoric crocodile found in Brazil allowed scientists to claim there was a "bridge" between South America and Madagascar about 70 million years ago. Scientists believe Africa broke away from Gondwana about 100 million years ago but a connection still existed between what is now South America, Antarctica, India, and Australia until about 70 million years ago. *© Sergio Moraes/Reuters/Corbis.*

energy for plate-tectonic processes begins as heat in Earth's interior. As this heat radiates into space through Earth's surface at a rate of about 0.07 watts per square meter, cooled upper mantle material slowly sinks and hotter material rises. The resulting rolling motion imparts motion to the plates.

Subduction zones are usually active earthquake zones; they are the only sites of deep earthquakes, whose sources range down to a depth of about 430 miles (700 km) in areas termed Benioff zones. Plate friction (drag) and mineral phase transitions—sudden collapses of molecular structure in large mineral bodies under heat and pressure—create the explosive forces observed in deep earthquakes. The release of forces due to sudden slippage of plates during subduction can be sudden and violent. Subduction zones also usually experience frequent shallow and intermediate-depth earthquakes. Undersea earthquakes can result in large waves known as tsunamis. A subduction earthquake under the Indian Ocean caused the Boxing Day tsunami of December 26, 2004, which killed almost a quarter of a million people, mostly along the coasts of India, Indonesia, Sri Lanka, and Thailand.

Because continental crusts do not subduct, a collision between continental crusts results in an uplift of both crusts with resultant mountain-building (orogeny). The formation and continued upward growth of the Himalayas at a rate of approximately one centimeter per year is a result of the collision of India with Asia.

■ Modern Cultural Connections

In the nineteenth and early twentieth centuries, several theories of crustal change and continental drift were current, including proposals that Earth as a whole was expanding or contracting. Several forms of the idea of continental drift were proposed early in the twentieth century, but the idea that continents might move was considered a crackpot notion by most geologists. This changed when technological advances made during World War II and throughout the 1950s allowed the

Cliffs of the Sierra Madre Oriental. Here, tectonic forces thrust up limestone strata to nearly vertical angles. © *Jonathan Blair/Corbis.*

discovery of prominent undersea ridges and rift valleys. The formation, structure, and dynamics of these ridges were ultimately explainable only by continental drift. In 1960, geologist and U.S. Navy Admiral Harry Hess (1906–1969) provided the missing explanatory mechanism for plate tectonics by suggesting that thermal convection currents in the athenosphere drive plate movements. Subsequently, geologists Drummond Matthews (1931–1997) and Fred Vine (1939–1988) confirmed Hess's assertions regarding seafloor spreading. Measurements of the magnetism of rocks in the Atlantic sea floor showed alternating north-south striping of magnetic field direction, with mirror-image striping on either side of the mid-Atlantic ridge. These stripes are a clear record of the growth of the sea floor in sheets spreading from the ridge; as rock cooled in the crust forming along the line of origin, it recorded the direction Earth's magnetic field happened to be pointing at the time. When Earth's magnetic field flipped, as it does occasionally, fresh crust on both sides of the ridge would record an opposite magnetic direction for awhile. As the crust grew and Earth's field flipped many times over tens of millions of years, two sets of stripes were formed, one to the east of the ridge and another to the west. No plausible explanation for the existence of these stripes exists, other than sea-floor spreading and movement of the continents.

The principal proofs of plate-tectonic theory lie in: (1) the geometric fit of the displaced continents (matching distant coastlines); (2) the similarity of rock ages and Paleozoic fossils in corresponding bands or zones in areas that were once adjacent, such as South America and West Africa; (3) the existence of ophiolite suites (slivers of oceanic floor containing fossils) found uplifted into the upper levels of mountain chains; (4) radiometric dating evidence, which shows that rock ages are similar

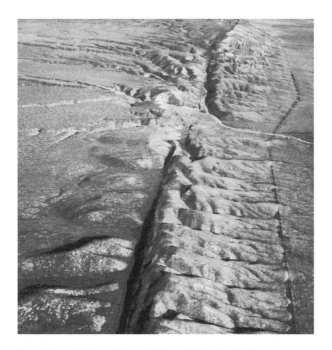

San Andreas Fault.   *Robert E. Wallace/U.S. Geological Survey.*

The deep-sea drilling vessel *Chikyu* is docked at Mitsubishi Heavy Industries' shipyard in Yokohama, suburban Tokyo, Japan, December 14, 2005. The vessel is part of a Japanese-led project to dig deeper into Earth's surface than ever before in the hopes of a breakthrough in detecting earthquakes. *STR/AFP/Getty Images.*

in equidistant bands symmetrically centered on the Mid-Atlantic ridge divergent boundary; and (5) paleomagnetic studies that show bands of magnetic orientation symmetrical to divergent boundaries (as described earlier for the Atlantic). These convergent and independent sources of evidence convinced working geologists in the 1960s and afterward that the motions described by plate tectonics are a reality.

Additional evidence is the fact that the age of the rocks making up the oceanic crust increases as their distance from the divergent boundary (i.e., mid-ocean ridge) increases. Rock-age studies also show that oceanic crust is young, containing no rocks formed more than 250 million years ago.

Understanding continental drift has indirect benefits for human societies. In particular, volcanoes, earthquakes, and tsunamis can cause large numbers of deaths and vast property loss: When such events can be predicted, their effects can be greatly reduced. Today, one reason that scientists study the nature of oceanic subduction is to be better able to forecast major eruptions and earthquakes. In California, for example, a network of ground-motion detectors tracks the slippage of the Pacific plate along the edge of the North American plate, which creates the San Andreas fault system.

Pseudoscientific beliefs about continental drift are held by a few people, such as that cultural similarities between South America and Africa (pyramid-building cultures arose in both) are explained by the fact that the two continents were once in contact. Such beliefs have no basis in fact: Human beings only evolved in the last million years or so, while it has been well over 100 million years since Africa was in contact with South America.

■ Primary Source Connection

Before comprehensive theories develop and scientists know how natural mechanisms work, there is a tendency to shape explanations and analyze data in terms of prevailing theory. For example, prior to the now well-tested and accepted theory of plate tectonics and evidence of glacial movements, there were various attempts to explain erratic boulders (rock not native to a particular area). In the example that follows, in 1839, English naturalist Charles Darwin (1809–1882), best known for his advance of evolutionary theory, compiled the following "Note on a rock seen on an iceberg in 61° south latitude" that was then published by the *Journal of the Royal Geographical Society of London* in

Volcanologist Christina Heliker shields herself from the heat as she takes a shovel-full of hot pahoehoe lava, during an eruption of the Kilauea Volcano. Understanding the causes of volcanic activity can help scientists better prepare for such events. © *Roger Ressmeyer/Corbis.*

which Darwin documented anecdotal observations that he thought might be relevant to developing an explanation of erratics.

NOTE ON A ROCK SEEN ON AN ICEBERG IN 61° SOUTH LATITUDE

HAVING been informed by Mr. Enderby, that a block of rock, embedded in ice, had been seen during the voyage of the schooner Eliza Scott in the Antarctic Seas, I procured through his means an interview with Mr. Macnab, one of the mates of the vessel, and I learnt from him the following facts:—On the 13th of March, when in lat. 61° S., and long. 103° 40′ E., a black spot was seen on a distant iceberg, which, when the vessel had run within a quarter mile of it, was clearly perceived to be an irregularly-shaped but angular fragment of dark-coloured rock. It was embedded in a perpendicular face

of ice, at least 20 feet above the level of the sea. That part which was visible, Mr. Macnab estimated at about 12 feet in height, and from 5 to 6 in width; the remainder (and from the dark colour of the surrounding ice, probably the greater part) of the stone was concealed. He made a rough sketch of it at the time, as represented at p. 524. The iceberg which carried this fragment was between 250 and 300 feet high.

Mr. Macnab informs me, that on one other occasion (about a week afterwards) he saw on the summit of a low, flat iceberg, a black mass, which he thinks, but will not positively assert, was a fragment of rock. He has repeatedly seen, at considerable heights on the bergs, both reddish-brown and blackish-brown ice. Mr. Macnab attributes this discolouration to the continued washing of the sea; and it seems probable that decayed ice, owing to its porous texture, would filter every impurity from the waves which broke over it.

Every fact on the transportation of fragments of rock by ice is of importance, as throwing light on the problem of 'erratic boulders,' which has so long perplexed geologists; and the case first described possesses in some respects peculiar interest. The part of the ocean, where the iceberg was seen, is 450 miles distant from *Sabrina* land (if such land exists), and 1400 miles from any certainly known land. The tract of sea, however, due S., has not been explored; but assuming that land, if it existed there, would have been seen at some leagues distance from a vessel, and considering the southerly course which the schooner *Eliza Scott* pursued immediately prior to meeting with the iceberg, and that of Cook in the year 1773, it is exceedingly improbable that any land will hereafter be discovered within 100 miles of this spot. The fragment of rock must, therefore, have travelled at least thus far from its parent source; and, from being deeply embedded, it probably sailed many miles farther on before it was dropped from the iceberg in the depths of the sea, or was stranded on some distant shore. In my Journal, during the voyage of H.M.S. *Beagle*, I have stated (p. 282), on the authority of Captain Biscoe, that, during his several cruises in the Antarctic Seas, he never once saw a piece of rock in the ice. An iceberg, however, with a considerable block lying on it, was met with to the E. of South Shetland, by Mr. Sorrell (the former boatswain of the *Beagle*), when in a sealing vessel. The case, therefore, here recorded is the second; but it is in many respects much the most remarkable one. Almost every voyager in the Southern Ocean has described the extraordinary number of icebergs, their vast dimensions, and the low latitudes to which they are drifted: Horsburgh* has reported the case of several, which were seen by a ship in her passage from India, in lat. 35° 55′ S. If then but one iceberg in a thousand, or in ten thousand, transports its fragment, the bottom of the Antarctic Sea, and the shores of its islands, must already be scattered with masses of foreign rock,—the counterpart of the "erratic boulders" of the northern hemisphere.

C.R. Darwin

DARWIN, C.R., "NOTE ON A ROCK SEEN ON AN ICEBERG IN 61° SOUTH LATITUDE." *JOURNAL OF THE ROYAL GEOGRAPHICAL SOCIETY OF LONDON* 9 (MARCH 1839): 528–529.

SEE ALSO *Earth Science: Gradualism and Catastrophism.*

BIBLIOGRAPHY

Books

McPhee, John. *Basin and Range.* New York: Farrar, Straus, & Giroux, 1980.

Periodicals

Darwin, C.R., "Note on a Rock Seen on an Iceberg in 61° South Latitude." *Journal of the Royal Geographical Society of London* 9 (March 1839): 528–529.

Matthews, Drummond H., and Simon L. Klemperer. "Deep Sea Seismic Reflection Profiling." *Geology* 15 (1987): 195–198.

Vine, F.J. "Spreading of the Ocean Floor: New Evidence." *Science* 154 (1966): 1405–1515.

Web Sites

United States Geological Survey. "This Dynamic Earth: The Story of Plate Tectonics." March 27, 2007. http://pubs.usgs.gov/gip/dynamic/dynamic.html (accessed February 6, 2008).

K. Lee Lerner

Mathematics: Calculus

■ Introduction

The calculus is a set of powerful analytical techniques, especially differentiation and integration, that utilize the concepts of rate and limit to describe the properties of functions. The formal development of the calculus in the latter half of the seventeenth century, primarily through the independent work of English physicist and mathematician Sir Isaac Newton (1642–1727) and German mathematician Gottfried Wilhelm Leibniz (1646–1716), was the crowning mathematical achievement of the Scientific Revolution. The subsequent advance of the calculus influenced the whole course and scope of mathematical and scientific inquiry. Although the logical underpinnings of calculus were hotly debated early on, the techniques of calculus were quickly applied to a variety of problems in physics, astronomy, and engineering. By the end of the eighteenth century, calculus had proved a powerful tool that allowed mathematicians and scientists to construct accurate mathematical models of physical phenomena ranging from planetary motions to particle dynamics.

■ Historical Background and Scientific Foundations

Early Development

Important mathematical developments that laid the foundation for the calculus of Newton and Leibniz can be traced back to techniques advanced in ancient Greece and Rome. Most of these techniques were concerned with determining areas under curves and the volumes of curved shapes. Besides their mathematical utility, these advancements both reflected and challenged prevailing philosophical notions about the concept of infinitely divisible time and space. Greek philosopher and math-

ematician Zeno of Elea (c.495–c.430 BC) constructed a set of paradoxes that were fundamentally important in the development of mathematics, logic, and scientific thought. The most famous of these is the dichotomy (two-parts) paradox: If you wish to walk from your chair to the door, you first traverse half the distance. But when you have done so, half the distance still remains before you. When you have traversed half that remaining distance, half of it still remains undone—and so on, forever. To reach the doorway, you must traverse an infinite number of finite distance intervals in finite time.

Zeno argued that this was impossible, and that motion is therefore an illusion. Zeno's paradoxes reflected the idea that space and time could be infinitely subdivided into smaller and smaller portions; his paradoxes remained mathematically unsolvable until the concepts of continuity, limits, and infinite series were introduced in the development of the calculus. In the calculus, it is elementary to show that the sum of an infinite number of finite terms (such as the times it takes to make the ever-shorter journeys in Zeno's dichotomy paradox) can indeed be a finite number. You can reach the door. Motion is not an illusion. Yet Zeno's paradoxes were not foolish; they pointed to deep mathematical questions that were not resolved until many centuries later, and his claim that motion is impossible is no more counterintuitive than some claims made by modern physics, such as the relativistic revelation that there is no such thing as a universal simultaneous moment of time.

Greek astronomer, philosopher, and mathematician Eudoxus Of Cnidus (c.408–c.355 BC) developed a method of exhaustion that could be used to calculate the area and volume under certain curves and of solids (i.e., the cone and pyramid). In this context, "exhaustion" refers not to physical tiredness but to the adding up of smaller and smaller areas and volumes to draw closer to a true solution. Eudoxus's method relied, like the calculus, on the concept that time and space can be divided into infinitesimally small portions. The method of exhaustion pointed the way toward a primitive geometric form of what in calculus is known as integration.

Although other advances by classical mathematicians helped set the intellectual stage for the ultimate development of the calculus during the Scientific Revolution, ancient Greek mathematicians failed to find a common link between problems related to finding the area under curves and to problems requiring the determination of a tangent (a line touching a curve at only one point). That these processes are actually the inverse of each other—one equals the other in reverse—became the basis of the calculus eventually developed by Newton and Leibniz. Today, the reciprocal or symmetrical relationship between integration (area-finding) and differentiation (tangent-finding) is known as "the fundamental theorem of the calculus."

In the Middle Ages, philosophers and mathematicians continued to ponder kinematics (questions relating to motion). These inquiries led to early efforts to plot functions relating to time and velocity. In particular, the work of French bishop Nicholas Oresme (c.1325–1382) was an important milestone in the development of kinematics and geometry, especially Oresme's proof of the Merton theorem, which allowed for the calculation of the distance traveled by an object when uniformly accelerated (e.g., by acceleration due to gravity). Oresme's proof established that the sum of the distance traversed (i.e., the area under the velocity curve) by a body with variable velocity was the same as that traversed by a body with a uniform velocity equal to that of the varying-velocity body at the middle instant of whatever period was measured. From this it could be shown that the area under the location-versus-time curve was the sum of all distances covered by a series of instantaneous velocities. This work would later prove indispensable to the quantification of parabolic motion by Italian astronomer and physicist Galileo Galilei (1564–1642) and later influenced Newton's development of differentiation.

During the Renaissance in Western Europe, a rediscovery of ancient Greek and Roman mathematics spurred increased use of mathematical symbols, especially to denote algebraic concepts. The rise in symbolism also allowed the development of and increased application of the techniques of analytical geometry principally advanced by French philosopher and mathematician René Descartes (1596–1650) and French mathematician Pierre de Fermat (1601–1665). Beyond the practical utility of establishing that algebraic equations corresponded to curves, the work of Descartes and Fermat laid the geometrical basis for the calculus. In fact, Fermat's methodologies included concepts related to the determination of minimums and maximums for functions that are mirrored in modern mathematical methodology (e.g., setting the derivative or rate-of-change of a function to zero). Both Newton and Leibniz would rely heavily on the use of Cartesian algebra in the development of their calculus techniques.

Although many of the elements of the calculus were in place by the mid–1600s, recognition of the fundamental theorem relating differentiation and integration as inverse processes continued to elude mathematicians and scientists. Part of the difficulty was lingering philosophical resistance toward the philosophical ramifications of the limit and the infinitesimal—a resistance descended from Zeno's paradoxes. In one sense, the genius of Newton and Leibniz lay in their ability to put aside the philosophical and theological ramifications of the infinitesimal in favor of developing practical mathematics. Neither Newton nor Leibniz were seriously worried by the deeper philosophical issues regarding limits and infinitesimals. In this regard, Newton and Leibniz worked in the spirit of empiricism that grew during throughout the Scientific Revolution—although ultimately, mathematicians would address those deep issues in developing the calculus further and making it a part of the rigorous system of modern mathematics.

A bitter, angry controversy arose concerning whether Newton or Leibniz deserved credit for the calculus. This controversy was grounded in the actions of both men during the late 1600s. Historical documents established that Newton's unpublished formulations of the calculus came two decades before Leibniz's publications in 1684 and 1686. However, most scholars conclude that Leibniz developed his techniques independently. Also, although the mathematical outcomes were identical, the differences in symbolism and nomenclature used by Newton and Leibniz are evidence of independent development.

The controversy regarding credit for the origin of calculus quickly became more than a simple dispute between mathematicians. Supporters of Newton and Leibniz often argued along blatantly nationalistic lines; the feud itself had a profound influence on the subsequent development of the calculus and other branches of mathematical analysis in England and in Continental Europe. English mathematicians who relied on Newton's "fluxions" methods were divided from mathematicians in Europe who followed the notational conventions established by Leibniz. The publications and symbolism of Leibniz greatly influenced the mathematical work of Swiss mathematicians (and brothers) Jakob Bernoulli (1654–1705) and Johann Bernoulli (1667–1748).

The publications of Newton and Leibniz emphasized the utilitarian aspects of the calculus. Nevertheless, the nomenclatures and techniques developed by Newton and Leibniz also mirrored their own philosophical leanings. Newton developed the calculus as a practical tool with which to analyze planetary motion and the effects of gravity. Accordingly, he emphasized analysis, and his mathematical methods describe the effects of forces on motion in terms of infinitesimal changes with respect to time. Leibniz's calculus, on the other hand, was driven by the idea that incorporeal (non-material) entities were the driving basis of existence and the changes in the larger world experienced by mankind; consequently, he sought to derive integral methods by which discrete infinitesimal units could be summed to yield the area of a larger shape.

While philosophical debates regarding the underpinnings of the calculus simmered, the first calculus texts appeared before the end of the seventeenth century. The first textbook in calculus was published by French mathematician Guillaume François Antoine l'Hôpital (1661–1704), though modern scholars now credit much of the content to Johann Bernoulli. L'Hôpital's *Analyse des Infiniment Petits four l'intelligence des lignes courbes* (1696) helped bring the calculus into wider use throughout continental Europe.

Within a few decades the calculus was quickly embraced and applied to a wide range of practical problems in physics, astronomy, and mathematics. Why the calculus worked, however, remained a vexing question

IN CONTEXT: AN ANGRY GENIUS

The fame and status of English mathematician and physicist Isaac Newton (1642–1727), plus the fact that he corresponded with his great rival in the development of the calculus, German mathematician Gottfried Wilhelm Leibniz (1646–1716), resulted in a charge of plagiarism against Leibniz by members of the British Royal Society (the most prestigious scientific organization of its day). Supporters of Leibniz subsequently leveled similar charges against Newton. Leibniz petitioned the Royal Society for redress, but Newton, a high-ranking member of the Society, hand-picked the investigating committee and prepared reports on the controversy for committee members to sign. Not surprisingly, the committee ruled against Leibniz. Before the dispute was resolved, Leibniz died.

Newton's anger remained unabated after Leibniz's death. In many of Newton's papers, he continued to set out mathematical and personal criticisms of Leibniz. Another feud, with English scientist Robert Hooke (1635–1703), resulted in Newton purging from his master-work *Philosophiae Naturalis Principia Mathematica* (1687) any reference to Hooke's important contributions. Another dispute, between Newton and the British Royal Astronomer John Flamsteed (1646–1719), resulted in Flamsteed's name also being deleted from the *Principia*.

The feud over credit for the calculus adversely affected communications between English and European mathematicians, who took sides along national lines. English mathematicians used Newton's "fluxion" notations exclusively when doing the calculus; European (especially Swiss and French) mathematicians used only Leibniz's *dy/dx* notation. The latter is standard today.

that opened it to attack on philosophical and theological grounds. This school of critics—eventually to be led in eighteenth century England by the Irish Anglican bishop George Berkeley (1685–1753)—argued that the fundamental theorems of calculus derived from logical fallacies, and that the great accuracy of the calculus actually resulted from the mutual lucky cancellation of fundamental errors. This argument was incorrect, but such attacks upon the calculus resulted in increased rigor in mathematical analysis, which was ultimately beneficial for modern mathematics. Scholars took Berkeley's criticisms seriously and set out to support the logical foundations of calculus with well-reasoned rebuttals.

In the late 1700s, French mathematician Jean le Rond d'Alembert (1717–1783) published two influential articles, "Limite" and "Différentielle," that offered a strong rebuttal to Berkeley's arguments and defended the concepts of differentiation and infinitesimals by discussing the notion of the limit. The debates regarding

IN CONTEXT: CALCULUS HELPS THREAD A COSMIC NEEDLE

On June 28, 2004, the Cassini space probe, a robot craft about the size of a small bus, built jointly by the United States and Europe, arrived at Saturn after a seven-year journey. The plan was for it to be captured by Saturn's gravity and so become a permanent satellite, observing Saturn and its rings and moons for years to come. But to make the journey, Cassini had reached a speed of 53,000 miles per hour (many times faster than a rifle bullet), too fast to be captured by Saturn's gravity. At that speed it would swoop past Saturn and head out into deep space. Therefore, it was programmed to hit the brakes, to fire a rocket against its direction of travel as it approached its destination.

For objects moving in straight lines, changes in velocity can be calculated using basic algebra. Calculus is not needed. But Cassini was not moving in a straight line; it was falling through space on a curving path toward Saturn, being pulled more strongly by Saturn's gravity with every passing minute. To figure out when to start Cassini's rocket and how long to run it, the probe's human controllers on Earth had to use calculus. The effects of the important forces acting on Cassini—in particular, its own rocket motor and Saturn's gravity—had to be integrated over time. And the calculation—carried out using computers, not, for the most part, on paper—had to be extremely exact, or Cassini would be destroyed. To get deep enough into Saturn's gravitational field, Cassini would have to steer right through a relatively narrow gap in Saturn's rings called the Cassini division (named after the same Italian astronomer as the probe itself), a navigational feat comparable to threading a cosmic needle. If it missed the gap, the Cassini craft would have been destroyed by collision with the rings.

Not all of NASA's navigational calculations have been correct; in 1998, a space probe crashed into Mars because of a math mistake. But in Cassini's case, the calculations were correct. Cassini passed the rings safely, was captured by Saturn's gravity, and began its orbits of Saturn.

George Berkeley (1685–1753), an Irish bishop and philosopher who tried unsuccessfully to establish a college in the Bermudas and finally returned to London to become the Bishop of Cloyne. *Library of Congress.*

the logic of the calculus resulted in the introduction of new standards of rigor in mathematical analysis and laid the foundation for subsequent rise of pure mathematics in the nineteenth century.

The first texts in calculus actually appeared in the last years of the seventeenth century. The publications and symbolism of Leibniz greatly influenced the mathematical work of two brothers, Swiss mathematicians Jakob and Johann Bernoulli. Working separately, the Bernoulli brothers improved and made wide application of the calculus. Johann Bernoulli was the first to apply the term "integral" to a subset of calculus techniques allowing the determination of areas and volumes under curves. During his travels, Johann Bernoulli sparked intense interest in the calculus among French mathematicians, and his influence was critical to the widespread use of Leibniz-based methodologies and nomenclature.

Impact

Physicists and mathematicians seized upon the new set of analytical techniques comprising the calculus. Advancements in methodologies usually found quick application and, correspondingly, fruitful results fueled further research and advancements. Although the philosophical foundations of calculus remained in dispute, these arguments proved no hindrance to the application of calculus to problems of physics. The Bernoulli brothers, for example, quickly recognized the power of the calculus as a set of tools to be applied to a number of statistical and physical problems. Jakob Bernoulli's distribution theorem and theorems of probability and statistics, ultimately of great importance to the development of physics, incorporated calculus techniques. Johann Bernoulli's sons, Nicolaus Bernoulli (1695–1726), Daniel Bernoulli (1700–1782), and Johann Bernoulli II (1710–1790), all made contributions to the calculus.

In particular, Johann Bernoulli II used calculus methodologies to develop important formulae regarding the properties of fluids and hydrodynamics.

The application of calculus to probability theory resulted in probability integrals. The refinement made immediate and significant contributions to the advancement of probability theory based on the late-seventeenth-century work of French mathematician Abraham De Moivre (1667–1754).

English mathematician Brook Taylor (1685–1731) developed what was later termed the Taylor expansion theorem and the Taylor series. Taylor's work was subsequently used by Swiss mathematician Leonard Euler (1707–1783) in the extension of differential calculus and by French mathematician Joseph Louis Lagrange (1736–1813) in the development of his theory of functions.

Scottish mathematician Colin Maclaurin (1698–1746) advanced an expansion that was a special case of a Taylor expansion (where $x = 0$), now known as the Maclaurin series. More importantly, in the face of developing criticism from Bishop Berkeley regarding the logic of calculus, Maclaurin set out an important and influential defense of Newtonian fluxions and geometric analysis in his 1742 *Treatise on Fluxions.*

The application of the calculus to many areas of math and science was profoundly influenced by the work of Euler, a student of Johann Bernoulli. Euler was one of the most dedicated and productive mathematicians of the eighteenth century. Based on earlier work done by Newton and Jakob Bernoulli, in 1744 Euler developed an extension of calculus dealing with maxima and minima of definite integrals termed the calculus of variation (variational calculus). Among other applications, variational calculus techniques allow the determination of the shortest distance between two points on curved surfaces.

Euler also advanced the principle of least action formulated in 1746 by Pierre Louis Moreau de Maupertuis (1698–1759). In general, this principle asserts economy in nature (i.e., an avoidance in natural systems of unnecessary expenditures of energy). Accordingly, Euler asserted that natural motions must always be such that they make the calculation of a minimum possible (i.e., nature always points the way to a minimum). The principle of least action quickly became an influential scientific and philosophical principle destined to find expression in later centuries in various laws and principles, including Le Chatelier's principle regarding equilibrium reactions. The principle profoundly influenced nineteenth-century studies of thermodynamics.

On the heels of an influential publication covering algebra, trigonometry, and geometry (including the geometry of curved surfaces) Euler's 1755 publication, *Institutiones Calculi Differentialis* (Foundations of Differential Calculus), influenced the teaching of calculus for more than two centuries. Euler followed with three volumes published from 1768 to 1770, titled

Institutiones Calculi Integralis (Foundations of Integral Calculus), which presented his work on differential equations. Differential equations contain derivatives or differentials of a function. Partial differential equations contain partial derivatives of a function of more than one variable; ordinary differential equations contain no partial derivatives. The wave equation, for example, is a second-order differential equation important in the description of many physical phenomena including pressure waves (e.g., water and sound waves). Euler and d'Alembert offered different perspectives regarding whether solutions to the wave equation should be, as argued by d'Alembert, continuous (i.e. derived from a single equation) or, as asserted by Euler, discontinuous (having functions formed from many curves). The refinement of the wave equation was of great value to nineteenth century scientists investigating the properties of electricity and magnetism that resulted in Scottish physicist James Clerk Maxwell's (1831–1879) development of equations that accurately described the electromagnetic wave.

The disagreement between Euler and d'Alembert over the wave equation reflected the type of philosophical arguments and distinct views regarding the philosophical relationship of calculus to physical phenomena that developed during the eighteenth century.

Although both Newton and Leibniz developed techniques of differentiation and integration, the Newtonian tradition emphasized differentiation and the reduction to the infinitesimal. In contrast, the Leibniz tradition emphasized integration as a summation of infinitesimals.

A third view of the calculus, mostly reflected in the work and writings of Lagrange, was more abstractly algebraic and depended upon the concept of the infinite series (a sum of an infinite sequence of terms). These differences regarding a grand design for calculus were not trivial. According to the Newtonian view, calculus derived from analysis of the dynamics of bodies (e.g., kinematics, velocities, and accelerations). Just as the properties of a velocity curve relate distance to time, in accord with the Newtonian view, the elaborations of calculus advanced applications where changing properties or states could accurately be related to one another (e.g., in defining planetary orbits, etc.). Calculus derived from the Newtonian tradition was used to allow the analysis of phenomena by artificially breaking properties associated with that phenomena into increasingly smaller parts. In the Leibniz tradition, calculus allowed accurate explanation of phenomena as the summed interaction of naturally very small components.

Lagrange's analytic treatment of mechanics in his 1788 publication *Analytical Mechanics* (containing the Lagrange dynamics equations) placed important emphasis on the development of differential equations. Lagrange's work also profoundly influenced the work of another French mathematician, Pierre-Simon Laplace

Gottfried Wilhelm Leibniz (1646–1716). *Library of Congress.*

(1749–1827) who, near the end of the eighteenth century, began important and innovative work in celestial mechanics.

■ Modern Cultural Connections

Since the eighteenth century, the calculus has become ubiquitous in the physical and social sciences. It is employed daily in almost all forms of physical science and engineering, much of biology and medicine, and economics. It functions almost as a universal language in fields that involve applied mathematics and is as much a part of the fundamental toolbox of higher mathematics as arithmetic itself.

The calculus is a true tool; that is, it does not merely express knowledge that is obtained from other sources, but makes possible the discovery of new knowledge and the design of devices that could not be produced by any other means. It is essential in the design of transistors, engines, and chemical processes; weapons, games, and medicines; indeed, of that whole panoply of technological devices that shapes so much of the modern human environment, from two-stroke engines and electrical generators to cell phones, nuclear weapons, computers, and communications satellites. Without the calculus, none of our advanced technologies would be possible, and efforts to understand or predict global climate change and most other aspects of the physical world would be futile. The calculus is the indispensable grammar of modern science and technology.

The Calculus of War and Peace

Like all scientific knowledge, calculus can be applied not only to creation but destruction. For example, the calculus-based concept of *inertial guidance* has been developed by missile-makers to a fine art.

The first ballistic missiles used in war, the V-2 rockets produced by Nazi Germany near the end of World War II (1939–1945), were fired at London from mainland Europe. They were intended as terror or "vengeance" weapons and so only needed to explode somewhere over the city, not over particular military objectives. Yet to hit a large city such as London at such a distance, a V-2 missile needed a guidance system, a way of knowing where it was at every moment so that it could steer toward its target. It was not practical to steer by the stars or the sun, because these are hard to observe from a missile in supersonic flight and would require complex calculations. Nor was it practical to steer by sending radio signals to the missiles, for without advanced radar (not yet available) controllers on the ground would be just as ignorant of the missile's location as the missile itself. Besides, the enemy might learn to fake or jam control signals, that is, drown them out with radio noise.

The solution was inertial guidance, which exploits the calculus fact that (a) the time derivative of position is velocity and (b) the time derivative of velocity is acceleration. By the fundamental theorem of calculus, which says that integration and derivative are opposites, we know that we can follow the trail backwards: the integral of acceleration is velocity, and the integral of velocity is position.

What designers need a missile to know is its position. But position is hard to measure directly. You have to look out the window, identify landmarks (if any happen to be visible), and do some fast geometry, likewise with velocity. But acceleration is easy to measure, because every part of an object accelerated by a force experiences that force. In addition, unlike velocity or position, a force can be measured directly and locally, that is, without making observations of the outside world. Therefore, the V-2's engineers installed gyroscopes (spinning masses of metal) in their missile and used these to measure its accelerations. Lasers, semiconductors, and other gadgets have also been used since that time. Some are more expensive and accurate than others, but all do the same job: they measure accelerations. Any device that measures acceleration is called an accelerometer.

Thanks to its accelerometers, an inertial guidance system knows its own acceleration as a function of time. Acceleration can be written as a function of time, $a(t)$. This function is known by direct measurement by accelerometers. The integral of $a(t)$ gives *velocity* as a function of time: $\int a(t)\ dt = v(t)$. And the integral of $v(t)$ gives *distance* as a function of time, which reveals one's position at any given moment: $\int v(t)\ dt = x(t)$. The actual calculations used in inertial guidance systems are, of course, more complex, but the principle remains the same.

The bottom line for inertial guidance is that, given an accurate knowledge of its initial location and velocity, an inertial guidance system is completely independent of the outside world. It knows where it is, no matter where it goes, without ever having to make an observation.

The V-2 inertial guidance system was crude, but since World War II inertial guidance systems have become more accurate. In the early 1960s they were placed in the first intercontinental ballistic missiles (ICBMs), large missiles designed by the Soviet Union and the United States to fly to the far side of the planet in a few minutes and strike specific targets with nuclear warheads. They were also used in the Apollo moon rocket program and in nuclear submarines, which stay underwater for weeks or months without being able to make observations of the outside world. Inertial guidance systems are today not only in missiles but in tanks, some oceangoing ships, military helicopters, the Space Shuttle and other spacecraft, and commercial airliners making transoceanic journeys.

Calculus makes inertial guidance possible, but also, in a sense, limits its accuracy. The problem is called integration drift. Integration drift is a pesky result of the fact that small "biases" are, for various technical reasons, almost certain to creep into acceleration measurements. Any bias in these acceleration measurements, any unwanted, constant number that adds itself to all the measurements, will result in a position error that increases in proportion to the square of time. As a result, no inertial guidance system can go forever without taking an observation of the outside world to see where it really is. Increasingly, inertial guidance systems are designed to update themselves automatically by checking the Global Positioning System (GPS), a network of satellites that blanket the whole Earth with radio signals that can be used to determine a receiver's position accurately.

Today, inertial guidance systems have reached a very high degree of accuracy. A ballistic missile launched from a nuclear submarine, which may begin with a knowledge of its initial position and velocity, can be launched from a still-submerged submarine by compressed air, burst through the surface of the water, ignite its rocket, fly blind to the far side of the planet, and explode its nuclear warhead within a few yards of its target.

IN CONTEXT: EVERYDAY CALCULUS

Calculus has been used literally thousands of times in the design of every one of the electronic toys we take for granted—including MP3 players, TV screens, computers, cell phones, etc. For long-term information storage, most computers contain a "hard drive" (some simple "terminal" computers do not, and new generation computers may contain "flash" drive type storage similar in principle to USB memory drives). A hard drive contains a stack of thin discs coated with magnetic particles. A computer stores information in the form of binary digits ("bits" for short, 1s and 0s) on the surface of each disc by impressing or "writing" on it billions of tiny magnetic fields that point one way to signify "1," another way to signify "0." The bits are arranged in circular tracks. To read information off the spinning disk, sensors glide back and forth between the edge of the disc and its center to place themselves over selected tracks. The track spins under the sensor; the bits are read off one by one at high speed; and within a few seconds, the program appears. In designing a data-storage disc, engineers attempt to *optimize* the amount of data stored on the disk, that is, to store the most bits possible.

At first it may seem logical that the best way to store data would be to completely cover the disk's surface with tracks. Although logical, that turns out to be the wrong approach. For the sake of keeping the read-write mechanism simpler (and therefore cheaper), every circular track has to have the same number of bits. However, the smaller you make the radius of the innermost track, the fewer bits you can fit on it. But all the tracks on the disc must, as specified above, have the same number of bits, so if you make the innermost track too small, it will hold only a few bits, and so will all the other tracks, and you'll end up with an inefficient disc. On the other hand, if you make the innermost track too big, there won't be much room for additional tracks between the innermost track and the outer edge of the disc, and again your design will be inefficient.

Calculus is used to solve the problem. Using derivative calculus to find the value of the innermost radius permits engineers to place the maximum amount of data (in term of number of bits) on the disk.

SEE ALSO *Mathematics: Trigonometry.*

BIBLIOGRAPHY

Books

Boyer, Carl. *The History of the Calculus and Its Conceptual Development.* New York: Dover, 1959.

———. *A History of Mathematics.* 2nd ed. New York: John Wiley and Sons, 1991.

Edwards, C.H. *The Historical Development of the Calculus.* New York: Springer, 1979.

Hall, Rupert. *Philosophers at War: The Quarrel between Newton and Leibniz*. Cambridge, UK: Cambridge University Press, 1980.

Kline, M. *Mathematical Thought from Ancient to Modern Times*. New York: Oxford University Press, 1972.

Kuhn, Thomas S. *The Structure of Scientific Revolutions*. Chicago: University of Chicago Press, 1970.

K. Lee Lerner
Larry Gilman

Mathematics: Foundations of Mathematics

■ Introduction

Physical science is based on the direct or indirect observation of objects or events. Mathematics, however, is the study of non-material objects or relationships—sets, equations, lines, and the like—that do not exist outside of the human mind, at least, not according to many modern mathematicians and philosophers of mathematics. Several basic questions about mathematics can be asked: How can mathematicians and scientists be sure that mathematical theories are true? Why should mathematics say one thing rather than another, and why are mathematics so useful in describing the physical universe?

Part of the answer to the last question is that not all mathematics are, in fact, useful in describing the physical universe. Only some systems of geometry, for example, describe the geometry of the universe we observe. But other geometries are not false or incorrect simply because they describe nothing physically real. A system of mathematical relationships is considered valid or true if it is consistent with itself, that is, if it works on its own terms: it need not correspond to anything in the world. In mathematics, correctness is judged by purely mathematical standards, not by the experimental standards of physical science.

Historically, the first foundation of mathematics was common sense, the intuitive human sense of arithmetic and geometry as tested against experience. The earliest forms of mathematics were, therefore, (a) the counting or natural numbers (1, 2, 3, etc.) and their associated operations (addition, subtractions, etc.) and (b) Euclidean geometry, the geometry still used in daily life, engineering, and most science because it provides a close approximation to the geometry of much of the physical world.

However, as mathematics became more abstract and complex in recent centuries, mathematicians and philosophers began to question the nature of mathematics itself. How do we know that a given mathematical

system of statements is valid or true, whatever those terms might mean? Modern mathematicians, logicians, and philosophers have sought to answer this question by discerning a theoretical foundation or bedrock for explanation and understanding mathematical truth. The field of study termed "foundations of mathematics" examines the nature of mathematics and its methods using various theoretical tools, especially mathematics itself.

■ Historical Background and Scientific Foundations

The questions with which inquiry into the foundations of mathematics starts are philosophical: that is, they are addressed to the nature of the ideas on which other ideas depend. Typical examples are: What are mathematical objects (e.g., variables, matrices, equations)? How do we have knowledge of mathematical objects and on what grounds do we believe—if we do believe—that this constitutes true knowledge? What is a proof? When do we know that a mathematical statement is provable?

A theory that provides answers to such questions is called a foundational theory. Any such theory must be able to account for a large part, preferably all, of mathematics, starting from a small number of base assumptions and principles. No foundational theory produced as of the early 2000s was fully satisfactory on these terms or had convinced a large majority of mathematicians and philosophers of mathematics.

The aim of foundations of mathematics is to organize all aspects of mathematics in such a way that at the base are the most fundamental concepts, assumptions and principles, and all other aspects depend on this base. There then arises the question of why certain fundamental notions are accepted rather than others; this is a matter of philosophical inquiry. The tools of foundational mathematics are mostly those of mathematical logic, a closely related discipline and, as its name indicates, a form of mathematics.

Foundational Crises before the Twentieth Century

The problem of giving secure foundations to mathematics arose with particular vigor at the beginning of the twentieth century due to the discovery of various contradictions at the basis of Cantor's theory of infinite sets; this upheaval in mathematical thinking is known today as the foundational crisis. But mathematics, in its long history, had known other foundational crises connected with new discoveries or inventions that raised doubts about what had formerly been taken to be the unshakeable basis of mathematical thought.

The discovery of irrational numbers by ancient Greek mathematicians may have caused the first foundational crisis. As the story goes, around 500 BC, Hippasus of Metapontum, disciple of Pythagoras, produced a geometric proof that the square root of 2 is an irrational number—that is, a number that cannot be expressed as the ratio of two whole numbers or as a terminating or repeating decimal. Pythagoras is said to have discovered this fact earlier, but to have kept it secret; however, many such stories are mythical, since few contemporary records survive. Pythagoras believed that all things are numbers—that is, that numbers are the basis of all reality—and the existence of a number that could not be expressed as a ratio of two whole numbers was, supposedly, disturbing to him. When Hippasus revealed this secret knowledge of the Pythagoreans, his brethren, according to legend, threw him off a ship and drowned him.

Another crisis in mathematics happened around 1850 and concerned the discovery of non-Euclidean geometries. In his book *The Elements*, the Greek mathematician Euclid (325–265 BC) organized geometry by providing an axiomatic system, that is, a limited number of basic assumptions (axioms) from which he thought it possible to derive all other true geometric propositions. His system was based on five axioms, propositions regarded as self-evident, which he thought could not be proved from simpler ones. One of the five axioms, the fifth, was problematic in the sense that it seemed not so self-evident as the others. This is the parallel postulate, which says that if a line A intersects two other lines, forming two interior angles (angles facing each other) on the same side that sum to less than 180°, then the two lines, if extended indefinitely, must meet (intersect) on that side of A. In Euclidean geometry, it is assumed that on the other side of A the two lines will diverge, that is, get farther and farther apart and never intersect. Or, if the angles sum to 180° exactly, then the lines are parallel, and never intersect on either side of A. What is problematic with the parallel postulate is the involvement of the notion of infinity. The axiom posits the possibility of an intersection that may happen at an infinite distance and therefore cannot be observed. For this reason the fifth axiom was never considered, even by Euclid himself, as being as self-evident as the others.

Due to these considerations, mathematicians of the nineteenth century thought that Euclid's fifth postulate was not an axiom but a theorem, that is, a proposition that could be proved—in some way not yet discovered—from the other four axioms, which were considered self-evident. Several attempts were made to prove that the fifth postulate was a theorem, but all failed. The result of these efforts, however, was a breakthrough: it was realized that while the first four axioms plus the fifth form a logically consistent system, Euclidean geometry, the first four axioms plus a statement which contradicts Euclid's axiom can form a logically consistent alternative system, a non-Euclidean geometry. Today logicians express this fact by saying that the fifth postulate is independent of the other four postulates. The first person to realize this fact was German mathematician and scientist Johann Carl Friedrich Gauss (1777–1855), but he never published his work. Euclidean geometry was still assumed at that time to be the geometry of the real space, but mathematicians were realizing that non-Euclidean geometries could be constructed that were not self-contradictory.

The discovery of the possibility of non-Euclidean geometries, which contradicts human intuition (e.g., in a non-Euclidean geometry the shortest distance between two points may not be a straight line), diminished mathematicians's confidence in intuition—one's sense of rightness or obviousness—as a foundation of mathematical knowledge and so promoted the study of mathematics using formal logic. It was logic that had shown that geometries other than Euclidean geometry (which scientists and mathematicians still thought described

This hand-colored woodcut appeared in a section on mathematics in the book *Margarita Philosophica* (The Philosophical Pearl) by Gregor Reisch (1467–1525). First printed in 1503, this book was once considered the "tree of all knowledge." The illustration shows computations being done with the aid of counters. *Image Works.*

real space) were possible; now, more emphasis was put on the study of the properties of different axiomatic systems, giving birth to a variety of new geometries.

Until this time, Euclidean geometry had been given a special status among mathematical disciplines, as it seemed to be directly justified by intuition based on spatial experience. For this reason, concepts in mathematical analysis were given a geometrical interpretation. But in the nineteenth century, work by many mathematicians, including the German Karl Weierstrass

(1815–1897) showed that it was possible to interpret notions such as limit, integral, and derivative (from calculus) in terms of statements about the real numbers rather than geometrically. By the end of the century, it had also been shown that all real numbers could be represented as sums of rational numbers. As rational numbers were known to be representable as pairs of natural numbers, the problem that was left was to explain what natural numbers are. For some, this question was simply not answerable: natural or counting numbers were fundamental, not further reducible. This was the opinion, for example, of Leopold Kronecker (1823–1891) who said that "God created the whole numbers; everything else is the work of man." Some others, like German mathematician Richard Dedekind (1831–1916) and German mathematician and philosopher Gottlob Frege (1848–1925), strove to find a further reduction by logical means.

Birth of Modern Logic

Logic was already an old discipline, part of philosophy since its invention by Greek philosopher Aristotle (AD 384–322) as part of the science of valid inference. It remained mostly unchanged until the nineteenth century when, thanks to the discovery of non-Euclidean geometries, the great potentiality of logical reasoning became clear.

A major advancement of logic, one fated to greatly influence not only foundations of mathematics research but also mathematics and philosophy in general, was achieved by Frege. In 1879, Frege, whose work was not widely recognized during his life, published a pamphlet called "The Concept Script" (*Begriffschrift* in German). This work, in which he sought to describe what he saw as the necessary laws of thought, is generally regarded as the birth of modern mathematical logic. For Frege, logic had the same role as the microscope had for scientific research: it was a more refined, precise way of seeing. In studying the fundamental properties of mathematics one cannot use natural language: it is too imprecise. Frege therefore invented an artificial language, a new ideography (system of signs standing for ideas), with the intent of rigorously describing logical concepts. His hope was to avoid the disadvantages of natural languages, such as vagueness and ambiguity. These characteristics of natural language were handicaps if one wanted to express and study the links between formulas in a mathematical demonstration. In 1884, Frege published the *The Foundations of Arithmetic* (*Grundlagen der Arithmetik*) with the purpose of showing that arithmetic (considered the most fundamental part of mathematics) was founded on logic alone, that arithmetical truths did not need support from empirical facts, and that mathematical intuitions were not in need of empirical confirmation. His thesis that arithmetic could be reduced to logic came to be known as logicism. According to the logicist thesis, all the things that mathematics speaks about can be treated as purely logical entities.

In his attempt to demonstrate that mathematics can be treated as a branch of logic, Frege implicitly used the informal notion of class, which corresponds in more modern mathematics to the concept of set. The notion of class was considered by Frege a logical notion, not a mathematical one. Using classes he was able to provide a definition of the most fundamental notion of arithmetic, that of number. For Frege, numbers were (or could be reduced to) classes: for example, the number two is the class of all classes that have two elements; the number three is the class of all classes having only three elements, and so on.

In his essay "The Basic Laws of Arithmetic" (*Die Grundgesetze der Arithmetik*), published in two volumes in 1893 and 1903, Frege finally presented his reconstruction of arithmetics based on logic alone, setting out in a rigorous way logical laws and rules from which it was possible to demonstrate, step by step and without appeal to any extra assumptions, arithmetical truths.

Frege's logicist program, if successful, would have proved that the certainty of mathematical knowledge does not derive from intuition or from empirical fact.

IN CONTEXT: RUSSELL'S PARADOX

In 1901, British philosopher Bertrand Russell (1872–1970) discovered a simple contradiction in Gottlob Frege's (1848–1925) logical system, which attempted to show that all mathematics can be reduced to logic. Russell wrote to Frege on June 16, 1902, announcing the news. Frege promptly replied, relating the paradox to the system in his work *The Basic Laws of Arithmetic*. In particular, Frege diagnosed the problem as concerning his Law V.

Law V says that any two concepts *G* and *F* are identical if exactly the same objects fall under them. For example the number 2 is the only number falling under the concept "square root of 4" and under the concept "even prime number;" thus, the two concepts are identical.

Implicit in Law V was the assumption that what falls under a concept forms a set. This opens the possibility of assigning a set of objects to any linguistically defined concept: namely, the set of the things to which the concept is applicable. For example, the set 2, whose only element is the number 2, extends the concept "square root of 4."

What Russell's paradox reveals is that there is not a set corresponding to every linguistically defined concept.

The second volume of the Grundgesetze was already in press and Frege could only add an appendix in which he says that his Law V "is not so evident, as the others." That was a classic understatement: in fact, the flaw undermined his whole effort to establish mathematics entirely on a foundation of pure logic.

Despite Frege's failure to establish a logicist foundation for mathematics, his mathematical logic became a standard tool for foundational research, and its invention is today recognized as one of the main philosophical achievements of Western philosophy, not only because of its application to the foundations of mathematics but because it triggered a series of developments that led to the invention of modern digital computers.

This thesis contrasted with the opinion of almost all mathematicians and philosophers of that time, who, influenced by the thought of the German philosopher Immanuel Kant (1724–1804) believed that mathematics rested on intuitions of space (in the case of geometry) and of time (in the case of arithmetic).

Frege's program failed. Just before the publication of the second volume, British mathematician and philosopher Bertrand Russell (1872–1970) found a contradiction in his system, the flaw known today as Russell's paradox. The contradiction was due to by Frege's use of the informal notion of class.

The Birth of Set Theory

In his logical system, Frege used the notion of "class," what today is called "set" (a collection of things) and it was the way in which this notion was used that turned out to be paradoxical—an unavoidable flaw in Frege's effort to reduce mathematics to logic. Russell's paradox exposed a flaw in the naïve conception of class. The notion of set started to be used at that time also by other mathematicians, mainly as a tool to solve mathematical problems, as in the case of the German mathematician Georg Cantor (1845–1918). Cantor, beside employing sets for solving mathematical problems, initiated the study of set-theoretical concepts, giving birth to what is today known as set theory.

Cantor's set-theoretical language turned out to be a powerful tool, as many problems in mathematics could be formulated as problems involving sets. Yet Cantor's theories were also affected by the discovery of Russell's paradox. Both theories were based on the possibility connecting linguistic expressions with sets, while Russell's paradox shows that the notion of set being used by the theories is paradoxical (in some circumstances contradicts itself). Russell showed that a set of objects, no matter how abstract, does not exist for every set definable in words. For example, there is no set corresponding to the concept of "set of all sets" or to the concept of "set of all sets that have the property of not being members of themselves." Suppose such a set—let us call it *A*—is given. *A* contains all the sets that have the property of not being members of themselves. Russell asked: Is *A* an element of *A* (that is, is it an element of itself)? Both the assumption that *A* is a member of *A* and the assumption that *A* is not a member of *A* (the only two possible answers) lead to a contradiction: *A* is a member of itself if *A* has the relevant property of not being a member of itself (a contradiction), and if *A* is not a member of *A*, then *A* has the relevant property of not being a member of itself and is a member of itself—again, a contradiction.

Russell was not the first to discover the paradoxicality of the Cantorian and Fregean notion of set; Cantor himself was aware of the fact even before Russell's discovery. However, Russell's was the simplest of the paradoxes thus far formulated and the most difficult to resolve.

The lesson drawn was that the notion of set must be regulated by axioms that state which sets exist or can be constructed starting from previously given sets. In 1908, German mathematician Ernst Zermelo (1871–1953) gave for the first time an axiomatic system for Cantor's set theory. Later on, many other mathematicians extended Zermelo's system or gave different axiomatizations of set theory. German mathematician Abraham Fraenkel (1891–1965), Hungarian mathematician John von Neumann (1903–1957), Swiss mathematician Paul Bernays (1888–1977), and Austrian mathematician and logician Kurt Gödel (1906–1978) are all important figures in the development of axiomatic set theory.

Cantor's theory of sets introduced a new way of doing mathematics and a new philosophy of the infinite. One of the most fascinating ideas that Cantor introduced in mathematics, one that been important in foundational research, is that there are different orders of infinity: that is to say, not all infinite sets are equivalent in size. In 1874, Cantor published his proof that there are as many rational and algebraic numbers as natural numbers but that the set of real numbers is strictly bigger in size—although all these sets contain an infinite number of members. This was a striking result, as all infinite collections had previously been considered the same size. Even more striking than the result itself was the way Cantor arrived at his conclusion, that is, by means of showing that it was impossible to enumerate all real numbers. In short, it proved that something was true by showing that something else was impossible.

Kronecker, who was a member of the editorial staff of *Crelle's Journal*, where Cantor's proof was published, did not like the revolutionary new ideas contained in Cantor's article and tried to prevent Cantor's later work from being published in the journal. Kronecker, like many mathematicians of his time, only accepted mathematical objects that could be constructed starting from the numbers that he believed were intuitively given, namely the natural or counting numbers. Cantor's ideas were, in Kronecker's eyes, meaningless because they were about objects that for him did not exist.

In developing his theory of infinite cardinalities—orders of infinities, some larger than others—Cantor conjectured that the set of real numbers (the continuum) is the smallest possible set that is strictly bigger than the set of natural numbers. This is termed the continuum hypothesis. Cantor sought to prove that this conjecture was correct, but all his attempts were unsuccessful, and the question was left for future mathematicians to resolve.

A few decades later, using mathematical logic techniques, Gödel proved that the continuum hypothesis is consistent with the axioms of what is nowadays

GEORG CANTOR (1845–1918), CREATOR OF TRANSFINITE SET THEORY

Georg Ferdinand Ludwig Philipp Cantor (1845–1918) was born in St. Petersburg, Russia, and died in Halle, Germany. He started to study mathematics at the Polytechnicum in Zürich in 1863, but soon, after the death of his father, moved to Berlin where he attended the lectures of Weierstrass and Kronecker, among others. At the University of Berlin, he completed his dissertation on number theory in 1867. In 1869 he was appointed to the University of Halle, where he presented his thesis, again on number theory, and received his professorial teaching qualification.

At Halle, Cantor turned his attention to the branch of mathematics known as analysis, which deals with all questions relating to convergence and limits. He did so because his senior colleague Heine asked him to prove a difficult problem on which he had worked, unsuccessfully, himself: the uniqueness of representation of a function as a trigonometric series. The problem was solved by Cantor by April 1870. It was this work that prompted the discovery of transfinite numbers. In 1874, he published an article in *Crelle's Journal* that is considered to have founded set theory, and between 1879 and 1884 he published a series of six papers in *Mathematische Annalen* providing a basic introduction to set theory.

The year 1884 was one of crisis for Cantor, the year of his first serious mental breakdown. The crisis may have been due to his inability to prove the continuum hypothesis, or to stress: in any case, it lasted about one month. After that Cantor changed his attitude, becoming more interested in philosophy and seeking to teach philosophy instead of mathematics. Further breakdowns occurred, and in 1899 Cantor was again hospitalized for mental instability. The last period of his life he spent in a sanatorium.

considered the standard axiomatic system for set theory, namely ZFC (Zermelo-Fraenkel with axiom of choice). This means that if we add to ZFC as a further axiom the statement that the continuum hypothesis is true, we do not obtain any contradiction. Later on, American mathematician Paul Cohen (1934–) proved also that the negation of the continuum hypothesis plus ZFC constitutes a consistent mathematical system. Together these results mean that the question of the truth or falsity of the continuum hypothesis cannot be settled by means of the ZFC axioms.

The Foundational Crisis of the Twentieth Century

By the beginning of the twentieth century, the mathematical and philosophical landscape had been greatly changed by the invention of mathematical logic and set theory. These proposed two new paradigms of doing mathematics. Logic, no longer seen as a part of philosophy, became a standard tool in mathematical investigations. Mathematics, because of the introduction of set theoretical concepts and techniques, became more abstract and no longer dealt fundamentally with numbers. This innovation was accompanied by complex foundational crises driven by the paradoxes that affected the basic notion of set.

These paradoxes, far from halting the progress of mathematics, had the effect of promoting the growth of modern logic. The language and methods of logic began to be extensively employed to give a precise and economical shape to mathematical theories. The expression of large parts of mathematics in a formal, axiomatic framework had many advantages: it was possible to make explicit all assumptions involved in mathematical reasoning, diminishing and even dismissing completely the importance of intuitive processes of thought. Intuition was thus extruded from formal mathematics in the sense that it could not be used to justify mathematical concepts or to prove the truth of mathematical results.

Around 1900, Italian mathematician Giuseppe Peano (1858–1932) devised a system of five postulates from which the entire arithmetic of the natural numbers could be derived:

1. 0 is a natural number (i.e., a counting number: non-negative whole number).

2. Every natural number has a successor.

3. No natural number has 0 as its successor.

4. Distinct natural numbers have distinct successors (i.e., no two natural numbers have the same successor).

5. (Induction axiom) If a property holds for 0, and holds for the successor of every natural number for which it holds, then the property holds for all natural numbers.

According to Bertrand Russell, Peano's postulates implicitly define what we mean by a "natural number." However, French mathematician Henri Poincaré (1854–1912) maintained that Peano's axioms only defined natural numbers if they were consistent, that is if no self-contradictory statement of the form "*P* and not *P*" could be proved inside the system. If such a proof exists, then the axioms are inconsistent and cannot be said to define anything.

David Hilbert (1862–1943), the leading personality in mathematics in this period, formulated a foundational program aiming to show that Peano's system was free from contradictions. Hilbert posed the problem of proving the consistency of arithmetic. In his famous list of unsolved mathematical problems, put forth at the International Congress of Mathematicians in 1900 in Paris, the problem of the consistency of arithmetics was second.

Hilbert's Program and Gödel's Incompleteness Theorems

Some mathematicians felt that the new non-intuitive methods and the mathematics that were the outcome of their application were devoid of mathematical meaning. But for most mathematicians, especially for Hilbert, it would have been absurd to renounce the powerful methods made available by Cantor's invention. Still, the paradoxes were present. In 1922, Hilbert launched a foundational program, wanting, as he said, to "settle the question of foundations once and for all." Stressing the importance of the use of formal axiomatic systems, he hoped to establish the noncontradictoriness of the formalized mathematical theories. The peculiarity of his program, what has come to be known as formalism, was that it required that a proof of noncontradictoriness should be produced by methods involving only combinatory relations among the linguistic symbols used to express mathematical statements. Such a proof would have secured the result against the radical criticism of those who, like Dutch mathematician Luitzen Egbertus Jan Brouwer (1881–1966), required for mathematics a more concrete, computational meaning.

It was the young Gödel who proved that Hilbert's dream was not realizable, at least not in the form desired by Hilbert. In 1931, Gödel proved a pair of theorems, called the first and second incompleteness theorems, which together set a strict limit on the power of logic. The second theorem particularly affected Hilbert's program by stating that the noncontradictoriness of Peano's axiom system cannot be proved by using means whose strength is weaker or equivalent to the system itself. In other words, no nontrivial mathematical system of statements can completely prove its own correctness. This result had important foundational repercussions, as it showed that Hilbert's program could not be carried out in its original form.

■ Modern Cultural Connections

The effort to found mathematics on a solid basis has continued to this day, as mathematicians develop and extend the essential innovations of the beginning of the twentieth century. Today foundational studies overlap with mathematical logic as articulated in model theory, axiomatic set theory, proof theory, and recursion (or computability) theory. Foundational studies are also of great importance for philosophy. Different positions concerning the foundations of mathematics have been defended, among them logicism (carrying on Frege's foundational work) and formalism (carrying on Hilbert's). Besides these, we have Platonism, a widespread position among mathematicians and scientists according to which one should act as if mathematical entities enjoyed actual existence; intuitionism, introduced by Brouwer, in which it is maintained that the primary source of mathematical knowledge is our intuitive sense that mathematical objects are real; and mathematics-as-language, according to which mathematics is produced by the imagination and operates like a language. Today there is still no universal agreement on what mathematics is: mathematics remains mysterious, even as mathematicians continue to produce new proofs, invent new types of mathematics, and gain new insights into ancient problems. A pragmatic common-ground position might be that mathematics is simply what mathematicians do—although it may be more.

In a sense, foundations of mathematics remain irrelevant to modern science. Physicists, engineers, and other scientists who employ mathematics do not need to know what mathematics "is": they only need to know what kind of mathematics works for them and describes the behavior of the world. Nevertheless, physicists are often compelled to wonder what mathematics "is," a question called by Hungarian physicist Eugene Wigner (1902–1995) the "unreasonable effectiveness of mathematics in the natural sciences." Repeatedly, modern physicists have sought mathematical systems to describe subtleties of the physical world, especially in quantum physics and relativity, only to find that mathematicians had already produced such systems for purely mathematical reasons, without any regard to their possible uses in physics. For example, non-Euclidean geometries were developed before Albert Einstein (1879–1955) proposed, in his theory of general relativity (1915), that the geometry of the physical world is actually non-Euclidean (though it closely approximates Euclidean geometry over short distances).

Foundations of mathematics, although having no direct application in physics or technology, have driven progress in mathematics itself by forcing mathematicians to examine exactly what they are doing. The resulting advances in mathematics are, if history is any guide, likely to have tangible benefits for the physical sciences sooner or later. As has already been noted, the development of set theory was an essential precursor to the exploitation of the full potential of the modern digital computer. Simple mechanical calculators could be and were in fact constructed before the development of set theory, but modern computer science relies essentially on many branches of mathematics that draw upon set theory, number theory, computability theory, and the like.

Among the questions that are actively considered in foundational studies today, some, like the continuum hypothesis, were formulated by the founders of mathematical logic and set theory. However, the field of foundations of mathematics has expanded and diversified, taking in elements of formal philosophy and a number of mathematical fields. The foundational crisis of the early twentieth century has not resulted in a clear victory for any one of the contending points of view—logicism, Platonism, formalism, intuitionism, or mathematics-as-language; rather, these views have continued to contend, but with a lessened sense of urgency, as it has become clear that regardless of which of these views (if any) ever does prevail, mathematics itself will continue, both on its own terms and as essential tool for all other scientific disciplines.

SEE ALSO *Mathematics: The Specialization of Mathematics.*

BIBLIOGRAPHY

Books

Dauben, Joseph W. *Georg Cantor: His Mathematics and Philosophy of the Infinite.* Princeton, NJ: Princeton University Press, 1979.

Hatcher, William S. *Foundations of Mathematics.* Philadelphia: W.B. Saunders Company, 1968.

Mancosu, Paolo. *From Brouwer to Hilbert. The Debate on the Foundations of Mathematics in the 1920s.* Oxford: Oxford University Press, 1998.

Russell, Bertrand B. *The Autobiography of Bertrand Russell.* Crows Nest, New South Wales, Australia: Allen & Unwin, 1967.

Periodicals

Sarukkai, Sundar. "Revisiting the 'Unreasonable Effectiveness' of Mathematics." *Current Science* 88 (2005): 415–423.

Wigner, Eugene. "The Unreasonable Effectiveness of Mathematics in the Natural Sciences." *Communications in Pure and Applied Mathematics* 13 (1960): 1–14.

Giuseppina Ronzitti

Mathematics: Measurement

■ Introduction

All modern science relies on measurements of observable properties of specific events or objects. Typically, a single measurement yields a number associated with a unit: for example, the number-unit pair "10 meters" could be the result of a distance measurement. A number-unit pair produced by making a measurement is called a datum; when there is more than one datum, they are called data. The numerical part of a datum records unique information, while the unit tells what sort of phenomenon the information refers to: in the datum "10 meters," the "10" is information, while "meters" tells us how to interpret that information. ("10 feet" is interpreted differently from "10 meters"; the units tell us that it describes a shorter distance.)

Data produced by measurements usually have some degree of uncertainty. For example, if a ruler that is only accurate to the nearest centimeter is used to measure a distance, then it is not enough to report "10 meters"; a scientist must record that the distance is 10 meters give or take 1 centimeter.

Without precise measurements whose uncertainties are well-understood, science could not disentangle the web of cause and effect that is the physical world. And although science is far more than the mere accumulation of measurements, without measurement science could not exist. Millions of phenomena are measured throughout all scientific fields: the weight of a mouse, the response time of a human subject, the brightness of a star, the temperature of a breeze.

■ Historical Background and Scientific Foundations

The development of measurement systems is one of the oldest scientific concerns, but the earliest standardized

units were developed primarily for trade and architecture, not for the sake of scientific measurement. As early as 3000 BC, early civilizations had developed certain units as common reference points to provide clarity where exchange of goods or coordination of activities was desired. Three of the most fundamental areas that people began to measure were time, length, and weight.

Most early measurement systems grew out of pre-existing patterns in nature, such as the length of the day or the size of a human foot or arm as a measure for length. These standards of measure were inexact and varied both within and between cultures. Since the 1700s, measurements have become increasingly uniform and exact, with all nations across the globe agreeing (for scientific purposes) on a single measurement system, the metric system. As scientific instrumentation became more complex, allowing for measurements on subatomic and galactic scales, scientists began to make measurements with greater precision than thinkers of earlier ages could have foreseen and of phenomena they could not have imagined.

Ancient Astronomical Calendars

Time was the first quantity for which ancient cultures invented units of measure: all human peoples experience the seasons, lunar cycles, rising and setting sun, and wheeling stars, and are motivated to keep track of them. Archaeological evidence suggests that even before developing written languages, prehistoric people tracked the passage of time by making marks on pieces of wood, stone, or bone to record the appearance of each full moon. The Sumerians, who lived in the area of the Near East known to archaeologists as Mesopotamia (now part of Iraq), were probably the first civilization to use calendars. Noticing that the cycle of seasons repeated every 12 full moons or so, the Sumerians counted 12 lunar months as a year. Other ancient civilizations also developed lunar calendars as well, identifying anywhere from two to five seasons depending on regional climate.

Although the moon's visibility and predictable orbit around Earth made it a strong choice as a basis for marking time, lunar calendars were problematic. Because the full moon appears about every 29.5 days, 12 lunar months contain only 354 days. As a result, lunar calendars fall increasingly out of sync with the seasons as the years pass. Different cultures have found different ways to accommodate the accumulating inconsistency.

Noticing that certain stars or constellations appeared in the sky at certain times of the year, some ancient astronomers began to use star patterns to measure the passage of time. Notably, the ancient Egyptians replaced their lunar calendar with the world's first solar calendar around the year 2700 BC. The Egyptians began each year with the rising of the star Sirius (now commonly called the Dog Star), which always occurred soon before the annual flooding of the Nile River. The Egyptians divided the year into 12 months of 30 days each, then added five days at the end to bring the calendar year to 365 days.

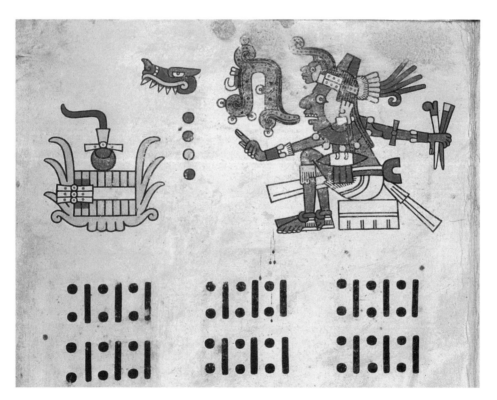

In the Codex Fejervary-Mayer (Mexico) a minor planetary deity is shown with bars and dots, the numerical system of the Aztecs (1350–1521 BC). Each bar represents five units and each dot one. When added together, these numbers probably represent the synodic period of the planet. *Werner Forman/Art Resource, NY.*

Ancient Devices for Measuring Time

The measurement of time intervals is an essential aspect of many scientific measurements, especially those involving events or processes. As we have seen, the notion of dividing time into intervals was almost inevitable, given the regular round of sky and seasonal phenomena that all cultures experience and their practical importance; however, such large, approximate units were not themselves useful for most scientific purposes, nor for many practical purposes, especially in cultures with complex economic affairs to coordinate. Devices for objectively marking the passage of smaller units of time—timepieces—were necessary. Around 3500 BC, the Egyptians began constructing obelisks that used the motion of the sun to tell time. An obelisk was a large pillar that cast a shadow during the day. Someone could estimate the time of day by looking at the base of the structure, which had markings to measure hours depending on where the obelisk's shadow fell at a time of day.

The hour was not yet an absolute unit of time, as it is today, but varied from day to day: each hour measured 1/12 the time from sunrise to sunset on that particular day. Because the amount of daylight varies from season to season, these shadow clocks actually measured "temporal hours," with longer hours in the summer and shorter hours during colder months. However, being near the equator, Egyptian sundials were fairly consistent year-round. Other civilizations in the Near East, Asia, and Europe later used sundials to tell time.

To tell time at night or during cloudy weather, the Egyptians invented the water clock around 1400 BC; the Chinese also used this type of clock. A water clock is a bucketlike device with a small hole near the base that allows water to escape. Markings on the inside of the bucket indicated the passing of units of time (such as hours). Because temperature changes cause the water to flow at different rates, water clocks could be imprecise, but they were independent of the seasons, clouds, and geographical location.

Water clocks were the first time-measuring devices that used an internal phenomenon (in this case, water flow) as a standard against which to mark the passage of time—the principle of all true clocks—and were used for many centuries. Italian physicist Galileo Galilei (1564–1642), widely considered the founder of modern physical science, used a type of water clock to measure time intervals while observing the motions of rolling balls and other objects in his laboratory. The equations of motion that he derived from these measurements are basic to modern physics. This illustrates the importance of measurement in the development of new theories in modern science, and the surprisingly important gains that can be made using even simple tools. In contemporary scientific developments, such simple measuring devices are no longer adequate, since extremely accurate

IN CONTEXT: THE PRIZE-WINNING CHRONOMETER

Early weight-driven and pendulum clocks were very inaccurate at sea due to temperature changes and the motion of ships. This was problematic because navigators relied on accurate timekeeping to locate their position at sea; the less accurate the clock, the farther off course a ship could go. In 1714, the British government offered a financial prize to whoever could develop an accurate method to determine a ship's longitude within 30 nautical miles at the end of a six-week voyage. To win the prize, the time-keeping device had to be accurate within three seconds per day, which was more accurate than any pendulum had been on shore. Over the next few years, numerous sailors tried to develop time-keeping devices, but none met the stiff criteria to win the competition.

After four unsuccessful attempts, during the 1720s carpenter and self-taught instrument-maker John Harrison (1693–1776) built a clock stable enough to withstand rough seas. Called a chronometer, Harrison's prize-winning invention lost only a few seconds in six weeks, earning him 20,000 British pounds (worth about $12 million today). Mechanically speaking, the device was essentially a well-made watch suspended in gimbals (rings connected by bearings) that remain horizontal even when a ship turns. Although advances in chronometry (time measurement) were motivated by economics and navigation, once available they quickly found use in the sciences, especially in astronomy.

measurements of hard-to-detect quantities are often required to distinguish between rival scientific theories: mechanical measuring devices of great subtlety are needed. The mechanical clock, developed relatively recently, is the direct technological descendent of such devices.

Mechanical Clocks and Time Measurement

Time measurement advanced greatly during the 1300s, with the invention of mechanical clocks. The earliest known mechanical clocks were in Milan, Italy, by 1309. Large and made of iron, these earliest clocks had no faces or hands, but relied on an attendant to strike a bell for the hour. In 1335, Milan had the first clocks that struck automatically. Clocks became fixtures in the church bell towers of European cities by the end of the Middle Ages, keeping everyone within earshot on the same time. These early clocks relied on a heavy weight attached to a chord tightly wound on a spool. As gravity caused the spool to unwind, it turned sets of gears that moved hands on the clock of the face. A counterweight at the other end of the rope triggered a hammer that

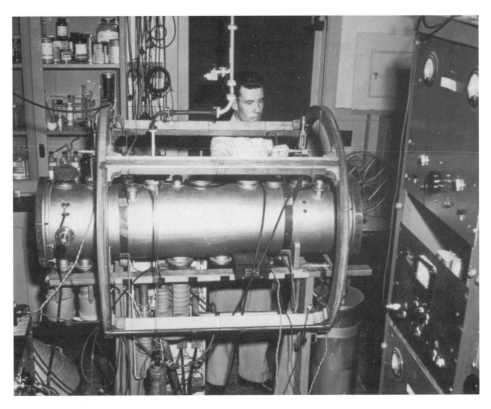

Atomic clock.   © *Corbis.*

struck a bell to ring when the clock reached the start of a new hour. Early clocks didn't keep very accurate time, losing as much as an hour a day and requiring regular adjustments, but the technology kept improving. The first household clocks appeared by 1400, and timepieces small and sturdy enough to easily carry on the person— pocket-watches—became possible a century later.

Increasingly accurate chronometry made fundamental scientific discoveries possible: for example, by the late 1600s clocks were accurate enough for astronomers to note that the movements of Jupiter's moon Io seemed to shift in time depending on where Earth was in its orbit. Danish astronomer Olaf Roemer (1644–1710) realized that the time-shifts could be explained by the extra time it took light, moving at a finite speed, to reach Earth from Jupiter and its moons when Earth's orbit took it farthest from Jupiter. By using the improved mechanical clocks of his day, Roemer was able to collect data on the time-shift of Io's eclipses that later astronomers could use to make a reasonable estimate of the speed of light.

Following the Babylonian mathematical tradition of dividing a circle into 360 degrees, early round-faced clocks—establishing a standard still followed today— divided hours into 60 minutes, and then into 60 seconds. This made for 3,600 seconds every hour, and 86,000 seconds in a 24-hour day. That method seemed to work until 1820, when a committee of French scien-

tists pointed out that because of Earth's elliptical orbit, the length of one day can vary by few thousandths of a second from one day to another, depending on the planet's position around the sun. Scientists therefore measured a year of days to find the average length, then defined a second as a specific fraction of that average day. This calculation turned the second into an exact scientific measurement in itself rather than a small slice of a larger measurement. The decision seemed to work until 1956, when scientists discovered that Earth's rotation is slowing down slightly each year by about 7.3 milliseconds. Meanwhile, the solar year is getting shorter by about 5.3 milliseconds each year.

The invention of the atomic clock during the mid–1900s greatly refined the measurement of time. The atomic clock depends on the vibrational frequency of certain atoms. With the heightened accuracy of such clocks, scientists can measure time by fractions of a second far too small for people to perceive in daily life. Once again, improvements in the ability to measure time gave scientists new power to test scientific theories: for example, several predictions of the theory of relativity about the nature of space and time have been verified using atomic clocks. These include, among many others, the prediction that a clock at the bottom of a tower will run slower than one at the top (because it is in a slightly more intense gravitational field) and that a clock that is moved about relative to a clock stationed on the ground will run more slowly than the stationary clock.

Origins of Distance and Weight Measurement

The measurement of distance and weight, along with the measurement of time, is one of the most fundamental tools of science. Around 2000 BC, the Sumerians began using a cubit as a standard measurement for length. A Sumerian cubit was the distance from a person's middle finger to the elbow, and a "foot" equaled two-thirds of a cubit. In the ancient Near Eastern city of Nippur, a copper bar was a standard measurement tool around 1900 BC. At 3.62 ft (110.35 cm) long (converting to today's measurements), the bar was divided into 4 "feet," each with 16 "inches."

Ancient standards for weight focused on small standards that were light enough for someone to carry. By 2400 BC, the Sumerian civilization was using a base unit of weight called a shekel, about 0.3 oz (8.36 grams). A larger unit, the mina, weighed 60 shekels. During this period, the Egyptians used balancing scales to develop standard measurements for weight. By hanging two pans on opposite ends of a hanging rod, the Egyptians could see if objects in the two pans weighed the same amount. The Egyptians placed seeds in these simple balances to weigh small items, such as precious gems and metals. The weight of one grain became the official measure (and is still used in the measurement of gunpowder in North America).

The Metric System

Many systems of units for weight, distance, and volume evolved in Europe and elsewhere. Over the centuries, a slow trend toward standardization was driven by the economic benefits of having a common system; this trend also, as we have seen in the case of time measurement, benefited science by making it possible for different researchers to compare results and to test each other's experiments based on written descriptions of what was done. The concept of the measurement unit slowly evolved from an improvised marketplace convenience to a system of universally acknowledged quantities allowing rigorous comparisons between observations widely separated in time and space. The peak of this process was reached with the construction of the metric system.

From the time of Charlemagne (Charles the Great; c.742–814), who ruled France from 771 to 814, numerous European leaders issued decrees that modified regional weights and measures. This legacy made work difficult for travelers and traders, who readily had to convert back and forth between different systems. Commercial trade, as well as scientific investigation, required

more uniformity. As a solution, mathematician Gabriel Mouton (1618–1694) proposed a standard unit of length based on the length of the meridian (imaginary curved line on the Earth's surface from the North Pole to the South), which astronomer Jean Picard (1620–1682) had determined to reasonable accuracy in 1670. In the late 1700s, as the Industrial Revolution increased the need for standard-sized parts, Mouton's idea caught on. In 1791, a commission of scientists led by astronomer Joseph Lagrange (1736–1813) tackled the issues. The result was the metric system. (The word derives from "metron," which means "to measure" in Greek.) Following Mouton's advice, the commission divided the meridian's length by 10,000,000 and called that distance a meter. The standard unit for weight, meanwhile, became the gram.

Based on multiples of 10, metric measures allowed for great ease in moving from larger and smaller units. For example, 1,000 times the weight of a gram is a kilogram; 1/100 the length of a meter is a centimeter. France began using the metric system in 1795, and several other European nations switched to metric in the decades that followed.

Over time, in response to scientific and industrial need for ever more precise and measurements, the metric system has been altered to make it more exact. For example, in the mid-twentieth century, officials at the International Bureau of Weights and Measures decided that defining the meter in terms of the size of Earth was too imprecise. The meter was temporarily redefined on the basis of the frequency of the light emitted by a particular isotope of the element krypton. In 1983, scientists devised a new basis for the meter from the speed of light in a vacuum. The length of a meter was redefined as the distance that light travels in a vacuum in 1/299,792,458 of a second, based on the 1967 definition of a second. This decision brought the measurement for time into the metric system—almost. Suggestions to develop a metric clock (with 10 hours in a day, 100 minutes in an hour, and 100 seconds in a minute) have never caught on: the Babylonian habit of using factors and multiples of 60 remains embedded in our most advanced chronometers.

■ Modern Cultural Connections

Measurement of fine variations in forces, radiation intensities, and time intervals continues to drive progress in some of the most fundamental questions facing modern physical science. For example, as of 2007 efforts were underway to measure the phenomenon of frame dragging predicted by the theory of general relativity. Frame dragging is a shift in the inertia properties of objects that are near to large, spinning masses (such as planets or stars), where inertia is the tendency of an object to remain in its state of motion unless acted upon by a force—a fundamental property of all matter. Like scores of earlier tests of relativity, these measurement studies (which use satellites orbiting Earth) seek to answer questions about the nature of space, time, and the properties of matter.

On a larger scale, cosmology (the study of the structure of the universe as a whole) is enlivened today by debates about the existence and nature of "dark matter" and "dark energy." In the late 1990s, measurements of light from very distant exploding stars showed that the acceleration of the universe is, contrary to scientists' expectations, accelerating: the energy driving this acceleration is termed "dark energy" because its nature remains mysterious. A number of measurements by satellite-based instruments are being planned that will resolve or at least constrain the nature of dark energy: these include measurements of the matter density of the universe (how much matter it contains, on average, in each unit of its volume) and fine variations in the cosmic microwave background radiation.

Nor are the questions being illuminated by measurement restricted to cosmic physical problems that most people cannot understand. Millions of separate measurements of temperature and the traces of ancient climate left in ancient ice layers or other materials have shown, in recent decades, that the climate of Earth is changing in response to greenhouse gases released by human activities. Although for some years the correctness of this view was disputed, it has been firmly established by an intense, international program of climate measurement, paleoclimatic proxy measurement, advancing knowledge of climate physics, and computer simulation, with effects on world political views and trends that are still only beginning to be felt.

So important is measurement to the conduct of modern science that a specialized field termed measurement theory has developed. Measurement theory systematically examines the ways in which numbers are assigned to phenomena during the making of scientific measurements. The effects of error and uncertainty are of particular interest in measurement theory, given that imperfection is inescapable in the making of measurements and that entire physical theories may stand or fall on fine variations in measured data.

SEE ALSO *Astronomy and Cosmology: Western and Non-Western Cultural Practices in Ancient Astronomy.*

BIBLIOGRAPHY

Books

Aveni, Anthony. *Empires of Time: Calendars, Clocks, and Cultures.* New York: Basic Books, 1989.

Barnett, Jo Ellen. *Time's Pendulum: From Sundials to Atomic Clocks, the Fascinating History of Timekeeping and How Our Discoveries Changed the World.* San Diego: Harcourt Brace & Company, 1998.

Bendick, Jeanne. *How Much and How Many? The Story of Weights and Measures.* Rev. ed. New York: Franklin Watts, 1989.

Blocksma, Mary. *Reading the Numbers: A Survival Guide to the Measurements, Numbers, and Sizes Encountered in Everyday Life.* New York: Viking, 1989.

Jones, Tony. *Splitting the Second: The Story of Atomic Time.* Bristol and Philadelphia: Institute of Physics Publishing, 2000.

Hebra, Alex. *Measure for Measure: The Story of Imperial, Metric, and Other Units.* Baltimore: Johns Hopkins University Press, 2003.

O'Malley, Michael. *Keeping Watch: A History of American Time.* New York: Viking, 1990.

Sobel, Dava. *Longitude: The True Story of a Lone Genius Who Solved the Greatest Scientific Problem of His Time.* New York: Walker and Company, 1995.

Steel, Duncan. *Marking Time: The Epic Quest to Invent the Perfect Calendar.* New York: John Wiley and Sons, 2000.

Zupko, Ronald Edward. *Revolution in Measurement: Western European Weights and Measures since the Age of Science.* Philadelphia: The American Philosophical Society, 1990.

Periodicals

Fernandez, M.P., and P.C. Fernandez. "Precision Timekeepers of Tokugawa Japan and the Evolution of the Japanese Domestic Clock." *Technology and Culture* 37 (1996): 221-248.

Web Sites

Astronomy Department, Cornell University, Ithaca, New York. "Timekeeping." http://curious.astro.cornell.edu/timekeeping.php (accessed May 8, 2008).

James Satter

Mathematics: Probability and Statistics

■ Introduction

Probability and statistics are the mathematical fields concerned with uncertainty and randomness. Since uncertainty is part of all measurement—no instrument is perfect—and all scientific knowledge is ultimately tested against measurements, probability and statistics necessarily pervade science. They are also found throughout engineering, economics, business, government, artificial intelligence, and medicine. Indeed, in today's world most people grapple with statistical concepts and claims in everyday life. For example, if the weather forecast predicts a 30% chance of rain, should I grab an umbrella as I go out? If a study reports that eating a certain fast food as often as I do increases the probability of my contracting heart disease by 10%, should I change my diet? What if the claimed risk increase is 80%?

Although human beings have always weighed probabilities intuitively when making decisions with uncertain outcomes, the application of mathematical methods to uncertainty is only a few centuries old. Early mathematicians did not see uncertainty as a natural topic for mathematics, which was thought of as the science of pure certainty, of absolute knowledge: The idea of deliberately introducing uncertainty into mathematics would probably have struck early philosophers as ridiculous or repulsive. Nevertheless, probability and statistics are today essential aspects of applied mathematics. For example, the behavior of the elementary particles of which the universe is composed is, according to the interpretation most common among physicists today, random in its essence and can only be described using probabilistic concepts.

■ Historical Background and Scientific Foundations

The Beginnings of Probability

The first mathematical treatments of probability arose from a desire to understand—and to triumph in—games of chance. Suppose, for example, that two gamblers play a dice-rolling game in which the rule is that the first player to win the game three times claims the wager. Now suppose the game is interrupted after it has begun but before one of the players has won three times. How should the wager be divided in order to fairly recognize the current positions of the two players? In other words, if the first player has won two games and the second player has won one game, what portion of the wager belongs to the first and what portion to the second?

This question was posed to French mathematician Blaise Pascal (1623–1662) by Antoine Gombaud, Chevalier de Méré (1607–1684), a fellow Frenchman and writer with a thirst for gambling. Similar problems had been addressed by various fifteenth and sixteenth century mathematicians, but without significant success. Pascal himself communicated the problem to another brilliant French mathematician, Pierre de Fermat (1601–1665). Both men proposed solutions to the problem and exchanged a series of philosophical letters during the summer of 1654. In this correspondence, Pascal and Fermat inaugurated the modern study of mathematical probability.

Engraving of French mathematician Pierre de Fermat (1601–1665). *SPL/Photo Researchers, Inc.*

A Dutch scientist and mathematician, Christiaan Huygens (1629–1695), learned of the Pascal-Fermat correspondence and published the first printed version of the new ideas developing in mathematical probability in a 1657 book called *De Ratiociniis in Ludo Aleae* (On Reasoning in Games of Chance).

Pascal also provided us with one of the most ambitious claims ever based mathematical probability—the claim that belief in God is mathematically justified. "Pascal's Wager," as his argument is called, is an early attempt to apply what mathematicians today call decision theory. Pascal claimed one could choose to either believe in God or not believe in God. Now, presumably God exists or does not. If one believes in God yet God does not exist, Pascal argued, no true loss is incurred. And if one does not believe in God and God does not exist, again nothing is lost. However, if one chooses to believe in God and God does exist, the rewards are infinite, and if one chooses not to believe in God yet God does exist, the penalty is immense. Pascal's conclusion: one should choose to believe in God.

It should be noted that Pascal's Wager is not the purely logical argument that it may seem at first sight. It depends, rather, on a number of assumptions about ethics, psychology, and theology, including the doctrine that God (if there is such a being) punishes disbelief, which not all religious groups affirm. In the case of Pascal's Wager, as with many arguments in the fields of game theory and decision-making that employ the mathematics of probability, the conclusion actually depends not only on mathematical reasoning but on implied claims about facts and values that may be questionable. In fact, so common is the cloaking of flawed arguments in the jargon and machinery of probability and statistics that the phrase "lies, damned lies, and statistics" has become a cliché. Experts and non-experts alike should scrutinize statistical arguments carefully, whether they happen to find their conclusions pleasing or not.

Basic Discoveries in the Theory of Probability

To modern persons, the probabilities associated with tossing a coin may seem obvious. If the coin is fair—not weighted to come up one way more often than the other—then the probability of getting "heads" on any given flip is ½ (0.5) and the probability of a "tails" is also 0.5. (Probabilities for given events are expressed as numbers between 0 and 1, with 0 denoting the probability of an impossible event and 1 the probability of a certain event.) Therefore, if one were to flip a fair coin 10 times one would expect to see about five heads and five tails. The word "about" is important: If you actually took a coin (presumed to be fair) out of your pocket and flipped it ten times, you would not think it bizarre to observe, say, six heads and four tails instead of five of each. If, on the other hand, you flipped a coin a million times and recorded about 600,000 heads and

400,000 tails, this would probably strike you as showing that the coin is not fair, even though the proportion of heads or tails to the total number of throws is the same as in the 10-flip case. Intuitively, we expect a fair coin to come up closer and closer to exactly half the more times it is flipped. The mathematical truth behind this intuition is termed the law of large numbers, which was first formalized by the Swiss mathematician Jacob Bernoulli (1654–1705). The law of large numbers essentially states that the probability of an event—such as recording a heads on a coin toss—is equal to the relative frequency of that event if the experiment is repeated a large number of times. Although this may seem obvious to many modern observers, understanding probability as an expression of relative frequencies over large numbers of trials was a breakthrough in the quest to create a mathematics of probability.

Bernoulli's statement of the law of large numbers was part of his attempt to define what he called "moral certainty." For Bernoulli, the achievement of moral certainty meant taking into account enough data or observations so that one could approach certainty as to probable outcomes. In his highly influential work, *Ars Conjectandi* (The Art of Conjecture), Bernoulli argued that if one were presented with an urn containing a large number of white pebbles and black pebbles, one could calculate the probable ratio of black pebbles to white pebbles by recording the color of pebbles drawn from the urn one at a time (with replacement of each pebble after it is drawn out and then restirring the urn). If the number of pebbles drawn were made large enough, the actual ratio of black to white pebbles could be approximated to any desired level of accuracy. Bernoulli chose to define a probability of at least 0.999 (one chance in a thousand of being in error) as the level required to conclude the outcome was correct with moral certainty. He proceeded to show mathematically how many trials were needed to produce such an outcome in a given situation. Today, the law of large numbers is central to many of the most important applications of probability theory.

Conditional Probability and Other Discoveries

Intuition can often be misleading in questions of probability. For example, if you have tossed a fair coin 6 times and it has come up heads every time—a very unlikely event—what is the chance, on the seventh throw, that it will come up heads yet again? The answer: 1/2. If the coin really is fair, then the chances of heads on any single toss are always 1/2, regardless of what low-probability series of throws may have come before. Another way of saying this is that the probability of getting heads on any given toss is not conditioned on the results of previous throws: It is independent of them. On the other hand, some probabilities are conditioned by other events: The probability that you will drown on

JACOB BERNOULLI (1654–1705)

Jacob Bernoulli was born in Basel, Switzerland in 1654. Although he was pressured by his father to study philosophy and theology, Jacob's true interest was in mathematics. After obtaining degrees in philosophy and theology from the University of Basel, he set out to study mathematics under some of the most important mathematicians in Europe. The University of Basel eventually offered Jacob the chair of mathematics, a position he would hold until his death in 1705.

Jacob was a member of perhaps the most remarkable mathematical family in history. He and his brother Johann made vital contributions to a wide array of mathematics in the late seventeenth and early eighteenth centuries. Three of Jacob's nephews were prominent mathematicians, as were two grand-nephews. The Bernoulli name was dominant in mathematics for over a century.

Jacob's work in the theory of probability provided a foundation for that discipline during its formative years. However, it is Bernoulli's work in calculus, another new (at the time) branch of mathematics, for which he is most often remembered today.

a given day is greater if you go swimming that day. The mathematical treatment of conditional probabilities was first developed by the British minister Thomas Bayes (1702–1761). Bayes published his work in 1764 in a paper called "Essay towards solving a problem in the doctrine of chances." Bayes's ideas have developed into the field known as Bayesian analysis.

Another important problem in mathematical probability was proposed and solved by the French naturalist Georges-Louis Leclerc (1707–1788, also known as the Comte de Buffon). Suppose a needle of a given length is tossed onto a floor marked by equidistant parallel lines, such as the cracks in a hardwood floor. What is the probability that the needle will come to rest across one of the cracks? The solution of this question turns out to be related to the geometrical constant π.

Buffon's needle problem is an example of a seemingly esoteric question that has implications for many practical problems. Each drop of the needle can be seen as the acquisition of a single random sample or snapshot of a landscape or space that one is searching for something (lines on the floor, submarines in the sea, astronomical objects in the sky, animals in a landscape, or other targets) but that is too large to search completely. The event of the needle's landing on or "finding" a line is comparable to the event of one's sample or snapshot detecting a target object in a large search space. Using geometrical probability methods descended from Buffon's needle problem, scientists can say how many

Swiss mathematician Jacob Bernoulli (1654–1705). © *Bettman/ Corbis.*

samples are needed to characterize the number of targets in a search space to any desired level of accuracy. (Whether it is practical to make the necessary measurements is another matter, to be decided case by case.)

Other concepts of probability arose in unusual places during the early development of the field. For example, the reliability of testimony and other legal questions was put to the test by the new theories developed in the eighteenth century. In France, the Marquis de Condorcet (1743–1794) discussed in his *Essay on the Application of Probability Analysis to Decisions Arrived at by Plurality of Votes* such topics as the mathematical expectation of receiving fair treatment in a courtroom. Such questions remain relevant today: statistics showing that black defendants are more likely than white defendants to receive the death penalty in U.S. courts for similar murders have been cited in recent years by persons arguing for a moratorium or ban on the death penalty.

Condorcet is also known for what is called the "Condorcet Paradox," which states that it is possible for a majority to prefer candidate A over candidate B, and candidate B over candidate C, but not prefer candidate A over candidate C. The paradox can arise because the majorities preferring A over B and B over C may not be made up of all the same individuals.

Statistics

Probability theory is the mathematical study of uncertainty and may refer to purely abstract quantities; statistics is the collection and analysis, using mathematical tools from probability theory, of numerical data about the real world. Modern statistics began to develop in the late seventeenth century, as governments, municipalities, and churches, especially in Britain, began collecting more data on the births, lives, and deaths of its citizens, and interested persons began analyzing this data in the attempt to find patterns and trends. One such man, John Graunt (1620–1674), a wealthy London cloth merchant, collected and organized mortality figures; in the process he founded the science of demography, the study of human populations using statistics. His book *Observations on the Bills of Mortality* (1662) was perhaps the first book on statistics. The importance of his book is that for the first time someone established the uniformity of certain social statistics when taken in very large numbers. His work represents a precursor to the modern life tables so vital to insurance calculations.

Another early application of statistics was found in the emerging field of actuarial science, the study of risks in the insurance business. In the seventeenth century, this application of statistics was developing into an important social and financial tool. Among the first to apply such methods to risk assessment was the Dutch political leader, Johan de Witt (1625–1672). In his book *A Treatise on Life Annuities*, De Witt laid out the fundamental idea of expectation in risk management. Today, actuaries (specialists in assessing insurance risks) employ advanced methods in statistics on behalf of insurance companies.

One of the most influential eighteenth-century writers on probability and statistics was Abraham de Moivre (1667–1754). De Moivre was born in France but fled to England as a young man due to religious persecution. De Moivre published two books on the subject of statistics, *The Doctrine of Chance* and *Annuities on Lives*, both of which went through many editions. In *The Doctrine of Chance*, de Moivre defined the crucial notion of statistical independence. Events are independent if the outcome of one event does not influence the likely outcome of the other. For example, successive flips of a coin are independent events, because each flip is a new experiment and is not influenced by earlier outcomes. But when pulling cards one by one from a deck, the results of earlier draws do influence what can happen in later draws: Later draws are, therefore, not independent of earlier draws. A person's chance of contracting lung cancer is independent of the color of their shoes, but may be dependent on whether they smoke. The concepts of dependence and independence are fundamental in modern probability and statistics.

De Moivre capitalized on his friendship with Edmund Halley (of Halley's Comet fame) to expand on Halley's work on mortality statistics and the calculation of annuities. Purchase of an annuity is the exchange of a lump sum of money by a person for a guarantee of regular payments until the person's death. The longer the person lives, the more payout they receive. This is a better deal for the buyer who receives the annuity and a worse one for the seller who pays it out. On average, a seller of annuities must pay out no more than they take in, or they will go out of business. Before de Moivre, pricing of annuities was at best a hit-or-miss proposition, with experience playing the most important role in determining value. De Moivre applied the new methods of probability to the calculation of annuities to ensure a fair price to the annuitant (the person paying for the annuity) and the seller of the annuity. His calculations resulted in tables showing the value of annuities for annuitant ages up to 86.

Perhaps de Moivre's most important contribution to statistics was his description of the properties of the normal curve, also called the bell curve because it is shaped like the cross-section of a bell. One of the most important distributions in statistics, it is used by statisticians for many purposes, such as employing samples to approximate the behavior of populations. More precisely, the normal curve can tell statisticians how much error to expect from their sample if they are using the sample to predict the behavior of a population.

Probability and Statistics in the Physical Sciences

One of the most important statistical techniques, the method of least squares, also happens to possess one of the more interesting and colorful histories. The method of least squares is a procedure by which a curve based on an equation—the curve may be as simple as a straight line, and often is—is fitted to a set of measured data points. Such fits are attempted when there is reason to believe that the physical or social process being measured by the data points is well-described by the theoretical curve, but the actual data have been scattered somewhat by random factors such as noise or random measurement error. The goal is to recover the best possible mathematical description or prediction of the underlying process. The best fit between curve and data is found when the curve is adjusted (say, the slope of the line is changed) until the total error or distance between all the data points and the points on the curve is least, that is, at a minimum. The method of least squares is one way of adjusting the curve shape to achieve this least error or best fit. German mathematician Carl Friedrich Gauss (1777–1855) discovered the method of least squares while still a teenager, but did not publish his results. The discovery resulted from his work in astronomy; Gauss wanted to come up with a system that accounted for various observational errors made in

the science. More than ten years after Gauss' discovery, another prominent mathematician, the Frenchman Adrien-Marie Legendre (1752–1833), independently discovered the method of least squares (also while working on astronomy) and promptly published. A bitter dispute arose between the two men over who should be given priority for the discovery.

Probability theory was shaped into something resembling its present-day form with the publication of *Théorie Analytique des Probabilités* (Analytical Theory of Probabilities, 1812) by French mathematician Pierre-Simon Laplace (1749–1827). Laplace's work incorporated most of the results in probability and statistics already known, as well as advances made by Laplace himself. Laplace's new results included a proof of the central limit theorem, which states that a random variable that is the sum of a large number of other random variables will always have a distribution of the particular type called "normal" (if the summed variables all have identical properties). Laplace's book, issued in many editions, influenced the study of probability throughout the nineteenth century.

Statistical methods eventually led to deep insights into formerly baffling physical problems. For example, three physicists, working independently in three different countries, discovered that statistical models could be used to predict the behavior of seemingly chaotic particles in samples of gas. These men, James Clerk Maxwell

James Clerk Maxwell (1831–1879), Scottish theoretical physicist. From Campbell & Garnett *The Life of James Clerk Maxwell*, 1882. *HIP/Art Resource, NY.*

IN CONTEXT: EUGENICS

Idiot, imbecile, degenerate, moron, dullard, feeble-minded—in the nineteenth and early twentieth century, all these words were attached by some scientists to people with actual or supposed limited mental abilities. Various scientists and social activists believed that the inheritance of inferior intelligence, along with flawed character traits believed to be inherited by criminals, prostitutes, illegitimate children, and the poor, was a major social problem. American sociologist Richard Dugdale (1841–1883) brought the problem to a head in when in 1875 he published a history of a New York family he called the Jukes. Dugdale studied the history of 709 members of the Juke family and found a high rate of criminal activity, prostitution, disease, and poverty. Dugdale and like-minded researchers assumed that such a pattern must be due to inherited genes, not inherited poverty. Similar research projects followed, with similar conclusions, supposedly based on scientific, statistical proofs. Entire families exhibited socially undesirable tendencies that cost governments millions of dollars in services.

Thus was born eugenics, widely considered a science in its day and advocated by many persons considering themselves enlightened or progressive. Eugenicists sought to improve the human gene pool by controlling human breeding. Forced sterilization of criminals and other persons deemed undesirable became the goal of many eugenicists. In the United States, many states passed bills legalizing such methods in the early twentieth century. The atrocities of the Nazis, who had admired the American sterilization laws and implemented far more violent and sweeping measures in the name of eugenics, caused eugenics to quickly fall out of favor in the 1930s and 1940s. The statistical claims of eugenics have been examined by latter-day scientists such as American evolutionary biologist Stephen Jay Gould (1941–2002) to expose the ways in which systematic errors in data collection and analysis, along with the framing of such work in untested assumptions about race, intelligence, gender, and criminality, enabled ideas with no basis in fact to achieve, for a while, the status of received scientific truths.

(1831–1879) of Scotland, Ludwig Boltzmann (1844–1906) of Austria, and Josiah Willard Gibbs (1839–1903) of the United States, made fundamental contributions to what would be called statistical mechanics. (Mechanics is the study of the motions of physical objects. Statistical mechanics is the study of the group behavior of large numbers of objects, such as atoms or molecules, moving randomly.)

Probability and Statistics in the Social Sciences

While physics in the nineteenth century was being revolutionized by statistical methods, statistical concepts were also being applied to the social sciences, where some of them had originated. Lambert Adolphe Jacques Quetelet (1796–1874), a Belgian scientist originally trained as an astronomer, introduced the science of "social physics" in his book *Sur l'homme et le développement de ses facultés*, (On Man and the Development of his Facilities, 1842). In this book, Quetelet applied statistical techniques to the study of various human traits (as measured in a large sample of Scottish soldiers) to arrive at what he called the "average man." These traits included not only physical measurements such as height and weight, but also other supposed characteristics relevant to society, such as criminal propensity. (The notion that a propensity to criminal behavior was innate in the individual was widespread among scientists during the nineteenth century.) Quetelet grouped the measurements of each trait around the average, or mean, value for that trait on a normal curve. He then treated deviations from the average as "errors."

One of the results of Quetelet's studies was a formula that is seen today in almost every doctor's office in the country—the Body Mass Index for measuring obesity by comparing weight and height to average figures. However, many medical scientists have argued recently that the Body Mass Index is not a useful tool for promoting human health because its assumptions about human variation, and its reliance on only two parameters (height and weight), are too simplistic.

Others also soon began to study the social sciences using statistical methods, sometimes with questionable results based on untested assumptions. For example, English scientist Francis Galton (1822–1911) applied statistical methods to human intelligence. Inspired by the work of his cousin, Charles Darwin, Galton became convinced that intelligence—conceived of as a single, measurable quantity inherent in each individual—was inherited. Galton therefore pursued the study of eugenics, the attempt to improve the human species by selective breeding (including, sometimes, sterilization of persons deemed unfit). Although eugenics was widely rejected after the horrors of Nazi Germany, which implemented eugenic ideas in extreme and horrific forms, Galton advanced the field of statistics with his development of the concepts of regression and correlation as well as his pioneering work in identifying people by their fingerprint patterns.

Galton's work inspired another English scientist, Karl Pearson (1857–1936, also a eugenicist). Pearson, a talented mathematician, became interested in the statistical laws of heredity and was instrumental in building the young science of biometrics, the application of statistical methods to individual human characteristics. He developed the idea of standard deviation, a central aspect of modern statistics, and he introduced the chi-squared statistic (named because its standard mathematical notation employs the Greek letter *chi*, pronounced "kie"), a tool for evaluating the standard deviation of a data set. The standard deviation of a collection of numbers is their average distance *from* the average—a measure of how spread-out the data are around their central average. In addition to many mathematical contributions to statistics, Pearson was Director of the Biometrics Laboratory at University College, London, and a co-founder of the journal *Biometrika*.

Karl Pearson's son, Egon Pearson (1895–1980), teamed with another mathematician, Jerzy Neyman (1894–1981), to lay the foundations of the modern method of hypothesis testing. A hypothesis test is a process by which the probability that a chosen hypothesis or proposed idea, termed the null hypothesis, is correct is determined. For instance, say a pharmaceutical company claims that its new drug increases the rate of recovery from a certain disease. Furthermore, assume that 40% of patients with this disease recover without the drug. In a test of the drug on a sample of patients with the disease, 46% recover. Obviously 46% is larger than 40%, but is the difference significant—that is, is it due to the drug? Might it be merely a coincidence whereby a few more patients in this sample just happened to get better than one would expect on average? How large does such a difference have to be before we are, say, 90% sure that it is a real difference, not a random one?

Hypothesis testing seeks to answer such questions. In this example, a hypothesis test would be used to determine whether the drug played a significant role in the recovery or whether the increase could be attributed to simple coincidence. Pearson and Neyman teamed up to introduce ideas such as the alternative hypothesis, type I and type II errors, and the critical region into the hypothesis testing process. Later, Neyman introduced the idea of a confidence interval, another process very familiar to modern students of statistics. Both hypothesis tests and confidence intervals may be used to compare means, proportions, or standard deviations in a wide array of applications in psychology, medicine, marketing, and countless other disciplines.

The early development of modern information technology was stimulated by the need to handle large quantities of statistical data. An American engineer, Herman Hollerith (1860–1929), while working for the United States Census Bureau, first began to think about building a machine that could mechanically tabulate the huge amounts of data collected during the census every ten years. At MIT, and later while employed by the United States Patent Office, Hollerith began to experiment with designs based on the punch-card system used with the Jacquard loom, an automated weaving machine used throughout much of the nineteenth century in textile production. Hollerith's machine won a competition to be used by the Census Office for the 1890 census. It was a smashing success, tabulating the census data in three months instead of the two years that would have been required for tabulation by hand. The company Hollerith formed to market his calculating machine, the Tabulating Machine Company, would, after several mergers and name changes, eventually become International Business Machines, known to everyone today as IBM. Until punch-cards were completely replaced by keyboards and video screens for computer programming in the early 1980s, the cards used to feed lines of program code to computers—one line per card—were known as Hollerith cards. Hollerith cards and machines supplied by U.S. companies were used by the Nazi regime to keep systematic track of people sent to extermination camps.

Twentieth Century Advances

Probability and statistics advanced greatly in the twentieth century. New theories and new applications continued to appear in disciplines as varied as physics, economics, and the social sciences. One of the most influential statisticians of the twentieth century was Englishman R.A. Fisher (1890–1962). Fisher continued the work of Galton, Karl Pearson, and others in biometrics. He was especially interested in the application of statistical processes to Mendelian genetics. Fisher's work was important in reconciling Mendel's genetics with Darwin's theory of natural selection, making possible what has been called the neo-Darwinian Synthesis. This is the union of genetics, mathematics, and traditional Darwinian theories of natural selection into a single theory of evolutionary change that has been highly successful in explaining the history of life. Fisher also produced groundbreaking results in sampling theory and in the theory of estimation. Among many important publications, Fisher's *Statistical Methods for Research Workers* (1925) provided a valuable guide in the application of statistics for researchers in various fields.

The Student's *t*-statistic, a tool used to estimate means for small samples, was employed by Fisher and others in the early twentieth century in the development of experimental design. The hypothesis-testing methods that came from Fisher's work have found application in almost every branch of science, from agriculture to zoology. Fisher realized that to make a claim about any matter of fact, one needs to be sufficiently sure that the result is not a chance occurrence. In one of Fisher's own examples, a lady claims that she can taste the difference in tea depending on whether the milk is added to the tea or whether the tea is added to milk. Fisher proposed an experiment in which the lady is given eight cups of tea, with four made each way. He then calculated the probability that the lady can guess correctly on each cup of tea without really discerning the difference. A definite probability of occurrence by chance can then be assigned to each possible outcome of the tea-tasting test. Although this example is trivial, similar insight is desired in real-world research; the application of such methods to studies of new medical tests and treatments is routine.

Specific mathematical questions that arose in the process of building the first atomic bomb in the 1940s led to the development of a statistical method called the Monte Carlo method. The Monte Carlo method is the use of random numbers to create a statistical experiment: that is, to see how some real-world process with random inputs (such as decaying radioactive atoms) might behave, one generates long lists of random numbers to specify those inputs, feeds them into equations representing the behavior of the system, and examines the system's behavior. The method was named after the roulette wheels in the famed Monte Carlo gambling center in Monaco. With the advent of electronic computers, the Monte Carlo method has become an important tool for physicists, engineers, mathematicians, climatologists, and many other professionals who deal with statistical phenomena.

The twentieth century also marked a time of renewed interest in the mathematical foundations of probability. One man, Russian mathematician Andrei Nikolaevich Kolmogorov (1903–1987), made the most significant contributions to this endeavor. In his book, *Foundations of the Theory of Probability*, Kolmogorov sought to make probability a rigorous, axiomatic branch of mathematics just as Euclid had done with geometry twenty-three centuries earlier. In doing so, Kolmogorov provided mathematically rigorous definitions of many statistical terms. His work also formed the foundation of the study of stochastic processes.

■ Modern Cultural Connections

Today, methods from probability and statistics are used in a wide array of settings. Psychologists use statistics to make inferences about individual and group behavior. Probability plays an important role in the study of evolutionary processes, both biological and social, as evolution is a mixture of random processes (genetic mutation) with nonrandom natural selection (survival of the fittest). Probability and statistics methods allow economists to build financial models and meteorologists to build weather models. Statistical analysis in medicine aids in diagnostics as well as in medical research.

With the birth of quantum physics in the twentieth century, probability became an integral part of the way scientists strive to understand and describe our physical world. According to the standard interpretation of quantum physics, events at the smallest size scales are truly random, not merely apparently random, because we cannot account for all the forces causing them (as is the case with rolling dice). Albert Einstein, reflecting on the uncertainty introduced into science by discoveries indicating the existence of a true or ontological randomness in nature, uttered his famous phrase, "God does not play dice"—though apparently, in some sense, God (or the underlying principles of the physical universe) does, at least at the quantum scale. However, the correctness of the standard interpretation is still debated in the scientific community.

One of the hottest new applications of statistics in the early twenty-first century was the field of data mining. Data mining combines statistics, artificial intelligence, database management, and other computer-related studies to search for useful patterns in large collections of data. Data mining has applications in business, economics, politics, and military intelligence, to name a few. For example, retailers have begun to use data mining to track customers' buying preferences and to use this information for more precise marketing strategies.

■ Primary Source Connection

Arguments about matters of public policy, such as the Iraq War, often lead to brawls over statistical claims. In this text, left-leaning media critics attack a statistical argument made by a prominent conservative.

ARE 2,000 U.S. DEATHS 'NEGLIGIBLE'?

On the October 13 broadcast of *Special Report*, the show he regularly hosts, 'Fox News Channel anchor Brit' Hume said of U.S. deaths in Iraq, "by historic standards, these casualties are negligible."

. . .

On August 26, 2003, Hume conjured up a bizarre mathematical formula to show that U.S. casualties were not a big deal:

"Two hundred seventy-seven U.S. soldiers have now died in Iraq, which means that statistically speaking U.S. soldiers have less of a chance of dying from all causes in Iraq than citizens have of being murdered in California, which is roughly the same geographical size. The most recent statistics indicate California has more than 2,300 homicides each year, which means about 6.6 murders each day. Meanwhile, U.S. troops have been in Iraq for 160 days, which means they're incurring about 1.7 deaths, including illness and accidents each day."

Hume's geographic comparison was meaningless, since the total population of California is far greater than the number of U.S. troops in Iraq—approximately 240 times greater. If Californians were being killed at the same rate that Hume cited for U.S. soldiers, there would be more than 400 murders per day, not six. When Washington Post reporter Howard Kurtz asked Hume about this point, Hume said: "Admittedly it was a crude comparison, but it was illustrative of something."

FAIR (FAIRNESS AND ACCURACY IN REPORTING),
"ARE 2,000 U.S. DEATHS 'NEGLIGIBLE'?" OCTOBER 25, 2005.
HTTP://WWW.FAIR.ORG/INDEX.PHP?PAGE=2706
(ACCESSED OCTOBER 12, 2007).

SEE ALSO *Mathematics: The Specialization of Mathematics; Mathematics: Trigonometry.*

BIBLIOGRAPHY

Books

Daston, Lorraine J. *Classical Probability in the Enlightenment.* Princeton, NJ: Princeton University Press, 1988.

David, F.N. *Games, Gods and Gambling.* London: Griffin, 1962.

Gigerenzer, Gerd, et al, eds. *The Empire of Chance: How Probability Changed Science and Everyday Life.* Cambridge: Cambridge University Press, 1989.

Gould, Stephen Jay. *The Mismeasure of Man.* New York: Norton, 1981.

Hacking, Ian. *The Emergence of Probability: A Philosophical Study of Early Ideas about Probability, Induction and Statistical Inference.* Cambridge, UK: Cambridge University Press, 1975.

———. *The Taming of Chance.* Cambridge, UK: Cambridge University Press, 1990.

Hald, Anders. *A History of Probability and Statistics and Their Applications Before 1750.* New York: Wiley, 1990.

Johnson, Norman Loyd, and Samuel Kotz, eds. *Leading Personalities in Statistical Sciences: From the Seventeenth Century to the Present.* New York: Wiley, 1997.

Krüger, Lorenz, Lorraine J. Daston, and Michael Heidelberger, eds. *The Probabilistic Revolution.* 2 vols. Cambridge, MA: MIT Press, 1987.

Owen, D.B. *On the History of Statistics and Probability.* New York: Dekker, 1976.

Stigler, Stephen M. *The History of Statistics: The Measurement of Uncertainty Before 1900.* Cambridge, MA: The Belknap Press of Harvard University Press, 1986.

Porter, Theodore M. *The Rise of Statistical Thinking, 1820–1900.* Princeton, NJ: Princeton University Press, 1986.

Periodicals

Box, Joan Fisher. "Gossett, Fisher and the *t* distribution." *American Statistician* 35 (1981): 61–66.

Cowan, R.S. "Francis Galton's Statistical Ideas: The Influence of Eugenics." *Isis* 63 (1972): 509–528.

Dutka, Jacques. "On Gauss' Priority in the Discovery of Least Squares." *Archive for History of Exact Sciences* 49 (1996): 355–370.

Ore, Øystein. "Pascal and the Invention of Probability Theory." *American Mathematical Monthly* 67 (1960): 409–419.

Van Brakel, J. "Some Remarks on the Prehistory of Statistical Probability." *Archive for the History of Exact Sciences* 16 (1976): 119–136.

Web Sites

FAIR (Fairness and Accuracy in Reporting), "Are 2,000 U.S. Deaths 'Negligible'?" October 25, 2005. http://www.fair.org/index.php?page=2706 (accessed October 12, 2007).

Todd Timmons

Mathematics: The Specialization of Mathematics

■ Introduction

Mathematics is the study of relationships among and operations on abstract objects that obey definite rules, including numbers, variables, functions, rules, spaces, shapes, and sets. In its ancient origins, mathematics was concerned solely with numbers and geometry (the properties of definite shapes), which arose from measurable and countable phenomena and could, in part, be applied directly to business and architecture. During the nineteenth century, an increasing number of mathematicians became fascinated by relationships of pure reason and by the deductions that could be drawn from those relationships, even where these results seemed to have no application to the real world. This formalization of symbolic logic and abstract reasoning allowed mathematicians to develop the definitions, relations, and theorems of pure mathematics, but also—unexpectedly—had the effect of advancing applied mathematics, namely, those mathematical methods useful to science, engineering, and economics.

In both pure and applied mathematics, nineteenth century mathematicians took on increasingly specialized roles corresponding to the rapid compartmentalization and specialization of mathematics in general. One no longer simply became a mathematician, but a specialist in some particular branch of mathematics. As in other fields, specialization in mathematics has allowed the individual worker to gain a deeper knowledge in some particular area, but at the expense of breadth: a quip often attributed to German biologist Konrad Lorenz (1903–1989) states, "A specialist knows more and more about less and less and finally knows everything about nothing." Overall, however, specialization in mathematics has been both unavoidable and beneficial.

Often, methods developed for the sake of specialized branches of pure mathematics have migrated to applied mathematics and become essential to new technologies and industries: for example, studies in abstract logic became, many decades after their original conception, the basis for modern digital computers.

■ Historical Background and Scientific Foundations

Starting in the late 1700s, the emerging European industrial revolution gave rise to new methods and knowledge in physics, astronomy, and engineering. These innovations created a spate of novel mathematical challenges. In the 1800s, mathematicians scrambled to invent and refine analytical methods to solve the seemingly endless list of questions and problems being raised by scientists and engineers. By the middle of the century, attention had begun to shift toward the operations of mathematical logic, considered on its own abstract terms. The result was an increased emphasis on the relationships and

WORDS TO KNOW

APPLIED MATHEMATICS: All mathematical knowledge used in physical science, economics, social science, and technology. Applied mathematics is often contrasted to pure mathematics, which is pursued without regard to any specific real-world use.

PURE MATHEMATICS: The pursuit and expansion of mathematical knowledge and technique without regard to whether the mathematics produced will ever have a technological application. Pure mathematics are often contrasted with applied mathematics. Often, pure mathematics have turned out to have unforeseen applications.

rules for evaluating axioms and postulates, which are the fundamental statements or claims on which structures of logical reasoning are raised.

Building on the calculus invented by innovators Sir Isaac Newton (1643–1727) of England and Gottfried Wilhelm von Leibniz (1646–1716) of Germany in the 1600s, nineteenth-century mathematicians extended the accuracy and precision of mathematical calculations. The increasingly complex and powerful structure of applied mathematics also depended on foundations laid by Swiss mathematician Leonhard Euler (1707–1783) in mechanics, differential and integral calculus, geometry, and algebra. Without all these tools, science and technology could not have continued to progress. The large strides in applied mathematics, however, left much theoretical and logical ground untouched. Mathematics relied on reasoning, but what was reason? How did one do it, and what were its rules, varieties, and limitations? After centuries of emphasis on practical applications, the nineteenth century was ripe for the development of a pure mathematics—a mathematics of pure reason.

With the nineteenth-century explosion in mathematical knowledge came, inevitably, specialization: the amount of total mathematical knowledge had become too large for any one person to master, the literature too large to read in a lifetime. Similar specialization was also occurring in the physical sciences. A result of the increasing specialization of mathematics was a schism between pure and applied mathematics. One definition of this division is that pure mathematics is advanced for theoretical reasons—it is concerned only with the mathematical validity of certain abstract, symbolic structures—while applied mathematics develops tools and techniques to solve problems in science, engineering, and economics. Such a simplistic definition, while containing much of the truth, omits the common history of and frequent crossovers between the two types of mathematics.

Starting in the nineteenth century, there evolved a profound difference in the methodology—the whole manner of proceeding—of pure and applied mathematics. In pure mathematics, deductions are valid if properly derived from given axioms and postulates. In the reasoning of applied mathematics, hypotheses (proposed explanations) are grounded in experimental evidence, cast in mathematical terms, and tested against further evidence. The choice of what mathematics to apply is based on observation, testing, and (often, in the earlier stages) speculation; some forms of mathematics will work, in practice, better than others, and one must find out by trial which these are. Pure mathematics is a matter of correctly following the laws of reasoning to produce a certain structure of mathematical relationships, whose interest is judged solely by its relationship to other mathematical structures. In applied mathematics, one seeks to discover which structure of

mathematical relationships can be used to describe a recurrent pattern of natural phenomena (that is, a law of nature).

The professional split between the pure and applied math did not occur all at once. For example, the theories of German mathematician and physicist Johann Carl Friedrich Gauss (1777–1855) embraced and embodied both pure and applied mathematical concepts. Gauss' practical mathematical discoveries included advances in the study of the shape of the Earth (geodesy), planetary orbits, and statistical methodology (i.e., least-squares methods). Gauss also advanced pure mathematics through seminal work in number theory, representations of complex numbers, quadratic reciprocity, and a proof of the fundamental theorem of algebra.

In 1847, English mathematician George Boole (1815–1864) published his *Mathematical Analysis of Logic*. He followed it in 1854 with another important publication, *The Laws of Thought*. Boole asserted that the development of symbolic logic and pure mathematical reasoning was being retarded by an overdependence on applied mathematics. He advocated the expression of logical propositions in symbols used to represent logical statements: logical propositions could then be proved or disproved just like any other mathematical statements. Boole championed pure mathematical reasoning by attempting to dissociate abstract mathematical operations from any specific applications.

Boole's publication drew a line in the sand for mathematicians and highlighted a trend away from Gauss-like mathematical universalism and toward increased specialization within the profession of mathematics. In particular, there was an increasing divergence between pure and applied mathematics. As a consequence of the popularity of formalism, such as was advocated by Boole, there also resulted an increasing number of mathematicians dedicated to pure mathematics, with minimal consequences—or so it appeared—for science and technology.

This divergence was not always viewed with favor. Although initially created as a subdiscipline of mathematics, the field of mathematical logic was widely ignored or held in disdain by many mathematicians. By the end of the century, however, symbolic logic progressed from academic obscurity to popular entertainment. The books of English mathematician Lewis Carroll (1831–1898) on logic, *The Game of Logic* (1887) and *Symbolic Logic* (1896), became popular topics of conversation and sources of entertainment both for scholars and laymen in Victorian England; his most famous work, *Alice in Wonderland* (1865), contained much logic-play and has been quoted by scientists and philosophers for over a century. Despite initial resistance, mathematical logic made strong inroads into philosophy after William Stanley Jevons (1835–1882) heralded its use in his widely read *Principles of Science* (1874). Jevon argued

William Stanley Jevons (1835–1882), English economist and logician, is credited with inventing one of the first computers, a logical machine that he exhibited to the Royal Society in 1870. *SPL/Photo Researchers, Inc.*

(correctly, as it turned out) that symbolic logic would be of importance to both philosophy and mathematics.

The increasing volume of work relating to number theory also lead to hierarchies with the emerging specialties of mathematics. Fueled by Gauss' work on the theory of numbers, algebraic theories of numbers took on a preeminent position in pure mathematics.

Some initially pure mathematical theory was met with outright derision and scorn. Among the most controversial of advances in mid-nineteenth century mathematics was the publication of non-Euclidean geometries by German mathematician Georg Friedrich Bernhard Riemann (1826–1866). Riemann asserted that Euclidian geometry—the ordinary geometry of triangles, lines, planes, and so forth, as taught in schools and used throughout architecture, surveying, and technology—was but one possible geometry, and that many others could be validly conceived. His expanded concepts of geometry treated the properties of curved space and seemed useless to nineteenth century Newtonian physicists. Many mathematicians also thought Riemann's conceptualizations bizarre.

Regardless, in the next century his theories proved of enormous consequence and value to the expansion of concepts of gravity and electromagnetism and of fundamental importance to the twentieth-century theoretical work of Albert Einstein (1879–1955) and others regarding the nature of space, time, and gravity. Topology as a specialization of mathematics (the study of the general properties of surfaces apart from changes in size or shape) was born in the advances of nineteenth century geometry and has also proved indispensable to modern physics.

Later in the nineteenth century, when Russian mathematician Georg Cantor (1845–1918) proposed his transfinite set theory, many thought it the height of abstraction. Advances in twentieth century physics, however, have also found use of Cantor's theories.

Yet pure and applied mathematics never became completely disconnected from each other: the challenges of real-world work have spurred developments in pure mathematics, while the abstract structures of pure mathematics have repeatedly been found necessary, sometimes many years after their development, for the description of the real world. Thus, not all developments in mathematics were polarized into the pure and applied camps. Early in the nineteenth century, the work of French mathematician Jean Baptiste Joseph Fourier (1768–1830) with mathematical analysis allowed him to establish what is known to modern mathematicians as the Fourier series, which is central to Fourier analysis, an important tool for both pure and applied mathematicians. And, late in the nineteenth century, group theory made possible the unification of many aspects of geometric and algebraic analysis.

Although there was an increasing trend toward specialization throughout the 1800s, near the end of the century French mathematician Jules Henri Poincaré (1854–1912) embodied Gauss's universalist spirit. Poincaré's work touched on almost all fields of mathematics. His insights provided significant advances in applied mathematics, physics, analysis, functions, differential equations, probability theory, topology, and the philosophical foundations of mathematics. Poincaré's studies of the chaotic behavior of systems subsequently provided the theoretical base for the continually evolving—and deeply practical—chaos theory of recent mathematics.

Mathematical rigor—the use of absolutely precise and unambiguous fundamental concepts and definitions—was required in the early part of the century by the extension and refinement of the calculus and was broadened near the end of the century by German mathematician Karl Theodor Wilhelm Weierstrass (1815–1897) to the types of analysis familiar to modern mathematicians (e.g., convergence series, theories using Abelian principles, periodic and elliptic functions, bilinear and quadratic forms).

The advancement of elliptic functions, principally through the work of Norwegian mathematician Neils Henrich Abel (1802–1829) and Prussian mathematician Karl Gustav Jacob Jacobi (1804–1851), provided mathematical precision in calculations required for discoveries in astronomy, physics, algebraic geometry and topology. In addition to their use in applied mathematics, however, the development of the theory of elliptic functions also spurred the study of functions of complex variables and provided a bridge between the widening chasm opening between pure and applied mathematics.

■ Modern Cultural Connections

Although there were subtle divisions of mathematics at the beginning of the nineteenth century, by the early twentieth century there were full and formal divisions of pure and applied mathematics and of subfields within each. University appointments and coursework syllabi began to reflect these divisions and an increasing number of professorial positions were designated for pure or applied mathematicians. Within each field, there was continuing sub-specialization as the amount of material to be mastered in each area grew over time. Yet the old arguments about the relative merits of pure and applied mathematics died out and remained, for the most part, dead: both types of mathematics offered ample challenge for the most brilliant minds.

One of the most dramatic examples of pure mathematics becoming essential in real life is the symbolic logic of Boole. Boole himself foresaw no practical uses for his methods, and his work accelerated the nineteenth-century trend toward a split between applied and pure mathematics—yet direct technological application of Boolean algebra in the digital computer has revolutionized warfare, personal life, finance, and science in the modern industrialized world. Boolean algebra is used by computer designers to specify the physical circuits that handle the billions of on-off signals that comprise the working information inside a digital computer. Also, as mentioned earlier, the non-Euclidean geometries of nineteenth-century pure mathematics have become everyday tools of modern physics.

Yet pure mathematicians do not pursue their craft for the sake of producing such tools: there is simply no way to tell which purely mathematical structures will someday find practical use and which will not. Instead, pure mathematicians pursue their theories with an interest that has often been compared to the creation of music or art. In doing so, they expand the range and rigor of human thought, expanding our intellectual possibilities: world-changing technological and scientific gains are a frequent side-effect of that drive.

SEE ALSO *Mathematics: Trigonometry.*

BIBLIOGRAPHY

Books

Boyer, Carl. *A History of Mathematics.* 2nd ed. New York: John Wiley and Sons, 1991.

Brooke, C., ed. *A History of the University of Cambridge.* Cambridge, UK: Cambridge University Press, 1988.

Carroll, Lewis. *The Game of Logic.* London: Macmillan, 1887. Reprinted in *Symbolic Logic and the Game of Logic.* New York: Dover, 1958.

Dauben, J.W. *The History of Mathematics from Antiquity to the Present.* New York: Garland Press, 1985.

Kline, M. *Mathematical Thought from Ancient to Modern Times.* New York: Oxford University Press, 1972.

K. Lee Lerner

Mathematics: Trigonometry

■ Introduction

Trigonometry is the branch of mathematics that deals with the properties of triangles and of the trigonometric functions, which were originally developed to describe certain properties of triangles but have applications in math and the sciences far beyond flat geometry.

For about 2,000 years, trigonometric functions were especially prominent in astronomy and geography, where they were first developed. Today they are part of the daily working language of scientists and engineers in a multitude of fields.

For centuries after their discovery, the trigonometric functions—of which the most familiar and fundamental are those known as the sine and cosine—were closely associated with the properties of triangles. Only in the 1600s did mathematicians free them from direct association with triangles and bring them into the family of general functions on a real axis. First, the Newton-Taylor expansion of smooth functions as infinite sums (series) of polynomials connected the trigonometric

functions to algebraic functions. In the eighteenth century Leonard Paul Euler (1707–1783) connected them to exponential functions using complex numbers. In the nineteenth century, French mathematician Joseph Fourier (1768–1830) developed his analysis, which became a powerful tool for analyzing signals in scientific and technological settings. Today trigonometric functions find application throughout much of higher and applied mathematics.

■ Historical Background and Scientific Foundations

The word "trigonometry" is from the Greek for "the geometry of triangles." However, the first applications of trigonometry probably did not originate with the Greeks—there are indications that Greeks learned some trigonometric ideas from the Egyptians and Babylonians. For example, it is possible that the Greek thinkers Pythagoras (c.582–c.500 BC) and Plato (c.427–c.347 BC) learned mathematics in Egypt. Alexander the Great's (356–323 BC) military exploits in 333–323 BC briefly united the Mediterranean and Middle East under Greek rule, and sharing of scientific ideas between different regions was more likely under these conditions. It was at about this time that the Greeks elevated science and mathematics to new levels of precision and abstraction. The *Elements* of Euclid (fl. third c. BC), written in the third century BC in Alexandria, Egypt, defined the field of geometry for the next 2,000 years.

Euclid developed the idea of similarity between triangles of different sizes but identical angles. This concept allowed measurement of the size of Earth using nothing but simple geometry and a vertical stick, a device called a gnomon (pronounced NO-mun). Using a gnomon, Eratosthenes (276–194 BC), measured the angular height of the sun in Alexandria on the day

The waters of an ancient well near Aswan, Egypt, mark the arrival of the summer solstice, with the sun almost directly overhead. In the third century BC in Alexandria, the Greek scholar Eratosthenes relied on both geometry and reports of the sun's annual appearance almost directly over Aswan to calculate Earth's circumference, arriving at an estimate close to the actual distance. *© Bob Sacha/Corbis.*

of the summer solstice (point on the calendar when the days cease to get longer and begin to get shorter again). Knowing that in another Egyptian city, Thebes, the sun stood directly overhead on the same day at the same hour, Eratosthenes could compute the angular difference between the two cities on the surface of the Earth, which Greek thinkers understood correctly to be spherical. (It is not true that Christopher Columbus [1451–1506] discovered, proved, or was the first to suggest that Earth is round.) Eratosthenes estimated that Earth's circumference is 50 times greater than the distance between Thebes and Alexandria—a reasonably correct value achieved using extremely simple tools.

Euclid did not introduce any of the specific functions associated with the angles, the trigonometric functions. (A function is a specific relationship between two or more sets of numbers.) Neither did the Greek scientist Archimedes (287–212 BC), who in his treatise *Measurement of a Circle* (c.240 BC) computed accurate bounds on π. In doing so, Archimedes computed estimates of the chords (circle-bridging line segments) associated with certain angles, namely those that are multiples of 3.75°. Today, his procedure would be viewed as the computation of the sines and tangents of those angles—two of the basic trigonometric functions.

In astronomy and physics, the Aristotelian idea that circular motion is more natural, perfect, and pervasive was dominant for over 1,500 years. Since the heavenly

bodies clearly do not move in simple circles, the idea of epicycles, circles moving along other circles, evolved. Hipparchus (190–120 BC), an astronomer of Rhodes (a Greek island), was the first to construct a viable astronomical system, around 160–140 BC. This might have been the earliest use of trigonometry in astronomy, for although Hipparchus never thought to introduce a specific function analogous to the modern sine [sin(x), read aloud as "sine x"], in his system the position of the sun was expressed by a formula that implicitly included sines.

Other early astronomers pushed trigonometry to greater sophistication. The *Almagest* of Claudius Ptolemy (AD c.90–c.168), composed around AD 150 in Alexandria, was the central textbook of ancient and medieval astronomers for about 1,400 years. It was supplanted only when Nicolaus Copernicus (1473–1543) wrote his *De Revolutionibus* (1543), which placed the sun, rather than Earth, at the center of the universe.

Despite the downfall of Rome and the advent of the Dark Ages in Europe, the *Almagest* was preserved by translation into Arabic (four different translations were made in ninth-century Baghdad). The first translation from Arabic back into Latin by Gerard of Cremona (c.1175) marked the scientific re-awakening of Europe.

In the *Almagest*, Ptolemy claimed to explain all of the phenomena in the sky, including the most intricate and puzzling then known, namely, the retrograde motions of the planets (that is, their apparent motion,

IN CONTEXT: THE TRIGONOMETRIC FUNCTIONS

Two basic mathematical relationships, the sine and cosine functions, arise in the study of triangles. Various simple ratios of these produce a family of secondary or derived functions, namely the tangent, cotangent, secant, cosecant. Together, these functions are known as the trigonometric functions.

Given a particular angle, θ (Greek letter theta), each trigonometric function specifies a number. For example, the sine of 45°, written sin(45°), is the square root of 2 ($\cong$.7071); the sine of 90° is 1; and so on.

The sine of an angle is found by assuming that the angle is inside a right triangle (a triangle having one right or 90° angle). The longest side of any right triangle is the side facing (across from) its right angle: this side is called the hypotenuse. The sine of any angle in a right triangle is the length of the side facing that angle divided by the length of the hypotenuse. The sine of a 90° angle, for example, is 1, because it is given by the length of the hypotenuse divided by itself.

The cosine of an angle is defined as the length of the side of the triangle adjacent to the angle divided by the length of the hypotenuse.

Both the sine and cosine, plotted in Cartesian coordinates with the angle θ on the horizontal axis and the magnitude of the function on the vertical axis, appear as wavy, up-and-down lines that repeat forever to left and right.

All other trigonometric functions are formed as simple ratios of the sine and cosine. The tangent is sine/cosine, the cotangent is cosine/sine, the secant is 1/cosine, and the cosecant is 1/sine. These functions are a useful shorthand but contain no information that is not found in the sine and cosine. Indeed, since $\sin(\theta) = \cos(90° - \theta)$ and $\cos(\theta) = \sin(90° - \theta)$, either of the sine or cosine alone would suffice to write down all other trigonometric functions. Such notation would, however, be obscure and inconvenient.

In modern mathematics, definitions of the sine and cosine that do not depend on the properties of triangles are sometimes used so that these functions can be extended to complex numbers.

---

at times, from east to west relative to the fixed stars, opposite to the west-to-east motions of all other heavenly bodies). The *Almagest* explicitly introduced the sine function, though under a different name. Ptolemy sought to use sine functions to characterize the error introduced in estimating the size of Earth using the gnomon by the fuzziness of the gnomon's shadow, which introduces angular uncertainty into observations of the sun's position.

Muslim mathematicians made further progress in trigonometry during the long sleep of European science. Their main goal was to find Kibla—the direction to Mecca from any locality on Earth—so that Muslims everywhere could face Mecca during prayer. In the eleventh century, Muslim scientists discovered a number of cosine and sine theorems. Later Muslim mathematics concentrated on producing more precise Kibla tables.

The introduction of the tangent function (as well as cotangent, secant, and cosecant) is attributed to Muslim mathematician Abu'l-Wafa (959–988), a member of the Caliph's court in Baghdad. Though these functions are simply different ratios of sine and cosine functions, their development allowed the use of a shorter, more useful algebraic notation that was important in the development of mathematics in medieval Europe.

The need of medieval European traders for convenient, accurate maps dictated further development of trigonometry. Since Earth is round, while maps on paper must be flat, mapmaking requires mathematical

understanding of the projection of the surface of a sphere onto a plane. Around 1569, Flemish cartographer Gerardus Mercator (1512–1594) found a convenient way to represent part of the globe on a flat map using logarithms and trigonometric functions.

The Mercator map projection, with its straight meridians and parallels, was widely used for navigation and is still one of the most common projections. The projection has its advantages, but latitude (distance from the equator) is distorted, with distortion increasing with distance from the poles until it approaches infinity at the poles. A Mercator map thus shows more or less accurate proportions near the equator, but is less accurate near the North and South Poles. For example, on a Mercator projection Greenland looks as big as South America, though it is in fact only a tenth as large.

The Mercator projection was the first case in the history of applied mathematics where the logarithm (inverse of the exponential function) and trigonometric functions intermingled. Previously, they had seemed to have little relationship.

In the seventeenth century, progress in trigonometry continued to be driven by astronomy. German astronomer Johannes Kepler (1571–1630) was the first to suggest, in his 1609 *Astronomia Nova*, that all planets move in elliptical orbits, rather than in circles (or circles moving in circles). His basic equation to relate the position of the sun to that of any given planet (e.g., Earth) included a sine function.

IN CONTEXT: TRIGONOMETRY AND NAVIGATION

Pilots, mariners, and mountaineers all use trigonometric concepts to find their way from one point to another. In navigation, positive angles are measured clockwise from North and are known as azimuths. Azimuths convey direction, so they can range from 0° to 360°, and the word azimuth is sometimes used synonymously with the word heading. The azimuth of a line running from south to north is 0° and the azimuth of a line running from north to south is 180°. This distinction is critical in navigation. In other applications, it may not be critical to distinguish the direction. For example, it does not matter whether the boundary of a country runs from north to south or south to north.

In navigation involving travel over large distances, Earth's curvature becomes important and spherical rather than plane; trigonometry must be used. Coordinates for navigation over long distances are given in terms of latitude and longitude, which are angular measurements. Trigonometry is used to calculate the distance between the starting and ending points of a journey, taking into account that the path follows the surface of a sphere and not a straight line. The latitude and longitude of waypoints along a journey can also be calculated using trigonometry.

Navigation on Earth is complicated by the fact that the North Magnetic Pole, to which compass needles are attracted, does not coincide with the North Geographic Pole. The North Magnetic Pole is located in far northern Canada. For very approximate navigation, for example if a hiker wants to know if she is generally headed north or south, the fact that the geographic and magnetic poles are different does not make much difference. For any kind of precise navigation or mapmaking, however,

the difference is important. The difference between true north, which is the direction to the North Geographic Pole, and magnetic north, which is the direction to the North Magnetic Pole, is known as magnetic declination. It is shown as an angle on topographic maps and navigational charts.

Magnetic north is about 20° east of true north in the northwestern United States and about 20° west of true north in the northeastern United States. The line of zero declination runs through the midwestern part of the country. In other areas of the world, the magnetic declination can be as great as 90° east or west in the far southern hemisphere. The North Magnetic Pole moves from year to year as a consequence of Earth's rotation, so the magnetic declination also changes over time. Government agencies responsible for providing navigation aids track the movement of the North Magnetic Pole, and maps are continually revised to reflect changing declination. Measurements by the Canadian government show that the North Magnetic Pole moved an average of 25 mi (40 km) per year between 2001 and 2005.

A simple trigonometric calculation illustrates the error that can occur if magnetic declination is not taken into account. The distance off course will be the distance traveled multiplied by the sine of the magnetic declination. In an area where the magnetic declination is 20°, therefore, a sailor following a course due north would find herself 21.1 mi (34 km) off course at the end of a 62.1 mi (100 km) trip. The longer the distance traveled, the farther off course the traveler will be. If the magnetic declination is only 10°, however, the error will be 62.1 mi (100 km) × sin 10° = 10.6 mi (17 km).

The application of trigonometric functions to technological innovation also began in the seventeenth century. One of the interests of mid-seventeenth century mathematicians was to find the area under a curve; avorite curves were the cycloid—the curve described in the air by a point on a rim of a rolling circular wheel—and a more general curve, the trochoid (or prolate cycloid), described by a point located inside the wheel.

Consideration of the cycloid, which requires the use of trigonometric functions, gave the solution to the tautochrone problem. The tautochrone is a curve, which, if imagined as a stiff wire along which frictionless beads slide, allows all beads to reach the bottom of the curve at the same time regardless of their starting point. The Dutch astronomer Christiaan Huygens (1629–1695) proved in 1673 that the cycloid curve is the tautochrone curve. Huygens used this curve to construct the first pendulum clock and to describe methods for improving the accuracy of pendulum clocks at sea.

The relationship of trigonometric functions to polynomials and exponential functions was the next area of progress, and eventually had important implications for technology and science. Isaac Newton's disciple, English mathematician Brook Taylor (1684–1731), discovered a method of expanding any function in the vicinity of a point, later known as Taylor's Theorem. The resulting expansion is called a Taylor expansion. In mathematics, an expansion is an expression for a function that consists of a sum of terms, perhaps an infinite number of them. In particular, Taylor showed that trigonometric functions can be represented as infinite Taylor expansions: for example, $\sin(x)$ can be expressed as a sum of powers of x multiplied by various constants. Taylor series expansions enabled mathematicians to recognize the relationship between trigonometric functions and logarithms, which can also be expressed as Taylor series.

Developing the Newton-Leibniz calculus, Swiss mathematician Leonard Euler (1707–1783) discovered,

IN CONTEXT: VECTORS, FORCES, AND VELOCITIES

Vectors are quantities that have both direction and magnitude, for example the velocity of an automobile, airplane, or ship. The direction is the azimuth in which the vehicle is traveling and the magnitude is its speed. Using trigonometry, vectors can also be broken down into perpendicular components that can be added or subtracted. Take the example of a ferry that carries cars and trucks across a large river. If there are ferry docks directly across from each other on opposite banks of the river, the captain must steer the ferry upstream into the current in order to arrive at the other dock. Otherwise, the river current would push the ferry downstream and it would miss the dock. If the velocities of the river current and the ferry are known, then the captain can calculate the direction in which he must steer to end up at the other dock. The velocity of the river current forms one leg of a right triangle and the velocity of the ferry forms the hypotenuse (because the captain must point the ferry diagonally across the river to account for the current).

If the current is moving at 3.1 mph (5 km/hr) and the ferry can travel at 7.5 mph (12 km/hr), the angle at which the ferry needs to travel is found by calculating its sine. In this case, the sine of the unknown angle is 5/12 = 0.4167. The angle can then be determined by looking in a table of trignometric functions to find the angle that most closely matches the calculated value of 0.4167, by using a calculator to calculate the sines of different angles and comparing the results, or by using the arc sine (asin) function. Each of the trigonometric functions has an inverse function that allows the angle to be calculated from the value of the function. In this case, the answer is sin 0.4167 = 25°. In other words, the captain must point his ferry 25° upstream in order to account for the current and arrive at the dock directly across the river.

Another application of vectors and trigonometry involves weight and friction. Automobiles and trains rely on friction to move uphill or remain in place when parked, and friction is required in order to hold soil and rock in place on steep slopes. If there is not enough friction, cars will slide uncontrollably downhill, and landslides will occur. Even if a car is traveling downhill, friction is required to steer. In the simplest case, the traction of a vehicle or the resistance of a soil layer to landsliding depends on three things: the weight of the object, the coefficient of friction, and the steepness of the slope. The weight of the object is self-explanatory. The coefficient of friction is an experimentally measured value that depends on the two surfaces in contact with each other and, in some cases, temperature or the rate of movement. The value used before movement begins, for example between the tires of a parked car and the pavement or a soil layer that is in place, is known as the static coefficient of friction. Once the object begins moving, the coefficient of friction decreases and is known as the dynamic coefficient of friction. Some typical examples of coefficients of friction are 0.7 for tires on dry asphalt, 0.4 for tires on frosty roads, and about 0.2 for tires on ice. The coefficients of friction for soils involved in landslides can range from about 0.3 to 1.0, with most values around 0.6.

Because weight is a force that acts vertically downward, trigonometry must be used to calculate the components of weight that are acting parallel to the sloping surface. The frictional force resisting movement parallel to the slope is $\mu \times w \times \cos \theta$, where w is the weight, μ is the coefficient of friction, and θ is the slope angle. The component of the weight acting downslope is $w \times \sin \theta$. Division of the frictional resisting force by the gravitational driving force gives the expression $\mu / \tan \theta$. If the result is equal or greater than 1, the car or soil layer will not slide downhill. If it is less than 1, then downhill sliding is inevitable. If the coefficient of friction for tires on dry asphalt is 0.7, then parked cars will slide downhill if the slope is greater than 35°. If the road is covered with ice, however, the coefficient of friction is only 0.2 and cars will slide downhill on slopes greater than 11°.

using Taylor series, that a complex exponential can be expressed as a sum of two trigonometric functions, one multiplied by the square root of −1. This discovery opened the way to many results in mathematics and technology. In particular, the use of Euler's identity allows the handling of periodic phenomena—events that repeat regularly in time, like the voltage of ordinary alternating-current power—using complex exponentials, which, despite their formidable-sounding name, are easier to manipulate mathematically than sinusoids. Today, complex exponentials are a standard tool in such fields as electrical engineering, where sinusoidal signals are commonplace.

The last major innovations in the fundamental theory of trigonometric functions came in the nineteenth century; since that time, there has been little mathematical innovation in this area. In the late nineteenth century, the development of non-Euclidean geometries—which would be essential to the development of general relativity theory in the early twentieth century—triggered the exploration of non-Euclidean trigonometries, in which, for example, the sum of the interior angles of a triangle is not 180°, as in ordinary plane geometry. In 1822, Fourier discovered that any periodic function (one that repeats itself exactly over a fixed time period), no matter how complex its shape or even if it contains sudden jumps or discontinuities, can be written as a superposition or adding-up of sine functions shifted and amplified by different amounts. That is, any periodic waveform or function can be built by piling up appropriately shifted and scaled sinusoids. A sum of sinusoids representing a particular periodic function is called a Fourier series.

Trigonometry has played a role in many computations over the years. For example, this 50-ft (15-m) Michaelson Pease Interferometer above the Mt. Wilson Observatory in California (pictured in 1930 with Dr. G.F. Pease) measured stars as large as 400,000,000 mi (643,737,600 km) in diameter. The instrument measured the heavenly bodies in fractional degrees of an arc, and from these figures, the diameter in miles was computed by means of trigonometry. *© Bettmann/Corbis.*

■ Modern Cultural Connections

Trigonometric functions are well-established mathematical tools that permeate the working equations of mathematics, social and physical science, and technology. There is no controversy about their basic nature or validity.

Fourier's insight—that one realm of functions can be mapped back and forth to another—extended in the nineteenth century to non-periodic signals (in the Fourier transform) and in twentieth century to discrete functions consisting of lists of numbers (in the discrete Fourier transform) and has found application throughout science and technology. It is by using such transform techniques—rooted directly in trigonometric functions and their close relatives, complex exponentials—that scientists can examine the frequency spectra of data gathered over time, such as audio signals or geological markers of ancient climate variations. Digital filtering of sampled audio signals often relies on the discrete Fourier transform and related techniques.

Significant developments in the application of sinusoidal functions have occurred at least as recently as the 1980s, when modern wavelet transforms were developed. These are now commonly applied in signal processing (for example, in removing noise from audio and video recordings). In some applications, wavelet techniques are superior to Fourier analysis, for example, chemists rely on trigonometry to analyze unknown substances using methods such as Fourier transform spectroscopy.

Trigonometric principles and calculations have applications in virtually every discipline of science and engineering, so their importance will continue to increase as technology continues to grow in importance, especially in fields such as computer and global positioning system technologies.

Computer Technologies

Both two- and three-dimensional computer graphics applications make heavy use of trigonometric relationships and formulae. Rotating an object in two dimensions, for example a spinning object in a video game or the text in an illustration, requires calculation of the sine and cosine of the angle through which the object is being rotated. Graphics objects are typically defined using points for which x and y coordinates are known. In some cases, the points may represent the ends of lines or the vertices of polygons. Many computer programs that allow users to rotate objects require the user to enter an angle of rotation or use a graphics tool that allows for freehand rotation in real time. Each time a new angle is entered or the mouse is moved to rotate an object on the screen, the new coordinates of each point must be quickly calculated.

Rotation of graphical objects in three dimensions is much more complicated than it is in two dimensions. This is because instead of one angle of rotation, three angles must be given. Although there are several different conventions that can be used to specify the three angles of rotation, the one that is most understandable to many people is based on roll, pitch, and yaw. These terms were originally used to describe the three kinds of rotation experienced in a ship as it moves across the sea and were adopted to describe the motion of aircraft in the twentieth century. Roll refers to the side-to-side rotation of a ship or aircraft around horizontal axis. An aircraft is rolling if one of its wings is going up as the other goes down. Pitch refers to the upward or downward rotation of the bow of a ship or the nose of an aircraft. As the bow or nose goes up, the stern or tail goes down and vice versa. The final component of three-dimensional rotation is yaw, which refers to the side-to-side rotation of the nose or bow around a vertical axis. Just as in two-dimensional graphics, the simulation of three-dimensional rotation by a computer program requires that trigonometric functions be calculated for each of the three angles and applied to each point or polygon vertex. Three dimensional graphics are also more complicated because the shape of each object being simulated must be projected onto a two dimensional computer monitor or other plane, just as Earth's spherical surface must be projected to make a map.

GPS Technology

Global positioning system (GPS) receivers embedded in cellular telephones, vehicles, and emergency transmitters will allow lost travelers to be located and criminal suspects to be tracked. Fast Fourier transforms will help to advance any kind of computation involving waveforms, including voice recognition technologies. Trigonometric calculations related to navigation will also become even more important as global air travel increases and the responsibility for air traffic control is increasingly shifted from humans to computers and GPS technology.

Even when GPS receivers are used to determine the locations of unknown points, the locations of known points are used to increase accuracy. This is done by placing one GPS receiver over a known point and using a second receiver at the point for which a location must be determined. In the United States, one of the known points might be a continuously operating reference station, or CORS, operated by the government and providing data to surveyors over the Internet. Data from the two receivers are combined, either in real time or afterwards by post-processing, to obtain a more accurate solution that can be accurate to a millimeter or so. Although it may not be obvious because the calculations are performed by microprocessors within the GPS receivers and on computers, they require extensive use of trigonometric functions and principles.

Once the locations of points or features are determined, they must be plotted on a map in order to be visualized. If the area of concern is relatively small, the map can be constructed using an orthogonal grid system of perpendicular lines measuring the north-south and east-west distance from an arbitrary point. If the area to be mapped is large, however, then trigonometry must be used to project the nearly spherical surface of Earth onto a flat plane. Over the centuries, cartographers and mathematicians have developed many specialized projections involving trigonometric functions. Some are designed so that angles measured on the flat map are identical to those measured on a round globe, some are designed so that straight-line paths on the globe are preserved as straight lines on the planar map.

SEE ALSO *Earth Science: Geodesy.*

BIBLIOGRAPHY

Books

Evans, James. *The History and Practice of Ancient Astronomy.* Oxford, UK: Oxford University Press: 1998.

Monmonier, Mark. *Rhumb Lines and Map Wars: A Social History of the Mercator Projection.* Chicago: University of Chicago Press, 2004.

Neugebauer, Otto. *A History of Ancient and Medieval Astronomy.* New York: Springer-Verlag, 1975.

Westfall, R.S. *Never at Rest: A Biography of Isaac Newton.* Cambridge University Press, 1980.

Periodicals

Duke, D. "The Equant in India: The Mathematical Basis of Indian Planetary Models." *Archive for History of Exact Sciences* 59 (2005): 563–576.

Ari Belenkiy

Physics: Aristotelian Physics

■ Introduction

No other philosopher had such a deep and long-standing impact on Western science as the Greek philosopher Aristotle (384–322 BC). In the fourth century BC he developed a fully comprehensive worldview that would, with only a few modifications, stand for about 2,000 years. Rather than merely collect isolated facts, he posed fundamental questions about nature and the methods needed to study it. Physics in the Aristotelian sense was a fundamental understanding of matter, change, causality, time, and space, all of which had to be consistent with logic and experience. From this he derived a cosmology that allowed him to explain all phenomena from everyday life to astronomy, including both natural phenomena and technology.

■ Historical Background and Scientific Foundations

Aristotle was born in Stagira, on the Chalcidic peninsula of Macedonia; his father, Nichomachus, was physician to King Amyntas III of Macedonia. Aristotle lived in a time of extreme political turbulence that deeply influenced his life. When the 17-year old Macedonian moved to Athens to enroll at the famous Academy of Plato (c.428–c.348 BC), the city-state had lost its former political hegemony, but still had an international reputation in education. When Amyntas's son, King Philip II, began to conquer the Greek states in 359 BC, it spawned a wave of anti-Macedonian sentiment that almost certainly made Aristotle's life difficult. When his patron Plato died in 347 BC and Athens declared war against Macedonia, Aristotle left for the city of Assos in Asia Minor (modern-day Turkey), where he led a group of philosophers.

Around 343 BC, Philip brought Aristotle to his court as a tutor for his son Alexander (356–323 BC), a brilliant young man who would go on to conquer the world's largest empire to date, ranging from Greece eastwards to India and southwards to Egypt. Under Alexander the Great's rule, Aristotle returned to Athens at the age of 49 to found a new school called the Lyceum. When Alexander died only 13 years later and his huge empire fell apart, Aristotle left for Chalcis, where he died shortly afterward.

Aristotle's intellectual work was truly encyclopedic. His writings cover fields as diverse as logic, epistemology, metaphysics, rhetoric, physics, chemistry, biology, psychology, political studies, ethics, and literature studies; many of these disciplines, most notably logic and biology, claim Aristotle as their founding figure. Even in mathematics, which Aristotle conspicuously neglected (although it was then a major topic at Plato's Academy), he influenced Euclid's (c.325–c.265 BC) geometry through his axiomatic approach to logic. Moreover, Aristotle's general approach became the standard scientific method for about 2,000 years.

Unlike other philosophers, who presented their views in aphorisms or narratives, Aristotle developed a systematic approach. For each issue, he first collected all the views and arguments by his predecessors, which makes his work a rich source for historical studies. Then he clarified the meaning of all pertinent concepts and analyzed the various views and their conflicts. To resolve a fundamental issue, Aristotle drew on different sources. Were the views in accordance with available empirical data? Were the arguments sound? Did the views appeal to common sense? Finally, did they fit with knowledge previously established by the same method? Working incrementally through the entire realm of knowledge with this method, Aristotle built a stable philosophical system that covered almost every discipline and stood for about two millennia with only slight modifications.

WORDS TO KNOW

ALCHEMY: The study of the reactions of chemicals in pre-modern times. It was often, but not always, directed by the goal of making gold. In a general sense, alchemy is perceived as the transmutation (or, transformation) of a common substance to something rare and valuable. Medieval alchemists are often portrayed as little more than quacks attempting to make gold from lead. This depiction is not entirely correct. To be sure, there were such characters, but for real alchemists, called adepts, the field was an almost divine mixture of science, mystery, and philosophy.

CALCINATION: An old term used to describe the process of heating metals and other materials in air.

COSMOLOGY: The study of the universe as a whole, its nature, and the relations between its various parts.

ELEMENT: Substance consisting of only one type of atom.

ETHER: Also spelled aether; the medium that was once believed to fill space and to be responsible for carrying light and other electromagnetic waves.

GEOCENTRIC: A geocentric model of the solar system places a stationary Earth at the center of the solar system, with the sun and planets orbiting Earth.

HELIOCENTRIC: A heliocentric model of the solar system places the sun at the center, with the planets and Earth orbiting around it.

MATTER: Anything that has mass and takes up space.

MECHANICAL PHILOSOPHY: Mechanical philosophy was a school of thought that prospered in the 1600s and asserted that the universe consists entirely of atoms in motion obeying mechanical laws (where the term "mechanical" refers to the motions and interactions of solid objects, not necessarily to the behavior of artificial machines).

MONISM: Any philosophy that denies the existence of duality or division in some aspect of being, such as between mind and body or nature and God. The term can refer to a philosophy or theology that denies the reality of all distinctions whatever.

Bust of ancient Greek philosopher Aristotle (384–322 BC). © *Bettman/Corbis.*

craft like carpentry, and optics was either a theory about visual sensation, or a branch of mathematical geometry. For Aristotle and his followers, mathematics was clearly distinct from physics, because it described nature in purely numerical terms. The task of physics was to explain nature.

Aristotle's approach is still appealing today because of his straightforward reasoning. For him, explaining nature meant answering "why" questions, insisting that scientists have fulfilled their duty only if all questions have been answered satisfactorily. He observed that people ask four different types of why questions, each of which requires an answer that reflects a distinct cause. Consider an example that covers the four different causes: "Why does a knife cut meat?" If you respond that the knife is made of iron, which is harder than meat, you are referring to the material cause. Arguing that the knife has a sharp blade provides the form cause. If you explain the mechanism by which the knife takes the meat apart, you give the efficient cause. And if you say that the knife can cut meat because that is the purpose for which it has been made, you provide the final cause. For a satisfying answer, you must refer to all four causes, although their relative importance may differ from case to case.

Of course, a meat-cutting knife is not an example of physics (the study of nature) in the ancient sense,

The Causality of Nature

The English term "physics" is derived from the Greek *physike episteme*, the knowledge and study of nature, or *physis*. Even in the early nineteenth century, physics was a blanket term for natural philosophy, which covered all scientific disciplines. In antiquity, however, the fields of modern physics (e.g., electricity, magnetism, and thermodynamics) were either undeveloped or misunderstood. For instance, mechanics was considered a because knives are artifacts and not natural things. Aristotle, however, was convinced that we ask the same four kinds of why questions for both. In particular, unlike modern physicists, he thought that scientists must not forget the final cause to provide satisfying answers. For instance, a blooming flower could not sufficiently be explained simply by explaining the mechanism that makes the flower bloom. A satisfying answer, according to Aristotle, must refer to the bloom's purpose—enabling the flower to reproduce—which he thought was embedded in the flower like an unfolding program. Moreover, the flower's proper form develops only in the state of blooming, and this is not only part of our concept of flowers, it is also part of the flower itself throughout its development.

Aristotle defined natural things as those that develop and are what they are only by virtue of causes internal to them. Artifacts, in contrast, are made by humans according to human goals, which are external to the objects. Examples of natural things are stars, animals, plants, stones, clouds, and basic materials; examples of artifacts are houses, furniture, cloth, and tools. However, the distinction is not a simple one. For instance, when a rotting chair looses its original form, it is still an artifact insofar as it is a piece of furniture, but it becomes a natural thing, a piece of matter, insofar as rotting is a natural process determined by its basic material properties. A hedge is natural insofar as it is a plant that grows according to its own principles, but artificial insofar as humans have shaped it into a certain form for human ends. Hence, the world cannot simply be divided into natural and artificial things—the division depends on how we perceive them.

The Dynamics of Nature

Aristotle was convinced that nature is essentially dynamic, and that natural things are under continuous development. Thus, understanding a natural thing requires two perspectives: we need to know 1) what the thing is composed of, and 2) how and why the thing alters. In response to the first question, Aristotle developed a metaphysical scheme that shaped his entire philosophy: Every real thing, whether natural or artificial, is composed of matter and form. For instance, a brick consists of clay shaped like a rectangle. Rectangular forms that are not materialized, as in geometry, are not real things but simply mathematical ideas. On the other hand, real things can be the material of which other real things consist if they are arranged in a certain form. For instance, bricks are the material for building houses and houses are the material of cities. Aristotle used this scheme to build up the entire cosmos.

To understand the dynamics of natural things, Aristotle distinguished between four kinds of processes. First, a thing can simply move in space without being changed. Second, it can grow or shrink, i.e. increase or decrease in size, without changing its characteristics.

Third, it can undergo qualitative changes—such as when a tadpole is transformed into a frog—without losing its identity. Finally, it can undergo substantial change when it emerges out of or turns into something entirely different; this occurs, for instance, when an animal dies and decomposes into basic materials or when basic materials undergo chemical transformation. Once we have identified the kind of change—spatial, quantitative, qualitative, or substantial—we can investigate its cause, which for Aristotle are both the efficient and final causes.

In every change, Aristotle believed, something must persist throughout the process. While this is obvious with spatial and quantitative changes, it is more difficult to identify in qualitative and, particularly, substantial changes. According to Aristotle, the matter of each real thing persists even though its form changes. For instance, when we form a mug from a lump of clay, the clay persists and gradually changes its form, going from a lump to a mug. Since some forms can't be made from clay (spider webs, for instance), matter and form are related. Thus, clay has the potential to assume the form of a mug, but not that of a spider web. This is more important for natural processes, where the causes of change are internal. For instance, a tadpole has the hidden potential to assume the form of a frog instead of a bird or something else. Therefore, Aristotle also described any process as a change from potentiality (a potential frog) to reality (a real frog).

Furthermore, Aristotle thought that change always requires some interaction between the changing thing and its cause, and that the change ends when the interaction stops, an idea that was revised in early modern mechanics. For instance, if we heat water with fire, fire acts on water because water is susceptible to the action of fire; as soon as we stop heating it, the water cools down. Similarly, if a change is driven by a final cause, the object of change needs to be susceptible to this final cause and stop changing as soon as the final cause is removed.

The Elements of Nature before Aristotle

One of Aristotle's most persistent contributions to science, and indeed the core of his physics, was his theory of the elements, which endured until the end of the eighteenth century and the dawn of the chemical revolution. Apart from astronomy, the theory of the elements was the core of ancient natural philosophy. It explained the plurality and change of all matter, disciplines now called chemistry and particle physics. Unlike today's scientists, however, ancient philosophers rarely conducted experiments, but searched instead for rational systems that were in accordance with all available and observable data. Before we deal with Aristotle's solution, we will briefly look at those of his predecessors.

Little is known of the pre-Socratic philosophers; only indirect reports and a few extant fragments remain, and those are difficult to understand. By the seventh century BC, Greek philosophers had broken with their religious traditions, rejecting the idea that natural phenomena were caused by supernatural forces. Instead they characterized the ultimate principles of nature by their material properties. Many pre-Socratics were monists, who argued that a single material principle underlay the plurality and change of all matter. For Thales (c.624–c.546 BC) this principle was water, for Anaximenes (c.585–c.528 BC) it was air, and for Heraklitus (c.540–c.480 BC), fire.

Pluralists, like Anaxagoras (c.500–c.428 BC), assumed that the infinite plurality of things required infinite principles, and that any change is caused by the mixing and separating of the elements. Working from Pythagoras' (c.582–c.500 BC) idea that everything is founded in the dualism of opposing principles, Empedocles (c.495–c.435 BC) developed the first ancient synopsis on which Aristotle would later draw. He combined the earlier suggestions of water, air, and fire with earth into a system of four elements that interacted with each other by the opposing principles of attraction and repulsion to form the plurality of all things.

The most interesting account may be that of atomism, expressed by Democritus (c.460–c.370 BC) and based on the earlier ideas of Leukippus (fl. 5th c. BC). On one hand, atomism resembled Anaxagoras' pluralism, because Democritus claimed that there were an endless number of atoms that form the variety of things, and that all change is caused by their separation and mixing. On the other, ancient atomism was a dualistic doctrine, because its proper principles were matter and void. Thus, atoms (from Greek *atomos*, indivisible, uncut) were thought to be a certain distribution of matter and void, such that matter forms invisibly small regions of irregular shapes that persist through all changes in time.

Atomism remained a prominent but much-contested doctrine throughout its history. Its critics, first among them Aristotle, had many objections. Since matter, according to Democritus and unlike all the other philosophies of nature, had no material properties, it was unclear how it differed from void. When Democritus argued that matter was full whereas void was empty, critics objected that the empty void was not a principle of nature but merely nothing, and that claiming the existence of nothing was a contradiction. The debate continued up to early modern times as the question of whether or not vacuums could exist.

Others argued that there was no empirical evidence for the existence of atoms. Further, since matter had no material properties, every explanation of material properties based on supposed atomic shape was highly speculative. Indeed, Democritus and his followers arbitrarily claimed various shapes to explain differences in color, taste, or any other empirical properties. Finally, atomism was a difficult concept for many people to grasp.

The idea that matter could be indivisible and that it had no intrinsic properties were counterintuitive, because evidence suggested just the opposite.

Plato had developed his own version of atomism that drew on earlier Pythagorean ideas, some sophisticated mathematics, and the doctrine of Empedocles. Although it was esoteric even for contemporaries, it became influential because Plato set his theory in the form of a creation myth. In it, the divine creator builds the world according to geometrical ideas by shaping not matter but space. Empedocles' elements of fire, air, water, and earth consisted of four invisibly small regular polyhedra. These were not atoms but consisted of indivisible triangles of two different types. Plato selected their mathematical construction in such a way that several material changes (e.g., fire boils water to become air-like steam) could be explained by a quasi-geometrical mechanism. For instance, the sharp-edged tedrahedra of fire could split the blunt-edged dodecahedra of water into their composing triangles, which could then reassemble to form the octahedra of air.

Aristotle's Elements of Nature

Aristotle rejected both kinds of atomism, and argued that Plato's system confused mathematical ideas with real things. Instead, he added a new foundation to Empedocles' four elements. In Aristotle's view, the elements of nature must represent the fundamental characteristics of nature, i.e., they must bear the basic properties of matter that drive the dynamics of nature. Matter's basic characteristic is its tangibility, which for Aristotle included two tactile properties: matter is more or less dry (hard) or wet (soft) and more or less cold and hot.

To cover the whole realm of these two property dimensions, each element had one extreme property from each dimension, producing four pairs of properties to which Aristotle related Empedocles' four elements: dry and cold were the characteristics of earth, wet and cold those of water, wet and hot those of air, and dry and hot those of fire. Moreover, for Aristotle, hard and soft were passive properties, because they determined the malleability of materials, whereas hot and cold were active properties because they could act on other materials. For instance, water expands if it is heated by fire and shrinks if it is cooled. The two pairs of properties thus represented both the empirical characteristics of matter and the basic interactions between materials.

Aristotle used his theory of the elements to explain a wealth of natural phenomena ranging from chemistry, physics, and meteorology to biology and medicine. Moreover, his theory allowed him to write the first treatise on what we would call the chemical processing of materials, including metallurgy and cooking. He made no fundamental distinction between natural and technological phenomena, because the materials and their interactions were essentially the same in natural and artificial processes. Furthermore, the elements structured the entire world in two different approaches.

The Hierarchical Structure of the World

As in Plato's theory, Aristotle's elements could interact with and transform each other. When an excess of fire (hot and dry) acted on water (cold and wet) to neutralize the elemental property cold, water turned into a kind of air (hot and wet). For Aristotle, the elements were real things, although they occurred only in impure forms or mixtures. According to his metaphysical doctrine, both elements and real things must be composed of form and matter, and their elemental properties were their specific form.

In elemental transformation, his paradigm for substantial change, the elemental form was replaced. His theory of substantial change required that a primary matter, devoid of any qualities, persisted through the change. Since primary matter had no qualities and no form, it was not a real thing but only the bearer of elemental properties and the substratum of substantial change that united the physical world. Nonetheless Aristotle's primary matter would later inspire numerous misunderstandings, particularly among alchemists in their experimental search for the basic principle of matter.

Starting with the elements composed of primary matter and their specific form of elemental properties, Aristotle developed a hierarchy of the physical world in which each step provided the matter for the next. For basic compounds, the elements served as matter and their composition as specific form. In the next step, heterogeneous compounds such as wood could be combined and structured to form parts of living beings, like a human arm or the trunk of a tree. If combined and organized according to certain forms and ends, they would form a living being. For Aristotle this required at least a "vegetative soul" to serve as the organizing principle and to control the metabolism. Animals differed from plants by an additional higher-order soul that allowed living beings to move and feel. Humans were endowed with an additional "intellectual soul" that enabled them to organize their life according to ideas and goals.

The inorganic world, including air, water, and earth, was spatially and chronologically structured to form regular and periodical phenomena like the weather and the seasons, for which Aristotle identified the sun, the moon, and the stars as their structuring and moving principle. Finally, since for Aristotle every movement must have a cause, he postulated gods as the ultimate cause of the regular motion of the stars. Like primary matter, these gods were not real things composed of matter and form. Rather, like the human intellect, which can organize real events through its nonmaterial existence and activity, the gods were pure form and so-called "unmoved movers." Entities of complete independence and modesty, they also served as models for human beings.

IN CONTEXT: ARISTOTLE'S COSMOS

For the foundation of his work, Aristotle turned to the ideas of his predecessor Empedocles (c.495–c.435 BC). Empedocles thought that change was the result of the interaction of four elements: fire, water, earth, and air. The forces of love and hate acted upon these elements, and their interaction first led to minerals, then plants, then animals from a long series of trial-and-error interactions.

Aristotle also thought that all matter in the terrestrial part of the universe, comprised of the area below the sphere of the moon to Earth (sublunar region), was made by interactions between the four elements. Earth, which was made of the heaviest earthly element, rested immovably in its natural place, at the center of the sublunar region. The watery sphere surrounded Earth, but the boundaries between the earth and water were irregular, because the higher parts of the land projected above the oceans that surround our globe. The sphere of the air was next. Above it, but below the moon, was the sphere of fire, which was the lightest of the elements and the transition to the eternal realms of the planets. This sublunar realm was a realm of change and corruption.

In the area of the universe beyond the moon, everything was made of a fifth element he called the ether. Planetary orbits were solid crystalline spheres made of the ether, to which the perfectly polished planetary bodies, also made of the ether, were firmly attached. The heavens were immutable.

The forces behind the movement of the elements were ultimately due to an eternal being called the Prime Mover. Taking a cue from Empedocles, Aristotle's Prime Mover functioned as an object of love and desire for the soul that animated the body of the outermost sphere of fixed stars, the *primum mobile*. The *primum mobile* rotated at an enormous speed every twenty-four hours and communicated this motion to the planetary orbital spheres. Later Christian commentators such as St. Thomas Aquinas (c.1225–1274) adapted Aristotle's idea of the Prime Mover to their conception of God, another reason for the durability of Aristotelian philosophy.

Anna Marie Eleanor Roos

The Cosmological Structure of the World

Aristotle viewed the cosmos as a series of spherical shells, each related to one element, around Earth. Ancient Greeks knew that the planet was a sphere, and even measured it with some precision. It consisted mainly of the element earth, with a surface largely covered by water, and its atmosphere was dominated by air. The lower atmosphere was filled with moisture—clouds and rain—owing to turbulence at the interface between the water and air shells that determined the weather. Above the atmosphere, the next sphere, reaching up to the height of the moon, was filled mainly with the element fire.

Aristotle saw ample empirical evidence of this shell model. In particular, earth was heavier than water, which in turn was much heavier than air; and flames obviously rose up into the air. In water, a stone sank down whereas a bubble of air rose up. Based on such empirical regularities he drew the general conclusion that each element tended to move to its specific shell, which he called its proper place. This theory could also explain any ordinary phenomenon on Earth that we now explain by the force of gravity.

Above the moon, things were obviously different, since the sun, stars, and planets appeared to move in semiregular circles around Earth, a motion that was, without the help of additional forces, impossible on Earth. Because of this, Aristotle postulated that the stars and their surroundings were composed of an entirely different matter, unknown to humans, which he called ether and which enabled circular rather than straight-line motion.

Aristotle's model was based on the theories the astronomer Eudoxos (c.395–342 BC), who developed a complex geometric model that explained the stars' irregular motions by the superposition of many regular circles. This geocentric cosmological model, with Earth at its center and all the celestial bodies moved around it in circular orbits, was later developed in greater detail by the Greek mathematician and astronomer Claudius Ptolemy (c.AD 90–c.168). As early as the third century BC, however, an astronomer from Aristotle's own school, Aristarchus of Samos (c.310–c.230 BC), suggested that the sun was at the center of the cosmos, with Earth moving around it. This heliocentric model, although known to many succeeding astronomers, did not gain acceptance until Polish astronomer Nicolaus Copernicus (1473–1543) developed it with a mathematical rigor that could explain the irregular orbits with greater precision.

Aristotle's cosmology would be incomplete without his views on time and space. If you ask, "What is in space beyond the sphere of the stars?" Aristotle would have responded that this question has no meaning because there is no space beyond the sphere of the stars. For him the entire cosmos was a huge but finite sphere composed of matter, with each element, including the ether, in its specific place. For Aristotle, space without matter did not exist, in either cosmology or in atomism.

Unlike space, however, he saw time as infinite, without beginning or end. The cosmos was eternal, without beginning or end, because both its emergence out of nothing and its vanishing into nothing violated the basic principles of his metaphysics of change. Moreover, owing to the regular movement of the stars, and ultimately to the eternal nature of the gods, neither radical nor evolutionary changes were possible. Indeed, Aristotle believed that biological species did not evolve but were stable, in the same way that minerals were. Even if, by some natural disaster, some species disappeared, the long-term balanced conditions on Earth would enable its reemergence.

Aristotle's Natural Philosophy in the Middle Ages

After the fall of the Roman empire in the fifth century AD, most of Aristotle's philosophy was lost to the West for centuries. His works were translated into Arabic in the Muslim world, where they helped form the basis of much Islamic science from the eighth century onward. As Europe made contact with the eastern world in the late Middle Ages, the writings of classical philosophers were rediscovered.

The earliest translations of Aristotle's scientific works from Arabic to Latin in the twelfth century shocked many medieval Christians. Until then only fragments of his logic had been known, but even that was enough to make him the unquestioned authority in all logical and philosophical matters. Now they learned that the revered philosopher had taught that the world was not created by God, as the Bible said, but eternal, without beginning and end. Moreover, Aristotle had defined gods as "unmoved movers" who guaranteed the eternal movements of the stars, but did not intervene in worldly events, effectively dismissing the idea of miracles or the role of angels.

The writings of the greatest Arabic Aristotlean scholars included those of Averroes (1126–1198), whose numerous commentaries on Aristotle's teachings were particularly influential. He posited that the human soul could not, according to Aristotle, survive physical death, which contradicted the Christian doctrine of the soul's immortality. Thus, Christian authorities' first reaction was to ban the teaching of Aristotle's science altogether on pain of death. However, the German philosopher Albertus Magnus (c.1200–1280), and particularly his pupil Thomas Aquinas (c.1225–1274), undertook enormous efforts to reconcile Aristotle's natural philosophy with Christian doctrine by writing voluminous commentaries that explained in great detail how Christians should interpret Aristotle's texts.

Thanks to these commentaries, Aristotle's revised natural philosophy moved into the core curricula of the newly established European universities, where it remained for at least four centuries. Furthermore, Aquinas's blend of Aristotelian and Christian views, which came to be known as Thomism, was made the official doctrine of natural philosophy and metaphysics by the Roman Catholic Church and has remained so up to today.

This theological assimilation gave Aristotle's natural philosophy an extraordinary status. On the one hand, any criticism or differing views were threatened by official sanctions, ranging from a ban on teaching

Portrait of Moorish philosopher and physician Averroes (1126–1198), twelfth century. Averroes adapted Aristotle's theories to Islamic theology, and much of our knowledge of the Classical philosophers comes through him. *NY Library Picture Collection.*

and publishing to excommunication and even death. The Italian philosopher Giordano Bruno (1548–1600), for example, was burned at the stake for his rejection of Aristotelean cosmology and his proposition of a heliocentric solar system and infinite universe. On the other, it closely related natural philosophy to theology, infusing all debates on natural philosophy, including attempts to overcome the Aristotelian system and to establish what we call modern science, with religion. Since Aristotelian natural philosophy was administered by the church, however, criticism grew with the Protestant Reformation.

Early Attempts to Overthrow the Aristotelian System

Apart from its Christian assimilation, Aristotle's natural philosophy was a strong system based on intertwined metaphysical principles that could not easily be altered. Radical changes were required to build a new system, but any such change was threatened with persecution. The French philosopher and mathematician René Descartes (1596–1650) solved this paradox by building a new system based on selected and remodeled Aristotelian principles. Where Aristotle had claimed four dif-

ferent causes in nature (formal, material, efficient, and final) with which scientists must explain natural phenomena, Descartes selected only the efficient cause.

The scientist's task, according to Descartes, was to explain all natural phenomena solely by its causal mechanism. Similarly, of Aristotle's four kinds of change (spatial, quantitative, qualitative, and substantial), Descartes choose only spatial motion, declaring that any qualitative or substantial change could ultimately be reduced to the motion and collision of particles in space. He remodeled Aristotle's principles of form and matter to become geometrical form and spatial extension, and characterized the elements by the geometrical form and size of their particles rather than the elemental qualities of hot, cold, wet, and dry. In the end, Descartes's universe strongly resembled ancient atomism, with invisible particles swirling around, but he rejected both the ideas of an empty space or vacuum and of indivisible particles.

Descartes's new emphasis, however, was the idea that the mechanism of any particle motion (and thus any natural phenomena) could be expressed mathematically. He developed a set of mathematical theories that would strongly influence English physicist and mathematician Isaac Newton's (1642–1727) later principles of mechanics. Indeed, Descartes, along with Italian mathematician and astronomer Galileo Galilei (1564–1642), formulated what we now call the principle of inertia, according to which a body set in motion tends to continue its motion in a straight line as long as no other external cause interferes.

This was an important departure from Aristotelian physics in two regards. First, Aristotle taught that motion or change continues only as long as the moving cause is effective; inertia required only a moving cause at the beginning of the motion. With respect to the entire universe, an initial impetus would suffice to cause all the succeeding dynamics of the universe. That idea was theologically appealing to Descartes and his followers of mechanical philosophy because it could convert Aristotle's "unmoved mover" into God, who started the dynamics of the universe at creation.

Second, since Descartes (unlike Galileo) claimed his principles were valid for all motions, he rejected the Aristotelian distinction between earthly and celestial physics. In particular, he dismissed the prominent idea that the natural motion of the stars was circular rather than straight and instead tried to explain the quasicircular movement of celestial bodies by gigantic vortices of celestial particles. This approach, despite its weaknesses, would later inspire Newton to unite earthly and celestial mechanics with the common force of gravitation.

The science of ballistics began to develop in the sixteenth century, following the model of the ancient mathematician and engineer Archimedes (c.287–212 BC), who used empirical measurements to solve engineering problems. In an effort to maximize the range of

René Descartes (1596–1650). *Library of Congress.*

inclined plain. The measurements confirmed his mathematical hypothesis, which came to be known as the law of free fall. It allowed Galileo to describe Tartaglia's trajectories as parabolic curves and to prove mathematically what Tartaglia had shown only by empirical tests: The maximum range of projectiles was achieved when the shot was made at a 45° angle. Newton integrated this into his laws of general mechanics, which combined celestial and ballistic motion in a uniform mathematical theory centered on the force of gravitation.

Mechanics was only a marginal part of Aristotle's comprehensive natural philosophy, because outside of astronomy it did not apply to natural phenomena. Although the rise of mathematically based mechanics by Descartes, Galileo, English natural philosopher Robert Boyle (1627–1691), Newton, and others is now called the scientific revolution, it did not touch on most topics covered by Aristotle. Indeed these subjects, the bulk of modern scientific disciplines, remained deeply influenced by Aristotle's philosophy for centuries. His biology in particular stood almost unmodified well into the nineteenth century, when it was finally eclipsed by Darwin's theories of evolution and natural selection.

Aristotle's theory of elements and compounds was the foundation of eighteenth-century chemistry, mineralogy, meteorology, geology, and medicine, even though in retrospect it more accurately covered what we today call thermodynamic phenomena, e.g. the boiling or freezing of water, rather than truly chemical transformations. Since it also claimed that one element could be transformed into another, it became the theoretical basis of alchemy, which proposed the very non-Aristotelian idea of studying nature by trying to transform it. Eventually, however, that became the approach of modern experimental laboratory science.

Despite their inability to transform base materials into gold, alchemists or "chymists," in sixteenth-century parlance, did create a plethora of new materials and chemical phenomena in their laboratories that defied Aristotelian explanation. For centuries Aristotelian elements had been supplemented only by additional "chymical principles" to account for such phenomena as burning, calcination, or acid-forming. It was not until the late eighteenth century that the theory of matter was put on a new, experimental basis. Instead of placing the elements in a metaphysical system, as Aristotle had done, French chemist Antoine Lavoisier (1743–1794) defined elements as any material of matter that resisted experimental efforts to take it apart.

projectiles, the Italian military engineer Niccolò Tartaglia (1499–1557) studied their paths. He was the first to analyze the curved trajectory as being simultaneously caused by the (artificial) impetus in the direction of the shot and the (natural) gravity down to the earth.

Once separated analytically, the two components of motion yielded to further empirical studies. The Dutch engineer Simon Stevin (1548–1620) dropped two lead projectiles of different size from the same height; he concluded that their velocity was the same regardless of their weight. This contradicted the physics of Aristotle, who had reached the opposite conclusion from the different velocities of, say, a piece of metal and a feather. Galileo, to whom Stevin's experiment has wrongly been attributed, further studied the motion of falling bodies by combining metaphysically inspired mathematical hypothesis with measurements.

Galileo reasoned that all natural motions must be mathematically simple. This meant that the simplest motion, with constant velocity or distance proportional to time, was reserved for celestial bodies. Freely falling bodies on Earth moved according to the second-simplest motion: constant acceleration or velocity proportional to time. Because sixteenth-century clocks were too inaccurate to prove his hypothesis, he modified the experiment to measure the time that a ball needed to roll down an

■ Modern Cultural Connections

Even though most of Aristotle's scientific answers are now outdated, his texts provide compelling reading. He poses "common sense" questions that provide a benchmark

IN CONTEXT: MODERN SCIENCE, OFTEN COUNTERINTUITIVE, DISPLACES SENSORY "COMMON SENSE"

On the surface, Aristotle's explanations agree with most basic sensory observations of motion, but there were some troubling exceptions that had important implications for the development of modern physics. If a rock is hurled from a catapult, it continues to travel even after it has left the arm of the machine and does not drop to the ground immediately, as Aristotelian physics would predict. Later commentators claimed that the air in front of the rock was disturbed by the motion of the rock and swirled behind the rock and pushed it along. Other physicists, such as John Philoponus (AD 490–570) modified Aristotle's theory with the concept of impetus. Philoponus claimed that a projectile moves on account of a force or impetus the mover gives it, which exhausts itself in the course of the movement. Impetus would keep a projectile moving after it left the catapult. Although erroneous, Philoponus' idea did demonstrate some of the inconsistencies of Aristotelian physics. Galileo later solved this problem by demonstrating that projectile or parabolic motion was the result of inertial and gravitational forces.

Aristotle also attempted to explain the speed of objects in free fall, claiming that the weight of the object divided by the resistance of the medium in which it traveled would result in its speed. This was why, he reasoned, a rock seemed to fall faster in the air than a feather, a claim that would not be disproved until the work of Galileo in the seventeenth century. Aristotle's concept of the speed of motion also did not explain free-fall acceleration, as the speed of objects in his scheme should be constant. Later commentators attempted to explain acceleration in Aristotelian terms as due to the increasing desire of the object to reach its natural place, a concept also shown to be false by Galileo.

According to his equation for speed, Aristotle also realized that if an object moved in a medium without resistance (no friction), it would go infinitely fast and move forever, a concept he rejected as absurd. The concept he rejected was the principle of inertia, later discovered by French philosopher and mathematician René Descartes (1596–1650) and comprising Newton's First Law—the property of an object to remain at constant velocity unless acted upon by outside forces. In a frictionless environment, a pushed object will move forever at a constant velocity.

Aristotle's concept of the natural circular motion of the planets also did not agree with empirical observation, as planets move irregularly in the sky in retrograde or looping fashion along the band of the zodiac. So influential however was Aristotle's cosmological system though, that the purpose of astronomy until the European Renaissance in the sixteenth century was to provide a mathematical theory that preserved the circle as a means of calculating planetary positions, yet explained observed deviations from those circular orbits. By using the devices of the epicycle and the equant point, Roman astronomer Ptolemy (c.AD 90–168) created a set of compounded circles to account for the irregularities in the apparent motions of all of the planets. Ptolemy's book *The Almagest* was the first mathematical treatise that systematically gave a complete and quantitative account of all the celestial motions, and it was based on Aristotle's concept of the heavens. *The Almagest* was not replaced until the work of Polish astronomer Nicolas Copernicus (1473–1543) which demonstrated the solar system was sun-centered, not Earth-centered. German astronomer Johannes Kepler's (1571–1630) first planetary law in the early seventeenth century also demonstrated that planets move in elliptical—not strictly circular—orbits.

Though sometimes internally inconsistent or incompatible with sense observation, Aristotle's physics remained the dominant paradigm in scientific thought for two thousand years.

AnnaMarie Eleanor Roos

of early human understanding of the Cosmos that are increasingly challenged by modern science (which is increasingly counterintuitive, especially in areas of quantum physics).

SEE ALSO *Astronomy and Cosmology: A Mechanistic Universe; Astronomy and Cosmology: Big Bang Theory and Modern Cosmology; Astronomy and Cosmology: Cosmology; Astronomy and Cosmology: Western and Non-Western Cultural Practices in Ancient Astronomy: ; Astronomy and Space Science: Astronomy Emerges from Astrology; Physics: Articulation of Classical Physical Law; Physics: Newtonian Physics.*

BIBLIOGRAPHY

Books

Aristotle. *On the Heavens*, translated by W.K.C. Guthrie. The Loeb Classical Library, Cambridge: Harvard University Press, 1960.

Aristotle. "Physics," *The Complete Works of Aristotle*, 2 vols. Edited by Jonathan Barnes. Princeton: Princeton University Press, 1984.

Aristotle. *Physics.* Introduction and commentary by William D. Ross. Oxford: Clarendon Press, 1998.

Barnes, Jonathan. *Aristotle.* Oxford: Oxford University Press, 1982.

Barnes, Jonathan, ed. *The Cambridge Companion to Aristotle*. Cambridge, Cambridge University Press, 1995.

Crombie, Alistair C. *The History of Science from Augustine to Galileo*. New York: Dover, 1995.

Grant, Edward. *Planets, Stars, and Orbs: The Medieval Cosmos, 1200–1687*. Cambridge: Cambridge University Press, 1994.

Guthrie, William K.C. *The Greek Philosophers from Thales to Aristotle*. London: Methuen, 1950.

Hall, Marie Boas. *The Scientific Renaissance 1450–1630*. New York: Dover, 1992.

Johnson, Francis. *Astronomical Thought in Renaissance England: A Study of the English Scientific Writings from 1500 to 1645*. New York: Octagon Books, 1968.

Koestler, Arthur. *The Sleepwalkers: A History of Man's Changing Vision of the Universe*. New York: Arkana, 1989.

Kuhn, Thomas S. *The Copernican Revolution: Planetary Astronomy in the Development of Western Thought*. Cambridge: Harvard University Press, 1966.

Lang, Helen S. *Aristotle's Physics and its Medieval Varieties*. SUNY Series in Ancient Greek Philosophy. Albany: State University of New York Press, 1992.

———. *The Order of Nature in Aristotle's Physics: Place and the Elements*. Cambridge, UK: Cambridge University Press, 1998.

Lloyd, G.E.R. *Aristotle: The Growth and Structure of His Thought*. Cambridge: Cambridge University Press, 1968.

Sachs, Joe. *Aristotle's Physics: A Guided Study*. Masterworks of Discovery. New Brunswick, NJ: Rutgers University Press, 1995.

Sambursky, Samuel. *The Physical World of Late Antiquity*. New York: Routledge and Kegan Paul, 1962.

Sarton, George. *A History of Science: Ancient Science through the Golden Age of Greece*. New York: Dover, 1995.

Solmsen, Friedrich R.H. *Aristotle's System of the Physical World: A Comparison with his Predecessors*. Cornell Studies in Classical Philosophy. Ithaca, NY: Cornell University Press, 1960.

Waterlow, Sarah. *Nature, Change, and Agency in Aristotle's Physics: A Philosophical Study*. Oxford: Clarendon, 1982.

Periodicals

Grant, Edward. "Aristotelianism and the Longevity of the Medieval World View," *History of Science* 16 (1978): 93–106.

Web Sites

Aristotle. *Physics*. The Internet Classics Archive. http://classics.mit.edu/Aristotle/physics.html (accessed August 28, 2007).

Joachim Schummer

Physics: Articulation of Classical Physical Law

◼ Introduction

A physical law is a description in words or mathematical symbols of a measurable, universally recurrent pattern in nature. For example, Newton's law of gravitation, $F = Gm_1m_2/r^2$, first published by English physicist Isaac Newton (1642–1727) in 1668, describes the force of gravity between any two objects. This physical law was one of the earliest to be discovered and is still one of the most useful. Though simple in form, it explains a great deal of what happens in the whole universe: why stars and planets and moons orbit each other as they do, why a free-flying ball or bullet travels in a parabolic path, and much more. Scientists have devised—and revised—scores of such laws over the last 400 years that now articulate modern science.

The creation of the first great system of physical laws, today called classical physics, began in seventeenth-century Europe and lasted until the end of the nineteenth century, when contradictions and inconsistencies in classical physics came to light. In the early years of the twentieth century a new physics was developed, one that extended the old body of physical laws with the advancement of relativity theory and quantum mechanics. During this period of radical revision, the very idea of a physical law changed; formerly, physicists had argued that the laws they discovered described reality perfectly, but today scientists view many laws as only approximately true. In fact, they spend much of their time trying to discover what revisions are necessary.

◼ Historical Background and Scientific Foundations

The physics that developed from the sixteenth to the early twentieth century is known as *classical physics*. This article restricts its attention the articulation of the laws of classical physics.

Milieu: The State of Science and Society

People have always observed regularities in nature and put their knowledge into words. For example, the phrase "What goes up, must come down" is a sort of common sense law of gravitation, an experience-based generalization about an important physical fact. Although it is not true under all conditions—an object that "goes up" fast enough need never come down again, but can go into orbit or leave the vicinity of Earth forever—such general understandings sufficed human society for tens of thousands of years. Such vague statements are sometimes called folk physics, naive physics, or intuitive physics. They do not predict any quantity that can be measured, such as time, velocity, or acceleration. They describe observation only in a general, and very localized, way.

In the European classical period and in the Middle Ages, intuitive physics was mingled with metaphysical thought—speculation about the nature of reality—to produce a scientific tradition that mixed religion, philosophy, mathematics, and trust in reason. Modern physics arose when this mixed medieval science began, in the late 1500s, to be systematically replaced by a new kind of science, one in which mathematical physical laws were tested by observation and experiment.

A scientific physical law is a statement that predicts how specific measurable quantities such as force, mass, speed, electric-field intensity, number of particles, time, or the like will behave in relation to each other. A law is usually stated as an equation, which is a mathematical expression with two terms or groups of terms separated by an equals sign—for example, Newton's law of gravitation as given above , or $F = ma$ (Newton's second law of motion, force equals mass times acceleration).

WORDS TO KNOW

COPENHAGEN INTERPRETATION: A way of viewing the equations of quantum mechanics. Its earliest proponents were Danish physicist Neils Bohr (1885–1962) and German physicist Werner Heisenberg (1901–1976), who worked together on quantum mechanics in Copenhagen, Denmark, in the late 1920s. According to the Copenhagen interpretation, the wave function that describes every particle's probability of being found in any given place is collapsed by measurement—that is, it is reduced from an infinite number of possibilities to a single, definite state. Other interpretations of quantum mechanics say that the wave function does not collapse, but continues to hold all its values in a state of superposition (all-at-onceness or layered being). On this theory, an infinite number of co-existing universes are being continually generated by all the particle interactions in the universe. The majority of modern physicists today affirm the Copenhagen interpretation, but a growing minority affirms the many-worlds interpretation.

ELECTROMAGNETISM: A form of magnetic energy produced by the flow of an electric current through a metal core. Also, the study of electric and magnetic fields and their interaction with electric charges and currents.

QUANTUM MECHANICS: A system of physical principles that arose in the early twentieth century to improve upon those developed earlier by Isaac Newton, specifically with respect to submicroscopic phenomena.

VARIABLE (MATHEMATICAL): In mathematics, a quantity that can stand for anywhere from two to an infinite number of specific numerical values. A single specific number is termed a constant; variables vary, constants are constant. Variables are most commonly represented by italicized letters of the Latin alphabet, such as x or y.

Once Newton's three laws of motion and the law of gravitation were formulated, any person trained to apply them could tell not only that a cannonball would come down after going up—which had always been obvious—but when it would come down and where, how fast it would be going, and at exactly what angle it would strike (allowing for some complication from air resistance). Moreover, the motions of the planets could be predicted with better accuracy than before. Once the study of moving objects and forces (mechanics) was brought under the sway of mathematical law in Newton's day, laws were also articulated in optics, chemistry, and other fields. New, powerful technologies were devised with the help of these precise laws, transforming industry, trade, and daily life. By providing new instruments for measurement and experimentation, the new

technology in turn transformed science itself. This cycle continues today.

The notion of mathematical physical law tested against observations may seem obvious today, but it only arrived after centuries of painstaking effort. Until the late 1500s, scientific thought was organized around ideas inherited from antiquity. Greek doctor and philosopher Aristotle (384–322 BC) taught that knowledge of the natural world should arise from experience, but did not wed mathematics to experience. The Greek mathematical physicist Archimedes (287–212 BC) elucidated mathematical laws for the simple machines (lever, pulley, etc.), yet Greece was conquered by Rome—Archimedes himself was reportedly killed by a Roman soldier. Roman culture was not, by and large, as interested in philosophy and mathematics as preceding Greek culture—and more emphasis was placed upon military and economic application.

Many of the writings of Aristotle and Archimedes were essentially lost to Europe after the fall of the Roman Empire, and the Arab civilization of the Middle East, which preserved copies, still did not develop a systematic, mathematical, experimental form of science based upon those writings.

One reason that the flourishing of science was delayed for so many centuries is that the way people looked at the world, their worldview, was not congenial to its invention. During the Middle Ages, which lasted roughly from the breakup of the Roman Empire in the AD 400s to the Protestant Reformation of the 1500s, the universe was seen primarily not as a machine to be described by mechanical laws but as a harmonious, meaningful whole, less like a clock than a living creature. All Earthly physical events were said to be driven ultimately by supernatural forces (a god or group of gods), also variously described as the First Cause or Prime (i.e., first) Mover. Following Aristotle, most medieval thinkers taught that objects moved as they did because it was more "natural" to them to be in some places rather than others. Flame and smoke, for example, rose because they sought their natural place higher up; stones fell because their natural place was lower down. Some philosophers, such as Jean Buridan of Paris (1300–1358), criticized this idea in the later Middle Ages, preparing the way for the new physics of the fifteenth and sixteenth centuries. However, knowledge of medicine, chemistry, physics, and the like was typically derived from revered books rather than from direct experimentation. Mathematics was carried on as a separate, abstract pursuit.

Medieval science was not entirely stagnant—new knowledge, especially mathematical and technological, continued to be gained—but by modern standards the rate of progress was very slow. Astronomy was one of the first areas in which breakthroughs were made. Long before the Scientific Revolution of the fifteenth and sixteenth centuries, the motions of the heavenly bodies had been observed carefully, and explanatory models had been created to account for them.

The first physical model to describe the heavenly motions with good accuracy was the Ptolemaic (pronounced tole-eh-MAY-ik) model. This model was the work of the Egyptian-Greek astronomer Claudius Ptolemy (TOLE-eh-mee, c.AD 90–c.168). According to Ptolemy's book *Almagest*, which remained the standard astronomical text of Europe for about 1,400 years, Earth is a sphere residing at the center of the universe. Around it are nested eight rotating, concentric, transparent ("crystal") spheres. From smallest to largest these are the spheres of the moon, the sun, Mercury, Venus, Mars, Jupiter, Saturn, and the fixed stars. No other planets were known until the discovery of Neptune by English astronomer Sir William Herschel (1738–1832) in 1781. The moon, sun, planets, and stars were supposed to be attached to these larger, invisible spheres much as a tack might be stuck into the surface of a soccer ball. The spheres would rotate independently around Earth, which is stationary. The rising and setting of the sun, for example, was supposedly due to the rotation of the sun's crystal sphere around Earth. Motions too complex to be accounted for in this way could be calculated using epicycles, which are hypothetical circular motions executed by a heavenly body on the surface of its Earth-centered crystal sphere. Circular motions were used because it was standard doctrine that perfection resides in the heavens, and that the circle is the most perfect geometrical shape—therefore, heavenly bodies must move in circles.

The Ptolemaic model did an excellent job of explaining the motions of the stars, planets, sun, and moon. Yet it suggested no physical reason why the spheres should move as they did, no mechanical explanation; the most popular motive power was angels. Moreover, the astronomical success of the Ptolemaic model did not help explain any non-astronomical phenomena, such as the way objects behave on Earth. It did not contain or imply any physical law: it was purely descriptive, and it could not be generalized. It was an intellectual triumph, but an isolated and a scientifically barren triumph.

In 1543, Polish astronomer Nicolaus Copernicus (1473–1543) published *De revolutionibus orbium coelestium* (*On the Revolution of the Celestial Spheres*). In it, he proposed the revolutionary idea that the sun, not Earth, was at the center of the universe. Scientists now know that the sun is also not at the center of the universe. The universe has no center. By displacing Earth from its central position, however, Copernicus began a process of remodeling that would eventually cast doubt on the centrality of human beings in the story of the cosmos. Perhaps, the new changes seemed to suggest, mechanism, not meaning, was the key to reality. Despite resistance from cultural religious authorities—Italian astronomer and physicist Galileo Galilei (1564–1642)

was threatened with torture in 1633 for affirming that Copernicanism was literally true—Copernicus's ideas gradually prevailed, helping set the stage for a new physics.

Danish astronomer Tycho Brahe (1546–1601), court astronomer and astrologer of Denmark, was born only a few years after Copernicus published his controversial book. He made precise measurements of the heavenly bodies' motions and tried to create a description of the cosmos that blended the Copernican system with the Ptolemaic. Soon, using Tycho's data, the best collected to that date, German astronomer and mathematician Johannes Kepler (1571–1630) made a discovery: The planets move not in circular orbits, as had been taught for thousands of years, but in elliptical orbits. (Elliptical orbits are ellipse-shaped. If the top is sliced off a circular cone at a slanting angle, the oval outline of the slice is an ellipse.) Moreover, Kepler described his discoveries in terms of three mathematical laws. These were some of the first physical laws of the new scientific age.

While Kepler was studying the planets, Galileo was studying objects on Earth. In particular, he measured the speeds of objects rolling down inclined planes and formulated mathematical expressions of laws describing his results. Galileo's laws are still used today in calculating the results of constant accelerations (changes in velocity), such as are experienced by objects falling freely under the influence of gravity.

As described later in this article, Newton managed to bridge the motion laws of Kepler, which applied only to the planets, and of Galileo, which applied only to everyday objects. With this unification, modern science came into being.

Cultural Changes

Two great social changes occurred just before modern science—essentially the project of describing all of nature in terms of mathematical physical laws—got under way. These were the replacement of feudalism with capitalism (a change sometimes called the Commercial Revolution) and the Protestant Reformation.

In the social and economic system of Europe in the Middle Ages, trade was primarily local, and seagoing navigation hugged the coasts. Money was used, but there was little industrialism, and the economy was basically agricultural. Then, around 1520, all this began to change. Globe-girdling voyages of discovery and the invasion of North and South America by England, France, and Spain increased long-distance trade. A European commercial culture arose with fundamentally different attitudes toward the world than had prevailed during the Middle Ages. Some modern cultural historians, such as Georg Simmel (1858–1918) and Morris Berman (1944–), have argued that the new money economy increased the importance of exact numerical

IN CONTEXT: NEWTON AND THE APPLE

There is a popular folk story that gravitation was discovered when English physicist Isaac Newton (1642–1727) was hit on the head by a falling apple. As with many such folk tales, although the anecdote is not exactly true, there is evidence that has origins in a real incident. A year before Newton's death, he told a man named William Stukeley a story in which Newton recalled his early thoughts about gravity and remembered pondering the fall of an apple. Newton claimed that at the time he wondered: "Why should it not go sideways, or upwards, but constantly to the earth's center? Assuredly, the reason is, that the earth draws it. There must be a drawing power...."

Other writers of Newton's day relate slightly different versions of the story, but the idea that Newton reasoned from apples to planets appears to be at least partially true. The story reminds us that science sometimes progresses by leaps of insight rather than by systematic reasoning and careful experimentation. However, it took Newton another 20 years to fully work out his theory of gravitation.

calculation, and that this predisposed people to expect exact numerical accounting in the physical cosmos also. Counting, weighing, and pragmatism (the philosophy of focusing on results rather than values or morality) were elevated to high status: nature began to be seen primarily as dead material to be manipulated rather than part of an organic chain of being stretching from the dust to the divine.

The new commercial economy also placed a new value on technical knowledge, which had hitherto been thought the proper concern of people in the trades, guilds, and working classes, not of philosophers and mathematicians. Early in the sixteenth century, the educated classes began to show new interest in the details of manufacture. Spanish scholar Joan Lluís Vives (1492–1540) published *De Disciplini Libri XX* ("Twenty books on disciplines"), in which he argued that a young nobleman's education should include some study of agriculture, textile manufacture, cookery, building, and navigation. German scholar Georg Agricola (1490–1555) visited mines and metallurgical workshops to study minerals and mining directly and published *De Natura Fossilium* in 1546, rejecting ancient authorities in favor of his own observations. His *De Re Metallica* (1556) described the practical processes of mining and metal making in great detail. Agricola is often characterized as the founder of the science of mineralogy. Many other treatises on the crafts of printing, papermaking, and the like were published during this period.

Also, the economy provided financial rewards for improvements in water-pumping, clock-making, and other technologies. Such improvements would, in turn, allow the manufacture of instruments to verify and to apply the new, quantitative, mathematical science of the seventeenth century and beyond. The scientific revolution could not have succeeded on the strength of either abstract thought without practical know-how, or know-how without rigorous abstraction.

At about the time of the Commercial Revolution, another great upheaval of European society was occurring: the Protestant Reformation. Although its cultural and historical origins are far more complex, the landmark beginning of the Reformation is traditionally dated to the posting by Martin Luther (1483–1546) of his ninety-five theses or propositions to the door of Castle Church, Wittenberg, Germany, on October 31, 1517. At the time both economic and religious thought was changing across Europe. A new emphasis on the testing of ancient authorities against individual conscience and reason was abroad. In this setting it would seem increasingly natural, even necessary, to test theoretical ideas against experiments, and to check reports of experiments by repeating them oneself.

The Revolution in Mechanics

"In the beginning," says historian of science Max von Laue (1879–1960), "was mechanics." Why? Because mechanics is the science of objects, motion, and forces, and objects can be touched, motions seen, forces felt. Mechanical processes such as falling and pushing can be measured using simple equipment and compared to mathematical theories. It was natural that the articulation of physical law would begin with mechanics rather than, say, electromagnetics.

The sixteenth century's success in wedding math to mechanics overturned the basically Aristotelian view of nature as organic and moved by propriety or "naturalness" rather than by forces. The success of mechanics seemed to prove that the universe is indeed a machine. The extension of mathematical law to other fields, including the study of heat, light, and electromagnetic forces, followed over the next two centuries.

The triumph of classical mechanics in the late 1500s and the 1600s was the application of the scientific method to long-pondered problems. The Greeks invented the beginnings of mathematical physics, and many philosophers of the Middle Ages, mostly famously English friar Roger Bacon (c.1214–c.1292), advised that theory be tested by experiment. Some medieval thinkers, such as William of Ockham (c.1288–1347), also an English Franciscan friar, questioned Aristotle's theories of motion and groped toward an understanding of acceleration (changing velocity) and inertia (the tendency of an object to maintain its state of motion unless acted upon by a force). Yet most medieval writers addressing physical science were content to quote alleged experimental results from ancient books rather than carrying out the experiments themselves, and thinkers such as Ockham did not succeed in producing an accurate science of mechanics. They were hampered by, among other things, their Aristotelian belief that a force of some kind, whether produced by surrounding air or by some power in the object, is needed to keep an object in motion. Medieval scientists taught that when a gun fired a projectile it endowed the projectile with a certain force called "impetus" that would push the projectile upward at an angle in a straight line until the impetus was exhausted, whereupon the projectile then dropped straight down upon the target. This is a commonsense view based on everyday experience: If you want to raise a brick, you have to apply force the whole time you are lifting it. If you let go, it falls straight to the ground.

Galileo rejected the impetus theory of motion. He understood that a force does not keep an object in motion: rather, an object stays in steady motion (that is, at rest, or moving in a straight line) unless a force acts upon it. Forces do not maintain states of motion; they change states of motion.

Galileo was the first scientist to systematically and thoroughly apply an experimental and mathematical working philosophy similar to that of modern science. Galileo, like many thinkers of the fourteenth and fifteenth century, was influenced by the Greek philosopher Plato (428–348 BC), who taught that the visible world is secondary to a higher realm of invisible, eternal forms. Galileo's version was that mathematical law is the ultimate reality behind the miscellaneous happenings of the physical world. Historian of science A.C. Crombie (1915–1996) described Galileo's view this way in 1959: "The object of science for Galileo was to explain the particular facts of observation by showing them to be consequences of ... general 'mathematical' laws, and to build up a whole system of such laws in which the more particular were consequences of the more general." This closely describes the whole project of modern science. "Facts of observation" include explanations both of naturally occurring events, such as the motions of the heavenly bodies—Galileo was a supporter of the new Copernican system—and of artificially arranged events or experiments, as Roger Bacon had advocated centuries earlier.

The ingredients of Galilean science had been present for many generations, but until the late sixteenth century, philosophical speculations about experimental method and actual physical investigations were rarely in the hands of the same people. Technicians, coinmakers, and the like worked out practical methods based on the observed behaviors of materials, while the book-learned handled mathematics and speculated on (but did not test) the behaviors of objects. For the two

IN CONTEXT: THE LAW OF GRAVITATION

Can a few rules, simple enough to be printed in large type on a credit card, describe how objects fall, why projectiles fly in parabolic arcs, the ocean tides, and the path of every object in the universe that moves freely through space, from atoms to galaxies? The answer is yes: all these events can be predicted to high accuracy using Isaac Newton's three laws of motion plus his law of gravitation, which describes the force of gravity pulling any two objects together. If the masses of the objects are symbolized as m_1 and m_2, then Newton's law of gravitation can be written as follows: $F = Gm_1m_2/r^2$.

Here F is the force felt by either object (both objects feel the same force, pointing toward the other object). G is a fixed number called the universal gravitational constant, and r is the distance between the two objects. Since the masses are on top of the fraction, adding mass to either object makes the force of gravity bigger. And since distance is on the bottom of the fraction, the larger it gets—that is, the farther away the two objects are from each other—the smaller the force of gravity gets. Newton's laws are still used today in all calculations of motion and gravity that do not involve speeds close to that of light or extreme gravity conditions such as those found near black holes. First published by Newton in 1687, the Law of Gravitation is one of the triumphs of classical physical law.

centuries prior to Galileo there is no record of anyone testing actual motions of earthly objects or heavily bodies against philosophical speculations about the nature of motion. Only when the three practices considered essential by Galileo—testing of theory against actual experiments, use of mathematics to state physical laws precisely, and the weaving together of laws into a unified system—were unified into a single method of interrogating nature did the basic practice of modern science come into being.

Newton completed the revolution that Galileo began. While he did important work in optics (the science of light), invented the first practical reflecting telescope in 1671, and published his major work on light, *Opticks*, in 1704, his truly revolutionary work was in mechanics. This culminated in 1687 with the publication of *Philosophiae Naturalis Principia Mathematica* (*Mathematical Principles of Natural Philosophy*). This book, usually referred to today simply as the *Principia*, is one of the important single works in the history of physics.

Focusing on Forces as Real Rather than as to Their Nature

Newton completed the Galilean revolution in mechanics partly by putting aside the vexing question of what, exactly, forces are. Instead, he followed the principle that it was sufficient to treat them as if they were real. As long as forces obeyed the equations written down for them—or rather, as long as equations could be devised that described what the forces did—it did not matter what caused forces: that question could be dealt with separately. Newton did express the opinion that some form of mechanical explanation for gravity would be found, particles of some sort bumping against other particles, as opposed to "action at a distance," which he condemned in a letter to scholar Richard Bentley (1662–1742) in 1692: "That gravity should be innate inherent & essential to matter so that one body may action upon another at a distant through a vacuum without the mediation of anything else by & through which their action or force may be conveyed from one to another is to me so great an absurdity that I believe no man who has in philosophical matters any competent faculty of thinking can ever fall into it."

The mathematics available to Newton at the beginning of his career could not calculate the effects of the new physical laws he sought, so he had to invent calculus, the mathematics of continuously varying and accumulating quantities. Calculus was invented independently at about the same time by German mathematician Gottfried Wilhelm von Leibniz (1646–1716). Calculus immediately began to be applied by mathematicians to all sorts of scientific questions, not only mechanical ones. Today, it is the universal language of science and technology. Other forms of higher mathematics are applied as needed in specific fields, but calculus is applied in virtually all fields of study concerned with the physical world. A version of Leibniz's notation (way of writing down calculus) is used today.

Newton's mechanics boil down to four elegant mathematical laws. The first three are the three laws of motion:

1. *An object retains its state of motion unless acted upon by a force.* The units of force used today are called newtons in honor of Newton's clarification of the concept of force.

2. *The total force acting on a body causes it to accelerate (change its velocity) to a degree that is proportional to the body's mass.* Stated as an equation, with F standing for force, m for mass, and a for acceleration, $F = ma$. Alternatively, $a = F/m$; that is, a heavier object (larger m) is accelerated less by a given force, and a larger force accelerates an object of given mass more quickly.

3. *Forces always occur in pairs that point in opposite directions*; also stated as *To every action there is an equal and opposite reaction.* For example, if a rocket motor pushes gases away from itself with a certain force, the gases also push the rocket motor away in the opposite direction with equal force. This is the principle of rocket propulsion.

4. The fourth basic law of Newtonian mechanics is Newton's law describing the gravitational attraction between two objects: $F = Gm_1m_2/r^2$. Here F is the force due to gravity, G is a fixed number called the universal gravitational constant, m_1 is the mass of one of the two objects, m_2 is the mass of the other, and r is the distance between them. By Newton's third law of motion, gravitation between two objects produces two equal and opposite forces of strength F, one pushing on each object.

The law of gravitation has several interesting consequences. Since the distance r is in the denominator, the force F gets smaller as r gets bigger; since r is squared, the force F decreases as the square of the distance. This means that if the distance is doubled, the gravitational force is cut to one fourth its original strength; if the distance is quadrupled, the force is cut to one sixteenth.

Newton was able to show mathematically that Kepler's laws of planetary motion are consequences of the same handful of mechanical laws that describe the motions of all physical objects, including falling apples. He thus forwarded the Galilean program of producing a system of mathematical physical laws in which the more specific laws were consequences of the more general laws. In this case, the Keplerian laws were specific to planetary motion; Newton showed that they were merely an application of the more general Newtonian laws to planetary motion.

Beyond Mechanics: Physical Law in the Eighteenth and Nineteenth Centuries

Over the next two centuries, the Newtonian revolution in mechanics was extended to other fields. The body of physical laws that has been created by this process is much too large to review here, but a few highlights can be noted.

The first century and a half after the revolution in mechanics saw little progress in our understanding of electricity and magnetism. Early on, English physician William Gilbert (1540–1603) made fundamental observations about these matters, arguing correctly that electricity and magnetism are not the same thing and proving by means of a physical model that Earth itself is a giant magnet. Gilbert coined the word "electricity." However, he was not able to formulate any mathematical, quantitative laws describing the properties of electricity or magnetism. The first law to accurately describe an electrical phenomenon was discovered in 1785 by French scientist Charles Augustin de Coulomb (1736–1806). Coulomb's law for the force between two charged objects was remarkably similar to Newton's law of gravitation: $F = Kc_1c_2/r^2$.

Here F is the force of electrical attraction or repulsion; K is a fixed number called the electrostatic constant; c_1 is the charge on one of the objects; c_2 is the charge on the other object; and r is the distance between them. Coulomb showed that the force arising from electrical charge behaves much like the gravitational force arising from mass, with a few important differences: first, the electrical force can be either attractive or repulsive. Like charges (two positives or two negatives) repel, and unlike charges (one positive and one negative) attract. Second, an electric field can be blocked by a conductor, while gravitational fields cannot be screened or blocked by any barrier.

Coulomb's law was only the beginning of the discovery of the laws governing electricity and magnetism. Electricity and magnetism have a complex relationship to each other: electric charges produce electric fields, much as masses produce gravitational fields, but changing or moving electric charges also produces a magnetic field, and changing or moving magnetic fields produces an electric field. A pair of electric and magnetic fields that are changing together, each generating the other, propagate through space as a wave—an electromagnetic wave—without any charge being present at all. Light, radio waves, and X rays are all electromagnetic waves.

Electromagnetism—the mutual production of electric and magnetic fields create—was first publicized in 1820 by Danish physicist Hans Christian Oersted (1777–1851). A mathematical law describing the attractive or repulsive force between two current-carrying conductors (e.g., wires) was discovered by French physicist André-Marie Ampère (1775–1836) a few years later. The physical law that describes the generation of a magnetic field by a changing electric field was described by English physicist Michael Faraday (1791–1867) in 1831.

The unification of all the miscellaneous laws describing the relationships of electric and magnetic fields was accomplished by Scottish physicist James Clerk Maxwell (1831–1879) in 1864, with his publication of the group of four equations known in his honor today as Maxwell's Equations. These laws are as basic a contribution to the body of physical law as those made by Newton and, later, German-American physicist Albert Einstein (1879–1955). They form the working basis of all technology that relies on magnetic and electric fields.

Meanwhile, laws were also being articulated in other fields. In the 1660s, Henry Power (1623–1668) and Richard Towneley (1629–1707), both English scientists, discovered a mathematical law relating the pressure and volume of a gas: pV = constant (at a fixed temperature). This law is known today as Boyle's law. The study of heat (thermodynamics) had been given new urgency by the invention of the steam engine, first built in 1698 but not analyzed mathematically until the early 1800s. Almost all English and French physicists

Engraving of Hans Christian Oersted (1777–1851), Danish physicist, seen here with his assistant observing an experiment to demonstrate the effect of an electric current on a magnetic compass needle. This was the first demonstration of a connection between electricity and magnetism and in retrospect his experiment is seen as the foundation of the study of electromagnetism. His discovery was published in 1820. Oersted did not carry his experimentation with electromagnetism any further, but his discovery unleashed a whirl of activity in the field of physics. In 1934 a unit of magnetic field strength was named an oersted in his honor. *SPL/Photo Researchers, Inc.*

concerned themselves with the question of the steam engine and its principles in the late 1700s and early 1800s; most of the basic laws of thermodynamics were developed by studying the steam engine.

The new laws in various fields were built on the Galilean model—that is, they were mathematical and (usually) explainable in terms of more fundamental laws. For example, the gas laws were explained in terms of large numbers of particles (gas molecules) flying about in obedience to Newton's laws of motion. Summary laws or surface laws such as Boyle's gas law remain useful, however, even when they are explained in terms of more fundamental laws, because they are much easier to apply to practical problems.

Also developed in the nineteenth century were laws of acoustics (the science of sound), optics, energy

Portrait of André-Marie Ampère (1775–1836), French physicist, mathematician, and pioneer of electrodynamics. He showed in a classical experiment that two parallel wires carrying currents attract if the currents are in the same direction. *SPL/Photo Researchers, Inc.*

conservation, the kinetic theory of gases, and statistical mechanics. All these equations remain essential in modern physics and technology. So complete did the picture of physics painted by the new sciences and their laws seem, so great was the success of science during this period, that some thinkers declared that science itself was almost finished—that scientists were on the verge of having answered all questions. For example, Albert Michelson (1852–1931) said in 1894 (and again in 1898) that "While it is never safe to affirm that the future of Physical Science has no marvels in store even more astonishing than those of the past, it seems probable that most of the grand underlying principles have been firmly established and that further advances are to be sought chiefly in the rigorous application of these principles to all the phenomena which come under our notice."

Yet contradictions were arising, contradictions that forced the birth of a new physics in the early twentieth century.

The Limitations of Classical Physical Law

What the scientists of the seventeenth through the nineteenth centuries did not realize was that all the laws they were discovering were approximate. Newton's law of gravitation, $F = Gm_1m_2/r^2$, is usually so close to true as to appear perfectly exact; yet there are conditions, real-world conditions, under which it is inaccurate. New, more general, laws had to be discovered when the approximate nature of classical law was realized. It proved to be impossible to adjust the classical laws in such a way as to explain the constancy of the speed of light (which had been proved by the 1887 experiment by Michelson and Edward Morley (1838–1923) in the quantization of energy in photons, light's mixture of wave and particle properties, and other phenomena. Resolving these difficulties required the laws of relativity and quantum mechanics, which were developed in the early twentieth century. These laws, too, are approximate, but what scientists call their domain of application—the range of conditions under which they are correct within the limits of our ability to make measurements—is much broader than that of classical physics. The old laws are still good for most conditions and are mathematically simpler, so they are still used today throughout science and technology—but they are used with the knowledge that they are imperfect.

Two assumptions that had been made throughout all of classical physics finally had to be abandoned: absolute space and time, and determinism.

Absolute space and time are independent of each other and of the objects they contain, and look the same to all observers, regardless of the locations or states of motion of those observers. If time were absolute, one could, in principle, distribute accurate clocks set to a single, agreed-upon time throughout the universe, and they would continue to agree forever. This is our commonsense experience of time: When we speak of the accuracy of a clock, for example, we never feel obliged to condition our description on where the clock is located or how it is moving. Time passes at the same rate everywhere—or does it? At the beginning of the twentieth century, it was discovered that time is not absolute. Neither is space. Moreover, they are not independent of each other, but are intimately related. Yet they are not interchangeable, two forms of the same thing: as Einstein, discoverer of relativity, put it in 1921, time is equivalent to space "in respect to its *role* in the equations of physics, 'but' not with respect to its physical significance."

Determinism was the second basic assumption of classical physics to be called into question. Since Newton, scientists have assumed that all appearances of randomness in the universe, such as the unpredictability of rolled dice, are illusions: on this view, all outcomes are actually the result of cause-and-effect chains that could not have turned out any other way. Some events seemed random to us only because we did not have the knowledge and computational power to predict how conditions were bound to work themselves out according to rigid physical law. The idea of universal, absolute determinism was famously expressed by French mathematician, Pierre Simon de Laplace (1749–1827) who argued that one could "embrace in the same formula the movements of the greatest bodies of the universe and those of the lightest atom; for it, nothing would be uncertain and the future, as the past, would be present to its eyes."

Since the early twentieth century, the interpretation of quantum mechanics accepted by the great majority of physicists, the Copenhagen interpretation, has been that at the scale of the very small, the subatomic scale, events are truly random. For example, the time when a given unstable or radioactive atom breaks down—which is experimentally unpredictable—is not determined by tiny, hidden differences between that particular atom and others of its kind; it is truly random. Whether this interpretation will be replaced some day by a new, more subtle form of determinism has been debated for decades by physicists, but as of the early 2000s the Copenhagen interpretation still prevailed among the majority of specialists studying the question.

■ Modern Cultural Connections

Classical physical law greatly multiplied the ability to predict and control physical events, whether for profit, pleasure, war, or knowledge. It transformed the world by making possible the Industrial Revolution, and continues to transform it through that ongoing flood of new technologies which we now take for granted—at least, those of us who live in the industrialized parts of the world. Almost all modern technology is made possible by applying the classical laws of mechanics, thermodynamics, optics, electromagnetics, chemistry, and a few other disciplines. Without succinct, accurate, mathematical laws, whatever information science managed to accumulate would be useless—a heap of unconnected facts.

It should be noted that sciences such as geology, biology, and astronomy, which primarily work to produce factual explanations and histories—what makes the continents move, how does a new species of finch appear, how did the galaxies form, and so on—are just as valid, important, and scientific as physics or the other elemental physical sciences. Without the sciences that describe the world, physics would be crippled in its search for universal laws; without the laws provided by physics, the other sciences would be a jumble of disjointed facts. Today, for example, the existence of a form of gravitating yet invisible matter clustered around the galaxies, called dark matter, has been proved by applying the laws of Newtonian physics to data obtained by telescopes. Dark matter implies what scientists call "new physics"—that is, it is not predicted by the laws of quantum physics as they now exist. Some modification of those laws will

be needed to account for the observations that prove the existence of dark matter. Without known physical laws, the observations could not be made; once they are made, the observations require revision of the known physical laws.

The proliferation of technology, enabled by scientific knowledge of physical laws, has effected every aspect of modern life in industrialized societies. The rise of the automobile has affected work and courtship patterns; telecommunications have changed the way we socialize; nuclear and other weapons have enabled destruction on a scale not even imagined by earlier centuries; and the unintended side-effects of applied science are causing soil loss and changing the climate, both of which ultimately threaten human survival by threatening the agricultural basis of all human life. Yet even this list does not capture all the ways in which scientific law has impacted the modern world.

Social and Philosophical Implications

Soon after what even its contemporaries referred to as the "revolution" in science achieved by Newton and the other physicists, the new ideas were popularized to a wide public by writers such as English philosophers John Locke (1632–1704) and David Hume (1711–1776) and French philosophers Bernard le Bovier de Fontenelle (1657–1757) and François-Marie Voltaire (1694–1778). The result was a shift in the way people in European societies saw the nature of the world—a shift more pronounced in the more educated (especially the more scientifically educated), but affecting all levels of society.

Locke taught that all ideas arise from experience and that ideas cannot be innate in our minds or a priori (i.e., obvious on their own merits—from the Latin for "what is before"). Locke's philosophy, especially as expressed in his *Essay Concerning Human Understanding* (1690), was boosted by the prestige of Newton and in turn boosted the prestige of mathematical, experimental science. Locke claimed—what Newton himself, ironically, as a dedicated Puritan Christian, would have denied—that human beings have no special place in creation. Locke's impact was great both in England and in continental Europe. His emphasis on experience, limited knowledge, and tentative conclusions has become largely habitual in modern thought. Influenced by Locke and the new science, Hume argued for the total rejection of all beliefs that are not modeled on Newtonian physics: "If we take in our hand any volume; of divinity [theology] or school metaphysics, for instance; let us ask, *Does it contain any abstract reasoning concerning quantity or number?* No. *Does it contain any experimental reasoning concerning matter of fact and existence?* No. Commit it then to the flames: for it can contain nothing but sophistry and illusion" (*An Enquiry Concerning Human Understanding*, 1748).

IN CONTEXT: SCIENCE VS. RELIGION

Since the early days of the new science of mathematical law and universal determinism, some thinkers have protested against its spiritual effects. English poet William Blake (1757–1827) wrote in 1802, "May God us keep / From Single vision & Newton's sleep!" By this he meant, more or less, that the spiritual effect of the Newtonian worldview is deadly—that it empties the world, reducing beauty, meaning, and the like to the status of arbitrary experiences or illusions that are real only inside the individual human mind, which itself is reduced to irrelevance, a sort of bright shadow cast by the deterministic chemical-mechanical workings of the brain. Blake called this "Newton's sleep."

In France, Fontenelle popularized the new science in works such as *Conversations on the Plurality of Worlds* (1686), a work of popular astronomy. Voltaire, at about the same time as Fontenelle, touted the success of Newton in explaining the universe and argued for a semi-religious worldview in which reason reigned supreme and Christianity was rejected as superstition. Historian of science Herbert Butterfield (1900–1979) has said that Fontenelle, Voltaire, and the anti-religious, rationalistic writers that followed in their steps, the *philosophes*, not only spread the new scientific knowledge but performed a second function: "the translation of the scientific achievement into a new view of life and the universe." Religion was not eliminated, of course, but its plausibility and authority were weakened.

The view of the world as a machine with inherent properties that cannot be expressed in mathematical form—the view that arose and became commonplace in Europe in the seventeenth and eighteen centuries—has had profound impact on all aspects of culture, including religion. Qualities such as beauty, meaning, holiness, and value, which were once assumed to be inherent in the physical world, have been relocated into the subjective realm of personal feeling: "Beauty in things exists merely in the mind which contemplates them," wrote Hume. The position of religion as a source of knowledge about the world has been weakened by the all-embracing, illuminative powers of scientific laws: The universe has come to seem self-sufficient, self-explanatory.

Although the connection between science and diminished religious belief is not a logical one—science as such makes no statements, pro or con, about a God or gods, moral values, beauty, or anything else that cannot be measured—the rise of science as articulated in physical laws has made religious belief less plausible for some people. The more educated in science a person is, the less likely they are to have religious beliefs. In the United States, for example, the journal *Scientific American* reported in 1999 that belief in a prayer-answering supernatural god and personal immortality was affirmed by over 90% of the general public, only 50% of scientists with degrees at the B.S. level, and less than 10% of scientists in the elite National Academy of Sciences. The journal *Skeptic* conducted a poll in 1998 that found that 40% of scientists believe in a supernatural god, but that the rate was lowest (20%) for physicists—those scientists most directly concerned with discovering and testing fundamental, universal physical laws on the Galilean model.

There is no consensus among either experts or the general population on the significance of these changes—only on the fact that they have happened.

■ Primary Source Connection

The following essay was written by the American physicist Richard Feynman (1918–1988), published in *The Feynman Lectures on Physics*. Feynman won the Nobel Prize in Physics in 1965 for his contributions to the advancement of Quantum Electrodynamics (QED). He also illustrated the mathematical laws of subatomic particles in what later became known as Feynman diagrams. He participated in teams that developed the atomic bomb and was instrumental in determining that leaking O-rings were the cause of the space shuttle *Challenger* disaster. Feynman popularized the study of physics by writing entertaining accounts such as *Surely You're Joking, Mr. Feynman* and through his respected lectures. This essay from one of Feynman's introductory lectures explores teaching and learning physics through experimentation, approximation, and imagination.

ATOMS IN MOTION

This two-year course in physics is presented from the point of view that you, the reader, are going to be a physicist. This is not necessarily the case of course, but that is what every professor in every subject assumes! If you are going to be a physicist, you will have a lot to study: two hundred years of the most rapidly developing field of knowledge that there is. So much knowledge, in fact, that you might think that you cannot learn all of it in four years, and truly you cannot; you will have to go to graduate school too!

Surprisingly enough, in spite of the tremendous amount of work that has been done for all this time it is possible to condense the enormous mass of results to a large extent—that is, to find *laws* which summarize all our knowledge. Even so, the laws are so hard to grasp that it is unfair to you to start exploring this tremendous subject without some kind of map or outline of the relationship of one part of the subject of science to another. Following these preliminary remarks, the first three chapters will therefore outline the relation of physics to the rest of the sciences, the relations of the sciences to each other, and the meaning of science, to help us develop a "feel" for the subject.

You might ask why we cannot teach physics by just giving the basic laws on page one and then showing how they work in all possible circumstances, as we do in Euclidean geometry, where we state the axioms and then make all sorts of deductions. (So, not satisfied to learn physics in four years, you want to learn it in four minutes?) We cannot do it in this way for two reasons. First, we do not yet *know* all the basic laws: there is an expanding frontier of ignorance. Second, the correct statement of the laws of physics involves some very unfamiliar ideas which require advanced mathematics for their description. Therefore, one needs a considerable amount of preparatory training to learn what the *words* mean. No, it is not possible to do it that way. We can only do it piece by piece.

Each piece, or part, of the whole of nature is always merely an *approximation* to the complete truth, or the complete truth so far as we know it. In fact, everything we know is only some kind of approximation, because *we know that we do not know all the laws* as yet. Therefore, things must be learned only to be unlearned again, or more likely, to be corrected.

The principle of science, the definition, almost, is the following: *The test of all knowledge is experiment.* Experiment is the *sole judge* of scientific "truth." But what is the source of knowledge? Where do the laws that are to be tested come from? Experiment, itself, helps to produce these laws, in the sense that it gives us hints. But also needed is *imagination* to create from these hints the great generalizations—to guess at the wonderful, simple, but very strange patterns beneath them all, and then to experiment to check again whether we have made the right guess. This imagining process is so difficult that there is a division of labor in physics: there are *theoretical* physicists who imagine, deduce, and guess at new laws, but do not experiment; and then there are *experimental* physicists who experiment, imagine, deduce, and guess.

We said that the laws of nature are approximate: that we first find the "wrong" ones, and then we find the "right" ones. Now, how can an experiment be "wrong?"

First, in a trivial way: if something is wrong with the apparatus that you did not notice. But these things are easily fixed, and checked back and forth. So without snatching at such minor things, how *can* the results of an experiment be wrong? Only by being inaccurate. For example, the mass of an object never seems to change: a spinning top has the same weight as a still one. So a "law" was invented: mass is constant, independent of speed. That "law" is now found to be incorrect. Mass is found to increase with velocity, but appreciable increases require velocities near that of light. A *true* law is: if an object moves with a speed with less than one hundred miles a second the mass is constant to within one part in a million. In some such approximate form this is a correct law. So in practice one might think that the new law makes no significant difference. Well, yes and no. For ordinary speeds we can certainly forget it and use the simple constant-mass law as a good approximation. But for high speeds we are wrong, and the higher the speed, the more wrong we are.

Finally, and most interesting, *philosophically we are completely wrong* with the approximate law. Our entire picture of the world has to be altered even though the mass changes only by a little bit. This is a very peculiar thing about the philosophy, or the ideas, behind the laws. Even a very small effect sometimes requires profound changes in our ideas.

Now, what should we teach first? Should we teach the *correct* but unfamiliar law with its strange and difficult conceptual ideas, for example the theory of relativity, four-dimensional space-time, and so on? Or should we first teach the simple "constant-mass" law, which is only approximate, but does not involve such difficult ideas? The first is more exciting, more wonderful, and more fun, but the second is easier to get at first, and is a first step to a real understanding of the second idea. This point arises again and again in teaching physics. At different times we shall have to resolve it in different ways, but at each stage it is worth learning what is now known, how accurate it is, how it fits into everything else, and how it may be changed when we learn more.

Richard Feynman

FEYNMAN, RICHARD. "ATOMS IN MOTION." *THE FEYNMAN LECTURES ON PHYSICS*. BOSTON: ADDISON-WESLEY, 1963. REPRINTED WITH PERMISSION OF CALIFORNIA INSTITUTE OF TECHNOLOGY.

SEE ALSO *Astronomy and Cosmology: A Mechanistic Universe; Maxwell's Equations: Light and the Electromagnetic Spectrum; Physics: Aristotelian Physics; Physics: Fundamental Forces and the Synthesis of Theory; Physics: Heisenberg Uncertainty Principle; Physics: Newtonian Physics; Physics: Special and General Relativity; Physics: The Quantum Hypothesis; Science Philosophy and Practice: Postmodernism and the "Science Wars"; Science*

Philosophy and Practice: Pseudoscience and Popular Misconceptions; Science Philosophy and Practice: The Scientific Method.

BIBLIOGRAPHY

Books

Bell, Arthur. *Newtonian Science.* London: Edward Arnold Publishers Ltd., 1961.

Cohen, I. Bernard. *The Newtonian Revolution: With Illustrations of the Transformation of Scientific Ideas.* New York: Cambridge University Press, 1980.

Crombie, A.C. *Medieval and Early Modern Science, Vol. II: Science in the Later Middle Ages and Early Modern Times: XIII–XVII Centuries.* New York: Doubleday, 1959.

Deason, Gary B. "Reformation Theology and the Mechanistic Conception of Nature." In *God and Nature: Historical Essays on the Encounter between Christianity and Science.* Edited by David C. Lindberg and Ronald L. Numbers. Berkeley: University of California Press, 1986.

Feynman, Richard. "Atoms in Motion." *The Feynman Lectures on Physics.* Boston: Addison-Wesley, 1963.

———. *The Character of Physical Law.* Cambridge, MA: The M.I.T. Press, 1967.

Kuhn, Thomas. *The Essential Tension.* Chicago: University of Chicago Press, 1977.

Purrington, Robert D. *Physics in the Nineteenth Century.* New Brunswick, NJ: Rutgers University Press, 1997.

Von Laue, Max. *History of Physics.* New York: Academic Press Inc., 1950.

Periodicals

Badash, Lawrence. "The Completeness of Nineteenth-Century Science." *Isis,* Vol. 63, No. 1 (March 1972): 48–58.

Larson, Edward J., and Larry Witham. "Scientists and Religion in America." *Scientific American* (September 1999): 88–93.

Larry Gilman
K. Lee Lerner

Physics: Cosmic Rays

■ Introduction

Cosmic rays are naturally occurring high-energy particles—protons, helium nuclei, and electrons—that travel at nearly the speed of light. Some scientists argue that cosmic rays may cause cloud droplets to form in Earth's atmosphere, increasing cloudiness when the sun emits more cosmic rays or when the solar system passes through a part of the galaxy where cosmic rays are more abundant. Increased cloud cover might, in turn, affect Earth's climate. As of 2007 climate scientists disagreed about whether cosmic rays are a significant influence on Earth's climate. The large hadron collider, an experimental device under construction at the European Organization for Nuclear Research (CERN) particle accelerator laboratory in Switzerland, due to be completed in 2010, might decide the question.

■ Historical Background and Scientific Foundations

Scientists have long known that slight changes in the sun's energy output might affect Earth's climate. American astronomer Jack Eddy (1932–) pointed out in the 1970s that the Little Ice Age, a cold period in Earth's climate from 1500–1850, coincided with a historic low point in the number of sunspot numbers known as the Maunder minimum. In 1997, Danish scientists Henrik Svensmark (c.1958–) and Eigil Friis-Christensen noted that Earth's global cloudiness had decreased by 3% from 1987–1990, at the same time that cosmic rays had decreased by 3.5% due to the regular cycle of solar activity.

They proposed that cosmic rays might increase cloud cover in this way: Because cosmic rays have high energy, they can strip electrons from atoms when they strike Earth's atmosphere. These stripped atoms, now

WORDS TO KNOW

COSMIC RAY: The fast-moving particles (protons, electrons, and helium nuclei) coming from outer space and moving near the speed of light with which Earth is constantly struck. Most cosmic rays originate near extremely heavy objects (such as black holes and neutron stars) that are accelerating nearby matter. Most of the matter accelerated near these objects gets sucked in, but some gets slung out: some of these particles become cosmic rays.

GLOBAL WARMING: Warming of Earth's atmosphere that results from an increase in the concentration of gases that store heat, such as carbon dioxide.

MAUNDER MINIMUM: A historic dip in the number of sunspots from about 1645 to 1715. During this period the European climate was unusually cold, but this may have been a coincidence. The Maunder minimum is named after English astronomer Edward Maunder (1851–1928), who identified its occurrence from historical records in 1893.

MILKY WAY: The galaxy in which our solar system is located.

SUNSPOT: A region on the surface (photosphere) of the sun that is temporarily cool and dark compared to surrounding areas.

with positive electric charges, might cause water to condense out of humid air into cloud droplets. This, in turn, might increase the number of clouds, their density, or both.

The cosmic ray theory conflicts with another scientific hypothesis that global warming is caused by increased carbon dioxide in Earth's atmosphere, not by changing levels of solar activity. In 2003 two other scientists, Nir J. Shaviv and Ján Veizer, took the dispute to a new level by arguing that over the last 545 million years, about two-thirds of Earth's climate changes could be attributed to highs and lows in the number of cosmic rays that strike Earth from outside the solar system. In addition to fluctuations in cosmic ray flow from our own sun, cosmic ray changes have occurred throughout Earth's existence over millions of years as the solar system's orbital path around the center of the galaxy take it into and out of the Milky Way's spiral arms.

■ Modern Cultural Connections

Evidence from several independent sources indicates that changing solar energy output has affected Earth's climate in the past. Moreover, the solar hypothesis and the carbon dioxide hypothesis do not necessarily refute each other: Earth's climate might be influenced by multiple factors. As of 2007, however, disagreement about the climatic effects of solar rays was still a source of controversy.

In 2004 German scientist Stefan Rahmstorf (1960–) and a group of 10 other scientists from the United States, Switzerland, and other countries published a paper disputing Shaviv and Veizer's 2003 cosmic ray climate theory. In 2007 British scientists Mike Lockwood and Claus Fröhlich (1936–) published a study indicating that "the observed rapid rise in global mean temperatures seen after 1985 cannot be ascribed to solar variability," no matter what mechanism for solar influence on climate is invoked or how that influence might be amplified, as, for example, by cosmic rays. The solar influence on climate, they assert, has decreased over the last 20 years while temperature has increased.

In this handout from NASA, dated 2005, the mosaic image, one of the largest ever taken by NASA's Hubble Space Telescope of the Crab Nebula, shows a six-light-year-wide expanding remnant of a star's supernova explosion. Many researchers argue that cosmic rays are a primary source of supernova explosions. *Getty Images.*

Present-day climate change, the authors contend, is due to human activity, not solar influence.

Nevertheless, the possibility of a cosmic-ray connection to climate during other periods, whether slight or great, remains open. At the CERN particle accelerator laboratory in Geneva, an experiment called CLOUD (Cosmics Leaving Outdoor Droplets) is under construction. Due to be completed in 2010, CLOUD will enable scientists to observe the effects of artificial cosmic rays on air and water vapor and may settle the question of whether cosmic rays can influence cloud formation.

SEE ALSO *Earth Science: Atmospheric Science; Earth Science: Climate Change.*

BIBLIOGRAPHY

Periodicals

Kanipe, Jeff. "Climate Change: A Cosmic Connection." *Nature* 443 (September 14, 2006): 141–143.

Rahmstorf, Stefan, et al. "Cosmic Rays, Carbon Dioxide, and Climate." *Eos: Transactions of the American Geophysical Union* 53 (2004): 38–40.

Schiermeier, Quirin. "No Solar Hiding Place for Greenhouse Skeptics." *Nature* 448 (July 5, 2007): 8–9.

Shaviv, Nir J., and Ján Veizer. "Celestial Driver of Phanerozoic Climate?" *GSA Today* (July 2003): 4–10.

Web Sites

American Geophysical Union. "Cosmic Rays Are Not the Cause of Climate Change, Scientists Say." http://www.agu.org/sci_soc/prrl/prrl0405.html (accessed August 8, 2007).

Proceedings of the Royal Society. "Recent Opposite Directed Trends in Solar Climate Forcings and the Global Mean Surface Air Temperature." http://www.pubs.royalsoc.ac.uk/media/proceedings_a/rspa20071880.pdf (access November 6, 2007).

Larry Gilman

Physics: Fundamental Forces and the Synthesis of Theory

■ Introduction

Since the beginnings of modern science in the 1500s and 1600s, physicists have tried to unify their theories, that is, to explain the known properties of energy and matter with fewer theories. For example, in the seventeenth century English physicist and mathematician Isaac Newton (1642–1727) unified Italian astronomer and physicist Galileo Galilei's (1564–1642) equations for the movements of accelerating everyday objects with German astronomer and mathematician Johannes Kepler's (1571–1630) equations describing the motions of the planets. Newton showed that his own three laws of motion, plus his law of universal gravitation, could explain everything that the other two theories had explained and more.

In the nineteenth century, Scottish physicist James Clerk Maxwell (1831–1879) showed that electric and magnetic forces could be described by a single, unified, mathematically precise theory. In the twentieth century, still more ambitious attempts at unification began as scientists sought to reconcile the theory of relativity with quantum physics and to explain the four fundamental forces (gravity, electromagnetic, weak, and strong) with a single, unified theory. The ultimate goal, which may still be decades away or permanently unattainable, is a Theory of Everything—or a finished description of the fundamental nature of Nature.

■ Historical Background and Scientific Foundations

The drive for unified physical theory has been at the center of Western science since the Middle Ages. Historians of science speculate that the nature of Christian religious doctrine, which framed European philosophical and scientific thought for over a thousand years, encouraged the assumption—by no means universal in human cultures—that nature is ruled by laws and that these laws can be comprehended by human beings.

One of the basic concerns of Western physics since the time of the ancient Greeks has been force. The existence of forces—pushes and pulls—has always been intuitively obvious, but it was not until Newton that the concept received a strict, mathematical definition. Newton defined the total force acting on an object as the time rate of change of that object's momentum (the product of its mass and velocity). In other words, a force is something that can make an object speed up, slow down, or change direction.

Newton used his definition of force, along with his law of universal gravitation, to show that the same gravitational force that governs falling apples and other objects here on Earth also guides the planets around the sun. In doing so he achieved the first great unification in modern physics, the synthesis or bringing-together of planetary mechanics with everyday mechanics.

Scientists in Newton's day already knew of the attractive and repulsive forces of electrical charges and magnets, but had no useful theory about them; magnetic and electric forces seemed to be two quite different things. In the early nineteenth century, English scientist Michael Faraday (1791–1867) and French scientist André-Marie Ampère (1775–1836) showed that moving electric charges produce magnetic fields. This proved that the two kinds of force, electrical and magnetic, were closely related. Faraday attempted to show that electromagnetism and gravity are also related, but failed.

Building on the discoveries of Faraday and others, Maxwell, in a series of papers published in the 1850s and 1860s, showed that a single group of equations, known today as Maxwell's equations, could describe all known features of the electrical and magnetic forces. He had achieved the first true unification of two distinct forces.

WORDS TO KNOW

ELECTROMAGNETIC FORCE: Electrically charged particles experience a force when they are in the presence of an electrical field, whether they are moving or stationary; they also experience a force in the presence of a magnetic field, if they are in motion. These forces are manifestations of a single fundamental force, the electromagnetic force.

ELECTROWEAK FORCE: In physics, the single unified force of which the electromagnetic force and weak interaction are both manifestations. The electromagnetic force is caused by an electromagnetic field; the weak interaction is involved in certain types of radioactive decay (e.g., beta decay of neutrons).

FORCE: Any external agent that causes a change in the motion of a free body or that causes stress in a fixed body.

GENERAL RELATIVITY: In 1915, German physicist Albert Einstein (1879–1955) announced a mathematical theory that described the nature of gravitation by appeal to the curvature of space and time. This theory, general relativity, has been one of the most successful scientific theories of all time and has passed many rigorous observational tests by successfully predicting the precise behaviors of various experimental setups and distant astronomical objects.

GRAND UNIFIED THEORIES (GUTS): Systems of physical laws that describe the strong nuclear, electromagnetic, and weak nuclear forces in terms of a single underlying field. All such theories omit the force of gravity; it has proved difficult to formulate a theory uniting gravity to the other forces. A Theory of Everything (TOE) did not yet exist as of 2008.

MAGNETIC FIELDS: Quantities occupying every point of space in the region of a magnet or a moving electric charge. Stationary magnets or moving electric charges will experience a magnetic force—a physical push or pull—inside a magnetic field.

QUANTUM MECHANICS: A system of physical principles that arose in the early twentieth century to improve upon those developed earlier by Isaac Newton, specifically with respect to submicroscopic phenomena.

SPECIAL RELATIVITY: The theory proposed in 1905 by German physicist Albert Einstein (1879–1955) that describes the effects of straight-line, unaccelerated motion on time, distance, and mass. According to the theory, which has been validated by thousands of experiments, observers in constant, straight-line motion with respect to each other see that the other's mass is increased, their size in the direction of travel is decreased, and their clocks run slower. These effects are real, not illusory. The equivalence of mass and energy, which powers nuclear weapons and stars, follows from the theory. It is incorporated into the more complex and far-reaching theory of general relativity, proposed by Einstein in 1915 and also validated by many experiments.

STANDARD MODEL: In particle physics, the theory that describes the fundamental particles that make up ordinary matter and the particles that are exchanged between them to create all of the fundamental forces except gravity. Gravity is not described by the Standard Model but by the theory of general relativity; scientists hope to someday modify the two so that they can be united into a single theory that describes all known phenomena.

THEORIES OF EVERYTHING (TOES): Theories of Everything are sought-after systems of physical law that would incorporate both quantum mechanics and general relativity into a single, coherent system of explanations. Presently, these two physical theories work extremely well in their own domains of application, but neither can be extended to all physical conditions.

UNIFICATION: In physics, the discovery of a single, consistent theoretical framework for describing phenomena that formerly had to be described by separate theories.

Next, German–American physicist Albert Einstein (1879–1955) showed that space and time, previously thought to be absolutely different things, were closely related. His theory unifying space and time, Special Relativity, appeared in 1905; soon afterward he showed that the force of gravity could be understood in terms of the curvature of space and time by objects. This theory, General Relativity, he perfected in 1915. Einstein shared Faraday's dream of unifying the gravitational force with electromagnetics; he spent the last thirty years of his life trying to work out a theory that would do so. His efforts were premature, however, and did not succeed.

Even as Einstein went to work on a unified field theory, which he was never to find, physicists already knew of two other apparently fundamental forces. The first was the strong or nuclear force. The need for such a force became clear with the discovery of the neutron 1932 by English physicist James Chadwick (1891–1974). Physicists then knew that the nucleus of the atom must be made up of protons, which are positively charged, and neutrons, which have no electrical charge. Because like charges repel, the protons in the nucleus would fly apart, pushed by repulsive forces of between 50 and 100 pounds acting on each particle, if they were not held together by some even stronger force—the strong force. The strong force is so called because at short distances it is stronger than any of the other forces; it drops off quickly with distance, however, so that outside the atomic nucleus the behavior of matter is dominated by electromagnetic and gravitational forces.

The existence of a fourth force, called the weak force, was proposed in 1934 by Italian physicist Enrico Fermi (1901–1954) to explain beta decay, a type of radioactive decay in which an atomic nucleus breaks up into smaller pieces, giving off a high-speed electron (beta particle) in the process.

Thus, by the 1930s four fundamental or basic forces were known. For about 40 years, these remained separate in physical theory. Finally, in the 1970s, Pakistani physicist Abdus Salam (1926–1996), American physicist Sheldon Glashow (1932–), and American physicist Steven Weinberg (1933–) succeeded in describing the electromagnetic force and the weak force as manifestations of a single, underlying force, the electroweak force. They were jointly awarded a Nobel Prize in physics in 1979 for their work.

In 1978, physicists coined the term Grand Unified Theories (GUTs) to describe physical theories seeking to unify the electroweak force with the strong force. Several such theories have been devised in the decades since, but as of 2007 none had been experimentally proven to be correct. Also since the 1970s, physicists have been trying to devise what are sometimes called Theories of Everything (TOEs), theories that would unify the strong and electroweak force with the weakest of all the forces, gravitation. As of 2007, these efforts had not yet succeeded.

The Fundamental Forces and the Standard Model

The four fundamental forces, as currently described, are as follows:

- *Strong force.* The strong force acts only on the particles called quarks, gluons, and hadrons. Its range is very short (about 10^{-15} meter). It glues together atomic nuclei despite the mutual repulsion of their protons.
- *Weak force.* The weak force acts on quarks and leptons. (Electrons are a type of lepton.) It is one ten-trillionth as strong as the strong force, hence its name, and acts at an even shorter range (about 10^{-18} meter). It is involved in some radioactive processes.
- *Electromagnetic force.* The electromagnetic force acts on all electrically charged particles, including electrons and protons. It is the force that attaches atoms to each other in molecules and so makes chemistry and biology possible. It is relatively strong—one percent as strong as the strong force—and can act over distances far greater than the width of an atom's nucleus.
- *Gravitation.* Gravitation is by far the weakest of all the fundamental forces; even the "weak" force is about 10^{26} times stronger than gravity. However,

gravity always acts positively, it acts at long range, and it acts on all particles, so its effects are important. Gravity shapes planets and stars into globes and ignites nuclear reactions in the hearts of stars through pressure-generated heat.

As of the early 2000s, all definite knowledge about the fundamental forces and particles had been encapsulated in a scheme known as the Standard Model, a sort of parts list for the universe. The Standard Model was developed in the 1970s and has been adjusted repeatedly since. In modern physics, all forces other than gravity are described as exchanges of particles, so the Standard Model is a list both of the particles that make up matter and of the particles that mediate the exchange of forces. To see how forces might actually consist of particle exchanges, imagine two people standing on wheeled stools throwing a heavy ball to each other: the people would be gradually pushed apart.

The Standard Model has been extended to account for the unification of the electromagnetic and weak forces as the electroweak force, but it does not account for gravity. Several Theories of Everything that would explain all fundamental particles and interactions have been proposed, but testing them experimentally has proved difficult. Gravity remains the greatest challenge. It is elegantly described by the theory of general relativity, but reconciling relativity with quantum mechanics—the physics of the very small, of which the Standard Model is one product—has not yet been possible.

■ Modern Cultural Connections

The drive toward unification in physics has produced a series of revolutions in our understanding of the physical world. These revolutions are not of merely theoretical interest: Newton's unification allows us to build and control satellites or space probes; Maxwell's unification of the electric and magnetic fields is fundamental to radio and modern electronics; and Einstein's unification of matter and energy provided the initial insights into the capacity of nuclear weapons and nuclear power.

For over half a century, the push for final unification in physics has motivated large bodies of fresh theoretical and experimental work. The primary candidate for total unification today is M-theory, which posits 11 dimensions (10 of space, one of time), and its associated string theories, which explain matter in terms of tiny, vibrating stringlike entities. M-theory has not yet been tested experimentally, and it is not clear whether it even can be.

Although earlier unifications in physics have affected society by making possible new technologies, it is harder to pinpoint the social consequences of recent attempts toward unification of the fundamental forces. These have not yet given rise to new technologies and may never do so. The social effects of the effort toward the synthesis of theory in physics are therefore not tangible or technological, at least so far. Success would tend to bear out Western civilization's ancient faith in the rationality of the universe and of the capacity of the human mind to understand that rationality. How it would ultimately impact art, philosophy, politics, or other non-technical realms is speculation.

■ Primary Source Connection

The Superconducting Super Collider (SSC) was a particle accelerator that was partially constructed near Waxahachie, Texas. Plans for the SSC called for an underground, ring-shaped tunnel complex with a 54 mile (87 km) circumference and an energy of 20 TeV per beam. Researchers hoped the SSC would help particles physicists observe elementary particles, like the Higgs boson, which had never been observed but was predicted by the Standard Model.

The project quickly outgrew its initial budget estimates and became mired in political controversy over its ultimate utility and cost. When Congress killed the SCC project in 1993, many researchers mourned the lost opportunity to study fundamental questions about the universe and sharply criticized politicians and the media for not understanding the scientific importance of the SSC. The SCC was never completed, leaving over 14 miles of tunnels and 17 large shafts in the ground at the time the project was abandoned.

SSC, R.I.P.

The good news for Superconducting Super Collider fans in 1993 was that Congress voted to protect the $640 million that had been earmarked for the SSC in this fiscal year. The bad news was that the money is to be spent ripping the project apart. "The SSC as we know it is dead," said Louisiana senator J. Bennett Johnston, a longtime SSC supporter, after the fatal vote in the House of Representatives. "It cannot be revived."

The SSC was arguably the most ambitious and certainly one of the most costly pure-science projects ever undertaken. The 54-mile underground tunnel circling Waxahachie, Texas—now one-quarter complete—was to hold 10,000 superconducting magnets capable of accelerating two small clouds of protons toward each other at nearly the speed of light. The energy of the resulting collision would have rivaled the energy of the universe immediately after the big bang and created a shower of exotic elementary particles. Physicists expected that one of those particles would be the Higgs boson—which, if it exists, would be the key to understanding the origin of mass. No existing accelerator packs enough of a punch to make a Higgs, and most physicists had pinned their hopes on the SSC.

IN CONTEXT: DARK MATTER, DARK ENERGY, AND NEW PHYSICS

In 1933, Swiss astronomer Fred Zwicky (1898–1974) noticed that the spiral shapes of the galaxies could be explained by the gravitation of their visible stars and gas; there must be some form of invisible or "dark" matter clumped around the galaxies that exerts gravitational force. In the 1980s and 1990s, observations from space satellites confirmed Zwicky's idea. Not only is there dark matter out there, the new data showed, but it outweighs all the ordinary, visible matter in the universe by a factor of at least 5. Our own galaxy, the Milky Way, has about 10 times as much dark matter as visible matter (stars, gas clouds, planets).

Clearly, dark matter exists and is very important. Yet nobody knows what it is. As of 2007, the most popular theory was that it is a form of non-baryonic cold matter. Baryons are protons and neutrons, which form the nuclei of ordinary atoms; in physics, "cold" simply means moving at speeds not close to the speed of light. On this theory, dark matter consists of large numbers of some unknown type of particle drifting in space.

Adding one mystery to another, in 1998 observations by scientists at Lawrence Berkeley National Laboratory in California and others showed for the first time that the expansion of the universe is accelerating—getting a little faster all the time. The only possible explanation is an invisible energy field pervading the cosmos, nicknamed dark energy.

Neither dark matter nor dark energy can be accounted for by existing physics. Until the mysteries of gravitation, dark matter, and dark energy are solved, it is highly unlikely that any unifying physical theory will explain the three still-fundamental forces—the strong force, electroweak force, and gravity—as aspects of a single, underlying force.

News of the SSC's death left much of the physics community stunned and disheartened—and not just the 150 physicists who were directly employed by the project but the thousands who work on particle physics at universities. Many of these physicists believe that only the SSC would have been able to provide the experimental data they needed to move forward. "It's tragic," says Lisa Randall, a particle physics theorist at MIT. "We were just at the point at which we were hoping to answer some of the fundamental questions that have been on our agenda for years, and now it's possible we'll never have those answers."

Europe's Large Hadron Collider, currently on the drawing board, will produce about a third of the energy of the SSC; some put its chances of finding the Higgs at one in three. Thus unless the European project gets lucky, physicists might have to give up—for the time

IN CONTEXT: THE DOUBTERS

Some scientists have taken a dim view of the quest for total unification in physics, arguing that it has sucked resources from other work and may not even be attainable. In the early 1990s, Howard Georgi (1947–), one of the first physicists to propose a Grand Unified Theory uniting the electroweak force with the strong force and co-proposer of the Supersymmetric Standard Model in 1981, wrote: "The legacy of grand unification, which in my view is very bad for the field of particle physics, is that it is considered reasonable—and even fashionable—for someone who calls him or herself a particle theorist to spend full time speculating about the world at distances much smaller than anything that we will ever be able to study in the laboratory."

While most scientists continue to support the quest for a theory that would unify all forces and account for all particles, there remain dissidents, including Georgi, who argue that a Theory of Everything may never be found.

being, at least—on unraveling this outstanding mystery: Why does matter have mass?

Some say most of the collateral damage from the SSC's demise will be felt by the upcoming generation of would-be physicists. "A lot of bright people are drawn into studying physics by the dream of discovering the fundamental laws of nature, even if most of them ultimately end up working on something else," says Steven Giddings, a theorist who currently works on black holes at the University of California at Santa Barbara. "If when I was in college I had gotten the message that our society lacks the will to pursue these fundamental questions anymore, I might well have gone to law school."

Still, not all physicists and certainly not all scientists mourn the SSC. Some had come to resent the mammoth project's drain on overall science funds, and some even questioned its chances of success. Some wondered whether the results SSC was after were really that much more important than research in solid-state physics, nuclear physics, astrophysics, or geophysics—projects with much lower price tags. The $640 million that will be spent on the SSC in this fiscal year, for instance, is more than the National Science Foundation expects to spend this year on all the Earth sciences and astronomy combined.

Critics of the project even included some particle physicists. "There is a group of people who felt the SSC was too big and too dependent on brute force," says Richard Blankenbecler, head of the theory group at the Stanford Linear Accelerator Center. "But at the time it was designed, there weren't any competing ideas, and people were just too impatient to get at the data." Blankenbecler adds that a new linear collider under development

at SLAC "may well end up a smaller, cheaper way of answering some of the same questions. People in this field are clever. They will come up with new ideas and bounce back."

But to its supporters, the death of the SSC was a shattering event. Some of them saw the project's demise as a gloomy portent for the future of science in general, even though funding for basic research in this country is at an all-time high. Giddings points out that the cost of the SSC would have been about $4 for every person in the United States over each of the next eight years. "That's less than the cost of a movie, and a lot less than the cost of a subscription to DISCOVER," he says. "How much would you pay to know what the world is made of?"

David H. Freedman

FREEDMAN, DAVID H. "SSC, R.I.P." *DISCOVER* 15, NO. 1 (JANUARY 1994): 101.

SEE ALSO *Physics: Heisenberg Uncertainty Principle; Physics: Nuclear Physics; Physics: QED Gauge Theory and Renormalization; Physics: Radioactivity; Physics: Special and General Relativity; Physics: The Inner World: The Search for Subatomic Particles; Physics: The Quantum Hypothesis; Physics: The Standard Model, String Theory, and Emerging Models of Fundamental Physics; Physics: Thermodynamics; Physics: Wave-Particle Duality.*

BIBLIOGRAPHY

Books

Hawking, Stephen W. *A Brief History of Time: From the Big Bang to Black Holes.* New York: Bantam Books, 1988.

Lincoln, Don. *Understanding the Universe: From Quarks to the Cosmos.* Singapore: World Scientific, 2004.

Oerter, Robert. *The Theory of Almost Everything.* New York: Pi Press, 2005.

Periodicals

Feng, Jonathan. "Searching for Gravity's Hidden Strength." *Science* 302 (October 31, 2003): 795–797.

Freedman, David H. "SSC, R.I.P." *Discover* 15, no. 1 (January 1994): 101.

Kane, Gordon. "The Dawn of Physics Beyond the Standard Model." *Scientific American* (June 2003): 68–75.

Salam, Abdus. "Einstein's Last Dream: The Space-Time Unification of Fundamental Forces." *Resonance* (December 2005): 246–251.

Weinberg, Steven. "A Unified Physics by 2050?" *Scientific American* (December 1999): 68–75.

Larry Gilman

Physics: Heisenberg Uncertainty Principle

■ Introduction

The Heisenberg uncertainty principle, also known as the principle of indeterminacy or the Heisenberg position-momentum uncertainty relation, is a quantum mechanics theory that limits just how precisely physicists can measure atomic particles like electrons. The uncertainty principle forces them instead to make probable estimates for pairs of intricately linked physical characteristics at any given time. These pairs of characteristics, such as position and momentum or time and energy, are known as conjugates.

It is possible to determine, for instance, the position of a particle precisely if the particle's momentum is not specified at the same time. However, if one measures position accurately then one must sacrifice an accurate knowledge of momentum. Thus, it is impossible to know the exact position and momentum of a particle at any given point in time.

■ Historical Background and Scientific Foundations

Werner Karl Heisenberg (1901–1976) attended Ludwig Maximilian University in Munich, Germany. In 1920 he went on to the University of Munich to study mathematics and physics from Arnold Sommerfeld (1868–1951), Wilhelm Wien (1864–1928), and Ernst Pringsheim (1859–1917) and, later, at the University of Göttingen from Max Born (1882–1970), James Franck (1882–1964), and David Hilbert (1862–1943).

At this time the leading atomic theory was that of Danish physicist Niels Bohr (1885–1962), Sommerfeld, and their colleagues. The Bohr-Sommerfeld quantum theory describes an atom as a tiny positive nucleus surrounded by negative electrons that orbit like planets around the sun. The theory was accurate when describing a hydrogen atom with one orbiting electron; it was unable to describe more complicated atoms, however.

In the early 1920s it became apparent that the theory did not satisfy all areas of physics, especially those of spectroscopy (the study of light emitted and absorbed by atoms), optics (the nature of light, and whether it consists of particles or waves), and atomic and molecular physics (the properties of atoms and molecules).

After Heisenberg was awarded his doctorate in 1923 from the University of Munich, he became a teaching assistant for Max Born. During this time he began to develop a theory of quantum mechanics, basing his work on matrix equations developed in the nineteenth century by the British mathematician Arthur Cayley (1821–1895).

Heisenberg was awarded the Nobel Prize for physics in 1932 for his theory of quantum mechanics, published in 1925; his principle of uncertainty, announced in 1927; and for the discovery of different (allotropic) forms of hydrogen. Today, the Heisenberg uncertainty principle is a central principle of modern physics.

The Science

In the everyday world measurements are easy to make. If you want to measure the dimensions of a table, a tape measure will be sufficient. You will not need to consider quantum forces because the table is so big when compared to these small and immaterial forces. When measuring small particles like atoms, however, quantum forces cannot be ignored.

Just how small is an atom? Its diameter is about 10^{-11} meters, or 1 hundred-billionth of a meter. The atom's nucleus has a diameter of about 10^{-15} of a meter, or 1 million-billionth of a meter. The width of a human hair, in comparison, is about 1 million carbon atoms, while an average human cell contains about 100 trillion atoms.

WORDS TO KNOW

ATOM: The smallest particle in which an element can exist.

CONJUGATE: In mathematics, the conjugate of a complex number is another number that has the same real part and a negative part of equal magnitude but opposite sign. The word "conjugate" has other technical uses in physics, chemistry, biology, and mathematics.

ELECTRON: A subatomic particle having a negative charge of −1.

MATRIX MECHANICS: A form of quantum mechanics. It provides a mathematical description of the physical laws governing the behavior of atomic- and subatomic-size objects. It was proposed in 1925 by German physicists Max Born (1882–1970), Werner Heisenberg (1901–1976), and Pascual Jordan (1902–1980).

MOMENTUM: The mass of a moving object multiplied by its velocity.

NUCLEUS: Any dense central structure can be termed a nucleus. In physics, the nucleus of the atom is the tiny, dense cluster of protons and neutrons that contains most of the mass of the atom and that (by the number of protons it contains) defines the chemical identity of the atom. In astronomy, the large, dense cluster of stars at the center of a galaxy is the galactic nucleus. In biology, the nucleus is a membrane-bounded organelle, found in eukaryotic cells, that contains the chromosomes and nucleolus. Intact eukaryotic cells are comprised of a nucleus and cytoplasm. A nuclear envelope encloses chromatin, the nucleolus, and a matrix, which together fill the nuclear space.

PHOTON: Smallest individual unit of electromagnetic radiation (light energy). These light particles are emitted by an atom as excess energy when that atom returns from an excited state (high energy) to its normal state.

PLANCK'S CONSTANT: A fixed number (constant) that appears in quantum physics: any photon can only carry some integer multiple (i.e., once, twice, three times, N times) of the fixed energy unit or quantum $E = hv$, where E is energy, v is frequency, and h is Planck's constant, whose value is approximately 6.626×10^{-34} J·s. German physicist Max Planck (1858–1947) described the value of this constant in 1899 and in 1918 was awarded the Nobel Prize in Physics.

QUANTUM MECHANICS: A system of physical principles that arose in the early twentieth century to improve upon those developed earlier by Isaac Newton, specifically with respect to submicroscopic phenomena.

Atoms are the smallest parts of an element that can be divided but still retain their unique identifying properties. They consist of a dense, positively charged nucleus surrounded by one or more negatively charged electrons. Measuring them accurately is not easy. To measure an electron, for instance, you can't use a tape measure. For this you need a high-powered atomic force microscope that uses light or some other form of electromagnetic radiation.

Electromagnetic radiation is waves of energy produced by the acceleration or oscillation (movement back and forth) of an electric charge. With both electric and magnetic components, electromagnetic radiation ranges from extremely high frequencies (short wavelengths) to extremely low frequencies (long wavelengths). From high to low frequency, the forms of electromagnetic radiation are gamma rays, x rays, ultraviolet radiation, visible light, infrared radiation, microwaves, and radio waves.

Electromagnetic radiation is made up of photons, which have enough energy and momentum that when they strike an electron they change its direction. Atomic force microscopes measure the exchange of energy that occurs when the two collide, and this, Heisenberg discovered, changes the physical characteristics of the electron being observed. Thus one may be able to get an accurate measurement of momentum, but acquire only a poor measurement of position.

The uncertainty principle explains the problem. Heisenberg stated that the process of measuring the position x of a particle disturbs the particle's momentum p, so that $Dx \times Dp$ is greater than or equal to h, where Dx is the uncertainty (difference) of the position of a particle along a spatial dimension, Dp is the uncertainty of the momentum, and h is Planck's constant.

Thus determining the exact position and momentum of a moving particle at any given point in time is limited to the precision of each measurement—with the product of these uncertainties equaling Planck's constant or greater. The more precise the measurement of position, the less precise the measurement of momentum, and vice versa. In large-scale measurements (such as tables), these errors are negligible, but they cannot be ignored in minute-scale measurements (such as atoms).

Heisenberg's theory is based on measurable radiation emitted by the atom, but other atomic or subatomic particles cannot be measured precisely. For example, Heisenberg stated that it is impossible to assign a position to an electron in three-dimensional space and in one-dimensional time. Heisenberg did not consider space and time to be two separate entities but one combined entity called space-time. Similarly, speed and position are not two separate things when dealing with atoms. They are inversely related, so that one can only be determined at the expense of the other one.

In accordance with his uncertainty principle, Heisenberg stated that numbers—ordinarily used to represent physical quantities—cannot be used to represent quantities involving miniscule materials such as atomic and subatomic particles. Instead, he used math-

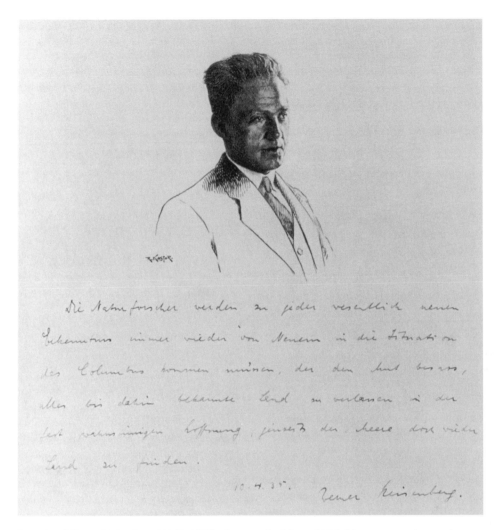

Portrait of Werner Heisenberg (1901–1976), German physicist and Nobel laureate, a major creative figure in the field of quantum mechanics. His papers on matrix mechanics (non-relativistic quantum mechanics) eventually led to his 1932 Nobel Prize. He is best known, however, for his uncertainty principle. Heisenberg led Nazi leader Adolf Hitler's A-bomb project, which he later claimed to have effectively stalled. *Library of Congress/Photo Researchers, Inc.*

ematical structures called matrices. In 1925, matrix mechanics—developed by Heisenberg, Max Born, and Pascual Jordan (1902–1980)—became the first complete formulation of quantum mechanics to describe the behaviors of subatomic particles.

Influences on Science and Society

The uncertainty principle has had a great impact on fundamental physics, chemistry, and other sciences. It has also been highly controversial. In fact, when confronted with Heisenberg's theory, Einstein famously scoffed, "God does not play dice with the universe," meaning that scientists should, instead, find logical explanations for behaviors found within the subatomic world.

The uncertainty principle involves four variables: the electron's position, momentum, energy, and time. When Heisenberg was developing his theory, scientists

held that the precision of any measurement was limited only by the accuracy of the measuring instrument.

Heisenberg, however, showed that at the atomic level and below an instrument's precision is irrelevant because quantum mechanics limits the precision possible when two canonically conjugate (specifically paired, inversely linked) variables are measured at the same time. For instance, momentum and position are canonic conjugates, as are energy and time. By proposing that certain aspects of subatomic reality are unknown, accepting and defining the bounds of uncertainty, Heisenberg freed scientists to use statistics to determine subatomic processes for the first time.

Although some physicists initially interpreted the uncertainty principle as a violation of fundamental physical laws, others realized its importance. Ultimately, it became the fundamental premise of quantum mechanics, which explained particles at atomic and subatomic levels, something that had been impossible with classical mechanics and electromagnetism.

■ Modern Cultural Connections

The Heisenberg uncertainty principle contradicts much earlier scientific dogma. Before its introduction, the central ideas of physics and chemistry were that (1) all matter consists of discrete particles with properties of gravitational mass and electrical charge, (2) light is a continuous electromagnetic wave that travels at a constant speed, (3) continuous electromagnetic fields are created by discrete charged particles, and (4) local interactions of electromagnetic charges are limited by the velocity of electromagnetic waves.

However, the quantum theory alters these ideas by showing that (1) all matter, besides being composed of discrete particles, also has wave properties, (2) light has discrete particle properties, (3) discrete statistical fields exist, and (4) instantaneous nonlocal matter interactions exist, and are called instant action at a distance.

The uncertainty principle also implies that space-time is smoothly curved in the macroscopic world. However, when viewed on increasingly smaller scales, quantum fluctuations occur as observations approach the Planck scale (10^{-33} cm). These effects are observed when theoretical physicists attempt to join the equations of general relativity (which involves the force of gravity) with those of quantum mechanics (which concerns the strong force, the weak force, and the electromagnetic force). So far such attempts have not produced a unified field theory that would unite all of the fundamental forces between elementary particles.

The uncertainty principle states that well-defined atomic orbits are not possible, the uncertainty in the electron's position is a basic atomic property, and observing any physical characteristic produces a disturbance—an interaction between the observer and the observed—that cannot be eliminated by more accurate measuring devices. The uncertainty principle also says the observer must be regarded as an inherent part of the measurement.

Quantum mechanics has improved on classical mechanics as the theoretical basis of all of physics. For most macroscopic systems, classical mechanics approximates most physical measurements, but in the atomic realm, only quantum mechanics—based on the Heisenberg uncertainty principle—can be used.

SEE ALSO *Physics: Maxwell's Equations, Light and the Electromagnetic Spectrum; Physics: The Quantum Hypothesis; Physics: The Standard Model, String Theory, and Emerging Models of Fundamental Physics; Physics: Wave-Particle Duality.*

BIBLIOGRAPHY

Books

Cassidy, David Charles. *Uncertainty: The Life and Science of Werner Heisenberg.* New York: W.H. Freeman and Company, 1992.

Duck, Ian. *One Hundred Years of Planck's Quantum.* River Edge, NJ: World Scientific, 2000.

Heisenberg, Werner. *The Physical Principles of the Quantum Theory.* Translated by Carl Eckart and Frank C. Hoyt. New York City: Dover, 1930.

Lindley, David. *Einstein, Heisenberg, Bohr and the Struggle for the Soul of Science.* New York: Doubleday, 2007.

Mehra, Jaqdish. *The Golden Age of Theoretical Physics.* River Edge, NJ: World Scientific, 2001.

Web Sites

American Institute of Physics. "Center for History of Physics; Niels Bohr Library and Archives." http://www.aip.org/history/ (accessed April 15, 2008).

———. "Quantum Mechanics 1925–1927: The Uncertainty Principle." http://www.aip.org/history/heisenberg/p08.htm (accessed April 15, 2008).

Fitzpatrick, Richard. University of Texas at Austin. "Heisenberg's Uncertainty Principle." December 12, 2006. http://farside.ph.utexas.edu/teaching/qmech/lectures/node26.html (accessed April 15, 2008).

New York Times. "New Detector May Test Heisenberg's Uncertainty Principle." http://query.nytimes.com/gst/fullpage.html?res=9C06E0DE163FF931A15754C0A9659C8B63 (accessed April 15, 2008).

———. "Professor Werner Heisenberg: A Pioneer of Quantum Mechanics." http://www-groups.dcs.st-and.ac.uk/~history/Obits/Heisenberg.html (accessed April 15, 2008).

Nobel Foundation. "Werner Heisenberg: The Nobel Prize in Physics 1932." http://nobelprize.org/nobel_prizes/physics/laureates/1932/heisenberg-bio.html (accessed April 15, 2008).

William Arthur Atkins

Physics: Lasers

■ Introduction

Laser is an acronym that stands for light amplification by stimulated emission of radiation. Lasers operate in the infrared, visible, or ultraviolet regions of the electromagnetic spectrum. Who deserves credit for the invention of the laser? The answer is not a simple one. Bell Laboratories' Web site states categorically that it was invented there. Others dispute that.

The first working laser saw the light of day on July 7, 1960. At first the invention seemed to be a solution looking for applications. In fact, for the first couple of years lasers seemed able to do little more than blaze holes in razor blades for TV commercials. Then, as the advantages of the strange new light produced by lasers became clearer, the devices seemed to find application everywhere.

Today the number and variety of lasers is astonishing, and they are used in such varied fields as medicine, science, industrial production, the home, office, communications, and the military.

The main reason for this wide use is that laser light has very special qualities: It is very pure in color, it can be very intense, and it can be directed with great precision. What makes all of this possible? Because of the way it is produced, it has a quality called coherence, which means that its waves remain in phase (in step) as they are produced.

■ Historical Background and Scientific Foundations

Stimulated Emission

The laser and its parent, the maser (microwave amplification by stimulated emission of radiation), can be traced back 90 years to their theoretical beginnings.

The German-born American physicist Albert Einstein (1879–1955), although best known for his work in relativity theory, also did important work on that other important twentieth-century scientific masterpiece, quantum theory. In one paper, published in 1916, he showed that under certain conditions, controlled emission of light energy could be obtained from an atom.

When an atom or molecule has somehow had its energy level raised by the input of energy, it can release this stored energy in one of two ways. It might be released in the form of a photon of light energy at some moment and in a direction that cannot be known in advance. This is spontaneous emission. It can also be stimulated by subjecting the particle to a small "shot" of electromagnetic radiation of the proper frequency. This is stimulated emission.

Einstein wrote that when an incoming shot of energy caused an electron to drop from a higher to a lower orbit, the electron emitted another photon. In other words, the energy of the emitted photon would be added to that of the photon that stimulated the action in the first place. Here, potentially, was light amplification. But what is special about the laser principle is that the emitted photon would be of the same frequency, in the same direction, and in step with (having the same phase as) the one that hit it.

Some two decades later, from 1939–1940, Russian physicist Valentin A. Fabrikant (1907–1991) began to think about what conditions would be needed to produce amplification of light in this way. At about the same time, American physicist Charles H. Townes (1915–), who had a PhD in physics from the California Institute of Technology, joined Bell Laboratories. Although Townes had worked in several theoretical fields, World War II was approaching, and he was assigned to work on a radar-directed bombing system. The system used microwave frequencies with a wavelength of several centimeters.

After the war, his work with microwaves fit in nicely with a growing interest in molecular (microwave) spectroscopy, and in 1948 he moved to Columbia University. Townes took his interest in microwave spectroscopy with him, and in 1949 began working with American physicist Arthur Schawlow (1921–1999), a newly minted physics PhD from the University of Toronto who had joined Columbia University on a fellowship.

Townes and Schawlow studied the idea that as the wavelength of the microwaves used in spectroscopy grew shorter, the radiation's interaction with molecules became stronger, leading to a more sensitive spectroscopic device. But generating the smaller wavelengths was a problem. Townes had the idea that the desired frequencies could be generated somehow by the use of gaseous molecules.

Townes's interest in millimeter waves—the next major step down in wavelength from microwaves—was for more powerful spectroscopes. The military, however, was funding some of the lab's research and had other interests. Microwaves, used in radar, for example, find wide application in communications, and this explained the military's interest: The higher the frequency of

electromagnetic radiation, the greater its information-carrying capacity. Systems utilizing millimeter waves—one frequency step shorter than microwaves—would be a huge advance, but researchers were stymied by the idea of building the required resonating cavity in millimeter proportions. To get around this, Townes, along with a couple of his more advanced students, began to work on a unit for microwaves that had a resonating cavity on the order of a few centimeters.

On April 26, 1951, after much thought, he figured out how to do this. Normally, more of the molecules in any substance are in low-energy states than in high ones. Townes wanted to change the natural balance and create a situation with an abnormally large number of high-energy molecules, then stimulate them to emit their energy by nudging them with microwaves. Here was amplification.

He could even take some of the emitted radiation and feed it back into the device to stimulate additional molecules, thereby creating an oscillator. This feedback arrangement, he knew, could be carried out in a cavity, which would resonate (just like an organ pipe and acoustic waves) at the proper frequency. The resonator would be a box whose dimensions were comparable with the wavelength of the radiation, that is, a few centimeters on a side.

On the back of an envelope he figured out some of the basic requirements. Three years and many experiments later, the maser was a reality. The original maser was a small metal box into which excited ammonia molecules were added. Townes later wrote: "The idea that I added ... was to use a resonant cavity so that the signal would go repeatedly through the gas, bouncing back and forth, picking up energy each time. The process would provide effectively infinite amplification."

When microwaves were beamed into the excited ammonia, the box emitted a pure, strong beam of high-frequency microwaves, far more frequency coherent (in step) than any that had ever been achieved before. The output of an ammonia maser is stable to one part in 100 billion, making it an extremely accurate atomic clock; masers' amplifying properties have also been found to be very useful for magnifying faint radio signals from space and for satellite communications.

Townes chose to work with ammonia gas because he knew that it was possible to separate the low-energy molecules from the high and to get the excited molecules into the cavity without too much trouble. This procedure for getting the majority of atoms or molecules in a container into a higher energy state is called population inversion and is basic to the operation of both masers and lasers.

It's worth noting that two Russians, Nikolay G. Basov (1922–2001) and A.M. Prokhorov (1916–2002), were working along similar lines independently of Townes. In 1952 they presented a paper at an All-Union

(U.S.S.R.) Conference, in which they discussed the possibility of constructing a molecular generator, that is, a maser. Their further publications, in 1954 and 1955, were in many respects similar to Townes's, and even showed a new way to obtain the active atomic systems for a maser.

Thus on October 29, 1964, when the Nobel Prize in Physics was awarded for fundamental work in the field of quantum electronics, which led to construction of oscillators and amplifiers based on the "aser" principle, it was awarded not only to Townes, but to Basov and Prokhorov as well.

A Laser Is Born

Following the maser development, there was much speculation about the possibility of extending the principle to the optical region. The difficulty, of course, was that optical wavelengths are so tiny—about one ten-thousandth of that of microwaves. The maser depended on a physical resonator a few centimeters or even millimeters in length. But at millimeter wavelengths, such resonators are already so small that they are hard to make accurately. Making a box one ten-thousandth that size was out of the question. Another approach was necessary.

In 1958 Townes and Schawlow (who was by then working at Bell Telephone Laboratories and had become his brother-in-law) outlined the theory and proposed a structure for an optical maser. They suggested that resonance could be obtained by making the waves travel back and forth along a relatively long thin column of amplifying substance that had parallel reflectors, that is, mirrors facing inward, at each end. One of the mirrors would be only partially reflective, to permit the beam to emerge when it grew to a certain strength. They dubbed this device, which would have a working medium of potassium vapor, an optical maser and published their theory in the respected journal *Physical Review* on December 15, 1958.

But according to another researcher, history has gotten things all wrong. According to Gordon Gould (1920–2005), a graduate student at Columbia who was working down the hall from Professor Townes at the time, the basic insights that made it all possible should have been credited to him.

The First Working Laser

After Townes and Schawlow's theory of the optical maser was published on December 15, 1958, the race to build the first actual device began in earnest. The clear winner, in 1960, was American physicist Dr. Theodore H. Maiman (1907–2007), then at Hughes Aircraft Company. Curiously, the active substance he used was neither the potassium vapor design suggested by Townes and Schawlow nor the gas laser suggested by Gould. Rather it was a single ruby crystal, with the ends ground flat and silvered.

Ruby is an aluminum oxide in which a small fraction of the aluminum atoms in the molecular structure, or lattice, have been replaced with chromium atoms. These atoms absorb green and blue light, imparting a red color. The chromium atoms can be boosted from their ground state into excited states when they absorb green or blue light. This process, by which population inversion is achieved, is called pumping.

Pumping in a crystal laser is generally achieved by placing the ruby rod within a spiral flash lamp. When the lamp is flashed, a bright beam of red light emerges from one end of the ruby, which has been only partially silvered. The duration of this flash of red light is quite brief, lasting only some 300 millionths of a second, but it is very intense. In the early lasers, such a flash reached a peak power of 10,000 watts.

The outside world seemed to have little understanding of the significance of Maiman's accomplishment. Maiman detailed his work for quick publication in *Physical Review Letters,* but passed it through the Hughes Patent Office first. The patent office cleared the paper but didn't think the report of his work was important enough to warrant filing for patent protection. Thus Hughes lost any claim to foreign patent rights. Second, the editor at *Physical Review Letters* sent Maiman a curt letter of rejection. Maiman's paper was subsequently published in the British science journal *Nature* on August 6, 1960.

A Multitude of Lasers

Although the researchers at Bell Laboratories were disappointed that Maiman had reached the goal of building the optical maser first, they pointed out, happily, that Maiman's laser was "only" a pulsed laser. In other words, light energy was pumped in and a bullet of energy sped out from it. Then the whole process had to be repeated. Pulsed operation is fine for spot welding and for operations such as radar-type range finding, where pulses of energy are used anyway. But even though lasers permit use of optical wavelengths with their much greater carrying capacity, pulsed systems would not be useful for communications (or for other applications to be discussed later). Continuous-wave (CW) operation remained a major goal.

In addition, solid crystals are difficult to manufacture. Hence, it was natural for laser pioneers to look hopefully at gas lasers, which would theoretically be easier to make, once the proper conditions were satisfied. Simply fill a glass tube with the proper gas and seal it. But other advantages would accrue. For one thing the relatively sparse population of emitting atoms in a gas provides an almost ideally homogeneous medium. That is, the emitting atoms (corresponding to chromium in the ruby crystal) are not "contaminated" by

GORDON GOULD (1920–2005)

Gordon Gould (1920–2005) was born in New York City. Even as a child, his heroes were inventors like Alexander Graham Bell (1847–1942), Thomas Edison (1847–1941), and Guglielmo Marconi (1874–1937). Knowing that to invent anything worthwhile he would have to understand the physics behind it, he concentrated on this subject throughout his school years. In 1957 he was working toward a PhD at Columbia University, an important center of physics. Charles Townes (1915–), inventor of the maser, was teaching there, and they occasionally discussed their work. Gould was doing research on optical and microwave spectroscopy, but was beginning to think about lasers. There is considerable disagreement between the two men as to how much each told the other about his laser ideas.

On the night of November 9, 1957, at the age of 37, Gould came up with his concept for the laser. The maser, Townes's invention, amplified microwaves. A laser would be much more powerful, since a photon of light by its very nature has a hundred thousand times more energy than a single unit of microwave energy. Gould—at least in his own estimation—had come up with one of the most important inventions of the twentieth century. Yet he was to embark on a comedy of errors that would prevent him from profiting from that invention for almost 30 years.

Knowing that he had something important in hand, he walked away from his almost-won doctorate. He consulted a patent attorney but, misconstruing the attorney's advice, thought he had to build a prototype before he could file for a patent. Therefore he spent most of 1958 trying to refine and improve his device and did not file until 1959. In the meantime, other laser researchers, including Townes and Schawlow, had already filed for the same or similar devices. What Gould did do, fortunately for him, was write down and diagram a careful description of his ideas in a school notebook, which he then took to a corner candy store, where he had the pages notarized. The date was November 13, 1957. Gould was the first to use the term laser.

In 1958 Gould joined a newly formed New York company called TRG, where he felt there was a better chance that he could develop his laser. The company applied for and won a contract from ARPA, the Department of Defense's Advanced Research Projects Agency, to support this work, but they were not successful in developing a working laser. Later Gould became a professor at Polytechnic Institute of Brooklyn, and in 1973 he helped found the industrial firm Optelicom, Inc.

Over the years, however, he also entered upon a long and very expensive patent war. Along the way he enlisted the aid of a series of partners who helped in his battles. It was a war fought in many courtrooms—against competing scientists, against firms that he claimed were illegally using "his" designs without paying licensing fees, and against the Patent Office itself. He insisted that Townes had used his ideas, while Townes insisted that whenever they talked before that November date, Gould would rush back to his office and write down what they had discussed. Gould says their discussions merely alerted him to the fact that Townes was thinking along similar lines to his.

At first things looked bleak for Gould. In March 1964, for instance, the Patent Office ruled against him and in favor of Bell Laboratories, upholding the optical maser patent awarded to Schawlow and Townes on March 22, 1960. Gould and his partners turned to the courts, suing the Patent Office in federal court when his own patents were disallowed. After a great deal of expensive litigation and more battles with the Patent Office, things began to turn in his direction. In May 1977 Gould's application for an optically pumped amplifier was allowed. After some other successes, he was awarded patents for a gas-discharge amplifier which, with the optically pumped amplifier, covered 80 percent of the lasers made in the United States. He was granted other patents as well.

The delay, curiously, had made the patents far more lucrative than they would have been if they had been issued immediately. In other words, if they had been issued when applied for in 1959, they would have run their 17-year course before laser applications exploded into the huge industry they became in later years.

the lattice or host atoms. Since only active atoms need be used, the frequency coherence of a gas laser would probably be even better than that of the crystal laser, they reasoned.

Less than a year after the development of Maiman's ruby laser, Iranian-born physicist Ali Javan (1926–) at Bell Laboratories proposed a gas laser employing a mixture of helium and neon. This was an ingeniously contrived partnership whereby one gas did the energizing, and the other did the amplifying. Javan's laser provided the first continuous output, generally referred to as CW, or continuous wave, operation.

Gas lasers now utilize many different gases for different wavelength outputs and powers and provide the "purest" light of all. An additional advantage is that the optical pumping light can be dispensed with; an input signal of radio waves of the proper frequency can do the job.

Power and Efficiency

The two units generally used to specify the power output of a laser are watts and joules. The watt, the rate at which (electrical) work is being done, is the more familiar unit—we need only think of a 15- or a 150-watt bulb to get an idea of its magnitude. The joule is a unit of energy and can be thought of as the total capacity to do work. One joule is equivalent to 1 watt-second, or 1 watt applied for 1 second. But it can also mean a

10-watt burst of laser light lasting 0.1 second, or a billion watts lasting a billionth of a second.

In the early years, crystal lasers were the most powerful, but other materials, such as liquids and specially prepared glass, have now been made that provide solid competition. In 2003 the U.S. National Ignition Facility Project produced 10.4 kilojoules of ultraviolet light in a neodymium glass-based laser beam, setting a world record for laser performance.

One of the least satisfactory aspects of the laser has been its notoriously low efficiency. For a while the best that could be achieved was about 1%, that is, 100 watts of light had to be put in to get one watt of coherent light out. In gas lasers the efficiency was even lower, ranging from 0.01 to 0.1%.

In gas lasers this was no great problem, since high power was not the objective. But with the high-power solid lasers, pumping power could be a major undertaking. A high-powered laser pump built by Westinghouse Research Laboratories could handle 70,000 joules. In more familiar terms, the peak power input while the pump is on is about 100,000,000 watts. For a brief instant this is roughly equal to the electrical power needs of a city of 100,000 people.

Two developments changed the efficiency levels. First, the carbon dioxide (CO_2) gas laser is quite efficient, with the figure having passed 15%. CO_2 lasers, producing either pulses or continuous waves, can put out powerful beams and have found wide use in many applications. Special-purpose devices have been constructed that produce 20 megawatts CW.

The second is the injection or semiconductor laser, in which efficiencies of more than 40% have been achieved. In a current program supported by the U.S. Defense Advanced Research Projects Agency, Alfalight, a semiconductor laser manufacturer, has demonstrated 65% efficiency; the program is shooting for 80% in three years. Unless unforeseen difficulties arise, this figure is expected to continue to rise to a theoretical maximum of 100%.

In the semiconductor laser all the functions of the laser have been packed into a tiny semiconductor crystal. In this case, electrons and "holes" (vacancies in the crystal structure that act like positive charges) accomplish the job done by excited atoms in the other types of lasers. Although the device itself is about the size of this letter "o," it is self-contained and can convert electric current directly into laser light. This has made possible a vast field of experiment and improvement in the world of lasers.

■ Modern Cultural Connections

Today lasers are in action practically everywhere we turn. In the home, for example, we have laser-based CD

IN CONTEXT: LASING—A NEW WORD

The all-important rod in the ruby laser is formed as a single large crystal. Because it must be free of extraneous material, it is grown artificially. That is, the crystal is formed as it is pulled slowly from a "melt," after which it is ground to size and polished.

In operation, the crystal rod contains many atoms in the ground state and a few in an excited state. When the pumping lamp flashes, it raises most of the atoms to the excited state, creating the required population inversion. Lasing begins when an excited atom spontaneously emits a photon parallel to the axis of the crystal. Photons emitted in other directions merely pass out through the transparent sides of the crystal. The emitted photon stimulates another atom in its path to contribute a second photon, in step, and in the same direction.

This process continues as the photons are reflected back and forth between the ends of the crystal. (We might think of lone soldiers falling into step with a column of marching men.) The beam continues to build; when amplification is great enough, the beam flashes out through the partially silvered side of the rod—a narrow, parallel, concentrated, coherent beam of light.

---

and DVD players, CD-ROM drives, pointers, laser thermometers for cooking, levels and "tape" measures for the craftsman, and even a laser guide that helps guide a driver into the proper parking spot in a garage.

Laser applications can, in fact, be divided into two broad categories: (1) commercial, industrial, military, home, and medical uses; and (2) scientific research. In the first group, lasers are used to do something that has been done in another way (but perhaps not as well). For example, one of the first medical applications was in eye surgery, for "welding" a detached retina back into place. The laser is particularly useful here because laser light can penetrate transparent objects such as the eye's lens without harming it, and can do the needed surgery ("stitching") on the retina.

The device used on the eye may have a power of only 5 watts, which is less than a typical night light. But the concentration of the beam is such that it can be focused down to the size of a single cell. Lasers have been replacing the eye surgeon's blade in treating farsightedness and astigmatism by reshaping the cornea to alter the way the eye refracts light.

In another medical application that puts the laser's unique qualities to work, urologists use its concentrated power to blast kidney stones while they watch the process through the same fiber-optic cable that is carrying the laser shots.

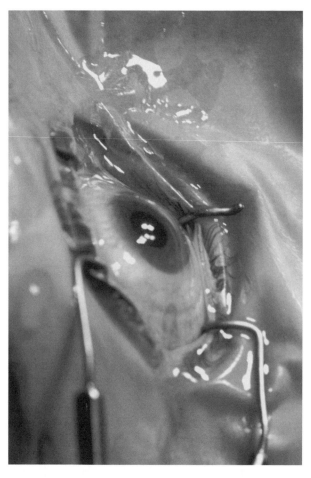

A patient undergoes an eye operation with laser at Istanbul Cerrahi Hastanesi (Istanbul Operation Hospital). *Mustafa Ozer/ AFP/Getty Images.*

The laser's unusual features have also led to its wide use in cosmetic surgery to treat everything from sagging eyelids to varicose veins, and for all kinds of skin rejuvenation, including treatments for wrinkles and problems with acne, skin texture, and discoloration. When used in general surgery, lasers can perform nearly bloodless cutting by cauterizing (sealing off blood vessels) as they cut. They are being used for pain relief, in place of radiotherapy for cancer treatment.

Highly accurate laser tracking and measurement systems have been developed for precision manufacturing. Laser light's high intensity can cut or penetrate almost anything, including the hardest known material, diamond. An additional advantage is that laser drills do not get dull with use.

Scientific Research

In the second category, scientific research, the narrowness of the laser beam has made it ideal for applications requiring accurate alignment. Perhaps the ultimate such application is the 2-mile-long (3.2-km) linear accelerator built in 1966 by Stanford University for what was at the time the United States Atomic Energy Commission. Only a laser beam could accomplish the incredible task of keeping the 7/8-inch (2.2-cm) bore of the accelerator straight along its full 2-mile (3.2 km) length. A remote monitoring scheme, based on the same laser beam, told operators when a section shifted out of line (due, for example, to small earth movements) by more than about 1/32 inch (0.79 cm) and identified the section.

The accelerator is housed at the Stanford Linear Accelerator Center (SLAC) in California. Established in 1962, it remains one of the world's leading research laboratories and is currently building what will be the world's first x-ray laser. This system will use the last 0.6 mile (1 km) of the 2-mile (3.2 Km) SLAC accelerator to generate the needed input energy. The device will open up a world of new applications.

In scientific research the laser has proved its value over and over. Pulses of laser light can be powerful, but they can also be short. Laser pulses lasting a few femtoseconds (one quadrillionth [10^{-15}] of a second) were produced in the 1980s. Current work at the University of Munich and the Max Planck Institute for Quantum Optics in Garching, Germany, has brought this down to attoseconds (one quintillionth [10^{-18}] of a second). This makes it possible to probe atomic and subatomic electron processes. For the first time researchers can "freeze" the motion of electrons in the length and timescales of atoms.

In both scientific and medical research, the laser is often combined with some existing kind of equipment, where its special features provide an advantage. For example, researchers at Stanford University hope to treat cancer by creating carbon nanotube messengers that, combined with other substances, will latch on to cancer cells. The objective is to turn a near-infrared laser into a cancer weapon; its light would pass through normal human tissue without harming it, but the 150-nanometer-long nanotubes would strongly absorb the radiation and turn it into heat that, they hope, will destroy the cancer cell.

The most exciting probability of all, however, is that lasers will undoubtedly change your life in ways that we cannot even conceive at this time.

■ Primary Source Connection

One of the many ways in which lasers are now being used is by the U.S. military. Though some critics argue that lasting eye damage can occur, as Will Knight reports in the *New Scientist,* the U.S. Department of Defense would like to begin using them in Iraq.

US MILITARY SETS LASER PHASRs TO STUN

The US government has unveiled a "non-lethal" laser rifle designed to dazzle enemy personnel without causing them permanent harm. But the device will require close scrutiny to ensure compliance with a United Nations protocol on blinding laser weapons.

The Personnel Halting and Stimulation Response (PHASR) rifle was developed at the Air Force Research Laboratory in New Mexico, U.S., and two prototypes have been delivered to military bases in Texas and Virginia for further testing.

The U.S. Department of Defense (DoD) believes the weapon could be used, for example, to temporarily blind suspects who drive through a roadblock. However, the DoD has yet to reveal details of how the laser works and has yet to respond to New Scientist's requests for further information.

Laser weapons capable of blinding enemies have been developed in the past but were banned under a 1995 UN convention called the Protocol on Blinding Laser Weapons. The wording of this protocol, however, does not prohibit lasers that temporarily dazzle a foe.

Permanent harm

"In the past, the problem with lasers of this type has been that they often permanently blind human targets," says Tobias Feakin, an expert at Bradford University's Non-Lethal Weapons Research Project in the UK.

But he says newer systems may avoid this problem by using less powerful laser beams. "This new wave of low-intensity laser weapons do not have a permanently damaging effect, apparently," he told New Scientist.

Several laboratories across the world are working on such weapons. But even low power laser systems can cause eye damage if they are used at close quarters or for extended periods.

The PHASR may attempt to address safety concerns by automatically sensing its distance from a target. The limited information released by the DoD includes mention of an "eye-safe range finder", which may mean the laser's power is adjusted depending on the distance to the target. The system is also said to incorporate a "two wavelength laser system", which may be designed to counter goggles that can filter out certain wavelengths of laser light.

Pulsing green light

Neil Davison, another expert at Bradford University, says the situation in Iraq may encourage the U.S. to push for the development of less-than-lethal laser weapons. "They already use bright white lights at vehicle checkpoints in Iraq to dazzle drivers who are approaching too fast," he says.

LE Systems, based in Connecticut, U.S., for example, makes the Laser Dazzler, which resembles an ordinary torch and emits a low power pulsing green laser light. The company says this device has been tested extensively and been shown to cause no lasting eye damage.

The possibility of causing lasting eye damage can be reduced by diffusing the laser beam or rapidly moving it across the target with a series of mirrors.

And the same U.S. military research lab developed another laser weapon more than a decade ago, called the Sabre 203. This device attached beneath the barrel of a normal rifle and emitted a low-power laser light over a range of 300 metres. It was used by U.S. forces in Somalia in 1995 but later shelved because of concerns over safety and effectiveness.

Will Knight

KNIGHT, WILL. "US MILITARY SETS LASER PHASRS TO STUN." *NEW SCIENTIST.* HTTP://WWW.NEWSCIENTIST. COM/ARTICLE/DN8275.HTML (ACCESSED NOVEMBER 14, 2007).

SEE ALSO *Chemistry: Molecular Structure and Stereochemistry; Chemistry: States of Matter: Solids, Liquids, Gases, and Plasma; Physics: Fundamental Forces and the Synthesis of Theory; Physics: Semiconductors; Physics: Spectroscopy; Physics: The Inner World: The Search for Subatomic Particles; Physics: The Standard Model, String Theory, and Emerging Models of Fundamental Physics; Physics: Wave-Particle Duality.*

BIBLIOGRAPHY

Books

Bromberg, Joan Lisa. *The Laser in America, 1950–1970.* Cambridge, MA: MIT Press, 1991.

Maiman, Theodore. *The Laser Odyssey.* Blaine, WA: Laser Press, 2000.

Park, David Allen. *The Fire within the Eye; a Historical Essay on the Nature and Meaning of Light.* Princeton, NJ: Princeton University Press, 1997.

Taylor, Nick. *Laser: The Inventor, the Nobel Laureate, and the Thirty-Year Patent War.* New York: Simon & Schuster, 2000.

Townes, Charles H. *How the Laser Happened: Adventures of a Scientist.* New York: Oxford University Press, 1999.

Periodicals

"An Unexpectedly Bright Idea—Case History." *Economist* 375, no. 8430 (June 9, 2005): 26.

Goldwasser, Sam. "Laser Pointers and Diode Laser Modules." *Poptronics* 3, no. 4 (April 2002): 46–48, 53.

Hannon, Kerry. "Vindicated!" *Forbes* 140, no. 13 (December 14, 1987): 35–36.

Web Sites

American Heritage. "Amazing Light." http://www.americanheritage.com/articles/magazine/it/1992/4/1992_4_18.shtml (accessed May 16, 2006).

Argonne National Laboratory. Nuclear Engineering Division. "Laser Applications Laboratory." http://www.ne.anl.gov/facilities/lal/projects/industry/online.html (accessed November 14, 2007).

Caliber: Journals of the University of California Press: Historical Studies in the Physical and Biological Sciences. "Who Invented the Laser: An Analysis of the Early Patents." http://caliber.ucpress.net/doi/abs/10.1525/hsps.2003.34.1.115 (accessed May 1, 2006).

Inc. Magazine. "Patent Pending." http://pf.inc.com/magazine/19890301/5568.html (accessed November 14, 2007).

Knight, Will. "US Military Sets Laser PHASRs to Stun." *New Scientist.* http://www.newscientist.com/article/dn8275.html (accessed November 14, 2007).

Laser Focus World Magazine. "Photonic Industry News." http://lfw.pennnet.com/ (accessed November 14, 2007).

Sam's Laser FAQ. "Safety, Info, Links, Parts, Types, Drive, Construction: A Practical Guide to Lasers for Experimenters and Hobbyists." http://www.laserfaq.org/sam/lasersam.htm (accessed November 14, 2007).

Hal Hellman

Physics: Maxwell's Equations, Light and the Electromagnetic Spectrum

■ Introduction

In the nineteenth century, knowledge of electromagnetism—all those phenomena related to electrical charges, electric currents, and magnetism—moved rapidly from experimental novelty to practical use. At the start of the century, only gas and oil lamps might be found in homes and businesses, but by the end of the century electric light bulbs were common. By 1865, a telegraph cable connected the United States and England. By the end of the nineteenth century, high-energy electromagnetic radiation in the form of x rays was being used to diagnose injury, and radio waves had been discovered, enabling a series of communications revolutions in the early twentieth century (the latest of which is the spread of cellular phones). The mathematical clarification and unification by Scottish physicist James Clerk Maxwell (1831–1879) of the scientific understanding of electromagnetism prepared the way for the more comprehensive theories of relativity and quantum mechanics in the early twentieth century. Maxwell's mathematical expressions describing the relationship between electrical and magnetic phenomena are known today as Maxwell's equations.

■ Historical Background and Scientific Foundations

Developments Leading Up to Maxwell's Equations

In the late eighteenth and early nineteenth centuries, philosophical and religious ideas led many scientists to seek a common, fundamental source for the seemingly separate forces of nature (electricity, magnetism, gravitation, etc.). Also, scientists were still concerned with the profound philosophical and scientific questions posed by Isaac Newton's (1643–1727) *Optics* (1704) regarding the nature of light. A substance termed ether—imagined as something like an intangible, invisible gas permeating all space—which would support the passage of light waves, much as air supports the passage of sound waves, was thought necessary to explain the wave-like behaviors of light.

Clarification of the relationship between electricity and magnetism was hampered in the latter eighteenth century and early nineteenth century by a rift between the descriptions of nature used by mathematicians and by experimentalists. Advances in electromagnetic theory during the nineteenth century produced unification of these approaches. The culmination of this merger came with the development by Maxwell of a set of four compact equations that accurately described electromagnetic phenomena better than any previous model. Indeed, so accurate are Maxwell's equations that they are still used routinely throughout science and engineering, although in some phenomena studied by physicists they must be supplemented or replaced by the equations of relativity and quantum mechanics.

In developing his famous equations, Maxwell built upon the mathematical insights of the German mathematician Carl Fredrich Gauss (1777–1855), the reasonings and laboratory work of French scientist André Marie Ampère (1775–1836), the observations of Danish scientist Hans Christian Oersted (1777–1851), and a wealth of experimental evidence provided by English physicist and chemist Micheal Faraday (1791–1867). Like most scientific leaps, including Albert Einstein's (1879–1955) a generation later, Maxwell's was made possible by a large amount of labor by earlier colleagues and contemporaries.

Simple observations by American statesman and scientist Benjamin Franklin (1706–1790) and other scientists in the mid-eighteenth century continued to intrigue scientists and inventors. As the nineteenth century progressed, electricity came to play an important role as increasingly technology-dependent industrial countries in Europe and North America developed machines with which to meet the needs of their rapidly expanding urban societies.

In 1820, Oersted demonstrated that magnetism has a relationship to electricity by placing a wire near a magnetic compass. When an electric current was passed through the wire, the compass needle showed a deflection (that is, moved from its rest position): This proved the presence of a magnetic field in addition to that of Earth. A year later, inspired by Oersted's demonstrations, Faraday proved his genius in the practical world of laboratory work by building a device he termed a "rotator," now credited as the first electric motor. Faraday's apparatus consisted of a wire carrying an electrical current and rotating about a magnet. Later, Faraday clearly demonstrated the converse phenomenon, the induction of current by magnets being rotated about the wire. The first practical electric motors were all designed according to principles discovered by Faraday, although it was another half century before their widespread application in industry, electric cars, elevators, and appliances. Faraday's method of producing electric current with moving magnets and fixed coils—exploiting the phenomenon of electromagnetic induction—is still used by virtually all modern power generators.

Faraday's 1831 publication of his work on electromagnetic induction was the first in a series of papers eventually collected as *Experimental Researches in Electricity* (1844, 1847). This work became a standard reference on electricity for nineteenth-century scientists and is credited with inspiring and guiding inventors such as Thomas Edison (1847–1931), the American who invented the light bulb, motion picture, and phonograph record.

During the last decades of the nineteenth century, electric motors powered an increasing number of time- and labor-saving devices, ranging from industrial hoists to personal sewing machines. Electric motors proved safer to manage and more productive than steam or internal-combustion engines. The need for production and distribution of electrical power spawned the construction of dynamos, central power stations, and elaborate electrical transmission systems.

Ampère, a professor of mechanics at the Ecole Polytechnique in Paris, France, was also influenced by Oersted's observations. Ampère's effect on the theoretical development of electromagnetic theory was as transformative as the influence of Newton on the scientific understanding of gravity. He deepened and clarified the relationship between electrical and magnetic phenomena through a series of brilliantly devised experiments that demonstrated the fundamental principles of electrodynamics (the effects generated by moving electrical charges and magnetic fields). Although Ampère made a number of experimental discoveries, it was his mathematical brilliance that was most important in laying the foundation for later electromagnetic theory. Ampère translated the electromagnetic phenomena observed by Faraday and other experimentalists into the language of mathematics, making possible precise treatment of and further insights into those phenomena.

The culminating fusion of nineteenth-century experimentation in and mathematical abstraction of electromagnetism came with the development of Maxwell's equations of the electromagnetic field. These equations were more than mere mathematical interpretations of experimental results; Maxwell's precise expressions predicted new phenomena (i.e., radio waves) and created the necessary background for quantum and relativity theory. Several major early-twentieth-century physicists, such as Max Planck (1858–1947), Einstein, and Neils Bohr (1885–1962), later credited Maxwell with laying the foundations of modern physics.

Maxwell first published his electromagnetic field equations in 1864. In 1873, his book *Electricity and Magnetism* fully articulated the laws of electromagnetism. Maxwell's statements about the propagation through space of electric and magnetic fields offered the first working theory of electromagnetic waves. They allowed the unification of known electrical and magnetic phenomena into the electromagnetic spectrum. Visible light, radio waves, x rays, and other phenomena are now known to be electromagnetic waves such as were described by Maxwell, differing only in their rate of vibration (frequency).

Illustration of generating artificial lightning in Nikola Tesla's laboratory. Tesla, an inventor, physicist, and electrical and mechanical engineer, used Maxwell's equations in his research in electromagnetism and waves, which led to his multiple contributions in various fields, including wireless transmissions.   © *Bettman/Corbis.*

The Equations

Although proof of the existence of the electron (the subatomic particle that carries negative electrical charge) did not come until the end of the nineteenth century, and knowledge of the proton (the particle that carries positive electrical charge) was not known until 1918, Maxwell's equations established that an electric charge is a source of an electric field, and that electric lines of force begin and end on electric charges (though they may also exist in the absence of electric charges, if a changing magnetic field is present).

Because they are mathematically elegant (concise) a full understanding of Maxwell's equations require an advanced level of mathematical sophistication (above the level of beginning calculus). In addition, several forms of Maxwell's equations are in common use; although they look different, they all say essentially the same thing.

In summary, Maxwell's first equation, also known as Gauss's law, relates the electric field (E) that passes through a particular surface area (A) (e.g., a sphere) to a charge (Q) within that surface. Maxwell's second equation

asserts that there are no "magnetic charges" and that, therefore, magnetic lines of force must always form closed loops. Maxwell's third equation, also known as Ampère's law, states that a magnetic field (B) can be induced by a circular loop by charges moving through a wire (i.e., an electric current designated as I) or by a changing electric field (electric flux). Maxwell's fourth equation, also known as Faraday's law, states that voltage is generated in a conductor as it passes through a magnetic field or as it cuts through magnetic lines of force (or, as also commonly stated, "cuts magnetic field lines").

■ Modern Cultural Connections

Prior to Maxwell's equations, scientists taught that all waves require a medium of propagation—some kind of flexible substance, whether solid, liquid, or gaseous—which could be set in motion in a way that transferred energy from one point to another. Maxwell's equations suggested that electromagnetic waves might not require such a medium: A changing electrical field and a changing magnetic field, paired together, might produce each other and move forward in space at the same time, constituting a wave without a medium—an electromagnetic wave. That the ether—a proposed transmissive medium of electromagnetic waves—does not exist was proved in 1887 by an ingenious experiment designed by Albert Michelson (1852–1931) and Edward Morley (1838–1923).

With his electromagnetic field equations, Maxwell was able to calculate the speed of propagation of an electromagnetic wave. This speed arises out of the equations themselves, and is the same for all electromagnetic waves. Maxwell's value for the speed of electromagnetic propagation fit well with experimental measurements of the speed of light, and Maxwell and other scientists realized that visible light must be electromagnetic in nature, part of a range or spectrum of such waves differing only in frequency of vibration. Exploration of the electromagnetic spectrum resulted in both theoretical and practical advances. For example, German physicist Heinrich Rudolph Hertz (1857–1894) demonstrated the existence of radio waves in 1888. Near the end of the nineteenth century, the discovery by German physicist Wilhelm Röntgen (1845–1923) of high-frequency electromagnetic radiation in the form of x rays found its first practical medical uses.

Maxwell's mathematical unification of electromagnetism thus laid the foundation of a series of technological, social, scientific, medical, and military revolutions based on wireless (radio) communications. They also made possible the development of relativity and quantum theory in the early decades of the twentieth century.

Maxwell's equations remain a powerful tool for understanding electromagnetic fields and waves. Although under certain conditions (for example, extremely high gravity, small size-scale, and velocities close to the speed of light) they must be replaced or modified by the equations of relativity and quantum mechanics, they are the working description of electromagnetic phenomenon used daily by most engineers and scientists. They are used, for example, in the design of electrical transmission lines and of radio, television, microwave, and other antennae.

Maxwell was the first scientist to produce a unification theory—a set of equations revealing that two or more apparently distinct forces, in this case the electric and magnetic forces, can be treated as manifestations of a single underlying phenomenon. The next unification of this type was not achieved until the 1960s, when scientists developed equations showing that electricity, magnetism, and the weak nuclear force could all be treated in a unified way. Today, the search for a unified theory of all forces continues.

SEE ALSO *Physics: Fundamental Forces and the Synthesis of Theory.*

BIBLIOGRAPHY

Books

Goldman, Martin. *The Demon in the Aether: The Story of James Clerk Maxwell.* Edinburgh, UK: Paul Harris Publishing, 1983.

Tolstoy, Ivan. *James Clerk Maxwell: A Biography.* New York: Harper & Brothers, 1987.

Whittaker, Edmund. *A History of the Theories of Aether and Electricity.* Chicago: University of Chicago Press, 1981.

K. Lee Lerner

Physics: Microscopy

■ Introduction

Though not a scientific field per se, microscopy (the use of one of many types of microscopes) is an important technological tool within many scientific disciplines. By allowing scientists to view samples of various materials at magnification, microscopy exposes details that were previously invisible. The use of microscopes has led to the discovery of the cell and the organelles within it, both crucial understandings in the field of biology, as well as many other discoveries.

Since its invention about 400 years ago, the microscope has undergone many changes and improvements, especially in professional models. Many of the most powerful microscopes no longer use visible light to form their images but employ electrons to create high-resolution images of very tiny objects, enabling researchers to study their structure. On the other hand, student-model microscopes have changed very little since their invention. Their continued use in education shows the power of this relatively simple tool.

■ Historical Background and Scientific Foundations

The most fundamental part of a microscope is a magnifying lens. Though rare, lenses of this type were produced in ancient times. During excavations in Nineveh, a city in ancient Mesopotamia, a quartz-crystal lens dating from around 640 BC was unearthed. Similar lenses have been found on the island of Crete and other ancient Greek sites. Some of these lenses could provide magnification up to 20 times, though most were only useful for much lesser magnification. The Greeks also used "burning glasses" to concentrate the sun's rays, and both Greeks and Romans used glass spheres filled with water as magnifiers. Engravers probably used lenses to create fine carvings, and philosophers such as Pliny the Elder (AD 23–79) were interested in them as objects of scientific curiosity.

During the Renaissance significant advances were made in the field of optics, when technology became reliable enough to make lenses that corrected vision. One of the most famous scientists who worked with optics was Italian physicist, astronomer, and mathematician Galileo Galilei (1564–1642). Galileo is especially known for his telescopes, which he built and used to study the solar system. His observations of Jupiter led to the discovery of several of its moons. Galileo also made an early compound microscope that he shared with his colleagues in the Academia dei Lincei, one of the earliest scientific academies.

Some confusion exists about the date of the microscope's invention and its inventor. Many credit Dutch opticians Hans Jansen and his son Zacharias (1585–1632), placing their invention around 1600; others cite another Dutch spectacle maker, Hans Lippershey (c.1570–c.1619). The Jansens' machine was a compound microscope, with a tube connecting a lens at the eyepiece to another near the sample. Both Galileo's device and the Jansens' were based on the design of telescopes, another early object of scientific curiosity.

The microscope allowed many kinds of scientists to study their fields in more depth than ever before. One of these was Marcello Malpighi (1628–1694), an Italian physiologist and one of the first to study microscopic anatomy. Malpighi questioned the fundamental principals of medicine in his day. He used his microscopic observations to prove, among other things, that blood was continuously circulated, not turned to flesh when it reached the edges of the body. He also studied the tissues of the brain, liver, kidneys, and other organs, giving the first descriptions of their microscopic features.

WORDS TO KNOW

COMPOUND MICROSCOPE: A multiple-lens microscope (two or more lenses housed in a long tube).

DEPTH OF FIELD: In photography, depth of field refers to the range of distance from the lens in which objects are in focus. A shallow depth of field means that only objects at or near a specific distance are in focus; a deep depth of field means that objects over a wide range of distances are in focus.

LENS: An almost clear, biconvex structure in the eye that, along with the cornea, helps to focus light onto the retina. It can become infected with inflammation, for instance, when contact lenses are improperly used.

MICROBIOLOGY: Branch of biology dealing with microscopic forms of life.

NANOTECHNOLOGY: Nanotechnology describes device components that are generally less than 100 nanometers (1/1,000,000 of a millimeter), thus making them on a molecular scale.

RESOLUTION: The ability of a sensor to detect objects of a specified size. The resolution of a satellite sensor or the images that it produces refers to the smallest object that can be detected.

SCANNING PROBE MICROSCOPE: Devices that produce highly magnified images of objects by moving a finely pointed probe near their surface without touching. The name of the device refers to the fact that it scans the probe back and forth over a sample to build up an image. Some scanning probe microscopes can image features as small as a single atom.

Today, Malpighi is considered to be a key pioneer of histology, the study of tissues, and is also very important to early work in embryology.

One of the most influential early microscopists was Robert Hooke (1635–1703), an English scientist and mathematician who made important contributions in many scientific fields. Hooke modified the Jansen compound microscope to suit his own uses and to make it more powerful. He also devised a system to illuminate his subjects for better viewing. By slicing cork very thinly, he discovered and named the plant cell. Hooke also made observations on many other kinds of plants and animals, publishing them in his seminal book called *Micrographia* (Small drawings) in 1665. It contained his observations on many different organisms, including insects, sponges, and bird feathers. Hooke also used his position as curator of experiments for the Royal Society of London to demonstrate his microscopic findings before many important scientists of the day.

Another giant of microscopy was Antonie van Leeuwenhoek (1632–1723), an uneducated Dutch cloth trader. Descended from basket makers and brewers, van Leeuwenhoek seems an unlikely man to revolutionize the science of biology. After studying Hooke's *Micrographia*, van Leeuwenhoek became fascinated with microscopy. He was familiar with magnifiers from his work in the cloth trade, where they were used to examine goods. He decided to build his own microscope, which he manufactured himself. Instead of using Hooke's popular compound design, van Leeuwenhoek constructed what amounted to extremely powerful magnifying glasses with powers of up to 200 times. His simple devices were easy and quick to make, though not always easy to use, and he made up to 500 of them during his life.

Using his magnifiers, van Leeuwenhoek began to explore the world around him on a microscopic level. His investigations of common substances such as pond water, tooth plaque, blood, semen, animal tissue, plant sections, minerals, crystals, and fossils led to his many discoveries. In the course of his investigations, van Leeuwenhoek discovered bacteria, protists, the algae *Spirogyra*, blood cells, sperm cells, and microscopic worms. He became a member of the Royal Society of London, and, though he never visited it, his correspondence laid

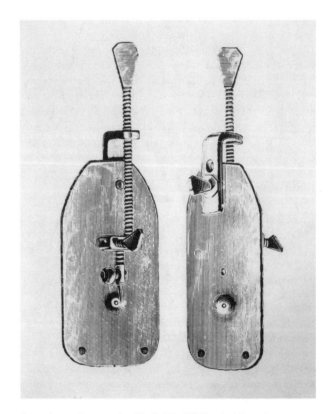

Antonie van Leeuwenhoek's (1632–1723) primitive microscope, actually a lens mounted between two metal plates. *© Bettmann/ Corbis.*

MARCELLO MALPIGHI (1628–1694)

In the second half of the seventeenth century, Italian physician Marcello Malpighi (1628–1694) used the newly invented microscope to make a number of important discoveries about living tissues and structures, earning himself enduring recognition as a founder of scientific microscopy, histology (the study of tissues), embryology, and the science of plant anatomy.

Early in his medical career, Malpighi became absorbed in using the microscope to study a wide range of living tissue—animal, insect, and plant. At the time, this was an entirely new field of scientific investigation. Malpighi soon made a profoundly important discovery. Microscopically examining a frog's lungs, he was able for the first time to describe the lung's structure accurately—thin air sacs surrounded by a network of tiny blood vessels. This explained how air (oxygen) is able to diffuse into the blood vessels, a key to understanding the process of respiration. It also provided the one missing piece of evidence to confirm William Harvey's (1578–1657) revolutionary theory of blood circulation: Malpighi had discovered the capillaries, the microscopic connecting link between the veins and arteries that Harvey—with no microscope available—had only been able to postulate. Malpighi published his findings about the lungs in 1661.

Malpighi used the microscope to make an impressive number of other important observations, all "firsts." He observed a "host of red atoms" in the blood—the red blood corpuscles. He described the papillae of the tongue and skin—the receptors of the senses of taste and touch. He identified the *rete mucosum*, the Malpighian layer, of the skin. He found that the nerves and spinal column both consisted of bundles of fibers. He clearly described the structure of the kidney and suggested its function as a urine producer. He identified the spleen as an organ, not a gland; structures in both the kidney and spleen are named after him. He demonstrated that bile is secreted in the liver, not the gall bladder. In showing bile to be a uniform color, he disproved a 2,000-year-old idea that bile was yellow and black. He described glandular adenopathy, a syndrome rediscovered by Thomas Hodgkin (1798–1866) and given that man's name 200 years later.

Malpighi also conducted groundbreaking research in plant and insect microscopy. His extensive studies of the silkworm were the first full examination of insect structure. His detailed observations of chick embryos laid the foundation for microscopic embryology. His botanical investigations established the science of plant anatomy. The variety of Malpighi's microscopic discoveries piqued the interest of countless other researchers and firmly established microscopy as a science.

out many of his significant discoveries. Amazingly, in 1981 several of van Leeuwenhoek's original samples were discovered in the Royal Society's archives by microscopist Brian J. Ford (1939–), whose reanalysis of the original specimens revealed the excellent quality of van Leeuwenhoek's work and once again confirmed its importance to science as a whole.

With improved lenses and mechanical fittings, the compound microscope became the most powerful and widely used magnifying instrument. The development of the stage and slides that rest upon it permitted better examination of samples. Staining also improved visibility. Today microscopy remains a useful tool within many scientific and technical fields. Three major types of microscopy are used to examine increasingly minute samples, providing more information than ever before.

The most recognizable type of microscope is the optical compound microscope. It developed directly from the earliest compound microscopes and shares many of the same features. The most significant development for this type of microscope was the addition of an artificial light source, which allowed consistent observation of samples. Different light sources can be used at different wavelengths or from various angles to illuminate indistinct structures. New ways of viewing

optical microscope images have also been developed. Stereomicroscopes create two images from slightly different angles and transmit them through two eyepieces, producing a three-dimensional view of the subject. Cameras can also be used to view the sample, creating photographic or digital images.

Electron microscopes use electrons, not light, to visualize samples, permitting much greater magnification. There are two major types. Transmission electron microscopes (TEM) work in a way similar to optical microscopes. They pass a beam of electrons through a sample and use a detector to form the image. The sample must be sliced into an extremely thin section for the beam to pass through it. Advanced-transmission electron microscopes have been able to deliver magnification of up to 50 million times, allowing even individual atoms to be visualized. This gives it an important role in nanotechnology development. The other type of electron microscope, the scanning electron microscope (SEM), is responsible for many dramatic and captivating images, as well as important discoveries. Its electron beam scans a sample's surface, producing an image with great depth of field (large portions of the image are in focus) giving it an almost three-dimensional appearance.

Transmission electron microscopes (TEM) and scanning electron microscopes (SEM) offer important

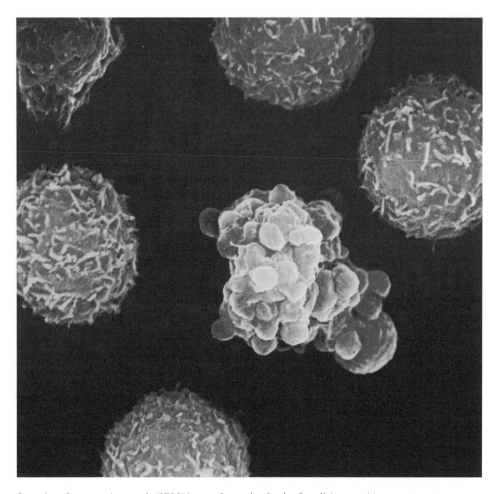

Scanning electron micrograph (SEM) image shows the death of a cell (apoptosis). *Gopal Murti/ Phototake.*

variations on basic electron microscopy. The TEM transmits electrons through an extremely thin sample. The electrons scatter as they collide with the atoms in the sample and form an image on a photographic film below the sample. This process is similar to a medical x ray, where x rays (very short wavelength light) are transmitted through the body and form an image on photographic film behind the body.

By contrast, the SEM reflects a narrow beam of electrons off the surface of a sample and detects the reflected electrons. To image a certain area of the sample, the electron beam is scanned in a back and forth motion parallel to the sample surface, similar to the process of mowing a square section of lawn.

The chief differences between the two microscopes are that the TEM gives a two-dimensional picture of the interior of the sample while the SEM gives a three-dimensional picture of the surface of the sample. Images produced by SEM are familiar to the public, as in television commercials showing pollen grains or dust mites.

In the early 1980s, a new technique in microscopy was developed which did not involve beams of electrons or light to produce an image. Instead, a small metal tip is scanned very close to the surface of a sample and a tiny electric current is measured as the tip passes over the atoms on the surface. Some probes are so sharp that the tip is composed of a single atom. The first scanning probe microscope, called a scanning tunneling microscope (STM), used a quantum property of electrons to examine a sample. Now many different types are specialized to gather different types of data, not just produce images, although they can be used to do so. Scanning probe devices are not exactly microscopes—the image is created not by "looking" at the surface, but by "feeling" it with the probe. When a metal tip is brought close to the sample surface, the electrons that surround the atoms on the surface can actually "tunnel through" the air gap and produce a current through the tip. This physical phenomenon is called tunneling and is one of the amazing results of quantum physics. If such a phenomenon could occur with large objects, it would be possible for a baseball to tunnel through a brick wall with no damage to either. The current of electrons that tunnels through the air gap is very much dependent on the width of the gap; therefore the current will rise and fall in succession with the atoms on the surface. This current is then amplified and fed into a computer to produce a three dimensional image of the atoms on the surface.

Without the need for complicated magnetic lenses and electron beams, the STM is far less complex than the electron microscope. The tiny tunneling current can be simply amplified through electronic circuitry similar to circuitry that is used in other electronic equipment, such as a stereo. In addition, the sample preparation is usually less tedious. Many samples can be imaged in air with essentially no preparation. For more sensitive samples that react with air, imaging is done in vacuum. A requirement for the STM is that the samples be electrically conducting, such as a metal.

Modern Cultural Connections

Some of the first scientific investigations using microscopes were in the fields of biology and the life sciences. Hooke and van Leeuwenhoek's discoveries have led to even more important advances within life sciences. Hooke's discovery of the cell was made possible by the

microscope, as was the later discovery of the organelles within it. Observing these structures and learning their functions have revealed the inner workings of the cell: metabolism and reproduction. Through the use of the microscope, a vast number of different cell and tissue types have been identified. By understanding their structures and roles within the body, scientists have learned more about the way life functions.

In zoology the microscope has been invaluable for studying both extremely small organisms and the tiny features present on creatures of all sizes. In *Micrographia,* Hooke observed a flea and drew pictures of it in minute detail. Insects are often studied by microscopy today, including some species that may be too small to study otherwise. Insects' tiny scales and hairs, complicated eye structure, and different body plans have been revealed by microscopy. Microscopy also revealed the hollow structure of polar bears' hair shafts, which provide extra insulation in their northern habitats. Botanists use microscopes to study plants in many of the same ways zoologists study animals. Both microscopic plants and tiny features of larger ones can be seen with a microscope.

The entire field of microbiology has been made possible by the use microscopes, which have been used to observe bacteria, viruses, fungi, protists, and algae; all are visible only with the aid of a microscope. Studying microorganisms has led to cures for disease, improved

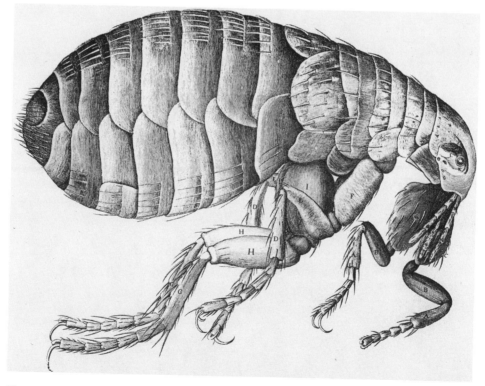

Illustration of a flea, a wingless, bloodsucking, parasitic insect from 1665 as depicted in *Micrographia* by Robert Hooke (1635–1703). *HIP/Art Resource, NY.*

industrial products, and new ways to decompose dangerous waste. The field still holds much promise, as biologists project that only a tiny percentage of existing microorganisms has been discovered and studied. Doctors and medical technicians use microscopes to study many of the same organisms and tissues as other scientists, hoping to cure disease.

Microscopy has become important in many technological fields as well. The minute analysis of materials has led to stronger, lighter, and more durable products. The production of microchips would not be possible without the ability to analyze and build them on a minute scale. The most powerful scanning probe microscopes can already image single atoms, allowing continuing advances in nanotechnology.

The microscope's lasting impact is its ability to reveal features that were previously hidden by their miniscule size. Studying these features has produced a better understanding of the basic function of things as diverse as bacteria and ball bearings. By exploring the tiniest, most basic elements of the world around us, we can better understand its function as a whole.

SEE ALSO *Biology: Cell Biology; Biomedicine and Health: Bacteriology; Biomedicine and Health: The Germ Theory of Disease; Biomedicine and Health: Prions and Koch's Postulates; Biomedicine and Health: Virology.*

BIBLIOGRAPHY

Periodicals

Sines, George, and Yannis A. Sakellarakis. "Lenses in Antiquity." *American Journal of Archaeology* 91 (1987): 191–196.

Web Sites

Florida State University. Molecular Expressions: Exploring the World of Optics and Microscopy. "Introduction to Microscopy." March 15, 2005. http://micro.magnet.fsu.edu/primer/index.html (accessed February 2, 2008).

Microscopy–UK. An Introduction to Microscopy. "The History of the Microscope." http://www. microscopy-uk.org.uk/index.html?http://www. microscopy-uk.org.uk/intro/histo.html (accessed February 2, 2008).

University of Dayton. "The History of the Microscope." http://campus.udayton.edu/~hume/Microscope/microscope.htm (accessed February 2, 2008).

University of Nebraska–Lincoln. Electron Microscopy. "What Are Electron Microscopes?" http://www.unl.edu/CMRAcfem/em.htm (accessed February 2, 2008).

Kenneth T. LaPensee

Physics: Newtonian Physics

■ Introduction

Newtonian physics, also called Newtonian or classical mechanics, is the description of mechanical events—those that involve forces acting on matter—using the laws of motion and gravitation formulated in the late seventeenth century by English physicist Sir Isaac Newton (1642–1727). Several ideas developed by later scientists, especially the concept of energy (which was not defined scientifically until the late 1700s), are also part of the physics now termed Newtonian.

Newtonian physics can explain the structure of much of the visible universe with high accuracy. Although scientists have known since the early twentieth century that it is a less accurate description of the physical world than relativity theory and quantum physics, corrections required for objects larger than atoms that move significantly slower than light are negligible. Since Newtonian physics is also mathematically simple, it remains the standard for calculating the motions of almost all objects from machine parts, fluids, and bullets to spacecraft, planets, and galaxies.

■ Historical Background and Scientific Foundations

Although Newton redefined the basic concepts of mechanics and devised his laws of motion and universal gravitation in the late 1600s, he based his work on important scientific discoveries about matter and motion that had already been established. Without these earlier achievements, he could not have produced the four laws that are the foundation of Newtonian physics: his three laws of motion and his law of universal gravitation.

From the at least the fourth century BC until Newton's time, European scientific thought was modeled largely on the theories of ancient Greek thinkers such as Plato (c.428–348 BC) and Aristotle (384–322 BC). So great was Aristotle's influence, in fact, that the world-view held by most European scholars until the seventeenth century is termed Aristotelian. This did not rule out the investigation of events using experiment and mathematics, which are now the heart of the scientific method, but it did not particularly encourage them either. This is because Aristotelians saw the universe and everything in it primarily in terms of their meaning, rather than cause and effect.

Aristotelians also inherited flawed assumptions about specific physical questions. For example, if an object were in motion, they assumed something must keep it in motion, whether a mysterious quality in the object called impetus, the surrounding air, or something else. For many centuries this mindset confounded efforts to unravel the physics of motion.

Despite many preconceptions, the Middle Ages and the Renaissance did produce some significant scientific developments. Manual workers such as joiners, builders, navigators, and shipbuilders accumulated knowledge about practical methods and materials. Scholars advanced knowledge in several branches of mathematics, recovering the long-forgotten or poorly copied works of the Greek mathematicians Euclid (born c.300 BC) and Archimedes (287–212) and making new discoveries of their own. In the 1500s, stimulated by existing Arab work, algebra was developed by Italian mathematicians such as Niccolò Fontana Tartaglia (1499–1557). The word "algebra" itself is from the Arabic *al-jabr*, meaning "reunion." Many tools, physical and intellectual, had to be in place before Galileo Galilei (1564–1642), Newton, and the other founders of modern physical science could achieve their triumphs in the sixteenth and seventeenth centuries.

The way was partly prepared for the new way of thinking that Newton and others called "experimental philosophy" or "mechanical philosophy" by French philosopher and mathematician René Descartes (1596–1650). Descartes assumed, like most earlier thinkers, that the universe can be explained by top-down reasoning from general first principles, with little or no need for particular experiments; famously, he thought that all knowledge could proceed from the logical statement "I think, therefore I am" (*Cogito ergo sum* in the original Latin).

Descartes reasoned wrongly that neither atoms nor a vacuum could exist. Yet he prefigured the modern scientific approach by seeking a comprehensive, mechanical, rational interpretation of nature. In particular, he proposed that the motions of the planets could be accounted for by a vortex or swirl of "subtle matter"—matter not perceptible to the senses—stirred throughout the solar system by the rotation of the sun on its axis. The sun, he theorized, like a spinning whisk at the center of a large bowl of cream, set the subtle matter twirling around it; since the twirling would naturally diminish with distance, this, according to Descartes, explained why planets more distant from the sun move more slowly than those that are closer.

Descarte's vortex theory of planetary motion was popular in Europe at the time Newton published his own theory of the solar system in *Philosophiae Naturalis Principia Mathematica* (Mathematical Principles of Natural Philosophy), usually referred to simply as the *Principia* (1687). Newton's mechanics explained both earthly and planetary motion, signaling the downfall of Descartes's vortex theory and his entire approach to knowledge. Experiment combined with mathematics, rather than top-down philosophical speculations, would define all serious attempts to understand the physical world from that time onward.

It was no accident that the motions of the planets concerned both Descartes and Newton. Before the advent of Newtonian physics, observational astronomy was the only science with mathematically precise knowledge or predictive power. Chemistry consisted mostly of unconnected bits of practical knowledge accumulated by trial and error. Modern concepts of the elements did not begin to develop until English scientist Robert Boyle's (1627–1691) experiments disproved Plato's theory that all matter is composed of four elements—earth, air, fire, and water—in 1660.

The existence of microorganisms was not known until 1676, when Dutchman Antoni van Leeuwenhoek (1632–1723) built the first microscope. Medicine, too, was rudimentary in Newton's day; the fact that the heart pumps blood through the body, for example, only became widely accepted after 1616, when this theory was published in works by English doctor William Harvey (1578–1657).

Social Milieu

The 1600s were a time of upheaval in all aspects of European society, including religion, science, politics, commerce, and the arts. The Protestant Reformation, starting in the early 1500s, had split the centuries-old religious consensus of Europe along approximately geographic lines, with Protestant countries to the North and Roman Catholic countries to the South. The Reformation called ancient patterns of thought into question and triggered wars that plagued the continent for decades.

The Commercial Revolution, which ran from the early 1500s to about 1650, also helped break up old patterns of thought and motivate new discoveries in science and technology. Techniques for long-distance ocean navigation were needed, stimulating new precision work in astronomy and clock making.

England, Newton's native country, suffered especially brutal upheavals. From 1642 to 1651, a bloody

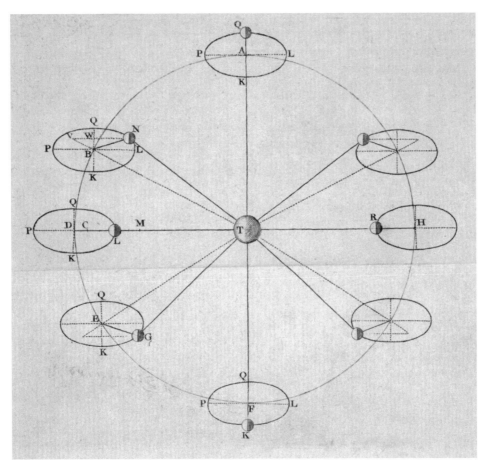

Newton's rendering of planetary motion, predicted by his laws of classical mechanics, as seen in his work, *The Mathematical Principles of Natural Philosophy.* © *Corbis.*

civil war pitted the Puritans (Calvinist Protestants) against the Anglicans (members of the English state church). After the Puritans beheaded King Charles I (1600–1649) in 1649, Oliver Cromwell (1599–1658) became first chairman of the Council of State, then Lord Protector from 1653 until his death in 1658. The British monarchy was restored under Charles II in 1660.

A few years later, during an outbreak of plague, young Isaac Newton—a Puritan who was also fascinated not only by science but alchemy and the biblical book of Revelation—took refuge from pestilence in his mother's country house. There, in a space of 18 months (1665–1666), he conceived the basic elements of a new physics: the three laws of motion, the law of universal gravitation, and calculus. He also did extensive work in optics, though he did not have the revolutionary effect there that he did in mechanics.

The Science

Medieval astronomy was based on Claudius Ptolemy's (AD c.90–c.168) *Mathematike syntaxis* (Mathematical

collection), better known in the West by a shortened form of its Arabic title, the *Almagest.* According to Ptolemy, the planets were embedded in vast crystal spheres centered around Earth and moving in changeless, perfect circles. Their motion was imparted by supernatural means from the outermost sphere of all, that of the fixed stars. This model was challenged by Nicolaus Copernicus (1473–1543) in the 1500s. In 1543 he published *De revolutionibus orbium coelestium* (Six books concerning the revolutions of the heavenly orbs), in which he proposed that the sun, not Earth, is at the center of the universe.

Copernicus's revision of the universe prompted a wave of new astronomical work. Tycho Brahe (1546–1601) made naked-eye observations of the motions of the sun, moon, and planets that were the most accurate to that date. After Brahe's death, his assistant Johannes Kepler (1571–1630), an advocate of the Copernican system, tried to fit Brahe's precise new observational data to equations describing planetary orbits, beginning with the planet Mars. At first he assumed a circular

IN CONTEXT: GALILEO AND THE CHURCH

Galileo Galilei (1564–1642) was an Italian physicist who perfected the modern scientific method. His work on accelerated motion was essential groundwork for Newtonian physics. Unfortunately, Galileo's defense of Copernican (or heliocentric) astronomy—the view that Earth rotates around the sun, not the other way around—ran afoul of established religious doctrine. The Catholic Church, which taught that Earth is stationary, declared in 1616 that heliocentrism was "false and altogether contrary to Scripture."

In 1633 the elderly Galileo was brought before the Inquisition and found guilty of heresy (preaching incorrect belief) and shown the instruments of torture that would be used on him if he did not retract his statements. Under duress, Galileo publicly retracted his belief in heliocentrism and spent the rest of his life under house arrest. Blind and disappointed, he died in 1642, the same year Isaac Newton was born. Because of Galileo's conviction, scientists were fearful of speaking truthfully in Southern Europe for decades afterward, and most of the work in the Scientific Revolution was thereafter done in England and Northern Europe.

The church eventually admitted its mistake, but not until many years later. In 1822 the church lifted its ban on books teaching the view that Earth goes around the sun; in 1981 Pope John Paul II (1920–2005) convened a new commission to study the Galileo case. In 1992 the commission declared that the case had been marked by "tragic mutual incomprehension." This has not been enough for some, including a former director of the Vatican observatory (from 1978–2006), priest George Coyne (1933–), who would have liked a more thorough admission of responsibility for Galileo's persecution and a true apology.

orbit for Mars, as all had done before him, but he could not make the observational data fit. Eventually he found that the best fit was given not by a circle but by an ellipse (a curve like the outline of an egg).

Kepler was the first to describe the motions of the planets in terms of mathematical laws. He stated three, two of which involved time as a variable. Using time to describe the world mathematically was a significant advance for physics; the European scientific tradition inherited from the Greeks was primarily static (motionless) and geometrical. Its attention went primarily to the shapes of curves and rarely used mathematics to describe dynamic (time-dependent) processes.

Kepler published two of his laws in 1609 and the third in 1619. They were purely descriptive, that is, they offered no explanation of why the planets acted as specified, nor did they describe how any other objects (such as falling apples) might move.

After Brahe and Kepler, Galileo laid crucial groundwork for Newtonian physics. He mistakenly rejected Kepler's proof that the planets moved in elliptical orbits, but conducted precise experiments in the laboratory to characterize the movements of accelerating bodies—objects that are changing the direction or rapidity of their motion. Like Kepler, he searched for mathematical laws to describe the way physical systems change over time.

Galileo concluded that the distance covered by a steadily accelerating object is proportional to the square of the time it has been accelerating. He also discovered that objects accelerate steadily under the influence of gravity, which he treated as a constant force unaffected by distance (which it is, approximately, near Earth's surface). He found that objects accelerate with equal speed regardless of their weight—that is, a heavier ball does not fall faster than a light ball of the same size. Perhaps most fundamentally, he found that objects tend to maintain their straight-line motion unless acted upon by a force. This overthrew the Aristotelian view that a force is needed to maintain an object's state of motion.

With Galileo's physics and Kepler's astronomy in place, the stage was set for Newton's triumph.

Newton's Physics

Newton's influence is due mostly to his major work, *Philosophiae Naturalis Principia Mathematica*, published in 1687 and best known by the shortened form of its Latin title, *Principia*. This work was produced partly at the urging of Newton's friend, English astronomer Edmond Halley (1656–1742), who also financed the project, helping to produce one of the most important works in the history of science.

Of all the scientists working in his day, only Newton conceived that there could be a single universal system of mechanics—that is, a physics that would describe both earthly and celestial motions at the same time. In the *Principia,* Newton established such a physics with his three laws of motion and his law of gravitation. Elaborating these laws and unifying them with a rigorous idea of "energy" in the late eighteenth century produced a system, Newtonian (or mechanical) physics, that is still used today for everything from engineering design to the analysis of galactic motion.

Newton's three laws of motion are as follows:

1. An object remains at rest or moves in a straight line at a constant speed unless acted upon by a nonzero total force.

2. A force acting on a body causes it to accelerate (change its state of motion) to a degree that is proportional to the body's mass.

Stated as an equation, writing F for force, m for mass, and a for acceleration, we have $F = ma$. In

Isaac Newton (1642–1727), English physicist and mathematician.   © *Bettmann/Corbis*.

other words, an object's velocity and momentum changes with time in proportion to the force acting on it.

3. Forces occur in pairs pointing in opposite directions.

This law is most often stated as: For every action there is an equal and opposite reaction. For example, when a gun fires, the force acting on the bullet as it accelerates through the barrel is equal to the recoil of the gun acting on the shooter's hand or shoulder.

The fourth basic law of Newtonian physics is the law of universal gravity: $F = Gm_1m_2/r_2$. Here F is gravitational pull, G is the universal gravitational constant (a fixed number, $G = 6.6742 \times 10^{-11}$ m^{3}kg^{-1}s^{-2}), m_1 is the mass of one object, m_2 is the mass of the other object, and r is the distance between the centers of the two objects. Larger masses mean larger gravitational force,

Хронологипi

IN CONTEXT: SURPRISE: NEWTON WAS RIGHT!

Even though three centuries have passed since Isaac Newton published his theory of gravitation in 1687, scientists are still testing it. Newton's law says that the gravitational attraction between any two objects decreases with the square of the distance between them; doubling the distance means one fourth the force. This type of relationship, called an inverse-square law, is accurate at the scale of baseballs, planets, or galaxies, but, according to quantum physics, should fail when objects are close together. Masses separated by as little as the width of a human hair (56 millionths of a meter or micrometers, μm) should, according to some theories, experience measurably less gravitational attraction than Newtonian law predicts.

In 2006, a group of physicists led by D.J. Kapner tested Newton's law of gravitation by measuring the gravitational pull between a pair of small, spinning metal disks as little as 56 μm apart. They found that Newton's law was still valid even at this distance.

This simple result—derived from an experimental setup that could fit inside a soda can—may have consequences for our understanding of the whole universe. In the 1990s astronomers discovered that 70% of all the energy and mass in the universe consists of a mysterious "dark energy" of a still unknown nature. Some theories attempting to explain dark energy, such as string theory, make predictions about the force of gravity. At least one such theory—the fat-graviton theory—was ruled out by the recent spinning-disk experiment. In coming years, even more sensitive tests of Newton's law will be made to further constrain physicists' attempts to explain the nature of universe.

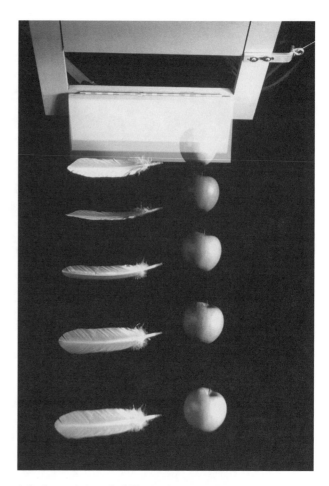

A feather and an apple falling at the same rate in a vacuum chamber. This experiment demonstrates the validity of Isaac Newton's (1642–1727) ideas on gravity and inertia. *© Jim Sugar/ Corbis.*

bigger *F*; more distance between the masses, bigger *r*, means less.

Newton showed that these four laws could account simultaneously for Galileo's laboratory results—the behavior of everyday objects in the vicinity of Earth's surface—and for the motions of the planets. A single set of laws, compact enough to jot down on a card, could describe the motions of all the stars, planets, and moons in the universe. Moreover, to achieve this astonishing result Newton had had to invent a new branch of mathematics to handle quantities that change with time in an unsteady way. He called it the method of fluxions, but today it is known as calculus, the basic mathematical language of all physical science.

Impacts and Issues

Most branches of science were in a disorderly state when Newton published his *Principia* in 1687. Newtonian mechanics was the first scientific discipline to achieve apparent perfection: Newton's laws passed every experimental test, explained the tides and other vexing problems of astronomy, and defeated the Cartesian theory of vortices.

But it was also controversial. Critics, such as German scientist Gottfried Leibniz (1646–1716), who invented calculus independently of Newton, attacked Newton's theory of gravitation as mystical or useless: how could one mass act on another instantaneously across a distance, without being in direct or at least indirect contact with it? The new theory did not *explain* anything. Newton was disturbed by this problem too, saying that he found what was called "action at a distance" implausible. Yet he defended his method, writing in 1715: "His [Leibniz's] arguments against me are founded upon metaphysical & precarious hypotheses & therefore do not affect me: for I meddle only with experimental Philosophy." It did not matter, Newton maintained, whether he had explained gravity or not: an

PHILOSOPHIÆ
NATURALIS
PRINCIPIA
MATHEMATICA.

Autore *J S. NEWTO N, Trin. Coll. Cantab. Soc.* Matheseos Professore *Lucasiano*, & Societatis Regalis Sodali.

IMPRIMATUR·
S. P E P Y S, *Reg. Soc.* P R Æ S E S.
Julii 5. 1686.

L O N D I N I,

Jussu *Societatis Regiæ* ac Typis *Josephi Streater.* Prostat apud plures Bibliopolas. *Anno* MDCLXXXVII.

Facsimile of Isaac Newton's (1642–1727) *Philosophiae Naturalis Principia Mathematica.* © Bettman/Corbis.

explanation was desirable, but could not be had until an accurate description of what gravity does was available.

Leibniz also criticized Newton's assumptions about absolute space, which, the latter stated, if it existed, would be flat everywhere (i.e., obey the laws of Euclidean geometry) and infinite in all directions. Leibniz's philosophical objection was vindicated over 200 years later when German physicist Albert Einstein (1879–1955) showed that the idea of absolute Newtonian space had to be abandoned and replaced by curved, relative space.

Despite its defects, Newtonian physics revolutionized science. Mathematically stated theories tested against physical observations became the standard in almost all fields of scientific thought. Banished forever was medieval reliance on authoritative books and the Cartesian reliance on reason unchecked by observations.

Impact on Society

The *Principia*, written in Latin and intensely mathematical, was not read widely. Its intellectual impact on society was mediated by authors who took up the cause of Newton's "experimental Philosophy" and wrote for the general public. The most influential of these was French thinker Voltaire, born François-Marie Arouet (1694–1778), who explained Newton's science accu-

rately in nontechnical terms in his book *Eléments de la philosophie de Newton* (Elements of Newton's philosophy, 1738).

For Voltaire and similar thinkers, Newton's triumph in mechanics proved that science would eventually explain everything, including human actions, in terms of rigid cause-and-effect (deterministic) laws: "It would be very singular," Voltaire wrote, "that all nature, all the planets, should obey eternal laws, and that there should be a little animal, five feet high, who in contempt of these laws, could act as he pleased, solely according to his caprice."

Armed with the ironclad credibility of the new science, these writers began to attack traditional religion. Their skepticism contributed to a decline in religious belief in industrial societies that has continued steadily to the present day and is especially true in Europe. In the United States, the general populace remains almost universally religious, but scientists express lower rates of religious belief in polls, with physicists—Newton's intellectual heirs—being the most nonreligious group (only about 20% of physicists believed in God as of 1998). This is ironic, given that Newton himself was a devout Christian.

Some of Newton's contemporaries, such as Irish philosopher Bishop Berkeley (1685–1753), attacked the new materialism. However, such holdouts fought a losing battle, and, during the eighteenth century, the intellectual climate in England and northern Europe became predominantly pro-scientific and deterministic.

The new science also transformed the world in economic, military, and other matters. Newtonian physics did not make these changes alone: chemistry, medicine, mathematics, electromagnetics, and other scientific fields were also crucial. Other forms of science took up the Galilean and Newtonian methods of mathematical law-testing. Together, the new science unleashed a flood of new technologies that drove the Industrial Revolution starting in the late 1700s and has continued to the present.

■ Modern Cultural Connections

Newtonian physics continues to be applied in every area of science and technology where force, motion, and gravitation must be reckoned with. However, today's physicists, unlike Newton, know that his laws do not work in all circumstances. The behaviors of objects traveling near the speed of light, or interacting on the size scale of subatomic particles, are not described accurately by Newtonian physics. Relativity theory and quantum physics are required in such non-Newtonian realms, and even these theories have limits.

Despite the advent of later, more complete theories, scientists continue to study Newtonian physics. As

IN CONTEXT: THE ULTIMATE NEWTONIAN MACHINE

A spacecraft is the ultimate Newtonian machine because it relies for propulsion on rockets, which are the most straightforward possible application of Newton's second law of motion, the principle that every force acting on some object is paired with an equal and opposite force acting on some other object. Gases exiting a rocket push against the rocket's combustion chamber, and the combustion chamber pushes with equal and opposite force against the gases. The gases fly off in one direction, the chamber (with rocket attached) in the opposite direction.

A spacecraft that has left the atmosphere is governed only by the forces exerted by its rockets—Newton's second law—and the force of gravity, described by Newton's law of universal gravitation. Newton's laws therefore account for almost everything that affects the path of a spacecraft in flight. During the 1968 journey of *Apollo 8*, which circled the moon, a child on Earth wondered aloud who was driving the spaceship. When the question was relayed to him by radio, astronaut Bill Anders replied, "I think Isaac Newton is doing most of the driving right now."

Spacecraft and rocketry have had a profound impact on modern society. Space probes have greatly multiplied our knowledge of the planets and more distant universes, satellites have transformed communications, and ballistic missiles—equipped with a non-Newtonian invention, the nuclear bomb—have made it physically possible to destroy most of the human race in a few minutes.

described in the sidebars, scientists have measured the force of gravity at distances as small as the width of a human hair to see if Newton's law holds for objects so close together. Their results, published in 2007, showed that Newtonian physics still held, even at such small distances. This ruled out of some of the most promising theories put forward to explain the mysterious fact (discovered in the late 1990s) that the universe is not only expanding, but expanding more quickly all the time.

In the early 1980s research began to focus on the possibility that Newtonian physics may not be correct even in non-quantum, non-relativistic realms—situations in which it has always been believed to be essentially perfect. A new type of physics, called modified Newtonian dynamics (MOND), first proposed by Israeli physicist Moti Milgrom in 1983, suggests that Newton's second law should be modified for small accelerations. Since small accelerations are common in astronomical settings, MOND would account for a number of observations that other theories describe as the result of an unknown, invisible form of matter called dark matter. MOND is controversial; some physicists support it, but most are convinced that some form of dark matter accounts for those observations. In the scientific style established by Newton and Galileo, the question will eventually be settled by comparing the predictions of rival theories to actual observations.

SEE ALSO *Astronomy and Cosmology: A Mechanistic Universe; Physics: Aristotelian Physics; Physics: Articulation of Classical Physical Law.*

BIBLIOGRAPHY

Books

Bell, Arthur. *Newtonian Science.* London: Edward Arnold Publishers, Ltd., 1961.

Cohen, I. Bernard. *The Newtonian Revolution: With Illustrations of the Transformation of Scientific Ideas.* New York: Cambridge University Press, 1980.

Crombie, A.C. *Medieval and Early Modern Science, Vol. II: Science in the Later Middle Ages and Early Modern Times: XIII–XVII Centuries.* New York: Doubleday, 1959.

Laue, Max von. *History of Physics.* Translated by Ralph Oesper. New York: Academic Press Inc., 1950.

Periodicals

Ball, Philip. "A Jump That Would Prove Newton Wrong." *Nature.* 446 (2007): 357.

Ignatiev, A. Yu. "Is Violation of Newton's Second Law Possible?" *Physical Review Letters* 98 (2007): 101101.

Kapner, D.J., et al. "Tests of the Gravitational Inverse-Square Law below the Dark-Energy Length Scale." *Physical Review Letters* 98, no. 2 (2007).

Speake, Clive. "Gravity Passes a Little Test." *Nature* 446 (March 1, 2007): 31–32.

Larry Gilman
Paul Davies
K. Lee Lerner

Physics: Nuclear Physics

■ Introduction

Nuclear physics became a scientific discipline and the atomic nucleus a subject of inquiry in the period between the discovery of radioactivity in 1896 and the identification of the neutron in 1932. Since then nuclear physics has spawned many fundamental topics of research, as well as a number of applications with formidable social and political implications. These applications have commanded so much attention that they tend to overshadow the scientific areas of the discipline.

■ Historical Background and Scientific Foundations

In 1886, while investigating the phosphorescent properties of uranium salts, Henri Becquerel (1852–1908) accidentally discovered that they emitted penetrating radiation that could expose (leave images on) photographic plates. Other researchers quickly pursued this finding. Marie and Pierre Curie (1867–1934 and 1859–1906), working at the Sorbonne in Paris, who coined the term "radioactivity," also discovered new radioactive elements, including radium and polonium.

Ernest Rutherford (1871–1937) found that uranium emitted two types of radiation: one that he called alpha radiation was rapidly absorbed; a second, much more penetrating type, he called beta. Eventually, by observing the mysterious alpha particles' behavior in electric and magnetic fields, Rutherford concluded that they were positively charged helium ions. In similar experiments, Walter Kaufmann (1871–1947) determined that beta radiation was composed of high-energy electrons. A third class of radioactivity, gamma radiation, was discovered by Paul Villard (1860–1934) and eventually shown to be high-energy electromagnetic waves.

Rutherford's most famous and important series of experiments, conducted in 1911 at Manchester University with Hans Geiger (1882–1945) and Ernest Marsden (1889–1970), led to his discovery of the positively charged atomic nucleus. J.J. Thomson (1856–1940) had identified the negatively charged electron in 1897, fueling speculation about the atom's internal structure, especially regarding the distribution of the positive charge. Thomson had proposed a model in which the atom's positive charge was distributed evenly in a sphere, with a diameter on the order of 10^{-10} m. The negative electrons, he thought, rested within the positively charged blob, somewhat like raisins in a pudding (hence the name "plum pudding model"). Others, such as the Japanese physicist Hantaro Nagaoka (1865–1950) speculated that the atom resembled a tiny planetary system, with the electrons rotating around a small positively charged center in much the same way as the planets rotate around the sun.

To test these hypotheses, Rutherford and his team shone a beam of alpha particles (emanating from a radium source) onto a thin gold foil target, then counted them in painstaking detail by visually observing their faint scintillations on zinc sulfide screens. The team found that most alpha particles went straight through the foil with little deflection. Less than 1% deflected to surprisingly large angles (in excess of 90 degrees). This led Rutherford to conclude that Nagaoka's guess was correct. Because only a small number of incident alpha particles were scattered at large angles, the positive charge had to be a concentrated core at the center of the atom (on the order of 10^{-14} m). At first, Rutherford referred to this core as the "charged center," but later he used the term "nucleus."

In 1919 Rutherford became the head of the Cavendish Laboratory at Cambridge University, a post he held until his death in 1937. One of his most important experiments in these years involved bombarding

WORDS TO KNOW

BETA DECAY: Process by which a neutron in an atomic nucleus breaks apart into a proton and an electron.

CRITICAL MASS: The minimum amount of fissionable uranium or plutonium that is necessary to maintain a chain reaction.

ELECTRON: A subatomic particle having a negative charge of −1.

FISSION: Splitting or breaking apart. In biology, the division of an organism into two or more parts, which each produce a new organism. Nuclear fission is a process in which the nucleus of an atom splits, usually into two daughter nuclei, with the transformation of tremendous levels of nuclear energy into heat and light.

FUSION: The process stars use to produce energy to support themselves against their own gravity. Nuclear fusion is the process by which two light atomic nuclei combine to form one heavier atomic nucleus. As an example, a proton (the nucleus of a hydrogen atom) and a neutron will, under the proper circumstances, combine to form a deuteron (the nucleus of an atom of "heavy" hydrogen). In general, the mass of the heavier product nucleus is less than the total mass of the two lighter nuclei. Nuclear fusion is the initial driving process of nucleosynthesis.

GAMMA RAY: Short-wavelength, high-energy radiation formed either by the decay of radioactive elements or by nuclear reactions.

ISOTOPE: A form of a chemical element distinguished by the number of neutrons in its nucleus, e.g., ^{233}U and ^{235}U are two isotopes of uranium; both have 92 protons, but ^{233}U has 141 neutrons and ^{235}U has 143 neutrons.

PROTON: Particle found in a nucleus with a positive charge. The number of these gives the atomic number.

QUARK: A type of elementary particle.

STRONG FORCE: In atomic physics, one of the four fundamental forces of nature (electromagnetic, weak, strong, gravitational). The strong force is repulsive at extremely close range (much smaller than an atomic nucleus) and attractive at longer ranges. Within the atomic nucleus, the attraction of the strong force between protons is stronger than the repulsive electromagnetic force between them (like charges repel); this is why atomic nuclei are not blown apart by the mutual electromagnetic repulsion of their positively charged protons. On the other hand, protons and neutrons are kept from simply collapsing into each other by the repulsive effect of the strong force at extremely short range. At longer-than-nuclear ranges, the Coulomb force of electromagnetic attraction or repulsion is far stronger than the strong nuclear force, so it dominates outside the nucleus.

WEAK INTERACTION: In physics, the electroweak force is the single unified force of which the electromagnetic force and weak interaction are both manifestations. The weak interaction is involved in certain types of radioactive decay (e.g., beta decay of neutrons).

nitrogen with a beam of alpha particles from a radium source. This transmuted the nitrogen nuclei and produced a stream of unknown particles. To identify them, Rutherford passed them through various absorbing screens to measure either their penetration range or their deflection as they passed through or deflected from magnetic fields of known strength. He identified the unknown particles as hydrogen nuclei—to which he gave the special name "proton"—and determined that the nuclear reaction had transmuted nitrogen into oxygen. A year later, Rutherford postulated the existence of a particle with the same mass as the proton but with a neutral charge. He also suggested that a nucleus might be formed with one proton and one of these "neutral protons," forming the nucleus of an atom that would behave chemically like hydrogen (since it had a charge of one unit) but have double the mass.

In 1932 Rutherford's speculations bore fruit when the neutral proton was discovered at Cambridge by James Chadwick (1891–1974), who found that beryllium could be made to emit an unknown and highly penetrating radiation. Trying to identify this radiation, Chadwick pursued a variation of Rutherford's disintegration experiments, bombarding beryllium with alpha particles from a polonium source. This increased the intensity of the unknown radiation. He placed an ionization chamber in front of the beryllium sample to detect the penetrating radiation, and found that when thick pieces of lead were put between the beryllium and the ionization chamber, the number of counts per minute did not change significantly. When paraffin wax was placed in the gap, however, there was a huge increase in the count rate.

By measuring the range of this ionizing radiation, Chadwick concluded that it was composed of protons. Although not detected directly, the unknown radiation still knocked hydrogen nuclei (protons) first out of the air and then out of the paraffin. Chadwick deduced that it had the same mass as protons—and therefore that he had found Rutherford's neutral proton, or "neutron." At almost the same time, Columbia University chemist Harold Urey (1893–1981) and his collaborators announced their isolation of Rutherford's "heavy hydrogen nucleus" with a charge of one unit and a mass of two units. They gave the name "deuterium" to this hydrogen isotope.

Rutherford's proton and Chadwick's neutron made the new results of nuclear physics easier to interpret. We can use one of Rutherford's 1919 transmutation experiments as an example. When an alpha particle (2 protons, 2 neutrons) strikes a nitrogen nucleus (7 protons, 7 neutrons), it produces an oxygen nucleus (8 protons, 9 neutrons) and a hydrogen nucleus (1 proton). The notation adopted to represent this is shown in Equation 1, where the lower number is the atomic number (the number of protons) and the upper number is the mass number (the number of protons plus neutrons, collectively referred to as "nucleons").

Disintegration experiments such as Rutherford's and Chadwick's created the need for an artificial laboratory radiation source, one that could produce greater numbers of particles at varying energies. A number of researchers built early accelerators, including Robert Van de Graaff (1901–1967) at Princeton University, but the first team to disintegrate nuclei with an artificially produced beam was that of John D. Cockcroft (1897–1967) and Ernest T.S. Walton (1903–1995) working at the Cavendish Laboratory under Rutherford. To accelerate protons, Cockcroft and Walton developed a voltage-multiplying circuit and fed its output into a linear discharge tube. In 1932 they accelerated a beam of protons and disintegrated lithium nuclei into two alpha particles. (See Equation 2.) With the basic model of the nucleus established and the development of laboratory equipment such as ionization chambers and accelerators, nuclear physics had become an established field of study.

The Forces of Nature

To the two fundamental forces of classical physics (gravitation and electromagnetism), nuclear physics added the strong force, which holds together the constituents of the atomic nucleus, and the weak interaction, which is responsible for certain processes such as beta decay. By 1920 Rutherford was convinced that there must be some sort of a strong nuclear force that overwhelmed the repulsive Coulomb (electric) force between the protons of the nucleus. However, because beta decay was one of the earliest topics studied by the nuclear physics community, the weak interaction attracted greater initial attention.

Beta decay presented a difficult problem. At first scientists did not realize that the electrons were being released from the nucleus and not from the atomic orbitals. Even when this became clear, however, the electrons' energies were hard to understand. According to energy-mass conservation, the initial mass of the decaying nucleus should equal the sum of the mass of the final nucleus, the mass of any particles released, and the kinetic energies. The alpha radiation energy spectra showed well defined peaks at the expected energies. Sharp peaks were also seen in spectra of gamma radiation. In 1914 Chadwick had found that beta decay appeared to violate mass-energy conservation. These spectra did not have well defined energy peaks but instead showed long peaks that extended from zero up to the energy expected from mass-energy conservation.

While Niels Bohr (1885–1962) accepted this possibility, Wolfgang Pauli (1900–1958) sought in 1930 to avoid this conclusion by postulating that a hitherto unobserved particle was also emitted during beta decay, which accounted for missing energy. To account for the fact that the particle had never been detected, Pauli supposed that it had no charge and no mass. For two years, this hypothetical particle was known as "Pauli's neutron," until Chadwick identified the massive neutron and Enrico Fermi (1901–1954) renamed Pauli's neutron the "neutrino."

Although the neutrino was not conclusively observed until 1955 by Frederick Reines (1918–1998) and Clyde L. Cowan Jr. (1919–1974) at the Savannah River Atomic Energy Plant in South Carolina, many nuclear scientists accepted its existence since it could maintain mass-energy conservation in beta decay. With Chadwick's neutron and Pauli's neutrino, beta decay could be seen as the spontaneous decay of a neutron within a nucleus into a proton, a beta particle (electron), and a neutrino. (See Equation 3.) In present language, the neutrino in this reaction would be an "anti-neutrino.")

The weak interaction was introduced by Enrico Fermi to explain the interaction of particles during the beta decay process. During its development, nuclear physics used the new quantum mechanics, in which particles were shown to have wave properties and were represented mathematically with wave functions. In a 1933 paper Fermi theorized that the probability of beta decay in a given nucleus was proportional to the product of

$$_2^4\alpha + {}_7^{14}N \rightarrow {}_1^1p + {}_8^{17}O$$

Equation 1. *Cengage Learning, Gale.*

$$_1^1p + {}_3^7Li \rightarrow {}_2^4He + {}_2^4He$$

Equation 2. *Cengage Learning, Gale.*

$$n \rightarrow p + e^- + n$$

Equation 3. *Cengage Learning, Gale.*

four wave functions, one for each particle involved in the process (neutron, proton, electron, and neutrino). By fitting beta-decay data he was able to determine the proportionality value's constant, finding very low values (on the order of 10^{-14} in dimensionless units); hence the name "weak interaction."

Fermi's success with beta decay (and some early work by Heisenberg on the strong force), led Japanese physicist Hideki Yukawa (1907–1981) to suggest an influential model for the strong force in 1935. In 1927 Walter Heitler (1904–1981) and Fritz London (1900–1954) had modeled the sudden increase in the force that occurs when two atoms come into contact and exchange electrons between orbitals. From this, Yukawa postulated that the force between nucleons was the exchange of a heretofore unidentified particle. He estimated that to have the correct magnitude and range, this unknown particle would have a mass greater than the electron but less than the proton (it is now called a meson, from the Greek *mesos,* "middle") and have an associated coupling constant that was much larger than the proportionality constant in Fermi's model of the weak interaction.

In 1937 C. Anderson (1905–1991) and Seth Neddermeyer (1907–1988) at Caltech found a new particle in earthbound cosmic rays that had the same mass as Yukawa's unknown particle. Further experiments, however, showed that the Anderson-Neddermeyer particle interacted much more weakly with nucleons than Yukawa's theory required. Ten years later, the English cosmic ray researcher Cecil Frank Powell (1903–1969) and coworkers at the University of Bristol used high-altitude balloons to observe cosmic rays in the upper atmosphere; they found that the Anderson-Neddermeyer particle was a decay product of a somewhat heavier meson that interacted strongly with nucleons. This p-meson, as it was called, was recognized as the one that Yukawa had posited, while the Anderson-Neddermeyer particle was named the m-meson.

Fission and Fusion

The idea of mass-energy equivalence, encapsulated in Einstein's famous equation $E = mc^2$ suggested that energy would be released in nuclear reactions when the mass of the products was less than the mass of the reactants. When John Cockcroft (1897–1967) and Ernest Walton (1903–1995) checked this carefully, they found that the mass of the incident proton and the target lithium plus the kinetic energy of the incident proton exactly balanced the (lesser) mass of the product alpha particles plus their (greater) kinetic energy. Such results gave added credence to speculation that nuclear reactions might be harnessed for power generation or a new type of bomb. In 1939 the Hungarian-born American physicist Leo Szilárd (1898–1964), working with Fermi, proved that a nuclear chain reaction was possible.

The discovery of nuclear fission began in the work of chemist Otto Hahn (1879–1968) and physicist Lise Meitner (1878–1968) at the University of Berlin. In their experiments, they bombarded uranium with neutrons to produce new isotopes. These new isotopes, close to uranium on the periodic table, were also unstable and decayed along chains involving beta and alpha decay. Usually, Hahn and Meitner identified the new isotopes by measuring their radioactive properties and isolated them chemically. In 1938 Meitner, an Austrian Jew, left Nazi Germany for Sweden, but Hahn continued his research with physicist Fritz Strassmann (1902–1980). One of the unknown nuclei produced in their neutron-irradiated samples of uranium appeared to be an isotope of radium. This was confusing since it did not seem to fit into any of the decay chains they'd seen. Hahn and Strassmann came to the unexpected conclusion that the unknown substance was not radium but barium. This was surprising, since barium was nowhere near uranium on the periodic table. Hahn and Strassmann speculated

Dr. Lise Meitner (1878–1968), Austrian-Swedish physicist and mathematician, shown in 1949. Her work with radioactive isotopes helped the development of the atomic bomb.
© Bettmann/Corbis.

that they had stumbled upon a reaction in which uranium broke up into two pieces.

Hahn immediately wrote to Meitner in Sweden about this finding. Fortuitously, her nephew Otto Frisch (1904–1979), who worked at Niels Bohr's institute in Copenhagen, was visiting. Frisch and Meitner realized that if uranium absorbed a neutron as it was being bombarded, a highly unstable nucleus would result, due partly to the Coulomb repulsion between its protons. Using Bohr's earlier model, which viewed the nucleus as a large drop of liquid, absorbing a neutron would cause the nucleus to become unstable. This vibrating liquid drop would eventually break up into two smaller drops of approximately equal mass (e.g., barium and krypton). The drops (nuclear fragments) would then be repelled by their respective positive charges. This would produce significantly more energy than the two products simply added together. Meitner and Frisch called this process fission.

Upon his return to Copenhagen, Frisch conducted an experiment in which he bombarded uranium with neutrons and found huge bursts of energy. The reaction created a sensation in the physics community, especially when it was confirmed that, in addition to barium, a krypton nucleus and three neutrons were formed. (See Equation 4.)

$$^{1}_{0}n + {}^{235}_{92}U \rightarrow {}^{91}_{36}Kr + {}^{142}_{56}Ba + 3({}^{1}_{0}n)$$

Equation 4. *Cengage Learning, Gale.*

It was clear that the three neutrons might be used to trigger further fission reactions. Suddenly, it appeared that Szilárd's idea of a chain reaction might be feasible.

The fact that fission had been discovered within Nazi Germany sent chills throughout the physics community. In August 1939 Szilárd and a number of other concerned physicists convinced Einstein (who by now had left Germany and was living in the United States) to send a letter to President Franklin Roosevelt, warning of the danger of allowing the Germans to build such a bomb. During the first three years of the war, the American research effort on a fission bomb was small; it picked up in late 1942, when it became clear that the weapon was a practical eventuality.

The Manhattan Project was organized under the Army Corps of Engineers at a number of installations, each with its own specialized scientific or technical problem. At the University of Chicago, Fermi and co-workers built the first nuclear reactor and demonstrated a chain reaction on December 2, 1942. In Hanford,

Scientists at a weekly meeting for the Manhattan Project during the development of the world's first atomic bomb, Los Alamos, New Mexico, in 1945. Such meetings made possible a free exchange of ideas throughout the project. Seated in the front row are (left to right) Norris Bradbury, John Manley, Enrico Fermi, and J.M.B. Kellogg. In the second row are J. Robert Oppenheimer (in dark suit) with Richard Feynman on his left. *Los Alamos National Laboratory/Photo Researchers, Inc.*

Washington, radiochemist Glenn Seaborg (1912–1999) and coworkers labored to find a chemical means of separating plutonium (a second material capable of undergoing nuclear fission), from samples of uranium. A large facility in Oak Ridge, Tennessee, used a process called gaseous diffusion to separate the uranium isotope useful for building a fission bomb, ^{235}U, from natural uranium samples, which contained large amounts of ^{238}U.

Finally, the facility at Los Alamos, New Mexico, was given the job of computing how large a mass of fuel was needed to start a chain reaction (a quantity called critical mass) and building the actual bombs. The Los Alamos effort, headed by J. Robert Oppenheimer (1904–1967), found two ways to assemble the nuclear fuel: the relatively simple gun method, in which two chunks of uranium were held at either end of a sort of gun barrel and then pushed together at the last moment; and the more difficult implosion method, in which plutonium was stored in a thin spherical shell and then pushed to its center by conventional explosives.

By the time the weapons were ready, the war in Europe had ended, but Japan fought on with relentless ferocity. The battle of Okinawa in March 1945 had killed over 12,000 American soldiers, 66,000 Japanese troops, and 150,000 Okinawan civilians—a chilling indication of the casualties that would result from a planned invasion of the Japanese mainland. To force an end to the war, despite an eleventh-hour effort by many scientists to convince the Truman administration not to use the bombs without warning, on August 6, 1945, the United States dropped a gun-type uranium bomb nicknamed "Little Boy" on Hiroshima, Japan. The bomb delivered the equivalent of 12.5 kilotons of TNT, laying waste to the city and killing about 70,000 people immediately. Three days later, the United States dropped "Fat Man," an implosion-type plutonium bomb on Nagasaki, with similar results. In response, a portion of the American physics community, including members of the Manhattan Project, founded the Federation of Atomic Scientists (now renamed the Federation of American Scientists) to press for arms control.

A mushroom cloud from an atomic bomb rises over Nagasaki, Japan, in August 1945, in the last days of World War II, shortly before the surrender of Japan. © *Nagasaki Atomic Bomb Museum/epa/Corbis.*

Immediately after the war, Congress established the Atomic Energy Commission (AEC, now the Department of Energy). The AEC developed and maintained America's nuclear weapons and also explored peacetime uses for nuclear physics, including power generation and medical applications. The AEC gave grants to universities which, in turn, operated the national laboratories. Lewis Strauss (1896–1974), the AEC's second chair, encouraged the development of fission energy, reinforced by the Eisenhower administration's 1954 Atoms for Peace program. At the first International Conference on the Peaceful Uses of Atomic Energy in Geneva, in 1955, the Oak Ridge National Laboratory transported and set up a working light-water reactor (using regular water and not "heavy water" made with deuterium), which lent the United States added scientific prestige on the international stage. A reorganization of the AEC in 1954 made it legal for the commission to enter into contracts with private companies; the first American nuclear power plant, built by the AEC but run by the Duquesne Light Company, began operation in 1957 in Shippingport, Pennsylvania.

In contrast to fission, fusion reactions had been known to the physics community before 1939. In 1920 the chemist Francis Aston (1877–1945) made precision measurements of the masses of light atoms; this showed that the total mass of four hydrogen atoms was higher than a single alpha particle. The British astrophysicist Sir Arthur Eddington (1882–1944) immediately realized that Aston's measurements, along with Einstein's mass-energy formula, implied that the fusion of hydrogen into helium might account for the great energy produced by the sun and stars. Because two protons had to overcome the Coulomb force repulsion before they could feel the short-ranged nuclear attraction, fusion reactions could only occur at the high temperatures and densities found within stars.

During the next two decades, George Gamow (1904–1968), Robert Atkinson (1898–1982), Fritz Houtermans (1903–1966), and others explored the reaction rates and energy release of fusion reactions in the stars. In 1939 physicist Hans Bethe (1906–2005) published a seminal paper in which he specified two processes by which fusion reactions created energy. The first process, the proton-proton chain, was found in smaller stars like the sun, while the second process, the carbon-nitrogen-oxygen cycle, was important in more massive stars.

After World War II it was clear that a bomb based on fusion reactions might be possible. Certain scientists, such as Oppenheimer, opposed the development of the fusion bomb. However, the burgeoning Cold War, punctuated by the first successful Soviet fission bomb test in 1949, gave the day to scientists such as Edward Teller (1908–2003), and AEC chair Strauss. A new national laboratory, Lawrence Livermore, in California, was devoted to its development. The hydrogen bomb,

$$\,_1^2\text{H} + \,_1^3\text{H} \longrightarrow \,_2^4\text{He} + \,_0^1\text{n}$$

Equation 5. *Cengage Learning, Gale.*

or H bomb as the new weapon became known, used a small fission bomb as a sort of fuse to ignite the fusion fuel. One such reaction was the fusion of deuterium and tritium. (See Equation 5.) The first successful fusion device, code named "Mike," was too large to fit into any military delivery vehicle and was detonated on November 1, 1952, in the Enewetak atoll in the central Pacific, unleashing a 10-megaton equivalent of TNT.

One great dream of the postwar years was to harness fusion reactions for power generation—if only the Coulomb repulsion between nuclei could be overcome by means other than a fission bomb. Calculations showed that creating a beam of particles to collide with a target would use more energy to produce than would be gained from fusion reactions. Attention therefore turned to heating gases to high temperatures, in the hopes that thermal energy would enable fusion reactions while the ionized fuel or "plasma" was confined for a short time in carefully designed magnetic field configurations.

Secret research programs on controlled thermonuclear fusion began at British national laboratories during the late 1940s and in the United States and the U.S.S.R. in the early 1950s. In mid-1958 the three countries agreed to declassify their fusion programs in time for the Second International Conference on the Peaceful Uses of Atomic Energy. Strauss pushed the fusion effort as a sort of crash program, modeled after the Manhattan Project, and hoped for a working fusion reactor at the conference. The Project Sherwood scientists—a part of the Atoms for Peace project that focused on fusion reactors—worked until the last possible moment but ultimately disappointed Strauss. In the coming years, practical fusion energy continued to be elusive. The fusion research community obviously has some intersection with nuclear physics but, because most of the fusion devices involve the confinement of hot ionized gases, much of its expertise lies in plasma physics.

Models of the Nucleus and Nuclear Reactions

In the early years of nuclear physics, experimentalists took the lead, but in the 1930s, theorists became increasingly important. As experimentalists continued to amass results, it became clear that the nucleus could behave in astonishingly complex ways. Theorists sought to impose order on the data by proposing models of the nucleus and of the reactions that the nuclei entered into.

The study of the internal structure of the nucleus began with the collation of experimental results. In

IN CONTEXT: GETTING MORE FROM LESS

If all protons and neutrons have the same mass, and if the numbers of protons and neutrons are conserved in many nuclear reactions, how can the mass of the products be less than the reactants?

When nucleons (protons or neutrons) are in bound systems, they lose some of their mass to the binding energies (due to the attractive nuclear force) that holds the nucleus together. There are two ways that the total mass diminishes and releases energy fission reactions: by taking apart nuclei, such as uranium, to give products such as barium and krypton; or with fusion reactions, by assembling nuclei such as deuterium and tritium into nuclei such as helium.

1948 the physical chemist turned nuclear physicist Maria Goeppert Mayer (1906–1972), working at the University of Chicago, noticed that certain nuclei have many more stable isotopes than might be expected. She also noticed the dependence of certain nuclear properties on the number of neutrons. For example, the binding energy of the last neutron placed in a nucleus was especially high in certain cases, and this corresponded to those nuclei with many stable isotopes. Mayer used her knowledge of atomic physics to make an analogy between it and a new model for the nucleus. She suggested that nucleons are in a nuclear potential in much the same way that electrons are in a Coulomb potential, and that nucleons fill energy shells of the nucleus in a way similar to how electrons fill shells of the atom. Mayer called the special numbers of neutrons that created particularly stable nuclei "magic numbers" (2, 8, 20, 28 …) and likened them to the closed electron shells of atomic physics. She constructed a nuclear shell model by assuming a simple form for the shared nuclear potential and assigning each of the nucleons in the nucleus an orbital and spin angular momenta. While the results were not as pretty as the ability of atomic physics to reproduce the layout of the periodic table, the nuclear shell model did make sense of a great deal of nuclear data.

With the publication of Mayer's work, a great deal of research turned to the new field of nuclear structure. Following the analogy with atomic physics, nuclei needed to be characterized in both their ground state and many excited states. Many excited states of nuclei could be described with the single-particle approximation of the shell model. Others, however, reasserted the need to see the nucleons of the nucleus as part of a collective motion; excited nuclei could rotate and vibrate in complex ways, not unlike the way that nuclei deformed in the fission process. In 1953 Danish physicists Aage

Bohr (1922–; the son of Niels Bohr) and Ben Mottelson (1926–) found a way to make the single-particle and collective models of the nucleus consistent with one another and to apply them to the available data. Their efforts resulted in a two-volume work, *Nuclear Structure,* which defined the field for many years.

The study of nuclear reactions is important to the determination of the ground state and excited states of nuclei as well as the modeling of nuclear properties such as a cross section (a measure of the degree to which a reaction occurs). Nuclear reactions often are studied in scattering experiments, in which a beam of projectiles hits a target and initiates a nuclear reaction, after which the reaction products move off at different angles relative to the original beam. There are two general categories of nuclear reactions: compound-nucleus reactions and direct reactions.

The first contribution to a theory of the compound nucleus was made by Niels Bohr in 1936. Bohr imagined that the projectile is entirely captured by the target nucleus, forming a new, compound nucleus. Because the energy of the projectile is soon shared with all of the nucleons, the compound nucleus is an excited system. Therefore, it is short-lived (on the order of 10–16 seconds), and breaks up into reaction products. States of the compound nucleus could be identified when the incoming particles were at an energy such that the "waves" of those particles resonated with the state of interest.

Experimental results showed such resonances clearly. When researchers graphed the cross section against the energy of the projectiles, large peaks were found at well-defined energies. Bohr's model for compound nucleus reactions was strengthened by the work of Eugene Wigner (1902–1995) and Gregory Breit (1899–1981) at Princeton University, who produced a general mathematical formalism for resonant systems that could be used to analyze compound nucleus reactions. The combined effect of Bohr's intuitive model and the Breit-Wigner formalism was galvanizing. The nuclear physics community enthusiastically conducted new experiments identifying ground and excited states of many compound nuclei.

In contrast to compound nucleus reactions, in a direct reaction the projectile does not enter into the target nucleus but instead only grazes the surface. The projectile might scatter elastically or interact with nucleons at the target's surface. In either case, the time for a direct reaction is much smaller than that for a compound reaction, on the order of about 10^{-22} seconds. Another contrast is found in the pattern made by the scattered particles. Compound nucleus reactions scatter fairly evenly to all angles since the compound nucleus has no "memory" of the incoming projectile after it is formed and so decays in any direction. However, in a direct reaction, the target nucleus "remembers" the initial beam direction;

its scattering patterns show high cross sections for small angles near the original beam direction.

The theoretical analysis of direct reactions was pioneered by Stuart T. Butler (1926–1982) at the University of Birmingham, Australia. In 1951 Butler developed mathematical formulae to analyze the stripping reaction, in which a projectile has a nucleon stripped off it as it passes the target. Consider, for example, a deuteron (deuterium nucleus) projectile glancing off of a target such that it loses its neutron, leaving only the proton to scatter away. Butler's analysis was able to deduce the final states (ground state and excited states) in which the final nucleus was left. For stripping reactions in which a single nucleon was stripped off, the identified states were often similar to those of the ideal shell model.

High-Energy Physics

By the 1950s, high-energy or particle physics had established itself as a new area of study and independent research. The field grew from elements of cosmic-ray research and experiments using newer, larger accelerators. By 1932 Carl Anderson (1905–1991) of Caltech had identified the electron's antiparticle, dubbed the positron, in the course of cosmic-ray studies. This discovery was a ringing confirmation of British physicist Paul Dirac's (1902–1984) combination of quantum mechanics and Einstein's special theory of relativity, which had predicted the existence of antiparticles in 1928. Identification of many other particles and their antiparticles followed. The m-meson was discovered by Anderson and Seth Neddermeyer (1907–1988) in 1937 and the p-meson by Powell and coworkers ten years later. In 1947 George Rochester (1908–2001) and C.C. Butler at the University of Manchester identified the K-meson or kaon, a new type of particle with a quizzical new quantum number called "strangeness."

The development of larger accelerators was pursued by many researchers, but the most famous is probably Ernest Lawrence (1901–1958). In 1928 Lawrence invented the cyclotron, a device that was able to accelerate charged particles by making them circle within a magnetic

IN CONTEXT: WOMEN IN NUCLEAR PHYSICS

Until the closing decades of the twentieth century, women were poorly represented in physics—and nuclear physics was no exception. While women faced an uphill battle in this field, however, they made significant contributions nonetheless.

Lise Meitner (1878–1968) was raised in an Austrian Jewish family. She became the first woman to earn a doctorate in physics at the University of Vienna—Austrian restrictions on women's education had precluded even her entry into the university before 1901—working with the theoretician Ludwig Boltzmann (1844–1906). Upon her graduation, she won a position at the University of Berlin, working with Max Planck (1858–1947), despite his disapproval of women in academia. When, in collaboration with Otto Hahn (1879–1968) and Fritz Strassmann (1902–1980), their discovery of fission earned the two men the 1945 Nobel Prize in chemistry, Meitner was ignored.

Historian Ruth Sime highlights a number of reasons why Meitner was not given her due credit, including the prize committee's favoring of chemistry over physics, anti-Semitism in Germany, Meitner's exile to Sweden, and bias against women scientists. Meitner's 1938 flight from Nazi Germany not only separated her from Hahn and Strassmann but also left her with difficult working conditions in Sweden, where she was given little more than laboratory space. Hahn's first paper on fission in 1939 did not even mention Meitner, perhaps out of fear that his German colleagues would be more prone to reject his results if they involved a collaborator who was a Jew and a woman. After the war, Hahn may have tried to rationalize Meitner's exclusion, perhaps worried about his own reputation and the rebirth of German science. Meitner's contribution was not recognized until near the end of her life when the Atomic Energy Commission gave her, Hahn, and Strassman the 1966 Enrico Fermi Award for their discovery of fission.

Maria Goeppert Mayer's (1906–1972) story demonstrates how difficult it was for many women to secure a position, despite producing significant work, and even being married to a fellow scientist. Goeppert Mayer was born in 1906 into a German academic family and received an excellent physics education, interacting with such luminaries as Niels Bohr (1885–1962) and Enrico Fermi (1901–1954). After marrying the physical chemist Joseph E. Mayer in 1930, she embarked on a long search for employment. She and her husband first moved to Baltimore, where Joseph took a position at Johns Hopkins.

In Baltimore and later in New York after her husband changed jobs, Goeppert Mayer kept working and publishing—specializing in chemical physics—despite the fact that she never received a professional position or salary. Her persistent work eventually paid off to some degree. In 1946, after she and her husband moved to Chicago, Goeppert Mayer was given two positions, one at the University of Chicago and another at the nearby Argonne National Laboratory. Although these positions gave her office space, they still did not include a salary. After doing her fundamental work on the nuclear shell model and being elected to the National Academy of Sciences, she and her husband moved to the University of California at San Diego in 1956, where she was finally given a professorship. In 1963 she shared the Nobel Prize in Physics with J. Hans D. Jensen for their work on the nuclear shell model. She was only the second woman to have won this prize (after Marie Curie).

field by having an oscillating electric field that delivered small "kicks" during each revolution. Lawrence continued to build larger and larger accelerators and in 1936 was given a separate laboratory at the University of California at Berkeley called the Radiation Laboratory. In 1955 Owen Chamberlain (1920–2006) and coworkers identified the antiproton at the Radiation Laboratory using the Bevatron accelerator.

After Lawrence's death in 1958, larger and larger machines were built. During the 1960s the laboratories of the European Organization of Nuclear Research (CERN) became among the most important in the world for both nuclear and high-energy physics. In 1967 the National Accelerator Lab was founded in Chicago; it was renamed Fermilab seven years later.

By 1960 high-energy physicists had identified a bewildering array of particles and antiparticles. In 1961, American Murray Gell-Mann (1929–) at Caltech and the Israeli physicist Yuval Ne'eman (1925–2006) independently came up with a new scheme to classify existing particles and, in the case of the omega minus particle, to predict the existence of a missing particle that was later identified experimentally. Gell-Mann referred to this classification scheme as "the eightfold way" since the particles were organized into groups of eight. Three years later, Gell-Mann and, independently, the Russian-born American physicist George Zweig (1937–), suggested the eightfold way could be explained by the fact that hadrons—particles such as protons and neutrons, which experience the strong force—were composed of subparticles, which he named "quarks."

Perhaps the greatest accomplishment of high-energy physics has been the so-called Standard Model, developed during the 1970s, which joins three of the four forces of nature into one consistent theory. The gluon was proposed as the particle that mediated the strong force between quarks and offered a more fundamental understanding of the strong force than the Yukawa theory of pion exchange. The electromagnetic interactions (between charged particles such as electrons) are mediated by bits of electromagnetic energy called photons.

The Standard Model predicted that the weak interaction was mediated by relatively massive particles called the W and Z bosons, first identified in 1983 by two research groups at the CERN. Efforts to consolidate the Standard Model have focused on the Higgs particle, which is related to the mediating particles for all three of the forces addressed by the Standard Model. Identifying the Higgs particle experimentally will require collisions at energies that are higher than those produced by any presently existing accelerator. In 1993 Congress canceled construction of the so-called superconducting supercollider (SSC). The high-energy community has therefore turned its attention to a machine at CERN, the large hadron collider (LHC), which produces proton-proton collisions and was scheduled to begin operation in May 2008.

■ Modern Cultural Connections

Nuclear physics continues to be an area of vital research. As in so many other scientific and technical fields, the computing revolution opened new doors. In studies of the strong force, for example, the "few nucleon" community takes precision measurements of three-nucleon systems, such as a nucleon scattering from a deuteron, and compares them to the results of computer simulations. The computer simulations begin with detailed representations of the basic nucleon-nucleon (strong) force and calculate each nucleon's interactions separately. Assuming that the three-nucleon calculation is trustworthy, these precision measurements will allow the basic nucleon-nucleon force to be even better understood.

Studies of the weak interaction continue also. In 1965 American physicist Ray Davis Jr. (1914–2006) and coworkers at Brookhaven National Laboratory set up a giant neutrino detector in an abandoned gold mine at Homestake, South Dakota. According to knowledge of the fusion reactions in the sun, a certain flux of neutrinos was expected at Earth's surface. However, Davis and coworkers found only about half of this flux. This "missing solar neutrino problem" was solved by the discovery that there are actually three different types of neutrinos; all are unstable and oscillate from one type to the other as they travel through space (as the result of having a tiny but non-zero mass). To test these ideas and to study neutrino oscillations, new underground experiments are being proposed for a still-to-be-built National Underground Science Laboratory, one location for which could be Homestake.

As particle physics has continued in its inexorable march to higher energies and bigger machines, nuclear physics has moved into abandoned energy fields with a somewhat different focus. One of the newer nuclear physics facilities, the relativistic heavy ion collider (RHIC) was completed in 1999 at the Brookhaven National Laboratory. The RHIC accelerates protons, deuterons, copper nuclei, gold nuclei, and lead nuclei to over 99.99% the speed of light, creating dense high-temperature nuclear matter. Under such conditions, nucleons lose their individual identity, creating, for a brief moment, a sort of "nuclear plasma" made from quarks and gluons. Such nuclear matter is thought to have existed moments after the beginning of the universe—the so-called big bang.

The military applications of nuclear physics continue to be important. Stockpile stewardship is the effort to insure that the U.S. arsenal of nuclear weapons is reliable, if and when it is needed. In 1963 the United States and the U.S.S.R. signed the Limited Test Ban Treaty, banning nuclear tests above ground, underwater, and in space. This left only underground tests as a means of checking nuclear devices but, in 1992, President George H.W. Bush put a moratorium on these. Since that time, the stockpile stewardship program has used computer simulations, along with the laboratory measurement of reaction cross sections and other nuclear properties. Untested new hybrid weapons, however, designed to be more stable and less vulnerable to both terrorists and accidental deployment, may force the United States to resume underground tests.

After September 11, 2001, concern over the possibility of terrorist attacks gave nuclear military applications renewed attention. One area of research centers on the detection of radioactive materials that might be concealed in baggage. New detection systems are being developed to identify the characteristic radiation of different materials in the relatively short time available at airport check-ins and highway ports of entry. New weapons are also being sought. In recent years, the Pentagon has encouraged the development of relatively small nuclear weapons that could be used in a battlefield situation. One class of tactical nuclear weapon, the so-called "bunker buster," is designed to penetrate the earth and destroy bunkers of chemical and biological weapons. Some members of the physics community

have voiced concerns about the fallout associated with such weapons; supporters claim that newer weapons use far less radioactive material, and that the weapon's subterranean target would limit fallout.

Neither fusion nor fission energy has lived up to the enthusiasm first generated in the 1950s. None of the many fusion experiments that have been attempted have even reached breakeven, the point at which fusion reactions produce energy as great as that which heated the fuel in the first place. The largest experiment currently being pursued, the international tokomak experimental reactor, that will be built in Cadarache, France, hopes to reach breakeven but will not begin operation until about 2016.

Fission power has met with considerable public opposition. The accident at Three Mile Island, Pennsylvania in 1979, and the more serious disaster in Chernobyl, Ukraine, seven years later, combined with concerns about storing spent nuclear waste, turned the American public against the idea of new nuclear power plants. Renewed concerns about energy shortages however, have brought renewed attention to nuclear power. The problem of storing nuclear waste might possibly be solved by the proposed facility at Yucca Mountain, Nevada, a ridge of volcanic rock chosen by the Department of Energy as a repository, though this site has been controversial. Meanwhile, problems with older plants might have been answered by new nuclear technology, such as

Steam rises from the cooling towers of PECO's nuclear power plant near Limerick, Pennsylvania. Though nuclear fission reactors efficiently generate electricity without creating air pollution, critics state that the possibility of catastrophic accidents and long-term storage of radioactive wastes outweigh the benefits. *George Widman/AP Images.*

gas-cooled reactors (including the so-called pebble bed reactor). Of course, questions and concerns remain with all of these developments.

SEE ALSO   *Physics: Radioactivity.*

BIBLIOGRAPHY

Books

Andrade, E.N. da Costa. *Rutherford and the Nature of the Atom.* Garden City, NY: New York Doubleday, 1964.

Badash, Lawrence. *Scientists and the Development of Nuclear Weapons: From Fission to the Limited Test Ban Treaty, 1939–1963.* Atlantic Highlands, NJ: Humanities Press, 1995.

Brown, L.M., and H. Rechenberg. *The Origin of the Concept of Nuclear Forces.* Bristol, UK: Institute of Physics, 1996.

Pais, Abraham. *Inward Bound: Of Matter and Forces in the Physical World.* Oxford, UK: Oxford University Press, 1986.

Segrè, Emilio. *From X-Rays to Quarks: Modern Physicists and Their Discoveries.* San Francisco: W.H. Freeman, 1980.

Sime, Ruth Lewin. *Lise Meitner: A Life in Physics.* Berkeley, CA: University of California Press, 1996.

Periodicals

Hinchliffe, Ian, and Marco Battaglia. "A TeV Linear Collider." *Physics Today* 57, no. 9 (2004): 49–55.

Marcus, Gail H., and Alan E. Levin. "New Designs for the Nuclear Renaissance." *Physics Today* 55, no. 4 (2002): 36–41.

McDonald, Joseph C., Bert M. Coursey, and Michael Carter. "Detecting Illicit Radioactive Sources." *Physics Today* 57, no. 11 (2004): 36–41.

Nelson, Robert W. "Nuclear Bunker Busters, Mini-Nukes, and the U.S. Nuclear Stockpile." *Physics Today* 56, no. 11 (2003): 32–37.

Web Sites

American Institute of Physics. Center for History of Physics. "The Discovery of Fission." <http//www.aip.org/history/mod/fission/fission1/01.html> (accessed November 5, 2007).

U.S. Department of Energy. Office of Nuclear Physics. "DOE/NSF Nuclear Science Advisory Committee. Opportunities in Nuclear Science." <http//www.er.doe.gov/np/nsac/nsac.html> (accessed November 5, 2007).

G.J. Weisel

Physics: Optics

■ Introduction

Optics is the branch of physics concerned with the nature and uses of light. Especially through systems made of lenses—pieces of glass or plastic shaped to alter the light passing through them—optics have made possible photography; the discovery of microorganisms through microscopes; the correction of some vision disorders by eyeglasses and contact lenses; the study of the distant universe through telescopes; movie and slide projection; and thousands of industrial processes, including the manufacture of computer microchips using optics to outline billions of transistors on tiny tiles of silicon. Lenses, mirrors, lasers, and optical fibers are used in data transmission, storage, and retrieval; in surgery; weapons targeting; document scanning and printing; and for many other purposes. Optical systems have been essential to scientific and technological progress.

■ Historical Background and Scientific Foundations

The earliest known theories about the nature of light were made by the ancient Greeks. Greek philosophers reasoned from the casting of shadows by solid objects that light must travel in straight lines. They also knew that light rays are reflected from a surface at the same angle that they strike it. The followers of Democritus (c.460–370 BC), who taught that the world was made of atoms and void (emptiness), speculated that light consisted of streams of particles and that visual sensation is caused when these particles strike the eye. Pythagoras (c.575–500 BC) explained that light is not the cause of visual sensation, but that "seeing rays" are emitted from the eye, rays which interact with light somehow at the surface of the object being viewed or in some other way.

Aristotle (384–322 BC) denied Democritus's theory of atoms and void, and proposed a quite different theory of light, namely, that light is an alteration of the materials between the source of illumination and the eye.

Greeks of the fifth century BC were able to make glass and understood the ability of lenses to concentrate sunlight to burning intensity. Later, the Romans also used lenses as burning-glasses, though they do not seem to have understood their full potential.

WORDS TO KNOW

CAMERA OBSCURA: A dark room or box with a light-admitting hole that projects an image of the scene outside.

LASER: Acronym for light amplification by stimulated emission of radiation; a device that uses the movement of atoms and molecules to produce intense light with a precisely defined wavelength.

LENS: In optics, a transparent material (such as glass or plastic) with two opposing surfaces that focuses rays of light to create images. These surfaces can either both be curved or have one curved and one straight surface. A lens can be used alone or in combination with other lenses, in applications such as eyeglasses and telescopes. In biology, an almost clear, biconvex structure in the eye that, along with the cornea, helps to focus light onto the retina. It can become infected with inflammation, for instance, when contact lenses are improperly used.

MYOPIA: A synonym for near-sightedness. It is caused by elongation of the eyeball in the direction of vision so that images of far-away objects are focused inside the eye rather than on the retina (the light-sensitive surface at the back of the eye's interior).

Fresco of "Burning Mirrors," Stanzio della Mattematica (1587–1609), attributed to Giulio Parigi.
Giulio Parigi/Getty Images.

Growth in optics was slow throughout the Middle Ages (AD 900–1400). Arab scientists such as Ibn Sahl (c.940–1000) and Ibn al Haitham (963–1039) published treatises on mirrors and lenses that would eventually influence European students of optics. Various properties of lenses and prisms were observed by Arab and European scientists in the Middle Ages—English scientist and Franciscan monk Roger Bacon (1215–1294) was the first person to write down observations on the magnifying properties of lenses and also wrote a book attempting to explain the nature of rainbows—but no general agreement was reached on the nature of light itself, and, apart from the use of lenses as burning-glasses to start fires, few practical uses were found for optics. This changed with the invention of eyeglasses in Italy around 1280. Although expensive, eyeglasses gradually became common in some places. In the 1400s, the ability of concave lenses to correct nearsightedness was discovered by German churchman and scientist Nicholas of Cusa (1401–1464). The making and wearing of eyeglasses spread awareness of lenses and the ability to craft them over all Europe, preparing the way for other advances in optics.

In the 1400s, the device known as the *camera obscura*, Latin for "dark chamber" and origin of the English word "camera," became popular in Europe. A camera obscura is a darkened room pierced by a pinhole or a larger hole in which a convex lens is inserted. The lens or pinhole projects an image of the outer world on the opposite interior wall of the dark room, where it can be traced on paper or simply viewed. Modern cameras are miniature camera obscura, where instead of the image

being traced by hand, it is recorded by an array of electronic components or a sheet of film coated with light-sensitive chemicals.

The rise of modern, scientific optics occurred gradually during the late 1500s and early 1600s, a period when all areas of physical science were undergoing revolutionary changes. In 1600, a Dutch maker of eyeglasses, probably Zacharias Janssen (1580–1638), invented the compound microscope, a combination of two or more magnifying lenses that allow the inspection of details smaller than the unaided human eye can see. Microscopes were steadily improved during the 1600s, achieving higher and higher magnifications and revealing the existence of a hitherto unknown micro-world. In 1675, Dutch scientist Antoni van Leeuwenhoek (1632–1723) discovered the existence of microorganisms.

The refracting telescope was invented in 1608, also by Dutch makers of eyeglasses. New scientific discoveries followed rapidly. The first person to turn the telescope on the heavens was Italian scientist Galileo Galilei (1564–1642), who had heard about the new invention and in 1609 built one of his own. On the basis of his observations, Galileo advocated the Copernican view that Earth moves around the sun. The official view of the church at that time was that Earth is stationary, and Galileo was forced to officially retract his support of Copernicanism.

The seventeenth century also was the time of many discoveries of optical phenomena. Dutch scientist Willebrord Snell (1580–1626) discovered the law of refraction, today known as Snell's law. This mathematical law relates the angle through which a light ray is bent when

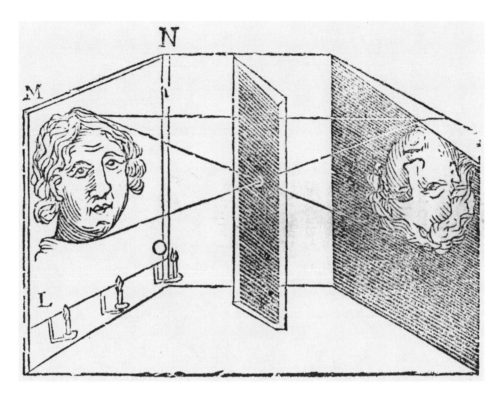

Sixteenth century print depicting a camera obscura developed by Giambattista della Porta.
© *The Print Collector/Alamy.*

passing from one medium into another (for example, from air into glass) to the index of refraction of each material. A transparent material's index of refraction is its tendency to slow light down.

With Snell's law in hand, scientists could make fast progress in geometrical optics, the mathematical description of the paths that light rays take through systems of lenses and mirrors. Geometrical optics are needed for the scientific design of microscopes, telescopes, eyeglasses, and other optical devices. Double refraction and polarization were discovered in the late 1600s, and in 1676 Danish astronomer Olaus (or Ole) Romer (1644–1710) made observations of the moons of Jupiter that allowed the first realistic estimates of the velocity of light. Romer's view that light travels from point to point at a finite velocity, rather than instantaneously, remained controversial for another 40 years but was verified by other scientists in the early 1700s.

Knowledge of how light behaved grew quickly in the seventeenth and eighteenth centuries, but the nature of light itself was still a mystery. Aristotle's views were still influential in many universities, and the atomistic view of Democritus (light consists of "corpuscles," or particles) was revived. An influential theory of light was put forward by French philosopher and mathematician René Descartes (1596–1650). Descartes theorized that all space is filled with subtle (hard to detect) material

of several types, and that some of this material swirls in tiny vortices or whirlpools. Descartes' theory was influential but was soon eclipsed by the work of English physicist Isaac Newton (1643–1724), who not only revolutionized physics with his laws of motion and gravitation but also did important work in optics.

Newton was a champion of the corpuscular, or particle, theory. He examined the properties of colors and prisms, proving that white light consists of a variety of colored lights that can be separated, whose colors cannot be altered, and which can be recombined into white light. So influential was Newton that the wave theory of light, championed by English scientist Robert Hooke (1635–1703) and Dutch astronomer Christiaan Huygens (1629–1695), was not widely accepted by scientists until the nineteenth century, even though it could describe a wider range of optical phenomena than Newton's particle theory.

In the nineteenth century rapid progress was made in all fields of physics, including optics. In 1801, English scientist Thomas Young (1773–1829) discovered the phenomenon of interference. This occurs when the peaks and troughs of waves, including light waves, add to or subtract from each other, causing areas of heightened brightness or darkness. Electricity and magnetism were unified optics in the work of Scottish physicist James Clerk Maxwell (1831–1879), who showed that light can

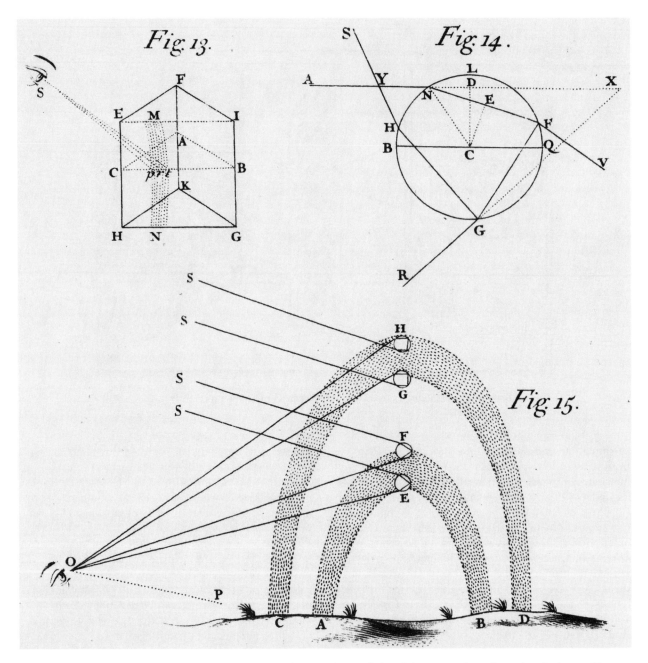

Diagram from Isaac Newton's (1642–1727) *Opticks*, 1704, showing a beam of white light passing through a series of prisms and lenses that split it into a colored spectrum, recombine it into a single beam, then split it once more. *©The Print Collector/Alamy.*

be explained as a transverse electromagnetic wave. That is, a ray of light can be viewed as an electrical field paired with a magnetic field, both fields pointing crosswise (transverse) to the direction of the ray. Each field produces the other by changing constantly in strength.

Not only did Maxwell explain the nature of light, he predicted that other types of electromagnetic waves would be discovered. His prediction came true in 1887, when radio waves were first studied by German physicist Heinrich Hertz (1857–1894). From that time onward,

visible light was understood as only one narrow band of the electromagnetic spectrum, that is, all possible electromagnetic waves ordered by frequency (speed of vibration) from lowest to highest.

Hertz did not foresee any use for the phenomenon of invisible electromagnetic waves: "It's of no use whatsoever," he told his students. But in 1896, Italian physicist Guglielmo Marconi (1874–1937) set up in England the world's first radio broadcasting station. Marconi's primitive station did not transmit voice or music, but bursts of radio energy that could be used to telegraph messages one coded letter at a time.

The camera was invented in the 1820s. The camera obscura of studio art, miniaturized and mounted on a tripod, was united with light-sensitive chemicals layered on glass plates to produce permanent photographs. Although for several decades cameras were used by only of a few hobbyists, they were mass-marketed starting with Kodak's one-dollar Brownie in 1900. (One dollar in 1900 was the equivalent of about $24 in 2007.) In the 1990s, the mass-marketed digital camera appeared, further expanding the ubiquity of the optically formed and technologically captured image. Today, imaging systems are built into cellular phones, personal digital organizers, and other multipurpose digital devices, and millions of people never leave home without one.

In the late nineteenth century, the triumph of the wave theory of light seemed complete. But German physicist Max Planck (1858–1947) showed in 1901 that the spectrum (mixture of intensities and frequencies) of electromagnetic radiation emitted by a perfectly black object (black body) can be best explained by assuming that energy is transferred only in chunks or fixed quantities (quanta). A few years later, German-American physicist Albert Einstein (1879–1955) showed that Planck's energy quanta, combined with the assumption that light is composed of tiny particles called photons, can explain certain features of the photoelectric effect, which occurs when light knocks electrons out of metal surfaces.

In the early twentieth century, light drove the discovery of new physics in two ways. First, through the theories of Planck and Einstein just described and through the work of many other people, light was basic to the development of quantum mechanics—the new theory of matter's properties at very small size scales. Today, light and all material particles are known to have the property of wave-particle duality. That is, a light wave is neither a wave nor a particle, but shows the characteristics of one or the other, depending on circumstances.

Secondly, through Einstein's theories of special relativity (1905) and general relativity (1915), light became essential to our understanding of the nature of time, space, matter, and energy, and thus of the size, shape, and history of the universe. The speed of light in a vacuum is the fundamental limiting velocity of the universe; no material object can be made to move as fast as or faster than light.

Science's fundamental understanding of light has not changed much over the last century, but the physics and technology of light have been intensely studied. A few notable developments are as follows:

- The laser was invented by 1958. The laser (short for "light amplification by stimulated emission of radiation") produces a beam of extraordinarily pure light. Laser light is pure in the sense that it is nearly coherent (all its waves march together) and monochromatic (all its energy is at a single wavelength or color). Laser light travels in tight beams that can be made extremely powerful. Among the hundreds of areas in which laser optics are used are eye surgery, optical-disc writing and readback, holograms, welding, nuclear fusion, distance measurement, entertainment, microscopy, weaponry, and spectroscopy.
- The first true optical fibers, invented in 1950, were thin rods or fibers of glass used in medicine

IN CONTEXT: THE IMPOSSIBLE PERFECT LENS

For almost two centuries, it was taught that there was a fundamental limit to the amount of detail that a magnifying lens could bring out: no detail smaller than the wavelength of the light being magnified could be seen.

But in 1968, Russian physicist Victor Veselago (1929–) predicted that materials with a negative refractive index might exist—materials that bend light in the reverse direction compared to ordinary lens materials such as glass. And in 2000, English physicist John Pendry (1944–) predicted that by using a material with a negative refractive index of –1, a "perfect lens" might be constructed—a lens that could focus an image with perfect resolution. Lenses with lesser negative refractive indices wouldn't produce perfect images but would still surpass the wavelength limit on clarity.

Scientists set out to build a material having a negative refractive index for microwaves, a type of radio wave. (Radio waves differ from visible light only in having fewer vibrations per second.) In 2001, a team of physicists at the University of California, San Diego, claimed to have built a metamaterial—a screen of tiny electronic components—that acted on microwaves with a negative refractive index.

Some physicists disputed this claim. They argued that Pendry and Veselago had done their math wrong. Negative refraction was impossible: it would violate Special Relativity and even causality, the rule that causes must follow effects, not precede them. They said that the wavelength barrier on resolution would never be broken.

In 2003 and 2004, microwave lenses were built that actually demonstrated negative refraction and surpassed the wavelength limit. Perfect lenses for microwaves or light are still a long way off, but we now know that they are theoretically possible.

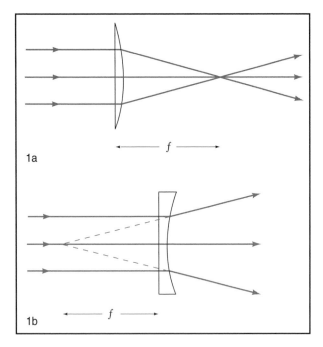

Convergence and divergence of lenses.   *Hans & Cassidy/The Gale Group.*

to allow doctors to peer inside the body. In 1956 the term "fiber optics" was coined and research began on the use of optical fibers to transmit information. After the perfection of low-loss (highly transparent) optical fibers in the 1970s, the modern system of high-speed digital data transmission through optical fibers began to come into being. Optical fiber was first deployed widely in the telephone system in the 1980s and today is the backbone of the long-distance data transmission network through which most telephone and Internet traffic passes.

The Science

Contemporary optics is usually divided into two general fields: geometrical optics and physical optics. Physical optics, in turn, includes the study of wave optics, light as an electromagnetic phenomenon, color, the interactions of light with matter, quantum optics, and relativistic optics.

Geometrical optics deals with the behavior of light when its wave nature and quantum nature can be ignored. Wherever the wavelength of light is much smaller than the objects it is interacting with, the light can be treated as an affair of raylike lines traveling from one point to another. Telescopes, microscopes, and cameras are all designed using geometrical optics.

Wave optics is concerned with light as a wave phenomenon. Interference is a wave phenomenon, and is used in holography to make three-dimensional images

recorded and projected using laser light. In astronomy, the wave-optical technique known as interferometry is used to add information about light or radio waves recorded at two or more widely separated points in order to craft higher-resolution images.

Physical optics is also concerned with the light phenomena of scattering, polarization, and spectra. Material objects both emit and absorb light in different patterns: each substance, in fact, absorbs or emits some colors (wavelengths) more strongly than others, producing a spectrum as telltale as a fingerprint. By comparing these telltale spectra with light received from distant stars and other objects, scientists can study the chemical composition of objects much too far away to ever sample directly. Spectra can also be used at closer range to examine the composition of objects small or large.

Quantum optics, in which the properties of individual photons (light particles) are significant, is a particularly active field in modern physics. Lasers are used in most quantum optics work because researchers can so precisely control the properties of the light they produce.

Impact on Science

Starting in the seventeenth century, optics enabled the range of scientific observation to explode outward from the realm of everyday experience, down into the microscopic and outward into the cosmic. Since the late nineteenth century, specialized optical systems for making images of many types have become essential in almost all scientific fields.

Not only has light been of use in research and practical technology, it has been an important challenge to scientific understanding. The question of light's nature also prompted development of both relativity and quantum mechanics at the beginning of the twentieth century, revolutionizing physics.

Impact on Society

The impact of optics on art, ideas, and human relations has been profound. During the Renaissance, many books on the uses of geometry and geometrical optics in art were written by learned people. Artists reconceived the task of painting as recording how things looked, guided by perspective, an aspect of geometric optics.

Optics has had great influence on daily life through technology. The billions of components that are built on a silicon chip to make a microprocessor are projected onto the raw chip's surface using techniques ultimately derived from the printing of photographs. The discovery of microorganisms using microscopes has made possible our modern knowledge of infectious diseases. Optical systems for examining the inside of the body through natural openings or small incisions have made surgery safer and more effective. Laser light techniques are used to cure many eyesight defects.

Perhaps the most far-reaching cultural influence of optics has been through photography. Visual reports of the larger universe from telescopes and space probes have radically revised our notions of the universe and our place in it. We have seen through photographs that we share space with trillions of other stars spread across unthinkable distances. Our view of earthly life has also been changed. Many painters have responded to photography by mimicking it (photorealism) or declining to represent objects at all (abstraction). American scholar Susan Sontag (1933–2004) famously described the psychological changes wrought by photography in her book *On Photography* (1977). "In teaching us a new visual code," Sontag wrote, "photographs alter and enlarge our notions of what is worth looking at and what we have a right to observe." The greatest change wrought by photographs, according to Sontag, is that they "give us the sense that we can hold the whole world in our heads—as an anthology of images."

■ Modern Cultural Connections

Optics continues to be an active field; new discoveries are frequent. The substitution of photons for electrons in computing devices would allow greater speed for less power, so several schemes for optical computing are being studied. Some would produce optical equivalents of the logic circuits found in conventional, electronic devices. Another approach is optical quantum computing, which combines optical methods with the quantum-information approach to computing. In some optical quantum computing experiments, a single atom confined in a microscopic chamber has been used as a source of photons with perfectly uniform properties. Single photons can be manipulated, as single atoms are in other quantum-computing schemes, to store "qubits" (short for "quantum bits") that can be used to perform computations with far greater speed than ordinary computers.

Culturally, the rapid spread of the digital camera and personal computer since the 1980s have made it possible for people in industrialized societies to capture, edit, and share images more quickly and cheaply and in far greater numbers than the film camera ever did. Millions of people are now posting and viewing online stills and videos of themselves (and of other people, not always with permission), with poorly understood social consequences. Thanks to the Internet video sharing site YouTube and similar venues, the production and distribution of video materials is in the process of being democratized, as millions of amateurs realize they do not have to submit their ideas to editorial selection. This will surely have ongoing effects on the visual arts.

The near-universal presence of imaging systems thanks to cellular telephones and other devices has made

IN CONTEXT: DOING THE WAVE

Under many conditions, light can be viewed as a wave moving through space. Unlike a wave in water or air, light is not a wave *in* something. It is an electromagnetic wave, that is, an electric field paired with an electric field, each reversing direction over and over again as the wave moves forward through space.

All electromagnetic waves have certain properties. First, they all move at the same speed in a vacuum, traditionally called the speed of light or c. The speed of light in a vacuum is 670,635,728 miles per hour (1,079,252,849 kilometers per hour). In substances such as water or glass, it moves more slowly. In fact, it can be slowed to a crawl in the laboratory.

Secondly, every electromagnetic wave, including every light wave, has a particular frequency. This is the number of times that the fields making up the wave reverse direction each second. We perceive visible light at shorter wavelengths as bluer and at longer wavelengths as redder.

Thirdly, every light wave has a fixed wavelength, the distance between any two neighboring peaks (or troughs) in the wave. Each frequency corresponds to one particular wavelength: a longer wavelength means a lower frequency and a shorter wavelength means a higher frequency. Shorter wavelengths tend to be more penetrating: The only difference between a beam of light that can be stopped by a piece of cardboard and a beam of X rays that can penetrate the body is that the X rays have a much higher frequency.

This is only the beginning, not the end, of a full description of the properties of light.

it more difficult to control the visual record of events: the photos of prisoner abuse by U.S. soldiers at the Abu Ghraib prison in Iraq released in 2004 and the illicit cell-phone video of the hanging of former Iraqi dictator Saddam Hussein in 2006, which showed executioners mocking the condemned on the scaffold, are prominent examples. The popularization of instant, digital imaging systems has provided many people with nearly unfettered and unfiltered access to world events. However, computerized imaging systems can also be used to watch people in public spaces or protect property. Some find these uses of imaging systems an invasion of privacy, while others find them a useful tool in preventing crime.

■ Primary Source Connection

The invisibility cloak is a staple utility of fantasy literature and science fiction novels. However, new research in optics is paving the way for invisibility devices. Newly

Over half the population of the industrialized world wears corrective optics—glasses or contact lenses. Some people have concluded that eyeglasses must be injuring the human genetic heritage. Their logic goes like this: In primitive times, people with bad vision would have died young. Natural selection would have screened out genes for badly formed eyes. Since the invention of glasses, however, people with bad eyes have been able to survive and pass on their bad genes. This idea was recently stated in scientific language by several scientists in the journal *Trends in Genetics* in 1998: "Poor vision would normally put one at a tremendous selective disadvantage, but the modern contrivance of corrective lenses has facilitated the maintenance of relevant myopia genes, and has led to a general weakening of visual capacities."

Yet myopia, or nearsightedness, one of the most common vision problems, is not simply caused by genes: rather, individuals (more in some ethnic groups) have genes that make them more vulnerable to developing the condition when they engage in much close-focusing work, such as reading, in childhood. Until civilization comes along and invents fine print, our genetic predisposition to myopia isn't activated. What's more, as *The Quarterly Review in Biology* said in 1991, "Corrective lenses were invented far too recently to have allowed a substantial increase in genes that cause myopia."

This might sound like a case of two groups of scientists contradicting each other—*Trends in Genetics* versus *The Quarterly Review in Biology*—but it's not. The only reference cited by the *Trends in Genetics* authors for their claim that glasses have "facilitated the maintenance of relevant myopia genes" is the paper in *The Quarterly Review in Biology* already quoted—which says the exact opposite.

Bottom line: Glasses have not injured the human gene pool, but preconceived ideas can cause intellectual myopia.

developed structures that bend, refract, and deflect light may help make targets "invisible" to observers. The following article notes that such a device, if ever fully developed, could have practical and military uses.

INVISIBILITY CLOAKS POSSIBLE, STUDY SAYS

Rarely, if ever, does physics news pique the interest of Pentagon brass, Harry Potter fans, and aspiring Romulans—those cloaking-device-wielding Star Trek baddies.

But a paper in tomorrow's issue of the journal Science might. In it researchers lay out design specs for materials that they say will be able to bend electromagnetic radiation around space of any size and shape.

The translation for Star Trek fans: Invisibility shields may not be science fiction for much longer.

The theoretical breakthrough is made possible by novel substances called metamaterials.

Invented six years ago, the man-made materials are embedded with networks of exceptionally tiny metal wires and loops.

The structures refract, or bend, different types of electromagnetic radiation—such as radar, microwaves, or visible light—in ways natural substances can't.

"[Metamaterials] have the power to control light in an unprecedented way," said Sir John Pendry, a theoretical physicist at England's Imperial College London.

"They can actually keep it out of a volume of space, but they can do so without you noticing that there's been a local disturbance in the light."

Theoretical Proof

The new study is by Pendry and physicists David R. Smith and David Schurig of Duke University in Durham, North Carolina.

The report explains not only how an invisibility cloak might work but also how to make one ... in theory, at least.

While their study did not produce cloaking devices, the team offers mathematical proof that the materials work, as well as technical requirements for their creation.

The underlying idea, Pendry said, is that "you can take either rays of light or an electric field or a magnetic field, and you can move the field lines wherever you want."

"So in the specific instance of cloaking, you take the rays of light, and you just move them out of the area that you don't want them to go in. ... Then you return them back to 'their' original path."

Schurig likens the effect to a rock in a stream. The rock symbolizes a metamaterial cloaking shell. The water plays the role of electromagnetic radiation flowing around the cloaking shell.

"Downstream you can't necessarily tell that there was an object distorting the flow," he said, adding that, even from the side, the disturbance is hard to discern.

In theory, planes, tanks, cars, and even entire buildings could be concealed.

"There's no limit on what you put inside," Schurig said. "If you build a cloak with a certain hold volume, you can swap things in and out of there, and it doesn't matter what they are."

But there are some catches—money, for starters.

While the raw materials (copper wire, for example) are relatively cheap, metamaterials are, for now, labor intensive and therefore expensive to manufacture.

Currently, a lab's typical output in a single go might fill a coffee cup.

Knights and Wizards

So far researchers have only developed metamaterials that divert radar and microwaves—rather than light waves, which are the key to invisibility.

While that's good news for Air Force generals who want to conceal warplanes, it's bad news for wannabe wizards hoping for a magic cloak.

Metamaterials that control visible light are particularly elusive in large part because the required matrix of metal loops and wires must be "nanosize," or exceptionally small.

That's not to say the stuff can't be manufactured. But so far no one has figured out how, says Gennady Shvets, a physicist at the University of Texas at Austin, who studies metamaterials of optical frequencies.

Of the study, Shvets said, "It was not a result that could be achieved by brute force but required some ingenuity.... I think it's great."

Pendry, the lead study author, points out another limitation. "You can't design a cloak, even in theory, that's perfect at every frequency" of electromagnetic radiation.

But the physicist, who earned a knighthood for his earlier work with metamaterials, says the cloaks should be able to work over a range of frequencies.

"There is, in fact, a trade-off between how thick you let me make the cloak and how much bandwidth I can give you," he said.

An invisibility cloak, for example, would need to be quite thick in order to bend the rainbow of colors, or wavelengths, that make up the spectrum of visible light—a broadband cloak.

"If you let me make a very thick cloak with lots of design flexibility, I can give you a broadband cloak. If you say, 'Well, I want it to be really thin,' then the more narrowband it has to be."

Stealth capabilities may get all the attention, but the researchers say there are many other applications.

"What we have here is a completely new way of controlling light and electric fields," Pendry said.

"We've thought of a few simple things, like cloaking or excluding magnetic fields. But I'd be very surprised

if those are the most important things you could do with it."

Smith, one of the Duke physicists, co-developed the first metamaterial while at the University of California, San Diego. He agrees with Pendry's optimistic forecast.

"This is just the start of what I think amounts to a lot of interesting things to come."

Sean Markey

MARKEY, SEAN. "INVISIBILITY CLOAKS POSSIBLE, STUDY SAYS." *NATIONAL GEOGRAPHIC NEWS* (MAY 25, 2006).

SEE ALSO *Physics: Articulation of Classical Physical Law; Physics: Maxwell's Equations, Light, and the Electromagnetic Spectrum; Physics: Optics; Physics: Spectroscopy; Physics: Wave-Particle Duality.*

BIBLIOGRAPHY

Books

Pedrotti, F., L.M. Pedrotti, and L.S. Pedrotti. *Introduction to Optics.* San Francisco: Benjamin Cummings, 2006.

Ronchi, Vasco. *The Nature of Light: An Historical Survey.* Cambridge, MA: Harvard University Press, 1970.

Sontag, Susan. *On Photography.* New York: Picador, 2001.

Steffens, Henry John. *The Development of Newtonian Optics in England.* New York: Science History Publications, 1977.

Periodicals

Cotter, D., et al. "Nonlinear Optics for High-Speed Digital Information Processing." *Science* 286 (1999): 1,523–1,528.

Knight, J.C. and P.St.J. Russell. "New Ways to Guide Light." *Science* 296 (2002): 276–277.

Krieger, Kim. "Lens Once Deemed Impossible Now Rules the Waves." *Science* 303 (2004): 1,597.

Markey, Sean. "Invisibility Cloaks Possible, Study Says." *National Geographic News* (May 25, 2006).

Schork N.J., L.R. Cardon, and X. Xu. "The Future of Genetic Epidemiology." *Trends in Genetics* 14; 7 (2004): 266-272.

Walmsley, Ian A., and Michael G. Raymer. "Toward Quantum-Information Processing with Photons." *Science* 307 (2005): 1,733–1,734.

Larry Gilman

Physics: QED Gauge Theory and Renormalization

■ Introduction

Quantum electrodynamics (QED) is a mathematical theory describing the interaction of electromagnetic radiation with matter. The development of QED theory was essential in the refinement of quantum field theory in the early twentieth century. QED is a fundamental scientific theory that accounts for most observed particle phenomena with extremely high precision. QED can be characterized as an extension of quantum theory to include special relativity. It is termed a gauge-invariant theory because its predictions are not affected by variations in space or time.

The practical value of QED is that it allows physicists to precisely describe the absorption and emission of light by atoms. QED also makes accurate predictions regarding the interactions between photons and charged atomic particles such as electrons. These abilities allow scientists to predict the properties of some molecules, to design highly sensitive sensors of electromagnetic fields, and to build single-atom memory-storage devices that will probably be used in future computer systems based on quantum principles.

■ Historical Background and Scientific Foundations

During the first half of the twentieth century, physicists struggled to reconcile Scottish physicist James Clerk Maxwell's (1831–1879) equations regarding electromagnetism with the emerging quantum and relativistic theories advanced by German physicist Maxwell Planck (1858–1947), Danish physicist Niels Bohr (1885–1962), German-American physicist Albert Einstein (1879–1955), and others. Prior to World War II, British physicist Paul Dirac (1902–1984), German physicist Werner Heisenberg (1901–1976), and Austrian-born American physicist Wolfgang Pauli (1900–1958) made significant independent contributions to the mathematical foundations of QED.

WORDS TO KNOW

ELECTROMAGNETISM: A form of magnetic energy produced by the flow of an electric current through a metal core. Also, the study of electric and magnetic fields and their interaction with electric charges and currents.

PHOTON: Smallest individual unit of electromagnetic radiation (light energy). These light particles are emitted by an atom as excess energy when that atom returns from an excited state (high energy) to its normal state.

RENORMALIZATION: A mathematical method in quantum physics first proposed in 1947 to deal with apparently absurd infinities in the equations of the theory known as quantum electrodynamics. In renormalization, infinities of opposite sign (positive and negative) are allowed to cancel each other out, producing a finite result that matches observation.

VIRTUAL PARTICLE: The Heisenberg uncertainty principle in quantum mechanics, named after its discoverer, German physicist Werner Heisenberg (1901–1976), states that it is impossible to know both the momentum and location of a particle at the same time with perfect precision. One of the consequences of this principle is that particles are permitted to pop into and out of existence for brief periods without violating the conservation of matter. These particles are termed virtual particles to distinguish them from permanent particles. Exchanges of virtual particles mediate the electromagnetic, weak, and strong forces.

U.S. physicist Julian Schwinger (1918–1994), famous for unifying the fundamental theories of electromagnetism and quantum mechanics in a theory of quantum electrodynamics (QED). *SPL/Photo Researchers, Inc.*

Working with QED was difficult at first, however, because infinite values kept appearing in the mathematical calculations (e.g., for emission rates or determinations of mass). Infinite values do not correspond to anything that can be meaningfully observed and may even point to places where a physical theory predicts the impossible and so fails. Thus, early predictions of particle behavior based on QED often failed to match experimental observations. Later, however, QED was rescued by a mathematical procedure termed renormalization, which allows positive infinities to cancel out negative infinities, and by other advances developed independently by American physicists Richard Feynman (1918–1988) and Julian Schwinger (1918–1994) and Japanese physicist Shin'ichiro Tomonaga (1906–1979).

The use of renormalization initially allowed QED theorists to use measured values of mass and charge in QED calculations. The result made QED a highly reliable theory with regard to its ability to predict and reflect the observed interactions of electrons and photons. QED theory was, however, revolutionary in theoretical physics because of the nature and methodology of its predictions. QED reflected a growing awareness of limitations on the ability to make predictions regarding behaviors of subatomic particles. Instead of making predictions based on mechanistic cause-and-effect interactions, QED relies on the probabilities associated with the quantum properties and behavior of subatomic particles: this allows the calculation of probabilities regarding outcomes of subatomic interactions. No one

outcome is necessarily determined: rather, there is a range of possible outcomes, having different, definite probabilities.

Renormalization was long viewed, even by Feynman and others, as a sort of sleight-of-hand trick to save QED from some hidden flaw. Upon receiving his Nobel Prize in 1965, Feynman said he thought that "the renormalization theory is simply a way to sweep the difficulties of the divergences of electrodynamics under the rug," though he could not, he added, be sure. In the end, his hunch turned out to be wrong: today, renormalization is considered a valid method in several areas of quantum physics.

As quarks, gluons, and other subatomic particles became known to physicists in the early- and mid-twentieth century, QED became increasingly important in explaining the structure, properties, and reactions of these particles. QED, also known as the quantum theory of light, eventually became one of the most precise, accurate, and well-tested theories in all science. QED-based predictions of the mass of some subatomic particles, for example, offer results accurate to six or more significant figures—one part in a million or better.

QED describes the phenomenon of light in ways that are counterintuitive (not typical of everyday experience) because it treats quantum properties of light. Considered as a quantum phenomenon rather than as an electromagnetic wave, light has certain properties that are conserved and that occur in discrete amounts called quanta. According to QED theory, light exists in a state of

American atomic physicist Richard Phillips Feynman (1918–1988). *Physics Today Collection/AIP/Photo Researchers, Inc.*

of travel) of the charged particles as they absorb or emit virtual photons. Only in this covert or hidden state, flashing between particles, do photons act as mediators of force: otherwise, they travel freely through space as light or other forms of radiation.

As virtual particles, photons are cloaked from observation and measurement: that is, they can only be detected by their effects. The naked transformation of a virtual particle to a real particle would violate the laws of physics specifying the conservation of energy and momentum (i.e., energy and momentum would emerge from below the threshold of quantum uncertainty and become a permanent part of the universe: this is not possible). Photons themselves are electrically neutral and only under special circumstances and as a result of specific interactions that preserve conservation do virtual photons become real photons observable as light.

Among other phenomena, QED accounts for the interactions of electrons, positrons (the positively charged antiparticle to the electron), and photons. In electron-positron fields, electron-positron pairs come into existence as photons interact with these fields. According to QED, the process also operates in reverse to allow photons to create a particle and its antiparticle (e.g., an electron and a positron).

QED's concept of forces such as electromagnetism arising from the exchange of virtual particles may carry profound implications regarding the advancement of theories relating to the strong, electroweak, and gravitational forces. Some physicists assert that if a unified theory can be found—a mathematical description of all known physical forces as aspects of a single, underlying force—it will rest on the foundations and methodologies established during the development of QED theory.

■ Modern Cultural Connections

Because QED is highly mathematical, only specialists with advanced, specialized schooling ever encounter its equations directly. Like some other areas of advanced physics, its cultural impact is primarily indirect.

In the early twentieth century, no practical uses for much of quantum physics, including QED, were yet imaginable: physicists pursued their understanding of these fundamental matters both from pure curiosity and from a historically founded confidence that uses would be found for such knowledge someday. In the early twenty-first century, as precise manipulation of single atoms and photons became commonplace in laboratories, QED became the basis of new sensors and computing devices. This new technological revolution—still in its early stages—depended on the refinement since the 1960s of techniques for building extremely small structures connected to complex electronic circuits on tiny chips of silicon. Such a circuit-on-a-chip is termed

particle-wave dualism (i.e., the electromagnetic wave has both particle and wave-like properties). Electromagnetism—the phenomena of electrical charge and magnetism on which much of our technology depends—results from the quantum properties of the photon, the fundamental particle responsible for the transmission or propagation of electromagnetic radiation. Unlike the particles of everyday experience, photons can also exist as virtual particles that are constantly exchanged between charged particles: indeed, all the fundamental forces of nature are understood, in quantum mechanics, as exchanges of virtual particles. The relatively familiar forces of electricity and magnetism arise from the exchange of these virtual photons between charged particles. The exchange of these virtual photons is the explanatory domain of QED. According to QED theory, virtual photons are passed back and forth between the charged particles somewhat like basketball players passing a ball between them as they run down the court. The force caused by the exchange of virtual photons results from changes to the velocity (speed and/or direction an integrated circuit because its parts are all integral to (a seamless part of) a single crystal of a semiconducting substance such as silicon or gallium arsenide.

In the late 1980s and early 1990s, integrated-circuit techniques allowed the creation of tiny superconducting (zero-electrical-resistance) cavities on chips. Methods for manipulating single atoms using lasers allowed researchers to place single atoms inside such cavities, where their interactions with photons—as described by QED—could be carefully observed and controlled. This field became known as cavity quantum electrodynamics (cavity QED or c-QED). Super-sensitive optical and microwave detectors, capable of detecting electromagnetic fields consisting of only a few photons, have been developed using cavity QED. Perhaps more importantly, in the early 2000s researchers were busily engaged in demonstrating how, using cavity QED methods, quantum information (in the form of quantum bits or "cubits") could be stored in single atoms. In 2007, researchers reported that they could imprint the state of a single photon on a single atom, force the atom to retain that information for a significant period of time, and then release the information at will in the form of a new photon. Such technologies will probably be needed in the development of future computer chips that greatly exceed today's most powerful computers in memory capacity and number of calculations per second. Just as today's

computer technology has had profound effects on society, future leaps to quantum computers with hundreds or thousands of times the power of today's machines are likely to have further effects, some unforeseen.

SEE ALSO *Physics: The Quantum Hypothesis.*

BIBLIOGRAPHY

Books

Feynman, Richard. *QED: The Strange Theory of Light and Matter.* Princeton University Press, 2006.
_____. *Quantum Electrodynamics.* Boulder, CO: Westview Press, 1998.

Periodicals

Haroche, Serge, and Jean-Michel Raimond. "Cavity Quantum Electrodynamics." *Scientific American* (April 1993): 54–62.

Web Sites

Feynman, Richard. "The Development of the Space-Time View of Quantum Electrodynamics." Nobel Prize.org. December 11, 1965. http://nobelprize.org/nobel_prizes/physics/laureates/1965/feynman-lecture.html (accessed January 23, 2008).

K. Lee Lerner

Physics: Radioactivity

■ Introduction

Radioactivity is the spontaneous breakup of the nuclei of unstable atoms, which releases radiation in the form of fast-moving particles or high-energy electromagnetic waves (gamma rays). Since the discovery of radioactivity in 1895, radiation from radioactive substances and other sources has been used for medical, military, and technological purposes. Radioactive materials are used to image the inside of the human body, to treat cancer, to power nuclear weapons and nuclear power plants, to trace chemical reactions and drug metabolism, and to determine the ages of ancient organic materials and of Earth itself (about 4.5 billion years). Radiation, regardless of its source, has harmful health effects (though in medical applications these may be outweighed by positive health effects): at low doses it may cause cancer or heritable mutations, and at high doses it sickens or kills through direct damage to body chemistry. This article considers both radioactive elements and radiation from other sources, such as x rays.

■ Historical Background and Scientific Foundations

The Discovery of X Rays

The study of radioactivity began with the accidental discovery of x rays by German physicist Wilhelm Conrad Röntgen (1845–1923) in 1895. For decades, physicists had experimented with current flow between electrodes (charged pieces of metal) inside partially airless glass tubes (cathode ray tubes, named for discharges from their positively charged electrodes, cathodes). Röntgen wanted to investigate cathode rays emitted when the pressure in the glass tube was very low. Much to his surprise, he discovered during an experiment with a cardboard-shrouded tube that, on the other side of his laboratory, a fluorescent-coated screen used to detect cathode rays began to glow. It was so far away that Röntgen doubted the fluorescence was caused by cathode rays (now known to be electrons in flight). Instead, he suspected that the fluorescence was caused by a new kind of ray.

As he investigated further, Röntgen discovered that the mysterious new rays penetrated most materials, but to different degrees. Sheets of paper were transparent to them; so were a thousand-page book and a piece of wood; aluminum, less so. The rays also blackened unexposed light-sensitive photographic plates, but thin plates of lead stopped them completely. Flesh was mostly transparent to the rays, bones mostly opaque. In his first communication about the rays, Röntgen included various photographs, including one showing the bones in his wife's hand. Röntgen called the new rays "X-rays." It took the research of many physicists over the next two decades to verify that x rays are electromagnetic waves of high energy and short wavelength—between 0.01 and 10 nanometers (nm, a billionth of a meter).

Scientists around the world immediately began to explore Röntgen's startling discovery. In February 1896, only two months after Röntgen published his first findings, research was so intense that the prestigious journal *Nature* had to declare that it could not keep pace with all the communications it was receiving on the topic.

X Rays in Medicine

Röntgen's photograph of the bones in a living human hand was splashed across the front pages of newspapers around the world, creating a public sensation. The medical profession quickly recognized x rays' potential. Because Röntgen's photograph depicted a hand clearly wearing a ring on the fourth finger, doctors saw the potential for locating metal objects (such as bullets or shrapnel fragments) in wounds; x-rays were first used for this purpose within a few weeks of their discovery. Industrial and commercial applications were also quickly found. The shoe-fitting fluoroscope, for example, which used x-rays to show the bones and soft tissues of a foot inside a shoe, became widely used in America, Europe, and Australia from the early 1920s until the end of the 1950s. This was an early example of the tragic misapplication of a poorly-understood technology: shoe-fitting x-ray machines exposed many thousands of people to cancer-causing x rays at doses that would today be considered dangerously high, while not providing any medical benefit. Indeed, they did not even improve service in shoe-stores significantly, but served primarily as a gimmick to impress customers. Children using the machine were exposed in a few seconds to as much radiation as a present-day industrial worker is allowed to receive in a year—and children are more vulnerable to radiation than adults. Moreover, the machines remained in use long after the harmful effects of radiation began to be understood, not being phased out in the United Kingdom until the mid-1970s.

Doctors soon discovered that when x rays were aimed at the human body, the skin often reddened in response, as if it had been sunburned. Since this was the same effect produced by ultraviolet light, which was used to treat various skin conditions, they assumed that x rays would have the same beneficial effects. X ray treatments were quickly introduced. When hair loss was noticed after treatment with radiation, many physicians began to use x rays as a depilatory (hair-removing) treatment. Doctors also experimented with x ray therapy on cancer patients, and early reports of successful healing led to a veritable boom in x ray therapy in the first years of the twentieth century. Unfortunately, the indiscriminate use of high-dose x rays probably caused far more cancer than it cured during this period.

X Rays' Harmful Effects

X rays' harmful effects quickly became apparent, however. At first the burns, rashes, dermatitis, and ulceration associated with x ray use were ascribed to apparatus malfunction, but Elihu Thomson (1853–1937), an engineer working for an x-ray machine manufacturer, doubted this. To prove that the x rays, and not his products, were responsible, he conducted a series of experiments on himself. By irradiating one of his fingers, he showed that x rays could produce severe, painful burns. He concluded that exposure to x rays beyond a certain limit would cause harm and warned his colleagues not to prolong exposure. U.S. inventor Thomas Edison (1847–1931) abandoned x ray research in 1903 after one of his assistants contracted fatal cancer in his hands by testing x ray tubes with them. By experimenting on guinea pigs, other scientists showed that x rays

---

WORDS TO KNOW

FISSION: Splitting or breaking apart. In biology, the division of an organism into two or more parts, which each produce a new organism. Nuclear fission is a process in which the nucleus of an atom splits, usually into two daughter nuclei, with the transformation of tremendous levels of nuclear energy into heat and light.

HALF-LIFE: The time it takes for half of the original atoms of a radioactive element to be transformed into the daughter product.

NEUTRON: A subatomic particle with a mass of about one atomic mass unit and no electrical charge that is found in the nucleus of an atom.

NUCLEUS: Any dense central structure can be termed a nucleus. In physics, the nucleus of the atom is the tiny, dense cluster of protons and neutrons that contains most of the mass of the atom and that (by the number of protons it contains) defines the chemical identity of the atom. In astronomy, the large, dense cluster of stars at the center of a galaxy is the galactic nucleus. In biology, the nucleus is a membrane-bounded organelle, found in eukaryotic cells, that contains the chromosomes and nucleolus. Intact eukaryotic cells are comprised of a nucleus and cytoplasm. A nuclear envelope encloses chromatin, the nucleolus, and a matrix, which together fill the nuclear space.

PROTON: Particle found in a nucleus with a positive charge. Number of these gives atomic number.

RADIATION: Energy transmitted in the form of electromagnetic waves or subatomic particles.

RADIOACTIVITY: The property possessed by some elements of spontaneously emitting energy in the form of particles or waves by disintegration of their atomic nuclei.

---

could blind, burn, cause abortions, or even kill; in light of these findings, shields began to be used to protect both patients and x ray operators. The growing recognition of x rays' harmful effects showed the need for safety guidelines.

The Discovery of Radioactive Elements

When Röntgen's discovery was first discussed at the French Academy of Sciences, members suggested that since the tubes emitting x rays were fluorescent, other fluorescent bodies might also emit the new rays. To test this hypothesis, the French physicist Henri Becquerel (1852–1908) conducted an experiment in which he placed sheets of uranium salt on a photographic plate that was wrapped in heavy black paper. This was placed in the sun for several hours, until the uranium salt

Scientist Marie Curie works in a laboratory. Curie, along with her husband, Pierre, first isolated the two highly radioactive elements-radium and plutonium-from uranium ore. *AP Images.*

became fluorescent. Upon developing the photographic plate afterward, Becquerel discovered silhouettes of the mineral crystals and concluded that the fluorescent uranium salt emitted radiation that penetrated paper.

In the days following his initial experiment the sun appeared only intermittently. While waiting for better weather, he placed his wrapped photographic plates and uranium crystals in a drawer. After a few days he developed the photographic plates, expecting to find only very weak images. But much to his surprise the images were quite intense. After experimenting with various uranium salts and other fluorescent minerals for several weeks, he concluded that the rays were emitted by the uranium. We now know that it was merely a coincidence that uranium salts emit visible light rays (fluoresce) under some conditions as well as emitting penetrating radiation: The fluorescence has nothing to do with the penetrating radiation.

Expanding on Becquerel's work, Marie Curie (1867–1934) discovered in 1898 that x rays were also emitted by thorium. Curie and her husband, Pierre Curie (1859–1906), also found that the mineral pitchblende was much more radioactive than its uranium content would indicate. They discovered that it contained another element even more radioactive than uranium, which they named polonium, in honor of Poland, Marie Curie's native country. A few months later they found that pitchblende contained yet another highly radioactive element, a substance they named radium. Marie Curie eventually died from a blood disease almost certainly caused by radiation exposure. In 1899, the French physicist and chemist André-Louis Debierne (1874–1949) discovered a third radioactive substance, actinium.

Early Investigations

The early study of radioactivity was experimental, focusing on the collection and classification of data as scien-

HENRI BECQUEREL 1852–1908)

Henri Becquerel (1852–1908) was born into a scientific family. Both his father, Alexandre Edmond Becquerel (1820–1891), and his grandfather, Antoine César Becquerel (1788–1878), were professors of physics at the Museum of Natural History in Paris. Each also served as president of the French Academy of Sciences.

Henri was educated at the École Polytechnique and the École des Ponts et Chaussées (Bridges and Highways School). Even before graduating he was teaching at the École Polytechnique; after graduation he became a government engineer in the Bridges and Highways Department. He also became an assistant at the Museum of Natural History, dividing his professional life among three institutions.

When the possibility of radiation from fluorescent bodies became a topic of discussion in the French Academy, Henri was

43 years old, settled and established, beyond the dogged pursuit and hard work of basic research. The study of fluorescence in uranium compounds fascinated both him and his father, however. Working with an inherited collection of uranium salts, he immediately took up this new field of investigation.

His discovery that uranium salts emitted x rays did not create the same excitement as Röntgen's discovery. During this time scientists studied numerous kinds of radiation—cathode rays, canal rays, x rays, and others; those discovered by Becquerel did not seem especially important at first. This changed when scientists learned that they were emitted not only from uranium, but from several different elements as well. In 1903 he shared the Nobel Prize for physics with Marie and Pierre Curie for their discovery of radioactivity. The International System unit of radioactivity, the becquerel, is named in his honor.

tists tried to answer many questions: What were these new rays? Were they emitted by all elements, or only by some? Was their activity affected by chemical processes, or by physical changes such as temperature? How did it fit into the periodic table of the elements?

During his first experiments, Becquerel noticed that the new rays would cause nearby electrically charged materials to lose their electrical charges. Unlike photographic plates, which provided only qualitative measurements, the discharge of an electroscope could quantify the radiation's intensity. He discovered that although the rays could penetrate paper, sheets of aluminum or copper decreased their intensity. In a series of similar absorption experiments, the New Zealand–born British physicist Ernest Rutherford (1871–1937) showed that the rays emitted from uranium contained at least two distinct types of radiation. The first, which was readily absorbed, he termed alpha radiation; the other, more penetrating type he termed beta radiation. In 1900, the French physicist Paul Villard (1860–1934) found a third type that was even more penetrating than beta radiation and was termed gamma radiation. Alpha (α), beta (β), and gamma (γ) are the first three letters of the Greek alphabet and are often used in science to denote quantities of interest.

Rutherford conducted a series of experiments in which he measured the decay rate of various radioactive substances. He found that they all decayed according to an exponential law, $I = I_o e^{-Lt}$—that is, that the radioactivity of a sample always decreased by the same percentage over an equal interval of time. The amount of time it took a sample's radioactivity to decrease by half (50%) Rutherford dubbed the "half-life" (i.e., after that much time, half the life of the sample's radioactivity was over). He also found that different substances have different half-lives, ranging from many thousands of years to a few seconds. These differences could be used to distinguish radioactive substances.

Discovering Transmutation

In 1900, Rutherford discovered that the element thorium, in addition to being radioactive, also emitted highly radioactive particles that traveled about the laboratory like a gas. Not knowing what it was, Rutherford called it "thorium emanation." With the help of a young British chemist and future Nobel laureate named Frederick Soddy (1877–1956), Rutherford hypothesized that thorium was transmuting into the gaseous element argon. This hypothesis was strengthened in 1902 when Rutherford showed that the emission of alpha rays changed the elemental identity of the source atom. It now seemed certain that one chemical element could, in fact, transmute into another.

Radiation's various components emerged gradually. By 1902 it had become clear that beta rays are, in fact, fast-moving electrons. In 1905, British physical chemist William Ramsay (1852–1916) and Soddy discovered that helium was produced by the radioactive decay of radium, and in 1908 Rutherford and the German physicist Hans Geiger (1882–1945) concluded that alpha rays consist of helium atoms that have gained a positive charge (i.e., lost their electrons: thus, an alpha particle is a fast-moving helium nucleus). Gamma rays were finally established as the same type of radiation as x rays, only with even higher energy. Despite this progress in establishing experimental data, scientists found it difficult to create an atomic model that could explain radioactive decay.

Harmful and Therapeutic Effects

It soon became clear that radioactive substances and x rays affected the skin in similar ways. In June 1901 Becquerel found that after carrying a tube of radium in his shirt pocket for a couple of hours, he developed a skin burn that took almost two months to heal. Pierre Curie attached a piece of radioactive material to his arm for 10 hours, resulting in a wound that took about four months to heal and left a heavy scar.

When the doctor who treated Becquerel's burn noticed that it resembled an x-ray burn, he suggested that radium might have therapeutic effects similar to those of x rays. Radium treatments for cancer were quickly introduced, and a variety of applicators were developed to

IN CONTEXT: N RAYS

As new kinds of radiation were discovered, claims also emerged for rays that turned out not to exist. In 1903 the noted French physicist René-Prosper Blondlot (1849–1930), working at the University of Nancy, France, thought he had discovered a new kind of ray that was emitted not only from discharge tubes, but also from gas burners, metals in states of strain, and from stretched muscles and the human nervous system! He called them n rays, for the University of Nancy.

After his discovery, several other scientists claimed also to have observed n rays from various animal and vegetable substances. However, these results could not be reproduced, and many in the scientific community began to doubt their existence. When American physicist Robert W. Wood (1868–1955) visited Blondlot's lab to investigate these findings, he secretly removed and replaced parts of the apparatus, effectively rendering the experiment invalid. Blondlot and his colleagues, however, still claimed to observe n rays—solely because they expected to. This episode became a classic warning against matching results to expectations. In 1904, the editors of the scientific journal *Nature* concluded that n rays were an illusion.

use the element in body cavities where x-ray treatment was difficult.

Popular hope for miraculous effects from radioactivity soared during the 1920s and 1930s. Inhaled radon gas, for example, was thought to have stimulating and restorative powers. So was the drinking of water from decorated urns in which radioactive substances were submerged. Likewise, since it was believed that low levels of radiation would kill germs and stimulate growth, radioactive materials were often used in products such as beauty creams. An unknown but certainly large number of cancers were caused by these useless exposures to radiation.

Induced Radioactivity

A new realm opened up in the years prior to World War II (1939–1945) when researchers discovered how to induce radioactivity in heavy elements by bombarding them with neutrons. Within just a few years, research on induced radioactivity lead to the discovery of nuclear fission, quickly followed by speculations about how the energy released during nuclear fission could be utilized.

When research on induced radioactivity began, however, no one considered the possibility that a nucleus

could be split. In 1934 the husband and wife team of Irène and Frédéric Curie-Joliot (1897–1956 and 1900–1958) discovered that when bombarding light elements with alpha particles, they would transmute into isotopes of known elements; contrary to the known, stable elements, however, these new isotopes were radioactive.

Because of the alpha particles' positive charge, which caused it to be repelled by atomic nuclei (which are all positively charged, the heavier the nucleus the more the charge), Curie and Joliot were able to induce radioactivity only in relatively light elements. But with the discovery of the electrically neutral neutron, Italian physicist Enrico Fermi (1901–1954) hypothesized that neutron bombardment might lead to the activation of heavy elements. To test their theory, Fermi and his collaborators began a series of experiments in which various elements were bombarded with neutrons. They found that for a large number of elements of any atomic weight, neutron bombardment produced unstable elements that disintegrated through the emission of beta particles.

Creating Elements Heavier than Uranium

In beta decay, the parent nucleus emits an electron and an antineutrino while transforming a neutron into a proton. The daughter nucleus is therefore lighter and of a higher atomic number than the parent. (The atomic number of an element is the number of protons in its nucleus.) During their neutron-bombardment experiments, Fermi's team turned their attention to heavy nuclei, especially uranium, which was element number 92 and the last element in the periodic table as it was then known. They wondered whether new elements with higher atomic numbers than uranium could be produced. This question opened the possibility that the list of chemical elements was not exhaustive—that new elements could be produced artificially.

Fermi's team bombarded uranium with neutrons and identified several beta-emitting products, one of which they found to be distinct from all elements heavier than lead. They concluded that it might be a new element, one with a higher atomic number than uranium that would surpass it in the periodic table; Fermi's team called it a "transuranium" element.

This was a daring hypothesis, since the last known gaps in the periodic table were just being filled. Many scientists accepted Fermi's contention that this was a new element, but the German chemist Ida Noddack (1896–1978) objected, saying that no conclusion could be drawn until the substance had been compared to all known elements, not only to those between lead and uranium. Fermi's team had not done this because they expected that parent and daughter nuclei would be close to each other in the periodic table. Noddack argued, however, that since nothing was known about neutron-

IN CONTEXT: THE CURIE FAMILY

Marie Curie (1867–1934) was born Maria Sklodowska in Poland. Her father taught mathematics and physics in a secondary school and her mother, who died when Marie was 11, managed a boarding school for girls. When she was 18, Marie worked as a governess to support her younger sister Bronia's medical education in France. In 1891 Marie moved to Paris to study physics and mathematics at the Sorbonne. She graduated in 1894, and in the same year she met Pierre Curie (1859–1906), a newly minted doctorate in physical chemistry. They married in 1895 and had two daughters, Irène (1897–1956) and Ève. Pierre was appointed professor of physics at the Sorbonne in 1904, and when he died two years later, the chair was bestowed on Marie, who became the first women to teach at the Sorbonne.

With Henri Becquerel (1852–1908), Marie and Pierre Curie shared the 1903 Nobel Prize for physics for their discovery of radioactivity. Marie Curie received a second Nobel Prize for chemistry in 1911 for her discovery of the elements radium and polonium. Her daughter Irène Curie-Joliot and son-in-law Frédéric Joliot-Curie (1900–1958) won the 1935 Nobel Prize for chemistry for their synthesis of new radioactive elements. Pierre Curie died in 1906 in a traffic accident; Marie died in 1934 from aplastic anemia, a disease certainly caused by her exposure to radiation.

induced transmutations, it was possible that the nucleus would simply fragment into much lighter elements than the original uranium. Her objections received scant attention, partly because her idea of the nucleus breaking apart seemed merely speculative and partly because she had made controversial claims of her own that had undermined her scientific credibility. However, on this point her guesses were approaching an important truth: atomic nuclei can fission (split apart).

Fermi's 1934 announcement piqued the interest of other scientists. The Curie-Joliots began similar experiments in Paris. In Berlin the physicist Lise Meitner (1878–1968) and chemists Otto Hahn (1879–1968) and Fritz Strassmann (1902–1980) competed to find new transuranic elements. Both groups conducted numerous experiments. Each produced results that were interpreted as transuranic, but most were complex products that required a variety of new hypotheses to be understood.

In two papers published in 1938, the Paris group claimed to have produced yet another transuranic element by neutron bombardment of uranium. They had difficulty identifying it, however, as it behaved more like a lanthanide (rare earth element) than any of the transuranics. The Berlin team immediately began to test the French results, but Meitner's participation was compromised when she had to flee Nazi Germany. Meitner was Jewish and had been protected up to that point by her Austrian citizenship. After Germany annexed Austria in 1938, however, she was subject to persecution by the Nuremberg Laws.

The Discovery of Nuclear Fission

Hahn and Strassmann stayed in Berlin, but corresponded with Meitner about their ongoing research. They found that the element produced by the Paris group was a mixture of several isotopes. In a famous experiment conducted on December 17, 1938, they were able to confirm the presence of barium, a lighter element than uranium. In a series of letters to Meitner, Hahn concluded that although it seemed impossible according to the laws of physics, as a chemist he had to conclude that their results indicated that the nucleus had been divided. He asked Meitner if she could figure out any other explanation.

Over Christmas, Meitner discussed Hahn and Strassman's results with her nephew, physicist Otto Frisch (1904–1979). Together they deduced that violent oscillations of the nucleus could indeed split it; within weeks of Hahn and Strassman's experiment, they published their discovery of nuclear fission.

This appeared to invalidate all findings of transuranic elements. As they continued to correspond in January and February 1939, Hahn was reluctant to relinquish the possibility of transuranic elements; Meitner,

ENRICO FERMI (1901–1954)

Enrico Fermi (1901–1954) was born in Rome. His aptitude for mathematics and physics, recognized and encouraged even in grammar school, earned him a university scholarship. After earning his doctorate degree in physics from the University of Pavia in 1922, he first went to Göttingen to work with Max Born (1882–1970) and later to Leiden, the Netherlands, to work with Paul Ehrenfest (1880–1933).

In 1927, Fermi became professor of theoretical physics at the University of Rome. When the fascist Mussolini government enacted anti-Jewish laws similar to Germany's Nuremberg Laws, Fermi, his Jewish wife Laura, and their two children emigrated to the United States in 1938. There Fermi joined the group of immigrant physicists, including Albert Einstein (1879–1955) who urged the Roosevelt administration to begin developing an atomic bomb to counter the ominous possibility that Germany would do so first.

Fermi went to Washington to present the idea in more detail to a group of officers. The initial meeting created little interest, but the growing Nazi threat eventually moved even Einstein, antiwar by conviction, to encourage work on a nuclear weapon. When the Manhattan Project was launched in 1942, Fermi became one of the lead physicists on the project.

on the other hand, wanted to reinterpret all previous results in the light of their new discovery. Meitner's logic triumphed, and by the beginning of 1939 they had retracted all their previous results.

With the discovery of fission, a number of questions surfaced. Splitting a heavy nucleus into two light nuclei released both neutrons and energy. If the released neutrons could cause further nuclei to split, a continuous chain reaction might occur that would release an enormous amount of energy. Within a year after the discovery of fission, research papers were discussing how the energy content of atomic nuclei could be made technically useful.

Nuclear Fission in Wartime

These speculations took on a new dimension with the outbreak of World War II (1939–1945), as scientists realized that a chain nuclear reaction could be used in an immensely powerful new weapon. Physicists who had fled from Europe to the United States were so worried about the prospects of a German atomic bomb that they urged government to keep all uranium research secret. From 1940 onward, British and American physicists agreed to stop publishing their work on nuclear energy. They could submit papers to the journals, but they would not be published until it was considered safe.

A number of physicists were so fearful that Germany would develop a nuclear weapon that they urged Albert Einstein (1879–1955) to share their concerns with President Roosevelt in the summer of 1939. In a series of letters between 1939 and 1940, Einstein warned the president that secret German nuclear research had ominous implications, and he encouraged the United States to take action to prevent them from developing such a bomb first. Although American physicists, chemists, and engineers began to work on uranium chain reactions, not until the Japanese attack on Pearl Harbor in December 1941 did their work begin to expand into the large-scale program needed to design and build an atom bomb. Roosevelt authorized the massive funds necessary to launch the Manhattan Project in 1942. (Neither Germany nor Japan, as it turned out, ever had a serious atomic-bomb program.)

In the spring of 1943, a huge research laboratory was set up for the Manhattan Project at Los Alamos, New Mexico. One problem the team faced was how to assemble a critical mass of fissionable material that would allow a chain reaction to occur. A critical mass of a radioactive material is a properly-shaped sample of material heavy enough, that is, containing enough atoms close enough together, to sustain a chain reaction driven by its own radioactivity. In a chain reaction, the neutrons emitted by some atoms will trigger the breakup of other atoms. In a bomb, the result is the breakup of so many atoms so rapidly as to constitute an explosion.

In the end, two methods of producing a critical mass at the desired moment were pursued: In the gun-type (uranium) bomb, one subcritical mass was shot into another to suddenly produce a critical mass and thus an explosion. In the implosion-type (plutonium) bomb, a spherical subcritical mass was surrounded by a chemical explosive. When detonated, the chemical explosive caused the plutonium to implode into a critical mass. The United States dropped "Little Boy," the gun-type uranium bomb, on the Japanese city of Hiroshima on August 6, 1945, killing approximately 140,000 people; it dropped the implosion-type "Fat Man" bomb on the city of Nagasaki on August 9, 1945, killing approximately 74,000 people.

After the end of World War II, nuclear reactors were developed to generate nuclear energy for civilian purposes, especially the generation of electricity. The nuclear power industry grew rapidly in the late 1950s as commercial power plants were built in the United States and other countries, subsidized by governments. Unexpectedly high costs, slower-than-forecast increases in demand for electricity, and (to a lesser extent) concerns about reactor safety and radioactive waste disposal later slowed this growth in the United States and most other countries; some, like Austria and Denmark, prohibited the establishment of nuclear power plants altogether.

A cleanup worker sprays numbers on the lids of drums containing radioactive debris. *AP Images/The Ottawa Daily Times.*

Radioactive Elements as Biological Tracers

Irene and Frédéric Joliot-Curie won the 1935 Nobel Prize for their synthesis of new radioactive isotopes of light elements. In his Nobel lecture, Frédéric Joliot-Curie noted that these isotopes could be used to detect physical and chemical processes in the body. The use of radioactive isotopes as medical tracers was pioneered by the Hungarian-born Swedish physicist Georg von Hevesy (1885–1966) when he lived in a boarding house. Suspicious that leftovers from the lodgers' plates were being recycled at later meals, he deposited a tiny amount of a radioactive isotope on a piece of meat that he left on his plate. The next day he brought an electroscope to the table and discovered that a supposedly fresh meat-containing dish was indeed radioactive.

By the same token, radioactive isotopes can be used to trace an element's progress through a body and its various physiological processes. With the use of the radioactive tracers, Hevesy revealed the principle of metabolic turnover, that is, that a substance administered to a living organism would leave the organism again, and that this turnover could be expressed by a decaying exponential function with a characteristic biological half-life for each element.

An important part of this discovery was the realization that some elements deposit selectively in different parts of the body, leading to searches for radioactive "magic bullets" to treat cancer and other diseases localized in particular organs. One of the most useful elements for this kind of treatment is iodine, which, when ingested, travels through the bloodstream to the thyroid gland. Patients with hyperthyroidism and some types of thyroid cancer could now be treated by administering a dose of radioactive iodine; this would concentrate in the thyroid gland where it would destroy the cells. However, most diseases are not treatable by this means.

Modern Cultural Connections

The application of radioactivity to the creation of nuclear weapons transformed the physical and psychological terms of modern life. With the proliferation of nuclear weapons in the 1950s and beyond—over 20,000 still remain in national arsenals, about 95% in the United States and Russia—the possibility that most of the human race could be extinguished in a sudden, self-inflicted disaster became real for the first time in history and has remained so. For the early nuclear nations, security seemed to inescapably depend on the possession of many thousands of nuclear weapons, carried at first by airplanes and later mounted on long-range missiles. Thus, since the end of World War II, whole societies have committed their sense of security to the maintenance of nuclear-weapons systems that can kill hundreds of millions or billions of people in a matter of minutes, and for half a century most persons have lived with the awareness (sometimes acute, sometimes muted) that not only one's own life but the lives of everybody that one knows could be ended suddenly, possibly even by accident, without any meaningful prospect of continuation in future generations.

In reaction to this nightmarish situation, early eagerness to declare that "the atom" (i.e., radioactive phenomena) has a peaceful, salvific (redeeming) side was strong. Promoters of nuclear power and medical technologies sometimes promised a utopian future where all problems would be solved by The Atom. In the first few decades after the development of nuclear power and weapons, public enthusiasm for peaceful nuclear energy was high. In the 1970s and beyond, public opinion in the United States and elsewhere became divided, with significant minorities or, in some countries, large majorities opposing the development of nuclear power and continued reliance on nuclear weapons.

The ability of low levels of radiation to cause cancer and genetic mutations in offspring has been a key part of the public debate over nuclear power. Radiation that separates electrons from atoms, a process termed ionization, is called ionizing radiation. X rays and the forms of radiation emitted by radioactive elements are ionizing. In large doses, ionizing radiation can ionize so many atoms in a cell that the resulting chemical reactions cause the cell's death. Or, ionizing radiation can cause cancer or heritable mutations by damaging the DNA (deoxyribonucleic acid) molecules present in almost all cells; these molecules contain instructions for day-to-day cell biochemistry and for producing offspring. Changes to DNA can make a cell cancerous or be inherited as mutations (usually harmful).

In the years immediately following Word War II, most analysis of long-term radiation effects on health was based on studies of the populations in Hiroshima and Nagasaki, who were exposed to a sudden burst of ionizing radiation from the atomic bomb. Other data were obtained from laboratory experiments on animals and from the results of human exposure to medical radiation. The Chernobyl disaster of April 1986 produced important data on effects of low-level radiation. The disaster occurred when operators shut off cooling systems for a power-generating nuclear reactor in Chernobyl, Ukraine. The reactor core melted down, causing steam explosions that blew open the containment building. The burning reactor then spewed radioactive materials into the atmosphere for days, out of control, dispersing large amounts of radioactive material, which was carried over much of Europe and Scandinavia by prevailing winds. The city of Chernobyl has been abandoned ever since the accident.

Among the approximately 6.6 million people exposed to Chernobyl, the World Health Organization estimated that as many as 50,000 new cases of thyroid cancer will develop. Such diagnoses have risen at least tenfold among Ukrainian children who were exposed to Chernobyl's radiation. Those who were under four years of age when the disaster occurred have a nearly 40% risk of developing the disease; those who were under two tend to develop particularly virulent forms. Because children affected by Chernobyl also produce more antithyroid antibodies than others, they may be at greater risk for hypothyroidism in later life.

In the early decades of scientific research on radiation and radioactivity, most scientists assumed that a tolerance dose could be established below which exposure to radiation was completely safe; the countervailing theory was that even low doses would cause harm (e.g., halving a dose would cause half the harm, all the way down to zero). These two schools of thought about radiation risk were at odds throughout the twentieth century. In 2006, after exhaustive review, the U.S. National Research Council Committees on the Biological Effects of Ionizing Radiation, a government body, stated definitively that the linear model (harm is proportional to dose down to zero) is best supported by data.

Between 1989–1990, a notable increase in genetic malformations in animals born in Russia near the Chernobyl nuclear disaster was noted, in particular in calves and pigs. The following year, almost 400 deformed animals were born, but they only lived for a few hours. In 1990, Igor Kostin took photographs of these mutations, including this eight-legged foal, and sent them to Mikhail Gorbachev; he received no reply from the Soviet leader.   *© Igor Kostin/Sygma/Corbis.*

Debate over the advisability of continuing or increasing reliance on nuclear power turns partly on the question of whether the large inventories of radioactive substances generated by nuclear power plants can be contained. Advocates of nuclear power point out that accidents such as Chernobyl have been rare, given the number of operating nuclear power plants. In addition they maintain that air pollution generated by burning coal to make electricity has killed many times more people than nuclear power. Opponents of nuclear power argue that reactors remain dangerous; that increased reliance on nuclear power would entail dependence on breeder reactors that are even more dangerous; that terrorists or military attackers could cause Chernobyl-like disasters by bombing even safe reactors; and that the containment of nuclear waste over hundreds or thousands of years cannot be guaranteed, subjecting hundreds of future generations to unknown risk for the sake of a short-term benefit (electricity).

Both opponents and proponents of nuclear power agree that the technology needed to produce nuclear power can be modified or exploited to produce nuclear weapons: for example, in the early 2000s, there was intense international debate over whether Iran's nuclear-power program was being secretly exploited by that country to produce nuclear weapons, as the supposedly peaceful reactor programs of Israel, India, North Korea, and Pakistan had previously been exploited. International inspectors of Iran's nuclear facilities had, as of early 2008, failed to produce definite physical evidence of a bomb program, though the controversy about their potential to do so continued.

In response to scientific evidence that even low levels of radiation cause harm, medical equipment that uses x rays and other forms of ionizing radiation has been reengineered over the decades to minimize the received dose. Mammography (breast x rays), for example, first developed around 1950, was improved in the 1970s to lower the radiation dose. The dose received by healthy tissue during radiation therapy has also diminished greatly. Stereotactic radiation therapy, in which radiation beams are focused at a localized tumor from hundreds of different angles for a short period of time, can deliver a precise, significant dose of radiation to a small tumor, while the surrounding tissue receives comparatively small doses.

■ Primary Source Connection

The disposal of radioactive wastes is one of the most controversial problems plaguing most forms of nuclear technology. In this article by Demetria Kalodimos, the discovery that radioactive waste is being placed in a landfill meant for household trash in Murfreesboro, Tennessee, is discussed.

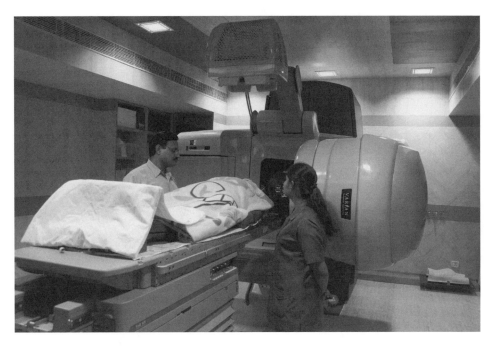

Indian technical medical staff in Ahmedadad, India, attend to a patient treated with South Asia's first Image Guided Radiation Therapy linear accelerator in November 2006.   *AFP/Getty Images.*

EXPERT: RUNOFF AT LANDFILL TESTS "VERY HIGH" FOR RADIOACTIVITY: SOME NUMBERS TWICE WHAT EPA ALLOWS

MURFREESBORO, Tenn.—The first test results are in for radioactivity at Murfreesboro's Middle Point landfill.

Channel 4 uncovered a little known state program that allows low-level radioactive waste from all over the country to be buried along with household trash at the landfill.

It's been happening for nearly 20 years, but the state has never required testing to monitor the effects of the dumping.

On a rainy Sunday afternoon, Channel 4 rode to the top of the Middle Point landfill to see what a state scientist could measure with a hand-held radiation detector.

Somewhere in the mountain of dirt and garbage, millions of pounds of low-level radioactive waste have been buried along with the household trash.

"If we stood around here all year, or anywhere else, with 11 micro r per hour, we would get about 96 millirem for the year," said state radiological inspector Billy Freeman.

The numbers sound complicated, but they're no higher or lower than you'd expect to find in this area from decaying rock, soil, the sun, what the scientists call "natural background."

The state said the added risk, even from tons of processed radioactive waste at the landfill, is minimal.

"(It is) almost inconsequential. It would be 1 percent of a member of the public's limit. One millirem per year is an inconsequential dose. Is it a realistic dose? Yes, it is a very realistic dose and that's why we chose it," said Freeman.

But even the state's radiation expert admits Channel 4's walking tour was less than realistic in terms of what might be going on deep under our feet.

"We did a rough and dirty walkover with a portable survey instrument. It's a very sensitive survey instrument, and I trust its readings. It's a calibrated device, so these are accurate readings. If your question is what is two feet down, six feet down, 100 feet down, in no way did our survey today give you an estimation of that, no," said Freeman.

This brings Channel 4 to another set of tests.

In the nearly 20 years Tennessee has allowed treated low-level radioactive waste to be dumped here, there has never been a requirement to test the air, water or soil for radioactivity.

The very first tests on the liquid runoff from the landfill, what's called leachate, paint a potentially troubling picture.

"The readings are very, very, very high," said Dan Hirsch, Expert of Nuclear Policy at the University of California at Santa Cruz.

Gross Alpha radiation in the leachate measured 82. The EPA standard for drinking water is 15.

Gross Beta, the leachate, measures 3,395. This is 68 times higher than the maximum allowed in drinking water.

Tritium, a radioactive element that attaches itself easily to water, measured at more than 38,000. This number is nearly twice what the Environmental Protection Agency allows.

It is true that no one drinks leachate, so is drinking water a fair comparison?

"It is relevant because landfills do leak," said Mark Quarles, ground water expert.

A professor of nuclear policy compared the numbers to a 2002 survey of 50 landfills in California.

"The gross beta readings are just astronomical. I've not seen radiation readings that high for leachate. The monitoring that we had in California suggested we had a problem. Although the highest reading we had was eight times lower than the reading reported in Tennessee," said Hirsch.

They also looked at waste-water sludge, comparing Middle Point's to a landfill that's not taking radioactive waste in Clarksville.

"When you look at the sludge of Murfreesboro to Clarksville, the gross Beta radiation is 9.5 times higher than that of Clarksville. The Tritium is 139 times higher than that of Clarksville," said Quarles.

Tennessee officials said a lot of everyday, careless disposal could be to blame.

Tritium, for example, is found in old illuminated exit signs, the kind that light up without a power source and quite often end up in landfills.

But Tritium, or Hydrogen 3, was also in loads of contaminated soil taken from Middle Point landfill by officials from the University of California at Los Angeles almost six years ago at the rate of 400 tons per month.

What's causing the problem? Is there a problem?

Without baseline testing, it may be impossible to know.

"Now we're so far into it they don't have the baseline, and the way you're supposed to do it is you're supposed to sample for all these constituents before you place the waste," said Quarles.

The state is still waiting for results of testing at the four other landfills in Tennessee that are accepting low-level treated radioactive waste.

A state advisory board will meet Thursday to discuss more testing and the 60 day moratorium the legislature has put on the dumping program.

Demetria Kalodimos

KALODIMOS, DEMETRIA. WSMV-TV. "EXPERT: RUNOFF AT LANDFILL TESTS 'VERY HIGH' FOR RADIOACTIVITY," JULY 4, 2007. HTTP://WWW.WSMV.COM/NEWS/13620876/DETAIL. HTML (ACCESSED OCTOBER 4, 2007).

SEE ALSO *Physics: Nuclear Physics; Physics: The Inner World: The Search for Subatomic Particles.*

BIBLIOGRAPHY

Books

Kevles, Bettyann Holtzmann. *Naked to the Bone: Medical Imaging in the Twentieth Century.* New Brunswick, NJ: Rutgers University Press, 1997.

Kragh, Helge. *Quantum Generations: A History of Physics in the Twentieth Century.* Princeton, NJ: Princeton University Press, 1999.

Romer, A., ed. *The Discovery of Radioactivity and Transmutation.* New York: Dover, 1964.

Walker, J. Samuel. *Permissible Dose: A History of Radiation Protection in the Twentieth Century.* Berkeley: University of California Press, 2000.

Web Sites

American Institute of Physics. "The Discovery of Fission." http://www.aip.org/history/mod (accessed May 8, 2008).

American Institute of Physics. "Marie Curie and the Science of Radioactivity." http://www.aip.org/history/curie (accessed May 8, 2008).

U.S. Department of Energy. "Human Radiation Experiments." http://www.eh.doe.gov/ohre/index.html (accessed May 8, 2008).

U.S. Environmental Protection Agency. "History of Radiation Protection." http://www.epa.gov/rpdweb00/understand/history.html (accessed May 8, 2008).

U.S. National Research Council. "BEIR VI: Health Risks from Exposure to Low Levels of Ionizing Radiation." http://dels.nas.edu/dels/rpt_briefs/beir_vii_final.pdf (accessed February 5, 2008).

Hanne Andersen

Physics: Semiconductors

■ Introduction

Semiconductors are materials that form crystal solids (such as silicon and gallium arsenide) and have properties between those of insulators and conductors. A typical semiconductor's conductivity (the degree to which it conducts electricity) depends on imperfections in the material (such as lattice defects), added impurities, and external conditions (such as applied electric fields). Man's increased ability to control the degree of conduction in semiconductors lies at the heart of the information revolution of the late twentieth and early twenty-first centuries.

■ Historical Background and Scientific Foundations

Early Work

Although the term semiconductor was first used around 1911, the properties of materials that we now call semiconductors had been studied beginning about 80 years earlier. The English physicist and chemist Michael Faraday (1791–1867) did the earliest known work during the 1830s with his research on silver sulfide, the conductivity of which increased as it was heated (exactly the opposite of most other materials such as metals). In 1873 the English engineer Willoughby Smith (1828–1891) published a paper on what we now call photoconductivity, showing that selenium's conductivity increased when light shined on it. Three years later, professor of natural philosophy William Grylls Adams (1836–1915) and his student Richard Evans Day discovered the photoelectric effect when they reported that selenium also produced electricity when light was shined on it. The German physicist Karl Ferdinand Braun (1850–1918) published a series of papers on the electrical behavior of the junctions that were formed when thin metal wires were pressed against the surfaces of metallic sulfides. He found that the current running through the devices depended on the polarity of the voltage across them, making him the discoverer of the rectifying contact, or diode.

The search for semiconducting materials grew during the early twentieth century. After testing thousands of materials, American engineer Greenleaf Whittier Pickard (1877–1956) found a way to detect radio signals using a slender silicon carbide wire, which he called a "cat's whisker," pressed against the surface of a silicon crystal (similar to the devices tested by Ferdinand Braun). In 1906 Pickard patented his crystal detector, which became a basic component of early radio sets. By conducting current more in one direction than in the other, the crystal detector converted the alternating current originating from a radio antenna to the simpler signals required for the listener's headphones. Nine years later, American physicist Manson Benedicks developed a crystal device based not on silicon, but on germanium. Neither Pickard nor Benedicks could explain why their devices worked, but produced solid empirical evidence of their behavior.

Although popular with radio hobbyists, crystal detectors were finicky and unreliable. In 1904, two years before Pickard's patent, the English electrical engineer and physicist John Ambrose Fleming (1849–1945) invented the thermionic valve, an electronic radio wave rectifier. Building on American inventor Thomas Edison's (1847–1931) invention of the light bulb, Fleming placed two electrodes inside a partially evacuated glass (vacuum) tube. By carefully shaping the electrodes—the negative cathode and the positive anode—he found that the electricity entering the tube as alternating current was "rectified" or converted into direct current. This thermionic or Fleming valve was later applied to diodes.

In 1907 American inventor Lee De Forest (1873–1961) invented the audion, a new type of vacuum tube,

WORDS TO KNOW

BAND THEORY: In quantum physics, the physical theory that describes what bands are available in a substance and how electrons will occupy those bands. A band is a specific energy level that an electron may possess in a given atom or molecule. Band values vary from substance to substance and are often numerous.

DEPLETION REGION: In integrated electronic circuitry, devices that control the flow of current in response to a control voltage (transistors), much as a valve controls the flow of water through a faucet, are made by mixing contaminant substances into parts of a solid crystal of silicon or other semiconducting material. These contaminants either donate extra electrons to the crystal matrix (add a negative charge) or are short of electrons (add a positive charge). Where a portion of the crystal with extra electrons touches a portion lacking electrons, some of the extra electrons are pulled over from the negatively-charged area to the positively-charged area; this continues until electrical forces are balanced. When a balanced state is achieved, an in-between layer exists that is neither positively nor negatively charged and so lacks carriers for current. This area, depleted of charge carriers, is termed a depletion region.

DIODE: An electronic device that permits current to flow through itself in one direction but not in the other.

DOPING: The act of adding impurities (dopants) to change semiconductor properties. In athletics, the illegal use of steroids, drugs, hormones, or techniques such as blood doping (injecting previously drawn red blood cells to increase oxygen) in order to enhance athletic performance.

FIELD EFFECT TRANSISTOR (FET): A transistor is an electronic device that controls a current flow in response to a separate electrical signal, like a valve regulating the flow of water through a faucet. A field-effect transistor (FET) accomplishes this by using an electrical field to pinch a conducting volume of semiconductor crystal; the stronger the field, the narrower the conducting zone and the higher its resistance (i.e., less current flows). Most of the transistors used in modern integrated circuits such as microprocessor chips are FETs.

HOLE: In the physics of solid materials (solid-state physics), a place in a crystalline array of atoms where an electron is missing, creating a localized positive charge. In semiconductors, holes can travel through the crystal as electrons shift, moving the gap from one position to the next. A moving hole behaves exactly as if it were a positively-charged particle or charge carrier.

JUNCTION TRANSISTOR: A transistor is an electronic device that controls a current flow in response to a separate electrical signal, like a valve regulating the flow of water through a faucet. A junction transistor, also known as a sandwich transistor, consists of three pieces or regions of semicon-

ducting crystal arranged in a sandwich. The outer layers consist of crystal that has been doped (slightly contaminated with atoms that donate electrons, endowing the crystal with negative charges); the middle layer consists of crystal that has been doped with atoms that lack electrons, endowing the crystal with positive charges. (A sandwich can also be made with positive and negative in reversed roles.) The ability of the junction transistor arises from the properties of the junctions between the positively and negatively doped silicon, hence the name.

MEAN FREE PATH: The mean free path of a free-moving particle is the average distance it travels before striking an obstacle (e.g., another particle). The mean free path is longer in less-dense media such as diffuse gases.

PERIODIC POTENTIAL: Periodic potential refers to the regular pattern of electric field potential inside a crystal, where atoms are built up in recurring (periodic) structures like a grid or lattice.

PHOTOELECTRIC EFFECT: The phenomenon in which light falling upon certain metals stimulates the emission of electrons and changes light into electricity.

P-N JUNCTION: In electronics, devices of microscopic size can be made by adding small amounts of dopant or contaminant substances to portions of a semiconducting crystal, usually silicon. Some dopants lack electrons and so contribute positive charges to the crystal; others have extra electrons and so contribute negative charges. Positive dopants are termed p-type, negative dopants n-type. Where a p-type region abuts or contacts an n-type region, a p-n junction is created. Electrons from the n-type side of the junction are drawn to the positive charges on the p-type side until electrostatic forces balance out and prevent further charge movement. This creates a slab-shaped region of electric neutrality along the p-n junction. Since this region is depleted of charge carriers, it is termed a depletion zone and cannot carry current. Placing a voltage across the p-n junction that is positive on the p side narrows the depletion zone and allows current to flow through it: applying voltage that is negative on the p side widens the depletion zone so that current cannot flow. A p-n junction thus allows one-directional current flow. Used in this fashion, a p-n junction is termed a diode.

SEMICONDUCTOR: Substance, such as silicon or germanium, whose ability to carry electrical current is lower than that of a conductor (like metal) and higher than that of insulators (like rubber).

SOLID STATE PHYSICS: All physics devoted to the properties of solids—substances that retain a fixed shape unless subjected to sufficient force—fall under the heading of solid-state physics.

by inserting a third electrode called the grid between the two electrodes of Fleming's rectifier. De Forest found that applying current to the grid both rectified and amplified the voltage running between the electrodes. This provided a cost-effective way to magnify voice transmissions via radio. AT&T also modified the audion in 1914 for use as a signal amplifier for long-distance telephone lines.

Although the redesigned audion was a success in the transcontinental phone line, it was clear that vacuum tubes were prone to failure. AT&T hoped that a new device might be developed that was based on semiconductors. This was the start of a commitment to industrial research and development that would, some 40 years later, lead to the invention of the transistor.

Semiconductors and Quantum Mechanics

Until the 1920s, work on semiconductors was primarily empirical and practical. After the discoveries of Austrian physicist Erwin Schroedinger (1887–1961) and German physicist Werner Heisenberg (1901–1976), quantum mechanics was applied to the developing field of solid-state physics (of which semiconductor work was one part). The first move toward a quantum theory of solids was an analysis of electrons in metals. German physicist Arnold Sommerfeld (1868–1951), working at the University of Munich, started from a classical model in which the electrons are seen as a noninteracting gas, and added so-called Fermi-Dirac statistics, in which no two electrons could share the same quantum states.

Although Sommerfeld's theory improved agreement with experimental data, there were still notable failings, including its inability to explain how electrons could have a large mean free path in crystals (that is, have such long distances between collisions). In 1928 one of Heisenberg's students, the Swiss-born American theorist Felix Bloch (1905–1983), answered this question in his doctoral thesis, by viewing the metal as a three-dimensional lattice. After Bloch used a periodic potential to represent the atoms making up the crystal lattice, he was able to assume that the function describing the location of the electrons took a form reflecting the pattern of the lattice potential. A long mean free path for electrons followed naturally from this assumption. More important, instead of finding discrete energy levels as in individual atoms, Bloch found that electrons in metals had a band, or range, of allowed energies.

Bloch's use of quantum theory made it possible to predict the conductivity of conductors, but it could not tell the difference between insulators, conductors, and semiconductors. This was accomplished by the English physicist Alan Wilson (1939–), whose papers on band theory almost single-handedly made the study of semiconductors, and solid-state physics in general, recognized fields of study. In addition to Bloch's formalism, Wilson used the ideas of German-born British physicist

Rudolf Peierls (1907–1995), who found that vacancies in almost-filled electron bands could be considered "holes" that behaved as though they were positively charged carriers. He also found that electron bands were often discontinuous in solids, where the bands of allowed energies were broken up by regions of forbidden energies, or "band gaps."

Wilson combined these ideas and found a simple and convincing explanation for the difference between types of materials. Conductors were materials in which the electrons filled to a level that was still within a continuous band of energies and were therefore free to move from atom to atom. Insulators were materials in which the electrons filled one band of energies, after which there was a large energy gap to the next available band. Even if an insulator were heated to high temperatures, the electrons would not be able to jump across the gap into the next band where they could move around the solid. Semiconductors were simply insulators with a smaller energy gap. When heated, electrons could jump out of the valence band and into the conduction band. This produced a free electron in the conduction band and a hole, or empty space in the valence band.

Pure (or instrinsic) semiconductors could be controlled by adding impurities that had atomic levels that fell between the valence and conduction bands, a process called doping. This could be done in two ways: Adding impurities that contain one more electron than the original semiconductor results in an n-type material (such as phosphorus-doped silicon). The extra electrons or majority carriers in the conduction band are available for conduction; the holes are minority carriers. Adding impurities that contain one less electron than the original semiconductor results in a p-type material (such as boron-doped silicon). In this case, the holes in the valence band are the majority carriers available for conduction.

Semiconductor theory also benefited from the study of rectifying contacts. One of the most important contributions came from German theorist Walter Schottky (1886–1976) in 1939, working at the German electronics firm of Siemens. He studied industrial rectifiers used in power applications that were based on junctions between copper and cuprous oxide, a p-type semiconductor. After a discussion with Peirels, Schottky noted that a small potential difference occurs at the surface of the semiconductor, and its free electrons are partly swept out. This leaves a depletion region of low conductivity at the surface, where no carriers were available for conduction. After calculating the depletion region's shape and width, Schottky found that if a positive bias (positive voltage) is applied to the semiconductor side, then the depletion region is eliminated and a current flows. If positive bias is applied to the metal side, however, then the normal potential difference between the materials increases, along with the width of the depletion region, and no current flows.

New Applications

These breakthroughs in solid-state theory enabled the development of devices that would have been beyond imagining in the late-nineteenth and early twentieth centuries. One of the most important uses of this science in World War II came from the development of radar. In 1917 Nikola Tesla (1856–1943) pioneered the first use of electromagnetic energy to detect objects at distance; by the mid-1930s an international effort had emerged, led mainly by British scientist Robert Watson-Watt (1892–1973), who patented the first workable system in 1935. In 1940 the United States agreed to assist the British war effort by establishing a sort of Manhattan Project for radar, the Radiation Laboratory (Rad Lab) at the Massachusetts Institute of Technology.

The British shared a new source of electromagnetic radiation with their American allies, a device called a cavity magnetron that used a rectifier, usually based on semiconductor materials, particularly silicon. The Rad Lab enlisted American scientist Frederick Seitz (1911–) and his solid-state physics group at the University of Pennsylvania to produce purer initial samples of silicon and control subsequent doping. Seitz, who had published an influential textbook summarizing then-current knowledge in solid-state physics, now led his research team in experiments that purified samples of silicon through repeated melting, then carefully doped them with boron.

Other new developments were encouraged by the war effort but arrived too late to affect its outcome. One group associated with the Rad Lab at Purdue University was headed by Austrian-born American physicist Karl Lark-Horovitz (1892–1958), who produced high-purity germanium samples doped with tin. The resulting rectifiers were able to withstand about ten times the voltage of silicon devices. The germanium rectifiers came off the Western Electric production lines in early 1945, not quite in time to affect the deployment of wartime radar systems, but making a considerable impact in general electronics.

Another wartime development was a new type of semiconductor diode discovered by American scientist Russell Ohl (1898–1987) at Bell Laboratories in 1939. Ohl accidentally found a junction between p-type and n-type regions while testing a silicon sample. Upon further investigation, he found that this p-n junction acted as a rectifier (current passed when a positive potential was applied to the p-type side of the diode but not when it was applied to the n-type side), and that it produced a relatively large voltage when light was shined on it via the photovoltaic effect, a type of photoelectric effect.

After the war, Bell Labs continued to discover devices that could replace the expensive and unreliable tube-based amplifiers. American scientist Walter Brattain (1902–1987) won a position in the division along with John Bardeen (1908–1981), who left academia for industrial research and development. The two worked with American engineer William Shockley (1910–1989), one of the division's directors.

The group first attempted to realize Shockley's concept of a field-effect transistor (FET), in which two electrodes (later named a source and a drain) were placed at either end of a piece of semiconducting material and an external electrode (the gate) was placed above. Shockley hoped that appropriate voltage on the external electrode would produce a charge layer to connect the input and output electrodes. Controlling current flow in this way made the device similar in operation to a triode vacuum tube.

Unfortunately, the first attempts in 1945 failed. Bardeen and Brattain suspected that rogue quantum states at the surface of the semiconductor were trapping charge carriers and forming an electric shield that canceled the external electrode's influence beyond the surface. To check this hypothesis, they conducted a number of experiments, eventually fixing the problem by placing an electrolyte (e.g., water) between the external electrode and the semiconductor.

This led the team to a new semiconductor design that used a slab of n-type germanium and three contacts: one to the bottom of the slab and two to the top. Bardeen and Brattain found that the first versions of this device gave small but promising amplification. At this point they made a crucial and unexpected discovery: When the two surface contacts were sufficiently close together, the operation of the device improved significantly.

Bardeen and Brattain then built yet another device using two rectifying surface contacts positioned close together with a clever spring-loaded assembly. On December 16, 1947, this point-contact transistor gave a strong amplification. The team determined that when one rectifying point contact (the emitter) was given forward bias (in this case, positive voltage), it injected holes into the n-type germanium, which were then drawn to the other point contact (the collector), since it was reversed biased (negative) voltage. Because the holes, the minority carriers inside n-type material, were as important as the electrons themselves (the majority carriers), the device was referred to as a bipolar transistor.

Shockley had not been greatly involved in the point-contact transistor and was eager to make his own contribution. In fact, a difference of opinion regarding the interpretation of the point-contact transistor led him to an even better transistor design. While Bardeen and Brattain thought that the minority carriers were confined to the transistor's surface and could not travel through the bulk of the semiconductor, Shockley took the opposite view, proposing a transistor that would involve the transport of minority carriers within the bulk.

This bipolar junction or npn transistor consisted of three layers of semiconductor; the outer two (emitter and collector) were n-type; the middle layer (the base) was p-type. Unlike the point-contact transistor, in which the emitter and collector were placed very close together at the surface of a large base, in the junction transistor the base had to be made very slender to allow the emitter and collector to be close enough together. The work of American physical chemist Gordon Teal (1907–2003) and American engineer Morgan Sparks (1916–2008) on growing germanium p-n junctions made such slender base regions technically possible; the new junction transistor was conclusively demonstrated on April 12, 1950. Two years later Henry Theuerer's work on the growing highly purified silicon junctions made it possible to construct silicon junction transistors.

Integrated Circuits

During the 1950s and 1960s, miniaturization in electronics was motivated by consumer products such as radios and hearing aids, military applications such as missile systems, and many areas of scientific research. The idea for a monolithic integrated circuit, one that combined many components in a single part, spurred much research. The idea was first realized in working devices by two people at roughly the same time. At Texas Instruments, American engineer Jack Kilby (1923–2005) recognized in September 1958 that resistors, capacitors, and transistors could all be made out of silicon and combined on a single part. However, because Texas Instruments had not yet developed the silicon technology that he needed, Kilby's first all-in-one circuits were germanium based, including an oscillator and a switching circuit called a flip-flop.

A few months later, at the start-up company Fairchild Semiconductor, American engineer Robert Noyce (1927–1990), came up with an even better approach to monolithic integration. In 1958 Fairchild began to manufacture discrete silicon transistors with the "mesa" approach pioneered at Bell Labs: Each thin silicon wafer was defined with a carefully applied wax patch. The exposed area was then subjected to acid etches which left raised "mesa" surfaces where the wax was. Soon, instead of wax patches, Fairchild turned to photolithography, in which a photosensitive material was applied to the silicon wafer and exposed to an image of the circuit. A weak acid then uncovered the desired areas, which were then subject to etching and eventual doping in diffusion ovens.

Another Fairchild scientist, Jean Hoerni (1924–1997), introduced the concept of planar manufacturing, in which many devices were constructed on a silicon wafer and connected by metal strips, again, using photolithography. During various steps, the wafers were given coverings of silicon dioxide to help with the etching and diffusion and give a protective coating to the

IN CONTEXT: OPERATION OF THE JUNCTION TRANSISTOR

When voltage is applied to an npn transistor, extra electrons in the n-type semiconductor recombine and cancel an equal number of holes in the p-type regions near the np boundaries. This forms a small depletion region at each boundary in which no charge carriers are available. When the transistor is biased (put under voltage) in a simple amplification circuit, the so-called common-emitter configuration, the voltage between the collector and emitter is much greater than the voltage between the base and the emitter. Therefore, the base-collector junction is strongly reverse biased, giving it a large associated depletion region. Because the base-emitter junction is slightly forward biased, the n-type emitter injects electrons into the p-type base. The electrons are then pulled across the depletion zone by the positively biased collector. Very small changes in the base-emitter current can result in large changes in the collector-emitter current. The gain factor is computed as the ratio of the collector current to the base current.

finished product. Fairchild's new technology worked wonderfully for the construction of integrated circuits; the company won a patent for it in 1961.

Integrated circuit manufacturing technologies led to the development of new devices, the most important of which was actually an old one: Shockley's 1945 idea of a field-effect transistor (FET). In 1960 Korean-born American physicist Dawon Kahng (1931–1992) and Egyptian-born Martin Atalla of Bell Labs developed a variation on Shockley's idea in which the drain and source electrodes were connected or not connected depending on whether the voltage on the gate electrode produced an inversion layer of minority carriers. For example, when a p-type substrate is subjected to a positive gate voltage, then an inversion layer of electrons connects the n-type source and drain. The new planar technology made it relatively easy to manufacture a metal-oxide-semiconductor FET (or MOSFET), especially since the silicon surfaces that had given Shockley so much trouble could now be rendered harmless with the "passivation" of a final silicon dioxide covering. The MOSFET, in turn, enabled a new surge of miniaturization and, because it was cheaper to fabricate, all but replaced junction transistors (though they were still manufactured for special applications).

Modern Cultural Connections

Miniaturization was crucial to the computing revolution that occurred in the years following the MOSFET's

IN CONTEXT: MATERIALS SCIENCE AND SEMICONDUCTORS

Research on semiconductors has interacted powerfully with materials science, the development of new solids for practical applications. After the pioneering work of researchers like Frederick Seitz (1911–) and Russell Ohl (1898–1987) before and during World War II, the development of the transistor and microelectronics has further heightened this interaction.

In 1948 William Shockley's (1910–1989) idea of a junction or npn transistor presented a serious technical problem. Although Russell Ohl had found a pn junction in semiconducting material, no one had yet found a reproducible way to create layers of differently doped semiconducting material. Despite initial disinterest from Bell Labs, Gordon Teal (1907–2003) and Morgan Sparks (1916–2008) developed a method of "growing" crystals of germanium. A tiny seed of germanium crystal was slowly pulled out of a pool of molten germanium and, as this happened, different dopants were added to the melt, producing sections of n- or p-type semiconductors.

After the npn transistor emerged in 1950, further improvements suggested themselves. Minority carriers, on which the new devices depended, were easily lost as they passed through the base, both to undesired impurities (not the carefully added dopants) and lattice defects in the semiconductor. In addition, because of its material properties, silicon clearly made a better transistor than germanium. In 1952 Bell Labs began to expand its use of the "float-zone" technique developed by William Pfann

in 1950, in which a heated element was repeatedly run through a rod of germanium, removing unwanted impurities. Henry Theuerer applied this technique to the purification of silicon, and the manufacture of silicon junction transistors became possible. In 1954 Teal, now working for the newly established Texas Instruments company, announced the production of the silicon transistor.

Since then, new semiconductor materials have been found, one large class of which are compound semiconductors. During the 1950s, German theoretical and applied physicist Heinrich Welker (1912–1981) showed that materials such as gallium arsenide (GaAs), aluminium antimonide (AlSb), and indium phosphide (InP) could be classed as semiconductors, based on American theoretical physicist Linus Pauling's (1901–1994) theory of the chemical bond. Welker confirmed this in experiments that demonstrated the ability of these materials to form rectifying contacts.

Gallium arsenide has been shown to have certain advantages over silicon, including the ability to operate at higher speed and higher power, making it attractive for applications in military communications. Its high speed makes GaAs seem desirable for computing systems as well, but so far has proven too costly. The structure of its electron bands makes GaAs excellent for generating light, and it has been used in light-emitting diodes (LEDs) and solid-state lasers.

invention. Two of the most important new devices were developed by Intel, a company founded in 1968 by scientists including Robert Noyce (1927–1990) and Hungarian-born American chemical engineer Andrew Grove (1936–2002). In 1970 Intel produced the first random access memory (RAM) using silicon-based processing (the Intel 1103). One year later it manufactured the first microprocessor (the Intel 4004). During the next decade, Intel continued to make improvements in their products and, in 1981, benefited greatly when IBM chose Intel's 8088 microprocessor for its personal computer. Though not the first personal computer (Xerox and Commodore had introduced their own some years earlier), IBM's Intel-powered model dominated the market.

Since the invention of the integrated circuit computer parts have become denser by virtue of their smaller feature size. Microprocessors constructed in the early 1970s contained about 5,000 transistors and had feature sizes (silicon or metal lines) of about 10 microns. Intel's Pentium microprocessors in the early 2000s contained about 30 million transistors and had feature sizes below 0.2 microns. At present writing, nanotechnology, the design and manufacture of functional electronic systems at the molecular level, is being explored by academic researchers and industry executives alike. Some forecasters

suggest that nanolithography will produce integrated circuits with features below the 0.05 micrometer level. Others suggest that nanotechnology will take computer hardware away from semiconductor-based technologies altogether and toward entirely new methods of storing information.

SEE ALSO *Chemistry: Molecular Structure and Stereochemistry; Chemistry: States of Matter: Solids, Liquids, Gases, and Plasma.*

BIBLIOGRAPHY

Books

Hoddeson, Lillian, E. Braun, J. Teichmann, and S. Weart. *Out of the Crystal Maze: Chapters from the History of Solid-State Physics.* New York: Oxford University Press, 1992.

Hoddeson, Lillian, and Vicki Daitch. *True Genius: The Life and Science of John Bardeen, the Only Winner of Two Nobel Prizes in Physics.* Washington, DC: Joseph Henry Press, 2005.

Riordan, Michael, and Lillian Hoddeson. *Crystal Fire: The Birth of the Information Age.* New York: W.W. Norton, 1997.

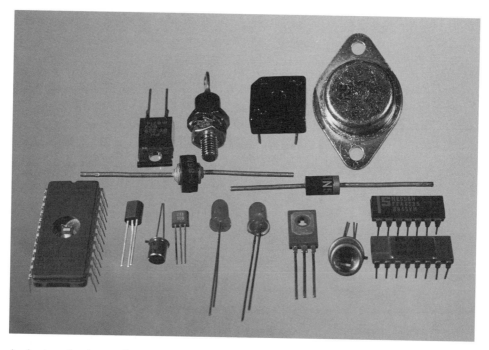

A selection of various applications of semiconductors, materials with properties between those of conductors and insulators. Combinations of different types of semiconductors allow the design of various types of electronic components to manipulate the flow of electric current. Semiconductor technology is crucial to the operation of technology such as computers. Components seen here include computer microchips (center left and center right), LEDs (light emitting diodes), as well as various other diodes, resistors and capacitors.   *Andrew Lambert Photography/Photo Researchers, Inc.*

Seitz, Frederick, and Norman G. Einspruch. *Electronic Genie: The Tangled History of Silicon.* Urbana and Chicago: University of Illinois Press, 1998.

Periodicals

Herring, Conyers. "Recollections from the Early Years of Solid-State Physics." *Physics Today* (April 1992): 26–33.

Jenkins, Tudor. "A Brief History of ... Semiconductors," *Physics Education* 40, no. 5 (2005): 430–439.

Noyce, Robert. "Microelectronics." *Scientific American* 237, no.3 (September 1977): 63–69.

Web Sites

Institute of Electrical and Electronics Engineers, IEEE Virtual Museum, Let's Get Small: The Shrinking World of Microelectronics. "Small Beginnings: From Tubes to Transistors." http://www.ieee-virtual-museum.org/exhibit/exhibit.php?id=159270 (accessed November 11, 2007).

Nobel Foundation. "Technology and Entrepreneurship in Silicon Valley." http://nobelprize.org/nobel_prizes/physics/articles/lecuyer/index.html (accessed November 11, 2007).

Gary J. Weisel

Physics: Special and General Relativity

■ Introduction

Even if you know very little about science you've probably heard of the theories of relativity—the words alone

WORDS TO KNOW

ACCELERATION: A change in the speed or the direction of an object's motion.

BLACK HOLE: An object so massive and compact that light is unable to escape its gravitational pull.

C: In physics, for over a century, the italicized Roman letter c, sometimes written c_0, has stood for the speed of light in a vacuum, namely 670,616,629 miles per hour (1,079,252,848 km/h) or, as more commonly expressed in science, 2.99×10^7 m/s. This speed is constant, that is, no matter how fast a source and observer are moving with respect to each other, the observer will always measure the speed of light as c. In media such as air or glass, light travels more slowly, but still has the property of being constant for all observers.

ENERGY: The capacity to effect change or do work.

INERTIAL REFERENCE FRAME: A reference frame that is at rest or moving at a constant speed.

LENGTH CONTRACTION: The apparent shortening of fast-moving objects as seen by a stationary observer.

LIGHT-YEAR: The distance traveled by light in one year, or a little more than 6 trillion miles (9.6 trillion km).

MASS: The amount of matter contained in an object.

SPACE-TIME: A four-dimensional union of time and space first postulated by Einstein in the theories of relativity.

TIME DILATION: The slowing down of time experienced by a stationary observer in a moving reference frame.

almost certainly make you think of Albert Einstein (1879–1955). And because Einstein is such an iconic genius, most people think that relativity is a very difficult mathematical concept that only a brilliant physicist could understand.

Einstein developed his theories between 1905 and 1911, but simpler ideas of relativity date back almost 400 years, to the time of Galileo Galilei (1564–1642). Everyone has a simple grasp of relativity that is based on everyday experience: moving around in cars, trains, and planes. The central question of any theory of relativity is simple: How does a person know if they are moving or standing still? And if they are moving, how do their observations of the world compare with those of a person who is not?

■ Historical Background and Scientific Foundations

Moving or Standing Still?

Answering the question "Am I moving or standing still?" is not as simple as it might seem. When we look around us, there are plenty of things we think are stationary—trees, buildings, the sidewalk. When we drive down the street, it is obvious to us that our car is moving, while the street and the buildings are not. But remember: We are on the surface of a moving planet. Earth's rotation spins us at nearly 1,000 miles per hour (1,609 km/hr), yet we don't notice this motion at all.

When we look up in the sky, we see the sun and moon rise and set, and the stars spin in the heavens. For thousands of years people assumed that Earth sat still while everything in the universe circled around it. But in 1543 Polish astronomer Nicolaus Copernicus (1473–1543) realized that stellar and planetary motions could be explained more simply and accurately if Earth moved as well. Earth, he believed, spins on its axis once per day and orbits the sun once per year; these combined motions explain everything from the rising and setting of the sun to the changing of the seasons to the intricate paths of the planets across the sky.

For many years Copernicus's ideas were of interest to only a handful of astronomers, who recognized their power for understanding and predicting the motions of the planets. To most people, however, it all seemed too outlandish to be true. The real uproar over a sun-centered universe erupted more than half-century later, when Galileo Galilei (1564–1642) provided solid proof that Copernicus was, in fact, correct. Only then did the intellectual and religious authorities see these new ideas as a threat. In Galileo's time, Copernicus's book was banned, teaching the sun-centered model was forbidden, and Galileo himself was tried for heresy. While these ideas seem unremarkable to us today, at the time they were truly revolutionary. Not only did they turn the known universe on its head, but they challenged accepted religious and intellectual orthodoxy as well.

One reason many people didn't take Copernicus's new theory seriously was because they found it impossible to believe that Earth was moving without being able to sense it. Wouldn't they be thrown off their feet by such rapid motion? Wouldn't there be a 1,000-mile-per-hour wind outside? If they jumped in the air, wouldn't they land hundreds of feet away as the planet moved out from underneath them?

Of course the answer to all of these questions is "no." In 1543 people had no experience that would allow them to understand what it feels like to move at hundreds of miles an hour. But today we do—if you've ever been on an airplane you've done so. And as long as the plane is in level steady flight it's easy to forget that you are moving at all. You can stand up, walk around the cabin, pour yourself a drink, all with no more difficulty than when you are standing still. The situation changes dramatically when the plane turns or climbs or experiences turbulence.

Galileo was the first to recognize this. A strong supporter of Copernican ideas, he went to great lengths to prove that the sun-centered model of the solar system was correct. To do so he developed an entirely new understanding of motion, starting with this simple fact: Moving at a constant speed feels just like sitting still.

Once we accept that both stationary and moving observers have a similar experience of the world, we can address the question of how their observations compare. A theory of relativity is basically a set of rules that allows us to compare the observations of two people who are moving at different speeds relative to one another. For example, imagine a car driving past a park where some kids are playing baseball. Say the pitcher is facing south, and the car is driving north at 40 miles per hour (64 km/h). The coach, standing behind the catcher with a radar gun, clocks the pitcher's throw at 50 mph

(80 km/h). The question that any theory of relativity attempts to answer is this: What speed would the driver of the moving car record for the pitch as they drove by?

The answer seems straightforward and obvious to most people. If the car was driving north at 40 mph and the ball is being thrown south at 50 mph, then relative to the driver, the ball is moving at 90 mph (145 km/h). If a wild pitch sent the ball flying into the car's windshield, the effect of the 50-mph pitch on the car would be the same as if it were sitting still and hit with a 90-mph Major League fastball.

This commonsense answer is the cornerstone of what physicists call classical or Galilean relativity. According to this theory, in order for two people who are moving relative to one another to compare their measurements of the speeds of various objects, all they have to do is add their own relative velocity to the speed measured by the other observer.

Relativity of Velocity

So what is the baseball's "real" velocity? Is it the speed measured relative to the players on the field? Or the speed measured relative to the car driving by? You may be tempted to answer that it is the baseball players who measure the actual speed of the ball, since they are the ones who are standing still. But this argument goes out the window when you realize that the baseball players are not standing still at all. They are standing on an Earth spinning at nearly 1,000 mph (1,609 km/h), and hurtling around the sun at more than 60,000 mph (96,561 km/h)! To an observer "at rest" on the moon or on Mars or at the center of the Milky Way, the ball could appear to be moving 500, or 5,000, or 500,000 mph. Which is the "true" velocity of the baseball? There is no way to tell. Or perhaps more accurately there simply is no true velocity. Every speed in the universe is relative to something else. This is the essence of any theory of "relativity."

This version of relativity was first expressed formally by Galileo in the early 1600s; it became an integral part of Isaac Newton's (1642–1727) more complete description of motion a generation later. Even today the ideas of classical relativity seem to work just fine for baseballs and cars, even planes and rockets. But our commonsense notion of how to add velocities together begins to fail when we look at the fastest object in the universe—light.

The Motion of Light

For centuries, scientists and philosophers had wondered about the nature of light, including how fast it moves—which is clearly very fast. When we turn on a lamp, the room fills with light immediately. We don't see it move outward gradually to the far corners of the room; it just seems to come on instantaneously. But is it really instantaneous or just very, very fast?

IN CONTEXT: GALILEO AND INERTIAL REFERENCE FRAMES

Galileo Galilei laid out his support for the new Copernican astronomical system in his work *Dialogue Concerning the Two Chief World Systems, Ptolemaic and Copernican,* published in 1632. To make his case, Galileo had to silence critics who objected to the idea of a moving Earth and insisted that surely they would be able to feel this motion. But Galileo knew that motion at a constant speed is undetectable. To prove to his detractors that Earth could be spinning rapidly and we would not feel it, Galileo compared our situation on the moving Earth to the following scenario:

> Shut yourself up with some friend in the main cabin below decks on some large ship, and have with you there some flies, butterflies, and other small flying animals. Have a large bowl of water with some fish in it; hang up a bottle that empties drop by drop into a wide vessel beneath it. With the ship standing still, observe carefully how the little animals fly with equal speed to all sides of the cabin. The fish swim indifferently in all directions; the drops fall into the vessel beneath; and, in throwing something to your friend, you need throw it no more strongly in one direction than another, the distances being equal; jumping with your feet together, you pass equal spaces in every direction. When you have observed all these things carefully (though there is no doubt that when the ship is standing still everything must happen in this way), have the ship proceed with any speed you like, so long as the motion is uniform and not fluctuating this way and that. You will discover not the least change in all the effects named, nor could

> you tell from any of them whether the ship was moving or standing still. In jumping, you will pass on the floor the same spaces as before, nor will you make larger jumps toward the stern than toward the prow even though the ship is moving quite rapidly, despite the fact that during the time that you are in the air the floor under you will be going in a direction opposite to your jump. In throwing something to your companion, you will need no more force to get it to him whether he is in the direction of the bow or the stern, with yourself situated opposite. The droplets will fall as before into the vessel beneath without dropping toward the stern, although while the drops are in the air the ship runs many spans. The fish in their water will swim toward the front of their bowl with no more effort than toward the back, and will go with equal ease to bait placed anywhere around the edges of the bowl. Finally the butterflies and flies will continue their flights indifferently toward every side, nor will it ever happen that they are concentrated toward the stern, as if tired out from keeping up with the course of the ship. The cause of all these correspondences of effects is the fact that the ship's motion is common to all the things contained in it.

A modern physicist would call Galileo's ship cabin an inertial frame of reference. According to both Galileo and Einstein, any observer in an inertial reference frame should see all of the laws of physics obeyed.

Galileo proposed an experiment to measure the speed of light by stationing two observers with lanterns atop two hills several miles apart. By flashing their lights at one another and awaiting a reply, the speed of light could be determined from the delay. But the speed of light is so fast that no delay was observed. Even a distance of a few miles was traversed in the blink of an eye. To measure the speed of light, much larger, interplanetary distances must be used. A beam of light travels from Earth to the moon in about 2 seconds and to the sun in about 8 minutes. Only by observing astronomical events separated by hundreds of millions of miles can we see the delays that Galileo tried in vain to measure.

These sorts of delays were first observed by Danish mathematician and astronomer Ole Rømer (1644–1710) in 1677. Rømer was observing the moons of Jupiter, discovered by Galileo only a generation before, and watching and timing them as they passed behind Jupiter itself. He found that if you measured the timing of an eclipse when Earth's orbit was near Jupiter, then used this information to predict an eclipse six months later, when Earth was far from Jupiter, your answer would be

off—it would happen 15 or 16 minutes later than you expected. Why? Rømer realized that this happened because the light had to travel further. From the amount of the delay and the size of Earth's orbit, Rømer was able to calculate the speed of light—around 186,000 miles per second (300,000 kilometers per second).

Light and the Ether

By 1700 it was clear that light moved very fast, but the question remained—what was light itself? What was it "made of?" Isaac Newton suggested that light was a stream of tiny particles. But others believed it was a kind of wave, like water or sound. Experiments in 1803 by Thomas Young (1773–1829) and others supported this view.

But if light were a wave, there had to be some substance that was doing the "waving." A wave is not an object, a wave is a vibration or disturbance in some other material. Ocean waves, for example, are a disturbance in water, and sound waves are a disturbance in air. If light waves traveled throughout the universe, from the sun to

Earth and out to the distant stars, there had to be some substance in space whose vibrations carried them. This invisible and purely hypothetical substance was named the "luminiferous æther," or ether.

If all of space really were filled with some sort of material, wouldn't there be some way to detect it? In 1887 Albert Michelson (1852–1931) and Edward Morley (1838–1923) devised an experiment to do so. Their experiment was not designed to detect the ether directly, but Earth's motion through it.

To understand the effect Michelson and Morley sought, imagine a fast ship moving across a rough ocean. As the ship crashes through the waves, an observer aboard the ship would see the waves passing by faster than if the ship were sitting still. Likewise, as Earth moves through space in its orbit around the sun, they presumed they would notice an effect on the speed of light—a sort of "ether wind"—caused by Earth's motion through the ether.

Michelson and Morley set up a very sensitive experiment consisting of an arrangement of mirrors that split a beam of light into two parts—one part traveling in the direction of Earth's motion through space, the other traveling perpendicular to it. By comparing the motion of these two beams of light, they hoped to detect the effect of the ether wind on one of the beams. The result? They found no effect. The speed of light was not affected at all by Earth's motion. This was a strange result indeed.

Light and Electromagnetism

While it took some time, the Michelson-Morley experiment eventually caused the "luminiferous ether" theory to fall out of favor. But if space was not filled with a light-carrying medium, then what was a light wave? Luckily, a new idea was waiting to take over. It had been discovered more than a decade earlier by James Clerk Maxwell (1831–1879).

Maxwell wasn't working on the problem of light at all. He was studying the properties of electricity and magnetism and the relationships between them. Scientists had known for some time that electric currents could create magnetic fields. (You may have done this yourself in science class by wrapping wire around a nail or a bolt and connecting it to a battery to make an electromagnet.) Researchers had also discovered that moving magnets could be used to generate electricity. Maxwell expressed these connections between electricity and magnetism in four simple equations, summarizing

Joos's interferometer, a device engineered by Zeiss and Scott, was designed and operated by the German physicist Georg Joos (1894–1959). Joos used it to carry out a Michelson-Morley experiment in 1930. The results showed that light does not travel faster or slower relative to a moving Earth. This led to Albert Einstein's postulate that the speed of light is the same for all moving objects, which was the basis of his 1905 theory of Special Relativity. *Volker Steger/Photo Researchers, Inc.*

everything that had been learned about them up to that point.

Maxwell then used his equations to study what would happen to an electrically charged particle moving back and forth in space. Since the particle has an electric charge, this would create an electric field in space around it. And since the particle is moving, the strength of its electric field would change from place to place over time. But Maxwell's equations told him that such a changing electric field would also create a magnetic field that would also change over time as the electron moved back and forth. Another equation told him that this changing magnetic field would create its own changing electric field in space.

Maxwell had stumbled on the remarkable possibility that an electric field might create a magnetic field that would create an electric field that would create a disturbance traveling through space—a disturbance that was made of electric and magnetic fields. The mathematical form of this disturbance was identical to the equation for a wave. When Maxwell calculated what speed his wave of electricity and magnetism would have as it moved through space, he made an even more remarkable discovery. Based on the numbers that were known to describe the strengths of electric and magnetic fields, this electromagnetic wave should move outwards in space at a speed of about 300,000 kilometers per second (186,000 miles per second): Exactly the speed of light!

The Science: Physics Faces a Dilemma

Maxwell's picture of light as an eletromagnetic wave was a huge success. It unified everything that was known about electricity, magnetism, and light at the time, and also predicted new phenomena such as radio waves and other electromagnetic radiation that wouldn't be discovered for decades to come. But one feature of Maxwell's discovery was unsettling. The equations predicted an exact value for the speed of light—a first in the history of physics. Never before had an equation in physics said—"the speed of some object is *this* particular number." According to Galileo and the classical theory of relativity, an object could not have one exact speed—it depends on who is measuring it. If Maxwell and his equations tell us the speed of light is 300,000 km/sec, it's fair to ask—as measured by whom?

Let's explore this using what Einstein called a *gedankenexperiment* a "thought experiment." Suppose a rocket was traveling through space at 1,000 km/s. (As you know, rockets can't travel anywhere near that fast. At the turn of the twentieth century, in fact, scientists could hardly have imagined a rocket at all! Einstein placed most of his own thought experiments aboard fast-moving trains.) Inside the rocket, the passenger points a flashlight from the rear of the ship to the front. The light travels at the speed of light according to the occupants of the ship. But what would someone outside of the ship see?

From an outsider's point of view, everything in the ship is moving at 1,000 km/s. When the flashlight is turned on, the light moves at the speed of light, 300,000 km/s. But since the flashlight is already moving, the light should appear to be moving 301,000 km/s to the observer. At least, this is what classical relativity tells us should happen. But if the outside observer tried to use Maxwell's equations to calculate the speed of the light, he or she would get an answer (300,000 km/s) that didn't match their observations (301,000 km/s). Something is wrong here.

So, as the twentieth century approached, physicists were faced with a problem. Everything they understood about motion told them that velocity is a relative concept, and that objects in the universe have no one true velocity. But Maxwell's equations seemed to disprove this. They predicted a single number for the speed of light that was based not on the motion of any particular observer, but on the properties of electric and magnetic fields. So who was right—Galileo or Maxwell? It was a difficult problem that puzzled some of the great scientists of the day. The solution to the dilemma would come from the virtually unknown, 26-year-old patent clerk named Albert Einstein.

Dr. Albert Einstein (1879–1955) writes out an equation for the density of the Milky Way on the blackboard at the Carnegie Institute, Mt. Wilson Observatory headquarters in Pasadena, California, on January 14, 1931. *AP Images.*

Einstein's Solution: Special Relativity

Einstein's solution was simple, but the implications were revolutionary. He began with a simple idea—that all moving observers should agree on the laws of physics. He then accepted Maxwell's equations and the speed of light they predicted as laws of physics. Then he imagined a situation very much like the rocket ship example above. How would observers inside and outside the ship measure the speed of the light?

The observers inside the rocket would see the light travel from the back to the front of the ship. They could divide the length of the ship by the time it took the light to travel and use that to calculate the speed. But an outside observer would see the light travel a longer distance—from the back of the ship to the front, which has moved some distance forwards from where it started. If the light moved a greater distance in the same time, it must have a greater speed. But this violates Einstein's principle that all observers must agree on the laws of physics. How do we fix our thought experiment to get both observers to measure the same speed for the light?

Einstein realized that the only way that two observers moving at different speeds could record the same speed for the light beam was if they recorded different values for the *time* that the light takes to travel the length of the ship. Even though this seems counterintuitive—how can two people watching the same thing happen disagree on how long it takes?—Einstein accepted this surprising conclusion and used it as the basis for constructing a new theory of relativity.

Until this point, scientists had always considered measurements of time and distance to be absolute and universal—something that all observers could agree on, no matter where they were or how fast they were moving. Issac Newton made this explicit in his *Principia Mathematica* in 1687:

> Absolute, true, and mathematical time, of itself, and from its own nature flows equably without regard to anything external, and by another name is called duration.... Absolute space, in its own nature, without regard to anything external, remains always similar and immovable.

This view agrees with our everyday understanding of time and space. But according to Einstein, space and time are not absolute. The only true absolutes in the universe are the laws of physics; the speed of light, according to Maxwell's equations, is one such law. Einstein built his whole theory of relativity around this idea—that all observers must agree on the laws of physics, including the speed of light, even if doing so means that they must disagree on such things as the flow of time and the measurement of space.

The implications of Einstein's idea are stunning. They suggest that when a person is moving very fast, he or she will experience time differently than an observer who is stationary. These effects are negligible when we travel at the relatively slow speeds of cars, trains, or planes, but if we were to travel faster and faster, approaching the speed of light, the effect would become more and more noticeable.

Imagine a spaceship that could travel very fast—say, 90% of the speed of light. If a traveler were to board this ship and take off into space very fast, he would notice nothing particularly strange from his point of view. Time would appear to flow normally, objects in the spaceship would appear just as they did before the trip began. But outside observers watching him zip past would notice some very strange things. First of all, were they to peek into the ship's windows as it streaked by, they would see that, although the ship itself is moving very quickly, everything inside seems to be moving in slow motion. Everything from the movements of the space traveler to the ticking of his watch would be slowed down to less than half its normal rate. Compared to the observers outside, inside the ship time has slowed down.

That's not all they would see. As the ship passed by the stationary observers, the ship and everything within it would have a sort of flattened appearance. This would not be apparent to the traveler inside the rocket, because the flattening would apply to everything around him.

IN CONTEXT: IT'S ALL RELATIVE

Einstein's special theory of relativity suggests that when a person moves very fast, time slows down for him, and measurements of distances in space are shortened along the direction in which he is traveling. But in any given situation, the question of *who* is moving complicates matters. If a rocket streaks past Earth at nearly the speed of light, people on Earth will see the rocket "squished" in the direction of motion, and observe that clocks aboard the rocket seem to be running slowly compared to our own. But remember that all velocities are relative, and that all observers have the right to consider themselves stationary (as long as they are in an inertial reference frame). This means that to an observer aboard the ship, the earthbound clocks seem to be running slowly, and Earth appears flattened like a pancake as it streaks by the ship's windows at nearly the speed of light!

Many people have misrepresented Einstein's contribution to physics by boiling it down to "everything is relative." But this is a gross oversimplification of Einstein's theories. In fact, while it is true that many measurements in Einstein's theory are relative, the most important aspect of the theory revolves around what is *not* relative—the laws of physics and the speed of light, which are the same for everyone everywhere in the universe.

According to Einstein, the traveler's measurements of space itself are altered due to his movement through it.

The Implications and Impact of Relativity

These two effects, called time dilation and length contraction are still surprising to us, since no one has ever traveled even close to the speed of light. One can only imagine how strange these ideas must have seemed in 1905. Here was a patent clerk from outside of the academic establishment, with no credentials to speak of, publishing a paper that challenged everything known about the nature of space and time. The reaction at first was similar to the reaction to Copernicus 350 years earlier—silence. It seemed as if scientists simply didn't know what to make of Einstein's ideas. But over the next few years, they increasingly realized the power of Einstein's new theory of relativity. Within a decade he was recognized as one of the brightest minds in physics.

Einstein's predictions about space and time were far ahead of their time. Today they are confirmed daily by physicists working with particle accelerators and observing high-energy particles that bombard Earth from space. These unstable subatomic particles can live for only a microsecond when sitting still, but can exist for several seconds or minutes when traveling close to the speed of light, when time slows down and stretches out their brief lifespans (when observed by the stationary scientists in their labs).

In 1971 J.C. Hafele and Richard E. Keating tested this concept of time dilation in an experiment that used very precise atomic clocks accurate to a few billionths of a second. They placed several of these sensitive clocks aboard airplanes and compared their times to an identical clock that remained on the ground. They found that time aboard the moving planes slowed down by a few hundred nanoseconds during their flights—just the amount that Einstein's equations predicted.

This slowing down of time aboard a moving airplane is obviously negligible at the speed of a commercial aircraft, which moves at a crawl compared to the speed of light. The few nanoseconds you gain during a long airplane flight don't amount to much no matter how often you fly. Even a professional airline pilot who flies back and forth across the Atlantic every day wouldn't extend his lifespan by so much as a microsecond. But if you could go faster and faster you would begin to notice dramatic effects on your experiences of space and time.

Suppose you built the very fast rocket we discussed earlier—one that could travel 270,000 km/s or 90% of the speed of light (even though no rocket or satellite built so far can travel faster than 62.14 miles/s, or 100 km/s). Now imagine you wanted to travel to a distant star, say Vega, which lies about 20 light-years from Earth. The trip would take you just over 20 years—22.2 to be exact. But that's only as measured by a stationary

observer on Earth. Because of time dilation, time aboard your rocket would move more slowly, and from your point of view, the trip would take less than 10 years!

The Ultimate Speed Limit

What would happen if you traveled faster than the speed of light? Would time stop? Would it flow backwards? We don't have to worry about such questions, fortunately, because Einstein's relativity equations make such situations impossible. Only an object with no mass, such as light, can travel at the speed of light. Material objects can approach the speed of light very closely, but never exceed it. As an object gets closer and closer to the speed of light, c, it takes more and more energy to speed up. A rocket can travel 90% or 99% or even 99.99% of the speed of light, but gaining that last little bit of speed would take an infinite amount of energy! The effect is almost as if the rocket gets heavier and heavier (and harder and harder to push) the closer it gets to c. This new relationship between force, mass, and acceleration revealed yet another revolutionary implication of Einstein's theory—an unexpected connection between energy and mass.

Energy and Mass: $E = mc^2$

What Einstein's theory did for our understanding of space and time, it also did for our understanding of matter and energy. For most of the 200 years before Einstein, chemists and physicists had gone to great lengths to prove some important facts about the changes that occur in nature. During the 1700s chemists painstakingly demonstrated that whenever a chemical change occurs in nature, matter is conserved—that is, not created or lost. The chemical elements or atoms that make up the substance are merely rearranged and recombined to make new compounds. If you were to carefully weigh the reactants that go into a chemical reaction, and then weigh the products of that reaction, you would get the same result.

During the 1800s physicists discovered much the same thing about energy. Whenever a process occurs in nature that transforms energy from one kind to another, the total amount of energy in the universe remains constant. For example, when you drive a car, the chemical energy in the gasoline is converted to kinetic energy—the energy of the motion of the car. When you put on the brakes, the friction in the brake pads converts the kinetic energy of the car into heat. If you were to add up the total amount of chemical energy that went into the car, it would be equal to the total amount of heat energy that comes out of the car. Energy, like matter, can never be created or destroyed.

These two laws of nature—the law of conservation of matter and the law of conservation of energy—were fundamental to our understanding of the world at the turn of the twentieth century. But just as Einstein showed that space and time are really two aspects of a single

ALBERT EINSTEIN (1879–1955)

Albert Einstein was born in Ulm, Württemberg, Germany, on March 14, 1879. After dropping out of high school, at 17 he enrolled at the Swiss Federal Polytechnic School in Zurich. He graduated in 1901, but, unable to find a teaching post, was forced to accept a position as a technical assistant in the Swiss patent office. During his years there, Einstein continued to work on physics in his spare time. He married Mileva Maric, a fellow student of mathematics and physics, in 1903. Two years later he earned his doctoral degree, but not before he had already written a handful of papers that would change physics, and our understanding of the universe, forever.

Nineteen hundred five is often referred to as Einstein's "miracle year." During this single year, he published not only his revolutionary work on relativity, but a paper on the photoelectric effect (an important milestone in the development of quantum mechanics) and papers on the size and motion of molecules. In fact, it was his work on the photoelectric effect, and not his papers on relativity, which was specifically cited when Einstein was awarded the Nobel Prize in physics 16 years later.

In 1912 Einstein returned to the Polytechnic School in Zurich as a full professor. By this time he and Mileva had two sons, Hans Albert and Eduard, and Einstein himself had made substantial progress in expanding his original work on relativity to include accelerated motion and gravity. This work was finally published in 1916 as *Die grundlagen der allgemeinen Relativitätstheorie* (Foundation of the general theory of relativity). By this time Einstein

had moved to Germany, accepting positions at the Kaiser Wilhelm Physical Institute and University of Berlin.

As Einstein's theories were confirmed by experiment in 1919, he rose to worldwide fame, earning the Nobel Prize for physics in 1921. But he was also Jewish, and in Nazi Germany fame was not enough to protect him from the rising tide of anti-Semitism. He fled Germany for the United States in 1933, landing at Princeton's prestigious Institute for Advanced Study.

In 1939 Einstein wrote a now-famous letter to President Franklin D. Roosevelt that warned of German experiments that had split the uranium atom, unleashing the potential of devastatingly destructive power. Ironically, this research had been made possible by Einstein's own theories and his now-famous equation $E = mc^2$. Forgoing his normally pacifist principles, Einstein urged Roosevelt to work on an atomic bomb if only to ensure that the Germans did not acquire such a weapon first.

While Einstein's work made the atomic bomb possible, he never worked directly on its development. In fact, his political views made the government view him with suspicion; he was deemed a security risk and denied the clearance that would have allowed him to work on the Manhattan Project. Those working on the project were even forbidden to discuss their research with him. When World War II ended in 1945 with the bombings of Hiroshima and Nagasaki, Japan, and the nuclear tension of the Cold War began, Einstein became more politically active and outspoken than ever, devoting much of his later life to the cause of peace and the fight against the proliferation of nuclear weapons.

entity called space-time, he demonstrated that there is a fundamental connection between matter and energy.

Einstein's most famous equation (perhaps the most famous in all of science), $E = mc^2$ suggested that matter could be destroyed and converted to energy. Likewise, energy could vanish and become bound up in matter. To Einstein, mass was simply another kind of energy.

How much energy is produced when mass is converted into energy via $E = mc^2$? A lot. The speed of light, c, is a very big number, and squaring it gives us a bigger number still. To find out how much energy is contained in an object, we multiply its mass by the speed of light (the speed of light = 299,792,458 m/s, and so the squared value is approximately 8.99×10^{16} m² / s²). If the book you are holding in your hands right now were converted entirely into energy, it could supply the electrical power needs of a large city like New York or Los Angeles for several months.

The implications of this discovery were far-reaching. The enormous amounts of energy locked inside of matter made a vast resource of previously unimagined energy available. Scientists soon realized that converting matter into energy by nuclear fusion was the fuel that

powered the sun and the distant stars. By 1939 they began to look for ways to harness this power through nuclear fission—the splitting of atomic nuclei, converting a tiny fraction of the mass of a uranium atom into energy. This is the process that underlies both nuclear weapons and atomic power plants.

General Relativity

The final task facing Einstein in his development of relativity was to find a complete treatment of all kinds of motion—not just inertial reference frames and objects moving at constant speeds, but reference frames that change speeds and direction and objects that move under the influence of external forces, such as gravity. In fact, Einstein recognized a symmetry between acceleration through space and gravity that led him to formulate a whole new theory about how gravity works.

Imagine a passenger inside a closed automobile with all of the widows painted black. As far as he can tell the car is sitting perfectly still. All of a sudden his head snaps back and he feels pushed back into the seat. What does he think just happened? Since most people know what it

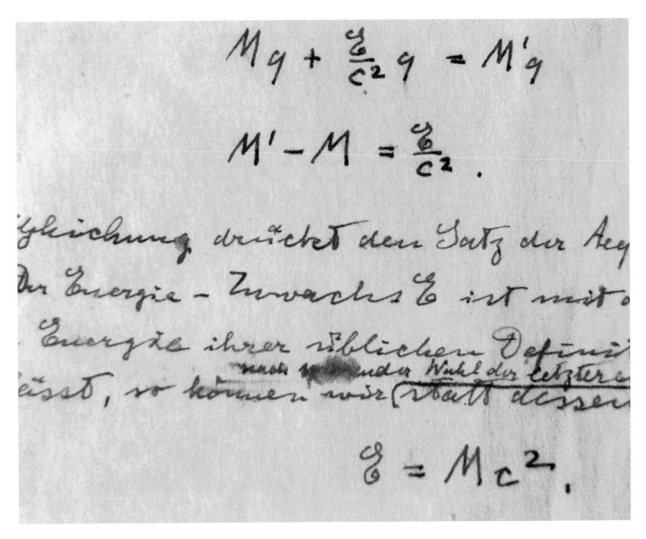

A page from a paper written in 1946 in which Albert Einstein (1879–1955) explained how he derived the formula E=mc², a consequence of his Special Theory of Relativity, first published in 1905. *Josh Reynolds/AP Images.*

feels like to be in a moving car, they would probably assume that the car had just accelerated forward abruptly. But Einstein recognized another possibility. If a very massive object like a large comet or asteroid passed behind the car, the passenger would feel its gravitational pull tugging it backward. With the windows blacked out, Einstein claimed that there was no way to tell the difference between the feeling of accelerating forward and the sensation of gravity pulling backward. Acceleration and gravity are, in some sense, equivalent.

The same effect occurs inside an elevator. When the elevator accelerates upwards, you briefly feel heavier. If you were to take a bathroom scale with you into an elevator and stand on it, you would notice that the scale reads a slightly heavier weight when the elevator begins moving up. If you didn't know the elevator was moving, you might imagine that for a split second that Earth's gravitational pull had increased slightly, temporarily increasing your weight.

Einstein considered this apparent equivalence between acceleration and gravity to be of great significance. Since he believed that observers in different reference frames should agree on all fundamental laws of physics, this thought experiment suggested to him that an observer accelerating through space and an observer sitting still in a gravitational field should be able to use the same equations to describe all their observations about the world. And since we already know from special relativity that moving through space affects measurements of space and time, Einstein argued that gravity must also affect space and time. In fact, he developed a whole new theory of gravitational attraction based on the idea that gravity was nothing more than a kind of curvature of space-time itself.

In the old picture of gravity developed by Isaac Newton, gravity was a force of attraction between all objects with mass. Newton provided no explanation for why two massive objects attracted one another, the force

This image, released by NASA Wednesday, April 19, 2006, shows a visualization of what Einstein envisioned in his theory of relativity, creating a three-dimensional simulation of merging black holes. It was the largest astrophysical calculation ever performed on a NASA supercomputer and provides the foundation to explore the universe in an entirely new way, through the detection of gravitational waves. *AP Images/NASA.*

realize that they are incomplete. While Einstein's theories are very good at describing space, time, and gravity for very large-scale objects like stars and galaxies, they fail when applied to problems of curved space and matter on the scale of atoms and smaller particles. In the realm of quantum mechanics, the science of the tiniest particles of matter, the intrusion of relativity can wreak havoc—producing nonsensical results, wrong answers, and infinities.

Much like the conflict between the theories of Galileo and Maxwell a century ago, scientists face a similar dilemma as they try to reconcile relativity with theories of subatomic particles. New approaches with exotic-sounding names like "string theory" and "supergravity" are taking Einstein's ideas to places he would never have imagined—expanding space-time into a universe with 11 dimensions. But while scientific progress may revise and extend the ideas of relativity, the profound changes in our concepts of space, time, and matter wrought by Einstein's theories cannot be undone.

of gravity was simply there. It was invisible, infinite in range, and more than a little mysterious. But Einstein's picture eliminated the need for an invisible force. In the general theory of relativity, a massive object causes space-time to bend or curve in its vicinity, and other objects appear to be attracted to it as they travel through this curved space. Imagine placing two heavy bowling balls on a trampoline. The bowling balls will bend the trampoline's surface, and if they are close enough together, they will roll toward each other. But there is no mysterious invisible "force" between the bowling balls, they are simply rolling along the curved surface of the trampoline.

The general theory of relativity predicted entirely new phenomena—from black holes to twisted space to ripples in the fabric of the cosmos. The first such prediction—that light should be bent by gravity when it passed a massive object—was tested and confirmed in 1919. But many of general relativity's predictions are still being examined and tested.

■ Modern Cultural Connections

Einstein's ideas, while they can seem unbelievable and strange, have become an indispensable part of our understanding of how the universe works. Scientists also

■ Primary Source Connection

The following article was written by Peter N. Spotts, a science and technology writer for the *Christian Science Monitor.* Founded in 1908, the *Christian Science Monitor* is an international newspaper based in Boston, Massachusetts. The article describes how the American physicist Albert Einstein introduced, and then later disregarded, a mathematical cosmological constant that scientists now say provides a clue to understanding the structure and future of the universe.

EINSTEIN WAS RIGHT! (FOR THE WRONG REASON)

Michael Levi is finding a golden opportunity in Albert Einstein's "greatest blunder."

In 1917, Einstein pulled a fudge factor out of thin air to coax his equations on general relativity into describing a universe astronomers actually saw.

Three years ago, two teams of astronomers startled the scientific world with evidence that this figment of Einstein's chalk board—which the legendary physicist later repudiated—has a measurable impact on the universe.

Now, what Einstein called a cosmological constant appears to be "the largest form of energy in the universe," Dr. Levi, a physicist, says. And, he laments, "we know nothing about it."

That is about to change.

On June 30, the National Aeronautics and Space Administration is slated to launch a spacecraft that will map the afterglow of the big bang in unprecedented detail. Data from the craft are expected to carry further clues about the "dark energy" Einstein's fudge factor describes.

Meanwhile, at the Lawrence Berkeley National Laboratory in Berkeley, Calif., Levi and colleague Saul Perlmutter are designing a space-based telescope that will use light from exploding stars in distant galaxies to trace the history of this energy's influence on the cosmos.

These projects are two among several aimed at unraveling the mysteries of dark energy, which has earned that moniker less for its lack of luminosity than for the ignorance surrounding it, some researchers say.

Fundamentally bizarre phenomenon

Indeed, the very idea of dark energy in the cosmos "is so bizarre from a fundamental-physics point of view that in their heart of hearts, people are still extremely skeptical," says Scott Dodelson, an astrophysicist and theoretician at the Fermi National Accelerator Laboratory in Batavia, Ill. "They say: 'You've got to prove it to me again and again—and you've got to prove it in different ways—or I won't believe it.'"

For his part, Einstein invoked dark energy out of expedience.

As he watched his equations on general relativity unfold, they led to the conclusion that the fabric of space-time is not holding steady, but could either expand or contract, depending on the universe's shape and on how densely it is packed with matter. Above a certain threshold, a densely packed universe would collapse under the pull of its combined gravity. Below that threshold, gravity grip loosens and the universe expands.

At the time, however, astronomers saw no large-scale motion. So Einstein added the cosmological constant, which applied brakes to the universe his equations described. When Edwin Hubble published his evidence in 1925 that the universe was expanding, and distant objects were receding from us at faster rates than close ones, Einstein tossed the cosmological constant out the window.

"Einstein was pulling something out of a hat," says Sean Carroll, an assistant professor of physics at the University of Chicago. "He thought of it as an extra term in his equations changing the response of spacetime to ordinary matter."

But, he adds, "We know now that the term he added is precisely equivalent in all ways" to a form of energy that has come to be known as vacuum energy.

Take a volume of space, he continues, strip it of every form of matter, and general relativity still allows the vacuum to contain energy. Its density would be constant: The amount of vacuum energy in a cubic inch of space in our solar system would match that of a cubic inch of space billions of light-years away. Depending on the value assigned to it, this energy could either retard or accelerate the expansion of the cosmos.

Many cosmologists hold that the universe got its kickstart with the sudden release of vast amounts of pent-up vacuum energy. During the universe's first billion trillion trillionth of a second, it burst from subatomic size to nearly its current volume. The result was the big bang, the primordial explosion that scientists say gave birth to the universe.

Yet only in 1998 were astronomers able to spot what looked to be the action of vacuum energy on the universe.

Two teams—one led by Dr. Perlmutter, the other by Brian Schmidt, with the Mt. Stromlo Siding Springs Observatories in Australia—reported that light from distant supernovae was dimmer than inflation theories said it should be. Currently, inflation holds that all the "stuff" in the universe is just dense enough to prevent collapse, but that expansion will slow without ever reaching a stop.

Yet light measurements taken from the supernovae, roughly 7 billion light-years away, indicated that the galaxies hosting the supernovae were farther away then they should have been if the universe was decelerating.

Last month, Lawrence Berkeley's Peter Nugent and Adam Riess of the Space Telescope Science Institute bolstered the 1998 results with data from a supernova more distant yet, which does fit with the expected rate of deceleration.

One explanation, researchers say, is that scientists have now bracketed the period in the universe's history when matter thinned sufficiently for gravity to give way to the universe's residual vacuum energy, which counteracts gravity and pushes clusters and superclusters of galaxies away from each other.

The most recent observation virtually eliminates objections that the 1998 data might merely be showing the effect of dust obscuring the supernovae, or other measurement errors, says Dr. Nugent, who also is a member of Perlmutter's team.

The Hubble Space Telescope result Dr. Riess and Nugent reported also "gives us a look at the epoch when gravity was dominant," Nugent says.

This distance scale is likely to be one of the most fruitful for followup studies of vacuum energy—if that's what it is—LBL's Levi says, because farther out, when the universe was younger and smaller, matter's higher density would give gravity the advantage over the much weaker vacuum energy. Any closer than about 7 billion light-years, and the vacuum energy's effect would be swamped by "local" regions of space where matter is relatively dense.

This rough distance is the target region for SNAP, a space telescope Perlmutter and Levi have proposed to tease more information out of supernovae. Known as type-1A supernovae, these stellar explosions yield consistent levels and changes of brightness as they evolve.

But type-1A explosions are rare, so researchers say they must gather repeat images of thousands of galaxies in less than two weeks to ensure they can spot an explosion soon enough to track it through its entire cycle.

With initial funding from the Department of Energy, the team is designing a 1.8-meter orbiting telescope with a million-pixel camera to fill that role. If all goes well, Levi estimates that the telescope could be ready for launch in 2008.

Next month, NASA is scheduled to launch the Microwave Anisotropy Probe, a spacecraft that will map the microwave background radiation from the big bang with extreme accuracy.

Tiny changes in the density of the radiation are thought to be the seeds from which galaxies and larger cosmic structures evolved. Buried in those fingerprints of the early universe are signatures that will help cosmologists refine their estimates of the universe's density and the share of that density that different forms of "stuff" account for, says MAP lead scientist Charles Bennett, with the Goddard Space Flight Center in Greenbelt, Md.

Density Defines Future of the Cosmos

Each cosmological theory predicts a certain pattern in the radiation, he says, turning density measurements into "a powerful tool" for pointing toward the correct theory.

Recent microwave background measurements from balloon-borne instruments in Antarctica have provided stunning confirmations of the inflation theories, researchers say. They yield a "flat" universe in which 5 percent of its density consists of matter and forms of energy humans can detect, 25 percent dark matter, which is inferred from the movement of galaxies, clusters, and super clusters. The remaining 65 percent consists of vacuum energy or its equivalent.

For Dodelson, it's an amazing time for astrophysics. "The time scale for change in cosmology is typically 500 years," he says.

"Five years ago, if someone told you there's a cosmological constant, you'd say he's crazy. Today it's just the

reverse. It might take 100 years to figure out what this stuff is. But it's remarkable we're living at this time."

Peter N. Spotts

SPOTTS, PETER N. "EINSTEIN WAS RIGHT! (FOR THE WRONG REASON.)" *CHRISTIAN SCIENCE MONITOR* (MAY 10, 2001).

SEE ALSO *Physics: Aristotelian Physics; Physics: Articulation of Classical Physical Law; Physics: Maxwell's Equations, Light and the Electromagnetic Spectrum; Physics: Newtonian Physics; Physics: Nuclear Physics; Physics: The Quantum Hypothesis;*

BIBLIOGRAPHY

Books

Bodanis, David. *E = mc²: A Biography of the World's Most Famous Equation.* New York: Berkley Trade, 2001.
Fölsing, Albrecht *Albert Einstein: A Biography.* New York: Penguin Books, 1998.
Galilei, Galileo. *Dialogue Concerning the Two Chief World Systems: Ptolemaic & Copernican.* Translated by Stillman Drake. Berkeley: University of California Press, 1953.
Greene, Brian. *The Elegant Universe: Superstrings, Hidden Dimensions, and the Quest for the Ultimate Theory.* New York: Vintage, 2000.
Newton, Isaac. *The Principia: Mathematical Principles of Natural Philosophy.* Translated by I. Bernard Cohen and Anne Whitman. Berkeley: University of California Press, 1999.
Wolfson, Richard. *Simply Einstein: Relativity Demystified.* New York: Norton & Company, 2003.

Periodicals

Spotts, Peter N. "Einstein was Right! (For the Wrong Reason.)" *Christian Science Monitor* (May 10, 2001).

Web Sites

American Institute of Physics. "World Year of Physics 2005: Einstein in the 21st Century." http://www.physics2005.org/ (accessed March 24, 2007).
Hebrew University of Jerusalem, and California Institute of Technology. "The Einstein Archives Online." http://www.alberteinstein.info/ (accessed March 24, 2007).
NOVA. "Einstein's Big Idea." http://www.pbs.org/wgbh/nova/einstein/ (accessed March 24, 2007).

David L. Morgan

Physics: Spectroscopy

Spectroscopy is the process of exposing a substance to some form of radiation, or gathering radiation emitted by the object, and then analyzing that radiation to gather information about the properties of the substance. Different kinds of radiation provide different kinds of information and are used to analyze different kinds of substances.

Spectroscopy's many guises have proved indispensable for research and analysis in all areas of chemistry as well as a variety of other scientific and technical areas. In analytical applications, spectroscopy can fingerprint (uniquely identify) a material by disclosing what elements it contains, and in what proportions. Usually it is not necessary to touch the object being studied, giving spectroscopy wide use in astronomy and astrophysics. At the same time, this remarkable scientific technique can also provide information on atoms and molecules and on even the tinier particles that make up the atom.

■ Historical Background and Scientific Foundations

Early Approaches

Although the fame of English physicist and mathematician Isaac Newton (1642–1727) rests on his contributions to calculus and the theory of gravitation, it was his work in optics that first made his reputation. In one of his early experiments, done in the mid-1660s, he cut a small hole in a window shade and held a glass prism in front of the beam of sunlight that streamed through it. Instead of a simple white spot on the opposite wall, he saw a narrow band of rainbow color—a spectrum.

Newton was not the first to produce a spectrum with a prism. Sixteen centuries earlier, Roman statesman and philosopher Seneca (4 BC–AD 65) observed that sunlight passed through an angular piece of glass revealed all the colors of the rainbow. But Seneca thought the glass had added the colors. Newton was the first to understand that the colors were contained within the light itself. He showed this by placing a second prism opposite the first and recombining the colors to produce white light.

The relationship of these facts to spectroscopy may be shown by a musical analogy. A person with a trained ear can listen to a chord (several notes played at once) and say immediately what notes it consists of. Each note in the chord can be compared with a color. If we mix several colors together, we can make another color, which we might call a color chord. But while the musical version is easily perceived using the human senses to be a mixture of specific notes, the color version is not: it just looks like a new color. The most experienced color expert can't tell what colors went into it simply by looking. What is needed is an instrument that has the capacity to analyze or separate out the different colors in the mixture (elements in the color's spectrum). The spectroscope is an instrument that does this job.

The Basic Optical Spectroscope

A spectroscope permits the visual observation of spectra. Developed in the late nineteenth century, the term derives from two root words: the Latin *spectrum*, meaning "image" or "appearance," and the Greek *skopein*, "to view."

In a simple spectroscope, light rays enter through a slit and are forced to be parallel by a collimating lens. A prism then disperses the light, which is then focused by a lens. In this system red is dispersed the least, violet most. Instruments that record spectral images on

WORDS TO KNOW

ABSORPTION SPECTRUM: The spectrum formed when light passes through a cool gas.

BAND SPECTRUM: A representation of the mixture of different frequencies of light (or other vibrations) from a given source. All substances are made of atoms, and light is emitted from atoms when their electrons are excited (raised to higher energy levels) by receiving energy. When excited electrons lapse to lower energy levels, they emit light at particular frequencies, giving rise to spectral lines. In some substances, especially molecules, these lines can be so closely spaced as to appear like a continuous band rather than a series of spectral spikes. These spectra are termed band spectra.

DIFFRACTION GRATING: A device consisting of a surface into which are etched very fine, closely spaced grooves that cause different wavelengths of light to reflect or refract (bend) by different amounts.

ELECTROMAGNETIC RADIATION: Radiant energy that travels at the speed of light, produced by periodic variations of electric and magnetic field intensities; examples include visible light, radio waves, x rays, and gamma rays.

EMISSION LINE: An atom or molecule emits photons (light particles) when its electrons jump from higher to lower electron levels. These photons, depending on the magnitude of the jumps possible in that atom or molecule, have certain specific energies. When displayed as a spectrum, these emissions, at certain favored frequencies, appear as spikes or vertical lines—emission lines.

FRAUNHOFER LINE: Light from the sun, displayed as a spectrum (a plot showing intensity or brightness over a range of frequencies), features a number of narrow notches. These notches or lines, named after German physicist Joseph von Fraunhofer (1787–1826), who discovered them in 1802, are frequencies of comparatively low intensity light that atoms in the atmosphere of the sun absorb. In effect, they are colors of light that are emitted from the sun's interior but that the sun's atmosphere filters out before they can be radiated into space.

FREQUENCY: The number of waves that pass a given point in a certain length of time.

PHOTON: Smallest individual unit of electromagnetic radiation (light energy). These light particles are emitted by an atom as excess energy when that atom returns from an excited state (high energy) to its normal state.

PRISM: A triangular or wedge-shaped block of glass that breaks up light into its constituent colors.

RADIATION: Energy sent out in the form of waves or particles.

REFRACTION: The deflection from a straight line of a light ray or other energy beam when passing from one optical medium, (such as air) to another (such as glass) in which its velocity is different.

SPECTRAL LINE: A spectrum is a graph of the frequencies found in a mixture of waves, such as light waves. The graph line is higher at frequencies where more energy is present; when energy is concentrated in a narrow range of frequencies, the line draws a spike or vertical line. For example, a pure blue light appears as a spectral spike or line in the blue part of the visible light spectrum. Alternatively, lines can appear as narrow notches in an otherwise continuous-appearing spectrum. These notches are called absorption lines because they usually represent frequencies that have been absorbed after light was emitted by some source, such as the sun.

SPECTROSCOPE: An instrument that performs analysis by producing a spectrum from which information can be extracted.

SPECTRUM: The dispersion of light or other type of energy into its components.

WAVELENGTH: The distance from the crest of one wave to the crest of the next.

photographic plates or in computer memory, spectrographs, are found on most large telescopes.

If, instead of using an eyepiece or recording device, we use a photoelectric cell to measure how much of each color appears, the instrument is called a spectrometer (from the Greek *metron*, "measure"). In this case, however, the prism is rotated so that only a narrow portion of the spectrum is received by the detector at any time. In some cases, the whole operation is done automatically, and the results are fed to a recording spectrophotometer, which prints a complete record of the spectrum. Such devices are used widely in analytical labs to both identify and quantify substances in solution. The first spectrophotometers appeared in the 1870s and were used to measure hemoglobin and other substances in blood and urine.

The Science

Like all waves, light has both a frequency and a wavelength. Wavelength is the distance from one crest of a wave to the next. Frequency is the number of waves that pass a given point in a certain length of time.

The rates of vibration of visible light waves are all near 100 trillion (10^{14}) per second. Visible light's

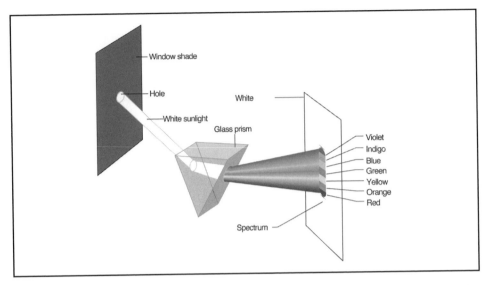

Diagram of visible light spectrum, shining through a glass prism and forming a rainbow. *Illustration by Hans & Cassidy. Cengage Learning, Gale.*

wavelength is very small, in the range of 0.00002 inches (0.508 μm). When working with visible light, scientists use the nanometer (nm, 10^{-9} m) or the angstrom (Å 10^{-10} m). The wavelength of violet light, for example, ranges from about 380 to 440 nm, while red light ranges from about 620 to 760 nm.

Continuous Spectra and Dark Lines

Any substance will emit visible light when heated to a high enough temperature (if it does not evaporate first). For example, the heating coils in an electric heater, when heated to about 1,400°F (760°C), emit light containing a range of colors that combine into a reddish glow. The tungsten filament of an electric bulb incandesces at about 4,200°F (2,316°C), emitting an intense yellowish-white light. An electric arc, which is hotter still—about 6,300°F (3,482°C)—is even more intense and is bluish in color.

All substances at temperatures above absolute zero, in fact, emit electromagnetic radiation (light) at all times, but objects that have not been raised to the relatively high temperatures mentioned in the previous paragraph do not emit visible light. Rather, they emit infrared light—literally, "below-red" light, so-called because its frequencies are lower than that of the lowest-frequency red light visible to the human eye. Spectroscopy can be performed on infrared light, but substances are often heated for spectroscopy because they emit radiation across a wider range of frequencies at higher temperatures, supplying more information. Also, astronomical objects such as stars are generally at high temperatures by their very nature. On the other hand, heating to high temperature may break up the very molecules that the

spectroscopist wishes to analyze (e.g., ink-soaked paper fibers in a criminal investigation). In these cases, the use of infrared spectroscopy is necessary.

When the light from a substance is passed through a spectroscope, the result is a continuous spectrum—a rainbow, in effect, whether it is a rainbow of visible colors or not. The detailed structure of this spectrum or rainbow—made apparent by using appropriate instruments—reveals much information about the substance that emitted the light. The insights gained from the unique qualities of heated substances are the basis for spectroscopy's explanatory power.

In 1802, well over a century after Newton's experiments, the English chemist and physicist William Hyde Wollaston (1766–1828) noticed that the sun's apparently continuous spectrum was crossed by a number of black lines, but he did not know what to make of them. About twelve years later, these dark lines were rediscovered by German physicist Joseph von Fraunhofer (1787–1826). Using better equipment than Wollaston's, he found that the spectrum was interrupted by "a large number of strong and weak lines." He realized that, contrary to Wollaston's theory, these could not be divisions between colors, since some of them clustered near the centers of colors. He still didn't understand their cause, but he put them to work as landmarks or points of reference that could be used to describe precisely any desired part of the spectrum.

Diffraction Grating

What made this discovery possible was Fraunhofer's development of a different kind of dispersion device, the diffraction grating, a transparent or opaque plate

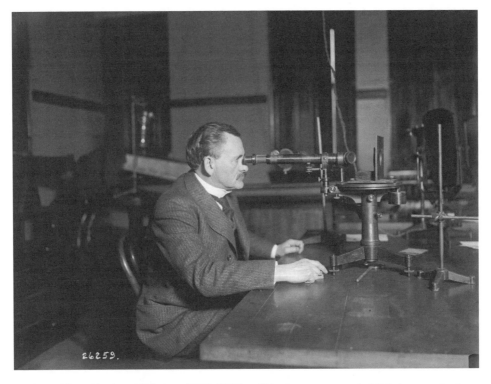

Professor Albert Abraham Michelson (1852–1931) and his giant spectrograph. Michelson was the first American to win a Nobel prize in physics. © *Bettmann/Corbis.*

inscribed with thousands of fine lines per inch. These lines break up the light rays as a prism does, but much more evenly. The result is a spectrum that displays a definite geometrical relationship between the position of the spectral line and its corresponding wavelength (e.g., a separation of 1 nm per mm on a photographic plate). Grating spectroscopes made direct determinations of absolute wavelength possible. This proved particularly useful later in the century, when scientists began to use spectroscopy to unravel the properties of atoms and molecules.

Also, diffraction gratings allow much greater dispersion of light; that is, the wavelengths are spread further apart. This facilitates observation and was perhaps the most important factor in Fraunhofer's work. Prism spectroscopes are commonly used in schools and demonstrations because of their simplicity, low cost, and impressively bright spectra.

■ Introduction

Chemical Analysis

Fraunhofer had an even greater scientific discovery within his grasp, but did not perceive that under the proper circumstances each element produced its own characteristic set of lines and that these lines are as

unique to it as fingerprints are to humans. That insight would be achieved by others.

Correlating spectral lines with distinct elements was finally accomplished by the German physicist Gustav R. Kirchhoff (1824–1887) and German chemist Robert W. Bunsen (inventor of the Bunsen burner, 1811–1899;) in 1859, with the spectroscope once again playing a major role. When gases are heated, the light they emit is quite different from that of a heated solid. If one introduces into the barely visible flame of a Bunsen burner a small amount of table salt, for instance, the flame turns a brilliant yellow. And if we look at this flame through a spectroscope, we do not see the usual rainbow-colored band, but a pair of closely spaced narrow yellow lines on a generally dark background. These bright emission lines, called D-lines, are produced by the sodium in the salt. Appearing at 588.99 nm and 589.59 nm, they are the same lines that Fraunhofer had seen.

In an related experiment, Kirchoff shone sunlight through a sodium flame and, using his spectroscope, observed two dark lines on a bright background exactly where the bright sodium lines were in the earlier experiment. He concluded that the gases in the sodium flame absorbed the D-line radiation from the sun, producing what is now known as an absorption spectrum. Kirchoff realized after further experiments that all the other Fraunhofer lines were also absorption lines. That

JOSEF VON FRAUNHOFER
(1787–1826)

Josef von Fraunhofer (1787–1826) was born in Straubing, Bavaria, now part of Germany. The son of a poor glazier (a craftsman who makes and works with glass), Fraunhofer exemplified the ideal combination of artisan and mathematically inclined theoretician. He acquired this expertise mainly through apprenticeships with glass and mirror makers, combined with independent study. By 1809 he had become a manager in an optical firm in Munich; he was made director of the glassmaking section in 1811; and in 1819, thanks to his optical research, he was elected to the Bavarian Academy of Sciences.

While trying to produce an improved telescope for his firm, Fraunhofer came across the lines seen earlier by Wollaston. But where Wollaston had detected only seven lines, Fraunhofer found hundreds. Using his new equipment, he went on to measure and map the position of the most prominent lines, which he indicated by capital letters, still known as Fraunhofer lines.

He also performed the following experiment: He separated the opening of his spectroscope into two halves. One was lighted by the sun and the other with a sodium flame. A dark pair of lines in the sun's spectrum matched a brilliant pair of lines in the sodium flame.

Fraunhofer did not pursue what now seems such a significant match, and so the origin of the lines remained a puzzle. Nevertheless, his pioneering work with the spectroscope was of great importance. After his untimely death from tuberculosis at 39, his tombstone was engraved with *Approximavit sidera*, Latin for "He approached the stars."

is, gases in the sun's atmosphere absorbed some of the radiation coming from the sun's interior. He concluded from certain Fraunhofer lines in sunlight that sodium and calcium were present in the sun's atmosphere, while lithium, another alkali metal, appeared to be absent.

Carrying this line of reasoning further, Kirchoff and Bunsen compared the solar lines with the spectra of known elements and found a number present in the sun, the most abundant of which was hydrogen. Thus was born an exciting new form of chemical analysis. In 1868 the English astronomer Sir Joseph Norman Lockyer (1836–1920) found a group of lines in the sun's spectrum that did not match those of any element found on Earth. Lockyer correctly attributed his discovery to an unknown solar element, which he called helium (from the Greek *helios*, meaning "sun"). His fellow scientists dismissed the idea, which was hardly surprising, since spectroscopy was still a new science; it seemed presumptuous to posit the existence of a new element merely on the basis of some lines in a spectroscope.

It took almost 30 years for helium to be found on Earth, but Lockyer lived long enough to see himself proved right, and, as is often the case in science, his discovery had other ramifications. The helium line was a prominent one because the sun contains a good deal of helium. In other words, there is a relationship between the amount of an element in a substance and the density of its spectral lines. Very accurate instruments, called densitometers, can measure the density of these lines. Comparing these results leads to quantitative analysis, that is, the determination of how much of a substance is present in a material.

Putting Spectroscopy to Work

During this early period of spectroscopy, interest was largely confined to the chemical and physical properties of matter and with correlating and charting the lines of various elements. Some 30 new elements were discovered in this way. A few were even named for the brilliant spectral lines they provided: thallium (green), rubidium (red), cesium (blue), and indium (indigo).

The Electromagnetic Spectrum

Once the spectrum of visible light was understood, it became reasonable to ask, "What happens at the two ends of the visible spectrum?" In 1800 the English astronomer Sir William Herschel (1738–1822) performed a most intriguing experiment: He took the temperature of the visible spectrum by placing a sensitive thermometer in the colors of light spread out by a prism. He found that the temperature rose as the wavelength grew longer (toward the red); even more interestingly, it continued to rise beyond the red, where no color could be seen.

Although the real meaning of this experiment only became apparent later, Herschel was the first to find a relationship between light and radiant heat (the kind we get from the sun or an outdoor fire). Both are examples of electromagnetic radiation. These heat rays came to be called infrared, from the Latin *infra*, meaning "below"; their wavelength varies from the red end of the spectrum at 760 nm all the way up to 10^6 nm.

Only a year later the German physicist Johann Ritter (1776–1810) found waves that were shorter than visible light. These were named ultraviolet (from the Latin *ultra*, for "beyond") and range from the violet down to about 2.5 nm. As with visible light, there are no sharp divisions between the various forms.

Since then, many kinds of electromagnetic radiation have been observed, ranging from the tiny high-frequency gamma rays at about 0.1 nm to extremely long low-frequency radio waves that range up to 10^{11} nm. Correlations between wavelengths and the matter that absorbs them make them useful as experimental tools to probe the properties of matter. Gamma rays, for instance, are absorbed by atomic nuclei, so they are used in nuclear physics; x rays are absorbed by the individual atoms of a gas or solid, so they are used to study atomic and solid-state phenomena, as well as for medical imaging; and wavelengths in the ultraviolet range (about 380 nm) are useful for studying viruses. Astronomers use a variety of wavelengths for their studies, e.g., longer-wavelength studies are useful for probing the cooler structures in the universe, such as dark cold molecular clouds, while high-energy gamma- and x-ray instruments are used to study remnants from massive dying stars and accretion of matter around black holes.

The Atom

Beginning around 1890 the power of spectroscopy helped unlock the secrets of the atom itself. We know now that atoms consist of a nucleus made up of protons and neutrons, plus surrounding electrons. The line spectrum of a heavy element like iron has many lines and at first glance would suggest that they are randomly arranged. This was found not to be the case.

Ironically, the path to understanding the atom was opened by hydrogen, the simplest element's line spectrum. It showed relatively few lines, arranged in an ordered pattern. Careful study revealed mathematical relationships for the lines in all the elements; materials scientists used this knowledge to learn more about how electrons act inside their atoms.

Scientists knew that a substance can only give off a spectrum when it has been given energy in some way, as by the hot flame of the Bunsen burner or the electrical input of a neon lamp. The excited substance then "de-excites" itself, but in a very special way. Because electrons' energy can only drop to a set of specific energy levels, only discrete amounts of energy can be lost. This loss corresponds to the set of sharp lines that appear in a line spectrum. This stunning discovery led to the mathematical formulas that accurately predicted their positions.

In 1913 the Danish physicist Niels Bohr (1885–1962) used one of these formulas to deduce the first really useful picture of the atom. It's probably safe to say that most of the information that atomic scientists have obtained about the nature of the atom has been obtained using spectroscopy.

Band Spectra

In addition to continuous and line spectra, there is a third type: band spectra. High-powered instruments indicate that these bands actually consist of many narrow lines, numerous at the darker end and gradually fading away at the other. Band spectra are caused by changes in molecules, rather than individual atoms. In other words, by using spectroscopic techniques with other kinds of radiation, scientists have been able to learn about the remarkable flexibility of the molecular world—the vibrations, rotations, and other motions that make up the vast majority of the substances.

Because different molecules absorb energy at different wavelengths, different forms of radiation are used, depending on what kind of information is being sought. Infrared spectroscopy, for instance, is particularly useful for showing how the molecules in a sample vibrate; ultraviolet radiation, which has more energy, can provide information on the chemical bonds in a molecule.

Spectroscopy also refers to several fields in which electromagnetic radiation is not involved. Separation or dispersion spectroscopy separates a beam of particles by energy or some other property. Beta-ray (electron) spectroscopy, for example, measures the energies of electrons emitted from nuclei. Mass spectrometers sort charged particles by mass; mass spectrometry played an important part in the discovery of isotopes, which are used in medicine, industry, and various other branches of science and technology.

Other types of spectroscopy are fluorescence, microwave, magnetic, alpha scattering, activation analysis, Raman, and resonance. All differ, but utilize the same basic idea. For example, nuclear magnetic resonance spectroscopy, which depends on small energy transitions in molecules in a magnetic field, has become an important method of analysis in chemical laboratories.

■ Modern Cultural Connections

Various forms of spectroscopy have been basic to modern advances in science and technology, with all that the changes imply for cultural and economic activity. Geologists, including petroleum geologists, have found spectroscopy invaluable for studying earth and rock samples. Metallurgists use it to study the purity of metals and alloys. It has been used in medicine to determine the presence or lack of certain metals in blood. The spectroscope has been employed extensively in astronomy, where it provides information about the velocities, composition, magnetic characteristics, and temperatures of stars; the motions of stars orbiting in pairs; and so on. The twin rovers that were landed on Mars by NASA in 2004, and were still exploring that planet as of early 2008, are each equipped with two spectrometers to gather data on the composition of minerals on that planet, a miniature thermal-emission spectrometer that can image whole landscapes at once, and a Mossbaüer spectroscope for close-up work on individual objects.

The development of lasers in the 1960s had a powerful effect on the field, because the special qualities of laser light can greatly increase the resolution and sensitivity of conventional spectroscopic techniques. In the early 2000s, a 785-nm infrared laser was being tested for use in a spectroscopic system for possible rapid diagnosis of pancreatic cancer, for example.

In forensics (crime investigation), infrared spectroscopy advanced greatly from the early 1990s to the early 2000s with the growing availability of cheap optics and fast computers for data processing. Infrared spectroscopy is now routinely applied by forensics laboratories to drugs, paints, fibers, explosives, inks, document materials, blood, and soil. Materials from questionable documents, crime scenes, and many other sources are analyzed to determine their unique spectroscopic (and thus chemical) identities, producing hard scientific evidence that can stand up in court. In 1989, for example, an instant-win lottery scratch-card that appeared to be the winning card for a large sum of money was challenged because its number did not appear in computer records as a winning number. Infrared spectroscopy of the card proved that it was authentic, not a forgery—the materials of which it was made were the precise materials of all other authentic cards—and the winnings were paid out to the card's owner.

■ Primary Source Connection

In 2007, an Oklahoma teen gained national recognition for developing an accurate, low-cost spectrograph. Mary Masterman's invention may improve global access to inexpensive spectrograph technology and garnered her a coveted Intel Science Talent Search prize.

OKLA. TEEN WINS $100,000 SCIENCE PRIZE

OKLAHOMA CITY—A 17-year-old girl won a scholarship worth $100,000 for building an inexpensive yet accurate spectrograph that identifies the "fingerprints" of different molecules.

Mary Masterman, a senior at Westmoore High School in Oklahoma City, was named the winner Tuesday of the annual Intel Science Talent Search.

More than 1,700 high school seniors across the nation entered the contest, which is in its 66th year.

Spectrographs, which measure wave lengths, are used in research such as astronomy and medicine and in industry. For example, they can be used as a sensing device to look for explosives or drugs or to help determine how old an art work is through its pigments.

They can cost as much as $100,000, but Masterman's invention—made of lenses, a laser, aluminum tubing and a camera—cost less than $1,000, Intel said.

Masterman received the honor from Intel Corp. Chairman Craig Barrett during a banquet Tuesday night in Washington.

"It was a complete surprise," Masterman said. "I wasn't expecting it."

The 40 finalists spent the last week in Washington, where they exhibited their projects at the National Institute of Science and met government officials including Vice President Dick Cheney and U.S. Education Secretary Margaret Spellings.

Masterman said she has been interested in science "ever since I was little. I can't remember ever not being interested." She credits her parents with encouraging her.

She said she has not decided where she will attend college but would eventually like to become a physicist or chemist.

Among the former winners of the competition's top award are six Nobel Laureates, three National Medal of Science winners, 10 MacArthur Foundation Fellows and two Fields Medalists.

"You're not only dealing with the top young person in the science field in the country in Mary, but you're dealing with 40 finalists who are doing breaking-edge research in total," said Brenda Musilli, Intel's director of education. "It's really something that's hard to imagine, how a young person like Mary could even achieve this level of capability at such a young age."

Murray Evans

EVANS, MURRAY. "OKLA. TEEN WINS $100,000 SCIENCE PRIZE." ASSOCIATED PRESS NEWSWIRE (MARCH 14, 2007).

SEE ALSO *Astronomy and Space Science: Pulsars, Quasars, and Distant Questions; Physics: Lasers.*

BIBLIOGRAPHY

Books

Colthup, Norman B., Lawrence H. Daly, and Stephen E. Wiberley. *Introduction to Infrared and Raman Spectroscopy.* New York: Academic Press, 1975.

Duckworth, H.E., R.C. Barber, and V.S. Venkatasubramanian. 2nd ed. *Mass Spectroscopy.* Cambridge and New York: Cambridge University Press, 1986.

Hollas, J. Michael. *Modern Spectroscopy.* 4th ed. Hoboken, NJ: John Wiley & Sons, 2004.

Jackson, Myles W. *Spectrum of Belief: Joseph von Fraunhofer and the Craft of Precision Optics.* Cambridge, MA: MIT Press, 2000.

Jenkins, Reese V. "Fraunhofer, Joseph." *Dictionary of Scientific Biography*, Vol. V. New York: Charles Scribner's Sons, 1972.

Macomber, Roger S. *NMR Spectroscopy: Basic Principles and Applications.* San Diego: Harcourt Brace Jovanovich, 1988.

McGucken, William. *Nineteenth-Century Spectroscopy: Development of the Understanding of Spectra, 1802–1897.* Baltimore: Johns Hopkins University Press, 1969.

Newton, Isaac. "A Letter of Mr. Isaac Newton." Reprinted in *Isaac Newton's Papers & Letters On Natural Philosophy.* Edited by I. Bernard Cohen. Cambridge, MA: Harvard University Press, 1958.

———. *Opticks; or, A Treatise of the Reflections, Refractions, Inflections & Colours of Light.* 4th ed. London, 1730. Reprinted. New York: Dover Publications, Inc., 1952, 1979.

Periodicals

Evans, Murray. "Okla. Teen Wins $100,000 Science Prize." *Associated Press Newswire* (March 14, 2007).

Rinsler, M.G. "Spectroscopy, Colorimetry, and Biological Chemistry in the Nineteenth Century." *Journal of Clinical Pathology* 34, no. 3 (March 1981): 287–291.

Wilkinson, T.J., et al. "Physics and Forensics Synchrotron Radiation and Infrared Spectroscopy Are Joining Forces to Fight Crime." *Physics World* 15, part 3 (2002): 43–46.

Web Sites

College of Wooster (OH). "Spectroscopy. Introduction to Spectroscopy: What Is Spectroscopy?" http://www.wooster.edu/chemistry/is/brubaker/intro_spectroscopy.html (accessed April 15, 2008).

Imperial College, London. Department of Chemistry. "NMR Spectroscopy. Principles and Application." http://www.ch.ic.ac.uk/local/organic/nmr.html (accessed March 28, 2008).

Spectroscopy Magazine. http://www.spectroscopymag.com/spectroscopy (accessed March 6, 2008).

University of California at Irvine, Department of History. "Spectroscopy and the Rise of Astrophysics" http://eee.uci.edu/clients/bjbecker/astrophysics.html (accessed March 28, 2008).

Hal Hellman

Physics: The Bohr Model

■ Introduction

One early theory of the structure of the atom was the Bohr model, developed in 1913 by Danish physicist Niels Bohr (1885–1962). By using quantum theory, Bohr's model improved on the earlier atomic models of British physicists J.J. Thomson (1856–1940) and Ernest Rutherford (1871–1937), which were based on classical (Newtonian) physics.

While working on his doctoral dissertation at Copenhagen University, Bohr studied the theory of radiation being developed by German physicist Max Planck (1858–1947). After graduation, Bohr worked in England with Thomson and subsequently with Rutherford. It was while working in England that Bohr developed his model of atomic structure.

■ Historical Background and Scientific Foundations

Before Bohr, the foremost classical model of the atom was the Saturnian or planetary model. This theory was suggested by Japanese physicist Nagaoka Hantaro (1865–1950) in 1904 and further developed by Rutherford. It proposed that the atom is structured like the Saturnian ring system or the solar system. In those larger systems, small objects orbit a central, massive object: The planetary model suggested that in the atom, small, electrically negative electrons orbit a relatively massive, positively charged nucleus. However, this view was controversial, and in 1907, J.J. Thomson proposed what was called the "plum pudding" model. In this alternative theory, negative electrons are embedded in a uniform globe of positive charge, much like plums in a pudding.

The plum-pudding model was disproved by 1911, when Rutherford showed that alpha particles fired at atoms sometimes bounce right back the way they came, as if they had struck a massive obstacle in the atom—a nucleus. Some form of planetary model was necessary to explain the behavior of atoms.

The classical planetary model of the atom allowed electrons to orbit at any distance from the nucleus. This implied that when an atom—for example, a hydrogen atom—is heated, its electrons should move away from the nucleus, then move back toward it again, giving up energy in the form of electromagnetic radiation (light). This offered a mechanism for the emission of electromagnetic radiation by atoms. However, the model predicted that an atom should produce a continuous spectrum of colors (light at all different rates of vibration) as it cooled because an electron would gradually give up its energy as it spiraled back closer to the nucleus. Yet spectroscopic observations showed that hydrogen atoms produce only certain colors when they are heated, which causes their electrons to shift repeatedly and rapidly between lower- and higher-energy states. Also, the description of electromagnetic radiation by Scottish physicist James Clark Maxwell (1831–1879) predicted that an electron orbiting a nucleus according to Newton's laws, like a miniature planet, would continuously lose energy and eventually, unless resupplied with energy from some outside external source, fall into the nucleus. In other words, a classical planetary atom could not be stable, though in the real world, most atoms are perfectly stable.

To account for the observed properties of hydrogen, Bohr proposed a radical new form of the planetary model. In Bohr's model, electrons can exist only in certain orbits—at certain distances from the nucleus—and that, instead of traveling between orbits like tiny spacecraft traversing the solar system, electrons make instantaneous (quantum) leaps or jumps between allowed orbits.

WORDS TO KNOW

BOHR MODEL: Neils Bohr (1885–1962), a Danish physicist, proposed the Bohr model of the atom in 1913, at a time when physicists were still struggling to understand atomic structure. In the Bohr model of the atom, negatively charged electrons are bound to a positively charged (and much more massive) nucleus by electrostatic attraction, but prevented from falling into the nucleus by the fact that they can only exist in certain permitted energy states as described by quantum mechanics.

ELECTRON: A subatomic particle having a charge of −1.

NUCLEUS: Any dense central structure can be termed a nucleus. In physics, the nucleus of the atom is the tiny, dense cluster of protons and neutrons that contains most of the mass of the atom and that (by the number of protons it contains) defines the chemical identity of the atom. In astronomy,

the large, dense cluster of stars at the center of a galaxy is the galactic nucleus. In biology, the nucleus is a membrane-bounded organelle, found in eukaryotic cells, that contains the chromosomes and nucleolus. Intact eukaryotic cells are comprised of a nucleus and cytoplasm. A nuclear envelope encloses chromatin, the nucleolus, and a matrix, which together fill the nuclear space.

ORBITAL: In an atom or molecule, a pattern of electron density or probable location that may be occupied by an electron. An orbital is not a path traced in space by an electron, as an orbit is traced by a planet or satellite in space: This is why the term "orbital" rather than "orbit" is used.

PHOTON: Smallest individual unit of electromagnetic radiation (light energy). These light particles are emitted by an atom as excess energy when that atom returns from an excited state (high energy) to its normal state.

In the Bohr model, the lowest, most-stable energy level is the innermost orbit (which is called an "orbital" or "shell" to distinguish it from the orbits followed by large objects like planets and satellites). The orbitals in any atom are numbered from the inmost outward. The first orbital of the hydrogen atom is spherical and is assigned a principal quantum number (n) of $n = 1$. Additional orbital shells are assigned values $n = 2$, $n = 3$, $n = 4$, etc. The orbital shells are not spaced at equal distances from the nucleus; rather, the radius of each shell increases as the square of n. Increasing numbers of electrons can fit into these orbital shells according to the formula $2n^2$. Accordingly, the first shell can hold up to 2 electrons, the second shell ($n = 2$) up to 8 electrons, the third shell ($n = 3$) up to 18 electrons. Subshells or suborbitals (designated s, p, d, and f), having differing shapes and orientations, allow each element a unique electron configuration.

As electrons move away from the nucleus, they gain potential energy and become less stable, tending to fall back into lower-energy orbitals closer to the nucleus. An atom with all its electrons in their lowest allowable energy orbits is said to be in a ground state, while an atom with one or more electrons raised to higher-energy orbits is said to be in an excited state. Electrons in atoms may acquire energy from thermal collisions (running into other atoms), collisions with subatomic particles, or absorption of a photon (a quantum packet of light energy). Of all the photons that an atom encounters, only those carrying an amount of energy equal to the energy difference between two allowed electron orbits will be

absorbed. Atoms release energy by giving off photons as electrons return to lower-energy orbits.

The leaping of electrons between orbits in the Bohr model accounted for Planck's observation that atoms emit and absorb electromagnetic radiation only in certain fixed energy units called quanta, rather than in a smooth range of in-between values. Bohr's model also explained quantum aspects of the photoelectric effect described by Albert Einstein in 1905.

According to the Bohr model, when an electron is excited by energy it jumps from its ground state to an excited state (i.e., a higher energy orbital). The excited atom can then emit energy only in certain fixed quantities as electrons jump back to lower-energy orbits located closer to the nucleus. This energy is emitted as quanta of electromagnetic radiation (photons, particles of light) that have the same energy as the difference in energy between the orbits jumped by the electron. Electron movements between different orbitals explain certain features of electromagnetic spectra. For example, in hydrogen, when an electron returns to the second orbital ($n = 2$) from a higher orbital, it emits a photon with energy that corresponds to a particular color or spectral line found in the Balmer series of bright lines in the visible portion of the electromagnetic spectrum. The particular color of the photon emitted depends on which higher orbital the electron jumps from. When the electron returns all the way to the innermost orbital ($n = 1$), the photon emitted has more energy and helps form a line in the Lyman series of bright lines, which is found in the higher-energy, ultraviolet portion of the spectrum.

Niels Bohr (1885–1962), Danish physicist and Nobel Prize winner in 1922, sits at his desk in an undated photograph. © *Bettmann/Corbis.*

When the electron returns to the third quantum shell ($n = 3$), the photon emitted has less energy and helps form a line in the Paschen series, which is found in the lower-energy, infrared portion of the spectrum.

■ Modern Cultural Connections

Later, more mathematically complex models based on the work of French physicist Louis Victor de Broglie (1892–1987) and Austrian physicist Erwin Schrödinger (1887–1961) that took into account the particle-wave duality of electrons proved more useful to describe atoms with more than one electron. According to these later models, an electron does not move in a circular path around the nucleus like a planet or moon: Rather, it is smeared out in space so that there is some probability of finding it at any points in its orbital. An electron is therefore most accurately imagined as occupying a fuzzy shell centered on the nucleus. Such shells have a variety of shapes, from simple spheres, to two-lobed hourglass shapes, to complex, many-lobed shapes. Today's standard model of the atom, which incorporates the particles known as quarks (whose existence was first proposed in 1961), further refines the Bohr model so that its properties match all those that have been ob-

served so far in experiments. Despite the need for such improvements, Bohr's original, simpler model remains fundamental to the study of chemistry. For instance, the valence shell concept used to predict an element's reactive properties is derived from the Bohr model of the atom.

The Bohr model remains a landmark in scientific thought that poses profound questions for scientists and philosophers. Along with related claims about the physical world made by quantum theory, it has called common-sense concepts of space, time, and language into question. For example, the idea that electrons make quantum leaps from one orbit to another, as opposed to simply moving between orbits, seems counter-intuitive—that is, outside the range of human experience or imagination. How can an object move from A to B without occupying a continuous series of locations between A and B? Yet electrons and other subatomic particles do just this. Bohr once said, "Anyone who is not shocked by quantum theory has not understood it."

Because quantum physics describes an atomic world that does not obey the rules of everyday physical common sense, the details of atomic structure cannot be portrayed in drawings without leaving out some aspect of their nature. This is why almost all pictures or drawings showing the structure of the atom show a simplified

Computer artwork depicting an atom. In the center of the atom is the nucleus, made up of a tightly packed cluster of protons and neutrons, around which electrons orbit. This classical model of the atom is known as the Bohr Model, after Danish physicist Neils Bohr (1885–1862). *Scott Camazine/Photo Researchers, Inc.*

version of the Bohr model, with electrons orbiting the nucleus like tiny, round planets. These pictures can be an aid to elementary understanding of the atom, but do not show what atoms look like. In fact, in quantum physics, atoms do not look like anything. They do not have any appearance at all because they cannot interact with light as do larger objects. This fact arises from their very nature, not from a mere lack of appropriate lighting, so atoms cannot be thought of as invisible or hard to see: they simply lack any visual appearance whatever, seen or unseen.

The phrase "quantum leap" (or "quantum jump") has entered everyday English as referring to a very large, sudden change. In the physics of the Bohr model of the atom, however, this phrase refers to the shifting of an electron from one orbital to another without traveling through any intermediate point. Such instantaneous shifts can be made only by very small particles moving over extremely small distances—almost the opposite of the popular phrase's meaning. This is an example of how ideas from physics often enter popular culture in distorted forms. Another example is the saying, "Einstein showed that everything is relative." He did not: He showed that some things are relative and others are not relative.

Bohr received a Nobel Prize in 1922 for his work on quantum physics.

SEE ALSO *Physics: The Inner World: The Search for Subatomic Particles; Physics: The Quantum Hypothesis; Physics: Wave-Particle Duality.*

BIBLIOGRAPHY

Books

Lakhtakia, Akhlesh, ed. *Models and Modelers of Hydrogen: Thales, Thomson, Rutherford, Bohr, Sommerfeld, Goudsmit, Heisenberg, Schrodinger, Dirac, Sallhofer.* Singapore: World Scientific Publishing Company, 1996.

Bohr, Niels. *The Unity of Knowledge.* New York: Doubleday & Co., 1955.

Web Sites

Bohr, Niels. "Atomic Structure." *Nature.* http://dbhs. wvusd.k12.ca.us/webdocs/Chem-History/ Bohr-Nature–1921.html (accessed January 8, 2008).

———. "On the Constitution of Atoms and Molecules." *Philosophical Magazine.* July, 1913. http://dbhs.wvusd.k12.ca.us/webdocs/ Chem-History/Bohr/Bohr–1913a.html (accessed January 8, 2008).

K. Lee Lerner

Physics: The Inner World: The Search for Subatomic Particles

■ Introduction

Greek philosophers of the fifth century BC were among the first people to guess that the world is made up of invisibly small particles. They speculated that these ultimate building blocks were *atomos*, uncuttable, from which we derive our word "atom." As it turns out, the elementary units of chemistry are not *atomos* at all, but can be divided into smaller particles, namely protons, neutrons, and electrons. Even protons and neutrons are not elementary, but are made up of smaller particles called quarks. Quarks are presently classed as truly elementary or *atomos*. Electrons also appear to be truly elementary.

Protons, neutrons, electrons, and quarks are subatomic particles, meaning that they are below (sub) the atomic level. Hundreds of other subatomic particles are not found in atoms. Most of these appear only under certain rare conditions, then rapidly break down into other particles and energy.

The basic structure of the atom and the existence of subatomic particles was not known until the early twentieth century. The search for experimental evidence of certain elementary particles and their properties continues in the early 2000s, over a century after the discovery of the electron in 1897.

■ Historical Background and Scientific Foundations

Scientific and Cultural Preconceptions

Although ancient Indian thinkers such as Kanada (c.600 BC) proposed atomic theories of matter independently of the Greeks, modern scientific investigation of atoms and subatomic particles grew out of the European intellectual heritage. The philosopher Democritus (c.460–c.370 BC) was the first known Greek exponent of atomism, the belief that the material world is composed of atoms. Democritus had no experimental evidence for atoms—the technology of his day was too limited—but it struck him that atoms could explain many observations. The properties of various substances could, he suggested, be due to the properties of their atoms: sticky substances consisted of atoms covered with tiny hooks, liquids consisted of slippery, round atoms, different flavors were the results of differently shaped atoms coming in contact with the tongue, and so on. Between atoms, he argued, was void—nothingness, empty space. "There is," he taught, "nothing but atoms and the void." After Democritus, Greek philosopher Plato (c.427–c.347 BC) and Roman philosopher and poet Lucretius (99–55 BC) also propounded forms of atomism.

For over two thousand years, people did not know how to use atoms to explain observations in an exact way or even to verify that atoms are real. Until the thirteenth century, the dominant authority in European scientific thought was the Greek philosopher Aristotle (384–322 BC), who denied the possibility of both atoms and the void. In 1277, theologians of the Church condemned Aristotle's strict determinism and asserted that God could create anything, including atoms and void. This opened room for philosophical speculation about multiple worlds, infinity, gravity, projectile motion, and atoms. European scientists proceeded to speculate about atoms, more or less fruitlessly, for the next 500 years.

Only after the scientific revolution of the sixteenth and seventeenth centuries did atomism begin to be used for the quantitative explanation of natural phenomena. For example, English scientist John Dalton (1766–1844) proposed in 1805 that the chemical law of multiple proportions could be explained by the weights of different atoms. The law of multiple proportions is the observation that whenever two or more elements combine to

WORDS TO KNOW

CHARGE: Describes an object's ability to repel or attract other objects. Protons have positive charges while electrons have negative charges. Like charges repel each other, while opposite charges, such as protons and electrons, attract one another.

COLLIDER (PARTICLE): A particle collider is a machine that accelerates (speeds up) atomic nuclei or subatomic particles such as protons to velocities approaching the speed of light. Particles are made to travel in either a straight line (in a linear collider) or a circle (in a ring collider) and eventually to collide with other particles, whether in a stationary target or a beam of particles traveling in the opposite direction. The goal is to produce high-energy collisions that spew out new particles whose interactions with surrounding detectors can be recorded. These particle showers reveal information about the physics of fundamental particles.

ELEMENTARY PARTICLE: A subatomic particle that cannot be broken down into any simpler particle.

HIGGS BOSON: The only particle predicted by the current Standard Model of particle physics that has not yet (as of early 2008) been experimentally observed. The Higgs boson, if it exists, is responsible for endowing all matter with mass. Intense and expensive searches for evidence of the Higgs boson are ongoing at the CERN laboratory in Europe and elsewhere.

PHOTOELECTRIC EFFECT: The phenomenon in which light falling upon certain metals stimulates the emission of electrons and changes light into electricity.

PHOTONS: Photons are light particles. According to modern physics, it is not accurate to imagine any subatomic particle, including a photon, as a tiny, hard ball: rather, a particle also has wave properties and is spread out through space. Thus, light has wave properties, despite being conveyed by photons.

QUANTUM PHYSICS: Quantum physics, also called quantum mechanics, is the science of the behavior of matter and energy at any scale where their quantum nature is significant (which is usually, though not always, the atomic or subatomic scale). The quantum nature of matter and energy refers to the restriction of changes to definite, sudden steps rather than smooth passage through a range of in-between values. Quantum physics began to be developed in the early twentieth century and continues to be one of the most vigorously investigated areas of physics.

QUARKS: Subatomic particles that combine in three to form other particles (neutrons, protons, and others) and are never observed alone. Quarks come in six varieties or flavors—the term is arbitrary and has nothing to do with the sense of taste—namely, charm, strange, top, bottom, up, and down.

RADIOACTIVE DECAY: The predictable manner in which a population of atoms of a radioactive element spontaneously disintegrates over time.

STANDARD MODEL: In particle physics, the theory that describes the fundamental particles that make up ordinary matter and the particles that are exchanged between them to create all of the fundamental forces except gravity. Gravity is not described by the Standard Model but by the theory of general relativity; scientists hope to someday modify the two so that they can be united into a single theory that describes all known phenomena.

WAVE-PARTICLE DUALITY: The combination of wavelike and particle-like properties possessed by all physical objects, which is especially apparent for single subatomic particles such as electrons. Under some conditions, a particle behaves as a small object that can be definitively found in one place or another; under other conditions, a particle behaves as a wave that is spread throughout space and can add or subtract from similar waves in a way that would be impossible for a tiny, hard object that has a single, well-defined location at all times.

make a compound, the ratio of their combining weights is always a fraction of small whole numbers. For example, 8 grams of oxygen always combine with 1 gram of hydrogen (an 8:1 ratio) to produce 9 grams of water, while 16 grams of oxygen always combine with 1 gram of hydrogen (a 16:1 ratio) to produce 17 grams of hydrogen peroxide. Dalton proposed that this fact could be explained by supposing that compounds like water consist of clusters of atoms. (Today we call these clusters *molecules*.) Each particle of water, for example, consists of two hydrogen atoms (written H_2) attached to a single oxygen atom (O), which we write as H_2O. Since each

oxygen atom weighs about 16 times as much as a hydrogen atom, 1 gram of hydrogen has twice as many atoms in it as 8 grams of oxygen—just the right numbers to combine to make 9 grams of water, as recorded by the law of multiple proportions.

Dalton's theory explained many facts, but there was still no direct evidence for the existence of molecules or atoms and no suspicion of the existence of subatomic particles. The inner structure of the atom itself was to remain mysterious for another century.

In the late 1700s, a curious phenomenon was observed: flashes of light were produced by electrical

IN CONTEXT: "THE MOST INCREDIBLE EVENT"

English physicist Ernest Rutherford (1871–1937) discovered the structure of the atom in 1911. He and his colleagues at the Cavendish Laboratory of Cambridge University, England, were studying the properties of alpha particles, high-speed particles emitted like tiny bullets from exploding atoms. At Rutherford's suggestion, Hans Geiger (1882–1945) and Ernest Marsden (1889–1970) verified that about one alpha particle in 20,000 bounces directly back toward its source from a thin foil of gold. This was a surprise, because the favored model of the atom in those days was the plum-pudding model. This conceived the atom as a droplet of positively-charged goo containing negatively-charged electrons like plums stirred into a pudding. But an alpha particle would not be bounced backward by colliding with a droplet of goo, nor by colliding with an electron embedded in the goo (an alpha particle is thousands of times more massive than an electron). To observe alpha particles bouncing directly backward was, Rutherford wrote, "the most incredible event that has ever happened to me in my life. It was almost as incredible as if you fired a 15-inch [artillery] shell at a piece of tissue paper and it came back and hit you."

The explanation that Rutherford proposed was that the positively-charged mass of each atom is not spread out as a fluid, but concentrated in its center in a tiny lump about a thousandth the width of the atom. This central lump or nucleus is surrounded by electrons, which account for a tiny fraction of the atom's mass but for most of its size.

Rutherford's explanation was correct. An alpha particle, incidentally, is now known to consist of a two protons paired with two neutrons—the nucleus of a helium atom minus its usual complement of two electrons.

In 1887, German physicist Heinrich Hertz (1857–1894) observed the photoelectric effect, in which light knocks electrons free from a metal surface. In 1905, German-American physicist Albert Einstein (1879–1955) showed that the amounts of energy possessed by the electrons liberated by the photoelectric effect could only be explained by assuming that light also consists of particles. He called the light particle a photon. Photons have nonzero energy, but, unlike electrons, zero mass. Light also remained a wave, however, when considered from some points of view, and this double nature was referred to as wave-particle duality.

A fuller understanding of the atom and its constituent particles came with the discovery of the atomic nucleus by English physicist Ernest Rutherford (1871–1937) in 1911. Rutherford showed that almost all the mass of an atom is concentrated in a tiny lump at its center, the nucleus. Electrons surround the nucleus, accounting for little of the atom's weight but most of its size. Since electrons were known to be negatively charged and atoms normally have zero charge, Rutherford and his colleagues reasoned that the nucleus must be positively charged. In 1918, he proposed that the nucleus of the simplest and lightest of all atoms, the hydrogen atom, is an elementary particle. In 1920, he dubbed this hypothetical particle the *proton*. A proton is 1836 times as massive as an electron and carries an equal (but opposite) electrical charge.

Yet a puzzle remained: elements heavier than hydrogen did not seem to contain enough hydrogen nuclei (protons) to balance out the charges on their electrons. For example, a calcium atom weighs 40 times more than a hydrogen atom, but its nucleus carries an electrical charge only 20 times that of a hydrogen nucleus. Physicists proposed that these heavier nuclei actually did contain more protons—in the case of calcium, 40 of them—but also electrons to cancel out some of those positive electrical charges. A calcium nucleus, if this theory were correct, would consist of 40 protons (charge $+40e$, where e is the charge of a single electron) and 20 electrons (charge -20), and so have a net charge of $+40-20 = +20$. The atom as a whole would be electrically neutral—zero total charge—when this nucleus was surrounded by a cloud of 20 electrons.

This reasonable-sounding theory turned out to be wrong. Only with the discovery of a fourth subatomic particle in 1932 by English physicist James Chadwick (1891–1974), the neutron, was the mystery solved. Now the basic structure of matter could be explained in terms of three subatomic particles: Atoms consist of protons and neutrons (or, in the case of hydrogen, a lone proton) bound together into dense, massive nuclei and normally surrounded by clouds of electrons. The number of electrons in an atom usually equals the number of protons in the nucleus. When the number of electrons attached to an atom is greater or smaller than the number of protons

discharges in glass tubes out of which most of the air had been pumped. In the 1850s, it was discovered that when even more air was removed (by improved pumps), mysterious rays seemed to be emitted by a positively-charged piece of metal (cathode) at one end of the tube, travel the length of the tube, and cause a green glow on the glass at the far end. The rays were dubbed "cathode rays." In the 1890s, British physicist J.J. Thomson (1856–1940) began a careful analysis of cathode rays. He hypothesized that they consist of a stream of negatively-charged particles. The existence of such particles, *electrons*, had been suggested by Irish physicist G. Johnstone Stoney (1826–1911) in 1874. By measuring the bending of cathode rays by electric and magnetic fields, Thomson was able to calculate the mass-to-charge ratio of the electron, that is, how much electrical charge there is for a given amount of electron mass. The electron was thus the first subatomic particle to be studied.

in the nucleus, the atom bears a negative or positive electrical charge. Electrical currents consist of free electrons in motion. Chemical reactions directly involve only the electrons of atoms, never the nuclei. Photons are not a building block of atoms, but can be emitted by atoms. They explain light and other forms of electromagnetic radiation such as radio waves and x rays.

Yet it already was known that the world of subatomic particles could not really be this simple. Observations of radioactive decay, which occurs when individual atomic nuclei spontaneously explode, releasing photons and various high-speed particles, had convinced physicists that an electrically neutral particle with very small mass must also exist—the neutrino (literally, "little neutral one"). The existence of the neutrino was proposed by Austrian physicist Wolfgang Pauli (1900–1958) in 1930, but not proved experimentally until 1956. As of 2007, it was still not known whether neutrinos have zero mass, like the photon, or very small mass; however, it was known that over 50 trillion neutrinos pass through the human body every second. Few of these particles interact with the body's matter; they pass right through.

A new theory, known as quantum physics, was also lengthening the list of subatomic particles in the 1920s and 1930s. In 1930, British physicist Paul Dirac (1902–1984) proposed for theoretical reasons that a positively charged opposite or antiparticle of the electron must exist, the *positron*. The positron was detected experimentally in 1932. Antiprotons and antineutrinos were also proposed and, eventually, detected. When particles and their corresponding antiparticles come into contact, they are annihilated and all their mass is released as energy.

As quantum physics was refined, it predicted the existence of more and more subatomic particles and antiparticles. The pion was proposed in 1935 to account for the strong force holding protons and neutrons together in the nucleus (the strong force overpowers the electrical repulsion that would otherwise cause the protons to fly apart, since electrical charges of the same sign repel each other). The pion and muon were first observed in 1937. The W and Z particles were predicted theoretically in the 1940s, to be detected years later. Other particles were observed directly in cosmic-ray collisions in particle chambers. Cosmic rays are very high-energy particles that arrive from outer space; when they strike atoms, large amounts of energy are released, some of which takes the form of particles. Many of these particles are short-lived, breaking down quickly into other particles and photons. By observing the tracks made by these short-lived particles in liquid or vapor, their properties can be deduced.

Starting in the 1950s, physicists stopped relying solely on cosmic rays to produce such collisions. Instead, they began building large machines called colliders that would accelerate charged particles (usually protons) to high energies. Particles moving in opposite directions would be smashed head-on against each other to produce energetic collisions. By observing these collisions, physicists discovered scores of new subatomic particles.

IN CONTEXT: "ACCIDENTAL"

In 1938, during a search for the mediator of the strong force, physicists discovered the mu meson (muon). Although the muon was ultimately determined not to be the carrier of the strong force, the findings initiated a series of "accidental" discoveries made by particle physicists. Eleven years later, while studying cosmic ray showers, English physicist Cecil Powell (1903–1969), discovered another type of meson, the pi meson (pion), which was found to interact with protons and neutrons. His "accidental" finding also confirmed a bond between theoretical physicists and "experimental" physicists in that theoretical physicists' predictions of the existence of subatomic particles would spur technological breakthroughs that then allowed experimental physicists to confirm the existence of predicted particles. Powell was subsequently awarded the 1950 Nobel Laureate in physics for his work developing the photographic methods used in the study of nuclear processes.

As the list of particles grew, the question of what was really meant by an "elementary" particle arose. Could there really be so many "elementary" particles? Why wasn't Nature simpler? Could simplicity perhaps be restored by theorizing that some particles are composed of a smaller number of even more fundamental particles, just as the atom itself had proved not to be *atomos*, uncuttable? The answer was not known certainly until the existence of the particle called the quark was demonstrated in experiments carried out from 1967 to 1973. The discovery of the quark and its place in the outline of modern physics's theory of the elementary particles is known as the Standard Model.

Quarks

Most of the new particles discovered in the 1930s through the 1950s were of the type called hadrons. Protons and neutrons are hadrons; electrons belong to another, less numerous group called the leptons. In the early 1960s, American physicists Murray Gell-Mann (1929–) and George Zweig (1937–) sought to simplify the long, messy catalog of hadrons by proposing that they are not truly elementary but are composed of smaller building blocks called quarks. Each quark, Gell-Mann and Zweig predicted, would have a charge of $+2e/3$ or $-e/3$. Two quarks of the $+2e/3$ type ("up" quarks), combined with a single quark of the $-e/3$ type ("down" quark), would form a particle with a charge of $+1e$, namely, a proton. Neutrons and other hadrons would be explained as other combinations of various quarks. By this theory, photons and electrons would remain truly elementary, *atomos* or indivisible.

Much as Rutherford and his colleagues used alpha-particle scattering from gold atoms to prove the existence of the atomic nucleus, particle scattering was used again to prove the existence of quarks. This time, however, instead of firing protons at alpha particles (helium atom nuclei), scientists fired electrons at protons. The experiments were carried out from 1967 to 1973 in Menlo Park, California, at the Stanford Linear Accelerator, a U.S. government facility operated by Stanford University. Just as the scattering angles of alpha particles probed the structure of the atom, the scattering angles of electrons probed the structure of the proton. The existence of quarks was proved by data from these scattering experiments.

The Standard Model

In the 1960s and 1970s, even before the existence of quarks was experimentally proved, a structure of mathematical theory to describe the particle world began to be developed—the Standard Model of fundamental particles and interactions. The Standard Model accounted for all known subatomic particles and predicted the existence of some that had not yet been detected. The mathematics underlying the Standard Model are complex, but its basic claims are not hard to summarize. The Model says that all matter and forces can be described in terms of two basic types of particles, fermions and bosons.

Fermions make up matter, that is, all particles with mass. They are subdivided into six leptons (the electron, electron neutrino, muon, muon neutrino, tau, and tau neutrino) and six quarks (the up, down, charm, strange, top, and bottom quarks). All matter, including the protons, neutrons, and other hadrons, is made of fermions.

Bosons, the particles that mediate forces, are massless. There are five types of bosons, according to the Standard Model: the photon, the W boson, the Z boson, the gluon, and the Higgs boson.

As of early 2008, the only particle in the Standard Model that had not been definitely observed in the laboratory was the Higgs boson, although signs of the Higgs boson may have already been observed in particle collisions at the Tevatron particle accelerator at Fermilab in Batavia, Illinois. Physicists hoped that the Large Hadron Collider at the CERN laboratory in Switzerland, an underground ring-shaped structure 17 miles in circumference and costing over $8 billion dollars, would begin gathering data on the Higgs boson when it was completed, probably in 2008. The Higgs boson is of particular importance in the Standard Model not only because its existence is still unconfirmed, but because it is essential to giving mass and inertia to other particles.

The Standard Model has passed many experimental tests, but is known to be incomplete. For example, it takes no account of the force of gravity, which is extremely weak compared to other forces. For example, the electromagnetic repulsion between two electrons is about 10^{35} times stronger than their gravitational

View of the experimental underground hall of the DELPHI experiment at CERN (Conseil Europeen pour la Recherche Nucleaire), the world's largest particle physics center. *T. Borredon/Explorer/Photo Researchers, Inc.*

attraction. Another fault in the Standard Model is its apparent inability to account for dark matter. Dark matter is gravitating matter that clumps around the galaxies but does not interact with light and so cannot be observed directly. There are a number of competing theories about the nature of dark matter, which is estimated to be about 85% of all the matter in the universe, but the favored theory at this time is that it consists of some kind of non-baryonic subatomic particle not accounted for by the Standard Model. If this turns out to be correct, it means that the Standard Model describes only a small fraction of the matter in the universe—though it does seem to account for all of the familiar kinds of matter, such as what makes up the stars and planets.

■ Modern Cultural Connections

Since its inception, the search for ever more fundamental particles has been one of the archetypal successes of physics, confirmation of the assumption that the universe can be understood by the human mind.

Until the early twentieth century, particles were conceived of as tiny, hard spheres like miniature billiard-balls, assumed to obey Newton's three laws of motion. But the subatomic particles turned out to be much more subtle and strange. They are not tiny, hard spheres, though in some situations they seem to behave that way, but in fact have a deeply ambiguous nature, both particlelike and wavelike—the ambiguity called wave-particle duality. Furthermore, most physicists have for the past century affirmed what is known as the Copenhagen interpretation of quantum physics (because it was formulated in Copenhagen, Denmark, in 1927 by Niels Bohr [1885–1962] and Werner Heisenberg [1901–1976]), which asserts that the behavior of individual particles is truly random—that is, not bound to occur a certain way because of the previous history of the universe, but unpredictable in its essence. Particle physics thus overturned the simplistic materialism of the nineteenth century. And, thanks to other aspects of quantum physics, the observing mind itself no longer seemed independent of the world, a mere shadow cast by its machine-like motions, but, rather, fundamental to it in some way, at least according to some interpretations.

English physicist Sir James Jeans (1877–1946) said in 1930 that "the stream of knowledge is heading towards a non-mechanical reality; the universe begins to look more like a great thought than a great machine. Mind no longer appears to be an accidental intruder into the realm of matter ... we ought rather to hail it as the creator and governor of the realm of matter." This view has been echoed as recently as 2005 in the prestigious science journal *Nature*, in which prominent American physicist Richard Conn Henry (1940–) wrote that "The universe is immaterial—mental and spiritual." In

---

IN CONTEXT: LARGE AND SMALL

Cosmology is the study of the universe as a whole—its size, shape, and history—and the distribution of the matter and energy within it. It is science on the largest possible scale. Yet it is closely involved with the study of elementary particles—science on the smallest possible scale.

In 1989 the Cosmic Background Explorer (COBE) satellite was launched by NASA to observe the cosmic microwave background radiation. This radiation consists of radio waves that shine from every part of the sky; the waves are the lingering glow of the big bang, the explosion from which the universe was born. Although the microwave background radiation is almost uniform across all parts of the sky, it does vary slightly, and, away from the radio noise of Earth's surface, COBE could map those slight variations. It was the first satellite (but not the last—the Wilkinson Microwave Anisotropy Probe was launched in 2001 on a similar mission) to measure the blotchiness of the microwave glow.

The universe was smaller than an atom when the big bang began, so random interactions between subatomic particles spanned the universe for a brief time early in its history. These interactions are recorded across the sky as the blotchy variations in the cosmic background radiation. The patterns in the microwave background radiation, the largest structures in the universe, are imprints of the wavelike processes that correspond to individual subatomic particles.

---

the twentieth century, the influence of the new knowledge about particles and their strange properties was felt throughout, not only in science and technology but in philosophy, theology, and art.

However, the effects of the new knowledge were not entirely positive: Understanding the structure of the atomic nucleus made possible the atomic bomb. The splitting of atomic nuclei by collision with neutrons was first observed in 1938. About seven years later, on July 16, 1945, the United States exploded the world's first atomic bomb. Two decades later, tens of thousands of nuclear weapons had been built. About 27,000 remain in the world today.

Socially, the effect of improving scientific understanding of subatomic particles has been primarily through technology. A detailed understanding of photons, electrons, and the properties of atoms has been essential to the electronics revolution, in which the dimensions of electronic devices have been pushed farther down into the microscopic scale, even approaching the atomic scale. Scientists hope to use the quantum properties of individual atoms to produce future generations of computers far more powerful than those available international particle physics research. The vast CERN was built primarily to search for the Higgs boson, and, it was hoped, would reveal particle physics beyond the Standard Model by generating extremely high-energy particle collisions after its completion in 2008.

Because the Standard Model accounts for the strong, weak, and electromagnetic fields and forces (including the electroweak theory force), it is a powerful theoretical model. However, the discovery of subatomic particles as a means to understanding the forces ruling the quantum world has become inexorably entwined with relativity-dominated cosmological theory regarding the origin and structure of the universe. A Standard Model built on particle physics research may be a step toward a grand unification theory that can unite incompatible quantum and relativistic theories. Articulation of such a theory would have profound scientific and philosophical impact comparable to, and perhaps surpassing, Newton's theories of gravitation, Einstein's theories of relativity, and quantum theory.

■ Primary Source Connection

Groups of scientists, funded by governments to the tune of billions of dollars, jockey to be the first to discover the Higgs boson, the last particle predicted by the Standard Model of particle physics, which has not yet been definitely observed.

"QUANTUM SCOOP: THE HOLY GRAIL OF PARTICLE PHYSICS MAY ALREADY HAVE BEEN FOUND."

Some call the Higgs boson the Holy Grail of particle physics. As the only undetected element of the field's theoretical masterpiece—the "standard model"—the Higgs guarantees a Nobel Prize for the experimenters who find it first. Now the European Union has spent an estimated $8 billion to build the world's largest particle accelerator, the large hadron collider, to finally track it down.

So goes the reasoning, at least, of popular science writers. In the last month, *The New Yorker*, the *New York Times*, and the *Boston Globe*, among others, have run articles on the LHC, which will be capable of reaching energies seven times greater than any comparable device ever created. All of this coverage has focused on the Higgs.

But what if someone else has already found it?

A rumor flying around physics departments these last few weeks claims that physicists working at the Tevatron, an accelerator located outside of Chicago, have found something new. Originally passed by word of mouth and private e-mail, the rumor made it into the blogosphere May 28, with an anonymous comment on the blog of a particle physicist living in Venice, Italy. Since then, the rumor has spread.

Trails left by subatomic particles create delicate spirals over the surface of a bubble chamber detector screen at Fermilab in Batavia, Illinois. The bubble chamber was invented by Donald A. Glaser (1926–) in 1952 to detect electrically charged particles. Glaser was awarded the 1960 Nobel Prize in physics for the invention. *© Kevin Fleming/Corbis.*

today, a field called quantum computing. Applied particle physics in the form of nuclear weapons continues to be a concern worldwide, as the number of nations with the ability to build such weapons slowly increases. As of mid 2007, ten countries were known to possess nuclear weapons: China, France, India, Israel, North Korea, the Russian Federation, Pakistan, the United Kingdom, and the United States.

Scientists were still searching for the last particle predicted by the Standard Model, the Higgs boson, as of 2008. After a decade of controversy and the expenditure of over $200 million, construction on the United States Superconducting Supercollider was stopped in 1993 by the United States Congress. Then-president William Jefferson Clinton lobbied against the shutdown because such an act reduced the United States "position of leadership in basic science." The termination of the SSC project left *Organisation Europeene pour la Recherche Nucleaire* (CERN) in Geneva, Switzerland, as the major site of

This isn't the first time a story like this has circulated. Until the LHC opens, the Tevatron remains the largest accelerator in the world. Among its most significant past discoveries is another standard-model particle, the top quark. And in 2009, it will shut its doors forever. Like the LHC, the Tevatron was built with the Higgs in mind, and as time runs out for America's biggest atom smasher, some nervy experimentalists have jumped the gun. Last summer, two Tevatron groups released some suggestive, but fruitless, graphs (PDF), just before the International Conference on High Energy Physics; in January, a new crop of rumors emerged, which were reported in the *Economist* and *New Scientist* in March. These other rumors have described "bumps": anomalies in the data that suggest a new particle but are too small for a definitive identification.

The current rumor, which comes in time for the summer conference circuit, may be different. It claims an experiment at the Tevatron has found a peak twice as high as the previous rumors' bumps. And unlike the other rumors, this one includes details: the new particle's mass, for instance, which fits within theoretical bounds on the standard model Higgs. Some versions include a decay chain, which describes what the new particle turned into as the experiment progressed, and which may be consistent with the standard model's predictions.

Of course, the rumor also claims that no one associated with the experiment will confirm the new findings until they've had time to publish, likely within the next few weeks. And until they do, no one can be certain what the Tevatron has—or has not—found.

The hype surrounding the Higgs boson is well-deserved. The standard model, a unified view of physics first presented by John Iliopoulos in 1974, describes everything we know about the smallest building blocks of nature yet observed. It's the most accurate theory ever developed, in any field. And without the Higgs, it doesn't make much sense: Based purely on first principles, elementary particles should be massless. Some, like photons, do have zero mass; yet others are surprisingly heavy. Enter the Higgs, which would—in theory—interact with these latter particles to make the difference.

So, if the rumor is true and the standard model Higgs has been found at the Tevatron, the LHC is in big trouble: Immediately, its "guaranteed" success—the final particle of the standard model, not to mention a couple of Nobel Prizes for European scientists—is gone.

The irony is that things look just as bleak for the LHC if the rumor is false, and the Europeans end up finding the standard model Higgs themselves. Physicists have developed such a complete description of elementary particles that, once the final piece of the theory is in place, the chances that the LHC will find anything the standard model doesn't predict are almost negligible.

Particle theorists talk a big game. They get excited and tell reporters, not to mention government funding agencies, that the Higgs is just the beginning: The LHC, some say, may find examples of a class of particles indicative of a new fundamental property of nature, called supersymmetry. Others say there may be two or three particles, which together perform the job the standard model assigns to the Higgs. The truth is that these alternatives patch up the standard model, should something unexpected happen. If a Higgs-like particle is found, say, but it's too light to be the standard model boson. Or if it decays in a surprising way. In cases like these, the LHC could indeed produce dramatic new discoveries.

But what happens if the Higgs turns out to be just right? Well, then the standard model predicts that you'd need a machine roughly a quadrillion times more powerful than the LHC to find anything new. With current technology, this would mean an accelerator the circumference of the Milky Way. Though some theorists—proponents, for instance, of string theory—speculate about what such an accelerator might find, few other physicists take them seriously.

In fact, finding the "just right" Higgs would be bad news all around. Surely the European Union wants more for its $8 billion than a single particle. But more importantly, it would provide the final proof of the standard model, which happens to be clunky, boring, and infuriatingly silent on the Big Questions that the final theory of physics was supposed to answer. Questions like: Why is there something, rather than nothing? And where does gravity fit in? If the standard model turns out to be a complete description of particle behavior, as the discovery of the Higgs would suggest, these questions may never be answered.

That's why particle physicists, and the EU member states that have spent Nepal's annual GDP to build this accelerator, are hoping that no one, in Chicago or Switzerland, finds the Higgs. The future of high-energy physics lies with the small chance that the standard model is wrong, and something exotic happens at LHC energies. Something, I hope, that will help us understand the why questions that the standard model leaves wide open.

James Owen Weatherall

WEATHERAL, JAMES OWEN. "QUANTUM SCOOP: THE HOLY GRAIL OF PARTICLE PHYSICS MAY ALREADY HAVE BEEN FOUND." *SLATE* (JUNE 4, 2007). HTTP://WWW.SLATE.COM/ID/2167563/PAGENUM/ALL/#PAGE_START (ACCESSED OCTOBER 2, 2007)

■ Primary Source Connection

Scientists can use neutrinos (subatomic particles only weakly blocked or deflected by matter) to probe Earth's core. The following *New Scientist* magazine article by Celeste Biever describes how scientists in Japan have measured the earth's planetary radioactivity and how this measurement could provide clues to fundamental questions about how Earth formed.

Celeste Biever is a regular contributor to *New Scientist* and is a freelance science writer based in Boston.

FIRST MEASUREMENTS OF EARTH'S CORE RADIOACTIVITY

EARTH'S natural radioactivity has been measured for the first time. The measurement will help geologists find out to what extent nuclear decay is responsible for the immense quantity of heat generated by Earth.

Our planet's heat output drives the convection currents that churn liquid iron in the outer core, giving rise to Earth's magnetic field. Just where this heat comes from is a big question. Measurements of the temperature gradients across rocks in mines and boreholes have led geologists to estimate that the planet is internally generating between 30 and 44 terawatts of heat.

Some of this heat comes from the decay of radioactive elements. Based on studies of primitive meteorites known as carbonaceous chondrites, geologists have estimated Earth's uranium and thorium content and calculated that about 19 terawatts can be attributed to radioactivity. But until now there has been nothing definitive about exactly how much uranium there is in the planet, says geologist Bill McDonough of the University of Maryland in College Park. "There are fundamental uncertainties."

There is one way to lessen this uncertainty, and that is to look for antineutrinos. These particles are the antimatter equivalent of the uncharged, almost massless particles called neutrinos and are released when uranium and thorium decay to form lead. If antineutrinos are being created deep within the planet they should be detectable, because they can pass through almost all matter.

Now, the KamLAND antineutrino detector in Kamioka, Japan, has counted such antineutrinos. An international team of scientists analysed the data and found about 16.2 million antineutrinos per square centimetre per second streaming out from Earth's core. They calculate that the nuclear reactions creating these particles could be generating as much as 60 terawatts, but are most likely putting out about 24 terawatts (*Nature*, vol 436, p 499). "We have made the first measurements of the radioactivity of the whole of Earth," says John Learned, who heads the KamLAND group at the University of Hawaii in Manoa. The KamLAND group's finding is like unwrapping a birthday present, says McDonough.

With time, as more antineutrinos are detected, KamLAND may be able to determine once and for all whether radioactivity is entirely responsible for heating Earth or whether other sources, such as the crystallisation of liquid iron and nickel in the outer core, also play a sig-

nificant role. "[Detecting anti-neutrinos] is the way of the future in terms of hard numbers about the system," says McDonough.

Antineutrinos could also reveal the radioactive composition of the crust and mantle, which will give geologists clues as to when and how they formed. But to do that, they will have to be able to pin down exactly where the antineutrinos are coming from, and this will require a whole network of detectors. "We are heading towards doing neutrino tomography of the whole Earth," says Learned. "This is just the first step."

Celeste Biever

BIEVER, CELESTE. "FIRST MEASUREMENTS OF EARTH'S CORE RADIOACTIVITY." *NEW SCIENTIST* (JULY 27, 2005): 9.

SEE ALSO *Physics: Heisenberg Uncertainty Principle; Physics: The Standard Model, String Theory, and Emerging Models of Fundamental Physics; Physics: Wave-Particle Duality.*

BIBLIOGRAPHY

Books

Brown, Laurie M., and Lillian Hoddeson, eds. *The Birth of Particle Physics*. New York: Cambridge University Press, 1983.

Han, M.Y. *Quarks and Gluons: A Century of Particle Charges*. Singapore: World Scientific, 1999.

Lincoln, Don. *Understanding the Universe: From Quarks to the Cosmos*. Singapore: World Scientific, 2004.

Schumm, Bruce A. *Deep Down Things: The Breathtaking Beauty of Particle Physics*. Baltimore, MD: Johns Hopkins University Press, 2004.

Weinberg, Steven. *The Discovery of Subatomic Particles*. Cambridge, UK: Cambridge University Press, 2003.

Periodicals

Biever, Celeste. "First Measurements of Earth's Core Radioactivity." *New Scientist* (July 27, 2005): 9.

Weatheral, James Owen. "Quantum Scoop: The Holy Grail of Particle Physics May Already Have Been Found." *Slate* (June 4, 2007). http://www.slate.com/id/2167563/pagenum/all/#page_start (accessed October 2, 2007).

Web Sites

Lawrence Berkeley National Laboratory. "The Particle Adventure." 2002. http://particleadventure.org/ (accessed June 13, 2007).

Larry Gilman
Paul Davies
K. Lee Lerner

Physics: The Quantum Hypothesis

■ Introduction

The quantum hypothesis, first suggested by Max Planck (1858–1947) in 1900, postulates that light energy can only be emitted and absorbed in discrete bundles called quanta. Planck came up with the idea when attempting to explain blackbody radiation, work that provided the foundation for his quantum theory.

Planck found that the vibrational energy of atoms in a solid is not continuous but has only discrete (distinct) values. Light energy is determined by its frequency of vibration, f. Energy, E, is described by the equation: $E = nhf$, where n is an integer and h is Planck's constant, equal to 6.626068×10^{-34} Joule-second (J-s).

Planck published his theory "On the Law of Distribution of Energy in the Normal Spectrum" in the German journal *Annalen der Physikm*, stating: "Moreover, it is necessary to interpret U_N 'the total energy of a blackbody radiator' not as a continuous, infinitely divisible quantity, but as a discrete quantity composed of an integral number of finite equal parts."

■ Historical Background and Scientific Foundations

Early in his life as a physics and mathematics student and, later, as a theoretical physicist, Max Karl Ernst Ludwig Planck was especially interested in the law of the conservation of energy, also called the first law of thermodynamics. It states that the increase in the internal energy of a thermodynamic system is equal to the amount of heat energy input into the system, minus work done by the system on its exterior surroundings.

Planck was also interested in the entropy law, the second law of thermodynamics, which states that entropy (disorder) of an isolated system, which is not in equilibrium, is inclined to increase over time and eventually approach a maximum value at equilibrium. This became the topic of his doctoral dissertation at the University of Munich, leading to Planck's quantum hypothesis.

In the late 1800s, physicists were having problems validating several laws of Newtonian mechanics, particularly the second law of thermodynamics. Their primary question was whether it meant that entropy results from the motions of a collection of molecules, as stated by Austrian physicist Ludwig Boltzmann's (1844–1906). Statistical (probabilistic) interpretation was controlled by energy and related physical quantities, as declared by German chemist Wilhelm Ostwald's (1853–1932) absolute energy interpretation. The debate led Planck to develop quantum theory, for which he won the Nobel Prize in 1918.

The Science

Planck explained entropy using Scottish theoretical physicist James Clerk Maxwell's (1831–1879) work with electrodynamics. This describes microscopic oscillators (radiating atoms) that produce the heat radiation emitted by blackbodies. (A blackbody is defined theoretically as any object that absorbs all light falling upon it, reflects none of it, and thus appears black. However, when a blackbody is heated, it emits radiation "light.") Blackbody radiation is the amount of radiant (heat) energy emitted at various frequencies for specific temperatures of a blackbody.

Many physicists had tried to explain blackbody radiation. They all failed, however, until Planck published his historic theory in 1900. Using the Boltzmann equation, he suggested that the total energy from blackbody oscillators could be divided into finite parts through a process called quantization. Planck described energy as being made of a finite number of equal parts. He included the constant $h = 6.55 \times 10^{-27}$ erg-second

WORDS TO KNOW

ANTIPARTICLE: In particle physics, most fundamental particles having non-zero rest mass (e.g., neutron, proton, electron) have an antiparticle that is an almost perfect mirror-image of the particle. When a particle and its antiparticle meet, they annihilate each other, releasing all their energy in the form of photons. At the big bang, about 13.7 billion years ago, slightly more particles were produced than antiparticles. After all particle-antiparticle pairs annihilated each other, the remaining fraction of particles became all the ordinary matter seen in the universe today.

BLACKBODY: An ideal emitter that radiates energy at the maximum possible rate per unit area at each wavelength for any given temperature. A blackbody also absorbs all the radiant energy incident on it; i.e., no energy is reflected or transmitted.

BOSON: According to the modern Standard Model of particle physics, all particles are either bosons or fermions. All forces are explained as exchanges of virtual bosons between fermions. Bosons have integer spin values; photons are bosons (spin = 1). The only particle predicted by the Standard Model that had not been experimentally observed as of early 2008 was the Higgs boson.

ENTROPY: Measure of the disorder of a system.

PHOTON: Smallest individual unit of electromagnetic radiation (light energy). These light particles are emitted by an atom as excess energy when that atom returns from an excited state (high energy) to its normal state.

QUANTUM: Something that exists in discrete units.

SPACE-TIME: The four-dimensional realm of space (which has three dimensions, often termed length, width, height) plus time (a fourth dimension). Time and space are neither independent nor interchangeable, but form a stretchy background, realm, or setting in which physical events occur. The term "stretchy" here refers to the fact that motions in space give rise to differences in time flow for moving and stationary observers: There is no one, unique rate of time flow throughout space, and no universal present moment. In particular, as described by special relativity, any stationary observer of a moving clock will see that clock as running slower than one to which they are attached. Here, "stationary" is a relative term because an observer attached to the "moving" clock will see exactly the same situation in reverse: Either observer's point of view can just as well be termed stationary (or moving) as that of the other. General relativity describes how space-time is curved by matter, giving rise to gravitation.

WAVE-PARTICLE DUALITY: The combination of wavelike and particle-like properties possessed by all physical objects, which is especially apparent for single subatomic particles such as electrons. Under some conditions, a particle behaves as a small object that can be definitively found in one place or another; under other conditions, a particle behaves as a wave that is spread throughout space and can add or subtract from similar waves in a way that would be impossible for a tiny, hard object that has a single, well-defined location at all times.

(1 erg = 10^{-7} joule), which he called the quantum of action, known today as Planck's constant.

In 1900 Planck proposed that heat energy E is emitted only in definite amounts called quanta. Thus, his equation became $E = hf$, where $h = 6.626 \times 10^{-34}$ J-s and f = frequency. Planck maintained that only certain specific energies could appear, and they were limited to n whole-number multiples of hf. Thus, $E = nhf$.

With this equation, Planck was able to explain blackbody radiation, showing that the hotter an object gets, the more radiation it produces. Since a blackbody absorbs all radiation frequencies, it should, he believed, under physical principles held at that time, radiate equally at all frequencies. Planck found instead that blackbodies emit more energy at some frequencies, and fewer at others. He didn't understand what this meant. The radiating atom appeared to contain only discrete quanta of energy, not the continuous energy he expected. This went against classical physics.

Near the end of 1900 Planck convinced himself that the second law of thermodynamics was not an absolute law. In addition, he assumed that his formulas correctly showed that blackbody oscillators radiate in only discrete amounts of energy, or quanta, not continuously. He published his quantum hypothesis in December 1900.

When Planck introduced the idea of energy quanta to the scientific community, it is unclear whether he really understood the relevance of quantum discontinuity: He was mostly interested in the accuracy displayed by his new law and its constant. Later, he called his equation "a fortuitous guess."

Influences on Science and Society

Quantum theory, the first evidence that the tiny world of atoms could not be accurately described with classical physics, became the basis of quantum mechanics—the branch of physics that studies the emission and

Max Planck (1858–1947), circa 1918, was the winner of the 1918 Nobel prize for physics, and was responsible for development of quantum theory. *HIP/Art Resource, NY.*

absorption of energy and the motion of particles at the atomic level. Quantum theory revolutionized scientific thought with respect to atomic and subatomic processes. It is held in the same regard as Albert Einstein's (1879–1955) theories of relativity, which revolutionized scientific thought with respect to space and time.

Einstein used Planck's idea of light quanta in 1905 to explain photons and the photoelectric effect mathematically, the first scientific work utilizing quantum mechanics. In 1907 Einstein showed the quantum hypothesis's wide application by using it to interpret the temperature dependence of the specific heats of solids. Two years later, in 1909, he wrote on the quantization of light and wave fluctuations, describing wave-particle duality—the theory that objects exhibit properties both of waves and particles.

In 1913 Danish physicist Niels Bohr (1885–1962) was the first to use the quantum hypothesis to explain atomic structure and spectra. He showed the association between electrons' atomic energy levels and light frequencies emitted and absorbed by atoms. In addition, Bohr postulated that an atom would not emit radiation while it was in one of its stable states, but only when it

traveled between them. The frequency of this radiation would equal the difference in energy between those stable states, divided by Planck's constant. This showed that atoms could not absorb or emit radiation continuously, but only in finite steps called quantum jumps.

For about twenty years physicists worked on the mathematics of quantum theory—finally expressing it mathematically in the 1920s. In 1924, French physicist Louis de Broglie (1892–1987) proposed that all forms of radiation, not just light, exhibited wave-particle duality. He suggested that particles, such as electrons, exhibit wavelike properties in certain circumstances. The de Broglie wavelength is equal to Planck's constant divided by momentum (mass times velocity).

Following de Broglie's work, German physicist Werner Heisenberg (1901–1976), using matrix mechanics, and Austrian physicist Erwin Schrödinger (1887–1961), using wave mechanics, used the wave function to relate the probability of finding a particle at a given point in space and time. In the late 1920s both de Broglie and Schrödinger introduced the concept of standing waves to explain their existence only at discrete frequencies and, consequently, only in discrete energy states.

In 1927 Heisenberg announced his uncertainty principle, which provides an absolute limit on the accuracy of certain measurements. It states that the action of measuring the position x of a particle disturbs the particle's momentum p so that $Dx \times Dp$ is greater than or equal to h;, where Dx is the uncertainty (difference) of the position of a particle along a spatial dimension, Dp is the uncertainty of the momentum, and h is Planck's constant. This means that one cannot know position and momentum simultaneously at the atomic level because a photon being used to measure an electron will alter its position and momentum when the electron is bounced off of it.

Planck's quantum hypothesis spawned subfields of quantum mechanics, quantum chromodynamics (the study of the strong interaction), quantum electrodynamics (the study of the relativistic aspects of electromagnetism), quantum electronics (the study of quantum mechanics based on interactions of electrons and photons), quantum gravitation (the current attempt to unify quantum mechanics and general relativity), quantum statistics (the study of particles in statistical mechanics), and various quantum field theories.

■ Modern Cultural Connections

Before the quantum hypothesis was introduced, the central concepts of physics were: (1) all matter consists of discrete particles with properties of gravitational mass and electrical charge, (2) light was a continuous electromagnetic wave that travels at a constant speed, (3) continuous electromagnetic fields are created by discrete charged

In this quantum entanglement experiment in 2003, lasers are being used in an experiment to demonstrate the breakdown of causality in quantum physics. This is part of the work at the GAP (Group of Applied Physics), the University of Geneva, Switzerland. *Pascal Goetgheluck/Photo Researchers, Inc.*

particles, and (4) local interactions of electromagnetic charges are limited by the velocity of electromagnetic waves.

Quantum theory changed these to: (1) all matter is composed of discrete particles and also has wave properties, (2) light has discrete particle properties, (3) discrete statistical fields exist, and (4) instantaneous nonlocal matter interactions exist, called instant action at a distance.

Modern physics is founded on general relativity and quantum mechanics, two theories that have passed many comprehensive and strenuous tests for validity. They contradict each other, however, when physicists attempt to join equations of quantum mechanics (involving the strong, weak, and electromagnetic forces) with equations of general relativity (associated with the gravitational force).

Physicists are searching for what is called a grand unified theory, or "theory of everything"—that will combine the four fundamental forces of nature into one all-inclusive force. This would form a far more complete understanding of space-time (a four-dimensional system consisting of three spatial coordinates and one coordinate of time) than is currently available. So far, none has been found.

Quantum theory explains the dynamics seen in the subatomic world. It has also contributed greatly to other sciences, such as applied chemistry and nuclear physics, and various technologies, such as the laser, electron microscope, and computer. Its complexity makes it difficult to use, however, because it contains an extraordi-

narily large number of subatomic particles, and requires numerous constants for its various equations.

Specifically, the concept of the quantum—the smallest unit in which energy can be measured—led to the field of particle physics. A classification of elementary particles, called the standard model, is used within particle physics. It describes the strong force, the weak force, and the electromagnetic force using gluons; W−, W+, and Z bosons; and photons, respectively. The standard model also contains 24 fundamental particles (12 particles and 12 antiparticle pairs).

In conclusion, some historians have written that Planck himself did not think these quantum jumps actually existed and that his quantum hypothesis was contrary to laws of classical mechanics and classical electrodynamics. Others point to Einstein as the physicist who first recognized the essence of quantum theory, arguing that Einstein was the first to identify the quantum discontinuity. Credit for the beginnings of quantum theory is generally given to Planck, with Einstein cited as the first to apply it to scientific pursuits. In reality, a field as complex as quantum theory took many scientists to develop it into a major branch of physics.

■ Primary Source Connection

The following article was written by Rushworth M. Kidder, a senior columnist for the *Christian Science Monitor* until 1990, when he founded the Institute for Global

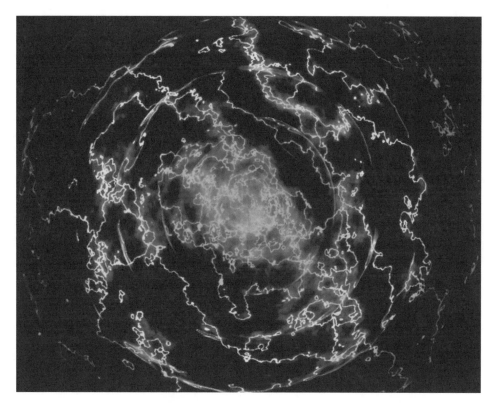

A quantum cloud of electrons surrounds a helium atom, showing electron density.   *Photo Researchers, Inc.*

Ethics. Kidder is the author of *Reinventing The Future—Global Goals For The 21st Century*. Founded in 1908, the *Christian Science Monitor* is an international newspaper based in Boston, Massachusetts. The article, originally published in 1988, describes how quantum mechanics has the potential to change the way the world culture is perceived, as well as how subatomic particles behave.

HOW MIGHT QUANTUM THINKING CHANGE US?

CAMDEN, MAINE—CALEB THOMPSON may not realize it, but he's preparing to encounter quantum mechanics—not in some far-off never-never land, but on an ordinary Tuesday here at Camden-Rockport High School. Surrounded by the intent faces of his lab partners in Mr. Bentley's second-period honors physics class, Caleb drops a steel ball bearing down a curved aluminum track. When it collides with a second ball bearing, both drop to the floor, landing on a carbon-paper-covered sheet of typing paper. Fellow student Peter Killoran, down on all fours with pencil in hand, labels the marks as the ball bearings drop.

"We're trying to prove that momentum is conserved," Peter explains. When enough marks have been recorded, he and his partners will compute distances and angles and analyze the vectors, or patterns. Their purpose: to demonstrate Newton's third law of motion, one of the landmarks of classical physics.

But why, in the age of quantum mechanics, string theory, and the search for a possible fifth and sixth force, are these students still learning about Newton's laws? Sitting at his cluttered desk at the front of the room, William N. Bentley explains.

"This is the bridge," says Bentley, who has taught the highly regarded physics course here for five years. "This is the first time [the students] consider subatomic particles."

The ball bearings, of course, are hardly subatomic. But their interactions, he notes, behave according to the same Newtonian laws that govern the collisions of everything from meteors to quarks. To bring home that point, Bentley will have his students perform another vector analysis in a few weeks. At that time, however, they'll do it on a photograph showing the tracks left by subatomic particles after they collide in a particle accelerator. And finally, at the very end of the school year, they'll arrive on the doorstep of quantum mechanics.

"All of a sudden," Bentley says, "we'll see the limitation of Newtonian mechanics."

Like millions of 20th-century high school students, Bentley's class has grown up in a century shaped by the

Newtonian world view. From their earliest years, in and out of science classes, these students have been taught that the material world has an objective reality. They've been told that its objects possess predictable attributes unaffected by observers. And they've been assured that those attributes can best be understood by reducing matter to its fundamental constituents.

Caleb and Peter may not go on to careers in physics. But one thing is sure. The world they will inhabit—the world of the 21st century—will be profoundly affected by quantum mechanics.

Not only will they benefit from the tangible results of this science—results that will presumably extend well beyond today's solid-state electronics, laser beams, superconductivity, holography, and other cutting-edge quantum technologies. They may also be influenced by the intangibles of this radically different way of comprehending the universe.

What will be the cultural impact of those intangibles? Will quantum mechanics, as it becomes more widely discussed, taught, and understood, change their world view? Will it, like the discoveries of Copernicus, Galileo, Newton, and Einstein, have a revolutionary effect on humanity's sense of itself and its universe? Will it produce what is often described as a "paradigm shift"—a fundamental change in the patterns of thinking that shape our vision of reality? Will quantum physics begin to produce a kind of quantum metaphysics? And if so, how might that metaphysics shift mankind's personal, social, political, and theological outlook?

Teasing out the meaning

THESE days, practicing physicists are divided about the relevance of such questions. Some stick to their mathematical formulas and their laboratory experiments, choosing not to think much about the philosophical ramifications. Others insist that the discoveries of quantum mechanics have been misappropriated—especially by proponents of Eastern religions eager to prove their own theses.

But still others feel that the weirdness of quantum mechanics cannot be brushed aside. They do worry, however, that generalizations about its deeper significance have been stretched too far, too fast. We simply don't know enough, they say, to justify some of the more heady assertions about matter, mind, and the universe.

Nobel laureate Murray Gell-Mann, professor of theoretical physics at the California Institute of Technology, is particularly wary. One of the most highly regarded thinkers about quantum matters, he decries the misuse of language among those who would tease out deep meanings from quantum physics.

Physicists, he observes, often turn to everyday language for analogies in their attempts to explain what they're encountering. That's fine, according to Dr. Gell-Mann. The problem arises, he says, when nonphysicists try to make a reality of what, for the physicists, is essentially just a metaphor.

Case in point: the idea that the fundamental particles have no size (or "extension") at all. "It's just words to say they don't have extension," says Gell-Mann.

Fellow Nobel Prize-winning physicist Steven Weinberg, who describes himself as "the most unphilosophical of physicists," agrees—and explains why. "One of the lessons of quantum mechanics," says Professor Weinberg from his office at the University of Texas, "is that the ideas that are useful in describing nature at the level of ordinary life may not even be meaningful when you get down to the level of the subatomic world."

"I don't think it's correct to say that particles have a definite size," he says, "And I don't think it's correct to say they don't."

The real problem, according to Nobel laureate Sheldon Glashow of Harvard University, lies in language itself, which "just doesn't work" to explain the complexities of quantum mechanics. "There's no reason to expect [our ordinary] language to have any relevance to the way things are," he says.

That, however, is the kind of puzzle that boggles the non-scientist's mind—and causes even the physicists interviewed for this series to admit, in many cases, that they themselves don't really understand quantum mechanics.

Most, however, would agree with the assessment of a Brandeis University historian of science, Samuel Schweber. He describes quantum mechanics as "a deep revolution." It is, he says, "so deep that—in some sense, in having affected so many different areas of thought and of intellectual life—we really have not assessed as yet the full impact of it."

If scientists, sociologists, and historians ever do assess that impact, what will they find? What are the philosophical elements of quantum physics that could reshape mankind's world view?

Classical physics assumes that, somewhere outside ourselves, there is a fixed and objective universe. It may be immeasurably large or small. It may be inextricably entangled and complex. But it's really there. All that keeps us from knowing it in detail are the limitations of our measuring devices and our computational abilities.

Quantum mechanics takes a different view. The material world is not lumpy but wavelike, not made of things as much as of fields. Moreover, its attributes seem to vary depending on the vantage point of the observer, somewhat like the location of a rainbow.

That does not mean, however, that quantum physics denies the existence of matter. "Nobody's saying there's not a world out there," says physicist John Ellis of the European Laboratory for Particle Physics (CERN) in Geneva. The well-defined laws governing that world may surprise us in their strangeness. But it's a world, all the same.

Princeton University physicist Robert Dicke agrees— although he emphasizes that reality depends not simply on an objective world but on our view of it. "Matter has a very real, solid existence," says Dr. Dicke, an emeritus professor who has taught quantum mechanics to generations of students, "as far as our view of it goes. But what is out there may not be nearly so well defined and solid and uniquely characterizable as our view of it."

What are the ramifications of this different sense of reality?

"I think that's what is new," says CERN physicist John Bell, "is the idea that what I call 'muddle' is permanently tolerable."

This "muddle," or confusion, he says, is not simply "a phase in the construction of the theory that is to be transcended and replaced by some deeper models in another phase of the theory.

"Somehow we have come to the end of the human capacity to form sharp pictures of what is going on," he adds, "and more and more we will have to rely on recipes that we don't understand."

That tolerance for "muddle," says Bell, is "the most characteristic feature of the orthodox school of quantum mechanics."

This lack of clarity, in fact, has its quantum counterpart in the electron. Once thought of as solid, definite objects, electrons are now pictured more as clouds or smears surrounding the proton. The result: At the quantum level, the edges of things are inherently fuzzy.

Is that true at human scale as well? It must be, argues Princeton University professor David Gross—since all the objects we encounter are made of atoms, and the outsides of all atoms consist of fuzzy clouds of electrons.

And as with objects, so with people. "You're a bit smeared out on the edges," he says with a chuckle, "but you don't notice it."

If matter is not quite the clear, basic, and all-determining thing that the classical world view suggests, then what is real? Some physicists point to something they describe—with hesitation, and sometimes even embarrassment—as mind or consciousness.

It's a concept that arises most prominently in discussions of the "observer-created reality"—where the presence of an observer is thought to cause matter to behave in ways it does not otherwise behave. Professor Gross puts the point most simply. "There isn't a real world out there that you can observe," he says, "without changing it."

That's a far cry, however, from saying that only the observer is real—or that everything is only as real as you make it. In fact, what physicists mean by an observer-created reality has to do with probabilities and measurement theory. The fact that the world shifts when it is measured means simply that you cannot make precise predictions about it. That's because the world is composed of probabilities rather than definite, fixed states.

Why does the world change when you observe or measure it? A physicist explains it by saying that a measurement causes this so-called "probability amplitude"— a mathematical way of picturing the probability as though it were a wavy line on a graph—to collapse into a much straighter line.

But measuring implies a measurer. Hence the observer becomes vitally important. As the presence of so many different interpretations of quantum physics suggests, however, the extent of the observer's importance remains a source of considerable discomfort for physicists.

"What is the meaning of the probability amplitude?" muses theoretical physicist Freeman Dyson of the Institute for Advanced Study in Princeton, N.J. "Does it just describe our state of ignorance, or does it describe something real?"

"Those questions really aren't answered," he continues. "And so I think in a certain sense that it's true to say that it has brought mind and consciousness back to the center of things—that you won't really understand quantum mechanics deeply unless you also understand the nature of mind."

Then is man, as the possessor of that mind, somehow restored to a central role in the universe—a role from which he was first banished when Copernicus found that the earth was not the center of the solar system? For most physicists, that's stretching things. Quantum cosmology, for example, points to what astrophysicist George Helou calls "the ongoing decentralization, a deemphasis of the importance of man." In an expanding universe, one can argue that every point is at the center— in only a tiny fraction of which stands an observer.

"We're not only not the center," says University of Chicago astronomer David Schramm. "We're just all part of the whole, riding with the wave of the big bang. We're certainly not [in] a significant role."

Questions about man's role

BUT questions about mankind's place in the universe continue to reverberate. "If it could ever be shown that the human mind is somehow a quantum mechanical effect like a semiconductor or a laser," says Nick Herbert, a physicist and science writer, "and that these concepts not only apply as metaphors but as direct descriptions of the way we are inside, then I think that would be a tremendous revolution."

"My guess is that the mind is somehow connected with these ideas," he adds.

Why can matter be described only in terms of probabilities? Because, physicists say, it's governed by random activity.

"Quantum mechanics asserts that at the very foundations of existence there is an essential randomness," says Rockefeller University physicist Heinz Pagels.

And that's a sharp break from classical physics.

In the world of Newtonian mechanics, the paths of particles and their velocities are the determining causes. Discover the initial paths and velocities of each particle, and you hold the key to every subsequent action and reaction in the universe: You could, in theory, explain everything.

But even Newton, who occupied himself with theological studies in his later years, recognized that an ultimate first cause must have produced those initial conditions. To be sure, actions might appear random. But trace them back far enough and you discover the will of the creator.

Quantum mechanics takes a different view. Trace actions back to their sources, and you discover not God but randomness. Even the big bang is viewed as a random event. There was a tiny but non-negligible probability that it might happen. And all of a sudden, in an entirely random way, it did.

Such probabilities challenge a commonplace of classical physics: the idea that the laws governing matter as we know it are the only conceivable laws. Some physicists point out that if a few of the basic constants of the physical universe—the ratio of the mass of the proton to that of the electron, for instance—were just slightly different, the universe would be a resoundingly different place.

Yet are those the only conceivable constants? That's a question that puzzles John Preskill, a California Institute of Technology physicist. "Is it possible," he asks, "that [such constants] really were, at some stage in evolution of the universe, sort of picked out from some probability distribution which was determined by the very early quantum state of the universe?" If so, then there is a probability that different constants could also have arisen. If the many-worlds thesis holds, in fact, they might already be in place in other universes.

So is there another universe out there where, say, the fine structure constant—a number based on electrical charge, Planck's constant, and the speed of light that is central to many calculations in physics—is something other than 1/137.0365? If there is, most physicists agree, it won't matter to us: We'll never be in a position to observe it.

In another way, however, randomness does affect us: when a particle, for no other reason than that it has a probability of doing so, suddenly breaks loose from an atom and flies away, as happens in radioactive decay. Once the particle flies away, its motions are governed by the strictly ordered laws that characterize the visible universe.

But how can such order grow up out of randomness? The analogy of an opinion poll helps. On a nationwide scale, one can discover, say, that 80 percent of the public approves of a particular presidential candidate. That fact tells you nothing about your next-door neighbor, however: He may be 100 percent against the candidate. And so on up and down your street: Ask resident after resident, and you get what may appear to be a chaotic distribution. Only when you take the nation as a whole, or at least a representative sample of it (as most polls do), can you get an accurate assessment.

In a similar way, randomness tells you nothing about how a particular particle will behave. It may have a 20 percent chance of suddenly disappearing from its present location and showing up elsewhere—even though it would have to pass through an apparently insurmountable "wall" of energy to get there.

That effect may be random. But it's by no means insignificant. It's known as "quantum tunneling," and it underlies the operation of semiconductors. It's the principle, in other words, upon which transistors and computer chips are constructed.

Why do some particles "tunnel" and others don't? The choice, it seems, is purely random. Yet the result, reined in by technology, produces the extreme precision of the modern-day computer.

If a single particle can tunnel through such a barrier, why can't a collection of them—a whole tennis ball, for example, or a whole human body?

Again, it's a matter of probabilities. "If I threw a ball against the wall," says astronomer Schramm, "the probability is much, much higher that the ball will bounce back than that the ball will go penetrating through the wall. But the probability of it going through the wall is not absolutely zero."

As with balls, so with people. "In principle," says Dr. Gross, gesturing across the dining room at the Princeton University Faculty Club, "there is some non-negligible probability that I'm located on the other side of the room. But that probability is so small that one would have to wait many lifetimes of the universe for me to jump over there and back again."

When Gross says "jump," he's choosing his words carefully. One of the basic ideas of quantum mechanics is that energy comes in unbreakable packets. A particle, in going from one state to another, does not move continuously between them. It's instantly translated, "jumping" to its next state. Hence the metaphor of a quantum leap. "Quantum mechanics means things are discrete," says Fermilab cosmologist Michael S. Turner, adding that "nothing in nature is continuous."

"If there's anything that mathematical logic says," asserts John A. Wheeler, a Princeton emeritus professor, "it's that you can't have a continuum. It says that that's an illusion, a myth, an idealization."

So are space and time also "discrete"—coming in chunks, rather than in a steady flow? Is life not a thread but a collection of blips?

If so, says Dr. Turner, then "we live on a lattice, not on a continuum. As you get down to the most fundamental level, you can either be here or there, but nowhere in between." Why? Because the ability to move from state to state is determined not only by the particle itself but by the parameters of the entire system.

Not surprisingly, then, various sciences are beginning to study systems in their entirety. These days, says Harvard physicist Roy Schwitters, "people are learning how to deal mathematically with assemblies of matter on a large scale: They can do physics in very complex and chaotic systems."

That has practical significance, he says, in allowing research into such things as superconductivity. But it also has metaphysical ramifications—as Stephen Toulmin, a Northwestern University historian of science, points out.

"The world as we in fact encounter it does not consist of atoms each of which behaves in an entirely independent way," he says. Instead, it's a world of "food chains, ecological systems, organs, organisms, families, communities." To understand the world properly requires what he calls "chain thinking."

It's no accident, says Dr. Toulmin, that the more interdisciplinary sciences like anthropology and ecology are currently held in much higher regard than in earlier periods, when classical physics reined. Why? Because, rather than breaking the world into isolated parts—looking for the basic building blocks of matter, for instance, as physics has done for so many decades—these sciences attempt to study a complex system as a whole.

What effect beyond the sciences?

WHAT effect could such thinking have beyond the sciences? One impact might be political. Historian Toulmin predicts that "the sense of every individual political nation as having its own absolute sovereignty will [give] way to a much deeper and more universal appreciation of the interdependence of the social units."

But how much do these ideas owe to quantum physics?

"The changes that are specifically associated with quantum mechanics are very difficult to disentangle from the larger changes of which they're a part," Toulmin says. "I don't want to talk about the wisdom of the East—I don't want to talk claptrap phrases about the end of linear causality. All the same, there is a general sense that we learn as much about the world from ecology as we do from breaking it up into tiny little bits."

That might be expected in studying biology, he observes. But Toulmin extends the idea to physics as well. "We understand how the actions of different parts of the universe are intelligible," he says, "only if we recognize their interdependence."

"The question is how we see all these different levels of organization in relation to one another—not how do we reduce them all to a single so-called fundamental level," he adds. "Fundamental always means totally fragmented."

For Toulmin, the distinction between "chain thinking" and "fundamentalism" even extends into religious phenomenon. "Fundamentalism as it exists in the religious life of America," he notes, "is based on the idea that every individual is saved separately."

By contrast, he says, "the general view in the history of Christianity was that it was always communities that were Christian—and it is only within the congregation that the individual has any hope of salvation."

So as scientific thought turns away from a preoccupation with fundamental particles and toward complex systems, will it provide a world view in which communities of all sorts—ecological, social, and religious—are seen to operate as wholes?

And will that view, ultimately, extend to the concept of mind itself—seen not as a collection of brain cells or even hemispheres, but as an integrated whole?

"I hope that some generation of humanity will find that the laws of physics and the laws of psychology begin to overlap," says Freeman Dyson, at Princeton. But, like so many physicists interviewed for this series, he urges caution. At present, he says, that possibility is remote.

"When I see people saying that this is already the case," notes Professor Dyson, "I see that they're using what for me are very flimsy arguments. And I react against that."

Where, then, does this leave Caleb and Peter and Mr. Bentley's ball bearings? These young men will be well into their chosen careers when quantum mechanics turns 100 years old in the year 2025. By then, it will have had a chance to sink more deeply into public thought. What will the world be like?

It may be a world where muddle is accepted as inherent in nature. If so, it could be a world which, in the terms of John Bell at CERN, will "encourage the people who are not very disciplined in thinking to feel that a kind of free fantasy is the right approach to intellectual affairs."

On the other hand, it could be a world in which rigid determinism of all sorts—hereditary, economic, racial, medical, educational—is replaced by an insistence that the observer, and especially the self, has a much larger role to play. If the universe—including, perhaps, the human body—is seen as not a ticking clock but a forum for limitless possibilities, so mankind may also be seen that way.

That could, on the one hand, lead to an intensity of individualism that would end in irresponsibility—a point raised by University of Texas physicist Joe Polchinsky. If the many-worlds interpretation takes hold, he says, "people may say, 'Well, it really doesn't matter what I do, since there's another universe where I made the opposite choice—so if I can't change anything, I may as well do what I please."

He characterizes such a view as "baloney"—a gross distortion of the implications of quantum mechanics. But he worries, nevertheless, about the cultural consequences of that distortion.

On the other hand, the perception of infinite possibilities could free mankind from the limits of fatalism and encourage greater individual responsibility. If reality, after all, is in some way shaped by the observer, the individual may discover that he or she is not simply a spectator but a participant.

Finally, if humanity is seen to be organized into complex systems, this fact could have profound effects on thinking. On one hand, it could engender despair over the loss of a private, exclusive, and ordered self. On the other hand, however, it could mean that wholeness and completeness, rather than analysis and fragmentation, would become the standard pattern for high-order thinking.

However the view changes, it may well have the effect of breaking the hammerlock of naive materialism. In the 21st century, says West German physicist Herwig Schopper, director general of CERN, "the changing world view will be in a way to abandon materialism and to go to what's more abstract."

For Caleb and Peter, the intellectual impact could be enormous—much like that described by Fermilab director Leon Lederman when he first encountered the world of quantum mechanics. "The goose pimples had goose pimples," recalls Dr. Lederman, sitting in his office beside a stuffed, two-foot-tall Albert Einstein doll.

But however heady the ideas, Caleb and Peter will also have to negotiate what will still appear to be a "real" world in the 21st century. It will still be world of cars as well as quarks, bridges as well as bosons.

"When I'm driving over a bridge," says Fermilab cosmologist Edward W. Kolb, "I don't worry about the probability that I'm going to tunnel through the bridge rather than drive over." However much you ponder quantum mechanics, he says, you still have an ordinary life to lead.

"You have to mow your lawn," he says with a sigh, "whether you think you know the origin of the universe or not."

Rushworth M. Kidder

KIDDER, RUSHWORTH M. "HOW MIGHT QUANTUM THINKING CHANGE US?" *CHRISTIAN SCIENCE MONITOR* (JUNE 16, 1988).

SEE ALSO *Physics: Heisenberg Uncertainty Principle; Physics: Maxwell's Equations, Light and the Electromagnetic Spectrum; Physics: The Standard Model, String Theory, and Emerging Models of Fundamental Physics; Physics: Wave-Particle Duality.*

BIBLIOGRAPHY

Books

Baggott, J.E. *Beyond Measure: Modern Physics, Philosophy, and the Meaning of Quantum Theory.* Oxford, UK: Oxford University Press, 2004.

Columbus, Frank, and Volodymyr Krasnoholovets, eds. *Developments in Quantum Physics.* Hauppauge, NY: Nova Science Publishers, 2004.

Isaacson, Walter. *Einstein, His Life and Universe.* New York: Simon & Schuster, 2007.

McMahon, David M. *Quantum Mechanics Demystified.* New York: McGraw-Hill, 2006.

Periodicals

Kidder, Rushworth M. "How Might Quantum Thinking Change Us?" *Christian Science Monitor* (June 16, 1988).

Web Sites

Oracle Education Foundation: ThinkQuest. "Quantum Mechanics." http://library.thinkquest.org/C004707/qm.php3?Java=No (accessed May 4, 2008).

PhysicsWeb: Institute of Physics Publishing. "Max Planck: The Reluctant Revolutionary." December 2000. http://physicsweb.org/articles/world/13/12/8 (accessed May 4, 2008).

Public Broadcasting Service: A Science Odyssey—People and Discoveries. "Max Planck: 1858–1947." http://www.pbs.org/wgbh/aso/databank/entries/bpplan.html (accessed May 4, 2008).

William Arthur Atkins

Physics: The Standard Model, String Theory, and Emerging Models of Fundamental Physics

■ Introduction

One of the crowning achievements of twentieth-century physics is the standard model of particle physics, an attempt to construct a complete description of all fundamental particles that exist in the universe and their interactions with one another. While the standard model represents a stunning success of the methods of modern physics and stands as a monument to the complex interplay between theory and experiment, it still leaves many questions about the nature of matter unanswered.

■ Historical Background and Scientific Foundations

One of the oldest goals of science has been to understand the fundamental structure of matter. This search began in the ancient world, when philosophers such as Democritus (460–370 BC), Epicurius (341–270 BC), and Lucretius (fl. first century BC) suggested that if you divide an object into smaller and smaller pieces, eventually you would arrive at an entity that could no longer be divided. These entities were called "atoms," from the Greek *atomos,* meaning indivisible.

The question of whether or not atoms were real remained a philosophical one for the next 1,500 years. But in the eighteenth century the emerging fields of physics and chemistry began to provide concrete evidence for their existence. Physicists began to realize that the properties of matter (such as heat) were the result of tiny moving particles. And chemists discovered a number of compounds that seemingly could not be broken down into other substances. The periodic table of the elements, first formulated by Dmitry Ivanovich Mendeleyev (1834–1907) between 1868 and 1870 was

basically a catalog of all of the atoms then known to exist in nature. In the early part of the twentieth century, this picture began to change as "indivisible" atoms were found to be composed of even smaller particles. The quest for the fundamental building blocks of matter was only beginning.

Parts of the Atom

The first fundamental particle that was identified as such was the electron. In 1897 British physicist Joseph John Thomson (1856–1940) was working with a recently invented device called a cathode ray tube—essentially a primitive version of the picture tubes used in early television sets. In studying the behavior of cathode rays, Thompson found that their behavior was that of a beam of tiny negatively charged particles. These "corpuscles" of electrical charge, as he called them, were much lighter than any known atom, and it seemed they could be generated from any kind of material. This led scientists to conclude that each atom had within it some number of these tiny charged particles—the first hint that atoms were neither elemental nor indivisible, but composed of even smaller particles.

If an atom included negative charges, it must also include positive charge, since matter is generally neutral. The earliest theory of atomic structure, held by Thompson and others, was the "plum pudding" model, which envisioned the atom as a uniform sphere of positive charge with tiny chunks of negative charge, electrons, embedded within it. This picture, however, was shattered in 1907 by New Zealand-born English physicist Ernest Rutherford (1871–1937).

In Rutherford's experiment, thin pieces of metallic foil were bombarded with alpha particles from a radioactive source. If the atom was a soft, uniform, positively charged sphere, most of the positively charged alpha

WORDS TO KNOW

ANTIPARTICLE: In particle physics, most fundamental particles having non-zero rest mass (e.g., neutron, proton, electron) have an antiparticle that is an almost perfect mirror-image of the particle. When a particle and its antiparticle meet, they annihilate each other, releasing all their energy in the form of photons. At the big bang, about 13.7 billion years ago, slightly more particles were produced than antiparticles. After all particle-antiparticle pairs annihilated each other, the remaining fraction of particles became all the ordinary matter seen in the universe today.

ATOM: The smallest particle in which an element can exist.

BARYON: Subatomic particles that participate in strong force interactions. Baryons are composed of three quarks (or three antiquarks).

BOSON: According to the modern Standard Model of particle physics, all particles are either bosons or fermions. All forces are explained as exchanges of virtual bosons between fermions. Bosons have integer spin values; photons are bosons (spin = 1). The only particle predicted by the Standard Model that had not been experimentally observed as of early 2008 was the Higgs boson.

ELECTRON: A subatomic particle having a charge of –1.

GLUON: The elementary particle thought to be responsible for carrying the strong force (which binds together protons and neutrons in the atomic nucleus).

HADRON: Subatomic particles that are affected by the strong nuclear force; they include both baryons (three-quark particles such as protons, neutrons, etc.) and mesons (two-quark particles such as pions).

KAON: Also known as a K-meson. A fundamental particle containing a single quark; it is a type of meson and is several times heavier than a pion.

LEPTONS: A group of subatomic particles not composed of quarks that includes electrons, muons, and tau particles.

MESON: Subatomic particles composed of a quark and an antiquark; they participate in strong force interactions.

NEUTRINO: A high-energy subatomic particle resulting from certain nuclear reactions that has no electrical charge and no mass, or such a small mass as to be undetectable.

NEUTRON: A subatomic particle with a mass of about one atomic mass unit and no electrical charge that is found in the nucleus of an atom.

NUCLEUS: Any dense central structure can be termed a nucleus. In physics, the nucleus of the atom is the tiny, dense cluster of protons and neutrons that contains most of the mass of the atom and that (by the number of protons it contains) defines the chemical identity of the atom. In astronomy, the large, dense cluster of stars at the center of a galaxy is the galactic nucleus. In biology, the nucleus is a membrane-bounded organelle, found in eukaryotic cells, that contains the chromosomes and nucleolus. Intact eukaryotic cells are comprised of a nucleus and cytoplasm. A nuclear envelope encloses chromatin, the nucleolus, and a matrix, which together fill the nuclear space.

PHOTON: Smallest individual unit of electromagnetic radiation (light energy). These light particles are emitted by an atom as excess energy when that atom returns from an excited state (high energy) to its normal state.

PION: A pion or pi meson is a type of fundamental particle containing two quarks. There are three pions; all are involved in carrying the strong nuclear force.

POSITRON: The antiparticle of the electron. It has the same mass and spin as the electron, but its charge, though equal in magnitude, is opposite in sign to that of the electron.

PROTON: Particle found in a nucleus with a positive charge. Number of these gives atomic number.

QUARK: A type of elementary particle.

STRONG FORCE: In atomic physics, one of the four fundamental forces of nature (electromagnetic, weak, strong, gravitational). The strong force is repulsive at extremely close range (much smaller than an atomic nucleus) and attractive at longer ranges. Within the atomic nucleus, the attraction of the strong force between protons is stronger than the repulsive electromagnetic force between them (like charges repel); this is why atomic nuclei are not blown apart by the mutual electromagnetic repulsion of their positively-charged protons. On the other hand, protons and neutrons are kept from simply collapsing into each other by the repulsive effect of the strong force at extremely short range. At longer-than-nuclear ranges, the Coulomb force of electromagnetic attraction or repulsion is far stronger than the strong nuclear force, so it dominates outside the nucleus.

WEAK FORCE: In physics, the electroweak force is the single unified force of which the electromagnetic force and weak interaction are both manifestations. The weak interaction is involved in certain types of radioactive decay (e.g., beta decay of neutrons).

particles would be only slightly deflected as they passed by or through the atoms. Instead, Rutherford and his collaborators, the German physicist Hans Geiger (1882–1945) and British-born New Zealand physicist Ernest Marsden (1889–1970) found that most of the alpha particles passed straight through the foil, while a very small number of them bounced almost straight back, as if they had encountered something solid. This led Rutherford to propose a new model for the atom in which the electrons moved about in a region of nearly empty space, while its positive charge was entirely concentrated in a tiny central region called the nucleus.

By 1918 subsequent experiments allowed Rutherford to identify the fundamental unit of positive charge as the proton. The other important component of atomic nuclei, the neutron, was discovered in 1932 by British scientist Sir James Chadwick (1891–1974). These three particles—the proton, the neutron, and the electron—gave scientists a more complete and fundamental description of the properties of the elements in the periodic table. These 90 or so elements were no longer thought to be elemental at all. Instead, each type of atom represented some combination of protons, neutrons, and electrons.

It was clear that this model of the atom, a tiny nucleus packed with positively charged protons, demanded a new kind of physical explanation. If the only force at work in the atom was that of electrical attraction and repulsion, the nucleus should fly apart due to the repulsive force between the protons. The existence of the nucleus suggested that some force bound protons and neutrons together.

Into the Nucleus

Experiments in the 1930s and 1940s used the first primitive particle accelerators to probe the structure of the nucleus and discover the force that held protons and neutrons together. But rather than simplifying the structure of matter, these experiments soon revealed a subatomic universe that was surprising and complex.

The first new particle, the muon, was discovered in 1936. Initially thought to be a new type of nuclear particle involved in the force between protons and neutrons, it was later found to have none of the properties that would be expected of such a particle, prompting the Hungarian-born American theoretical physicist Isidore Isaac Rabi (1898–1988) to ask jokingly, "Who ordered that?" In 1947 pions explained many of the processes taking place inside the atomic nucleus. That same year, however, the lambda was found to be heavier than the proton and neutron—another surprise. The kaon was discovered in 1949, delta particles in 1952. By the middle of the 1960s, more than a hundred new particles had been discovered. Rather than simplifying our understanding of matter, they revealed a vast and complicated array of previously unknown particles.

The Science: The Electron, Muon, and Tau

The muon, a particle with properties that are nearly identical to the electron, was discovered by American physicist Carl David Anderson (1905–1991) in 1936. With a mass more than 200 times greater than the electron, the muon is unstable, existing only for a few millionths of a second on average before decaying into an electron and neutrinos. But in every other respect—charge, spin, interactions with other particles—it behaves just like an electron. In fact, scientists have even created short-lived "muonic atoms" whose electrons were replaced by muons.

Another member of this group was discovered in the mid–1970s. The tau has a mass more than 3,700 times the electron. Its lifetime is also very short, typically less than a trillionth of a second. Along with the electron, the muon and tau make up a group of particles that physicists call leptons, from the Greek for small, thin, or delicate. Each lepton has an associated neutral partner, a particle with no electrical charge, called a neutrino.

The Neutrino

Neutrinos were first proposed as a solution to a nagging observation that plagued nuclear physicists in the very early part of the twentieth century. In 1911 Austrian-born physicist Lise Meitner (1878–1968) and German chemist Otto Hahn (1879–1968) were studying the radioactive process of beta decay, a process in which one of the neutrons in an atomic nucleus is transformed into a proton plus an electron, which is emitted from the nucleus as a particle of radioactivity. When the energy of the emitted electrons was measured, the process seemed to violate the law of conservation of energy because a tiny bit of energy seemed to be "lost" in the process. This was puzzling, since the idea that energy is conserved in all physical processes was—and still is—absolutely fundamental to our understanding of the universe.

A solution was proposed in 1930 by the Austrian-born physicist Wolfgang Pauli (1900–1958), who suggested that a small, light, neutral particle might also be produced in beta decay, and that this particle was escaping the experiment undetected, carrying with it some of the missing energy. Some years later, this hypothetical particle was named the neutrino by Italian-born American physicist Enrico Fermi (1901–1954), but more than two decades would pass before its existence was finally confirmed experimentally in 1956.

The neutrino interacts so weakly with ordinary matter that it typically passes through solid objects unaffected. Of the trillions of neutrinos that strike every square inch of Earth each second, most of them pass right through the entire planet and emerge from the ground on the other side; only about one out of every hundred billion is stopped by interacting with another particle along the way. Modern neutrino detectors, despite being much more sensitive than those used to discover neutrinos in the 1950s, still detect only a small handful every day. And yet their study is of great value to physicists and astronomers, giving them insight into the fundamental interactions of matter, as well as processes that occur in the core of the sun and distant exploding supernovas. Today, we know that there are three types of neutrinos. The original neutrino postulated by Pauli is known as the electron neutrino, but there two other types or "flavors": the muon neutrino, discovered in 1962, and the tau neutrino, discovered in 1975. Each is paired through its interactions with one of the charged leptons: the electron, muon, and tau.

One question about the neutrino that was unanswerable until very recently was whether or not they were massless, like a photon, or just very, very light. The main observable difference between a massless neutrino and one with mass is that the latter can change or "oscillate" from one flavor to another. Recent measurements suggest that some of the electron neutrinos produced in the sun's core transform into muon neutrinos during the 93-million-mile trip from the sun to Earth. This suggests that neutrinos do in fact have a small mass, although the experiments are not yet sensitive enough to measure it accurately.

Antiparticles

In 1927 British theoretical physicist Paul Dirac (1902–1984) was working to combine the quantum mechanical description of the electron with Albert Einstein's (1879–1955) theory of relativity. The result, now known as the Dirac equation, stated that the electron seemed to possess positive energy states in addition to its usual negative energy. This was eventually interpreted to mean that another particle should exist, identical to the electron in most respects, but with the opposite electrical charge. The positron was discovered in 1932 by Carl Anderson (1905–1991), who also discovered the muon.

We also know now that nearly all particles possess antiparticles. There are anti-muons and anti-taus, as well as antineutrinos and antiquarks. Some uncharged particles, however, such as the photon and the Z, are their own antiparticles.

The Rise of the Quark Model

Patterns in the masses, charges, and other properties of mesons (subatomic particles composed of a quark and an antiquark) and baryons (subatomic particles composed of three quarks, or three antiquarks) led physicists to speculate that the particles had some sort of underlying structure, not unlike the way the periodic table explains the elements as assemblages of protons, neutrons, and electrons. In 1961 American physicist Murray

Gell-Mann (1929–) proposed that the proton, neutron, and other heavy particles were made of smaller particles that he named quarks. Gell-Mann took the name from a line in James Joyce's novel *Finnegan's Wake:* "Three quarks for Muster Mark. Sure he hasn't got much of a bark. And sure any he has it's all beside the mark." Gell-Mann's model, which consisted of three quarks—the up quark, down quark, and strange quark—explained the number of hadrons (subatomic particles that are composed of two or more quarks or antiquarks, and that take part in the strong force interaction—both baryons and mesons are hadrons) perfectly, as well as the pattern of their masses, charges, and other properties.

At first only three quarks were necessary to explain all of the known heavy particles. These particles had fractional electrical charges compared to the electron and proton, and could be combined in groups of two or three to form all of the known baryons and mesons. For example, the up quark has a charge of $+\frac{2}{3}$, while the down quark has a charge of $-\frac{1}{3}$. A combination of two ups and a down gives a charge of $+\frac{2}{3} +\frac{2}{3} -\frac{1}{3} = 1$, which corresponds to a proton. A combination of two downs and an up, on the other hand, gives an electrical charge of $\frac{2}{3} -\frac{1}{3} -\frac{1}{3} = 0$. So two downs and an up make a neutron. Lighter mesons like the pion, kaon, and others were explained as a pair that consisted of a quark and an antiquark.

The quark model successfully explained the properties of all known hadrons in the 1960s, although some theoretical considerations suggested there could be more than three quarks. In 1974 the discovery of the J/Psi particle required the addition of a fourth quark to the model, the charm. This added some symmetry to the model, since now there were two quarks with a $+\frac{2}{3}$ charge (the up and the charm) and two with a $-\frac{1}{3}$ charge (the down and the strange). In 1977 a fifth quark was discovered—another $-\frac{1}{3}$ charged quark that was eventually named the bottom. Physicists were sure that a sixth would be found to complete the picture, a heavy quark with a $+\frac{2}{3}$ charge named the "top." Decades of searching finally bore fruit when the top quark was produced in 1995 at the powerful Tevatron particle accelerator at Fermilab near Chicago.

Standard Model Fermions

The six leptons (the electron, muon, tau and their partner neutrinos), six quarks (up, down, strange, charm, bottom, top), and their antiparticles are the fundamental constituents of all matter in what has come to be called the Standard Model of particle physics. These particles share one common property: They have the same quantum mechanical "spin," meaning they all carry a quantity of angular momentum equal to ½ Planck's constant. Physicists call this group of "spin ½" particles fermions. But fermions make up only half the particle physics picture.

Another group of particles in the standard model are related to the forces between particles of matter.

Exchange Forces and the Photon

The first force described by particle physicists was the electrical force, recognized since antiquity. The word "electron" comes from the Greek word for the substance amber, since rubbing amber with a cloth was known to generate "static electricity" that attracted other small light objects. Electrical forces, like gravity, were seen as mysterious invisible fields that acted through empty space. While the strength and other properties of such forces were well understood, what was lacking was an understanding of the mechanism of how they occurred.

Quantum theories helped shed some light on the mechanism of electrical interactions between particles. Since 1905 quantum theories had suggested that the interaction between particles and light happened in a discontinuous fashion, rather than a smooth continuous absorption of energy, as older theories of electromagnetism had suggested. Albert Einstein coined the term "photon" for a chunk of light energy, essentially a "particle of light." The emission and absorption of photons by electrons and other charged particles eventually led to the idea that the mechanism for electrical attraction and repulsion was an exchange of photons between the charged particles.

Two electrons exchanging photons with one another would move away from each other like two people on skateboards tossing a heavy object back and forth between them. These photons were called virtual photons since they existed only for a brief time before being reabsorbed by the other particle. The energy that creates virtual photons can be thought of as being borrowed for a brief time from the slight energy fluctuations allowed by the Heisenberg uncertainty principle, which states that it is not possible to know a particle's location and momentum precisely at the same time. A complete theory of the interactions between charged particles via photon exchange was finalized in the 1940s and given the name quantum electrodynamics or QED. The basic idea of particle exchange as the carrier of force between particles was useful for explaining other natural forces as well.

Quarks, Gluons, and the Strong Force

The idea that protons, neutrons, and other hadrons were made of more fundamental particles called quarks required an explanation of the force that binds quarks together. A theory of this interaction would have to explain some of the quarks' curious features, such as why they only appear in groups of two and three, and that no isolated quark had ever been detected. The force between quarks is known as the strong force, because it is obviously stronger than the force of electrical repulsion

that pushes apart the like-charged quarks inside a proton and neutron.

Since a typical hadron contains three quarks, the charge responsible for the strong force was also deduced to come in three varieties, instead of the two (positive and negative) that create the electrical force. The new strong charge was named "color charge," and its three varieties were called red, green, and blue because they combine to make a neutral combination, just like the three primary colors combine to make a colorless white light.

The particle that carries the strong force between quarks is called the gluon. Unlike the photon, which transmits electrical forces between electrical charges but has no electrical charge of its own, the gluon has color charge. And, just like the quarks, it feels the strong force. This is one reason that the strong force is so strong, and the reason that no one has ever seen an isolated quark. The strong force actually gets stronger with distance, so that the quarks in a proton are trapped for good, a feature of the theory known as confinement.

The complete theory of quarks, gluons, and the strong force, known as quantum chromodynamics or QCD, is the result of the work of a number of physicists in the 1960s and 1970s. QCD turns out to be very challenging mathematically, and while it is difficult to produce exact results for many problems, the theory has successfully described the interactions of quarks.

The Weak Force

The existence of the neutrino, a neutral particle that interacts with matter only very weakly, suggested that there was some new force at work between neutrinos and the other particles of ordinary matter. Dubbed the weak force, it is felt by all particles of matter, regardless of their charge, but is associated most closely with neutrinos, since it is the only force they feel (other than gravity, which is weaker still).

The weak force is carried by a trio of particles called weak bosons: the W+, W−, and Z⁰. These force-carrying particles are among the most massive in the standard model, and their large mass is responsible for the weakness and the short range of the weak force. The Heisenberg uncertainty principle dictates that the more energy and mass a force-carrying virtual particle has, the shorter its life span and the shorter the distance it can travel between particles.

The full theory behind the weak force arose in the mid to late 1960s through the independent work of American theoretical physicist Sheldon Glashow (1932–), Pakistani nuclear physicist Abdus Salam (1926–1996), and American nuclear physicist Steven Weinberg (1933–). The complete theory, called electroweak theory, explained the action of the weak force and predicted the existence of W and Z bosons, which were not discovered experimentally until 1983 at

One of the first observations of a W particle. Electronic display of the results of a high-energy proton-antiproton collision in the UA1 detector at CERN in 1982–1983. The discovery of the W particle, one of the carrier particles of the weak nuclear force, was announced at CERN, the European particle physics lab near Geneva, Switzerland, in January 1983. *CERN/Photo Researchers, Inc.*

European Organization for Nuclear Research (CERN) in Switzerland.

Besides governing any interaction involving neutrinos, the weak force plays a role in any process where a quark changes flavor from one type to another. For example, in beta decay (the process that led physicists to first postulate the existence of the neutrino) a neutron must change into a proton. This requires a down quark to change into an up quark. This can occur if the down quark emits a W−, changing its charge from −⅓ to +⅔. (The W− then decays into an electron and a neutrino, or, specifically, an antielectron neutrino.) This type of interaction occurs in any nuclear process in which neutrons turn into protons, or vise versa, including many types of radioactivity, and in the process of nuclear fusion.

Mass and the Higgs Boson

W and Z bosons are unique among force-carrying particles because they have mass, while the photon and gluon do not. One mechanism for generating a mass for particles is known as the Higgs mechanism, named after the British physicist Peter Higgs (1929–), who invented a mathematical method called spontaneous symmetry breaking that makes the mechanism possible. The method invokes a new sort of field, the Higgs field, which exists at every point in space. As particles move through space, their constant interactions with this Higgs field produce the sort of resistance to motion that

we typically associate with the property called mass, much like the resistance we feel when wading though deep water in a swimming pool.

The particle responsible for this interaction is known as the Higgs boson, the last major missing piece of the standard model puzzle. Although it is an essential part of the theory, no experiment has ever detected the Higgs boson directly. Based on the range of possible Higgs masses predicted by the standard model, the next generation of more-powerful particle accelerators, such as the large hadron collider, is expected to achieve the energies necessary to verify its existence.

Influences on Science and Society

The search for a theory of fundamental particles stretches from Thomson's discovery of the electron in 1897 to the discovery of the top quark in 1995 to the modern search for the Higgs boson—more than 100 years of experimental and theoretical work. Since 1950 more than twenty Nobel Prizes in physics have been awarded for work that is either directly or indirectly related to the development of the modern standard model. During this time, experiments that could be conducted by one or two scientists with some simple laboratory equipment have been replaced by "big science"—huge facilities costing billions of dollars with dozens of scientists and hundreds of technical and support personnel responsible to gather each new piece of data. This is true in all of the sciences, but it is particularly so in experimental particle physics, where new accelerators can cost billions of dollars. The standard model continues to drive inquiry in basic physics, not only because of its successes, but because of its failures as well. There are many things that the standard model does not do, or at least does not do to the satisfaction of many particle physicists.

What About Gravity?

One thing the standard model does not do is include a description of the force of gravity. As far as we can tell there are only four forces in nature, and the standard model deals with three of them: the electromagnetic force, the strong force, and the weak force. (Gravity is handled by Einstein's general theory of relativity, which describes forces in a very different way.) The standard model is a quantum field theory, that is, it explains forces in terms of exchanges of quanta of fields, such as the photon and the gluon. But general relativity considers gravity to be a result not of particle exchanges, but of the curvature of space-time.

Many physicists view this situation—where three of the forces of nature are treated with one set of mathematical tools, while a completely different approach is used for the fourth—to be unsatisfactory. They believe that there is a deeper theory, a "unified" theory, that would provide a single description of all the forces of nature. So far, attempts to create a fully unified theory of gravity and quantum physics have proven unsuccessful.

Supersymmetry

Another unexplained feature of the standard model is that, while fermions are free to interact in ways that allow them to transform from one type of matter particle into another, there are no interactions that transform fermions into bosons or vice versa. For a number of reasons, it is tempting for theoretical physicists to postulate a whole new set of particles and reactions that would bridge the gap between bosons and fermions and to construct a theory that was fully symmetrical in its treatment of the two classes of particles. This extension of the standard model, known as supersymmetry, postulates that every particle in the standard model has a supersymmetric partner on the other side of the fermion-boson divide. So spin ½ electrons and neutrinos have spin 1 partners called the selectron and the sneutrino. Likewise, the spin 1 superpartners of the quarks are known as squarks, and the photon and gluon have spin ½ partners as well—the photino and the gluino. None of these supersymmetric particles have ever been detected.

Supersymmetry may seem to add unnecessary complexity to the standard model by postulating additional particles for which there is no evidence. But the addition of the new supersymmetric particles actually solves a number of important theoretical and mathematical puzzles in the standard model, leading many physicists to trust that it is a correct theory that will be vindicated as the next generation of more powerful particle accelerators begins to produce the superpartner particles in the coming decades. Only time will tell.

Other Questions

Another weakness of the standard model is its inability to answer a number of questions. Why are there six leptons and six quarks? Why do the particles have the particular masses, charges, and other properties that they do? What determines the relative strengths of the various forces of nature? In addition, the number of free parameters in the standard model—numbers that are not predicted by the theory but must be determined experimentally and entered into equations by hand—is quite high. There is a sense among some physicists that a complete theory of fundamental particles should explain all these particle properties from basic principles, without reliance on parameters that just happen to have some particular value.

A related issue is that the number of fundamental particles seems to be quite large: six leptons, six quarks, and their antiparticles; the photon, the W, and the Z, and eight colors of gluons; and perhaps a supersymmetric partner for each of those. If the age-old quest to explain the structure of matter is an attempt to describe nature in terms of some very small number of basic constituents, the standard model does not seem to pass this test.

Unification and String Theory

The desire to unify our description of quantum forces with our understanding of space-time and gravity, and to simplify our picture of dozens of particles, antiparticles, and superpartner particles into a theory with some very small number of entities has led to a group of approaches known collectively as "string theory." The idea behind string theory is that the universe actually has more than three dimensions. Just like a piece of paper has a large width and height but a very small thickness, so the universe might have the three very large dimensions that we normally experience, but many other "thicknesses" in dimensions that are visible only at sizes at the scale of elementary particles. In string theory, fundamental particles are treated not as tiny zero-dimensional points but as loops or "strings" that wrap around these tiny hidden dimensions. The most commonly discussed string theories have six additional space dimensions, making the universe ten dimensional—with three large space dimensions, six small space dimensions, and one time dimension.

In string theory, there is only one fundamental object: the string. But depending how the string is situated or vibrating in the various dimensions, it can have different properties when viewed from our three-dimensional

Professor of Physics Andrew Strominger (right) and Cumrun Vafa (1960–), both of the Harvard University Department of Physics, specialize in the research of theoretical high-energy physics with emphasis on fundamental questions, as the nature of quantum gravity and the relation between geometry and quantum field theories. Known as "string theory," the idea has emerged as a strong candidate for a unified description of gravity and other forces. *© Rick Friedman/Corbis.*

vantage point. No physicist has yet figured out how to construct a specific version of string theory that describes the particular particles we find in the universe. The number of possible ways to deal with the extra dimensions is very large, perhaps as large as 10^{500} by some calculations. The astonishing number of different possible string theories suggests that finding the "right" one for the universe could be very difficult. But the ease with which string theory accommodates standard model physics (including supersymmetry and gravity) make it an intriguing avenue of research that has already given physicists some important insights into the problem of unification.

Not all physicists believe that string theory is the right approach to unification. Many feel that its large number of possible configurations means that we will never be able to find the correct set of string theory equations to describe the universe. Others doubt that string theory will ever be able to make any concrete predictions that are can be tested. If a theory never makes any testable predictions, then it is arguably not a scientific theory at all.

■ Modern Cultural Connections

The quest to probe the fundamental structure of the universe more and more deeply leads us to confront not only questions of how and when we know a theory is a good one, but forces us to examine our priorities as a society when it comes to investment in scientific research. While potential practical benefits could result from almost any scientific endeavor (PET scans and MRIs, for example, are two valuable medical diagnostic tools that owe their existence to our understanding of the properties of matter on the subatomic level), as we examine more and more esoteric phenomena at higher and higher energies, the motivation becomes the desire for knowledge for its own sake rather than the quest for new technologies or new applications. One can reasonably question whether such discoveries are worth the price tag, especially considering the rising cost of the experimental facilities needed to conduct research at the cutting edge of particle physics. In the United States, much funding for pure research of this nature comes from the federal government. Thus the question of what scientific research gets funded can become a political question as much as a scientific one.

An example of the conflict between pure science, government funding, and changing political landscapes can be found in the story of the superconducting super collider (SSC), proposed in 1983 as a powerful next-generation particle accelerator, designed to collide beams of protons at energies far exceeding the most powerful accelerators of the day. The goal of the SSC was to detect the Higgs boson and to explore the possibility of unknown phenomena beyond the standard model, such as supersymmetry. The project required a huge financial investment from the government. Initial estimates suggested that the project would cost more than $3 billion. Despite the cost, in 1987 the federal government decided to approve construction of the SSC at a location in Texas.

Construction began in 1991, but the project was never completed. A number of factors, including mounting budgets (the cost of the project ballooned to more than $11 billion) and changing administration priorities led Congress to shut down the SSC project in 1993. To Congress, the SSC represented a costly program that was an easy target for budget cuts. But physicists considered its cancellation a tragic blow to basic physics research. Despite such missteps, particle physics research continues, often with the help of international collaborations.

The most recent new particle accelerator, the large hadron collider at CERN (The European Organization for Nuclear Research), is scheduled to begin operation in 2008. And physicists are already planning more energetic machines, such as the international linear collider, a global effort scheduled for well into the 2010s. These new machines will probe the standard model at higher and higher energies and test its predictions to greater and greater accuracy. Physicists hope to see something new (and perhaps even unexpected) at these higher energies that would indicate new physical phenomena beyond the standard model—perhaps evidence of supersymmetry or extra dimensions, or even something that no one has thought of yet. But even if cracks do appear in the standard model, it will be impossible to deny its importance in shaping our view of the fundamental structure of matter, and the universe.

■ Primary Source Connection

The following article was written by Rushworth M. Kidder, a senior columnist for the *Christian Science Monitor* until 1990, when he founded the Institute for Global Ethics. Kidder is the author of *Reinventing The Future—Global Goals For The 21st Century*. Founded in 1908, the *Christian Science Monitor* is an international newspaper based in Boston, Massachusetts. In his 1988 article, Kidder describes how string theory seized the imagination of the physics community.

STARTLING STRINGS

HOLD a tiny rubber band 10 inches from your eye. It appears to be what it is—a closed loop of rubber. Put it 10 yards from your eye. Now it appears to be a dot, a single point whose features are indiscernible. Put it 10 miles from your eye. You see nothing at all. That rubber band, many physicists now agree, is like a fundamental particle of matter. The naked eye can't see such particles at all. They're simply too small.

How can you see them at closer range? A particle accelerator helps. Like a giant microscope, it uses high energies to let you "see" tiny particles. Even that, however, doesn't provide the high resolution needed to see the outlines of the rubber band. It sees the rubber band, all right—but as a dot, not a loop.

But what if you had an accelerator that generated the kind of energy available at the big bang? It would be like seeing a particle up close through a mammoth microscope.

And would those dots still be dots?

No, says string theory, a new and promising way of looking at matter that in the last five years has seized the imagination of the physics community. Those dots, say string theorists, only look like dots. In fact, they're really like rubber bands—or, more accurately, loops of string. What's more, the 60 or so particles (depending on who's counting) in the "particle zoo" that makes up matter turn out to be, in this theory, different manifestations of a single loop-shaped object.

Then why do all these particles—quarks and leptons, anti-quarks and anti-leptons, and all the rest—seem to be so different? According to an extension of string theory called superstring theory—which incorporates a new symmetry called "supersymmetry"—the differences arise because these loops vibrate in different ways. Set a loop oscillating in a certain way, and it will appear to have certain properties. Seen from a distance, it might appear to be a "charmed" quark. With a different oscillation, the same string might seem to be a muon. If we could only see it up close, however, it would reveal itself to be just another dancing, jiggling, rolling pattern being played on a one-size-fits-all loop.

If that sounds strange, imagine a blind Martian coming to earth and hearing a one-stringed violin. He could be forgiven for thinking that the instrument had dozens of strings—one for each note he hears. In fact, all those notes come from just the one string. The secret? Each note results from a different vibration of the string.

So promising is the superstring theory, physicists say, that it appears capable of embracing a vast array of phenomena and explaining them as parts of a splendid, well-balanced whole. Even the four forces of nature—gravity, electromagnetic force, the strong force that binds protons and neutrons into the nucleus, and the weak force responsible for radioactive decay—may prove to be different manifestations of a single force. Physicists refer to supersymmetry, only half jokingly, as the TOE—the theory of everything.

But the TOE is not without its problems. "Right now," says Rockefeller University physicist Heinz Pagels, "it's turned out to be a theory of nothing. By that I mean that, although it's extremely elegant both conceptually and mathematically, it has failed to make contact not only with experiment, but with the ordinary theories that we now know describe experiment."

The problem, in part, is one of scale. These strings or loops, according to the theory, are just 10 to the −33 centimeters long. That means that it would take a million of them—multiplied by a billion, then by another billion, then by still another billion—to add up to a centimeter. Even the mammoth superconducting supercollider—the SSC, currently proposed by the United States Department of Energy—will develop energies that will "see" only to the range of 10 to the −15 centimeters.

Moreover, such strings can't exist in our ordinary three-dimensional universe. "It appears that the superstring theory implies that space-time is 10 dimensional," says cosmologist Michael Turner. That means nine spatial dimensions and a 10th in time. In our everyday world, he notes, "we know of three spatial dimensions. That would say that we missed twice as many."

Where are they? The best explanation, apparently, is that they're somehow "folded up" within the three-dimensional world, like leaves waiting to mature.

And then there's the mathematics, widely described as immensely challenging. So complex is it, in fact, that superstring theory has had to await the discoveries of highly elaborate mathematics for it to progress. Result: There has been a plethora of solutions to these theoretical problems, many of which claim to be good descriptions of the universe.

Despite the obstacles, however, physicist John Schwarz—who, with British physicist Michael Green, is one of the founders of string theory—is convinced that there are now only three possible string theories that could be right.

"Once you've said which of those theories is the right one," he explains across the cluttered desk in his office at the California Institute of Technology, "you've given a completely unambiguous fundamental theory of nature. And now all you have to do to describe all of physics is to solve the equations."

"That's, of course, the hard part," he adds with a chuckle.

Rushworth M. Kidder

KIDDER, RUSHWORTH M. "STARTLING STRINGS." *CHRISTIAN SCIENCE MONITOR* (JUNE 16, 1988).

SEE ALSO *Physics: Heisenberg Uncertainty Principle; Physics: QED Gauge Theory and Renormalization; Physics: Radioactivity; Physics: Special and General Relativity; Physics: The Quantum Hypothesis.*

BIBLIOGRAPHY

Books

Greene, Brian. *The Elegant Universe: Superstrings, Hidden Dimensions, and the Quest for the Ultimate Theory.* New York: W.W. Norton, 1999.

Oerter, Robert. *The Theory of Almost Everything: The Standard Model, the Unsung Triumph of Modern Physics.* New York: Pi Press, 2006.

Periodicals

Kidder, Rushworth M. "Startling Strings." *Christian Science Monitor* (June 16, 1988).

Mervis, Jeffrey. "10 Years After the SSC: Scientists are Long Gone, but Bitter Memories Remain." *Science* 302, no. 5642 (October 3, 2003): 40–41.

Web Sites

Lawrence Berkeley National Laboratory, Particle Data Group. "The Particle Adventure." http://www.particleadventure.org (accessed November 10, 2007).

Nobel Foundation. "Solving the Mystery of the Missing Neutrinos." http://nobelprize.org/nobel_prizes/physics/articles/bahcall/index.html (accessed November 10, 2007).

David L. Morgan

Physics: Thermodynamics

■ Introduction

Beginning in the nineteenth century as the study of heat energy and transfer, thermodynamics established a new model for understanding the natural world. Instead of viewing the universe as a giant clockwork mechanism, it became more common to view it as a heat engine in which the conservation and transformation of energy determined the function of everything from atoms, molecules, and fields to chemical and geological phenomena, living processes, and possibly even the birth and death of the universe itself. Thermodynamics also plays an important role in the study of such current problems as global warming, pollution, and increased demands for efficient energy. Through its connection to the theory of information, thermodynamics is also a central part of computer science and globalized information systems such as the Internet.

■ Historical Background and Scientific Foundations

Early Theories of Heat

The ancient Greeks thought of heat as an actual substance. The philosopher Empedocles (c.490–430 BC) considered fire one of the four basic elements, along with water, earth, and air. By the fourth century BC Aristotle (384–322 BC) began to interpret heat as a quality that could be added to matter in different quantities. Warm bodies, he believed, had more of this quality than cold ones. Aristotle's ideas dominated natural philosophy throughout the Middle Ages, but by the end of this period scientists began to question the true nature of heat. Was it a separate material substance, as Empedocles claimed, or simply a state (condition) of ordinary matter?

By the seventeenth century several natural philosophers suggested that heat might be associated with some type of motion, but there was no consensus as to how this happened. Some believed that heat was the motion of the particles that made up ordinary matter, while others suggested that it was the motion of a subtle fluid-like material distinct from ordinary matter. Searching for the answer, natural philosophers made significant advances in the study of heat.

British natural philosopher and theologian Robert Boyle's (1627–1691) experimental studies of the relationship between heat and the volume of a gas led to what is now known as Boyle's law, which states that the pressure and volume of a gas varies directly with temperature. Galileo Galilei's (1564–1642) invention of the thermometer around 1592 allowed researchers to measure heat more precisely, but it also resulted in a certain confusion between temperature (the intensity of heat) and the total quantity of heat in a body. Edmond Halley (1656–1742), of Halley's comet fame, noted that the temperature of boiling water was the same as steam. Though it took additional heat to convert boiling water into steam, he didn't seem to understand the importance of this observation to a comprehensive theory of heat, however.

By the eighteenth century scientific ambiguity about the nature of heat had led to two distinct hypotheses: a material theory, in which heat was considered the motion of the ordinary particles of matter, and a material theory, eventually labeled caloric, in which heat was perceived as a subtle fluid that was distinct from ordinary matter but was attracted to and could combine with it. Many eighteenth-century advances in heat theory were discovered by chemists and physicians who supported the caloric theory. This led to the idea that heat was conserved, since it was believed that matter, even a subtle matter, could not be created or destroyed.

WORDS TO KNOW

BLACK HOLE: A single point of infinitely small space containing the mass and gravity of a collapsed massive star. The gravity is so strong that light can't escape.

CALORIC THEORY OF HEAT: In the 1780s, French scientist Antoine Laurent Lavoisier (1743–1794) proposed that heat is a substance resembling a fluid that can flow from one object to another. This fluid he named caloric. In the mid 1800s and beyond, this theory was replaced by the modern understanding of heat, which is that heat is the motion of large numbers of particles in a substance (atoms or molecules).

CARNOT CYCLE: A Carnot cycle—named after the French scientist who first identified such a cycle, Nicolas Carnot (1796–1832)—is a series of temperature-and-pressure states through which a system can be changed, returning eventually to its starting state. In moving through a Carnot cycle, a system does work on its environment, such as turning a shaft or moving a piston. All engines that turn heat into useful mechanical work are based on thermodynamic cycles like the Carnot cycle, but the Carnot cycle is the most efficient possible cycle and can only be approximated in a real machine.

ENERGY: The ability to do work or transfer heat.

ENTROPY: Measure of the disorder of a system.

FORCE: Any external agent that causes a change in the motion of a free body or that causes stress in a fixed body.

KINETIC ENERGY: The energy due to the motion of an object.

LATENT ENERGY: Also called latent internal energy or latent heat. The latent energy of a physical system is energy stored in the system by virtue of its phase (liquid, solid, or gas). For example, water contains latent energy that is released during freezing (i.e., freezing water releases latent energy as heat).

MATERIAL THEORY OF HEAT: Several material theories of heat were proposed in previous centuries. According to

these theories of heat, heat is a material substance or fluid; when one body is heated and another cooled, this fluid is supposed to flow from the hot to the cool body. Since the nineteenth century it has been known that heat is actually the rapid, random motion of atoms and molecules, not a substance in its own right.

MECHANICAL EQUIVALENT OF HEAT: The amount of work that must be done to produce a certain amount of heat. Mechanical work can be converted into heat by friction. Heat and mechanical work (force acting through distance) are interchangeable because heat consists of objects (particles) in motion and thus is essentially mechanical.

MECHANICAL THEORY OF HEAT: Developed in the mid-nineteenth century and refined since, the theory (system of explanations) stating that heat is not a substance or fluid in its own right but consists entirely of mechanical motions, namely, the motions of particles. In a solid, these motions consist of oscillations or vibrations around a fixed point; in a gas or liquid, they consist of rapid straight-line movement interrupted frequently by collisions with other particles.

MOTIVE POWER: The motive power of a device is the source of energy that makes it go. The term is usually applied to machines that actually move (e.g., vehicles).

SPECIFIC HEAT: Also known as specific heat capacity. The amount of heat energy required to raise one unit mass of that substance by a fixed unit of temperature (usually 1° Celsius). It takes more energy to heat substances that have high specific heats.

VIS VIVA: In the late 1600s, German scientist Gottfried Leibniz (1646–1716) proposed a theory of kinetic energy conservation that became known as *vis viva* (Latin for "living force"). The theory is now obsolete; kinetic energy is not, in fact, conserved, though momentum, a closely related quantity, is conserved.

The material theory, however, was less accepted until the theory of the conservation of energy emerged in the middle of the nineteenth century.

The idea that heat was always conserved led some researchers to distinguish between its intensity and its quantity. In the middle of the eighteenth century, Scottish chemist and physicist Joseph Black (1728–1799) discovered that heat had different effects on various substances. For example, a 1-pound (0.45-kg) iron bar at 212°F (100°C) had more ability to burn a person's hand than a 1-pound block of wood at the same temperature. From this Black discovered that different substances had different "heat capacities," a concept that he later refined into the concept of specific heat: the quan-

tity of heat, now measured in calories, required to raise one gram of water 1° Celsius (33.8°F).

Black's idea of heat capacity also led him to explain Halley's observation that steam and boiling water both existed at the same temperature. Black argued that the heat required to change a substance from one state (in this case a liquid) to another (a gas), is called latent heat. Steam carries this heat energy as it evaporates, then releases it as it condenses. This concept was successfully applied in the development of steam engines.

By the end of the eighteenth century, natural philosophers began to criticize the caloric theory of heat. One of the first to do so was Benjamin Thompson (1753–1814), an expatriate American who had

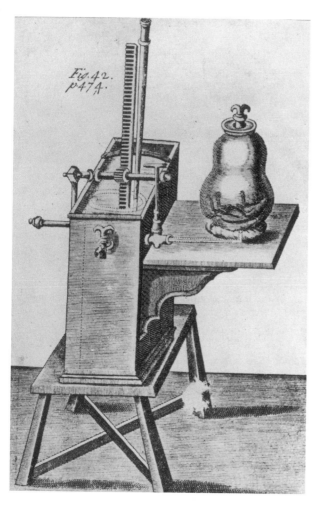

Fig. 42.
p. 474.

An illustration from 1725 showing an experiment with a vacuum pump invented by Robert Boyle (1627–1691), Irish-born English philosopher, naturalist and chemist. *Hulton Archive/Getty Images.*

During the early years of the nineteenth century, while chemists focused on the distribution of heat, physicists focused on its transmission or transfer. The leading figure in this effort was the French scientist Joseph Fourier (1768–1830). Ignoring the material vs. mechanical theories debate, Fourier used mathematics to develop differential equations that described heat flow through bodies of different shapes. His theories found almost immediate use among scientists studying the effects of solar and geothermal heat on Earth's surface.

The Steam Engine

Along with early theories of heat, the other important influence on thermodynamics was the invention of the atmospheric steam engine in 1712 by British engineer Thomas Newcomen (1663–1729). Working from earlier discoveries that the atmosphere exerted a force and that condensing steam created a vacuum, Newcomen invented an engine in which steam condensed inside a cylinder fitted with a piston to create a vacuum; atmospheric pressure then pushed the piston to the bottom of the cylinder. At this point in time, however, scientists thought that the steam engine's power source was atmospheric pressure, not heat.

One of the first to recognize the heat's primary role in the steam engine was James Watt (1736–1819), a Scottish inventor and instrument maker at Glasgow University. While attempting to get a small Newcomen engine to work, he realized that the steam had to reheat the cylinder each time after it condensed—an inefficient process. Using Black's concept of specific heat, Watt knew that the iron cylinder absorbed much of the heat on each alternate cycle. Watt added a second cylinder in which the steam could condense and cool, allowing the main cylinder to stay hot. The engine's power was now clearly associated with the flow of heat from the hot boiler to the cool condenser.

Watt's 1769 patent included another important improvement that influenced the development of thermodynamics. He realized that the pressure of the steam entering the cylinder was usually higher than atmospheric pressure; this allowed him to extract extra work from the steam by using its expansive power to push the piston for part of its cycle. Understanding the role of expansion in the cycle of a heat engine would prove a crucial element of the foundation of thermodynamics.

Until this point, the work done by a steam engine was calculated simply by multiplying the volume of the cylinder by the atmospheric pressure. Because Watt's engine allowed the steam to expand as it pushed against the piston, pressure dropped throughout the stroke; this made calculating the work performed difficult. In 1796 one of Watt's assistants attached a small pressure gauge to the main cylinder. As the piston moved and the pressure dropped, a pen attached to the gauge marked a pressure-volume (PV) curve on a piece of

supported the Royalist cause during the Revolutionary War. After spending time in England, where he was knighted by George III, he settled in Bavaria where he became minister of war and police and received the title Count von Rumford. While supervising cannon manufacture at the Munich arsenal, Rumford noticed that a seemingly inexhaustible amount of heat was generated as the boring tool rubbed against the inside of the cannon.

Since there seemed to be no limit to the amount of heat generated in this process, he argued that it was inconsistent with the caloric theory of heat, which postulated that only a fixed amount of heat existed in any material body. In a 1798 paper he argued that heat was a form of motion, since the motion of the boring tool was the only thing touching the cannon. Although it may seem obvious that Rumford's discovery played a crucial role in establishing a mechanical theory of heat, at the time most natural philosophers rejected his work and continued to support the caloric theory.

paper. The area under the curve delineated the work done, and PV indicator diagrams became a fundamental element of the new science of thermodynamics. Although Watt's invention of the separate condenser made the role of heat in the steam engine seem obvious, conventional wisdom still considered the steam engine primarily a pressure engine rather than a heat engine.

The Beginnings of Thermodynamics

In 1804 Arthur Woolf (1776–1837) patented a new high-pressure steam engine that was much more efficient than Watt's. After the Napoleonic Wars ended in 1815 one of Woolf's partners in France began building compound steam engines, which allowed steam to expand in two stages. French engineers, many with significant backgrounds in science, were intrigued by the new design and sought to understand the reasons behind its dramatically increased efficiency. The leading figure in this development was French physicist and engineer Nicholas Léonard Sadi Carnot (1796–1832), who had been trained at the École Polytechnique, France's leading engineering school. Carnot, a supporter of the mechanical theory of heat, also believed that heat was the universal motive force responsible for such phenomena

as wind and ocean currents. This led him to speculate that just as water could generate power by falling from a height, heat flow from a higher temperature to a lower one could also be used to generate power.

During the eighteenth century, scientists and engineers developed a theory for the motive power of water that led to the discovery of optimum conditions for its most efficient use. Drawing an analogy between water and heat, Carnot developed a similar set of conditions for the most efficient use of the motive power of heat. He demonstrated that an ideal engine operated in a cycle (now known as a Carnot cycle), in which some working substance such as steam or air underwent a series of expansions and compressions; during the cycle, heat flowed from a higher to a lower temperature, doing work in the process. In his pamphlet *Réflexions sur la puissance motrice du feu et sur les machines propres à développer cette puissance* (Reflections on the motive power of fire and on machines fitted to develop that power, 1824), Carnot used this analysis to draw several important conclusions:

First he argued that an engine that was more efficient than one following a Carnot cycle could be used to run a Carnot cycle in reverse; this would *accumulate* heat, which, in turn, could be used to run the more-efficient Carnot engine—theoretically resulting in perpetual motion. Since perpetual motion is impossible, he argued, an engine following a Carnot cycle must be the most efficient possible.

Second, because an engine following a Carnot cycle was the most efficient possible, the power produced by heat engines depended only on the temperature difference through which the heat fell, not upon the working substance that was used in an engine.

Most importantly, while Carnot based his reasoning on the material theory of heat, his conclusions turned out to be independent of any particular theory. Although his ideas would later become the fundamental elements of the new science of thermodynamics, they had little impact on most scientists and engineers until the late 1830s and early 1840s, when the French engineer Émile Clapeyron (1799–1864), produced a mathematical analysis of Carnot's work that was later translated into English and German.

Conservation of Energy

During the 1840s Carnot's theory ran into conflict with the new discovery of the mechanical equivalent of heat by English physicist James Prescott Joule (1818–1889). Carnot's theory had explained the work in a heat engine as the result of the flow of heat from a higher temperature to a lower one, but Joule's experiments showed heat could be transformed into mechanical work and vice versa. Joule's experiments were part of a movement in science that would lead to the reformulation of mechanical theories in terms of the new concept of energy.

IN CONTEXT: THE MOTIVE POWER OF HEAT

Sadi Carnot's (1796–1832) *Réflexions sur la puissance motrice du feu et sur les machines propres à développer cette puissance* (Reflections on the motive power of fire and on machines fitted to develop that power, 1824) is one of the fundamental works of thermodynamics. In it, he argues that heat is not only the source of the steam engine's power, but is the motive power behind all natural phenomena. As such, heat must be studied as a general phenomena and not one simply limited to steam engines. Although others had noted the role of heat in various geological and meteorological phenomena, Carnot was the first to see it as a universal motive force:

"Every one knows that heat can produce motion. That it possesses vast motive-power no one can doubt, in these days when the steam-engine is everywhere so well known.

To heat also are due the vast movements which take place on the earth. It causes the agitations of the atmosphere, the ascension of the clouds, the fall of rain and of meteors, the currents of water which channel the surface of the globe, and of which man has thus far employed but a small portion. Even earthquakes and volcanic eruptions are the result of heat."

SOURCE: *Carnot, Sadi. Reflections on the Motive Power of Fire. Edited by E. Mendoza, New York: Dover Publications, 1960.*

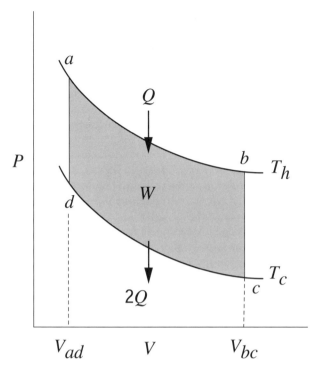

The Carnot cycle, showing conservation of heat. *Illustration by Argosy. Cengage Learning, Gale.*

British physicist James Prescott Joule (1818–1889). *Library of Congress.*

Joule grew up near the industrial city of Manchester, England, where his grandfather and father had established a large brewery. Joule spent two years studying with the famous meteorologist and chemist John Dalton (1766–1844) from whom he learned many of the skills of experimental chemistry. Although Manchester's manufacturing and railroads ran on steam power, by the 1830s Joule was studying the newly invented electric motor, which he believed could become a more economical power source than steam.

At the time, electric motors had only one-fifth the efficiency of the best steam engines. To understand this unfavorable comparison he conducted a series of experiments during the 1840s. Joule discovered that battery-generated electrical currents produced heat, a phenomenon that was clearly generated by chemical activity inside the battery. But electricity could also be generated by a magneto, in which a coil of wires was rotated in a magnetic field. This too generated heat. Since no chemical activity occurred, and no other part of the circuit was being cooled, Joule concluded in 1843 that heat was being generated by the mechanical effort used to rotate the coil of wires. He was able to demonstrate that 838 ft-lbs (116 kg/m) of work would raise 1 pound (0.45 kg) of water 1°F (0.55°C).

To demonstrate his belief that the mechanical equivalent of heat was universal and not peculiar to electric circuits, Joule carried out a series of further experiments that did not include electricity. In the most of

these, conducted in 1845, he placed a paddle wheel rotated by falling weights inside a water-filled cylinder. With this apparatus, he was able to calculate that the water's temperature increase was caused by the paddle wheel's rotation. In every experiment Joule found the mechanical equivalent of heat to be very close to what he measured in his electrical experiment.

Although his experiments demonstrated that mechanical work could be transformed into heat, Joule also believed that heat could be transformed into mechanical work. This led him to formulate a new conservation law. According to the material theory of heat, heat was always conserved, but the mechanical equivalent of heat meant that heat could appear in certain situations, as it did in Joule's various experiments, and that it could also disappear in others, as it did in a heat engine. Because its appearance and disappearance were always associated with doing or creating mechanical work, Joule proposed that while heat was not conserved, a quantity that natural philosophers had once labeled *vis viva*, or "living force," was. This quantity was mv^2, where m is an object's mass and v its velocity. Today, using the modern concept of energy developed in the nineteenth century, we would say that this quantity is twice the kinetic energy of an object with mass m and velocity v. Joule was mistaken

English physicist James Prescott Joule's (1818–1889) water friction apparatus. Joule proved that energy cannot be created or destroyed, but can only change in form. The paddle wheel was turned by falling weights possessing kinetic energy as they fell. The temperature of the water rose by an amount that depended on how far the weights fell. This showed that the kinetic energy of the weights had not existed as heat energy in the water. Joule's experiment also proved that heat is produced by motion, refuting the caloric theory. *© SSPL/The Image Works.*

in thinking that kinetic energy is conserved (neither diminished nor increased): Energy as such is conserved, but not any one of its forms (e.g., kinetic energy, potential energy, electrical energy, chemical energy, etc.).

The idea of the conservation of energy, which would later become the first law of thermodynamics, was based on the work of several scientists, including Joule, Julius Robert Mayer (1814–1878), and Hermann von Helmholtz (1821–1894). Mayer, a German physician serving aboard a ship in the East Indies in the early 1840s, noticed that the blood of patients in the tropics was brighter red than of those in Europe. He concluded that in a warm climate blood needed less oxidation to

maintain body temperature than in a cold climate. This led Mayer to argue that other physiological processes, such as muscular activity, could also be the result of the chemical combustion of food.

If this were true it meant that heat (food combustion), was being transformed into work (muscular activity). After returning to Germany in 1842 he broadened this idea to include inorganic processes as well. He noted that a gas expanding against some external pressure did work and, in the process, absorbed heat. Using published data on the specific heat of gases, Mayer was able to calculate the mechanical equivalent of heat (independent of Joule, who published his results shortly afterward).

This work led Mayer to put forward the idea that would later be labeled the conservation of energy. He published a paper in 1842 in which he concluded that forces could be converted into each other, but were indestructible, or conserved. At the time, "force" (*kraft*, in German), was defined as the ability to cause motion; it would later become the definition of energy. Therefore, with some hindsight, one can give Mayer credit for the discovery of the conservation of energy.

While Mayer proposed the idea of the conservation of energy and Joule provided experimental evidence that established the principle, Hermann von Helmholtz provided its mathematical foundation. Like Mayer, von Helmholtz's interest in physiology led him to the conservation of energy. Von Helmholtz also still used the German *kraft* for force, since the concept of energy had not yet emerged. Von Helmholtz rejected the widely held notion that heat produced by animals was based on the existence of some special *vis viva*, and argued instead that it could be explained by normal physical and chemical processes, such as the oxidation of food. From this he concluded that there must be some constant relationship between mechanical work and heat. To support his belief in the constancy, or conservation, of force in physiology, von Helmholtz put forward a mathematical proof that the conservation of force held for all natural processes.

In his 1847 paper "On the Conservation of Force," von Helmholtz showed that in any system of material objects governed by attractive and repulsive forces, the sum of the quantity known as *vis viva* plus something he called "tensional force" was a constant. Not until the concept of energy emerged in the 1850s did people recognize that *vis viva* and tensional force corresponded to kinetic and potential energy, respectively, and that his proof that the sum of the two were constant corresponded to a statement of the conservation of energy.

The First Law of Thermodynamics

Although Mayer, Joule, and von Helmholtz were major contributors to the discovery of the conservation of energy, its formulation as part of thermodynamics was primarily the result of the work of a group of British scientists and engineers, including William Thomson, later Baron Kelvin (1824–1907); William John Macquorn Rankine (1820–1872); and James Clerk Maxwell (1831–1879); along with the German scientist and engineer Rudolf Clausius (1822–1888). During the late 1840s Thomson realized that Carnot's theory of heat engines, which argued that work was produced by simply transferring heat from a higher temperature to a lower one, was contradicted by Joule's theory that heat was converted into work. Thomson also noted that irreversible phenomena, such as the conduction of heat through a solid, did not produce work; according to Joule's theory, however, the dissipated heat should be able to produce work.

At about the same time, Clausius wrote a paper in which he argued that Joule's and Carnot's theories could be reconciled if, during a Carnot cycle, only *some* of the heat was converted into work, the remainder was transferred from a higher temperature to a lower one, and there was some fixed relationship between the two processes.

During the first half of the 1850s Thomson and Rankine began to reformulate the laws of thermodynamics in terms of the new concept of energy. Although the term had a long history, it had been often used in vague and imprecise ways. Thomson and Rankine argued that it could be used as the basis for understanding all processes in natural philosophy, including mechanics, chemistry, electromagnetism, and thermodynamics. "Energy" came to be defined as the ability to do work.

Rankine further distinguished between the actual energy found in moving things, and potential energy, such as that stored in weights positioned at some height, electric charges, and certain types of chemical energy. (In 1862 Thomson and Scottish physicist and mathematician P.G. Tait [1831–1901] substituted the term kinetic energy for actual energy.) In 1853 Rankine reformulated von Helmholtz's conservation of force as the universal principle of the conservation of energy, which stated that the total energy in the universe was a constant. This became the first law of thermodynamics.

The Second Law of Thermodynamics

While the transformation of heat into work led to the formulation of the first law of thermodynamics, the problem of heat dissipation led to the second. As we have seen, Clausius argued that only a portion of heat in an engine was converted into work, while another portion was simply dissipated. In 1854 he began to reformulate his ideas, arguing that the dissipation of heat from a warmer body to a colder one had the "equivalence value" of the work required to move that heat back to the warmer body. In 1865 Clausius introduced the term "entropy," from the Greek word for transformation or change, to refer to the equivalence value of the transformation of heat.

Clausius and Rankine, who had developed a similar formula for the equivalence value of heat, demonstrated that for reversible processes, such as the Carnot cycle, the change of entropy is zero. Clausius also analyzed irreversible processes, such as those encountered in actual heat engines, and discovered that in those cases the entropy always increased. Using the new concept of entropy, Clausius was able to formulate the second law of thermodynamics, which stated that the total entropy in the universe is always increasing.

While the concept of entropy became fundamental to thermodynamics, there was a great deal of debate as to how it should be interpreted physically. In the 1860s Clausius suggested that entropy might be associated with the dispersal or rearrangement of the molecules that composed matter. Maxwell went further and suggested that if the molecules in a body or gas had a range of velocities, represented by something like a bell curve, the second law of thermodynamics, or the entropy principle, might be essentially a statistical law rather than one that could be explained in terms of the motions of individual molecules.

In an 1877 paper, Austrian physicist Ludwig Boltzmann (1844–1906) developed statistical mechanics, which states that the properties of a substance's atoms determine its larger physical properties. Boltzmann's statistical mechanics expressed entropy as proportional to the probability of finding a system in a state with a given distribution of molecular motions. Since disorderly states were more probable than orderly states, the second law of thermodynamics could be interpreted as simply saying that a system would tend to go from a less probable state (lower entropy and more order) to a more probable state (higher entropy and more disorder).

Thermodynamics and Science

The first and second laws of thermodynamics had emerged from the study of heat, but scientists soon recognized that energy and entropy were universal concepts that could be applied to a wide range of scientific topics. One of the first to do so was American scientist Josiah Willard Gibbs (1839–1903). During the 1870s he applied thermodynamics to the field of physical chemistry, where he used concepts such as energy and entropy to study mixtures of substances in different chemical phases, such as solid, liquid, or gas.

Gibbs introduced temperature-entropy and volume-entropy diagrams, using them to formulate his "phase rule" relating the degrees of freedom of a chemical system to the number of components and the number of phases in it. For example, in a system where water exists in all three phases (ice, water, and steam), there are no degrees of freedom, so with any temperature or pressure change one or more of the phases will disappear. Gibbs's work became so widely known that historian Henry Brooks Adams (1838–1918) attempted to apply the phase rule to the study of history, postulating that history went through phases, which he labeled religious, mechanical, electrical, and ethereal. Brooks posited that just as changes in temperature and pressure could change the phase of water from solid to liquid or to a vapor, social and political forces could change history from one phase to another.

When thermodynamics was applied to geology and biology, it conflicted with the theory of evolution. Starting with Earth's present temperature and applying the formula for the dissipation of heat, Thomson argued that Earth could not have been hospitable to life for a long enough period in the past for Darwinian natural selection to have taken place. Other critics argued that the emergence of higher forms of life by purely physical means was highly improbable and therefore violated the second law of thermodynamics.

By the twentieth century scientists discovered that the basis for this criticism was faulty. When heat generated by radioactive decay was later taken into account, Earth's age was much older than Thomson's estimate. Others argued that the second law applies to the total entropy in the universe; this meant that there could be pockets of decreasing entropy, represented by the evolution of higher forms of life, as long as this was offset by an increase of entropy in some other part of the universe.

Thermodynamics also played a significant role in the development of a new model of cosmology. An important implication of the second law was the "heat death" of the universe. If dissipation were a fundamental characteristic of nature—not just heat engines—then at some future point entropy would reach a maximum. With everything at the same temperature, there would be no possibility of extracting any work from such a system. Also, since maximum entropy would be the most probable state, the universe could not move to a more probable state. But, as James Jeans (1877–1946) would later note, it was the movement from less probable to more probable that defined "time's arrow," or the direction of the flow of time from past to future.

Therefore at a state of maximum entropy, or probability, the concept of the flow of time would become meaningless and the universe would effectively come to an end. This idea led many late-nineteenth century writers to begin to question the idea of human progress and raise the idea that Western civilization was in a decline. Such conclusions contributed to a general mood of pessimism at the end of the century. The universal application of the laws of thermodynamics also replaced the time-honored concept of a clock as a model for the universe with the concept of a heat engine.

By the end of the nineteenth century a number of scientists and philosophers, particularly in Germany, began to see the concept of energy as independent of any mechanical hypothesis. This group, led by the physical chemist Friedrich Wilhelm Ostwald (1853–1932), and

known as "energeticists," came to reject all material explanations of natural phenomena, including atomism, arguing instead that everything could be explained using the concept of energy. While most scientists continued to accept atomism, the energeticists' ideas influenced the philosophy of science, especially ideas put forward by the Austrian physicist Ernst Mach (1838–1916). Using Carnot's idea that the theory of heat engines was independent of any particular hypothesis about the nature of heat, Mach argued that thermodynamics should become the model for all scientific theories since it was independent of any hypothesis, such as atomism, that could not be confirmed by the senses and therefore might not be true.

Thermodynamics had a significant impact on twentieth century science. Applying its laws to electromagnetic radiation during the late nineteenth century led German physicist Max Planck (1858–1947) to suggest in 1900 that radiation was not emitted and absorbed in a continuous way but in discrete packages, called quanta. Shortly thereafter, Albert Einstein (1879–1955) applied Boltzmann's statistical mechanics to light, concluding that light exists as discrete particle-like packages of energy, later called photons. These discoveries became the basis for quantum mechanics during the early part of the twentieth century.

■ Modern Cultural Connections

In 1948 Claude Shannon (1916–2001), an American mathematician and electrical engineer working at Bell Labs, helped establish modern information theory by drawing on thermodynamics. He noted that the more disorder there was in an information system the less information it contained. This led him to see information as the negative of entropy and to develop a mathematical equation for it that was similar to Boltzmann's equation of entropy. This work has been one of the foundations of the modern information age.

About the same time, Ilya Prigogine (1917–2003), a Russian-born chemist living in Belgium, applied thermodynamics to systems that were very far from equilibrium, which he labeled dissipative. This work, which led him to become one of the founders of chaos theory, earned him the 1977 Nobel Prize for chemistry in 1977.

Thermodynamics' most recent role has been in the understanding of black holes. Between 1972 and 1974 British theoretical physicist Stephen Hawking (1942–) and Mexican-born theoretical physicist Jacob Bekenstein (1947–) used a combination of quantum mechanics, the general theory of relativity, and thermodynamics to predict that black holes would slowly radiate away. They discovered that in this process, the black hole's event horizon (a line that, if crossed, will cause an object to fall into a black hole) can never decrease. Since this was similar to the second law, in which entropy can never decrease, they proposed that black holes have entropy that is proportional to the area of the event horizon.

While many scientific theories were overthrown by the theory of relativity and quantum mechanics during the early twentieth century, the laws of thermodynamics remained unchanged and were recognized as universal laws essential to understanding a wide range of scientific and technological phenomena, many of which were far removed from their original study of heat and steam engines.

SEE ALSO *Physics: Fundamental Forces and the Synthesis of Theory; Physics: Wave-Particle Duality.*

BIBLIOGRAPHY

Books

Cardwell, D.S.L. *From Watt to Clausius: The Rise of Thermodynamics in the Early Industrial Age.* Ithaca, NY: Cornell University Press, 1971.

Carnot, Sadi. *Reflections on the Motive Power of Fire, by Sadi Carnot; and other Papers on the Second Law of Thermodynamics, by É Clapeyron and R. Clausius.*

E. Mendoza, ed. New York: Dover Publications, 1960.

Harman, P.M. *Energy, Force, and Matter: The Conceptual Development of Nineteenth-Century Physics.* Cambridge: Cambridge University Press, 1982.

Kuhn, Thomas S. "Energy Conservation as an Example of Simultaneous Discovery." In *Critical Problems in the History of Science.* Edited by Marshall Clagett. Madison: University of Wisconsin Press, 1959.

Layton, Edwin T., Jr., and John Lienhard, eds. *History of Heat Transfer.* New York: American Society of Mechanical Engineers, 1988.

Prigogine, Ilya, and Isabelle Stengers. *Order Out of Chaos: Man's New Dialogue with Nature.* Boulder, CO: New Science Library, 1984.

Smith, Crosbie. *The Science of Energy: A Cultural History of Energy Physics in Victorian Britain.* Chicago: University of Chicago Press, 1998.

Smith, Crosbie, and M. Norton Wise. *Energy and Empire: A Biographical Study of Lord Kelvin.* Cambridge: Cambridge University Press, 1989.

David F. Channell

Physics: Wave-Particle Duality

■ Introduction

Wave-particle duality is the ability of particles such as atoms, photons, and electrons to behave under some conditions like waves and under others like particles. The objects we call "particles" are actually neither waves nor particles but something else that the human mind is not capable of visualizing. Wave-particle duality has been observed in many experiments and is described by the mathematical laws of quantum mechanics.

Theoretically, duality is a property not only of atomic or subatomic particles but also of larger objects such as footballs, people, and planets. However, duality is harder to observe for larger objects and has not yet been measured for any object bigger than a molecule.

■ Historical Background and Scientific Foundations

In 1675, English physicist Isaac Newton (1642–1727) published an essay titled *Hypothesis of Light* to defend the theory that light consists of tiny particles. "Are not," he asked rhetorically, "the Rays of Light very small

Bodies emitted from shining substances?" He proposed that these tiny objects or corpuscles obeyed his three laws of motion, just like any other material objects. By proposing various attractive and repulsive forces, he could account for many facts of optics, such as the properties of lenses.

However, a rival school of thought held that light is really a kind of wave. Dutch scientist Christiaan Huygens (1629–1695) argued that light is a series of pulses in an ethereal or subtle medium. Huygens and others used the wave theory to explain reflection, refraction (bending of light in transparent media), and other properties of light.

Debate over the rival theories raged for over a century, yet the tide slowly turned in favor of the wave theory. In 1801 Thomas Young (1773–1829) discovered that light displays interference, a wave property that can also be seen in a basin of still water. Letting a drop fall anywhere on the surface of the water will cause waves to spread and to reflect from the sides of the basin. The waves reflected from different parts of the basin can be seen to pass through each other; moreover, wherever two peaks or valleys meet for a moment, they form a higher peak or a deeper valley, and where a peak and a valley of equal height pass through each other, they cancel out. Addition of peaks or valleys is called positive interference, while canceling of peaks and valleys is called negative interference.

By the mid 1800s, the wave theory of light had become so successful at explaining optics that it had triumphed. In 1864, Scottish physicist James Clerk Maxwell (1831–1879) showed that light can be described as an electromagnetic wave—that is, a time-varying electric field paired with a time-varying magnetic field, each generating the other as the two move forward. The nature of light seemed, to most scientists, a closed question. Light was clearly a wave.

Louis de Broglie (1892–1987), French physicist, was instrumental in showing that waves and particles can behave like each other at a quantum level (wave-particle duality). *Omikron/ Photo Researchers, Inc.*

IN CONTEXT: THE PHOTOELECTRIC EFFECT

The photoelectric effect occurs when light is shone on certain metals and, as a result, electrons fly off the surface of the metal. According to the wave theory of light, this occurs because light waves impart energy to the electrons, knocking them out of the metal much as ocean waves crashing on a rocky beach might fling rocks into the air. Bigger waves can throw rocks farther, so more intense (brighter) light should, similarly, produce electrons with more energy.

But the photoelectric effect doesn't work this way. For a given color (wavelength) of light, the maximum electron energy is the same for bright light or dim light. When the light is made more intense, more electrons are knocked out of the metal, but they do not have higher maximum energy. The only way to produce electrons with higher energy is to use light with a shorter wavelength (bluer light).

In 1905, German-American physicist Albert Einstein (1879–1955) showed that the photoelectric effect can be explained by assuming that light consists of tiny particles called photons. Each photon carries a fixed quantity (or "quantum") of energy and can deliver that energy to a single electron. The maximum energy of the kicked-out electrons is therefore limited by the energy in each photon, not by how many photons are arriving (the intensity or brightness of the light). And the energy in each photon is determined by the wavelength of the light.

In 1887, however, German physicist Heinrich Hertz (1857–1894) observed a phenomenon that would, in time, radically change the apparently invincible wave theory. This phenomenon was the photoelectric effect. In 1905, German-American physicist Albert Einstein (1879–1955) announced that the photoelectric effect can be explained by assuming that light consists of particles, which he called "photons." Yet all the evidence for light's wavelike nature remained, and Einstein did not deny it. How could light be both a wave and a particle, spread out in space but also (somehow) localized to a single point in space? It seemed a bizarre claim, like saying that something could be a musical sound and a bullet at the same time. Yet experiments showed that this was the case. Despite its seeming contradiction to the human experience of the every day world, wave-particle duality is a fact.

Einstein's work established wave-particle duality only for light, and it was not fully accepted by the physics community for years. Other physicists, including Neils Bohr (1885–1962) and Louis de Broglie (1892–1987), took the theory of wave-particle duality farther. Bohr applied Einstein's work to the spectra emitted by atoms, and in 1924 de Broglie predicted that all particles, not just photons, should display wave-particle duality. In particular, if de Broglie was right, electrons could be made to interfere with each other just like photons. By passing electrons through rows of atoms in a nickel crystal, physicists Clinton Davisson (1881–1958) and Lester Germer (1896–1971) proved in 1927 that de Broglie was indeed correct: Interference patterns can be created using electrons. Einstein had shown that waves are particles: Now it had been shown that particles are also waves.

■ Modern Cultural Connections

However counterintuitive (differing from normal perceptions and assumptions), the wave nature of light was well-established experimentally. It is, therefore, a human limitation that we cannot accurately describe the exact nature of light, but must instead characterize it in terms such as particles and waves, a situation which further

IN CONTEXT: EXPERIMENTAL EVIDENCE

The classic proof that particles have wave properties is the two-slit interference experiment. In this experiment, a barrier is set up with two slits cut in it so that particles shot towards the barrier might pass through either slit. On the far side of the barrier, arriving particles are counted by photographic film or some other method. An interference pattern is observed at the detector, stripes where particles are more (and less) likely to arrive. The areas of likeliest arrival are regions of positive interference between waves (particles) passing out of the slits.

Strangely, electrons, photons, and other particles not only interfere with each other when passing through an interferometer, but with themselves. That is, even when particles are shot at the barrier one at a time, the interference pattern built up on the far side over many trials is exactly the same as if they were fired in showers. This leads to the conclusion that each single particle passes, wavelike, through both slits at the same time—even though, if a detector is placed in either or both slits, the particle is always found to have chosen only one path or the other. At the same time, this act of detection destroys the interference pattern: It is as if trying to catch the particle in the act of being a wave forces it to stop being a wave. As Bohr put it, the types of properties—particle or wave—that we can attribute to a quantum system depend on the type of observation we choose to make of that system. For example, measuring the location of a particle collapses its wavelike properties.

It should be remembered that wave-particle duality is not an effect that appears only in experimental setups. All particles and waves, everywhere and at all times, possess wave-particle duality.

Arthur Compton (1892–1962), U.S. physicist. Compton won the Nobel Prize for Physics in 1927 for showing that photons have energy and momentum, and that in quantum mechanics, an object can behave as both a particle and a wave at the same time. This is known as wave-particle duality. He also studied cosmic rays, and NASA's Compton Gamma Ray Observatory satellite was named in his honor. *AIP/Photo Researchers, Inc.*

forces us to accept that light has characteristics of both a particles and waves, no matter how hard this situation might be to picture in the mind's eye.

The impact of wave-particle duality on physics, on technology via physics, and on society via technology has been profound. In 2007, physicist Alain Aspect (1947–) described wave-particle duality as "the main ingredient of the first quantum revolution," that is, that burst of new physics knowledge in the early twentieth century that led eventually to the transistor, the laser, and the other microelectronic devices that have made our information-dependent society possible.

Scientists continue to test wave-particle duality. For example, the de Broglie quantum theory asserts that even large objects like people and cars have wave-particle duality; however, to state a claim theoretically is not the same as experimentally proving it. Scientists have therefore sought to detect the wave-particle duality of larger and larger particles by forcing them to produce interference patterns just like photons or electrons. As of 2007, the largest particle whose wave-particle duality had been observed in the laboratory was the $C_{60}F_{48}$ molecule, 60 carbon atoms combined with 48 fluorine atoms (atomic weight 1,632).

In philosophy and theology, wave-particle duality has diminished the persuasiveness of what are sometimes called Either-Or arguments—arguments that rely on the common-sense assumption that a thing cannot have two natures at once or be in two places at once. Other influences by wave-particle duality on society are mostly indirect, mediated by the growth of information technologies. Quantum cryptography, quantum computing, and the engineering of nanometer-scale transistors are all technologies depending on acceptance and use of the nature of wave-particle duality.

■ Primary Source Connection

Something is counterintuitive if it is different from what everyday experience or common sense suggests.

Wave-particle duality is a part of the often counterintuitive quantum world and its increasingly broad application. It forces scientists into sometimes philosophically uncomfortable arguments in which acceptance of experimental results is paramount, and the inability to articulate or comprehend the exact state of nature provides insight on the limitations of human thought and perception.

In a heralded 1996 article, "Weirdness Makes Sense," published in the *New York Times*, Timothy Ferris (1944–), a professor emeritus of journalism at the University of California at Berkeley, wrote of the paradoxes of the quantum world. In the excerpt published below he articulated the difficulties in reconciling the quantum theory with the logic and language of everyday experience.

WEIRDNESS MAKES SENSE

If the next Einstein were born today, what might he or she be doing in the year 2022, having reached the age, 26, at which Einstein formulated the theory of relativity?

My suggestion would be—solving the problem of quantum weirdness. The term is scientific slang. It stands for a conundrum, more properly known as the "quantum observership" or "quantum measurement" problem, that has defied some of our century's strongest minds.

Quantum physics is a famously strange realm where matter, energy and knowledge are spooned out in indivisible units, the quanta—as if the world were a pub where you could quaff a pint of beer, or no beer, but never a half pint. In quantum physics we have learned to accept such unlikelihoods as "quantum leaps," in which particles vanish from one place and reappear in another—instantly. (Raymond Chiao of the University of California at Berkeley has recently measured photons, the carriers of light, quantum-leaping at velocities that would amount to twice the speed of light if they crossed the intervening space, which they don't.) But quantum physics itself is not the problem. It remains a highly successful branch of science that promises to cruise with flying colors through the centennial, in 2000, of Max Planck's discovery of the quantum principle. Weirdness arises when we try to reconcile some of the oddities of the quantum world with the dictates of common sense.

The history of that effort is littered with the bleached bones of mighty thinkers. Einstein pondered its paradoxes for decades and got next to nowhere. Niels Bohr didn't get much further. David Bohm, John Stewart Bell and Hugh Everett 3d labored mightily at it, and may someday be revered as pioneers, but all three went to their graves with little more than obscurity to reward their efforts. Quantum weirdness is so weird that just discussing it stands the normal rules of exposition on

their heads: to understand it is to become not enlightened but confused. As the physicist John Archibald Wheeler says, "The quantum is the greatest mystery we've got."

Some scientists dismiss quantum weirdness as a mere brain teaser. But it was a brain teaser that started Einstein on the road to relativity, when at age 16 he wondered what he would see if he observed an electromagnetic field while traveling at the velocity of light. Superconductivity was once a brain teaser. So were the mathematical oddities that led to digital computers. With recent improvements in technology, physicists are actually conducting the thought experiments that Einstein and Bohr could only imagine, and their efforts have already led to intriguing technological breakthroughs—a new kind of laser in Germany, a demonstration of quantum cryptography in England and the promise of powerful microscopes, miniature particle colliders and computers fast enough to break any code. Where such diamonds are strewn on the surface, it's reasonable to wonder what lies deeper down.

The essence of quantum weirdness can be summed up in the statement that quantum systems—typically, photons and electrons, things smaller than an atom—exhibit "nonlocal" behavior. In all previous scientific investigations, nature acts locally. For a cause here to produce an effect over there, an intervening mechanism must link the two. Such a mechanism is "local" in that you can identify it here and now; the waves that make a skiff bob in its moorings can be traced to a passing ship. If no waves or other mechanisms could be found connecting the ship to the skiff, we would have nonlocal behavior, which seems as inexplicable as if a car were to continue accelerating down the road after losing its drive shaft.

Newton worried that his theory of gravitation was nonlocal, since he never found a mechanism that could propagate gravitational force across space. Einstein cleared that up with the general theory of relativity, which revealed that gravitation results from the curvature of space. Changes in a gravitational field are conveyed, at the velocity of light, by ripples in space itself. As Einstein also found, neither this nor any other causal mechanism can work at faster than light speed. Yet quantum systems evidently behave in such a way that a cause here produces an effect over there instantaneously, with no discernable causal mechanism between the two points, and with insufficient time for such a mechanism, working at light speed, to have carried the news from one place to the other. Einstein regarded nonlocality as absurd. He called it "spooky action at a distance."

Quantum nonlocality first reared its weird head with Werner Heisenberg's uncertainty principle. Heisenberg discovered that certain kinds of information about quantum systems can be obtained only at the cost of forsaking other kinds of information. You can ascertain exactly where a particle is (its position) or exactly where it is going (its momentum), but not both. The more precisely you pinpoint its position, the less you can know about its momentum, and vice versa. It is as if you were handed an autographed baseball in a lightless room and told you could take but a single flash photograph of it. The photo might show the signature on the ball, or it might show the manufacturer's emblem embossed on the opposite side, or part of the signature and part of the emblem. But you could not learn everything about both the autograph and the emblem from a single photo. Bohr called these complementary aspects of the system. In real life, you could take the baseball out into the sunlight and examine it fully. But with quantum systems, what you see is all you get.

The central tenet of complementarity resides in the "wave-particle duality." Subatomic particles act like either waves or particles, depending on how they are examined. But common sense says they cannot be both at once, since waves and particles behave very differently. A wave is all over the place; a particle is in one place only.

The distinction is demonstrated by a simple test that the physicist Richard Feynman called "the experiment with the two holes." To run the experiment, punch two small holes in a sheet of steel, fire a stream of photons or other quanta at the sheet and record what comes through, using a detector of some sort on the far side of the steel sheet. (The detector can be something as simple as a sheet of photographic film.) When both holes are open, the detector records an interference pattern—the signature of interacting waves. Drop two stones in a pond and an interference pattern appears where the waves intersect. Wave peaks reinforce each other where they coincide, as do valleys, and where a wave peak intersects with a valley, the two cancel each other out. In the two-holes experiment, the interference pattern appears even if you send only a single photon through the apparatus: the photon finds its way through both holes and interferes with itself. Close one hole, however, and the photon's wavelike behavior disappears. Now it acts like a bullet: either it emerges from the single open hole to register a point impact on the detector, or it misses the hole and hits the steel sheet, and nothing comes through.

The weird thing is that the photon does this—responds to whether one or both holes are open—instantly, even if you wait until the last moment, just before it reaches the steel sheet, before deciding to close one hole or leave both open. It is as if the particle (or wave, whichever you prefer) were everywhere at once, feeling out the entire setup and responding to it instantaneously, everywhere.

Bohr explained the wave-particle duality by declaring that subatomic systems don't have either of their complementary states until they are observed. This view came to be known as the "Copenhagen" interpretation of quantum mechanics, named for the city where Bohr and his colleagues set up shop. It might be summarized as "Don't ask, don't tell." Are photons particles or waves? Don't ask! They are neither—or they are both. Their complementary states are only resolved, one way or the other, by their being observed.

In 1935 the Austrian physicist Erwin Schrodinger challenged Bohr's view of the role of observation in quantum systems by proposing a famous thought experiment, known ever since as "Schrodinger's cat." An unfortunate cat is placed in a sealed box with a quantum device that has a 50-50 chance of going to a particular state within, say, one hour. If the state is not achieved, nothing happens. If it is achieved, it explodes a cyanide capsule and kills the cat. At the end of the hour, but before we open the box, what has happened? If we accept the Copenhagen assertion that the system has no state until it is observed, we have to believe that the cat, until observed, remains in a "superposed" state of both dead and alive. If you object that the cat itself made the observation, then let's leave the cat out of it, run the experiment with an empty box, then clear the laboratory except for a graduate student who opens the box and will be killed if the cyanide has been released. As we stand outside the closed lab door, we reflect that according to Bohr the cyanide has been neither released nor unreleased until the student opens the box. The student's own observation, and not the prior state of the cyanide canister, will decide whether he lives or dies. Does anybody believe that the world really works this way?

To further highlight the weirdness of the Copenhagen view, Einstein and two of his Princeton colleagues constructed the E.P.R. (Einstein-Podolsky-Rosen) thought experiment. Let an atom spit out two particles, which then fly apart for an enormous distance. The physics equations tell us that the two particles must have opposite spin: if one is "spin up," the other must be "spin down." But according to the Copenhagenians, the particles have no spin state at all until observed. So now we observe one particle and find, say, that it is spin up. We have thus "resolved" the spin state of a particle that allegedly had no such state until we made the observation. But that means that the other particle, millions of miles away, suddenly "became" spin down. How did it know that its state was now supposed to be spin down?

Logic would seem to dictate that the particles actually had spin states all along. Einstein took that position, postulating that some underlying, causal agency must carry the spin. In other words, the spin state is not decided at the moment of observation after all, as Bohr claimed, but was in the particles to start with. But there is no evidence that any such mechanism exists. Theories like Einstein's are therefore known as "hidden variables" interpretations, since they propose the existence of something hidden in the quantum systems.

Since the Copenhagen view rejected hidden variables, the question remained at an impasse for decades. Which interpretation a given physicist preferred mainly depended on which seemed less repugnant, the hidden variables theory or spooky action at a distance.

Then the Irish physicist John Stewart Bell thought up an experiment that would decide whether hidden variables exist, as Einstein believed. The experiment, which involved obtaining the statistics of large numbers of photon interactions, was not technologically feasible when he proposed it in 1964. But in the 1970's Bell's experiment was performed, first by John Clauser and Stuart J. Freedman at Berkeley and later by Alain Aspect and his colleagues at the University of Paris. The verdict was clear: there are no hidden variables of the sort Einstein envisioned. Quantum physics really does exhibit nonlocality.

Some scientists reacted with a shrug of the shoulders. Quantum physics is simply like that, they asserted, and if it doesn't fit our notions of common sense, too bad. But quantum weirdness remains disturbing insofar as we wish to reconcile quantum theory with the logic and language of the wider world—something that Bohr himself held to be a fundamental imperative of the scientific enterprise. And as Feynman noted, quantum weirdness "is impossible, absolutely impossible to explain in any classical way. …"

Timothy Ferris

FERRIS, TIMOTHY. "WEIRDNESS MAKES SENSE"
NEW YORK TIMES (SEPTEMBER 29, 1996).

SEE ALSO *Physics: Heisenberg Uncertainty Principle; Physics: Maxwell's Equations, Light and the Electromagnetic Spectrum; Physics: Optics; Physics: QED Gauge Theory and Renormalization; Physics: The Quantum Hypothesis.*

BIBLIOGRAPHY

Books

Achinstein, Peter. *Particles and Waves.* New York: Oxford University Press, 1991.

de Broglie, Louis. *Physics and Microphysics.* Wakefield, MA: Pantheon Books, 1955.

de Broglie, Louis. *The Revolution in Physics: A Non-Mathematical Survey of Quanta.* New York: Noonday Press 1960.

Periodicals

Arndt, Markus, et al. "Wave-particle Duality of C_{60} Molecules." *Nature* 401 (1999): 680–682.

DeKieviet, Maarten, and Joerg Schmiedmayer. "Atom Waves in Passing." *Nature* 437 (2005): 1102.

Dowling, Jonathan P. "To Compute or Not to Compute?" *Nature* 439 (2006): 919–920.

Ferris, Timothy. "Weirdness Makes Sense." *New York Times* (September 29, 1996).

Rae, Alastair I. "Waves, Particles, and Fullerenes." *Nature* 401 (1999): 651–653.

Larry Gilman
Paul Davies

Science Philosophy and Practice: Ethical Principles for Medical Research Involving Human Subjects

■ Introduction

Medicine is both an art and a science. As an art, it creates mutually beneficial relationships between healers and patients, cures patients when possible, and comforts patients when death, loss of function, or deterioration of health is imminent. As a science, it rigorously studies biology, physiology, anatomy, mechanics, chemistry, and many other fields of inquiry in order to gain knowledge as precise as methods and conditions allow, then determines how to use this knowledge most effectively in the healing art. Teaching medicine properly involves imparting to students both this art and this science in depth. Medicine is thus frequently and aptly described as a "three-legged stool," consisting of patient care (the art), research (the science), and education. If any of these three legs should break, the stool would collapse. Each of these three aspects is equally important to maintaining the integrity of the whole medical profession.

Unfortunately, medical research cannot be conducted only in test tubes or petri dishes, under microscopes, or with laboratory animals. It sometimes also requires human experimentation. When it does, the absolute responsibility of the scientist is to ensure that the safety, autonomy, privacy, dignity, and values of each human subject are respected and protected to the greatest extent possible. The key to protecting human subjects of medical research encompasses the paired concepts of informed consent and informed refusal, usually called just "informed consent."

Informed consent recognizes the autonomy and essential human dignity of each subject or potential subject. By this principle, all potential human subjects or their guardians or legal representatives are completely free to decide without coercion or fear of retribution whether or not to participate in any scientist's experiment. No penalties shall arise if the decision is to refuse, even if the participant has second thoughts and walks away from the experiment in the middle or at any other time. The researcher's corresponding duty is to provide enough clear, honest, and easily understood information for each potential subject, guardian, or surrogate to make a well reasoned decision. Before this decision is reached, the scientist must explain the entire protocol of the envisioned experiment, including disclosure of all anticipated risks and benefits, in plain language. The potential subject must explicitly acknowledge understanding. Each prospective research subject has the right to all of this information.

■ Historical Background and Scientific Foundations

Modern medical research began in 1747 when Scottish naval surgeon James Lind (1716–1794) conducted the world's first controlled clinical trial. His aim was to discover the cause of and cure for scurvy, which was devastating the British Navy and merchant fleet. Aboard HMS *Salisbury*, Lind drafted 12 sailors with similar early symptoms of scurvy after about a month a sea. He divided them into six pairs to test six different possible remedies. To their daily rations he added, respectively, one quart of apple cider, a gargle of 25 drops of dilute sulfuric acid three times a day, two teaspoonfuls of vinegar three times a day, two oranges and a lemon, a half pint of sea water, and a strong herbal purgative three times a day. Only the pair that ate the citrus fruit showed significant improvement. Lind's success led the Royal Navy to order in 1795 that citrus juice be given to all sailors.

Lind informed the sailors of his plans, but did not ask their consent. He ordered them to participate, in accordance with contemporaneous British naval protocol. In his experiment, no further harm could come to any of the participants beyond what they were already likely to suffer from scurvy. Such lack of added risk in clinical trials is not always the case. Many clinical trials are downright dangerous. The general recognition that human research subjects have the right to be protected from these dangers was long in coming.

Acceptance of Lind's new empirical method of medical research was also slow. At the end of the eighteenth century most of the medical world was still accepting the wisdom of the ancients and other authorities, experimenting by trial and error, and taking advantage of unusual patients, especially those who owed their lives to the investigating physicians. Only rarely in this era was medical research on human subjects done as rigorous science, and even when it was, doctors enlisted participants by pressure, coercion, or guilt more often than by request.

English country doctor Edward Jenner (1749–1823) was an exception to this trend. In 1796 one of his patients, Gloucestershire milkmaid Sarah Nelmes, had cowpox. Local folklore claimed that people who got cowpox never got smallpox. Jenner wanted to test this hypothesis. With Nelmes's permission, he obtained fluid from her lesions. He then asked a local farmer, James Phipps, the father of an eight-year-old, for permission to use Phipps's son in a smallpox immunization experiment. Jenner explained his entire theory and obtained the farmer's free consent. He then inoculated the healthy boy with the dried cowpox fluid. Phipps got cowpox. After he recovered, Jenner inoculated him with dried fluid from a smallpox victim. Phipps did not get smallpox.

Edward Jenner (1749–1823). © *Bettmann/Corbis.*

WORDS TO KNOW

COMMON RULE: A set of ethical guidelines for medical research in the United States that has been adopted by 18 federal agencies. All universities and hospitals hoping to receive U.S. government funding must adhere to the Common Rule guidelines, and most research adheres to them even if it is not federally funded. The Common Rule forbids certain kinds of human research altogether and requires meaningful, informed consent from subjects (with full freedom to decline participation) for those kinds of research that are permitted.

HELSINKI DECLARATION: A set of ethical principles governing medical and scientific experimentation on human subjects; it was drafted by the World Medical Association and originally adopted in 1964.

IATROGENIC: (Pronounced eye-at-roh-GEN-ik.) Any infection, injury, or other disease condition caused by medical treatment.

INFORMED CONSENT: An ethical and informational process in which a person learns about a procedure or clinical trial, including potential risks or benefits, before deciding to voluntarily participate in a study or undergo a particular procedure.

INTERNAL REVIEW BOARD (IRB): Also known as an institutional review board. A committee of personnel in a hospital where research involving human subjects is performed. The IRB's mission is to make sure that such work strictly obeys laws governing research involving humans.

PLACEBO: A false treatment that gives a patient the sense that a medicine is working when, in actuality, no medicine was actually received.

VECTOR: Any agent, living or otherwise, that carries and transmits parasites and diseases. Also, an organism or chemical used to transport a gene into a new host cell.

Jenner repeated this experiment on several other Gloucestershire children, including his own son, always with informed consent, and always with the same outcome. He published his results in 1798 as *An Inquiry into the Causes and Effects of Variolæ Vaccinæ.* Even though his experiments were entirely successful and even though he had the support of his patients, both the established medical community and the church condemned his work as preposterous and unnatural. Until about 1802 he was generally ridiculed, but gradually his findings were accepted. By 1810 most English-speaking doctors vaccinated their patients. The University of Oxford awarded Jenner an honorary degree in 1813.

On the other side of the informed consent spectrum from Jenner was American army surgeon William

Beaumont (1785–1853). At Fort Mackinac, Michigan, in 1822 he saved the life of French-Canadian fur trader Alexis St. Martin (1794–1880), who had suffered a shotgun wound in his lower thorax and upper abdomen. The wound did not close properly. The inside of the stomach remained open, so that whatever St. Martin ate or drank would leak out unless he wore a plug, compress, and bandage.

Beaumont could have repaired the hole surgically, but he perceived a unique opportunity to study the physiology of digestion, which until then was very poorly understood. Seduced by the prospect of fame as a scientist, Beaumont decided not to close the hole. In 1823 he hired St. Martin as his personal servant, then in 1824, when St. Martin was fully recovered, he began performing physiological experiments through the hole that he had allowed to become permanent. As their relationship was both doctor/patient and employer/employee, one easily infers how Beaumont used his power to compel St. Martin to participate. Three times St. Martin ran away. Twice Beaumont tracked him down and convinced him to return for more experiments. The third time, in 1833, Beaumont could not persuade him. The experiments ended.

Beaumont's science was good. He published his first article about St. Martin's stomach in 1825 and his monumental book, *Experiments and Observations on the Gastric Juice, and the Physiology of Digestion*, in 1833. St. Martin was bitter for the rest of his life about Beaumont gaining prestige at St. Martin's expense and without his consent. Disabled by his injury, he had no dignified way to earn a living, but occasionally made money by exhibiting himself as a freak. He and his family subsisted in poverty. When he died, his wife allowed his body to decompose before she told anyone that he was dead. He had instructed her to do this so that the medical community could not preserve his stomach and continue to exploit him after death.

Early medical empiricism grew most significantly in France. French researchers such as Jean-Nicolas Corvisart (1755–1821), Xavier Bichat (1771–1802), Guillaume Dupuytren (1777–1835), René-Théophile-Hyacinthe Laënnec (1781–1826), François Magendie (1783–1855), Pierre-Charles-Alexandre Louis (1787–1872), and Jean Cruveilhier (1791–1874) obtained valuable results and exerted tremendous influence on medical research methodology throughout the Western world. By the middle of the nineteenth century, the typical medical publication title, "Observations on … ," had given way to "Researches on … " This shift indicated that meticulous observation, which physicians had always encouraged, was still honored, but that the emphasis was now on methodically designed empirical study. Especially in the work of Louis, the move toward rigorous empiricism was a major event in the history of medicine. Yet none of these scientists were noted for their consideration of their human subjects.

American military physician Walter Reed (1851–1902) was, for his era, unusually careful to obtain informed consent from his research subjects. In 1900 Surgeon General of the Army George Miller Sternberg (1838–1915) appointed Reed to lead the Yellow Fever Commission, charged with controlling that disease in Cuba. Reed decided to test the hypothesis of Cuban physician Carlos Juan Finlay (1833–1915) that yellow fever was transmitted by mosquitoes. He asked for volunteers. The first two were Private William Hanaford Dean (1877–1928) and a member of Reed's team, surgeon James Carroll (1854–1907), who allowed mosquitoes to bite them under controlled conditions. Both caught yellow fever but survived. Then another member of Reed's team, surgeon Jesse William Lazear (1866–1900), allowed himself to be bitten. He caught yellow fever and died.

Using Lazear's notebooks, Reed deduced that yellow fever was noncontagious and vector borne. Even though he did not know the identity of the pathogen, he correctly deduced its life cycle from human to mosquito and back to human. He then asked for more volunteers to test his conclusions. His commanding officer, Leonard Wood (1860–1927), authorized military funds to pay these volunteers, who included both local civilians and American soldiers.

The immediate benefits of Reed's yellow fever research are incalculable. In 1901 William Crawford Gorgas (1854–1920), the U.S. Army's chief sanitation officer for Cuba, instituted strict policies to destroy mosquitoes around Havana. Within three months that area was free of yellow fever. In 1904 Gorgas established similar but more extensive policies in Panama, where yellow fever and malaria were decimating workers and hindering construction of the canal. Yellow fever was eradicated from the Canal Zone by 1906 and the frequency of malaria there was greatly reduced by 1914.

Sometimes occasions for medical research occur fortuitously or unsystematically, rather than by design with specific ends in view. This was true in St. Martin's case and is particularly true in war, where plentiful presentations of unusual injuries provide military and naval surgeons with unique opportunities to advance medical, surgical, biomechanical, and even physiological knowledge, even while they are trying their utmost to save lives. Soldiers and sailors near death from battle wounds are in no condition to give or withhold consent. Their surgeons, with wide leeway to invent, experiment, and explore, naturally learn much from what would scarcely be encountered in civilian practice. This is different from St. Martin's case insofar as the military or naval patient's relationship with the particular military or naval doctor typically ends with discharge from the hospital.

The long-term consequences of military and naval surgery are often beneficial for generations of future patients. French surgeon Ambroise Paré (1510–1590) served in the war between King François I and the Duke of Savoy. At the battle of Turin in 1536, the Savoyards shot so many French soldiers that Paré ran out of oil to cauterize wounds. Until that time most European surgeons had followed the ancient Arabic practice of cauterizing puncture wounds with hot oil. In desperation, Paré just wrapped the newer wounds with bandages soaked in egg yolk, rose oil, and turpentine. He was soon surprised to learn that the bandaged patients healed better than the cauterized ones. He immediately abandoned cauterization. After he published his treatise on wounds in 1545, other surgeons throughout Europe abandoned it too. In 1552 King Henri II made Paré the royal physician.

Whatever new knowledge military and naval surgeons gain from their war service is theirs to use and publish as they see fit, regardless of how their patients feel about it. The United States Surgeon General's massive six-volume *Medical and Surgical History of the War of the Rebellion, 1861-65* (1870–1888) was the first official report of such new knowledge. The two world wars produced similar sets: *The Medical Department of the United States Army in the World War* (15 volumes, 1921–1929) and *The History of the Medical Department of the United States Army in World War II* (35 volumes, 1952–1976). In none of this published research was the consent of patients taken into account. All have substantially advanced medical science.

Scientific and Cultural Preconceptions

With regard to human research subjects, the standard presumption of doctors throughout the nineteenth century, and of military and naval doctors well into the twentieth century, was that consent, informed or not, was unnecessary. Patients should be willing to make sacrifices for the sake of science, and that was that. Society seldom questioned the paternalism or judgment of the medical establishment and generally believed that "the doctor knows best." Patient-initiated lawsuits for fraud, negligence, or malpractice were rare and typically unsuccessful.

Ironically, Germany, later infamous as the site of Nazi crimes against research subjects, was the first country to enact informed consent laws. In the 1890s in Breslau, Prussia, German dermatologist and bacteriologist Albert Neisser (1855–1916) studied syphilis and intentionally infected several prostitutes, but then claimed that they had contracted the disease while following their profession. He had neither obtained his subjects' permission nor told them the nature of his experiments. Reacting to Neisser's fiasco, the imperial Prussian government decreed in 1900 that medical research could not be conducted without each subject's consent. The decree had little effect, but was significantly strengthened by the Weimar government in 1931. This legal standard remained in effect in Germany even while the Nazis violated it in the 1930s and 1940s.

Twenty years after the Nazi biomedical research crimes were revealed, the West still remained falsely and complacently confident that such heartless travesties of science could only happen under military tyrannies, not in progressive democracies. Yet, while not nearly as bad as the Nazis, Western researchers continued routinely to violate the principle of informed consent. They harmed their human subjects and saw nothing wrong with doing so. The typical victims of such improper research were defenseless or marginalized populations. Three American instances of such abuse were the syphilis study of African-Americans in Tuskegee, Alabama, conducted by the federal government from 1932 to 1972; the hepatitis study of children at the Willowbrook State School for the Retarded, Staten Island, funded by New York State from 1955 to 1972; and the use of inmates at Holmesburg Prison, Philadelphia, as guinea pigs for chemicals and drugs from 1951 to 1974.

In Tuskegee, the United States Public Health Service (USPHS) ran a study to chart the full natural course of syphilis. Federal government physicians in the first few years of the study identified and enrolled 399 poor and undereducated African-American men who suffered from the disease. The USPHS lied to these victims about the nature of their ailment and the purpose of the study, purposefully withheld treatment from them, and compensated them only with trinkets, cigars, cigarettes, bus rides, funeral expenses, and tiny amounts of cash. Even after penicillin was determined to be an effective antisyphilis therapy in the 1940s, the USPHS refused to allow participants in the study to receive it or even learn about it. On May 16, 1997 President Bill Clinton offered a formal, written apology to the African-American community for the Tuskegee syphilis study.

Saul Krugman (1911–1995), later the discoverer of the hepatitis B vaccine and president of the American Pediatric Society, used mentally retarded children at Willowbrook as research subjects for experiments on hepatitis. He deliberately infected hundreds of newly admitted children, either without the informed consent of their parents or legal guardians or with consent obtained by deceit. Willowbrook was severely overcrowded, sanitation was minimal, and most of children there already had hepatitis when Krugman began his study. That was why he chose only new admittees as subjects. His team fed extracts of human feces to subjects in order to test the transmission of the disease.

Biomedical and other experiments on criminals were common in the United States until 1979, when the Belmont Report of the U.S. Department of Health, Education, and Welfare (DHEW) provided the spark to end the practice. Before then, any individual, corporation, organization, or government entity with a legitimate interest in testing drugs or chemicals on human subjects could enjoy nearly unimpeded access to prisoners. Nationwide, the most notorious of these

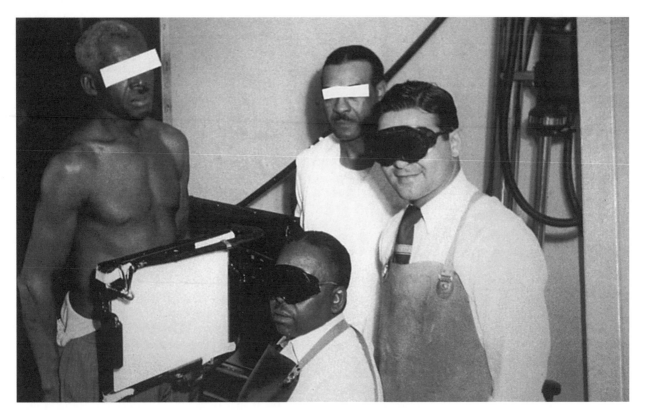

An unidentified man receives "treatment" during the Tuskegee syphilis study. From 1932 to 1972, the experiment studied the effects of untreated syphilis in almost 400 African-American patients, all of whom were told they were receiving beneficial medicine.   *© Corbis Sygma.*

facilities was the Philadelphia county prison at Holmesburg, Pennsylvania.

Instigated by dermatologist Albert M. Kligman (1916–), Holmesburg's research program was remarkable for its sheer size and range. Hundreds of researchers conducted experiments on thousands of poorly informed prisoners, who were deliberately kept ignorant about the nature of the substances they were testing and the possibly dangerous implications of their participation. Prisoners were paid for their time, but many felt compelled to cooperate with the researchers so that they could buy protection from homosexual rape, which was rampant in the cellblocks. Kligman exposed them to poison ivy, dioxin, radioactive materials, bizarre potions, and many other painful or hazardous substances. The army performed chemical warfare tests on inmates. Ortho Pharmaceutical Corporation developed its popular acne medication, Retin-A at Holmesburg in the 1960s without providing the prisoners with either treatment for side effects or relief from pain. Holmesburg experiments concerned not only pharmaceuticals and chemicals. In one protocol prisoners allowed their fingernails to be extracted for $150 each so that researchers could study the untreated wounds.

The first inkling of the principle of informed consent in America arose from a 1914 court case, *Schloen-*

dorff v. the Society of the New York Hospital. The plaintiff, as a patient under general anesthesia in 1908, had awakened to discover that she had undergone a surgical procedure which she had neither anticipated nor approved. The surgeon had removed a fibroid tumor, but she had agreed only to an exploratory examination. Gangrene in her left arm and other complications had set in, and two fingers had to be amputated. The New York State Court of Appeals Judge Benjamin Cardozo wrote: "... the wrong complained of is not merely negligence. It is trespass. Every human being of adult years and sound mind has a right to determine what shall be done with his own body; and a surgeon who performs an operation without his patient's consent commits an assault, for which he is liable in damages." Some lawyers complained that Cardozo was technically incorrect, as the offense in question was in fact battery, not assault. Yet the principle that Cardozo established in this decision remains a key element in American jurisprudential thinking about informed consent.

In 1935 a Michigan case, *Fortner v. Koch,* established biomedical research on humans as a social and scientific necessity conditional on strict legal regulation, such as informed consent. The term "informed consent," was known earlier in other contexts, but its use in American medical jurisprudence began when attorney

Paul G. Gebhard (1928–1998) employed it in his *amicus curiæ* brief from the American College of Surgeons for the 1957 malpractice case of *Salgo v. Leland Stanford Jr. University Board of Trustees.*

The Science

The only strictly scientific question relevant to the ethical conduct of medical research is methodological: Does obeying these principles make better science and, conversely, does flouting them make worse science? The statistical significance of the results of clinical trials improves proportionately to the number of subjects. Therefore one responsibility of physicians and scientists is to recruit as many subjects as possible for each study. Potential subjects, if they are to consent freely to the proposed experiment, must have faith in the process. If they do not trust the physician or the scientist, then they will not consent, the subject pool will be smaller, and the science will be poorer.

Cultivating this faith and trust is easier when the public has a high regard for the whole medical and bioscientific enterprise. Consistently encouraging informed consent and allowing informed refusal should be an integral part of each physician's or scientist's best effort to develop this good public image. Each encounter between physician and patient or between scientist and subject has the potential to affect public relations in general. Investigators should therefore be honest, respectful, plainspoken, and polite with their subjects, not only out of simple human decency, but also because to do so is good science.

In the landmark 1966 article "Ethics and Clinical Research," in the *New England Journal of Medicine,* Henry K. Beecher listed 22 examples, including Willowbrook, of physicians experimenting on patients for the sake of scientific knowledge and imaginary future patients instead of providing therapies for these present patients. Insofar as all physicians' first duty is to their present patients, such preference for scientific research over patient care violates the Hippocratic Oath, especially when patients are deceived or underinformed.

Beecher argued against the then-prevalent notion among scientists that bowing to the principle of informing potential subjects of likely risks would compromise science and retard progress. They believed that they had to deceive the public to gain enough subjects for research, that they were not acting unethically because scientific progress was a greater good than being frank with their subjects, and that science was so ultimately important that a few deaths or maimings along the way in its service would not matter. Within a few years after Beecher's article appeared, regulatory agencies, grant-funding organizations, and the general public, who had previously taken little notice of the ethical dimension of biomedical research, began to ferret out and stop unethical research protocols. His article did not trigger

---

IN CONTEXT: MENGELE AT AUSCHWITZ

Josef Mengele (1911–1979) was a wealthy Bavarian who joined the *Sturmabteilung* (SA, or storm troopers) early in the Nazi era. In 1935 the University of Munich awarded him a Ph.D. in physical anthropology for a dissertation on racial differences in the jaw. The University of Frankfurt gave him an M.D. in 1938 for similar research on the lip, jaw, and palate. That same year he joined the *Schutzstaffel* ("protective echelon," or SS), Heinrich Himmler's elite military unit. As a medical officer in the *Waffen SS,* the most dreaded wing of the SS, Mengele was ordered to Auschwitz in May 1943 as camp physician.

Mengele's natural cruelty had free rein at Auschwitz. He indulged his bizarre fascination with deformities, eye colors, dwarfs, giants, and other genetic differences, especially in twins. He was the absolute judge of life and death throughout the camp. The experiments that he and his staff conducted on those he selected not to die in the gas chambers were arbitrary, heartless, sexually perverted, and of no scientific merit. He did not use anesthesia for any kind of surgery. He did not care about the pain, infections, maimings, or deaths he caused among his captive human research subjects. Of about 3,000 twin children on whom he experimented, only about 200 survived.

Mengele escaped the Russian advance in January 1945. American soldiers captured him, but Mengele tricked his guards into releasing him after he obtained a false set of papers. He went into hiding , and four years later sneaked into South America. West Germany issued a warrant for his arrest in 1959, both universities revoked his degrees in 1964, and Israeli agents actively sought him, but he eluded capture and punishment for the rest of his life.

---

much outrage among the readers of the *New England Journal,* who were mostly physicians and bioscientists. But when the regular press discovered the Willowbrook study in the late 1960s, the public outcry succeeded in getting that research shut down.

Beecher claimed that most reasonable patients would not knowingly risk their lives or their good health for the sake of research. He further suggested that biomedical experiments that involved significant and obvious risks to their subjects were *ipso facto* unethical because he could confidently assume in these cases that informed consent had either not been obtained or had been obtained dishonestly. He questioned whether the results of such experiments should be allowed to be published, even if these results were solid or useful.

Influences on Science and Society

In reaction to documented abuses of human research subjects, the principle of informed consent began to be recognized in the late nineteenth century and was gradually codified in the twentieth as the key concept in biomedical research ethics, one that was enforced by governments and the biomedical enterprise itself. Thus, informed consent has played an increasingly important role in both research and clinical medicine, especially since the verdict in the Nazi "Doctors' Trial" was handed down in Nuremberg, Germany, in 1947. Developments beyond this include the 1964 Helsinki Declaration of the World Medical Association (WMA), 1974 U.S. National Research Act, 1979 Belmont Report, and the 1991 Common Rule of the U.S. Department of Health and Human Services (DHHS).

The international military tribunals in which the Allies tried Nazi war criminals after World War II included the Doctors' Trial from December 9, 1946 to August 20, 1947. The four judges, all Americans, acquitted seven and convicted 16 German physicians of torture, murder, or performing inhumane experiments of questionable scientific value on unwilling subjects. The court further found that the Nazi regime had

sanctioned cruel biomedical research on concentration camp prisoners as a matter of policy.

The most significant result of this trial was not the convictions but the "Nuremberg Code," the 10 ethical principles contained in the part of the verdict subtitled "Permissible Medical Experiments":

1. The voluntary consent of the human subject is absolutely essential.

2. The experiment should be such as to yield fruitful results for the good of society, unprocurable by other methods or means of study, and not random and unnecessary in nature.

3. The experiment should be so designed and based on the results of animal experimentation and a knowledge of the natural history of the disease or other problem under study that the anticipated results will justify the performance of the experiment.

4. The experiment should be so conducted as to avoid all unnecessary physical and mental suffering and injury.

5. No experiment should be conducted where there is an a priori reason to believe that death or disabling injury will occur; except, perhaps, in those experiments where the experimental physicians also serve as subjects.

6. The degree of risk to be taken should never exceed that determined by the humanitarian importance of the problem to be solved by the experiment.

7. Proper preparations should be made and adequate facilities provided to protect the experimental subject against even remote possibilities of injury, disability, or death.

8. The experiment should be conducted only by scientifically qualified persons. The highest degree of skill and care should be required through all stages of the experiment of those who conduct or engage in the experiment.

9. During the course of the experiment the human subject should be at liberty to bring the experiment to an end if he has reached the physical or mental state where continuation of the experiment seems to him to be impossible.

10. During the course of the experiment the scientist in charge must be prepared to terminate the experiment at any stage, if he has probable cause to believe, in the exercise of the good faith, superior skill and careful judgment required of him that a continuation of the experiment is likely to result in injury, disability, or death to the experimental subject.

Some of the hundreds of preserved brains of children killed under the Nazi euthanasia program of the physically or mentally handicapped are seen in this picture from Vienna. The children were murdered at a city hospital and their bodies used for decades for medical research. *Ronald Zak/AP Images.*

Witness J. Bzize from Poland shows the scars on her right leg, which were the result of medical experiments conducted while she was imprisoned at the Ravensbrueck concentration camp, shown during the "Doctors Trial" of the Nuremberg Trials. Bzize had been a member of the Polish underground movement and was imprisoned at the concentration camp on March 28, 1941. The Doctors Trial considered the fate of twenty-three German physicians who either participated in the Nazi program to euthanize persons or who conducted experiments on concentration camp prisoners without their consent. Sixteen of the doctors charged were found guilty. Seven were executed. *USHMM Photo Archives.*

The Nuremberg Code reaffirmed and reinterpreted the ancient principle of medical beneficence that is a central element in all codes of professional medical ethics, including the Hippocratic Oath. Just as physicians must always put the interests of their patients first, so biomedical researchers must put the interests of their human subjects first. Accordingly, the Nuremberg Code has been the basis of all subsequent bioscientific jurisprudence and regulation.

The first major document to serve as a corollary to the Nuremberg Code was the Helsinki Declaration. Its creator, the WMA, was founded in September 1947 specifically in response to the Doctors' Trial verdict. Its mission was to work with national medical associations across the world to develop and uphold high standards in medical ethics. At its Eighteenth General Assembly in 1964 it drafted a comprehensive policy statement designed to help prevent not only deliberate crimes such as those the Nazis committed, but also accidental biomedical innovation tragedies such as the thalidomide experiments of the 1950s, which caused thousands of unforeseen birth defects. Recognizing the need to keep the Helsinki Declaration current with situational changes in bioethics and emergent subtleties of legal interpretation, the WMA approved updates in 1975, 1983, 1989, 1996, 2000, 2002, and 2004.

Among the strengths of the Helsinki Declaration is its insistence that researchers evaluate and respect each subject's competence, or ability to think in general, and capacity, or ability to understand and decide a particular question. In this way the autonomy of the subject should be more strongly protected, not only during the informed consent process but also throughout the experiment itself. Along these same lines of legal and ethical reasoning, the National Bioethics Advisory Commission issued an important report, *Research Involving Persons with Mental Disorders That May Affect Decisionmaking Capacity*, in 1999.

Reacting to the negative publicity surrounding both Tuskegee and Willowbrook, in 1974 DHEW appointed the National Commission for the Protection of Human Subjects of Biomedical and Behavioral Research. Their comprehensive policy statement was named after Belmont House in Elkridge, Maryland, where they met from 1976 to 1978. The Belmont Report established strict ethical guidelines for federally funded research, strengthened the 1974 DHEW mandate that each entity receiving federal money for research must have an institutional review board (IRB) to approve each protocol, and inspired further regulation and legislation.

Until 1991 no uniform American policy existed to cover all federal agencies that award scientific research grants. The Department of Health and Human Services (DHHS), which succeeded DHEW in 1980, rectified this situation by writing the Federal Policy for the Protection of Human Subjects, known informally as the Common Rule because of its wide application. It emphasized the ascendancy of the individual's rights over the scientist's need to find humans on whom to experiment. IRBs gained additional powers and responsibilities under the Common Rule and the Office for Protection from Research Risks (OPRR), founded by DHEW in 1972 and renamed by DHHS in 2000 as the Office for Human Research Protections (OHRP).

Impact on Science

Science in general has lost much popular credibility since its heyday in the 1950s and 1960s, partially because the public has come to see it as unconcerned with human values. The Western debut of acupuncture in the early 1970s began a massive public shift toward preferring alternatives to established methods and products of rigorous Western biomedical science. This is largely due to a pervasive opinion that this science is cold and heartless while other medical systems are warm and responsive. At the dawn of the twenty-first century, science is not always seen as the ultimate deliverer of humankind from its earthly problems, but as part of these problems. In response to widespread reaction against the social optimism of science, slogans like DuPont's "Better Things for Better Living ... Through Chemistry" and General Electric's "Progress Is Our Most Important Product" disappeared from advertising rhetoric and "No Nukes!" replaced "Our Friend the Atom." Scientists began to realize that a more human face was needed to restore good public relations.

Internal Review Boards (IRBs) are part of a strategy to restore science's good name. One of their purposes is to ensure that decisions about nonscientists participating as subjects in scientific experiments are not made in coldly logical, fanatically motivated, or financially conflicted ways by the scientists themselves, but by a cross-section of the community, including nonscientists, presumably impartial. An IRB typically consists of a wide variety of well educated individuals. Each must have a minimum of five members, including at least one scientist, at least one non-scientist, and at least one person who is not in any way affiliated with the institution that is proposing the research protocol. It is also expected to reflect gender, ethnic, and racial balance.

IRB procedures, even when well followed, cannot prevent occasional tragedies in research. Healthy subjects sometimes die unexpectedly as a result of experimentation. In 1999 at the University of Pennsylvania, 18-year-old Jesse Gelsinger, who suffered from a rare metabolic disease, died during experimental gene therapy. In 2001 Ellen Roche, a healthy 24-year-old medical technician, volunteered for an asthma study at the Johns Hopkins University and within a month was dead from a reaction to an experimental medication. Such cases routinely receive immediate federal review, to try to prevent their recurrence and to remedy inadequacies in the protection of human research subjects. In both cases, federal authorities immediately stopped the experiments. In the Gelsinger case, they fined the university $517,000 for misstatements to the IRB, placed restrictions on the professional activities of the three researchers, and cited one of them, James M. Wilson, for conflict of interest because of his financial ties to Genovo Inc., and Targeted Genetics Corporation. In the Roche case, OHRP investigators determined that the IRB had failed to do its job properly, either because the research team did not give it enough accurate information or because of its own negligence.

American scientists often complain about IRBs for two basic reasons. First, the requirement demands so much red tape that the research schedule often suffers. Second, the nonscientific IRB members often do not understand specific scientific protocols well enough to make reasoned judgments about them. Yet, given the general recognition since the Common Rule was enacted that biomedical science conducted without the approval of its human subjects and their home communities is bad science, IRBs seem to have become a permanent fixture on the biomedical research landscape. Without having to write protocols for general inspection or undergo the scrutiny of reviewers before experiments are allowed to begin, scientists might be tempted to revert to pre-Nuremberg standards, thus further alienating the public and making subsequent recruitment of human subjects more difficult.

Impact on Society

There are two kinds of medical research—clinical and nonclinical. Correspondingly, there are two kinds of human medical research subjects—patients in experimental therapies supervised by a physician and otherwise healthy people supervised by a nonmedical provider, such as a PhD biochemist. Physicians may try to enroll their appropriate patients in clinical trials, but if they do so, they must be especially mindful of their supreme duty to serve each patient's best interest. The boundary between physician as clinical practitioner and physician as medical researcher or promoter of medical research is often blurry. Not all patients can benefit from experimental procedures, unusual protocols, or new drugs.

Physicians must simply treat such patients and not try to enroll them in studies, even if approved, safe, and germane research programs are readily accessible. Trying to persuade a patient to enter a clinical trial is ethical only if the prognosis is discouraging and if all available treatments are either unproven or insufficient. If the prognosis with an approved and proven treatment is good, the physician would act unethically even to suggest enrolling that patient in a trial.

Patients are naturally under stress. Physicians have a duty either to try to reduce this stress or at least not add to it. They act unethically when they try to persuade a patient to enter a research study knowing, or even suspecting—but not revealing—that so doing would likely increase that patient's stress.

Clinical researchers must minimize the possibility of iatrogenic (doctor-caused) injuries, illnesses, or adverse effects. Experimentation on a particular patient increases the danger for that patient, but may serve the greater good for many future patients. Because whether to allow such individual sacrifice must be each patient's free decision, the physician must tell the patient the whole truth. Part of proper informed consent procedure is that the physician must tell patients who may become research subjects whether or not they might receive placebos in lieu of actual therapy. Placebos, which are often used as controls in clinical trials, may put patients at undue risk. The Helsinki Declaration approves placebos only when no reliable therapy exists for the patient's condition.

Some physicians receive financial rewards, favored status, or other incentives from medical instrument or pharmaceutical companies for referring patients for clinical trials of experimental equipment or drugs. Conflicts of interest may arise for physicians who stand to gain materially for enrolling patients in trials of new therapies that may not be best for these patients.

Scientists investing in corporations that conduct research in which these scientists are even peripherally involved has been shown to compromise the integrity of the science. Entities paying for trials expect favorable, not objective or impartial, reports. Even payments to subjects might skew scientific results. Subjects should be compensated for their time and trouble, but payments should be neither so low as to belittle each subject's contribution nor so high as to affect the potential subject's judgment whether to participate or the actual subject's judgment whether to continue or withdraw.

■ Modern Cultural Connections

Ever since theologian Paul Ramsey (1913–1988) founded the consumer health movement in 1970 with his book, *The Patient as Person*, a major focus in the ethics of both clinical medicine and biomedical research

IN CONTEXT: JOHN MOORE'S SPLEEN

On October 5, 1976, Alaska pipeline engineer John Moore entered the University of California at Los Angeles (UCLA) Medical Center with an enlarged spleen. Three days later his physician, David W. Golde (1940–2004), diagnosed hairy cell leukemia and told Moore that his life would be in danger unless his spleen were removed. Moore consented to the surgery. Under Golde's instructions, surgeon Kenneth Fleming performed a successful splenectomy on October 20 and gave Moore's spleen to Golde. The spleen weighed 14 pounds, about 22 times its normal size. Golde and his research assistant, Shirley G. Quan, immediately recognized the enormous commerical potential of an immortal cell line cultured from this organ.

Moore was not told either that Golde had a financial interest in his body tissue or that Golde had kept his spleen. He naturally assumed that the spleen had been discarded like any other medical waste. Between November 1976 and September 1983, at Golde's insistence, Moore made increasingly expensive and inconvenient flights to Los Angeles, so that, as Golde told him, follow-up checks could be performed to ensure that Moore's cancer remained in remission. Eventually Moore became suspicious of Golde requiring him to come to California instead of allowing him to have these follow-ups done at a local medical facility. He and his lawyer did some sleuthing. They discovered that Golde and Quan had created the "Mo" cell line, published their research, and had obtained a lucrative patent. Moore sued.

In the case of *Moore v. Regents of the University of California*, the Supreme Court of California ruled in 1990 that Golde was at fault by deliberately concealing from Moore the financial implications of the splenectomy, but that, even if Golde had fully disclosed this information, Moore would still not be entitled to any share of the profits.

has been the tension between the traditional paternalism or condescension of the physician or scientist and the rightful autonomy of the patient or subject. Ramsey's legacy has been a dramatic upsurge in patient activism, which has also developed into research subject activism or at least assertiveness.

Religion is the basis of many of the claims of patients and subjects against physicians and scientists. Among such religious concerns are beliefs about transfusions, transplants, stem cells, embryos, DNA, autopsies, the preparation and burial or cremation of the dead, and the integrity of the body. Consistent with its key principle of protecting the autonomy of the individual, American law in each of these instances supports the demands of religion over those of science and personal beliefs over scientific evidence.

IN CONTEXT: THE MILGRAM EXPERIMENT

In 1961 social psychologist Stanley Milgram (1933–1984) began an experiment at Yale University to test the border between cruel obedience and compassionate disobedience. He solicited male subjects between 20 and 50 years old and paid them each $4.50. The experimenter tested the subjects in pairs. An apparently random, but actually rigged, drawing selected one as teacher and the other as learner. The teacher stayed in the room with the experimenter, while the learner went into another room. The teacher and the learner could communicate but could not see each other. The teacher then asked the learner a series of questions supplied by the experimenter. For each wrong answer the experimenter ordered the teacher to give the learner increasingly powerful electric shocks to a maximum of 450 volts.

No shocks actually occurred. The learner was in fact not a subject, but an accomplice of the experimenter. The fact that the experimenter lied to the subjects, putting them under emotional stress by making them believe that they were inflicting pain by following orders, created an ethical controversy. Yet such lies were necessary to perform the experiment at all, because this stress was precisely what Milgram was investigating. Except for the stress, there was no risk to any of the subjects, and they were all told the truth immediately after their sessions ended.

Milgram's results were important. He determined, and subsequent reproductions of his experiment have confirmed, that about 65% of ordinary people will obey commands even when doing so means violating their conscience or being sadistic. He thus advanced knowledge of the darker side of human nature and provided strong empirical evidence for German philosopher Hannah Arendt's (1906–1975) theory of the banality of evil. Both Milgram and Arendt were motivated by the question of whether Nazi war criminal Adolf Eichmann (1906–1962) was truly guilty or just a tool of his commanders.

In 2001 the Institute of Medicine, one of four independent components of the National Academies, published a landmark report, *Preserving Public Trust: Accreditation and Human Research Participant Protection Programs.* This report investigated the frequency and efficacy of government intervention in biomedical research programs that either IRBs or federal overseers determined were dangerous to human subjects. It offered new and stronger recommendations for accreditation, accountability, and initial protocol approval.

Society, government, and the biomedical community are obligated to protect vulnerable populations, such as children, prisoners, immigrants, researchers' employees, and physicians' patients, from exploitation. When patients or potential research subjects are in-

competent, incapacitated, or underage, the informed consent of the family or other legal surrogate is necessary to fulfill the ethical requirements. While American researchers no longer have access to prisoners in the United States, or relatively easy access to the American poor, they continue to have access to Third World populations, particularly in Africa, whom they can use as subjects. Just as American manufacturers open factories overseas in order to avoid expensive, unionized American labor, so American researchers conduct tests overseas in order to circumvent strict American regulations and to avoid having to use savvy Americans as subjects.

Gina Kolata's story in the November 22, 2006 *New York Times,* "Study Questions Need to Operate on Disk Injuries," probed research subject recruitment tactics and expressed misgivings about the ethics of a successful surgical test. Because of possible disservice to patients, some of the researchers themselves had doubted its ethics from the beginning, but participated anyway. A complicated but reliable and relatively risk-free operation to relieve pain from a herniated or ruptured vertebral disk had long been known. At the time of the experiment, about 300,000 Americans a year were electing to undergo this surgery. Some surgeons wondered whether the operation was necessary. They hypothesized that patients who either waited or did not have the surgery at all would not suffer beyond what they were already suffering from the condition. Over a period of two years, 2,000 patients for whom herniated disk surgery was indicated simply waited. Their outcomes were not appreciably different from those of patients who had chosen to have the surgery.

The informed consent movement since Nuremberg, and especially since Belmont, has reduced physician paternalism in clinical medicine and self-righteousness among medical research scientists. Decisions whether to treat and how to treat are no longer the doctor's, but the patient's, or ideally a consensus of both. Informed consent is what allows human research subjects to be active and voluntary participants rather than victims or unwilling tools of the scientific process. Since Nuremberg, the principle that neither physicians nor scientists may do anything to anyone's body without that person's freely given consent has been universally recognized as a basic human right.

■ Primary Source Connection

Congress passed the National Research Act of 1974 in the wake of negative publicity for both the government and the scientific community regarding the cruel treatment of some medical research subjects. Among the consequences of this act was the Department of Health, Education, and Welfare (DHEW) creating the National

Commission for the Protection of Human Subjects of Biomedical and Behavioral Research, which met at the Belmont House in Elkridge, Maryland, beginning in February 1976.

The Belmont Report prompted much new legislation and regulation in the decades following. In 1991 the Department of Health and Human Services (DHHS) codified the Federal Policy for the Protection of Human Subjects. This policy is called the "Common Rule" because, even though created by DHHS, it applies equally to all federal departments and agencies that grant money for scientific research that involves human subjects. Among other provisions, it strengthened the 1974 mandate that every institution receiving federal funds for research must have its own Institutional Review Board (IRB) to approve all research protocols. In 2000, the OPRR, now within DHHS, became the Office for Human Research Protections (OHRP).

Legal and ethical issues surrounding informed consent and informed refusal remain in a state of flux and continue to be a major concern of the federal government. Abuses, carelessness, and deaths still occur, even at the most prestigious American academic research institutions.

THE BELMONT REPORT: ETHICAL PRINCIPLES AND GUIDELINES FOR THE PROTECTION OF HUMAN SUBJECTS OF RESEARCH
Part B: Basic Ethical Principles

The expression "basic ethical principles" refers to those general judgments that serve as a basic justification for the many particular ethical prescriptions and evaluations of human actions. Three basic principles, among those generally accepted in our cultural tradition, are particularly relevant to the ethics of research involving human subjects: the principles of respect of persons, beneficence and justice.

1. **Respect for Persons.** — Respect for persons incorporates at least two ethical convictions: first, that individuals should be treated as autonomous agents, and second, that persons with diminished autonomy are entitled to protection. The principle of respect for persons thus divides into two separate moral requirements: the requirement to acknowledge autonomy and the requirement to protect those with diminished autonomy.

An autonomous person is an individual capable of deliberation about personal goals and of acting under the direction of such deliberation. To respect autonomy is to give weight to autonomous persons' considered opinions and choices while refraining from obstructing their actions unless they are clearly detrimental to others. To show lack of respect for an autonomous agent is to repudiate that person's considered judgments, to deny an individual the freedom to act on those considered judgments, or to withhold information necessary to make a considered judgment, when there are no compelling reasons to do so.

However, not every human being is capable of self-determination. The capacity for self-determination matures during an individual's life, and some individuals lose this capacity wholly or in part because of illness, mental disability, or circumstances that severely restrict liberty. Respect for the immature and the incapacitated may require protecting them as they mature or while they are incapacitated.

Some persons are in need of extensive protection, even to the point of excluding them from activities which may harm them; other persons require little protection beyond making sure they undertake activities freely and with awareness of possible adverse consequence. The extent of protection afforded should depend upon the risk of harm and the likelihood of benefit. The judgment that any individual lacks autonomy should be periodically reevaluated and will vary in different situations.

In most cases of research involving human subjects, respect for persons demands that subjects enter into the research voluntarily and with adequate information. In some situations, however, application of the principle is not obvious. The involvement of prisoners as subjects of research provides an instructive example. On the one hand, it would seem that the principle of respect for persons requires that prisoners not be deprived of the opportunity to volunteer for research. On the other hand, under prison conditions they may be subtly coerced or unduly influenced to engage in research activities for which they would not otherwise volunteer. Respect for persons would then dictate that prisoners be protected. Whether to allow prisoners to "volunteer" or to "protect" them presents a dilemma. Respecting persons, in most hard cases, is often a matter of balancing competing claims urged by the principle of respect itself.

2. **Beneficence.** — Persons are treated in an ethical manner not only by respecting their decisions and protecting them from harm, but also by making efforts to secure their well-being. Such treatment falls under the principle of beneficence. The term "beneficence" is often understood to cover acts of kindness or charity that go beyond strict obligation. In this document, beneficence is understood in a stronger sense, as an obligation. Two general rules have been formulated as complementary expressions of beneficent actions in this sense: (1) do not harm and (2) maximize possible benefits and minimize possible harms.

The Hippocratic maxim "do no harm" has long been a fundamental principle of medical ethics. Claude Bernard extended it to the realm of research, saying that one should not injure one person regardless of the benefits that might come to others. However, even avoiding harm requires learning what is harmful; and, in the process of obtaining this information, persons may be exposed to risk of harm. Further, the Hippocratic Oath requires physicians to benefit their patients "according to their best judgment." Learning what will in fact benefit may require exposing persons to risk. The problem posed by these imperatives is to decide when it is justifiable to seek certain benefits despite the risks involved, and when the benefits should be foregone because of the risks.

The obligations of beneficence affect both individual investigators and society at large, because they extend both to particular research projects and to the entire enterprise of research. In the case of particular projects, investigators and members of their institutions are obliged to give forethought to the maximization of benefits and the reduction of risk that might occur from the research investigation. In the case of scientific research in general, members of the larger society are obliged to recognize the longer term benefits and risks that may result from the improvement of knowledge and from the development of novel medical, psychotherapeutic, and social procedures.

The principle of beneficence often occupies a well-defined justifying role in many areas of research involving human subjects. An example is found in research involving children. Effective ways of treating childhood diseases and fostering healthy development are benefits that serve to justify research involving children—even when individual research subjects are not direct beneficiaries. Research also makes it possible to avoid the harm that may result from the application of previously accepted routine practices that on closer investigation turn out to be dangerous. But the role of the principle of beneficence is not always so unambiguous. A difficult ethical problem remains, for example, about research that presents more than minimal risk without immediate prospect of direct benefit to the children involved. Some have argued that such research is inadmissible, while others have pointed out that this limit would rule out much research promising great benefit to children in the future. Here again, as with all hard cases, the different claims covered by the principle of beneficence may come into conflict and force difficult choices.

3. **Justice.** — Who ought to receive the benefits of research and bear its burdens? This is a question of justice, in the sense of "fairness in distribution" or "what is deserved." An injustice occurs when some benefit to which a person is entitled is denied without good reason or when some burden is imposed unduly. Another way of conceiving the principle of justice is that equals ought to be treated equally. However, this statement requires explication. Who is equal and who is unequal? What considerations justify departure from equal distribution? Almost all commentators allow that distinctions based on experience, age, deprivation, competence, merit and position do sometimes constitute criteria justifying differential treatment for certain purposes. It is necessary, then, to explain in what respects people should be treated equally. There are several widely accepted formulations of just ways to distribute burdens and benefits. Each formulation mentions some relevant property on the basis of which burdens and benefits should be distributed. These formulations are (1) to each person an equal share, (2) to each person according to individual need, (3) to each person according to individual effort, (4) to each person according to societal contribution, and (5) to each person according to merit.

Questions of justice have long been associated with social practices such as punishment, taxation and political representation. Until recently these questions have not generally been associated with scientific research. However, they are foreshadowed even in the earliest reflections on the ethics of research involving human subjects. For example, during the nineteenth and early twentieth centuries the burdens of serving as research subjects fell largely upon poor ward patients, while the benefits of improved medical care flowed primarily to private patients. Subsequently, the exploitation of unwilling prisoners as research subjects in Nazi concentration camps was condemned as a particularly flagrant injustice. In this country, in the 1940's, the Tuskegee syphilis study used disadvantaged, rural black men to study the untreated course of a disease that is by no means confined to that population. These subjects were deprived of demonstrably effective treatment in order not to interrupt the project, long after such treatment became generally available.

Against this historical background, it can be seen how conceptions of justice are relevant to research involving human subjects. For example, the selection of research subjects needs to be scrutinized in order to determine whether some classes (e.g., welfare patients, particular racial and ethnic minorities, or persons confined to institutions) are being systematically selected simply because of their easy availability, their compromised position, or their manipulability, rather than for reasons directly related to the problem being studied. Finally, whenever research supported by public funds leads to the development of therapeutic devices and procedures, justice demands both that these not provide advantages only to those who can afford them and that such research should not unduly involve persons from groups unlikely to be among the beneficiaries of subsequent applications of the research.

Part C: Applications

Applications of the general principles to the conduct of research leads to consideration of the following requirements: informed consent, risk/benefit assessment, and the selection of subjects of research.

1. **Informed Consent.** — Respect for persons requires that subjects, to the degree that they are capable, be given the opportunity to choose what shall or shall not happen to them. This opportunity is provided when adequate standards for informed consent are satisfied.

While the importance of informed consent is unquestioned, controversy prevails over the nature and possibility of an informed consent. Nonetheless, there is widespread agreement that the consent process can be analyzed as containing three elements: information, comprehension and voluntariness.

Information. Most codes of research establish specific items for disclosure intended to assure that subjects are given sufficient information. These items generally include: the research procedure, their purposes, risks and anticipated benefits, alternative procedures (where therapy is involved), and a statement offering the subject the opportunity to ask questions and to withdraw at any time from the research. Additional items have been proposed, including how subjects are selected, the person responsible for the research, etc.

However, a simple listing of items does not answer the question of what the standard should be for judging how much and what sort of information should be provided. One standard frequently invoked in medical practice, namely the information commonly provided by practitioners in the field or in the locale, is inadequate since research takes place precisely when a common understanding does not exist. Another standard, currently popular in malpractice law, requires the practitioner to reveal the information that reasonable persons would wish to know in order to make a decision regarding their care. This, too, seems insufficient since the research subject, being in essence a volunteer, may wish to know considerably more about risks gratuitously undertaken than do patients who deliver themselves into the hand of a clinician for needed care. It may be that a standard of "the reasonable volunteer" should be proposed: the extent and nature of information should be such that persons, knowing that the procedure is neither necessary for their care nor perhaps fully understood, can decide whether they wish to participate in the furthering of knowledge. Even when some direct benefit to them is anticipated, the subjects should understand clearly the range of risk and the voluntary nature of participation.

A special problem of consent arises where informing subjects of some pertinent aspect of the research is likely to impair the validity of the research. In many cases, it is sufficient to indicate to subjects that they are being invited to participate in research of which some features will not be revealed until the research is concluded. In all cases of research involving incomplete disclosure, such research is justified only if it is clear that (1) incomplete disclosure is truly necessary to accomplish the goals of the research, (2) there are no undisclosed risks to subjects that are more than minimal, and (3) there is an adequate plan for debriefing subjects, when appropriate, and for dissemination of research results to them. Information about risks should never be withheld for the purpose of eliciting the cooperation of subjects, and truthful answers should always be given to direct questions about the research. Care should be taken to distinguish cases in which disclosure would destroy or invalidate the research from cases in which disclosure would simply inconvenience the investigator.

Comprehension. The manner and context in which information is conveyed is as important as the information itself. For example, presenting information in a disorganized and rapid fashion, allowing too little time for consideration or curtailing opportunities for questioning, all may adversely affect a subject's ability to make an informed choice.

Because the subject's ability to understand is a function of intelligence, rationality, maturity and language, it is necessary to adapt the presentation of the information to the subject's capacities. Investigators are responsible for ascertaining that the subject has comprehended the information. While there is always an obligation to ascertain that the information about risk to subjects is complete and adequately comprehended, when the risks are more serious, that obligation increases. On occasion, it may be suitable to give some oral or written tests of comprehension.

Special provision may need to be made when comprehension is severely limited—for example, by conditions of immaturity or mental disability. Each class of subjects that one might consider as incompetent (e.g., infants and young children, mentally disabled patients, the terminally ill and the comatose) should be considered on its own terms. Even for these persons, however, respect requires giving them the opportunity to choose to the extent they are able, whether or not to participate in research. The objections of these subjects to involvement should be honored, unless the research entails providing them a therapy unavailable elsewhere. Respect for persons also requires seeking the permission of other parties in order to protect the subjects from harm. Such persons are thus respected both by acknowledging their own wishes and by the use of third parties to protect them from harm.

The third parties chosen should be those who are most likely to understand the incompetent subject's situation and to act in that person's best interest. The person authorized to act on behalf of the subject should be given an opportunity to observe the research as it proceeds in order to be able to withdraw the subject from the research, if such action appears in the subject's best interest.

Voluntariness. An agreement to participate in research constitutes a valid consent only if voluntarily given. This element of informed consent requires conditions free of coercion and undue influence. Coercion occurs when an overt threat of harm is intentionally presented by one person to another in order to obtain compliance. Undue influence, by contrast, occurs through an offer of an excessive, unwarranted, inappropriate or improper reward or other overture in order to obtain compliance. Also, inducements that would ordinarily be acceptable may become undue influences if the subject is especially vulnerable.

Unjustifiable pressures usually occur when persons in positions of authority or commanding influence—especially where possible sanctions are involved—urge a course of action for a subject. A continuum of such influencing factors exists, however, and it is impossible to state precisely where justifiable persuasion ends and undue influence begins. But undue influence would include actions such as manipulating a person's choice through the controlling influence of a close relative and threatening to withdraw health services to which an individual would otherwise be entitled.

United States Department of Health, Education, and Welfare. National Commission for the Protection of Human Subjects of Biomedical and Behavioral Research

NATIONAL COMMISSION FOR THE PROTECTION OF HUMAN SUBJECTS OF BIOMEDICAL AND BEHAVIORAL RESEARCH. *THE BELMONT REPORT: ETHICAL PRINCIPLES AND GUIDELINES FOR THE PROTECTION OF HUMAN SUBJECTS OF RESEARCH*. WASHINGTON, D.C.: USGPO, 1979.

SEE ALSO *Biology: Sociobiology; Biomedicine and Health: Dissection and Vivisection.*

BIBLIOGRAPHY

Books

Annas, George J. *Law, Medicine, and the Market.* New York: Oxford University Press, 1998.

Berg, Jessica W., Paul S. Appelbaum, Charles W. Lidz, and Lisa S. Parker. *Informed Consent: Legal Theory and Clinical Practice.* New York: Oxford University Press, 2001.

Blass, Thomas. *The Man Who Shocked the World: The Life and Legacy of Stanley Milgram.* New York: Basic Books, 2004.

Brody, Baruch A. *The Ethics of Biomedical Research: An International Perspective.* New York: Oxford University Press, 1998.

Coleman, Carl H. *The Ethics and Regulation of Research with Human Subjects.* Newark, NJ: LexisNexis, 2005.

DeRenzo, Evan G., and Joel Moss. *Writing Clinical Research Protocols: Ethical Considerations.* Burlington, MA: Elsevier Academic, 2006.

Doyal, Len, and Jeffrey S. Tobias, eds. *Informed Consent in Medical Research.* London: BMJ Books, 2001.

Faden, Ruth R., Tom L. Beauchamp, and Nancy M.P. King. *A History and Theory of Informed Consent.* New York: Oxford University Press, 1986.

Freund, Paul A., ed. *Experimentation with Human Subjects.* New York: George Braziller, 1970.

Getz, Kenneth, and Deborah Borlitz. *Informed Consent: The Consumer's Guide to the Risks and Benefits of Volunteering for Clinical Trials.* Boston: CenterWatch, 2003.

Hornblum, Allen M. *Acres of Skin: Human Experiments at Holmesburg Prison: A True Story of Abuse and Exploitation in the Name of Medical Science.* New York: Routledge, 1999.

Iltis, Ana Smith, ed. *Research Ethics.* New York: Routledge, 2006.

Institute of Medicine. Committee on Assessing the System for Protecting Human Research Subjects. *Preserving Public Trust: Accreditation and Human Research Participant Protection Programs.* Washington, DC: National Academies Press, 2001.

Institute of Medicine. Committee on Ethical Considerations for Revisions to DHHS Regulations for Protection of Prisoners Involved in Research. *Ethical Considerations for Research Involving Prisoners.* Washington, DC: National Academies Press, 2007.

King, Nancy M.P., Gail E. Henderson, and Jane Stein, eds. *Beyond Regulations: Ethics in Human Subjects Research.* Chapel Hill: University of North Carolina Press, 1999.

Kodish, Eric, ed. *Ethics and Research with Children: A Case-Based Approach.* New York: Oxford University Press, 2005.

LaFleur, William R., Gernot Böhme, and Susumu Shimazono, eds. *Dark Medicine: Rationalizing Unethical Medical Research.* Bloomington: Indiana University Press, 2007.

Lavery, James V., Christine Grady, Elizabeth R. Wahl, and Ezekiel J. Emanuel, eds. *Ethical Issues in International Biomedical Research: A Casebook.* Oxford: Oxford University Press, 2007.

Macklin, Ruth. *Double Standards in Medical Research in Developing Countries.* Cambridge: Cambridge University Press, 2004.

Manson, Neil C., and Onora O'Neill. *Rethinking Informed Consent in Bioethics.* New York: Cambridge University Press, 2007.

Menikoff, Jerry, and Edward P. Richards. *What the Doctor Didn't Say: The Hidden Truth about Medical Research.* New York: Oxford University Press, 2006.

Milgram, Stanley. *Obedience to Authority: An Experimental View.* New York: HarperCollins, 2004.

Miller, Arthur G. *The Obedience Experiments: A Case Study of Controversy in Social Science.* New York: Praeger, 1986.

Moreno, Jonathan D. *Undue Risk: Secret State Experiments on Humans.* New York: Routledge, 2000.

National Bioethics Advisory Commission. *Research Involving Persons with Mental Disorders that May Affect Decisionmaking Capacity.* Springfield, VA: National Technical Information Service, 1999.

Quinn, Susan. *Human Trials: Scientists, Investors, and Patients in the Quest for a Cure.* Cambridge, MA: Perseus, 2001.

Rollin, Bernard E. *Science and Ethics.* New York: Cambridge University Press, 2006.

Schmidt, Ulf, ed. *Justice at Nuremberg: Leo Alexander and the Nazi Doctors' Trial.* New York: Palgrave Macmillan, 2004.

Shah, Sonia. *The Body Hunters: Testing New Drugs on the World's Poorest Patients.* New York: New Press, W.W. Norton, 2006.

Shamoo, Adil E., and David B. Resnik. *Responsible Conduct of Research.* Oxford: Oxford University Press, 2003.

Smith, Trevor. *Ethics in Medical Research: A Handbook of Good Practice.* Cambridge: Cambridge University Press, 1999.

Weindling, Paul Julian. *Nazi Medicine and the Nuremberg Trials: From Medical War Crimes to Informed Consent.* New York: Palgrave Macmillan, 2004.

Periodicals

Beecher, Henry K. "Ethics and Clinical Research." *New England Journal of Medicine* 274 (1966): 1354–1360.

Vollmann, Jochen, and Rolf Winau. "The Prussian Regulation of 1900: Early Ethical Standards for Human Experimentation in Germany." *IRB: Ethics and Human Research* 18, 4 (July-August 1996): 9–11.

Eric v.d. Luft

Science Philosophy and Practice: Lysenkoism: A Study in the Dangers of Political Intrusions into Science

■ Introduction

Lysenkoism was a pseudoscientific belief system associated with Soviet plant breeder Trofim Denisovich Lysenko (1898–1976). Lysenko rejected nearly a century of advances in genetics, the study of inherited characteristics in living things. His influence on science and agriculture in the Soviet Union in the 1930s through the early 1950s illustrates the disastrous consequences that may ensue when politics and ideology interfere with science. Lysenkoism became the Soviet government's official program and had major effects on government policies. It worsened food shortages in the Soviet Union, and real scientists were imprisoned for disagreeing with it. Some were killed.

WORDS TO KNOW

LAMARCKISM: The belief that acquired characteristics can be inherited, that is, that changes to an organism that happen during its life can be passed on to offspring.

LYSENKOISM: A type of pseudoscience that arose in the Soviet Union in the 1930s and destroyed Soviet biology for decades. Lysenkoists denounced modern evolutionary biology and genetics.

NATURAL SELECTION: Also known as survival of the fittest; the natural process by which those organisms best adapted to their environment survive and pass their traits to offspring.

■ Historical Background and Scientific Foundations

Ten years after Russia's 1917 revolution, which led to the formation of the Union of Soviet Socialist Republics (USSR), a plant breeder named Trofim Denisovich Lysenko observed that pea seeds germinated faster when maintained at low temperatures. Instead of reasoning that the plant's ability to respond flexibly to temperature variations was a natural characteristic—as further testing would have confirmed—Lysenko erroneously concluded that low temperatures forced seeds to alter their inherited characteristics.

Lysenko's erroneous conclusions were influenced by the teachings of Russian horticulturist I.V. Michurin (1855–1935), a holdover proponent of the Larmarckian theory of evolution by inheritance of acquired characteristics (the belief that species can evolve by individuals acquiring traits during their lifetime and then passing them on to offspring). Lamarckism had, at one time, been a legitimate scientific theory. In the late eighteenth and early nineteenth century, French anatomist Jean-Baptiste Lamarck (1744–1829) attempted to explain such adaptations as the long necks of giraffes by arguing that a giraffe, by stretching its neck to get leaves on high tree branches, actually made its neck lengthen, and that this individual would pass a longer neck on to its offspring. Thus, according to Lamarck, the extremely long necks of modern giraffes were the result of generation after generation of giraffes stretching their necks to reach higher for food. The evidence for Lamarckian evolution was once thought convincing: Even English naturalist Charles Darwin (1809–1882), discoverer of the real source of adaptive evolution (natural selection of random variations in inherited characteristics), taught that Lamarck's mechanism contributed to evolution.

Trofim Denisovich Lysenko (right), a Soviet geneticist and agronomist and President of the Lenin Academy of Agricultural Sciences, measures the growth of wheat in a collective farm field near Odessa in the Ukraine.   *© Hulton-Deutsch Collection/Corbis.*

Unfortunately for Lysenko and Soviet science, Lamarck's theory of evolution by the inheritance of acquired characteristics was incorrect. By the beginning of the twentieth century, scientists had already discarded Lamarckian evolution in favor of Darwin's concept of natural selection. Today, we know that the inheritable aspects of traits are determined by long, ladder-like molecules of DNA (deoxyribonucleic acid) found in the nucleus of almost every cell. These molecules of inheritance are not influenced by the use, disuse, or even loss of body parts. Darwinian natural selection explains the long necks of giraffes as the result of the greater feeding success, over many generations, of giraffes who happened, thanks to random changes in DNA, to have longer necks. Giraffes who, by chance, had shorter necks or other unfavorable characteristics have not left any offspring.

Despite the fact that Lamarck's theory of evolution by inheritance of acquired characteristics had already been widely discarded as a scientific hypothesis in the early twentieth century, a remarkable set of circumstances gave Lysenko the opportunity to sweep aside more than a hundred years of scientific investigation and advocate his own schemes for enhancing agricultural production. When Lysenko promised greater crop yields to government officials, a Soviet Central Committee, desperate to increase food production after famine in the early 1930s, listened with an attentive ear. Lysenko claimed that the spirit of Marxist theory (on which the Soviet Union was based) called for a theory of species formation which would entail "revolutionary leaps." Lysenko attacked Mendelian genetics and Darwinian evolution as a theory of "gradualism."

Lysenko constructed an elaborate hypothesis that came to be known as the theory of phasic development. One of Lysenko's ideas was to "toughen" seeds by treating them with heat and high humidity to increase their ability to germinate under harsh conditions. The desire to plant winter instead of spring forms of wheat was heightened by the need to expand Russian wheat production into areas climatically colder than traditional growing areas. The Nazi invasion during the Second World War made it critical to plant in colder, previously fallow eastern regions as the USSR was deprived of its Ukrainian breadbasket by Hitler's onslaught.

Faced with famine, Soviet agricultural planners became unconcerned with long-term scientific studies, making them vulnerable to Lysenko's unfounded claims. They believed what they wanted to believe, and looked no further into the validity of Lysenko's claims.

Lysenko ruled virtually supreme in Soviet science for years, extending his influence beyond agriculture to other areas of science. In 1940, Soviet dictator Joseph Stalin (1922–1953) appointed Lysenko Director of the Soviet Academy of Science's Institute of Genetics. In 1948, the Praesidium of the USSR Academy of Science passed a resolution virtually outlawing any biological work not based on Lysenko's ideas.

Although thousands of experiments carried out by geneticists all over the world failed to provide evidence for Lysenko's notions of transmutation of species—indeed, produced vast amounts of evidence against it—Lysenko's followers made increasingly grandiose claims regarding crop yields and the transformation of species. Not until 1953, following the death of Stalin, did the Soviet government publicity acknowledge that Soviet agriculture had failed to meet economic plan goals and thereby provide the food needed by the Soviet State.

The Lysenkoist episode reversed a longstanding tradition of Russian scientific progress. Despite the near-medieval conditions in which most of the population of Czarist Russia lived, the scientific achievements of pre-revolutionary Russia rivaled those of Europe and America. In fact, achievement in science had been one of the few avenues to wealth open to the non-nobility. The revolution had sought to maintain this tradition and win over the leaders of Russian science. From the earliest days, revolutionary leaders Lenin and Trotsky fought, even in the midst of famine and civil war, to make resources available for scientific research.

In the political storms that ravaged the Soviet Union following the rise of Stalin, including mass executions of dissidents and engineered famines in the Ukraine that killed millions, Lysenko's idea that all organisms, given the proper conditions, have the capacity to be or do anything seemed to have certain attractive parallels with the social philosophies of Karl Marx (and the twentieth century French philosopher Henri Bergson), who promoted the idea that man was largely a product of his own will. Enamored for ideological reasons with Lysenko's pseudoscientific claims, Stalin took matters one step further by personally attacking modern genetics as "counter-revolutionary" or "bourgeois" science. ("Bourgeois" is a French word meaning upper- and middle-class; in Soviet jargon, it was equivalent to "enemy of the revolution.") While the rest of the scientific community knew that evolution could not be understood without Mendelian genetics, Stalin used violence and political power to suppress scientific inquiry. Under Stalin, science was made to serve political ideology: Scientists were required to say the things that those in power wanted them to say, regardless of physical reality.

The victory of Stalin's faction within the ruling party changed the previously nurturing relationship between the Soviet State and science. Important developments in science (including what we would term today the social sciences) were terminated by state terror. During the 1930s and 1940s, scientists were routinely executed, imprisoned, or exiled. Soviet science was largely carried forward in specially-built labor camps, where scientists denounced publicly as "saboteurs" continued their work in isolation from the outside world.

Information on genetics was eliminated from Soviet biology textbooks as Lysenko attempted to reduce his conflict with classical geneticists to politics. He stated that there existed two class-based biologies: "bourgeois" (bad) and "socialist, dialectical materialist" (good). The entire agricultural research infrastructure of the Soviet Union—a country where millions teetered on the edge of starvation—was devoted to a disproved scientific hypothesis, and inventive methods were used to falsely "prove" that there was no famine and that crop yields were actually on the rise.

Soviet Central Committee support of Lysenko was critical to his success. It was known that Stalin clearly expressed his positive attitude toward the idea of the inheritance of acquired characteristics and his overall support of Lamarckism; in such an atmosphere, some of Lysenko's supporters even denied the existence of chromosomes (dark objects in cell nuclei containing heritable material; it was not known until the 1950s that this material is the molecule DNA). Genes were denounced as "bourgeois constructs." Under Lysenko, Mendelian genetics was branded "decadent," and scientists who rejected Lamarckism in favor of natural selection became "enemies of the Soviet people."

Some scientists resisted. Soviet geneticist Nikolay Ivanovich Vavilov (1887–1943) tried to expose Lysenko's claims as pseudoscientific. As a result, Vavilov was arrested in August 1940 and later died in a prison camp. Throughout Lysenko's reign there were widespread arrests of geneticists, who were denounced as "agents of international fascism." In fear for their lives, many Soviet scientists submitted. Some presented fraudulent data to support Lysenko, others destroyed evidence showing that he was wrong. Letters by scientists who had once advanced Mendelian genetics were made public in which they confessed the errors of their ways and extolled the wisdom of the Party.

Lysenko falsely predicted greater crop yields through hardening of seeds and a new system of crop rotation. His crop rotation method eventually led to soil depletion that required years of replenishment with mineral fertilizers. Under Lysenko's direction, hybrid corn programs based on successful U.S. models were ended and the research facilities destroyed because Lysenko opposed what he termed "inbreeding."

When Nikita Khrushchev (1894–1971) assumed the post of Soviet Premier following the death of Stalin in 1953, opposition to Lysenko began slowly to grow. Khrushchev eventually stated that under Lysenko "Soviet agricultural research spent over 30 years in darkness." In 1964, Lysenko's doctrines were officially discredited, and intensive efforts were made toward reestablishing Mendelian genetics and bringing Soviet agriculture, biology, and genetics into conformity with Western nations.

■ Modern Cultural Connections

The ultimate rejection of Lysenkoism was a victory for empirical evidence: Lysenkoism simply did not work. The only results it produced were agricultural disasters.

Lysenkoism has entered our cultural heritage in several ways. In the early 2000s, some scholars of the English author George Orwell (1903–1950) argued that he was inspired to write his seminal novel *1984*, about political control of people's perceptions of reality, after friends drew his attention to Lysenkoism. Today, the word "Lysenkoism" is often used when one person wishes to accuse another of distorting scientific facts to please political masters or further a non-scientific ideological agenda. For example, in the 1980s U.S. biologist Stephen Jay Gould (1941–2002) was accused of attacking the IQ theory of human intelligence for political reasons, and one of his fellow scientists accused him of "Neo-Lysenkoism." (The analogy was weak: Gould never sought to silence those who disagreed with him and had no form of government power backing him up.) Proponents of "intelligent design" and other forms of creationism sometimes accuse Lysenko's old foes, evolutionary biologists, of Lysenkoism: "Lysenkoism is now rearing its ugly head in the US," wrote Jonathan Wells in 2006, "as Darwinists use their government positions to destroy the careers of their critics." Critics of creationism have also compared creationists to Lysenkoists.

In the early 2000s, a number of scientists accused the presidential administration of George W. Bush (1946–) of Lysenkoism because it used its authority to prevent government-employed scientists from sharing mainstream scientific views on global climate change, stem cell research, and other issues.

SEE ALSO *Biology: Genetics; Science Philosophy and Practice: Pseudoscience and Popular Misconceptions.*

BIBLIOGRAPHY

Books

Graham, L. *Science in Russia and the Soviet Union.* Cambridge, UK: Cambridge University Press, 1993.

Joravsky, D. *The Lysenko Affair.* Cambridge, MA: Harvard University Press, 1970.

Lysenko, T.D. *Agrobiology.* Moscow: Foreign Language Press, 1954.

———. *Soviet Biology: A Report to the Lenin Academy of Agricultural Sciences.* New York: International Publishers, 1948.

Soyfer, Valery. *Lysenko and the Tragedy of Soviet Science.* New Brunswick, NJ: Rutgers University Press, 1994.

Periodicals

Darlington, C.D. "T.D. Lysenko (Obituary)." *Nature* 226 (1977): 287–288.

Web Sites

Horton, Scott. "The New Lysenkoism." *Harper's Magazine.* July 11, 2007. http://www.harpers.org/archive/2007/07/hbc-90000486 (accessed January 18, 2008).

The Editors of Scientific American. "Bush-League Lysenkoism." *Scientific American.* May 2004. http://www.sciam.com/article.cfm?chanID=sa006&colID=2&articleID=0001E02A-A14A-1084-983483414B7F0000 (accessed January 18, 2008).

K. Lee Lerner

Science Philosophy and Practice: Ockham's Razor

■ Introduction

Simplicity is a virtue in any explanation or description. Both understanding and communication are facilitated by keeping concepts and the words that express them as simple as possible without sacrificing accuracy or completeness. This principle has been known, practiced, and codified since ancient times. The best known version of the principle is attributed to English Franciscan nominalist philosopher William of Ockham (or Occam) (c.1288–1347). Though scientists more commonly call it Ockham's razor, it is properly called either the principle of parsimony or the principle of economy, because it shaves away superfluous concepts.

Ockham's razor has been the key principle of scientific method since the beginning of modern empiricism in the experiments of English polymath Francis Bacon (1561–1626) and especially since the standardization of the speculative–empirical cycle (observe, create hypothesis, test hypothesis, observe) in the twentieth century.

Probably the best example of the razor working to improve scientific understanding is the cosmological revolution wrought by Polish astronomer Nicolaus Copernicus (1473–1543). Before his time, the prevailing geocentric view of the universe was bolstered by notions of epicycles and other fanciful mathematical fabrications that "explained" retrograde planetary and lunar motions. By going against religious dogma and positing a heliocentric system, Copernicus was able to explain these motions with a relatively uncomplicated arrangement of ellipses.

■ Historical Background and Scientific Foundations

Greek philosopher Aristotle (384–322 BC) wrote in *De Caelo* that assumptions should be as few as possible, consistent with the known facts. In *Physics* he asserts that the universe has only one mover. In Book Alpha Major of the *Metaphysics* he refutes his teacher Plato's (c.427–c.347 BC) view that each thing on Earth has an ideal form in a supraterrestrial, intelligible realm. Aristotle's counterargument was that things are substances, and there is no reason to understand any substance as existing doubly, as both itself and its idea, when it could just as well be understood as existing singly.

The principle of parsimony was familiar in medieval scholastic philosophy before Ockham's birth. Like most Aristotelian ideas, it appears in the writings of Italian Dominican philosopher St. Thomas Aquinas (c.1225–1274), notably in *Summa Contra Gentiles*, in which he argues that because nature does not use two means when one would suffice, we generate superfluity if we perform any task in several ways when we could do it in one way. Moreover, he continued, if there are two or more metaphysical theories of a thing, at most only one of them could be correct.

The principle is also found in the texts of Scottish Franciscan philosopher John Duns Scotus (1266–1308), who may have been Ockham's teacher. The French Franciscan Archbishop of Rouen, Odo Rigaldus (1205–1275), wrote in *Commentary on Sentences* that to posit many entities when we could posit only one is vain.

Ockham wrote several statements of the principle. Its most common formulation, "Entities are not to be multiplied without necessity," is attributed to Ockham but does not appear in any of his surviving works. This version has not been found in any text earlier than the 1639 commentary on Duns Scotus by Irish Franciscan theologian John Ponce of Cork (1603–1670). Rather carelessly disregarding the actual history of the principle, Scottish philosopher Sir William Hamilton (1788–1856) coined the term "Ockham's razor" in 1852, though Hamilton did not give credit where credit is due. Ockham's razor would more fairly be called "Aristotle's razor."

The Principle of Sufficient Reason

The principle of sufficient reason (PSR) claims that everything has a reason or cause as to why it exists. Disproportionately associated with German rationalist philosophers Gottfried Wilhelm von Leibniz (1646–1716) and Christian Wolff (1679–1754), it appears also in the writings of Plato, Aristotle, Thomas Aquinas, and German philosopher Arthur Schopenhauer (1788–1860). It does not oppose Ockham's razor. The two principles complement each other. PSR supports not positing too few explanatory factors, while the razor supports not positing too many. The goal of investigation is to find exactly the right number.

Neither philosophy nor science can seriously challenge either PSR or the razor. They have been involved, at least implicitly, in every important philosophical or scientific contribution since the dawn of rigorous thought in ancient Greece. PSR has been invoked to discredit scientific theories, notably the idea of the spontaneous generation of life. To explain the origin of living beings, such as frogs, by saying that they just oozed out of swamp mud is simpler than talking about zygotes and tadpoles. But that explanation does not fit all the known facts. French biochemist Louis Pasteur (1822–1895) depended more on PSR than on the razor to disprove spontaneous generation.

Applying PSR adds concepts, but only necessary ones. Applying the razor removes concepts, but only unnecessary ones. The closer the harmony between these two opposite tendencies in any philosophical or scientific inquiry, the more defensible that inquiry is. Striking that balance requires both analytic subtlety and speculative comprehension.

■ Modern Cultural Connections

Ancient and medieval thinkers devised elaborate cosmologies and metaphysical systems to account for the structure and origin of Earth, the nature of things, and the birth, growth, and death of living beings. Many of these constructs have found their way into the doctrines of major world religions and are still difficult or impossible to reconcile with each other or with modern science. Others, such as the humoral theory of the elements and the flat Earth theory, were discredited as new facts came to light and as science applied Ockham's razor to these facts.

Since Bacon, the razor has been applied consistently and pervasively in natural science, but less so in other domains of thought, such as philosophy. It is conspicuously absent from theology. Many philosophers and scientists argue that there is no good argument for not applying it to all these domains because it enables beliefs to be based on facts and deductions rather than legends, traditions, prejudices, and intuitions.

WORDS TO KNOW

EMPIRICISM: The philosophical doctrine, first developed in the 1600s and 1700s, that all knowledge arises from sense impressions—things seen, heard, felt, and so on. It opposes the view that some knowledge is available to the mind a priori, that is, by the very nature of thought.

NOMINALIST: A person who affirms nominalism, the philosophical doctrine that universal concepts or abstractions (including mathematical abstractions) have no reality except as names (hence "nominalism," from the Latin *nomen*, for "name").

ONTOLOGICAL ARGUMENT: A philosophical argument for the existence of God. It was proposed in the Middle Ages and later famously articulated by French mathematician and philosopher René Descartes (1596–1650). "Ontological" means having to do with the nature of being. According to the ontological argument, God must be perfect, but a perfect being that did not exist would not be perfect; therefore, God must exist. Although forms of the ontological argument are still defended today, most philosophers and theologians have abandoned it as invalid because it attempts to establish a fact (God's existence) through appeal to a mere pattern of words. Persons responding to the ontological argument often point out that one could say "I have the idea of a perfect unicorn, and a perfect unicorn would have the attribute of existence, therefore my unicorn exists," even though there are, in fact, no unicorns.

POLYMATH: Any person who is highly capable in several fields of knowledge—for example, physics, mathematics, and music.

RATIONALIST PHILOSOPHERS: All philosophers that privilege or emphasize the use of reason in the quest for truth, while downplaying faith, experience, intuition, or other ways of knowing. Non-rationalist philosophers are not, in general, anti-reason, but see reason as one mode of mental activity leading to knowledge or wisdom, rather than as the only admissible one.

SUPERFLUOUS: Anything which is unnecessary or more than needed for some purpose.

The application of Ockham's razor to science is essentially a simple refusal to believe that which cannot be proved. Scientists form prudent theories from what the facts indicate, not from what either metaphysics or religion tells them the theories ought to be.

While its influence on science has been profound and positive, its influence on society has not been as significant. Many social problems can be traced to superstitions and other erroneous beliefs that the razor might eliminate if applied.

IN CONTEXT: PLATO'S BEARD

American logician Willard Van Orman Quine (1908–2000), following a line of argument begun by English analytic philosopher Bertrand Russell (1872–1970), wrote in *From a Logical Point of View* that the Platonic doctrine that non–being exists "might be nicknamed *Plato's beard*; historically it has proved tough, frequently dulling the edge of Occam's razor." Plato's beard is a paradox of reference. It allows meaningful discussion of entities that patently do not exist, such as unicorns, because these nonexistent entities "exist" as ideas.

At first glance, a sentence without a definite referent might seem nonsensical. Yet the sentence, "Unicorns cannot fly," makes sense even though the word "unicorn" does not denote anything real. We use Ockham's razor to excise unicorns from our world, which can well be explained without mentioning them. Yet, even though we can never observe them, we can still understand and employ descriptions of them as fictitious creatures with distinct properties, just as if they were real. Plato's beard becomes intellectually troublesome only if we believe in the actual existence of such fictions.

Ockham's razor can be applied to the phenomenon of crop circles, such as these found in a wheat field in Wiltshire, England. Although claims have been made that aliens create such circles, it has been shown that people can produce them. As such, Ockham's razor would favor the latter, simpler, proven opinion. *Robert Harding Picture Library Ltd/Alamy.*

The razor's limited social impact has resulted mostly from its inconsistent application. Ockham himself modified application of the principle by declaring that it did not apply to supernatural concepts of God or scripture. Instead of using it only to eliminate explanations not supported by experience or deductive reasoning, Ockham, serving his faith, used it also to eliminate explanations not in accord with his religious beliefs.

Despite the fact that both the ontological argument for the reality of a supernatural God and the five ways of Aquinas are consistent with the razor, if the razor were accepted universally, then it would directly challenge many superstitions, religious doctrines, and folk beliefs.

■ Primary Source Connection

What follows is a portion of a transcript of Dr. Anthony Garrett (1957–), a physicist at Cambridge University, explaining some modern applications of Ockham's razor during an April 2000 radio interview with Robyn Williams, host of the Canadian radio show *Ockham's Razor.*

PHYSICIST ANTHONY GARRETT EXPLAINS THE MEANING OF OCKHAM'S RAZOR

Anthony Garrett: … In the 20th century it (Ockham's Razor) has found mathematical expression within probability theory. So: is this mathematical version the final, precise version of the idea, settled for all time? No, it isn't. Mathematics is precise, and words have shades of meaning which make them ambiguous; mathematics and words are suited to different things. So today's version of the Razor in probability theory, corresponds to one interpretation of the words, but other interpretations could generate different mathematical realisations. The idea made precise in probability theory is a very important one though, relating theory and practice in some very basic areas of physics. Let me illustrate this.

The new mathematical razor applies whenever there is a set of numerical data that are polluted by noise, by processes we do not know the exact details of, and we are interested in whether there is a signal, hidden in that noise. An example that recurs as data become ever more accurate is whether there are further planets in our solar system. Look at the motion of the outermost planets such as Pluto. We know that Pluto is influenced principally by the gravitational field of the sun, round which it orbits; then by the gravitational fields of the other planets that we know about. But if you look closely enough at Pluto's motion you will still find small deviations from the motion predicted, even taking the known planets into account. The question is: do those deviations contain subtle evidence for any further unknown planets, or are they best explained as noise due to comets and meteors passing by Pluto, dust, and so on? If the answer is Yes, it is more likely that there is another planet, then where do the data suggest that this other planet is orbiting, and how big is it, what is its influence?

To see how this plugs into the idea of Ockham's Razor, imagine hunting for not just one extra planet, but two, three, four and so on. Obviously if we suppose there are enough extra planets we can fit the predicted orbit of Pluto to the observed fluctuations of the orbit very finely indeed. But the mind revolts at the idea of dozens of extra planets; it makes obvious better sense to suppose the fluctuations are due to irregular passing comets, dust, and so on. The physicist Richard Feynman once said that given enough parameters (that means planets in this case) he could fit an elephant to the curve. We call this phenomenon Overfit. On the other hand, with fewer extra planets we cannot fit the data so closely, which is obviously something that you want to do. So intuitively, there is going to be a trade–off between how well you can fit the data and how many extra planets you suppose there are. In other words, how complicated the theory is. This trade–off between goodness of fit to the observations and the simplicity of the theory you're using is made precise in the new mathematical razor. It allows us to say how probable it is that there is zero, one, two or more undetected planets; and if one or more, what is the best guess of where their orbits are and their masses. Both the number of extra planets, and their positions and masses, are chosen so as to allow the best fit to the data. Ockham's Razor is not just "choose the simplest theory that fits the facts," but "choose the simplest theory that fits the facts well," and there is a measurable trade–off, between goodness of fit and simplicity of the theory; a trade–off between flexibility and economy.

Ockham's Razor is also the motivation behind unification of physical theory. A good example of this came nearly 100 years ago. The German physicist Max Planck had invented an early version of the quantum theory that explained a baffling phenomenon: the speed of electrons that were thrown off when light is shone at a metal. His equations called for a new physical constant, a new constant of nature, whose value had to be found from the observations he made. But the same idea was then applied to explain the amount of radiation given off by a hot body, an electric fire, for example, and also to explain the wavelengths of light that are absorbed by hydrogen atoms. But of these further phenomena had been experimentally studied and each had required its own physical constant of nature to be set separately from the observations. The new idea related these two extra constants to Planck's and accurately gave their values. Three supposedly separate phenomena had been shown to have the same underlying explanation. The quantum idea was rapidly accepted in consequence.

My last example is from cosmology. When Einstein worked out his general theory of relativity and gravity

IN CONTEXT: OCKHAM'S RAZOR APPLIED TO ABSTRACT LANGUAGE

Ockham's razor implies that relationships between entities are not themselves to be treated as entities. Abstractions, the common aspects among several observed species or individuals, are not to be reified, personified, or otherwise granted entity status. They are not real things, but only ideas or interpretations in the mind of the observer.

Using the word "man" in the singular as a substantive without an article violates this implication of the razor. We would not say "Horse runs," "Snake creeps," or "Pig wallows," but we often say "Man thinks," "Man builds," or "Man cares." "Man" in this context does not refer to anything real. If we obeyed the razor in this regard, we would say "People think," "Humans build," or "Human beings care." As we typically say "Horses run," "Snakes creep," or "Pigs wallow," we acknowledge at least in those cases that we are referring to aggregates of individuals rather than to abstract, non–existent essences or universals.

There is no such thing as the universal "man." There are only individual human beings. Perhaps this is why the academic journal called *Man and World* was renamed *Continental Philosophy Review* in 1998. If philosophers and poets allow such titles as *An Essay on Man* rather than *An Essay on Humankind*, then, on the same grounds, ornithologists should be able to use, say, *An Essay on Bird* rather than *An Essay on the Class Aves*.

early in the 20th century, and improved on Newton's venerable theory, there was room for an arbitrary constant, known as a parameter, in his equations. To keep things simple he was tempted to put it to zero, but another consideration weighed even more heavily: he believed on philosophical grounds that the universe was unchanging on the large scale. He believed it was unchanging in how the great clusters of stars, called galaxies, relate to each other. This meant that his number could not be zero, for technical reasons.

But some years later, it was found that the galaxies were in fact all rushing away from one another. In Einstein's mind, an informal version of the Ockham analysis immediately took place and he reverted to the value zero for his number, which is called the 'cosmological constant' today. In this spirit, a translation of the Latin Ockham's Razor, 'entia non sunt multiplicanda praeter necessitatem' would be 'Parameters should not proliferate unnecessarily.' This particular plot has thickened though: the value of Einstein's cosmological constant is once again in question. Is it zero, or is it very small, and should be chosen so as to best fit the data?

We don't know yet. This is why these questions are exciting.

It is a long way from the modest 13th century village of Ockham to modern research laboratories with state–of–the–art technology, proving the secrets of elementary particles and cosmology. Our link is William, and the principle he wrote about which allows us to improve our answers to questions about the universe, according to the data that is coming out of those laboratories. I think he would be pleased.

Anthony Garrett

GARRET, ANTHONY. "OCKHAM'S RAZOR." INTERVIEW WITH ROBYN WILLIAMS ON *AUSTRALIAN BROADCASTING CORPORATION'S RADIO NATIONAL* (APRIL 16, 2000).

SEE ALSO *Science Philosophy and Practice: Postmodernism and the "Science Wars"; Science Philosophy and Practice: Pseudoscience and Popular Misconceptions; Science Philosophy and Practice: The Scientific Method.*

BIBLIOGRAPHY

Books

Baum, Eric B. *What Is Thought?*. Cambridge, MA: MIT Press, 2004.
Panaccio, Claude. *Ockham on Concepts*. Burlington, VT: Ashgate, 2004.
Pols, Edward. *The Recognition of Reason*. Carbondale, IL: Southern Illinois University Press, 1963.
Quine, Willard Van Orman. *From a Logical Point of View: 9 Logico–Philosophical Essays*. Cambridge, MA: Harvard University Press, 1971.
Rowland, Wade. *Ockham's Razor: The Search for Wonder in an Age of Doubt*. Toronto: Patrick Crean, 1999.
Rubenstein, Richard E. *Aristotle's Children: How Christians, Muslims, and Jews Rediscovered Ancient Wisdom and Illuminated the Dark Ages*. Orlando, FL: Harcourt, 2003.
Wassermann, Gerhard D. *From Occam's Razor to the Roots of Consciousness: 20 Essays on Philosophy, Philosophy of Science and Philosophy of Mind*. Brookfield, VT: Avebury, 1997.
Weinberg, Julius R. *A Short History of Medieval Philosophy*. Princeton: Princeton University Press, 1964.

Periodicals

Thorburn, William M. "The Myth of Occam's Razor." *Mind* 27, 3 (July 1918): 345–353.

Other

Garret, Anthony. "Ockham's Razor." Interview with Robyn Williams on *Australian Broadcasting Corporation's Radio National* (April 16, 2000).

Eric v.d. Luft

Science Philosophy and Practice: Postmodernism and the "Science Wars"

■ Introduction

In the second half of the twentieth century, a loosely-defined school of thought termed "postmodernism" gained many adherents in some university departments, especially those devoted to the non-scientific study of culture. Postmodernism is a style of thought that tends to reject the belief that any qualities possessed by human beings are inherent in their nature (an idea that postmodernists call "essentialism"). It also emphasizes that all belief systems, including science, are profoundly shaped by relationships of "power," namely, the power held by some human beings over others.

Some extremist postmodernists have suggested that science is a construct shaped entirely by power relationships, rather than a system of statements about a real physical world that is equally valid everywhere and at all times. In this view's most extreme form, DNA, gravity, and other entities discussed by science are not real, but are arbitrary products of a predominantly white, European, male subculture. Many scientists have reacted angrily to such claims.

Starting in 1996, the debate between scientists and postmodernists received a burst of public attention when American physicist Alan Sokal (1955–) published a lengthy, fake article entitled "Transgressing the Boundaries: Towards a Transformative Hermeneutics of Quantum Gravity" in the postmodernist journal *Social Text*. Sokal wrote deliberate nonsense dressed up in postmodern jargon to prove that it could be published as serious postmodernist scholarship. When the editors found out that they had been fooled by Sokal, they accused him of unethical deception and argued that he did not understand the ideas he was criticizing.

■ Historical Background and Scientific Foundations

Modernism is a style of thought, closely associated with art, literature, and architecture, that arose and flourished in the late nineteenth and early twentieth centuries. Modernists tended to assume that ongoing improvement of the human condition ("progress") was possible through logical thought, applied science, and the construction of rational, unornamented buildings. Following World War II (1939–1945), with its large-scale proof that progress is anything but inevitable—two of the most culturally and scientifically advanced countries on Earth, Germany and Japan, had shown themselves capable of savagery on a global scale, resulting in scores of millions of deaths—many thinkers reacted against the simple storyline assumed by modernism. In the 1950s, this reaction fed the movement known as existentialism. In the 1960s, many young thinkers and radicals in Western countries tended to reject the culture of the Establishment (the way of life and thought associated with the dominant military-political-cultural order), which included mainstream science. Starting around 1970, these tendencies took new form in the intellectual movement called postmodernism.

Postmodernist thinkers were at first concerned mostly with literature, but soon extended their criticism to all forms of culture, including beliefs about morals, gender, and science. Postmodernism's central assertion was that there is no such thing as "truth": all beliefs are constructs and might have been constructed differently.

In 1975, Austrian-American philosopher Paul Feyerabend (1924–1994) published *Against Method*, the first major work to extend the postmodernist critique to scientific thought itself, particularly the scientific method. Some of Feyerabend's claims—such as that working scientists do not, in fact, always operate according to some rigid recipe for producing new knowledge—are now commonplace among philosophers and sociologists of science; other claims, however, seemed to call the validity of science itself into question. For example, in his concluding chapter Feyerabend wrote that science is dominant because scientists have "*the power* to enforce their wishes … just as their ancestors used *their* power to force Christianity on the peoples they encountered during their conquests. Thus, while an American can now choose the religion he likes, he is still not permitted to demand that his children learn magic rather than science at school."

By the early 1990s, a number of postmodern theorists had made pronouncements about science that struck scientists as outrageous. For example, in 1990, Mark and Deborah Madsen wrote that to be postmodern, science must be "free from any dependence on

This "Doonesbury" cartoon strip from January 14, 2007 addresses "situational science"—when one's own beliefs are inconsistent with scientific evidence. *DOONESBURY © 2006 G.B. Trudeau. Reprinted with permission of UNIVERSAL PRESS SYNDICATE. All rights reserved.*

the concept of objective truth" (Madsen and Madsen, 1990). Scientists responded that this sort of thing was nonsense, and the "science wars," as the dispute was quickly dubbed, were on.

In 1996, a prominent postmodernist journal of culture studies, *Social Text*, announced that it would publish a special issue entitled "Science Wars" that would be devoted to the postmodernist view of science. One physicist, Alan Sokal (1955–), convinced that the more extreme postmodern views of science were meaningless, decided to submit a learned-sounding article full of deliberate nonsense. If the journal published the article, Sokal reasoned, it would prove that at least some prominent postmodernists could not tell the difference between sense and nonsense.

The article was accepted by *Social Text* and published. In it, Sokal pretended to argue for the extreme postmodern position that "physical 'reality,' no less than social 'reality,' is at bottom a social and linguistic construct, [and] that scientific 'knowledge,' far from being objective, reflects and encodes the dominant ideologies and power relations of the culture that produced it." Toward the end of the article, Sokal asserted that the content of science must be controlled by leftist or progressive political strategies in order to become "liberatory."

As Sokal himself described his parody article in the issue of *Lingua Franca*, released at the same time as the *Social Text* special issue on the "Science Wars," "[n]owhere in all of this is there anything resembling a logical sequence of thought: one finds only citations of authority, plays on words, strained analogies, and bald assertions."

Although scholars targeted by Sokal's attack were politically left-wing, he emphasized in *Lingua Franca* that he was not attacking left-wing politics as such: quite the opposite. He said that he considered himself a leftist or socialist and was grieved that postmodernists were undermining any possibility of radical social change by undermining the idea of an objective, real world about which it is possible to learn and which it is possible to change. His goal, he said, was "to defend the Left from a trendy segment of itself."

Sokal's act of intellectual guerrilla warfare received huge publicity, being covered by major newspapers in several countries. Dozens of editorials and letters to the editor followed, along with TV and radio appearances by persons with pro and con opinions. Although there is no way to measure the effect of such an event on intellectual trends, it is likely that the influence of the postmodern critique of science was diminished by the Sokal affair.

Defenders of postmodernism and *Social Text* argued that Sokal had acted unethically by violating academic trust and that he did not understand the postmodern critique of science. For example, American scholar of

law and literature Stanley Fish (1938–) replied in an editorial in the *New York Times* (May 21, 1996) that what "sociologists of science" really say is "that of course the world is real and independent of our observations but that accounts of the world are produced by observers and are therefore relative to their capacities, education, training, etc." The co-editors of *Social Text* wrote in a letter to the same paper that "disbelief in the existence of the physical universe" was of course "nonsense," and that what they really advocated was "questioning ... the scientific community's abuses of authority, its priestly organization and lack of accountability to the public" (May 23, 1996).

However, such defenses avoided mention of the many bizarre-sounding attacks on the idea of physical reality that Sokal had quoted from postmodernist literature. Instead, they fell back on defending an uncontroversial idea from mainstream sociology and science studies: the view that all knowledge is "socially constructed," namely, discovered or invented and then learned by human beings, who are always embedded in networks of human relationships (societies). However, Sokal was not attacking that view, which is accepted by most scientists. When mainstream sociologists say that a belief is "socially constructed," they do not mean that the belief is necessarily arbitrary or unreal: for example, we must be taught arithmetic socially (by other people), and the symbolic language we use to work arithmetic might have been constructed any number of different ways, but this does not mean that 2+2 could, in some other society, equal anything but 4.

What Sokal claimed was that postmodernism (not mainstream sociology) had reduced itself to chronic nonsense in its quest for ever-more-radical rhetoric. Various scholars came to Sokal's support with horror stories of scholarly conferences where they had been asked by fellow scholars if they really "believed in DNA" (Ehrenreich and McIntosh, 1997) or of classrooms where they were scolded by students for even allowing discussion of the idea that there are innate biological differences between men and women (Epstein, 1996).

■ Modern Cultural Connections

After 1996 and 1997, when the Science Wars were at their public height, those who favored postmodernism continued to do so, for the most part, as those who disliked postmodernism continued in their view. Few minds were changed, as is normal in such disputes. In the early 2000s a new twist appeared, namely, the appropriation by some creationists of aspects of the postmodern view of science. Creationists are persons who deny that natural processes such as evolution can account for the existence of life, or at least for some complex structures in living things.

In 2005, in the legal case *Tammy Kitzmiller, et al. v. Dover Area School District, et al.*, parents challenged a school-board rule that students in science classes must be read a statement calling evolution into doubt and recommending Intelligent Design (a form of creationism) as an alternative theory. At the trial, philosopher Steven Fuller (1959–) testified on behalf of the defendants from a point of view that some have characterized as postmodern. Fuller argued that Intelligent Design is scientific, not religious, and can therefore be legally taught in U.S. public-school classrooms, stating that scientific theories are socially produced and non-absolute. When asked during cross-examination, "If you contrast the higher-order claims made by evolutionary theorists with the claims made by intelligent design, do you see a comparative or a different situation with respect to testability?", Fuller, who is not himself an Intelligent Design creationist, replied: "Well, frankly ... the theoretical frameworks in which both evolutionary theory and intelligent design operate are largely both metaphysical," that is, not testable against a definite physical reality. Some advocates of Intelligent Design, such as U.S. law professor Philip Johnson (1940–), have explicitly advocated an approach to science based on a "relativist approach to knowledge claims," (PandasThumb.org, 2005) recalling some of the more extreme postmodern views of science.

■ Primary Source Connection

The following speech was delivered by Bertrand Russell, on receiving the Kalinga Prize for the Popularization of Science, at the United Nations Educational, Scientific and Cultural Organization (UNESCO) Headquarters on January 28, 1958. Bertrand Russell (1872–1970), a prolific English philosopher, historian, and author, also won the Nobel Prize for Literature in 1950.

THE DIVORCE BETWEEN SCIENCE AND 'CULTURE'

The modern tendency of treating science and humanities as separate specializations only started in the 19th century. Before that period, such as in the era of the Greek philosophers and the Renaissance, both fields were considered compatible and worth pursuing together by dedicated lovers of knowledge.

There was a time when scientists looked askance at attempts to make their work widely intelligible. But, in the world of the present day, such an attitude is no longer possible. The discoveries of modern science have put into the hands of governments unprecedented powers both for good and for evil. Unless the statesmen who wield these powers have at least an elementary understanding of their nature, it is scarcely likely that they will use them wisely. And, in democratic countries, it is not only statesmen, but the general public, to whom some degree of scientific understanding is necessary.

To insure wide diffusion of such understanding is by no means easy. Those who can act effectively as liaison officers between technical scientists and the public perform a work which is necessary, not only for human welfare, but even for bare survival of the human race. I think that a great deal more ought to be done in this direction in the education of those who do not intend to become scientific specialists. The Kalinga Prize is doing a great public service in encouraging those who attempt this difficult task.

In my own country, and to a lesser degree in other countries of the West, "culture" is viewed mainly, by an unfortunate impoverishment of the Renaissance tradition, as something concerned primarily with literature, history and art. A man is not considered uneducated if he knows nothing of the contributions of Galileo, Descartes and their successors. I am convinced that all higher education should involve a course in the history of science from the seventeenth century to the present day and a survey of modern scientific knowledge in so far as this can be conveyed without technicalities. While such knowledge remains confined to specialists, it is scarcely possible nowadays for nations to conduct their affairs with wisdom.

There are two very different ways of estimating any human achievement: you may estimate it by what you consider its intrinsic excellence; or you may estimate it by its causal efficiency in transforming human life and human institutions. I am not suggesting that one of these ways of estimating is preferable to the other. I am only concerned to point out that they give very different scales of importance. If Homer and Aeschylus had not existed, if Dante and Shakespeare had not written a line, if Bach and Beethoven had been silent, the daily life of most people in the present day would have been much what it is. But if Pythagoras and Galileo and James Watt had not existed, the daily life, not only of Western Europeans and Americans but of Indian, Russian and Chinese peasants, would be profoundly different from what it is. And these profound changes are still only beginning. They must affect the future even more than they have already affected the present.

At present, scientific technique advances like an army of tanks that have lost their drivers, blindly, ruthlessly, without goal or purpose. This is largely because the men who are concerned with human values and with making life worthy to be lived, are still living in imagination in the old pre-industrial world, the world that has been made familiar and comfortable by the literature of Greece and the pre-industrial achievements of the poets and artists and composers whose work we rightly admire.

The separation of science from "culture" is a modern phenomenon. Plato and Aristotle had a profound respect for what was known of science in their day. The Renaissance was as much concerned with the revival of science as with art and literature. Leonardo da Vinci devoted more of his energies to science than to painting. The Renaissance artists developed the geometrical theory of perspective. Throughout the eighteenth century a very great deal was done to diffuse understanding of the work of Newton and his contemporaries. But, from the early nineteenth century onwards, scientific concepts and scientific methods became increasingly abstruse and the attempt to make them generally intelligible came more and more to be regarded as hopeless. The modern theory and practice of nuclear physicists has made evident with dramatic suddenness that complete ignorance of the world of science is no longer compatible with survival.

Bertrand Russell

RUSSELL, BERTRAND. "THE DIVORCE BETWEEN SCIENCE AND 'CULTURE'." *UNESCO COURIER* (FEB 1996): 50 (1).

SEE ALSO *Science Philosophy and Practice: The Scientific Method.*

BIBLIOGRAPHY

Books

Feyerabend, Paul. *Against Method.* New York: Verso, 1993 (orig. 1975).

Lingua Franca, et al., eds. *The Sokal Hoax: The Sham that Shook the Academy.* Lincoln, NE: University of Nebraska Press, 2001.

Sokal, Alan, and Jean Bricmont. *Fashionable Nonsense: Postmodern Intellectuals' Abuse of Science.* New York: Picador, 1998.

Periodicals

Epstein, Barbara. "The Postmodernism Debate." *Z Magazine* (October 1996): 57–59.

Fish, Stanley. "Professor Sokal's Bad Joke." *The New York Times* (May 21, 1996).

Madsen, Mark, and Deborah Madsen. "Structuring Postmodern Science." *Science and Culture* 56 (1990): 467–472.

Russell, Bertrand. "The Divorce between Science and 'Culture'." *UNESCO Courier* 49, no. 2 (Feb 1996): 50.

Scott, Janny. "Postmodern Gravity Deconstructed, Slyly." *The New York Times* (May 18, 1996).

Sokal, Alan. "A Physicist Experiments with Cultural Studies." *Lingua Franca* (May/June 1996): 62–64.

Sokal, Alan. "Transgressing the Boundaries: Toward a Transformative Hermeneutics of Quantum Gravity." *Social Text* 46/47 (1996): 217–252.

Web Sites

Ehrenreich, Barbara, and Janet McIntosh. "The New Creationism: Biology Under Attack." *The Nation.* June 9, 1997. http://cogweb.ucla.edu/Debate/Ehrenreich.html (accessed January 29, 2008).

Matzke, Nick. "ID = Postmodern Creationism." August 4, 2005. http://www.pandasthumb.org/archives/2005/08/id-postmodern-c.html (accessed January 29, 2008).

Sokal, Alan. "Transgressing the Boundaries: Toward a Transformative Hermeneutics of Quantum Gravity." 1996. http://www.physics.nyu.edu/faculty/sokal/transgress_v2/transgress_v2_singlefile.html (accessed January 29, 2008).

Larry Gilman

Science Philosophy and Practice: Professionalization

■ Introduction

A profession is any field of study in which specialized knowledge is acquired after intensive academic instruction is applied with rigorous standards. Such fields include medicine, law, applied science research, teaching, and engineering. The concept of a profession has evolved over centuries, as training has progressed from informal and formal apprenticeships to highly structured academic training programs with meticulous licensing standards. Each discipline, particularly medicine and science, has structured the training and education programs deemed necessary to meet contemporary standards and legitimize their place in the highest ranks of society.

Current standards for physicians include medical training at the undergraduate and/or graduate level, passage of licensure examinations, internships, residencies, and continuing education. Science and engineering professionals must complete specific undergraduate and graduate training, pass licensure examinations in many fields, and experience continuing education as well. These standards vary from country to country and sometimes within a specific country.

WORDS TO KNOW

APPRENTICE: A person serving at low wages in order to learn a skill, trade, or craft from an established practitioner. Apprenticeship was widespread in the European economy during the later Middle Ages.

GUILD: In the economy of medieval Europe, an association of craftsmen (e.g., bakers, stonemasons) that controlled admission to the profession, prices, and other aspects of their trade.

JOURNEYMAN: In the European economy in the Middle Ages, a person who had completed their apprenticeship (learned the essentials of their specific trade) but had not yet achieved the status of a master craftsmen. Journeymen often traveled from job to job or worked briefly for various employers.

LICENSE: An official government permission to carry on a certain activity. While science is not a licensed profession, medicine is. To practice as a doctor in the United States, one must possess a license granted by the state in which one has attended medical school. In most other countries, medical licensure is granted by the central government.

MASTER: In a professional guild, the term "master" was used for those members with the highest skill level and seniority within the guild, above both journeymen and apprentices.

■ Historical Background and Scientific Foundations

Medical or scientific training, until the rise of formal education, took place through apprenticeships and individual experiments. Medical practitioners fell into four basic categories: midwives (exclusively women), physicians (university trained), surgeons, and apothecaries (both apprenticeship trained). Each practitioner provided a different service to the public and was trained by his or her predecessor. In many fields, fathers passed on knowledge to sons. Craftsman organizations date back thousands of years in India, ancient Greece had organizations segregated by skill, while the formal guild system began in Europe during the eleventh and twelfth centuries. Formal trade organizations formed in the Muslim world as early as the ninth century.

Structured guilds in Europe first appeared in Germany in the tenth century. While some, such as those in Florence, Italy, were created to help merchants gain economic advantage, craft guilds emerged as groupings of artisans designed to fix prices, set quality standards, and control the numbers of artisans producing a particular product or offering a service. Guild members frequently swore an oath when joining the organization to band together with fellow artisans with absolute loyalty.

Membership within guilds was divided into three levels: masters, journeymen, and apprentices. Guilds acted as social service organizations for members, ensuring pensions, aiding widows, and providing assistance when sales were low. Prohibiting new members from joining a guild was the primary method of control over a certain field. Apothecaries formed guilds in some areas in Europe in the seventeenth century, while Amsterdam had a surgeon's guild in the sixteenth century, and physician guilds persist into the twenty-first century as economic alliances. Midwife guilds are noted in the Old Testament of the Bible, in Exodus, although such guilds were less commonplace than artisan, physician, surgeon, or apothecary guilds.

Guilds enforced extended apprenticeships for members; because many practitioners passed their skills on to sons, the extended apprenticeship also served to maintain the father's authority and status as shop or service

Illustration shows a doctor and an apothecary working.
© *Bettmann/Corbis.*

master. The transition from apprentice to journeyman could take more than a decade in some fields; modern apprenticeships in trades take between three and five years, while current physician training has new medical school graduates undergoing an internship program of one year before moving on to residency, which can last several years more, depending on specialization.

Journeymen in guilds did not own their shops or practices, but were employees or contract workers. Beyond journeymen, masters owned their own shop—an apothecary might learn as an apprentice in his father's shop, rise to journeyman to work for other shops in the region, and then settle back to be the master of his father's store after his father's retirement or, more likely, death. The Royal College of Surgeons of Edinburgh was founded in 1505; the college began as a craft guild and accepted for membership only those persons who had apprenticed for more than six years with a master surgeon and who passed an extensive examination.

The rise of universities as training centers in the sciences and medicine developed during the Renaissance and through the Enlightenment. Medical training centers in the Middle East had been in place from the tenth century through the thirteenth and fourteenth centuries. In Europe, universities offered medical and scientific training programs that were increasingly formalized by the 1700s; the first medical college in the United States, the Academy and College of Philadelphia, opened in 1765, while the London Hospital Medical College in England opened in 1785.

Professional societies and universities developed contemporaneously; guilds in medicine and science were slowly replaced with associations and societies such as the Royal Medical Society in England, established in 1734, the Conseil de l'Ordre des Médecins, established in 1845 in France, the American Medical Association, established in 1847, and later, the German Bundesärztekammer in 1947. Less concerned with economic protectionism, these new professional societies offered guidelines and regulations for training in the respective fields, professional oversight of members to maintain rigorous standards, licensure and examination protocols, and censure for those members who did not uphold the ethics code determined for the profession by the society.

Such associations also served to limit the number of practitioners in the respective fields, to reduce competition and control price structures for services. These associations often worked in concert with university training programs, and the give and take between universities and professional associations persists into the twenty-first century. In the engineering profession, associations such as Skule, established in Canada in 1885, and Verein Deutscher Ingenieure in Germany, founded in 1856, marked divisions between different types of engineers, created educational frameworks to qualify for

Marie Sklodowska Curie (right) pictured with her husband, Pierre Curie, in their laboratory in France. Marie Curie was both a physicist and chemist and broke many barriers in the field of science, becoming the first female professor at the University of Paris and the first woman to win the Nobel Prize (which she eventually won twice, in two different fields). *AP Images.*

examinations, and structured examinations for each subfield within engineering, controlling the use of the various engineering titles (i.e., Professional or Chartered Engineer), and working with universities and engineering firms to establish internships and training guidelines.

Examinations were developed both by universities and professional societies to standardize the profession, select applicants for university training, and limit those who wished to be fully licensed and practice their skills. Licensure and certification examinations have taken many forms; in the United States, the Medical College Aptitude Test (MCAT) was first offered in 1928, largely to improve attrition rates in medical colleges, which ranged from 5% to as high as 50%. Within two decades the MCAT helped improve the pool of medical school candidates, and attrition rates settled to a steady 7%. The MCAT persists today as an examination used for qualification to enter medical school.

Both entrance examinations to begin study and examinations for licensure were shaped and driven by professional associations and university systems. While the Graduate Record Examination (GRE) is used for entrance into arts and sciences programs in the United States, and the MCAT is used for medical school, many licensure exams serve as exit exams, to ensure a certain

level of ability and knowledge in professionals in fields such as nursing (NCLEX-RN), engineering (NCEES), and medicine (state boards). The ability to work as a professional in these fields depends on passage of such licensure examinations; failure to pass means exclusion from the field.

■ Modern Cultural Connections

The status of women in the professions of science and medicine has been a source of contention since the development of guilds, associations, and courses of higher education. Women were routinely excluded from medical and scientific guilds, with the exception of midwifery; nonmedical guilds, such as textile guilds, occasionally permitted women to enroll as members, and women could at times inherit membership from their fathers or husbands, but such exceptions were rare. In general women were excluded in the belief that they would neglect their domestic roles if they became guild members.

In science and medicine, women began to make inroads by the mid-1800s. Elizabeth Blackwell (1821–1910) graduated in 1849 from medical school in Geneva, New York, the first woman ever to do so in the

United States. Marie Curie (1867–1934), a Polish immigrant to France, eventually became the first female professor at the Sorbonne and the only two-time Nobel sciences winner. As Donna M. Hughes notes in her article "Women, Science, and the Women's Movement" in *Sisterhood is Forever: The Women's Anthology for the New Millennium*, "Women were active in science and engineering in the late 1800s and early 1900s: their numbers rose with the suffrage wave of the Women's Movement ... Women's participation rose again during World War II, when men were called to military service—but after the men returned, women were demoted or dismissed."

Women's participation is lowest in the field of engineering, the highest in biological science; women's participation in graduate-level medicine and science was greatest in the 1880s to 1920s, and did not resume similar levels in the United States until the 1990s. Outright gender discrimination in college entrance policies, restrictions on research assistantships and internships, and a post-World War II (1939–1945) resurgence in identifying women with domestic roles contributed to the retraction of women in graduate-level science programs and careers.

Many female professional organizations arose to counteract these limitations; the American Association of University Women encompasses all female university professionals, while the Society of Women Engineers provides professional direction and networking for female engineers.

Professional science and engineering has been affected strongly by military and corporate research interests over the past 50 to 60 years. The defense industry, pharmaceutical companies, and biomedical engineering corporations work in tandem with many major research universities, offering substantial grant dollars and research funding for both applied and pure research. Professional associations have struggled with these outside interests, as the funders often place demands on the research methods, evaluation of outcomes, and funding allocation—all issues that provoke critics to charge that such interference reduces autonomy. As federal research dollars decrease, more university professors and professional scientists turn to outside sources, leaving professional organizations to find balance in regulating their profession in the face of possible conflicts of interest.

SEE ALSO *Science Philosophy and Practice: Research Funding and the Grant System; Science Philosophy and Practice: Science Communications and Peer Review; Science Philosophy and Practice: Scientific Academies, Institutes, Museums, and Societies.*

BIBLIOGRAPHY

Books

Bynum, William. *Science and the Practice of Medicine in the Nineteenth Century.* Cambridge: Cambridge University Press, 1994.

Conrad, Lawrence I., et al. *The Western Medical Tradition.* Cambridge: Cambridge University Press, 2006.

Duffy, John. *From Humors to Medical Science: A History of American Medicine.* Champlain, IL: University of Illinois Press, 1993.

Grant, Edward. *The Foundations of Modern Science in the Middle Ages.* New York: Cambridge University Press, 1996.

Kimball, Bruce A. *The "True Professional Ideal" in America: A History.* Cambridge, MA: Blackwell Publishers, 1992.

Lane, Joan. *A Social History of Medicine: Health, Healing, and Disease in England, 1750–1950.* New York: Routledge, 2001.

Morgan, Robin. *Sisterhood Is Forever: The Women's Anthology for the New Millennium.* New York: Washington Square Press, 2003.

Porter, Roy. *Disease, Medicine, and Society in England, 1550–1860.* Cambridge: Cambridge University Press, 1995.

Starr, Paul. *The Social Transformation of American Medicine.* New York: Basic Books, 1982.

Periodicals

Stevens, Rosemary A., Ph.D., M.P.H. "Themes in the History of Medical Professionalism." *Mount Sinai Journal of Medicine* 69 (2002): 357–362.

Waddington, Ivan. "Professionlisation." *British Medical Journal* 301 (1990): 688–690.

Melanie Barton Zoltán

Science Philosophy and Practice: Pseudoscience and Popular Misconceptions

■ Introduction

The prefix *pseudo-* means fake. Pseudoscience is fake science. Many people hold pseudoscientific beliefs. Some of the better-known of these include extrasensory perception (ESP), unidentified flying objects (UFOs), the planetary theories of American psychologist Immanuel Velikovsky (1895–1979), pyramid power, alien abductions, the Loch Ness Monster, Bigfoot, and creation science. Popular misconceptions about science, as distinct from pseudoscience, are commonplace but mistaken ideas about the physical world or about how science works. Such beliefs may be encouraged by pseudoscience or may arise from inadequate science education.

■ Historical Background and Scientific Foundations

Most people in modern society assume that science is the most reliable source of knowledge about the physical world. By claiming to be scientific or using scientific-sounding words, almost any claim can gain some believers. This is true even for claims based on incomplete information, flawed experiments, hallucinations, or private fantasy. People who lack critical-thinking skills or science education are often taken in by pseudoscience.

Today, science is a growing, interlocked system of strictly defined facts and explanations tested against observations. This form of science only came into being after the Scientific Revolution of the sixteenth and seventeenth centuries. When speaking of beliefs held before or even during that period, it is not historically meaningful to speak of "pseudoscience," because no distinction was yet possible. Until the eighteenth century, even leading scientists often held beliefs that today would be considered pseudoscientific.

For example, German astronomer and mathematician Johannes Kepler (1571–1630) was a founder of modern science who discovered the first mathematical laws of planetary motion. Yet Kepler believed that the orbits of the planets could be explained by a mystical relationship between the five Platonic solids (the cube, tetrahedron, and three other polyhedrons having identical faces). Kepler was also a professional astrologer (a person who predicts human affairs from the positions of the moon, sun, planets, and stars). Today, both astrology and the mystical significance of the Platonic solids are considered pseudoscience.

Another example is English physicist Sir Isaac Newton (1642–1727). Newton established the elements of modern mechanics, the science of objects in motion. Yet he spent at least as much time on alchemy, a mystical system devoted to transforming physical elements and the inward, spiritual self, using a combination of laboratory methods and magic spells.

Only with the development of strict scientific standards of experiment, verification, and publication over generations did a clear separation appear between science and pseudoscience. The word "pseudoscience" itself was not coined until 1844.

Some philosophers, such as Paul Feyerabend (1924–1994), have argued that it is not possible to define "science" clearly enough to distinguish scientific from pseudoscientific claims. However, most working scientists would agree that all real science has certain features:

1. *Falsifiability.* To be scientific, a claim must be capable of being disproved or falsified. For example, the statement that there is an elf in the room who cannot be seen, heard, smelt, felt, or observed by instruments cannot be proved false; such a statement is therefore not scientific. The claim that gravity bends light waves, on the other hand, can be tested by experiment. If it is false we

WORDS TO KNOW

ACUPUNCTURE: The Chinese practice of treating disease or pain by inserting very thin needles into specific sites in the body.

ALCHEMY: The study of the reactions of chemicals in pre-modern times. It was often, but not always, directed by the goal of making gold. In a general sense, alchemy is perceived as the transmutation (or, transformation) of a common substance to something rare and valuable. Medieval alchemists are often portrayed as little more than quacks attempting to make gold from lead. This depiction is not entirely correct. To be sure, there were such characters, but for real alchemists, called adepts, the field was an almost divine mixture of science, mystery, and philosophy.

ASTROLOGY: The practice of studying the apparent motions of the planets, moon, sun, and stars in order to draw conclusions about human character (supposedly affected by sky patterns at the time of one's birth) or about the future of a person's affairs. Astrology has been practiced since at least about 3000 BC; European, Indian, and Chinese astrological systems have all been developed. Although astrology is still widely popular, it is not a form of science.

CREATION SCIENCE: A form of creationism, the belief that life, or at least some life or features of life, were created miraculously rather than by natural evolutionary processes. Creation science was developed after the U.S. Supreme Court ruled in *Epperson v. Arkansas* (1968) that all U.S. state laws banning the teaching of evolutionary biology were unconstitutional. In response, creationists developed creation science, a pseudoscientific set of claims asserting that the Genesis story of creation, taken literally, is supported by scientific evidence (it is not). In 1987, the Supreme Court ruled in *Edwards v. Aguillard* that creation science is religious doctrine, not science, and cannot be taught in U.S. public schools because the First Amendment to the U.S. Constitution forbids the furtherance by government of any religion.

CULTURAL REVOLUTION: A period of government-mandated social upheaval in China from 1966 to 1976. Its proclaimed purpose was to solidify the Communist Revolution by purging it of old habits, old culture, old ideas, and old customs. Religion was persecuted; science and education were severely set back; and several million deaths were caused,

both by direct killings and disruption of agriculture and the economy.

FALSIFIABILITY: A claim is falsifiable if there is, at least in principle, some way of proving that the claim is false. For example, the claim that gravity obeys an inverse-square law is falsifiable (has falsifiability): One can conduct tests and see what kind of law gravitation obeys and whether it is in fact an inverse-square law. However, the claim that all life is a dream, or that God created the universe five minutes ago complete with all our memories of a deeper past, is not falsifiable: All possible evidence could be part of the dream or could have been created along with everything else. Most scientists agree that although falsifiability is not enough to make a claim scientific, to be scientific a claim must be falsifiable.

HOMEOPATHY: An alternative medical system invented by German doctor Christian Hahnemann in the early 1800s. In homeopathy, diseases are treated using extremely small doses of substances believed to cause symptoms resembling the disease itself. Some remedies marketed as homeopathic are actually herbal remedies (e.g., arnica gel for its anti-inflammatory properties) and contain significant quantities of their active ingredients; these remedies are not, strictly speaking, homeopathic. Homeopathy is pseudoscientific (i.e., appears scientific but is not): Controlled studies have shown that homeopathy is no better than the placebo effect, which is the tendency of patients to feel better if they believe they are receiving an effective treatment, regardless of the presence of any actual treatment.

LAMARCKISM: The belief that acquired characteristics can be inherited, that is, that changes to an organism that happen during its life can be passed on to offspring.

PLATONIC SOLIDS: The convex geometric shapes that have for their sides or facets identical polygons with sides of equal lengths. Polygons are flat shapes whose sides consist of straight line segments: Familiar polygons having sides with equal length are the square, pentagon, and equilateral triangle. Only five solid shapes having such shapes as sides or facets are possible, namely the five Platonic solids. The two most familiar Platonic solids are the cube and the tetrahedron (four-sided pyramid).

can prove that it is false. It is therefore a scientific claim. (It turns out to be true.) Yet falsifiability is not the only feature of a scientific claim; simply making up falsifiable statements is not a form of science. Some pseudoscientific claims can be disproven, in whole or part, but this alone does not mean they are scientific claims—even if the people making them happen to have academic degrees in

science. For example, if a person claims that they can fly by flapping their arms, their statement can be falsified but it is not science.

2. *Clear relationship to existing knowledge.* To be scientific, a statement must employ terms from existing scientific knowledge. For example, statements about "energy" that do not employ any

This anonymous seventeenth century painting depicts Michel Nostradamus (1503–1556), French physician and astrologer. Nostradamus established a reputation as a doctor by producing remarkable cures for victims of the plague in the south of France. In 1555 he published *Centuries*, a book of prophecies that gained him favor with the French court and continues to cause fascination to this day. *HIP/Art Resource, NY.*

scientific or testable definition of "energy" are pseudoscientific.

3. *Repeatability of observations.* Scientific claims must be based on observations that can be made by any properly equipped observer. It is a popular misconception that science can only study events that can be directly observed. However, scientists can often study events that are hidden from direct view or that happened in the past. They do so by reasoning from information that they gather in the present, such as fossils, DNA samples, rocks, or astronomical observations. What is essential for science is that these data can be checked by other scientists.

It should be noted that not all nonscientific beliefs are pseudoscientific. Beliefs about right and wrong, meaning, religion, and beauty, for example, are not generally considered pseudoscientific. Only when claims are made about the nature of the physical world without sound scientific basis does pseudoscience appear. For example, a person who claims that spirits make their car

go is making a pseudoscientific claim; science cannot consider supernatural explanations because there is no way to observe supernatural forces, if any such exist. We can, however, verify that the person's car will not run without gas in the tank.

A few beliefs that were once dismissed as pseudoscience have since been accepted by science. For example, in the 1700s and early 1800s, official scientific bodies such as the French Academy of Science maintained that reports of hot stones falling from the sky were a popular delusion. In 1808, U.S. President Thomas Jefferson (1743–1826) dismissed a report that two professors had verified a meteorite, writing that he "would rather believe that two Yankee professors would lie than believe that stones fall from heaven." However, meteorites are now known to be real (they are rocks that fall to earth from space). More recently, the theory of continental drift, first proposed in 1908, was not accepted by most geologists until the 1960s.

Acupuncture was long considered pseudoscience by conventional medical authorities in the West. However, numerous scientific studies over the last few decades have shown that acupuncture is partly effective for some medical conditions. In 1997, a panel of experts convened by the U.S. National Institutes of Health (NIH) to evaluate acupuncture stated that "there is sufficient evidence of acupuncture's value to expand its use into conventional medicine and to encourage further studies of its physiology and clinical value." There is a large body of scientific evidence that acupuncture does have some effectiveness for treating postoperative pain, chemotherapy nausea, and lower back pain; according to the NIH, "the data in support of acupuncture [for some conditions] are as strong as those for many accepted Western medical therapies." However, acupuncture has not been shown to be more effective than placebo in treating arthritis, depression, asthma, cancer, and nicotine addiction. While it may seem effective, that effectiveness is either psychosomatic or based upon different physiological principles than supposed by pre-scientific traditions. Belief that acupuncture can treat any medical problem at all is pseudoscientific.

Some claims that were once widely accepted as scientific have since been relegated to the status of pseudoscience. For example, claims that men are more intelligent than women and that darker-skinned groups are more childlike and more closely related to apes (hence less intelligent) than light-skinned groups were widely considered scientific throughout the nineteenth century and early twentieth century. These beliefs are now seen as pseudoscientific.

Science vs. Pseudoscience

A selection of common pseudoscience ideas follows, accompanied by brief explanations of why they are not scientific.

Velikovsky's planetary mechanics. In 1950, Immanuel Velikovsky published a book, *Worlds in Collision*, to propose a number of ideas about planetary mechanics. For example, he proposed that the planet Venus did not form out of the primordial disk of dust and gas orbiting the sun, but was ejected from Jupiter a few thousand years ago by an unspecified mechanism and passed through the inner solar system several times before settling into its present-day orbit. He claimed that the craters on the moon were produced not by meteorite impacts but by giant electrical sparks. Velikovsky's theories violate many laws of physics and have not contributed to modern planetary science. They have been refuted in detail by Carl Sagan (1934–1996) and many other scientists. However, his book has long been a best seller and remains in print.

UFOs. A 2001 Gallup poll of American adults found that 30% affirm the statement that "some of the unidentified flying objects that have been reported are really space vehicles from other civilizations," while 33% affirmed that "extraterrestrial beings have visited the earth at some time in the past." However, there is no scientific evidence for either of these statements. No unambiguous, clear video footage of an alien spacecraft has ever appeared, despite the proliferation in recent years of hundreds of millions of video cameras around the world; no space probe, satellite, or manned spacecraft has encountered any sign of nonhuman races in space; no ancient structure, such as the Great Pyramids of Egypt, contains alien technology that could not have been built by ancient human engineers, despite many assertions to the contrary in UFO literature. There is no scientific reason to believe that UFOs are real.

ESP. Belief in ESP (extrasensory perception) is also common; according to a 2001 National Science Foundation survey, 60% of U.S. adults believe that "some people possess psychic powers or ESP." Most people have personal experiences that seem to them to have an inescapably psychic or extrasensory dimension. However, over half a century of efforts to clearly identify psychic powers have failed. There has been no scientific proof that any person has been to be able to predict the future, view remote events, read minds, or move objects telekinetically.

Energy healing. In recent years, a wide range of alternative healing practices going by names like "energy work," "subtle energy medicine," and "vibrational healing" have arisen. These practices typically employ pseudoscientific language to promote themselves: for example, one practitioner promises to "treat people by using pure energy" utilizing "specialized forms of energy to positively affect the energetic systems that may be out of balance. This energy acts as a type of wave-guide to redirect or realign the subtle energies [of the body] that may be affected." Despite the use of technological-sounding words like *energy, energetic system, waveguide, redirect,*

> ## IN CONTEXT: LYSENKOISM
>
> In 1931, the government of the Soviet Union decreed that the state-run agencies devoted to agriculture and plant breeding must shorten the time for developing new cereal crop varieties for different climates from 10 years to 4 years. Trofim Lysenko (1898–1976) boasted that he could do it in only two and a half years using methods that defied mainstream scientific wisdom.
>
> Lysenko believed in a form of Lamarckism, an idea in nineteenth-century biology named after its most famous proponent, Jean-Baptiste Lamarck (1744–1829). According to Lamarckism, organisms can inherit traits imprinted on their ancestors by experiences: For example, giraffe's necks might get longer generation after generation as a result of stretching for leaves. (Natural selection explains that giraffes survived because they were long-necked and could easily reach an available food source while shorter-necked animals trying to reach the same food source would die out.) Charles Darwin himself, originator of the modern theory of evolution, believed that Lamarckism might contribute to evolutionary change. But long before 1931, Lamarckism had been discredited and become pseudoscience. Lysenko not only supported Lamarckism but denied modern Mendelian genetics, the standard explanation of how characteristics are passed from one generation to another by genes.
>
> Joseph Stalin (1879–1953), dictator of the Soviet Union, was pleased by Lysenko's extravagant promise. He publicly honored Lysenko and had him appointed director of the Institute of Genetics. Many geneticists (scientists who study heredity) who opposed Lysenko's pseudoscientific ideas were executed or sent to labor camps, where many died.
>
> By the 1970s, Lysenkoism had been replaced by mainstream biology, but only after great harm had been done to Soviet agriculture. Mao Zedong (1893–1976), dictator of Communist China from 1949 to 1976, was also a believer in Lysenkoism. Mao's nonscientific agricultural policies contributed to massive famines in China during the Cultural Revolution that killed an estimated 30 million people.

and *realign,* such methods use no measurable, physical form of "energy" and are pseudoscientific. People often feel better after receiving such treatments, but this is not scientific evidence that "energy" is being manipulated. Rather, they reflect the fact that people who believe that they have received any effective treatment tend to feel better even if they have received a placebo (a useless treatment), such as water pills. Other pseudoscientific healing techniques include homeopathy and Reike.

Creation Science and Intelligent Design

Creation science and intelligent design theory are pseudoscientific ideas that are particularly prominent today.

According to the scientific history of life on Earth, which is accepted as factual by approximately 99.9% of scientists who work in geology and the life sciences, the Earth is 4.5 billion years old. Life arose several billion years ago and has evolved since then into a wide variety of forms. All living things, including humans, are related to each other through common ancestors, just as cousins descend from common (shared) grandparents.

However, some people prefer a literal (exactly-what-it-sounds-like) interpretation of the biblical book of Genesis. These persons believe that Earth is about 10,000 years old rather than 4.5 billion years old, and that the ancestors of living species did not evolve but were brought suddenly into being by miracles—created. Hence, this belief is called creationism. According to a 1997 Gallup poll, about 44% of the U.S. public affirms the statement that "God created man pretty much in his present form at one time within the last 10,000 years."

In the late 1960s, after the U.S. Supreme Court decided in the case *Epperson v. Arkansas* that states cannot ban the teaching of evolution in high-school science classes, creationists developed a new form of creationism called "creation science." Creation science claims that the literal Genesis account, including a 10,000-year-old Earth and a Noachian flood, is supported by scientific evidence. The scientific community rejects this claim. In 1987 the U.S. Supreme Court ruled in *Edwards v. Aguillard* that creation science is religious, not scientific, and so cannot be taught in high school science classrooms.

In response to the Court's decision, a new form of creationism called "intelligent design" was developed, becoming popular in the 1990s. Intelligent design, which usually strips all references to religion from its arguments, argues that some aspects of living things are too complex to have evolved and therefore prove the existence of an intelligent designer. Advocates sometimes deny that this designer is necessarily God. Some intelligent design advocates are willing to admit the shared ancestry of all living things and the great age of Earth, but all insist that intervention by an intelligent designer must have occurred at some point in the history of life.

Many major scientific organizations, including the National Academy of Sciences and the American Association for the Advancement of Science, have issued statements warning that intelligent design is not science at all. In 1999 the United States National Academy of Sciences stated: "Creationism, intelligent design, and other claims of supernatural intervention in the origin of life or of species are not science because they are not testable by the methods of science."

Proponents of intelligent design accuse mainstream scientists of defending old ideas—in this case, evolution—for unscientific reasons. They point to the slow acceptance of continental drift and other cases where ideas originally considered pseudoscientific were eventually accepted as science and say that their theory is in the same position. What distinguishes authentic maverick or rebel science from pseudoscience, however, is that authentic science eventually produces fruitful research that can stand up to detailed review by independent experts. This is how continental drift theory became mainstream science. In over 15 years of existence, by contrast, intelligent design theory has not produced any original research. In 2005, in the case *Kitzmiller v. Dover Area School District,* a federal judge ruled that intelligent design's claims to be scientific (instead of religious) are a "sham," not sincere. Intelligent design, like creation science, is pseudoscience.

There is no authentic scientific debate about intelligent design or creation science. However, political controversy on the subject remains intense, especially in the United States. Both creationists and some antireligious philosophers such as Daniel Dennett (1942–) and Richard Dawkins (1941–) proclaim that religion and evolution are natural enemies, and that one must destroy or drive out the other. However, many large religious organizations—for example, the Catholic Church, the Episcopal Church, and non-Orthodox Jewish denominations—officially deny that there is conflict between their beliefs and evolutionary biology.

Characteristics of Pseudoscience

Pseudoscientific claims, though they vary widely in nature, tend to share typical features:

1. Supporters rely on weak evidence. Blurry photographs, uncheckable stories of personal experiences, poorly designed experiments, anecdotes, and faked artifacts are often used.

2. A pseudoscientific theory is often constructed so as to avoid falsification.

3. Alternative explanations such as fraud, coincidence, placebo effect, and self-deception are dismissed without serious consideration. Positive evidence is taken as proof; negative evidence is explained away.

4. Supporting evidence, if any, is drawn from out-of-date scientific sources; newer sources are ignored or downplayed.

5. Pseudoscientific research programs do not progress. This has been the case with Velikovsky's theories, ESP, UFOs, and intelligent design. No new knowledge is produced.

Popular Science Misconceptions

Popular science misconceptions are mistaken ideas about the physical world. Some appear in science textbooks

A private test of the "Keely Motor." This perpetual motion machine attracted considerable attention in the late 1870s. The wizard of "Etheric Force," John W. Keely professed to utilize a "free energy" on the perpetual motion principle. Investigation of his device after his death indicated that it was actuated by compressed air secretly conveyed in small tubes. © *Bettmann/Corbis.*

intended for elementary and high school students. The U.S. National Research Council identifies five forms of science misconception:

1. *preconceived notions,* for example, the idea that because water flows in rivers above ground, it must flow in rivers underground too;

2. *nonscientific beliefs from religious or mythical sources,* for example, the idea that the earth is only about 10,000 years old;

3. *conceptual misunderstandings* arising from inadequate science teaching, for example, persistence in the belief that objects slow down when not pushed by a continuous force;

4. *vernacular misconceptions* arising from confusion between everyday words and sound-alike words in science, such as "work" or "energy"; and

5. *factual misconceptions,* that is, false beliefs about particular matters of fact, such as the belief that the Gulf Stream is caused by the Mississippi River.

The following are a few popular misconceptions of the factual-misconception type:

There is no gravity in outer space. Fact: gravity is everywhere. We feel its force only when we resist it, such

IN CONTEXT: FALLEN STARS

Astrology is a method of calculating the supposed influence on human affairs of stars and planets. In its most common modern form, an astrologer uses a specialized sky diagram called a horoscope to calculate a natal chart for an individual person. The natal chart supposedly indicates that person's temperament and behavior and issues they are likely to encounter in life. A 2001 survey by the U.S. National Science Foundation found that 41% of the American public thinks that astrology is either "very scientific" or "sort of scientific."

Since astrology makes predictions about things that can be observed, namely personality and behavior, it can be partly falsified. The results of a particularly thorough test were published in the science journal *Nature* in 1985. Scientists performing the test worked with leading astrologers to design the study. Volunteers filled out a survey called the California Personality Inventory (CPI), a standard measure of personality traits. Astrologers created natal charts for the volunteers based on their birth dates and then attempted to match anonymous CPI profiles to their astrological predictions.

The performance of the astrologers was, in the language of the scientific report, "consistent with chance." That is, pure guessing would probably have done as well.

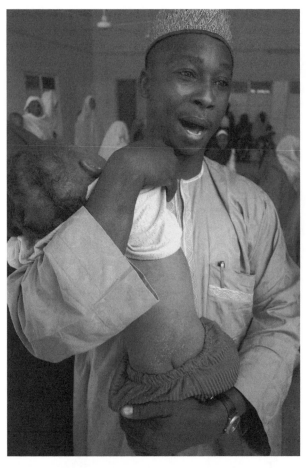

At a clinic in Nigeria in 2005, a man is shown carrying his one-year-old child, who is suffering from a skin condition caused by measles. Hundreds of children died from an upsurge in measles cases, despite a series of local vaccination campaigns aimed at combating the disease. Accusations by some Islamic preachers that the vaccines were part of an anti-Muslim American plot threatened efforts to combat a measles epidemic that killed hundreds of children, according to local health workers and parents in largely Muslim northern Nigeria.   *AP Images/George Osodi.*

as by standing on a solid surface. When we fall freely, such as when we jump off a diving board, we do not feel gravity. In space, objects are weightless because they are (in general) falling freely.

Oceans and lakes are blue because they reflect the sky. Fact: Water itself is blue. Small amounts of water only look clear because our eyes cannot detect their slight blueness. Large amounts of water are blue because they absorb enough of the red light passing through them so that the light which gets through looks blue to us. Impurities and reflections may add other colors to water, such as brown, gray, and green.

Dinosaurs are extinct because they grew too big for their brains. Fact: Dinosaurs were capable survivors that lasted for over 160 million years, about a hundred times longer than humans have been around. Not all dinosaurs were big or had unusually tiny brains. Dinosaurs

became extinct for a number of reasons—including the impact of a giant asteroid about 65 million years ago whose dust darkened the sky and killed off most plants—but stupidity was not one of them. Extinction is normal: 99.9% of all species that have ever lived are now extinct.

Evolution means that only the strong survive. Fact: Evolution has little to do with being stronger or winning fights. Any creature that has descendants is an evolutionary success, and the world is full of weak, slow, or small creatures that are producing lots of descendants.

Lightning never strikes twice in the same place. Fact: Lightning often strikes repeatedly in the same place. A metal tower at the top of a skyscraper may be struck dozens of times in a single thunderstorm.

Impact on Science

Pseudoscience can distract scientists from real work because scientists are often called upon to write books or attend public debates to answer pseudoscientific claims. Sufficiently sophisticated pseudoscience can even, occasionally, get published as science, causing confusion and wasting research time, money, and effort sorting out false leads. One such case was the polywater affair. In the 1960s, Russian scientists claimed to have observed a new form of liquid water that did not freeze or boil like ordinary water. Polywater was hailed by some top scientists as a major discovery and a number of research projects were undertaken to study its supposed properties. Finally, however, polywater was shown to be only impure water, not a new form of water.

Another way in which pseudoscience can harm science is by confusing public opinion. Since much scientific research relies on public funding, political opposition to authentic science, as in the case of evolution, has the potential to reduce the money available for good science.

■ Modern Cultural Connections

Scientific misconceptions and pseudoscientific beliefs range from harmless to deadly. At the harmless extreme, a person may choose to believe in ESP, UFOs, Velikovsky, creation science, or other pseudoscientific ideas while remaining emotionally balanced and able to participate in work, family, politics, and other aspects of normal life. In short, pseudoscience is not a form of mental illness. At the other extreme, people may forego lifesaving medical care or even commit suicide under the influence of pseudoscientific beliefs. For example, in 1997, 39 members of a UFO cult in California called Heaven's Gate committed group suicide in the belief that their minds would migrate from their bodies to a "higher evolutionary level" in association with the arrival of Comet Hale-Bopp.

Another example of dangerous pseudoscience has been popularized by the number-one bestselling book

Some people with terminal diseases seek out alternative forms of medical help. Here, an HIV-positive man consults a traditional healer in Matibi, Zimbabwe. The man is inhaling herbal smoke to invoke his ancestral spirits to help cure him. © *Gideon Mendel/Corbis.*

and movie *The Secret* (2006), by Rhonda Byrne, which proclaims that there is "universal law of attraction" by which people get what they desire. Note the similarity of the phrase "universal law of attraction" to "universal law of gravitation," which is real science language. This form of pseudoscience borrows terminology from mechanics, particle physics, neurology, and other science fields to bolster its claim that people create their own realities. Since the law of attraction is supposed to work for negative results as well as positive ones, advocates say that people with cancer "attract cancer into their lives" or "cause their own cancer" and have advised that positive thinking can be substituted for medical treatments. However, while mental attitudes can interact with many health conditions, negative thoughts do not cause cancer and positive thoughts do not cure it. Cancer is caused by a variety of changes to DNA (hereditary molecules) in some of the cells in a patient's body. It is treated by trying to kill the altered, cancerous cells while sparing normal cells.

Pseudoscientific beliefs have the sad effect of isolating people from the useful and beautiful system of modern scientific knowledge. Science links all phenomena into a single explanatory web of great subtlety. It continually tests its explanations for consistency with the real world and remains open to change based on real-world facts. It is the basis of all medicine and technology. Persons who are misinformed about science not only inhabit a shrunken, deformed intellectual world, but are less able to rise to science-related challenges facing the world such as global climate change and the evolution of new diseases.

SEE ALSO *Science Philosophy and Practice: Ockham's Razor; Science Philosophy and Practice: Postmodernism and the "Science Wars"; Science Philosophy and Practice: The Scientific Method.*

BIBLIOGRAPHY

Bauer, Henry H. *Science or Pseudoscience: Magnetic Healing, Psychic Phenomena, and Other Heterodoxies.* Chicago: University of Illinois Press, 2001.

Friedlander, Michael W. *At the Fringes of Science.* Boulder, CO: Westview Press, 1998.

Park, Robert L. *Voodoo Science: The Road from Foolishness to Fraud.* New York: Oxford University Press, 2000.

Shermer, Michael. *Why People Believe Weird Things.* New York: Henry Holt and Company, 2002.

Periodicals

Makgoba, M.W. "HIV/AIDS: The Peril of Pseudoscience." *Nature* 288 (2000): 1171.

Web Sites

National Science Foundation, National Science Board: Science and Engineering Indicators. "Science and Technology: Public Attitudes and Public Understanding—Science Fiction and Pseudoscience." April 2002. http://www.nsf.gov/statistics/seind02/c7/c7s5.htm#c7s5l2 (accessed May 3, 2008).

Larry Gilman

Science Philosophy and Practice: Research Funding and the Grant System

■ Introduction

Most modern scientific research is expensive: the days when one could discover fundamental laws of physics using home-made tools, as Italian physicist Galileo Galilei (1564–1642) did at the dawn of the Scientific Revolution, are gone. Money for research comes primarily from three sources: private foundations such as the Rockefeller Foundation or Gates Foundation, private corporations such as IBM, and government agencies such as the National Institutes of Health (NIH), the Pentagon, and the National Science Foundation (NSF), the last three being funding agencies of the U.S. government.

A sum of money given to fund specific research is called a grant. Scientists wanting grants must apply for them, explaining in writing, in detail, why the proposed research is important and convincing potential funders that they are qualified and equipped to do it. The num-ber of scientists seeking funding in any given field is always greater than the number that can be funded, so grant-seeking is competitive: This means that grant-givers must decide which research to fund and which research not to fund. The need to persuade funders to give money shapes the growth of modern scientific knowledge in ways that are controversial. Some persons argue that the system works well, others that some aspects and types of knowledge are neglected because funding institutions care less about them.

Today, science research, especially basic science research, is heavily dependent on grants from public and private institutions. By 2000, U.S. universities and colleges were receiving almost $20 billion annually for research programs. Most of this money was invested in science research. Grant-derived funding paid a range of costs associated with research, from test tubes to the salaries of technicians and professional investigators.

WORDS TO KNOW

GRANT PROPOSAL: A formal request for a sum of money (a grant) from a private foundation or a government agency. The authors must describe their qualifications, the question their research is designed to address and how it will address it, what results they think will probably be obtained, and what the importance of their proposed work will be. Usually more grant proposals are received than can be funded, so the process of obtaining grants is competitive.

TRIAGE: The process of allocating limited medical resources in a situation of overwhelming need; life-and-death cases likely to benefit from treatment are the first to be treated.

■ Historical Background and Scientific Foundations

Until recently, most scientific research was conducted by independently wealthy individuals in their spare time. Since the fundamental laws of mechanics, optics, electricity, and other aspects of science are discoverable using relatively simple, cheap equipment, this system worked well for centuries. By the early twentieth century, however, science was getting much more complicated—and expensive. More fundamental research in physics and other fields was being carried out in the laboratories funded by major universities. However, government funding of scientific research was still slight, and corporate funding of basic research was still unheard of.

World War II (1939–1945) changed this picture radically and permanently. Victory in that war was heavily influenced by technology: It became clear that governments needed weapons based on the newest science. Radar, for example, was secretly developed at the Massachusetts Institute of Technology (MIT) and in the United Kingdom, helping turn the tide of air war in Europe, while several governments raced to develop jet aircraft. The war project that depended most on basic scientific research was the drive to build an atomic bomb. The U.S. government began its massive, expensive, and intensely secret Manhattan Project in 1941 and exploded the world's first atomic bomb in 1945.

The scientists employed in the Manhattan Project were mostly not government scientists, but university professors and students hired into the program for the duration of the war. The Los Alamos Scientific Laboratory in Nevada, formed to carry on the Manhattan Project, was managed by the University of California starting in 1943 (and still is, in part). The spectacular success of the Project, along with other weapons made possible by scientific advances, made clear the relationship of national security to science, and the Manhattan Project created a precedent for government-university collaboration.

In the last year of the war, 1945, Vannevar Bush (1890–1974), Director of the Office of Scientific Research and Development, wrote a report for President Harry Truman (1884–1972) in which he urged strongly that the government fund all forms of scientific research, not only those with obvious, direct benefits for military technology. This was vital to long-term national survival, Bush argued: "We can no longer count on ravaged Europe as a source of fundamental knowledge."

The government took Bush's advice, establishing the NSF in 1950. The NSF quickly began giving grants, and by 2008 was funding about a fifth of all federally supported science research in U.S. universities. Other federal funding comes from a patchwork of agencies including the National Aeronautics and Space Administration (NASA), the NIH, the Pentagon, and others. After World War II the NIH expanded its grant-giving for biological and medical research from $4 million in 1947 to $1 billion in 1974 to about $28 billion in 2007. Similar systems were created in other industrialized countries: Only governments could supply the large sums of money needed to conduct modern, fundamental research in physics, chemistry, and biology. Because of its large population and industrial sophistication, the U.S. government has been by far the largest single funder of scientific research in the world.

Efforts to balance the federal budget in the 1980s (e.g., the Gramm-Rudman-Hollings Act of 1985) and the end of the Cold War in the early 1990s led to cutbacks in U.S. science funding in some areas. Funding for science research derived from both private and federal grants increased throughout the 1990s, but the rate of growth in competition for grant dollars greatly exceeded that rate of real growth in funds available.

Tight competition for federal funding caused universities to seek more funding from private companies, allowing industry to increasingly shape the course of research in the university system. In the early 2000s, the continued shifting of resources to military spending (by 2008, the U.S. military budget was almost half of total world military spending) and political pressure to reduce taxes led to renewed pressure on grant sources, increased competition in grant-seeking, and increasing control of industrial funders over some university research priorities. The cost of doing research continued to rise, so that a single grant dollar purchased less scientific knowledge with each passing year. From 2000 to 2006, although the NSF's grant budget grew by 44%, applications for funding grew even faster, resulting in greater competitiveness: 30% of grant applications to the NSF were funded in 2000, but only 21% in 2006.

While the grant system has funded most of the considerable scientific progress in the last half-century, it has also been criticized. Research that is likely to enhance technologies of destruction is favored by the military goals of much government grant funding, while research that is likely to yield patentable, profitable technologies is favored by industry. Forms of knowledge that do not yield profit, political payback, or weapons have been funded, but at much lower rates. In biology, for example, genetic engineering has been well-funded by industry because of its potential to produce patentable life-forms and high profits down the road: Ecology, paleontology, soil science, taxonomy, and other disciplines have been poorly funded.

Critics of the present grant system maintain that it has produced mediocre and misshapen science. Defenders argue that it has been effective. In the following section, some of these arguments are explored.

■ Modern Cultural Connections

Critics of the current grant process often focus on the problems with obtaining federal grants. They especially point out that grants are easier to obtain for studies involving the application of research rather than for basic science research. However, the existing system produces steady output of scientific innovation. There are many grants awarded by public and private sources that act to support good science at the most fundamental level. For example, some grants are designed solely to prepare undergraduate students for graduate education in science. Although certainly not basic-science research, these types of grants are important in the training of future scientists. In a sense, it is the most basic and most fundamental investment in science.

The present system also contains checks and balances that are intended to promote good science by

In a protest over low financing and back wages, Russian scientists carried a mock coffin with a sign saying "Russian Science" during a rally in downtown Moscow in 1996. Some 700 scientists of the Russian Academy of Sciences participated. *AP Images/Sergei Karpukhin.*

ensuring a distance between the research lab and the marketplace. Programs designed strictly for the marketplace are the antithesis of rigorous science, in which failure may be as informative as success: Recognizing this truth, many grant agencies (e.g., the NSF) explicitly refuse to grant money for the development of products for commercial markets.

It is true that over the last quarter century or so, there has been a shift away from basic science research to more applied science research within granting agencies such as the NIH. However, defenders of the grant system argue that this trend is balanced by the actions of other agencies to specifically encourage rigorous pursuit of basic science knowledge. For example, the NSF specifically discourages proposals involving particular medical goals (i.e., where the aim of the project may be the diagnosis or treatment for a particular disease or disease process).

The present grant system also seeks to provide special support for women, minority scientists, and scientists with disabilities. As with direct grants to students, grants to faculty at non-research colleges, with primarily undergraduate students, are designed solely to provide the most fundamental support of science in the development of the next generation of researchers. Grants can also be used to remedy a shortage of investigators in a particular area or research.

Most grant review processes seek to promote good science by allocating resources based upon the significance of the project (including potential impact on theory) and the capability and approach of the investigator or investigative team. Evaluating committees—especially when staffed with experts and functioning as designed—are able help fine-tune research proposals so that methodologies are well-integrated and appropriate to the hypothesis advanced. In cases of research involving human subjects or potentially dangerous research (e.g., genetic alteration of microorganisms), the grant-review process also provides some oversight of procedures that assure that research projects are conducted with due regard to ethical, legal, and safety considerations: Federal agencies do not grant funding to proposals that do not explain how they will meet certain ethical and safety standards.

Some critics of the existing funding system argue that although grants are designed to promote good science, the process has become so cumbersome, clogged, and confused that despite noble intent, it increasingly encourages mediocre, "safe" science.

Increasing competition and dependence on grants to fund increasingly complex and expensive research programs has exacerbated pre-existing weaknesses in strained grant evaluation systems. Moreover, the specific reforms designed to cope with increasing numbers

Demonstrators line a sidewalk outside Robert Wood Johnson University Hospital in New Brunswick, New Jersey, during a forum on the state's stem cell research initiative, in May 2004. New Jersey Governor James E. McGreevey signed an agreement at the forum to create the first state-supported stem-cell research institute, which would be constructed near the hospital. *AP Images/Daniel Hulshizer.*

of grant applications (e.g., triage and electronic submissions) are proving to have the unintended side effect of profoundly shaping the kinds of science research funded.

Grant awards, critics also argue, are rapidly becoming a contest of grantsmanship (the ability to write proposals and secure grants) rather than being decided on scientific merit. Emphasis on the form and procedures of the grant evaluation process, rather than on the substance of the science proposed, forces researchers away from the lab and into seminars on the craft of grant writing. Even more ominously for science, the investigators are forced, in many cases, to develop research proposals specially designed to please grant review committees. This impacts science research in several ways.

First, there is a loss of scientific diversity as proposals that have predictable outcomes are viewed as less risky investments of precious capital by grant evaluation committees. This drives research toward what critics of the current grant process term "safe science" and away from the types of risky research that are the likeliest path to more spectacular scientific insights and advances.

Second, as grantsmanship becomes increasingly important, new investigators fight an uphill battle to gain funding and build labs. Already several steps behind seasoned principal investigators who know how to craft strong proposals, new researchers often struggle along on smaller grants designed for new scientists. There is little funding of dissertation research: The NSF, for example, actively discourages graduate students from submitting grant proposals. Grants to scientists starting out on research programs are often insufficient. In fact, only

about one out of four researchers seeking initial NIH funding actually apply for the easier-to-obtain grants designed for researchers making their first application for funding as a principal investigator (leader of a research effort). More confining and debilitating to new researchers are early development grants that carry restrictive clauses that prohibit researchers from seeking other types of funding.

Actual funding reflects an increasingly brutal reality for investigators at all levels. The grant process is extremely competitive. Most grant proposals are not funded, and the percentage of proposed projects funded has steadily declined since the mid-1980s to current levels at which only about 10% of proposals, overall, are ultimately funded. In this environment, some scientists and their sponsoring institutions become proposal mills—often putting out a shotgun pattern of many proposals in hope that one or two may get funded. The time cost is a staggering drain on scientists and scientific research. Many investigators spend more time on the grant application process than on actual research.

Models of evaluation often work against more open-ended, basic-science-oriented research proposals. Basic science proposals usually contain a wider range of possible outcomes than do more narrowly focused goal-oriented projects (e.g., projects regarding a specific clinical application); reviewers tend to regard this unpredictability as a negative and so give a lower project priority to the proposal under review.

This emphasis on predictable outcomes may distort the research process itself: Biasing research toward a particular goal may tilt interpretation of data. It is a well-known axiom of science that researchers, regardless of discipline, often find the results they are looking for because even the most intellectually honest researchers are prone to shade and interpret—quite unconsciously and unintentionally—data that correspond to expected results. Indeed, this is the whole rationale behind the double-blind study that is standard in so much medical research: People conducting the research must be kept from knowing which medications are "supposed" to work, lest they unintentionally feed their expectations back into the research process.

In sum, critics of the present grant system argue that it encourages mediocre science because it encourages predictability. Researchers who fail to predict all possible outcomes for a project in their grant applications receive worse priority scores and so are less likely to be funded. Under these conditions, research tends to become an exercise in producing predicted results. This fundamentally reshapes the intent of research and results, creating a weak foundation upon which to build future applied research.

SEE ALSO *Science Philosophy and Practice: Science Communications and Peer Review.*

BIBLIOGRAPHY

Periodicals

Mervis, Jeffrey. "Grants Management: NSF Survey of Applicants Finds a System Teetering on the Brink." *Science* 317 (2007): 880–881.

Rajan, T.V. "Would Harvey, Sulston, and Darwin Get Funded Today?" *The Scientist* 13 (1999): 12.

Web Sites

Bush, Vannevar. "The Endless Frontier: A Report to the President by Vannevar Bush, Director of the Office of Scientific Research and Development." National Science Foundation. July 1945. http://www.nsf.gov/about/history/vbush1945.htm#ch3.8 (accessed January 22, 2008).

Brenda Wilmoth Lerner
K. Lee Lerner

Science Philosophy and Practice: Science Communications and Peer Review

■ Introduction

Peer review is the process by which scholarly articles and the research they contain are evaluated for accuracy and relevance by qualified experts. The scientific community employs peer review as a way to evaluate whether articles should be published in journals. A similar system is used to determine whether applications for research grants should be approved.

The strength of the peer review process lies in the experts, sometimes employed as journal editors, who evaluate the submitters' work. Expert evaluation encourages authors to meet commonly accepted standards of quality in their research and the conclusions they draw from it. The most effective evaluators are often those who are actively conducting research in the same field. However, to ensure neutrality, the identity of an article's author is often concealed during the peer review process. Some journals, such as the prestigious *Nature*, view peer review as a technique for improving articles and the journal as a whole.

■ Historical Background and Scientific Foundations

Today, peer review is considered essential to the credibility of an article. Just as universities and schools must be accredited to issue degrees that are respected, scientific journals must conduct peer review to ensure the merit of their articles. Though it is a requirement today, peer review was not formalized or used consistently until the middle of the twentieth century.

Medicine was the first field to employ systematic peer review and declare it to be an essential method of quality control. In the mid-ninth century, the Syrian physician Ishap bin Ali al-Rahwi (854–931) described a peer review system in which physicians were required

to keep notes on their patients and treatments. In the case of a death, a panel of other physicians would evaluate whether the treatment was correct and whether the treating physician could be sued.

As the sciences developed, peer review was present in an informal way through the relationships among individual scientists. Colleagues and friends evaluated each others' work as part of the collaborative process, sometimes in association with scientific academies. Scientific controversies were often vigorously debated among the members of academies, improving the state of science through the application of the scientific method. This system of peer review was not a formal evaluation of scientific papers, but an open dialogue between qualified experts.

In the early twentieth century, formal scientific journals were well established. Often under the authority of a single expert editor, articles would be accepted or rejected based on the editor's sole evaluation. The Nobel Prize–winning physicist Max Planck (1858–1947),

Martin Fleischmann, left, talks to reporters about cold fusion as University of Utah chemist B. Stanley Pons listens, in Los Angeles, May 9, 1989. The pair of scientists claimed they had achieved nuclear fusion at room temperature and held a press conference to announce their research rather than releasing it in a peer-reviewed journal, prompting criticism within the scientific community. *AP Images/Doug Pizac.*

editor of the journal *Annalen der Physik,* was a good example of this model. As one of the leading researchers of his day, Planck was active in the field of physics, constantly reading research papers, and acquainted with the different kinds of research going on at the time. However, the publishing industry could not always secure editors of such high quality. This led to the development of the more collaborative modern system of peer review.

■ Modern Cultural Connections

Today peer review is the standard by which scientific research is judged for quality. Research, articles, or claims that do not withstand the scrutiny of peer review are generally considered flawed or incorrect. Peer review varies among the many different scientific journals, but generally editors seek the advice of expert reviewers to better determine the quality of articles they consider for publication. In the classic blind review method, the reviewers' and authors' identities are concealed from one another to prevent conflicts of interest or other unethical behavior. The reviewers must read and provide comments to the editor in a timely manner to avoid unnecessary delay. This system is praised for excluding poor-quality work and for helping avoid duplicate publications of similar work.

Despite the system's broad acceptance, criticisms are numerous. Though peer review is widely believed

to improve the quality of published papers, there is no measurable evidence that yet confirms this claim. Also, the personalities and egos of the reviewers and authors can interfere with the neutrality of the process, particularly when they are rivals in the same field. Other critics allege that many reviewers have an unconscious bias toward orthodox or traditional thinking and wrongly reject unconventional approaches in research. Recently, major cases of fraud have called the effectiveness of peer review into question. One example is that of the South Korean stem cell researcher Hwang Woo-Suk (1953–), whose articles, published in the journal *Science,* contained large amounts of fabricated data. The inability to detect even large-scale fraud is a much-discussed weakness of modern peer review.

Editors try to minimize these weaknesses by introducing elements of openness to the review and selection of articles. Several well-known scientific journals, including *Nature* and the *Australian Medical Journal,* have experimented with the use of an online forum as part of their peer review processes. The system at *Nature* subjected approved articles to both traditional peer review and the new online approach simultaneously. Subscribers registered at the journal's Web site were invited to participate, limiting input from inexpert reviewers. The comments were moderated by the journal and reviewers were required to identify themselves. Though many people read the articles online, few actually commented

The journal *Science* is a leading scientific journal in which scientists publish research findings. Such publications subject scientific findings and arguments to intensive review by other scientists, a process termed peer review. *AP Images.*

on them, and even fewer of these comments were judged to be helpful to the editors or of similar quality to the traditional blind reviewers' comments. Because of these problems, editors at *Nature* abandoned this system. Other journals have had more success and continue to use online, open peer review to improve their articles.

This initial failure does not mean that openness in peer review or the use of the Internet to include more reviewers will ultimately be rejected. Serious problems have arisen in peer-reviewed journal articles despite the proper application of the current system. Members of the scientific community are trying to determine the best methods for improving peer review. Success, whether it is measured in increased openness, improved analysis, or other techniques, should improve science overall and help direct researchers to the best research methods.

■ Primary Source Connection

The Internet now serves as an efficient mechanism for the scientific community to share vital information. Several major universities are pondering and experi-

menting with the merits of open publication of scientific papers on the Web as an alternative to requiring that faculty strictly publish in often expensive traditional journals.

The Internet has also become an essential tool for scientists to communicate and share information when time is critical, especially during outbreaks of disease. In the following article, virologist Jack Woodall recounts the founding of ProMed mail, a Web-based forum where clinicians or researchers in the field can report data and observations of disease outbreaks from any spot on the globe to a centralized point where fellow experts can instantaneously review and respond. In 2003, the outbreak eventually identified as SARS was first reported to ProMED by a physician who learned from a colleague via e-mail that an unusual pneumonia was killing people and taxing the resources of hospitals in the Guangzhou region of China.

Woodall, who was an original founder of ProMed mail, continues to serve as its associate editor. Woodall is a graduate of Cambridge University in the United Kingdom, and he received his Ph.D. from London University. During a long and distinguished career, Woodall has served and worked at the East African Virus Research Institute in Uganda, the Belem Virus Laboratory in Brazil, and the Yale Arbovirus Research Unit in New Haven, Connecticut. He has also served as the head of the Arbovirus Laboratory for the New York State Health Department and as director of the Centers for Disease Control and Prevention's San Juan Laboratories in Puerto Rico. In 1981, he began work for the World Health Organization (WHO) and was a member of the WHO Gulf Emergency Task Force in support of the U.N. Special Commission (UNSCOM) in Iraq, and leader of the WHO delegation to the Third Review Conference on the Biological Weapons Convention. Until 2007, he served as visiting professor at the Institute of Medical Biochemistry and as director of the Nucleus for the Investigation of Emerging Infectious Diseases at the Federal University of Rio de Janeiro, Brazil.

In addition to his continued work with ProMED mail, Woodall is a member of the International Society for Infectious Diseases (ISID), and Web site editor and council member (ex officio) of the American Society of Tropical Medicine and Hygiene. He is a member of the Biological Weapons Working Group of the Center for Arms Control and Non-Proliferation and a board member for Sabin Vaccine Institute in Washington D.C.

PROMED MAIL

ProMED mail, to give it its baptismal name, or ProMED, as everyone now calls it, was a happy accident. Barbara Hatch Rosenberg of the Federation of American Scientists organized a meeting in 1993 in Geneva,

Switzerland, co-sponsored by the World Health Organization (WHO), to float the idea of a world-girdling chain of institutes capable of sending out teams to the site of any unusual disease outbreak in their neighborhood. The objective would be to determine whether it was of natural or unnatural origin. The conference itself was unusual in bringing together experts on not only human, but also animal and plant diseases, and on bioterrorism, which at the time was not high on anyone's priority list.

There were some 60 participants from 15 countries. The conclusion was that such a chain was highly desirable from the point of view of human health and food security. At a follow-up conference in the United States in 1994, further steps were outlined, and the late Dr. Robert Shope suggested the name ProMED, for Program for Monitoring Emerging Diseases. It was decided that an e-mail list be set up to enable discussion among the participating institutions. Charles Clemens of Satel-Life offered to host the e-mail list, and I offered to run it, with the assistance of Stephen Morse, then of Rockefeller University. I was working for the New York State Health Department in Albany, New York, at the time, and was one of only a few of the conference participants who had access to e-mail then. Thus, ProMED-mail, so called to distinguish it from its parent program, was launched in August 1994.

It turned out that absolutely no one in the program had anything to say to each other, so as we were supposed to be monitoring outbreaks, Steve and I started posting outbreak reports from the media. Then, in May 1995, the Ebola epidemic in Kikwit, Zaire, hit the media, and people surfing the Web discovered that ProMED was posting information about it. I well remember the thrill when our mailing list, which had begun with 40 members in seven countries, hit 250. Later we were written up in the *Wall Street Journal* and our numbers went overnight to 500. Today, in mid-2007, we stand at over 40,000 in at least 180 countries, with many more accessing our Web site at http://www.promedmail.org. And, thanks to foundations and donations, we still provide worldwide coverage, 7/365, without fee.

The uniqueness of ProMED is its stable of experts in the fields of clinical and veterinary medicine, microbiology, and plant pathology, all of whom serve on a part-time basis. It is the only free disease reporting system to cover human, livestock, wildlife, and food and feed crop diseases in one place, the latter because of the potential impact of animal and vegetable diseases on nutrition and therefore on human health. Since 2000, ProMED has been a program of the International Society of Infectious Diseases, which guarantees its freedom from political constraints that often cause delays in outbreak reporting. In fact, WHO [World Health Organization] has said that it uses ProMED reports to convince recalcitrant countries to report outbreaks officially, in view of the fact that a report has already appeared on ProMED.

During the anthrax-by-mail episode in the United States in 2001, the Science Advisor to the President told us that the White House's main sources of updates on the situation were CNN and ProMED. The Chief Veterinary Officers of Australia and New Zealand routinely copy livestock outbreak reports to ProMED at the same time as they send them to the World Animal Health Organization, and we get reports directly from hospitals and research institutes involved in outbreaks. We emphasize reports on outbreaks caused by select agents from the bioterrorism A list, such as anthrax and botulinum toxin, so that our readership understands that such natural outbreaks are not uncommon in some countries. We cover outbreaks due to biological toxins. Otherwise, we report on emerging diseases such as bird flu, using a rather broad definition of emerging that includes dengue but excludes most tuberculosis and HIV/AIDS reports.

ProMED has parallel lists in Spanish, Portuguese, and Russian, with a French version scheduled to launch shortly. These are not straight translations of the English reports, but are mainly reports of regional interest. Chinese and Japanese translations of many ProMED reports are found on the travel health Web sites of Hong Kong and Tokyo International Airport.

Jack Woodall

WOODALL, JACK. "PROMED MAIL." IN *INFECTIOUS DISEASES: IN CONTEXT*. EDITED BY BRENDA WILMOTH LERNER AND K. LEE LERNER. DETROIT: GALE, 2007.

SEE ALSO *Science Philosophy and Practice: Professionalization; Science Philosophy and Practice: Research Funding and the Grant System; Science Philosophy and Practice: Scientific Academies, Institutes, Museums, and Societies.*

BIBLIOGRAPHY

Books

Woodall, Jack. "ProMed Mail." In *Infectious Diseases: In Context*. Edited by Brenda Wilmoth Lerner and K. Lee Lerner. Detroit: Gale, 2007.

Periodicals

Gitanjali, B. "Peer Review—Process, Perspectives, and the Path Ahead." *Journal of Postgraduate Medicine* 47 (2001): 210–214.

Greaves, Sarah. "*Nature*'s Trial of Open Peer Review." *Nature* (December 2006).

Judson, Horace Freeland. "Structural Transformations of the Sciences and the End of Peer Review." *Journal of the American Medical Association* 272 (1994): 92–94.

Spier, Ray. "The History of the Peer-Review Process." *Trends in Biotechnology* 20 (August 8, 2002).

Web Sites

International Society of Infectious Diseases. "ProMED." http://www.promedmail.org (accessed February 5, 2008).

Nature. "Peer Review: Debate." December 2007. http://www.nature.com/nature/peerreview/debate/index.html (accessed January 2008).

Kenneth T. LaPensee

Science Philosophy and Practice: Scientific Academies, Institutes, Museums, and Societies

■ Introduction

Scientific academies, societies, or institutes are groups of intellectuals, usually researchers, who are interested in the same broad topics and share their studies and ideas with one another for the advancement of scientific knowledge. In the United States, the National Academy of Sciences conducts research and advises the different government departments to aid in decision-making. Many other countries follow the same approach with their own academies of science.

A few of the oldest national academies sprang from private scientific academies that worked for the advancement of a discipline or field, or for the advancement of science in general. Some of these private academies are very old, dating from the Renaissance or Enlightenment. Academies such as these contributed to many basic discoveries by encouraging both cooperation and competition between great minds. Today, public and private scientific academies continue this tradition by holding meetings, publishing journals, and generally encouraging the advancement of scientific thought and research.

■ Historical Background and Scientific Foundations

The term "academy" was first associated with learning and scholarship through the ancient Greek philosopher Plato (c.427–c.347 BC). Initially, the word was simply the name of his neighborhood, but it soon became associated with the school of philosophy that he oversaw in his home. In the ancient world, philosophy and science were not considered the separate disciplines they are today. At the time, philosophers wanted to explain the world as they saw it: The character of man, the nature of the gods, the composition of the universe, mathematics, geometry, ethics, rhetoric, and many other disciplines fell within the scope of their interest. Other philosophers, both before and after Plato, established schools in this model. Aristotle's philosophical school, the Lyceum, inspired the names of many academic institutions, particularly in Europe and Russia. Socrates (c.470–399 BC) is equally famous for shunning the idea of schools altogether and making teaching and philosophy the core of his interactions with other people.

Many philosophical academies continued in one form or another until the Roman Empire began to collapse in the fifth century AD. After the fall of Rome, much ancient learning was lost. During the Middle Ages, learning refocused around the Christian Church, particularly in monasteries. The Renaissance period, however, saw the rebirth of systematic scientific inquiry and the complete redefinition of artistic style. Science

French minister of finance Jean-Baptiste Colbert presents the members of the Royal Academy of Science, founded in 1667, to King Louis XIV. *Reunion des Musees Nationaux/Art Resource, NY.*

and art became intertwined with the rise of great masters such as Leonardo da Vinci (1452–1519). As much a scientist as he was an artist, da Vinci's wide-ranging interests mirrored the development of the age as a whole. He explored his interests in anatomy, engineering, and the natural world through observations recorded in his notebooks, but did not conduct many experiments to test theories.

It was during the later part of the Renaissance that the earliest "modern" scientific academies were founded. Academia Secretorum Naturæ, or the Academy of Nature's Mysteries, was founded by Giambattista della Porta (c.1535–1615) in Naples before 1580. Della Porta was interested in optics, mathematics, hydraulics, pharmacology, and meteorology, but also explored alchemy, astrology, and the occult. This led to the dissolution of his academy by Papal decree. He then joined the Accademia dei Lincei, or Academy of Lynxes (so named because the lynx is sharp-sighted), founded in Rome by Federico Cesi (1585–1630) in 1603. Although Cesi was a botanist, he invited scientists of all disciplines to join his group. Most prominent among the Lincei was Galileo Galilei (1564–1642), the astronomer, physicist, and mathematician who confirmed Copernicus' heliocentric theory of the solar system.

Upon Cesi's death, his academy disbanded. During the following centuries it was revived several times; its

current incarnation is designated as the national scientific academy of Italy. Other academies followed, notably in France, Germany, Russia, Ireland, England, and many other countries in Europe. Most were founded privately and eventually grew to receive the patronage of kings or universities.

■ Modern Cultural Connections

Today, scientific academies and institutes take many different forms. The National Academies of the United States are designed to advise the government on scientific, medical, or engineering matters. The National Research Council conducts research with the purpose of answering questions relevant to public policy. Members are elected to the National Academy of Science in recognition of their excellence in research. The governing board is selected from members of the academy.

Some private academies still exist, though in a much more organized form than in the past. Organizations such as the American Association for the Advancement of Science (AAAS) are as much involved with promoting science as they are with conducting research. Individual scientists still make up the membership of such organizations. They attend conferences, publish in the academies's journals, and serve on committees.

The International Solvay Institutes for Physics and Chemistry in Brussels held the first world physics conference in 1911 and began to host them every three years. The most famous conference was the fifth conference on electrons and photons in 1927. Attendees included Albert Einstein, Niels Bohr, and Werner Heisenberg. Of the 29 attendees, 17 were current or future Nobel Prize winners. *Science Source.*

Scientific academies are often interested in promoting scientific research and education, advising governments and universities on scientific policy, and encouraging the responsible use of science to solve problems. The term "scientific institute" often refers to an organization engaged in active research on a particular topic or group of topics. Sometimes education in this field is also an active part of an institute's mission.

Scientific museums may or may not be affiliated with scientific academies, institutes, or universities. For example, the Museum of Science in Boston was founded by a group of men who wanted to share their mutual interest in science, much like the earliest scientific academies. Many popular scientific museums focus on technology and its development over time. Sometimes, scientific museums focus on zoology, botany, paleontology, geology, and other sciences of the natural world. These museums are often known as natural history museums.

Modern scientific academies, institutes, and societies have developed from small, private groups exploring their interests to the vast institutions we know today. Some of the most fundamental discoveries in science were made possible because of the support of the earliest academies. Today, these organizations continue to encourage new discoveries, advocate the importance of science, and educate the public on scientific topics.

SEE ALSO *Science Philosophy and Practice: Professionalization; Science Philosophy and Practice: Research Funding and the Grant System; Science Philosophy and Practice: Science Communications and Peer Review.*

BIBLIOGRAPHY

Periodicals

Drenth, P.J.D. "Scientific Academies in International Conflict Resolution." *Technology in Society* 23, no. 3 (2001): 451–460.

Web Sites

National Academy of Sciences. "About the NAS." http://www.nasonline.org/site/PageServer?pagename=ABOUT_main_page (accessed January 12, 2008).

Rice University. Galileo Project. "Giambattista della Porta." 1995. http://galileo.rice.edu/Catalog/NewFiles/porta.html (accessed January 12, 2008).

University of St. Andrews. School of Mathematics and Statistics. "The Accademia Dei Lincei." August 2004. <http://www-groups.dcs.st-and.ac.uk/~history/Societies/Lincei.html> (accessed January 12, 2008).

Melanie Barton Zoltán

Science Philosophy and Practice: The Formulation and Impact of Naturalism, Reductionism, Determinism, and Positivism

■ Introduction

The "systems of thought" discussed below are not so much coherent systems as they are themes that appear and reappear with varying impact throughout the history of scientific thought. Before elements of the now standard "scientific method" were developed in practice by pioneers such as Italian astronomer and physicist Galileo Galilei (1564–1642), many philosophers attempted to reason a method for developing accurate information about the natural world.

From the days of Ancient Greece well into the Middle Ages, the quest for knowledge about the physical world occurred alongside what are now considered to be more metaphysical concerns. Metaphysics (roughly translated as "after physics") is a branch of philosophy that concerns itself with an understanding of being as a whole, the meaning of life, and humanity's place in the universe. In science's formative days, legitimate methods for obtaining this information ranged from direct observation to the alleged use of divine revelation. Thinkers like French philosopher René Descartes (1596–1650) outlined a system of scientific inquiry, extracting it from the larger realm of concerns and methods and limiting it to subjects about which it could develop reliable information. Naturalism, reductionism, determinism, and positivism hold an important place in the family of concepts that constitute the modern understanding of scientific thought.

■ Historical Background and Scientific Foundations

Systems of thought are relative to time and place, and scientific thought is affected by these systems in two different but interrelated ways. One is their relation to the field of metaphysics. Science, as a discipline, is not directly concerned with metaphysics. Conflicts arise, however, when the practice of science or its conclusions influence metaphysics indirectly. One of the earliest examples of this conflict can be found in the Copernican Revolution, where, through scientific observation, Polish astronomer Nicolaus Copernicus (1473–1543) drew the conclusion that Earth was not the center of the universe, seemingly devaluing the Christian church's metaphysical claim that Earth was superlative among God's creations (and therefore necessarily at its physical center). A recent example of this conflict is the debate over the teaching of evolution in schools.

The second way in which systems of thought impact scientific thinking is their direct comment on the process of rational thinking or science itself.

Naturalism

Naturalism, as it applies to metaphysics, is the idea that the physical world is all that exists and that it is orderly. This is a deceptively simple formulation, however, as it has been interpreted to mean radically different things. The Greek thinker Thales (c.624–c.546 BC) has been called the first naturalist, because he argued that the entire universe was composed of a physical substance—water. Later Greek thinkers Democritus (c.460–c.370 BC), Lucretius (99–55 BC), and others followed his lead, theorizing a world composed of tiny particles—atoms. The common element among these theories, and the reason they can be called naturalistic, is their exclusion of gods or other supernatural divine forces in their conception of the universe.

Naturalistic thinkers did not all profess atheism however. Naturalism also describes a method for acquiring knowledge that focuses on observation of physical evidence to explain natural phenomena and ignores the question of a supernatural god altogether. This was especially useful during the Middle Ages and Renaissance,

WORDS TO KNOW

CONUNDRUM: Any difficult question, especially one expressed as a verbal puzzle.

DETERMINISM: The notion that a known effect can be attributed with certainty to a known cause.

METAPHYSICS: The branch of philosophy that is concerned with basic concepts such as time, space, being, reality, and the like. Especially in the twentieth century it has interacted closely with physics, which is also concerned with such questions.

NATURALISM: Philosophical naturalism is the doctrine that nature is all that exists: there are no souls, spirits, or gods. Methodological naturalism is the practice of looking only for natural (physical) explanations for observable phenomena, while excluding the possibility that spirits, God, gods, or the like have caused phenomena. Methodological naturalism is basic to modern science, but does not require adherence to philosophical naturalism; that is, religious believers may, as scientists, be methodological naturalists without being philosophical naturalists.

NIHILISM: The attitude or doctrine that the universe as a whole, and human life in particular, are meaningless.

POSITIVISM: The philosophical doctrine that only science can produce valid knowledge. According to positivism, there is no other source of knowledge but science (e.g., religion, intuition, common sense, etc.). This view was first articulated by French philosopher August Comte (1798–1857).

REDUCTIONISM: A style of thought that can be contrasted to holism. Reductionism assumes that the properties of any complex system—a human brain, an ecosystem, or other—can be reduced (hence "reductionism") to the interactions of its component parts. Holism asserts that at least some system properties arise out of the system as a whole (hence "holism") and are not possessed by, or reducible to, the properties of the system's isolated, individual parts.

SCIENTIFIC METHOD: Collecting evidence meticulously and theorizing from it.

as methodological naturalism, in contrast to ontological or metaphysical naturalism. Naturalism's lasting impact on scientific thought has been to delineate science from everything that is not science through the preponderance of physical evidence in the acquisition of knowledge and the search for a physical cause to natural phenomena.

Reductionism

The first modern reference to reductionism can be found in Descartes' work *Rules for the Direction of the Mind*. In this work Descartes outlines principles for the process of "rational analysis," a method designed to produce an accurate understanding of the world. If naturalism provided one half of the conceptual weight of the scientific method, rational analysis provided the other. Rational analysis, especially when combined with experiment and observation, is considered the basis of the modern scientific method by some philosophers. Reductionism was essential to rational analysis and appears as a rule in Descartes' treatise. Reductionism is a method of analysis that understands a thing by reducing it to its parts. This method has led to our modern day understanding of the human body as a collection of organs, for example, or the atmosphere as a composition of different gases. Indeed, the drive to understand "the very simplest" elements of the universe has led scientists to discover atoms and subatomic particles.

Reductionism has a controversial edge with some philosophers when used to devalue any definition of "knowledge" that is not scientific knowledge. In practice, this type of reductionism, rather than maintaining science's characteristic indifference toward metaphysics, enters the debate by arguing that metaphysics, as a field of thought, is totally invalid. Whereas naturalism limits existence to the natural world and allows for interpretations of the divine, reductionism demands a categorical denial of anything divine or supernatural. In addition, it makes broader claims, particularly in relation to the options for the understanding of humanity's place in the world. For example, a reductionist is forced to account for human things like love, art, beauty, and friendship, in terms of subhuman chemical processes. This type of reductionism, though not properly scientific, is largely responsible for the sometime association of scientific thought with modern nihilism, the view that life is meaningless.

Determinism

Implicit in science's longing to understand the natural world is the desire to predict, and therefore control, events. The scientific method even enshrines this idea of predictability, insisting that valid results of an experiment must be duplicable. In order to do this, one must be able to properly identify the cause of an event.

when a powerful religious authority regarded scientific activity with suspicion. During this time naturalism as a methodology gained popularity, not as an affront to the Christian faith, but as an extension of it—a way to understand divine creation. Scientists who professed belief in Christianity asserted that God operated through natural processes, and thus their work, rather than challenging their faith, reinforced it. Today, this type of naturalism, grounded in religious faith, is known

The general secretary of the Science and Rationalists Association of India (left) speaks during a protest rally in India in October 2003. The rationalists demonstrated in Calcutta, the adopted home of Catholic nun and humanitarian Mother Teresa (1910–1997). The protesters rejected a purported miracle that the Vatican cited as grounds for Mother Teresa's beatification. *AP Images/Bikas Das.*

Then one must assume that every event does indeed have a cause. This idea immediately seems intuitive but also begs the question of a "first cause" for everything in the universe. Ironically, science's search for a first cause has thrown this intuitive truth into question, or at least rendered complex a seemingly simple proposition.

Determinism also questions the existence or non-existence of human free will. If all events are caused by some other event, humans would seem to be morally "off the hook," retaining no responsibility for their actions. This implication in particular has generated a great deal of resistance throughout history to the idea that nature operates according to set laws. Although this notion is generally accepted today, the conceptual paradox between natural laws and human free will remains, in many ways, unresolved.

Science has commented directly and indirectly on this debate since its inception. The idea that the universe operates according to laws—first introduced by the ancient Greek atomists and then formally in English physicist and mathematician Sir Isaac Newton's (1642–1727) *Principia* (1687)—suggests that events unfold in the universe with the predictability of balls in a game of billiards. The potential of these newly discovered laws was concisely described by French mathematician Pierre Simon de Laplace (1749–1827) in his seminal work *Celestial Mechanics* (1796). He argued that, with sufficient information about the present state of the universe, and with sufficient cognitive faculties, a man could predict

"the movements of the greatest bodies of the universe and those of the tiniest...."

The "classical" description of a universe governed by knowable laws, while a philosophical conundrum, was not seriously called into question until the twentieth century, by German–American physicist Albert Einstein's (1879–1955) Theory of Special Relativity and the advent of quantum mechanics. The former offered a radical interpretation of time and, in turn, causality. Quantum mechanics suggested that phenomena were the result of a combination of both determinacy and chance. The latter, now the basis of modern scientific advances, demonstrates that matter behaves very differently at the atomic level than at the level of objects. Systems viewed at the quantum level appear to behave randomly or chaotically, confounding attempts to predict their activity with Newtonian laws of physics. Quantum mechanics speaks not of determinism but rather of probabilities.

Positivism

The term "Positivism" was coined and developed by French philosopher Auguste Comte (1798–1857) in the mid–1800s. Positivism claims to be the application of scientific principles of inquiry to the great questions of philosophy and society. It was designed as an attempt to extend the reach of scientific thinking beyond traditional boundaries, while retaining a legitimately scientific process. It is both a worldview and a scientific methodology.

IN CONTEXT: CLASSICAL AND MEDIEVAL PHILOSOPHICAL CONFLICTS

The Western rediscovery of Aristotle's (384–322 BC) philosophy in the twelfth century brought the theologically dominated medieval world into sharp conflict with Aristotelian logic as contained in the *Book of Causes* and other works of Arabic scholars. Although the original author of the *Book of Causes* is unknown (some scholars argue that it originated in ninth- or tenth-century Baghdad, others argue that it was written in twelfth century Spain), the book elaborated upon Aristotelian concepts by asserting that from each cause there results a certain order to its effects. In contrast to medieval beliefs regarding miracles, the *Book of Causes* argued that a god could not do anything contrary to the order he had already established. Aristotle's concept of a Prime Mover as a being that remains unmoved, unchanging, and impersonal was also incompatible with the Christian concept of a god who regularly intervened in the affairs of people through miracles. In addition, Aristotle's argument that the universe was circular and eternal contrasted with the Christian doctrine of creation.

Most early medieval scholars rejected the eternity of the universe as philosophically absurd. Some, however, made tenuous connections to Aristotle's Prime Mover by asserting that God was ultimately the cause of all phenomena, and that God worked through natural mechanisms.

Positivism begins by asserting that the only real knowledge available to human beings is knowledge of "phenomena." Phenomena are understood to be objects and events as they are experienced in the world through the senses (the usual realm of scientific inquiry). This is contrasted with "noumena"—the essence or idea of a thing—as comprehended by the mind. Positivism echoes Descartes' reductionism and invokes naturalism's insistence on observation, but it makes far more ambitious claims than either of these systems. Comte suggested that scientific thinking is in fact the final stage in a historical progression of human thinking. Indeed, scientific thinking had the potential, when applied as positivistic philosophy, to "lay down a definite basis for the remodeling of society."

The concept of progress is implicit in positivism. Though a common notion today, progress was a new idea in Comte's time and shaped much of the thinking of the late 1800s. Throughout the Enlightenment, scientific advances challenged the concept of a static social structure ordained by a supernatural god. Instead, it seemed possible to create a new society free of the injustices of the past, and science seemed to be the per-

fect guiding process. The optimism of this era proved intoxicating and birthed widely different movements. Comte was also a pioneer in sociology, which takes the "illnesses" of society as its object of study.

■ Modern Cultural Connections

The essential tensions between science, philosophy, and religion have not lessened over the centuries. These tensions over what constitutes reason—and of the interplay of reason and faith—manifest in debates over a wide range of issues that cross many cultural lines in present day.

■ Primary Source Connection

In a November 2005 article, Associate Press reporter Nicole Winfield reported on an event typical of the continuing tensions between ideologies.

VATICAN: FAITHFUL SHOULD LISTEN TO SCIENCE

VATICAN CITY - A Vatican cardinal said Thursday the faithful should listen to what secular modern science has to offer, warning that religion risks turning into "fundamentalism" if it ignores scientific reason.

Cardinal Paul Poupard, who heads the Pontifical Council for Culture, made the comments at a news conference on a Vatican project to help end the "mutual prejudice" between religion and science that has long bedeviled the Roman Catholic Church and is part of the evolution debate in the United States.

The Vatican project was inspired by Pope John Paul II's 1992 declaration that the church's 17th-century denunciation of Galileo was an error resulting from "tragic mutual incomprehension." Galileo was condemned for supporting Nicolaus Copernicus' discovery that the Earth revolved around the sun; church teaching at the time placed Earth at the center of the universe.

"The permanent lesson that the Galileo case represents pushes us to keep alive the dialogue between the various disciplines, and in particular between theology and the natural sciences, if we want to prevent similar episodes from repeating themselves in the future," Poupard said.

But he said science, too, should listen to religion.

"We know where scientific reason can end up by itself: the atomic bomb and the possibility of cloning human beings are fruit of a reason that wants to free itself from every ethical or religious link," he said.

"But we also know the dangers of a religion that severs its links with reason and becomes prey to fundamentalism," he said.

"The faithful have the obligation to listen to that which secular modern science has to offer, just as we ask that knowledge of the faith be taken in consideration as an expert voice in humanity."

Poupard and others at the news conference were asked about the religion-science debate raging in the United States over evolution and "intelligent design."

Intelligent design's supporters argue that natural selection, an element of evolutionary theory, cannot fully explain the origin of life or the emergence of highly complex life forms.

Monsignor Gianfranco Basti, director of the Vatican project STOQ, or Science, Theology and Ontological Quest, reaffirmed John Paul's 1996 statement that evolution was "more than just a hypothesis."

"A hypothesis asks whether something is true or false," he said. "(Evolution) is more than a hypothesis because there is proof."

He was asked about comments made in July by Austrian Cardinal Christoph Schoenborn, who dismissed in a *New York Times* article the 1996 statement by John Paul as "rather vague and unimportant" and seemed to back intelligent design.

Basti concurred that John Paul's 1996 letter "is not a very clear expression from a definition point of view," but he said evolution was assuming ever more authority as scientific proof develops.

Poupard, for his part, stressed that what was important was that "the universe wasn't made by itself, but has a creator." But he added, "It's important for the faithful to know how science views things to understand better."

The Vatican project STOQ has organized academic courses and conferences on the relationship between science and religion and is hosting its first international conference on "the infinity in science, philosophy and theology," next week.

Nicole Winfield

WINFIELD, NICOLE. "VATICAN: FAITHFUL SHOULD LISTEN TO SCIENCE." *USA TODAY* (NOVEMBER 3, 2005).

SEE ALSO   *Science Philosophy and Practice: Ockham's Razor; Science Philosophy and Practice: Postmodernism and the "Science Wars"; Science Philosophy and Practice: Pseudoscience and Popular Misconceptions; Science Philosophy and Practice: The Scientific Method.*

BIBLIOGRAPHY

Books

Comte, Auguste. *A General View of Positivism.* London: Routledge and Sons, 1907.

Descartes, René. *Rules for the Direction of the Mind.* Indianapolis: Bobbs Merrill, 1964.

Frankl, Viktor E. "Reductionism and Nihilism." In *Beyond Reductionism: New Perspectives in the Life Sciences.* Edited by A. Koestler and J.R. Smythies. New York: Macmillan, 1970.

Laplace, Pierre-Simon. *Celestial Mechanics.* New York: Chelsea Publishing. 1966.

Lindburg, David C., and Ronald L. Numbers. *When Science and Christianity Meet.* Chicago: University of Chicago Press, 2003.

Mill, John Stuart. *Auguste Comte and Positivism.* London: N. Trübner & Co., 1866.

Russell, Bertrand, and Michael Ruse. *Religion and Science.* Oxford: Oxford University Press, 1961.

Wilson, Edward O. *Consilience: The Unity of Knowledge.* New York: Alfred A. Knopf, 1999.

Periodicals

Winfield, Nicole. "Vatican: Faithful Should Listen to Science." *USA Today* (November 3, 2005).

Web Sites

The New England Complex Systems Institute. "Concepts in Complex Systems: Reductionism." 2000. http://necsi.org/guide/concepts/reductionism.html (accessed December 11, 2006).

Angela Scobey

Science Philosophy and Practice: The Scientific Method

■ Introduction

The scientific method is the approach used by scientists in the discovery of new scientific knowledge. A simplified outline of this approach, reduced to the making of observations, the formation of hypotheses (possible cause-and-effect explanations), and the testing of hypotheses by further observations, is often taught to students as the scientific method. Although there is in fact no single scientific method, no universal way of conducting scientific research, all scientists' approaches to knowledge discovery have certain elements in common that distinguish science from other ways of knowing—the ways of knowing characteristic of religious belief, common sense, personal relationships, and so on. It is generally agreed among scientists and philosophers that scientific claims must be capable of being falsified by other scientists, must fit into some framework of explanatory ideas (a theory), and must make meaningful predictions about the observable universe. The making of observations, the formation of cause-and-effect explanations, and the comparison of these hypotheses to further observations are basic to the creation of new scientific knowledge, though they are far from the whole story and are rarely practiced in a straightforward, mechanical way.

■ Historical Background and Scientific Foundations

Scientific ways of thinking have developed over a period of about 2,000 years, starting from forms of common-sense problem-solving used by people in everyday life. Today, scientific knowledge is a large, ever-growing system to which individual scientists can add only by following some form of the procedure known as the scientific method.

The Greek philosophers of several centuries BC, especially Aristotle (384–322 BC), were among the first persons to think carefully about how we acquire knowledge about the natural world—scientific knowledge. Aristotle taught that science depends on two basic forms of reasoning, induction and deduction.

Induction is the inference (reasoning-out) of general principles from specific observations. For example, if we observe without exception that heavy objects fall straight downward when released, we may reason inductively that all objects have some property in common—mass—on which some force associated with the Earth acts—gravity. (Other inferences are possible from this simple observation, and the Greeks did not, in fact, reason in terms of "forces.") Further, careful observations of exactly how quickly objects fall may allow us to induce a strict, mathematical law describing how the force of gravity accelerates bodies. This was done by European physicists in the seventeenth century at the dawn of the Scientific Revolution.

Deduction, on the other hand, is the prediction of particular events or observations from general principles or laws: It is like induction working backwards. For example, once we have proposed a law of gravitation, we can deduce from it how a space probe should behave en route from Earth to Mars. If the probe behaves as predicted, the law is confirmed, at least thus far.

Both induction and deduction are part of the scientific method. From observations, laws may be produced using inductive reasoning. From these laws, predictions may be deduced. These predictions can be tested by arranging further observations. From these further observations, adjustments to the proposed scientific laws may be made.

A version of this scientific method was described by the Arab scientist Ibn al Haytham (965–1039) in the eleventh century. The English philosopher and

Franciscan monk Roger Bacon (1219–1294) proposed a version of al Haytham's method that even more closely prefigured the modern ideal: observe phenomena, propose a hypothesis to explain what is observed, make fresh observations to test the hypothesis, and publish your work so that others can check it. In the 1600s, physicists (scientists who study the fundamental laws governing all physical objects), including Isaac Newton (1643–1727), proposed further standards for scientific thought. In the nineteenth century, several philosophers refined these standards into a series of ideal steps that became known as the scientific method. In the twentieth century, the older, simplistic view of science as merely turning the crank on the scientific method was challenged by many philosophers and historians.

The classic scientific method still taught commonly to high-school and college students today is as follows:

1. Observe natural phenomena.

2. Propose a possible cause-and-effect explanation, a hypothesis, to account for the observations.

3. Use the hypothesis to predict phenomena not already observed.

4. Arrange observations of the predicted phenomena.

5. If the new observations do not agree with prediction, go back to step 2 and revise your hypothesis. Repeat these steps until your hypothesis accounts for all known observations.

Many philosophers, historians, and scientists have pointed out that in real life, science does not always follow the steps of the scientific method in an orderly way. Scientists often make intuitive guesses based on very slight observations; existing theories influence which observations are made out of the infinite number of observations that could be made, and anomalous observations that seem to conflict with an otherwise well-supported theory may be ignored or put aside for a time.

In the 1930s, American philosopher Karl Popper (1902–1994) proposed that the distinguishing mark of a truly scientific idea is that it is falsifiable—that is, it makes predictions that can be tested against observation (step 4 of the classic scientific method). In recent decades, other philosophers have pointed out that Popper's definition is inadequate: Simply making a falsifiable prediction is not the same thing as doing science. For example, a person might claim, on a whim, that cars run on air. This claim might be falsified by observing that cars need to be supplied with liquid fuel or electricity in order to run. However, even though it is falsifiable, the cars-run-on-air hypothesis is not scientific: It is merely a free-floating claim that does not follow from specific observations or from any coherent cause-and-effect theory. All scientific claims are falsifiable, but not all falsifiable claims are scientific.

WORDS TO KNOW

DEDUCTION: A form of logical reasoning. In deduction, one reasons from general laws or principles to particular facts. For example, knowing that a stone has been dropped at a certain time and place and knowing the law of gravitation, one can deduce when the stone will strike the ground. The complementary or reverse form of reasoning is induction, which reasons from particular observations to general principles. For example, after making measurements of how dropped stones accelerate, one might induce the law of gravitation.

FALSIFIABLE: A theory or claim is falsifiable if there are reasonably possible conditions under which the theory or claim can be proved false. To be scientific, an idea must be falsifiable (though it takes more than falsifiability to make an idea a scientific one).

HYPOTHESIS: An idea phrased in the form of a statement that can be tested by observation and/or experiment.

INDUCTION: A form of logical reasoning. In induction, one reasons from particular facts to general laws or principles. For example, by making measurements of how dropped stones accelerate, one might induce the law of gravitation. The complementary form of reasoning is deduction, which reasons from general principles to particular events. For example, knowing that a stone has been dropped at a certain time and place and knowing the law of gravitation, one can deduce when the stone will strike the ground.

THEORY: An explanation for some phenomenon that is based on observation, experimentation, and reasoning.

Explanations based on magic, gods, miracles, or other supernatural causes are never scientific, because if supernatural forces do exist, they might, in principle, cause anything at all to happen. Such forces are potentially compatible with all possible observations and so their existence cannot ultimately be tested. Science studies only natural explanations because they are the only explanations that can be ruled in or out by the scientific method.

■ Modern Cultural Connections

Not all scientists practice the scientific method in exactly the same way. For example, scholars often distinguish between the historical sciences and experimental sciences. Historical sciences, such as astronomy, geology, evolutionary biology, and astrophysics, seek to explain chains of events that occurred in the past. Experimental

IN CONTEXT: OBSERVATION VS. THEORY

In the late 1990s, some scientists observed an effect that was dubbed the Pioneer anomaly. The *Pioneer 10* and *Pioneer 11* space probes, launched by the United States in 1972 and 1973, respectively, to explore the outer solar system (they still sail through space) were observed to be traveling along paths that could not be exactly explained using existing gravitational laws. Scientists studied every cause for the Pioneer anomaly that they could think of—fuel leaks acting like weak rockets, resistance from dust and gas floating in space, measurement errors due to waves breaking on the shore miles from radar stations tracking the spacecraft, and many more—but could not account for the anomaly. The same effect has been observed, although with a wider margin for error, for two other probes flying through the outer solar system, *Ulysses* and *Galileo*.

Since no force could be identified that might cause the anomalous movement of these spacecraft, it appeared that these observations might contradict the prevailing theory of gravitation, general relativity. However, general relativity has been supported by many other observations, and no scientist has been willing to reject it based only on the Pioneer anomaly. This illustrates the principle that a few anomalous observations are not necessarily enough to overturn a well-supported scientific theory.

As of 2008, the Pioneer anomaly remained unexplained, while astronomers continued to make many other observations that confirmed predictions of general relativity.

Australian scientist Barry Marshall is shown at the Queen Elizabeth II hospital in Perth. Marshall and fellow Australian Robin Warren won the 2005 Nobel Prize in physiology or medicine for their work on how the bacterium helicobacter pylori plays a role in gastritis and peptic ulcer disease. Marshall utilized the scientific method when he served as a human subject during his research by infecting himself with the bacterium, developing the inflammation, and recovering the organism from the lining of his stomach. *© Tony McDonough/epa/Corbis.*

sciences, such as physics, test their hypotheses in controlled settings (e.g., laboratories). Both types of science are completely scientific and operate according to the basic principles of scientific method. The difference is that historical scientists usually predict new observations of naturally occurring evidence or events, rather than arranging experiments.

Confusion sometimes arises when people assume that making observations means performing experiments—that is, manipulating objects and forces in a laboratory to produce a certain outcome. For example, people who disbelieve the biological theory of evolution (in science, a "theory" is not a guess but any well-supported explanation for a body of facts) sometimes argue that because nobody was present to observe the evolution of life, and because evolution cannot be repeated in a laboratory, scientists' claims about evolution are a matter of faith, not science.

However, this objection is based on a misunderstanding of the scientific method. Observations do not need to be based on laboratory experiments to support

or contradict a scientific theory. For example, in the twentieth century, fossils gathered by paleontologists (scientists who study fossils) showed that the earliest known four-limbed land-dwelling animals appeared in the Late Devonian, about 360 million years ago. This corresponds to step 1 of the scientific method: Observation. Evolutionary theory predicts that such animals evolved from four-limbed fish living in shallow coastal waters just before the appearance of the first definitely land-dwelling animals, and that transitional (or, in-between) animals must have developed at that time. This corresponds to step 2: Create a Hypothesis. In the early 2000s, several paleontologists, reading in a geology textbook that shallow-shore rocks dating to about 375 million years ago are found on Ellesmere Island in northern Canada, reasoned that fossils of transitional animals should be found there. This corresponds to step 3: Make a Prediction Based on Hypothesis. These scientists arranged an expedition to Ellesmere Island—step 4: Arrange New Observations—and in 2004 discovered fossils of just such a transitional animal, now known as *Tiktaalik*. The discovery of *Tiktaalik* fossils was an observation that confirmed the hypothesis of evolution, as many other observations have done over the last century and a half.

■ Primary Source Connection

Two millennia of scientific discovery are summarized and ranked into a "best of" list in this article by *New York Times* reporter George Johnson. Johnson is also the author of *Miss Leavitt's Stars* and received a Templeton-Cambridge Journalism Fellowship in Science and Religion in 2005.

HERE THEY ARE, SCIENCE'S 10 MOST BEAUTIFUL EXPERIMENTS

Whether they are blasting apart subatomic particles in accelerators, sequencing the genome or analyzing the wobble of a distant star, the experiments that grab the world's attention often cost millions of dollars to execute and produce torrents of data to be processed over months by supercomputers. Some research groups have grown to the size of small companies.

But ultimately science comes down to the individual mind grappling with something mysterious. When Robert P. Crease, a member of the philosophy department at the State University of New York at Stony Brook and the historian at Brookhaven National Laboratory, recently asked physicists to nominate the most beautiful experiment of all time, the 10 winners were largely solo performances, involving at most a few assistants. Most of the experiments—which are listed in this month's *Physics World*—took place on tabletops and none required more computational power than that of a slide rule or calculator.

What they have in common is that they epitomize the elusive quality scientists call beauty. This is beauty in the classical sense: the logical simplicity of the apparatus, like the logical simplicity of the analysis, seems as inevitable and pure as the lines of a Greek monument. Confusion and ambiguity are momentarily swept aside, and something new about nature becomes clear.

The list in *Physics World* was ranked according to popularity, first place going to an experiment that vividly demonstrated the quantum nature of the physical world. But science is a cumulative enterprise—that is part of its beauty. Rearranged chronologically and annotated below, the winners provide a bird's-eye view of more than 2,000 years of discovery.

Eratosthenes' measurement of the Earth's circumference

At noon on the summer solstice in the Egyptian town now called Aswan, the sun hovers straight overhead: objects cast no shadow and sunlight falls directly down a deep well. When he read this fact, Eratosthenes, the librarian at Alexandria in the third century BC, realized he had the information he needed to estimate the circumference of the planet. On the same day and time, he measured shadows in Alexandria, finding that the solar

rays there had a bit of a slant, deviating from the vertical by about seven degrees.

The rest was just geometry. Assuming the earth is spherical, its circumference spans 360 degrees. So if the two cities are seven degrees apart, that would constitute seven-360ths of the full circle—about one-fiftieth. Estimating from travel time that the towns were 5,000 "stadia" apart, Eratosthenes concluded that the earth must be 50 times that size—250,000 stadia in girth. Scholars differ over the length of a Greek stadium, so it is impossible to know just how accurate he was. But by some reckonings, he was off by only about 5 percent. (Ranking: 7)

Galileo's experiment on falling objects

In the late 1500's, everyone knew that heavy objects fall faster than lighter ones. After all, Aristotle had said so. That an ancient Greek scholar still held such sway was a sign of how far science had declined during the dark ages.

Galileo Galilei, who held a chair in mathematics at the University of Pisa, was impudent enough to question the common knowledge. The story has become part of the folklore of science: he is reputed to have dropped two different weights from the town's Leaning Tower showing that they landed at the same time. His challenges to Aristotle may have cost Galileo his job, but he had demonstrated the importance of taking nature, not human authority, as the final arbiter in matters of science. (Ranking: 2)

Galileo's experiments with rolling balls down inclined planes

Galileo continued to refine his ideas about objects in motion. He took a board 12 cubits long and half a cubit wide (about 20 feet by 10 inches) and cut a groove, as straight and smooth as possible, down the center. He inclined the plane and rolled brass balls down it, timing their descent with a water clock—a large vessel that emptied through a thin tube into a glass. After each run he would weigh the water that had flowed out—his measurement of elapsed time—and compare it with the distance the ball had traveled.

Aristotle would have predicted that the velocity of a rolling ball was constant: double its time in transit and you would double the distance it traversed. Galileo was able to show that the distance is actually proportional to the square of the time: Double it and the ball would go four times as far. The reason is that it is being constantly accelerated by gravity. (Ranking: 8)

Newton's decomposition of sunlight with a prism

Isaac Newton was born the year Galileo died. He graduated from Trinity College, Cambridge, in 1665, then holed up at home for a couple of years waiting out the plague. He had no trouble keeping himself occupied.

The common wisdom held that white light is the purest form (Aristotle again) and that colored light must therefore have been altered somehow. To test this hypothesis, Newton shined a beam of sunlight through a glass prism and showed that it decomposed into a spectrum cast on the wall. People already knew about rainbows, of course, but they were considered to be little more than pretty aberrations. Actually, Newton concluded, it was these colors—red, orange, yellow, green, blue, indigo, violet and the gradations in between—that were fundamental. What seemed simple on the surface, a beam of white light, was, if one looked deeper, beautifully complex. (Ranking: 4)

Cavendish's torsion-bar experiment

Another of Newton's contributions was his theory of gravity, which holds that the strength of attraction between two objects increases with the square of their masses and decreases with the square of the distance between them. But how strong is gravity in the first place?

In the late 1700's an English scientist, Henry Cavendish, decided to find out. He took a six-foot wooden rod and attached small metal spheres to each end, like a dumbbell, then suspended it from a wire. Two 350-pound lead spheres placed nearby exerted just enough gravitational force to tug at the smaller balls, causing the dumbbell to move and the wire to twist. By mounting finely etched pieces of ivory on the end of each arm and in the sides of the case, he could measure the subtle displacement. To guard against the influence of air currents, the apparatus (called a torsion balance) was enclosed in a room and observed with telescopes mounted on each side.

The result was a remarkably accurate estimate of a parameter called the gravitational constant, and from that Cavendish was able to calculate the density and mass of the earth. Erastothenes had measured how far around the planet was. Cavendish had weighed it: 6.0×1024 kilograms, or about 13 trillion trillion pounds. (Ranking: 6)

Young's light-interference experiment

Newton wasn't always right. Through various arguments, he had moved the scientific mainstream toward the conviction that light consists exclusively of particles rather than waves. In 1803, Thomas Young, an English physician and physicist, put the idea to a test. He cut a hole in a window shutter, covered it with a thick piece of paper punctured with a tiny pinhole and used a mirror to divert the thin beam that came shining through. Then he took "a slip of a card, about one-thirtieth of an inch in breadth" and held it edgewise in the path of the beam, dividing it in two. The result was a shadow of alternating light and dark bands—a phenomenon that could be explained if the two beams were interacting like waves.

Bright bands appeared where two crests overlapped, reinforcing each other; dark bands marked where a crest lined up with a trough, neutralizing each other.

The demonstration was often repeated over the years using a card with two holes to divide the beam. These so-called double-slit experiments became the standard for determining wavelike motion—a fact that was to become especially important a century later when quantum theory began. (Ranking: 5)

Foucault's pendulum

Last year when scientists mounted a pendulum above the South Pole and watched it swing, they were replicating a celebrated demonstration performed in Paris in 1851. Using a steel wire 220 feet long, the French scientist Jean-Bernard-Léon Foucault suspended a 62-pound iron ball from the dome of the Panthéon and set it in motion, rocking back and forth. To mark its progress he attached a stylus to the ball and placed a ring of damp sand on the floor below.

The audience watched in awe as the pendulum inexplicably appeared to rotate, leaving a slightly different trace with each swing. Actually it was the floor of the Panthéon that was slowly moving, and Foucault had shown, more convincingly than ever, that the earth revolves on its axis. At the latitude of Paris, the pendulum's path would complete a full clockwise rotation every 30 hours; on the Southern Hemisphere it would rotate counterclockwise, and on the Equator it wouldn't revolve at all. At the South Pole, as the modern-day scientists confirmed, the period of rotation is 24 hours. (Ranking: 10)

Millikan's oil-drop experiment

Since ancient times, scientists had studied electricity—an intangible essence that came from the sky as lightning or could be produced simply by running a brush through your hair. In 1897 (in an experiment that could easily have made this list) the British physicist J. J. Thomson had established that electricity consisted of negatively charged particles—electrons. It was left to the American scientist Robert Millikan, in 1909, to measure their charge.

Using a perfume atomizer, he sprayed tiny drops of oil into a transparent chamber. At the top and bottom were metal plates hooked to a battery, making one positive and the other negative. Since each droplet picked up a slight charge of static electricity as it traveled through the air, the speed of its descent could be controlled by altering the voltage on the plates. (When this electrical force matched the force of gravity, a droplet—"like a brilliant star on a black background"—would hover in midair.)

Millikan observed one drop after another, varying the voltage and noting the effect. After many repetitions he concluded that charge could only assume certain fixed values. The smallest of these portions was none other than the charge of a single electron. (Ranking: 3)

Rutherford's discovery of the nucleus

When Ernest Rutherford was experimenting with radioactivity at the University of Manchester in 1911, atoms were generally believed to consist of large mushy blobs of positive electrical charge with electrons embedded inside—the "plum pudding" model. But when he and his assistants fired tiny positively charged projectiles, called alpha particles, at a thin foil of gold, they were surprised that a tiny percentage of them came bouncing back. It was as though bullets had ricocheted off Jell-O.

Rutherford calculated that actually atoms were not so mushy after all. Most of the mass must be concentrated in a tiny core, now called the nucleus, with the electrons hovering around it. With amendments from quantum theory, this image of the atom persists today. (Ranking: 9)

Young's double-slit experiment applied to the interference of single electrons

Neither Newton nor Young was quite right about the nature of light. Though it is not simply made of particles, neither can it be described purely as a wave. In the first five years of the 20th century, Max Planck and then Albert Einstein showed, respectively, that light is emitted and absorbed in packets—called photons. But other experiments continued to verify that light is also wavelike.

It took quantum theory, developed over the next few decades, to reconcile how both ideas could be true: photons and other subatomic particles—electrons, protons, and so forth—exhibit two complementary qualities; they are, as one physicist put it, "wavicles."

To explain the idea, to others and themselves, physicists often used a thought experiment, in which Young's double-slit demonstration is repeated with a beam of electrons instead of light. Obeying the laws of quantum mechanics, the stream of particles would split in two, and the smaller streams would interfere with each other, leaving the same kind of light- and dark-striped pattern as was cast by light. Particles would act like waves.

According to an accompanying article in *Physics Today*, by the magazine's editor, Peter Rodgers, it wasn't until

1961 that someone (Claus Jönsson of Tübingen) carried out the experiment in the real world.

By that time no one was really surprised by the outcome, and the report, like most, was absorbed anonymously into science. (Ranking: 1)

Correction: September 27, 2002, Friday. An article in *Science Times* on Tuesday about the experiments selected by physicists as the 10 most beautiful in history referred incorrectly at one point to the magazine edited by Peter Rodgers, which first printed the list. It is *Physics World*, not *Physics Today*.

A painting with the article, identified as an image of Henry Cavendish, was published in error. It showed another 18th-century scientist, Joseph Priestley, who did not figure in the list.

Correction: October 7, 2002, Monday. An article in *Science Times* on Sept. 24 about physicists' selections of the 10 most beautiful experiments misstated a portion of Newton's theory of gravity, cited in a discussion of Cavendish's torsion-bar experiment. Newton held that the strength of attraction between two objects increases with the product of their masses, not with the square of their masses.

George Johnson

JOHNSON, GEORGE. "HERE THEY ARE, SCIENCE'S 10 MOST BEAUTIFUL EXPERIMENTS." *NEW YORK TIMES* (SEPTEMBER 24, 2002).

SEE ALSO *Biology: Evolutionary Theory; Biology: Paleontology.*

BIBLIOGRAPHY

Books

Gauch, Hugh G., Jr. *Scientific Method in Practice.* Cambridge, UK: Cambridge University Press, 2002.

Periodicals

Cleland, Carol E. "Methodological and Epistemic Differences Between Historical Science and Experimental Science." *Philosophy of Science* 69 (2002): 474–495.

Johnson, George. "Here They Are, Science's 10 Most Beautiful Experiments." *New York Times* (September 24, 2002).

Web Sites

Wolfs, Frank. University of Rochester (New York). "Introduction to the Scientific Method." http://teacher.pas.rochester.edu/phy_labs/AppendixE/AppendixE.html (accessed February 6, 2008).

Larry Gilman